## ADVANCED AIRCRAFT FLIGHT PERFORMANCE

This book deals with aircraft flight performance. It focuses on commercial aircraft but also considers examples of high-performance military aircraft. The framework is a multi-disciplinary engineering analysis, fully supported by flight simulation, with software validation at several levels. The book covers topics such as geometrical configurations, configuration aerodynamics and determination of aerodynamic derivatives, weight engineering, propulsion systems (gas turbine engines and propellers), aircraft trim, flight envelopes, mission analysis, trajectory optimisation, aircraft noise, noise trajectories and analysis of environmental performance. A unique feature of this book is the discussion and analysis of the environmental performance of the aircraft, focusing on topics such as aircraft noise and carbon dioxide emissions.

Dr. Antonio Filippone's expertise is in the fields of computational and experimental aerodynamics, flight mechanics, energy conversion systems, propulsion systems, rotating machines (helicopter rotors, propellers, wind turbines), systems engineering, and design and optimisation. He has published more than eighty technical papers, ten book chapters, and two books, including *Flight Performance of Fixed and Rotary Wing Aircraft* (2006).

*Cambridge Aerospace Series*

*Editors*
Wei Shyy
and
Vigor Yang

1. J. M. Rolfe and K. J. Staples (eds.): *Flight Simulation*
2. P. Berlin: *The Geostationary Applications Satellite*
3. M. J. T. Smith: *Aircraft Noise*
4. N. X. Vinh: *Flight Mechanics of High-Performance Aircraft*
5. W. A. Mair and D. L. Birdsall: *Aircraft Performance*
6. M. J. Abzug and E. E. Larrabee: *Airplane Stability and Control*
7. M. J. Sidi: *Spacecraft Dynamics and Control*
8. J. D. Anderson: *A History of Aerodynamics*
9. A. M. Cruise, J. A. Bowles, C. V. Goodall, and T. J. Patrick: *Principles of Space Instrument Design*
10. G. A. Khoury (ed.): *Airship Technology*, Second Edition
11. J. P. Fielding: *Introduction to Aircraft Design*
12. J. G. Leishman: *Principles of Helicopter Aerodynamics*, Second Edition
13. J. Katz and A. Plotkin: *Low-Speed Aerodynamics*, Second Edition
14. M. J. Abzug and E. E. Larrabee: *Airplane Stability and Control: A History of the Technologies that Made Aviation Possible*, Second Edition
15. D. H. Hodges and G. A. Pierce: *Introduction to Structural Dynamics and Aeroelasticity*, Second Edition
16. W. Fehse: *Automatic Rendezvous and Docking of Spacecraft*
17. R. D. Flack: *Fundamentals of Jet Propulsion with Applications*
18. E. A. Baskharone: *Principles of Turbomachinery in Air-Breathing Engines*
19. D. D. Knight: *Numerical Methods for High-Speed Flows*
20. C. A. Wagner, T. Hüttl, and P. Sagaut (eds.): *Large-Eddy Simulation for Acoustics*
21. D. D. Joseph, T. Funada, and J. Wang: *Potential Flows of Viscous and Viscoelastic Fluids*
22. W. Shyy, Y. Lian, H. Liu, J. Tang, and D. Viieru: *Aerodynamics of Low Reynolds Number Flyers*
23. J. H. Saleh: *Analyses for Durability and System Design Lifetime*
24. B. K. Donaldson: *Analysis of Aircraft Structures*, Second Edition
25. C. Segal: *The Scramjet Engine: Processes and Characteristics*
26. J. F. Doyle: *Guided Explorations of the Mechanics of Solids and Structures*
27. A. K. Kundu: *Aircraft Design*
28. M. I. Friswell, J. E. T. Penny, S. D. Garvey, and A. W. Lees: *Dynamics of Rotating Machines*
29. B. A. Conway (ed.): *Spacecraft Trajectory Optimization*
30. R. J. Adrian and J. Westerweel: *Particle Image Velocimetry*
31. G. A. Flandro, H. M. McMahon, and R. L. Roach: *Basic Aerodynamics*
32. H. Babinsky and J. K. Harvey: *Shock Wave–Boundary-Layer Interactions*
33. C. K. W. Tam: *Computational Aeroacoustics: A Wave Number Approach*
34. A. Filippone: *Advanced Aircraft Flight Performance*

*Ignoranti quem portum petat nullus suus ventus est.*

No wind is favourable to a sailor who does not know at which port to land.

[Lucius A. Seneca (4 BC–AD 65), *Moral Letters to Lucilius* (letter 71)]

# Advanced Aircraft Flight Performance

**Antonio Filippone**
The University of Manchester

**CAMBRIDGE**
**UNIVERSITY PRESS**

32 Avenue of the Americas, New York NY 10013-2473, USA

Cambridge University Press is part of the University of Cambridge.

It furthers the University's mission by disseminating knowledge in the pursuit of education, learning and research at the highest international levels of excellence.

www.cambridge.org
Information on this title: www.cambridge.org/9781107024007



First published 2012

*A catalogue record for this publication is available from the British Library*

*Library of Congress Cataloguing in Publication data*

Filippone, Antonio, 1965–
Advanced aircraft flight performance / Antonio Filippone.
p. cm. – (Cambridge aerospace series ; 34)
Includes bibliographical references and index.
ISBN 978-1-107-02400-7 (hardback)
1. Airplanes – Performance. 2. Airplanes – Design and construction. I. Title.
TL671.4.F449 2012
629.132'3–dc23 2012015394

ISBN 978-1-107-02400-7 Hardback

# Contents

# Tables

# Preface

This book is a derivative of an earlier textbook on flight performance. This new work reflects my increased wisdom on the subject and represents an almost complete departure from closed-form solutions that are traditionally taught in under-graduate and post-graduate programs. Over the past several years, I have benefited from the experience of teaching a flight performance course to senior engineers from industry, government departments and academia. In the process, I learned a few new things that now find a place somewhere in the book.

There is an increase in numerical methods in all fields of engineering; nevertheless, flight performance has remarkably resisted change. Some closed-form solutions have been retained for those engineers who need a quick answer. The modern airplane is a complex engineering machine governed by systems, software and avionics. Primitive methods are still widely used, which are then applied to aircraft design and produce results of dubious accuracy that cannot be assessed. Worryingly, these methods are used in most "conceptual design" and "multi-disciplinary optimisation" methods. Now assume, more realistically, that you have been hired to provide flight prediction tools to an airline operator or a manufacturer of engines or airframes, a national or international aviation authority, an air traffic control organisation. Why should they trust your performance software? What is the risk of under-predicting the mission fuel for an intercontinental flight?

As we worried about conceptual design, the world has moved on. There is increased emphasis on airplane evolution and upgrading, which is now reflected in my thinking. At the same time, the environmental performance of the aircraft has become very prominent. Therefore, part of this book is devoted to a wide spectrum of environmental aspects of flight. My initial concerns have slowly shifted from noise to engine emissions. Noise disappears as the aircraft moves away from the receiver, although not many would like to agree. Exhaust gases remain with us for the next few generations. In particular, aircraft condensation trails are there to remind us that aviation is having a measurable impact on our skies. The lack of flexibility in aircraft levels, stepped cruise and descent, and the use of holding patterns in congested air space are all problems that need a solution in the coming years.

The book contains considerable advanced material across several disciplines, including aircraft noise, environmental performance, airframe-propulsion integration, thermo-structural performance and flight mechanics. I am conscious of the audacity of the task I have undertaken, but I am confident that this work meets the expectations of the aviation industry and the academic world.

I have developed some fully comprehensive flight codes. One code in particular, FLIGHT, to simulate aircraft performance and mission analysis of transport aircraft, contains most of the cross-disciplinary aspects of performance discussed in this book. In its present form it consists of about 160 KLOCS (thousand lines of code). Other codes discussed in the book include the propeller code, that is fully integrated with FLIGHT, as well as a supersonic flight performance code (SFLIGHT). Several block flowcharts have been included to help with the understanding of computer programs, numerical models, system analysis and flight performance. The following material is made available to readers:

- Computer code FLIGHT (demo version)
- Computer code Prop/FLIGHT (demo version)
- Computer code SFLIGHT (demo version)
- All charts and figures in any suitable graphical format

Separate technical documents will be issued to the readers wishing to work with these computer models.

Dr. Z. Mohammed-Kassim, my long-time associate, has actively contributed to the work on aircraft noise and to considerable code debugging. My doctoral student Nicholas Bojdo took great care in reading some chapters. I am indebted to my editor, Peter Gordon, who has been enthusiastic about my work from the beginning of the project to the end. The editorial and production work was efficiently managed by Peggy Rote at Aptara, Inc.

Finally, I thank my wife, Susan, for having the patience to tolerate my late nights at the desk, especially when I reached the *tunnel phase* of my work, that is, when I thought the book was finished but in fact there was no end in sight. A sabbatical leave from the University has allowed me to step up my efforts. I am grateful to the University, and the School, for the opportunity they have given me.

# Nomenclature

## Organisations

Below is a list of organisations that publish regularly documents (technical reports, papers, journals, regulations) as well as more general information of aviation.

| | |
|---|---|
| AAIB | Air Accidents Investigation Branch, United Kingdom (www.aaib.gov.uk) |
| AIAA | American Institute of Aeronautics & Astronautics (www.aiaa.org) |
| ANSI | American National Standards Institute (www.ansi.org) |
| ASTM | American Society for Testing and Materials (www.astm.org) |
| BTS | Bureau of Transportation Statistics, USA (www.bts.gov) |
| CAA | Civil Aviation Authority (www.caa.co.uk) |
| EASA | European Aviation Safety Agency (www.easa.eu.int) |
| ESDU | Engineering Data Unit (www.esdu.com) |
| FAA | Federal Aviation Administration (www.faa.gov) |
| FSF | Flight Safety Foundation (www.flightsafety.org) |
| IATA | International Air Transport Association (www.iata.org) |
| ICAO | International Civil Aviation Organisation (www.icao.int) |
| IPCC | Inter-governmental Panel for Climate Change (www.ipcc.ch) |
| Jane's | Jane's Information Systems (www.janes.com) |
| MIL | Military Standards (www.mil-standards.com) |
| NASA | National Administration for Space and Aeronautics (www.nasa.gov) |
| NATO | Advisory Group, Aerospace Research & Development (www.rta.nato.int) |
| NATS | National Air Traffic System, United Kingdom (www.nats.co.uk) |
| NTSB | National Transportation Safety Board, United States (www.ntsb.gov) |
| RAeS | The Royal Aeronautical Society (www.aerosociety.org) |
| SAE | Society of Automotive Engineers (www.sae.org) |
| SAWE | Society of Allied Weight Engineers (www.sawe.org) |

## Acronyms Used in This Book

| | |
|---|---|
| ACT | Additional Centre Tank |
| AEO | All Engines Operating |
| AF | Activity Factor |

| | |
|---|---|
| APU | Auxiliary Power Unit |
| ASDA | Accelerate-Stop Distance Available |
| ASI | Air Speed Indicator |
| ASK | Available Seat per Kilometre |
| ATC | Air Traffic Control |
| AUW | All-Up Weight |
| BFL | Balanced Field Length |
| BPR | By-pass Ratio |
| BRGW | Brake-Release Gross Weight |
| CAS | Calibrated Air Speed |
| CASK | Cost per Available Seat per Kilometre |
| CDA | Continuous Descent Approach |
| CG | Centre of Gravity |
| CTOL | Conventional Take-off and Landing |
| DOC | Direct Operating Costs |
| DOCG | Dry Operating Centre of Gravity |
| DOF | Degree of Freedom |
| DOW | Dry Operating Weight |
| EAS | Equivalent Air Speed |
| EBF | Externally Blown Flap |
| ECS | Environmental Conditioning System |
| EGT | Exhaust Gas Temperature |
| EPNdB | Effective Perceived Noise, in $dB$ |
| EPNL | Effective Perceived Noise Level |
| ETOPS | **E**xtended **T**win-Engine **OP**eration**S** |
| FADEC | Full Authority Digital Engine Control |
| FCA | Final Cruise Altitude |
| FCOM | Flight Crew Operating Manual |
| FDR | Flight Data Recorder |
| FL | Fuselage Line; Flight Level |
| FLS | Flight Level Separation |
| FMS | Flight Management System |
| GPS | Global Positioning System |
| GPU | Ground Power Unit |
| GRW | Gross Ramp Weight |
| GTOW | Gross Take-off Weight |
| IAS | Indicated Air Speed |
| ICA | Initial Cruise Altitude |
| ICW | Initial Cruise Weight |
| IDA | Initial Descent Altitude |
| IGE | In Ground Effect |
| ILS | Instrument Landing System |
| ISA | International Standard Atmosphere |
| KCAS | Calibrated Air Speed in knots |
| KEAS | Equivalent Air Speed in knots |
| KIAS | Indicated Air Speed in knots |

| | |
|---|---|
| KTAS | True Air Speed in knots |
| LRM | Long-Range Mach number |
| MAC | Mean Aerodynamic Chord |
| MBGW | Maximum Brake-Release Weight |
| MCP | Maximum Continuous Power |
| MEW | Manufacturer's Empty Weight |
| MIL | Military Standards (USA) |
| MLW | Maximum Landing Weight |
| MRM | Maximum-Range Mach number |
| MRW | Maximum Ramp Weight |
| MSP | Maximum Structural Payload |
| MTOP | Maximum Take-off Power |
| MTOW | Maximum Take-off Weight |
| MZFW | Maximum Zero-Fuel Weight |
| NADP | Noise Abatement Departure Procedure |
| OASPL | Overall Sound Pressure Level |
| OAT | Outside Air Temperature |
| ODE | Ordinary Differential Equation |
| OEI | One Engine Inoperative |
| OEW | Operating Empty Weight |
| OGE | Out of Ground Effect |
| OPR | Overall Pressure Ratio |
| PAX | Passengers |
| PNL | Perceived Noise Level |
| PNLT | Perceived Noise Level, Tone Corrected |
| PWL | One-third octave band Power Level |
| SAR | Specific Air Range |
| SAT | Static Air Temperature |
| SEL | Sound Exposure Level |
| SEP | Specific Excess Power |
| SFC | Specific Fuel Consumption |
| SHP | Shaft Horse Power |
| SI | International Units System |
| S/L | Sea Level |
| SPL | Sound Pressure Level |
| STOL | Short Take-off and Landing |
| TAS | True Air Speed |
| TAT | Total Air Temperature |
| TMA | Terminal Manoeuvre Area |
| TOCG | Take-off Centre of Gravity |
| TOD | Top Of Descent |
| TODA | Take-off Distance Available |
| TODR | Take-Off Distance Required |
| TOGA | Take-off and Go-Around |
| TORA | Take-off Distance Required |
| TORR | Take-Off Run Required |

| | |
|---|---|
| TOW | Take-off Weight |
| TSFC | Thrust-Specific Fuel Consumption |
| ULD | Unit Load Device |
| VMC | Minimum Control Speed |
| VMCA | Minimum Control Speed in Air |
| VMGC | Minimum Control Speed on the Ground |
| VMO | Maximum Operating Speed |
| VNE | Velocity Not to Exceed |
| WAT | Weight-Altitude-Temperature |
| WBM | Weight and Balance Manual |
| ZFCG | Zero-Fuel Centre of Gravity |
| ZFW | Zero-Fuel Weight |

The U.S. Department of Defense and NATO publish a dictionary of acronyms and aviation jargon. A detailed list of symbols follows each chapter.

# Technology Warning

This book makes reference to real flight vehicles in realistic flight conditions. The data used to model these vehicles have been extracted, elaborated, interpolated or otherwise inferred from documents available in the public domain. These documents are either published by the manufacturer or the operators, or both. They are supplemented with official data published by several aviation authorities at the national and international level. Many of these documents are freely available to the public in electronic format from the manufacturers, through their websites, or the websites of their customers, or by third parties. No commercial, sensitive or restricted data have been disclosed anywhere. All sources have been cited when appropriate. There is no implication that the data refer to any particular aircraft owned or operated by any organisation. The flight performance shown is often validated, but sometimes it is not. Whenever figures or tables report the term "simulated" or "validated", they refer to simulations carried out with the comprehensive performance code `FLIGHT` and its related software technology (available from the author).

Readers should be made aware that the statements made in this book are the author's own. Readers should use judgement before making technical, commercial, military, marketing or business decisions. The author cannot take responsibility for any action resulting in damage, accident or loss, as a consequence of statements made in this book. None of the graphs, figures and tables shown in this book can be used to make a final judgement on any airplane, any manufacturer, any flight, any service or any design. **Use of the graphs for flight planning is prohibited**. If you are in doubt, please consult the author, or use the performance codes from the aircraft manufacturers.

# 1 Prolegomena

Commercial aviation has grown to become the backbone of the modern transport system. Demand for commercial air travel has grown exponentially since the 1960s. The expansion of the aviation services is set to increase further. Even in the worst moments of recession, air transport has only suffered a blip in its expansion. Our global economy has become so dependent on air transport that any inconvenience caused by weather and external factors can cause mayhem. To understand the size of commercial aviation, reflect upon the fact that by the year 2000, there had been 35.14 million commercial departures worldwide, for a total of 18.14 million flight hours[1]. In some countries, air transport accounts for one quarter of all transported goods by value. It is estimated that there are about 50,000 airports and airfields around the world[2] and 18,000 commercial airplanes flying every day. A number of large airports are being constructed in some rapidly expanding regions. Modern airports are the size of a city: London Heathrow covers about 12 $km^2$; this is an area large enough to park about 1,800 Airbus A380s nose-to-tail. The support required by the aircraft is extraordinary and involves engineering, logistics, integrated transport systems, security systems, energy and people. However, at the heart of everything is the aircraft itself, interpreted as a flight system: this is the subject of our book. There will be only a superficial mention of various externalities, including air traffic control, queueing models, stack patterns, logistics, supply chains, and so on. More specifically, this book deals with the analysis, simulation and prediction of aircraft flight performance at several levels, including aerodynamics, weight performance, flight mechanics, aircraft noise and environmental emissions. Advanced aircraft performance analysis involves at least the following engineering activities:

- verification that an aircraft achieves its design targets
- efficient operation of existing aircraft or fleet
- selection of a new aircraft
- modification and upgrade of the flight envelope
- upgrading and extension of the mission profile
- aircraft and engines certification process
- environmental analysis, including emissions on the ground and in flight

- aircraft noise emissions
- design of a new aircraft with and without optimisation methods

Note that in this list of specifications, only the last one involves the design of a brand-new aircraft, something often referred to as "conceptual design". Safety in aviation dominates over everything else. Thus, health monitoring, adverse weather events and human factors become an integral part of operating an aircraft and maintaining safety records.

The methods used for the evaluation of aircraft performance are based on theoretical analysis and flight testing. The latter method is made possible by accurate measurement techniques, including navigation instruments.

*Performance flight testing* involves the calibration of instruments and static tests on the ground, testing at all the important conditions, gathering of data, data analysis, calibration with simulation models, determination of charts for the certification and the flight manual.

*Performance analysis* is based on the elaboration of flight data, either from flight testing or from flight data recorders. It is normally done at the operational level by commercial airlines in the attempt to determine how the airplane performs over time, in comparison with other airplanes and in comparison with the manufacturer's claims. The complexity of the modern vehicle and the variability of all external factors contribute to changes in performance that often do not correspond to the technical specifications. Most manufacturers have their own flight performance codes tuned for their own aircraft.

*Performance prediction* is at the base of any concrete aircraft design and operation. The estimation of weights, range and power plant size requires the calculation of basic aircraft performance from a few input data. In this case the approximation is generally good enough for parameter estimation and design. A manufacturer claims (typically) that: its airplane has 20% lower direct operating costs (including −30% fuel consumption) than competitor X. How are we going to find out that this is the case? Do we buy one of its airplanes and one from competitor X? Do we spend 1,000 hours in flight testing and then decide? In the field of aircraft performance there is no laboratory experiment that can be used for verification.

## 1.1 Performance Parameters

Most of the aircraft flight parameters are stored by the Flight Data Recorder (FDR), commonly called the *black box* (although its colour is often orange). The set of data recorded is now standardised and contains all of the parameters that may be useful for accident investigation, systems monitoring, flight trajectory analysis and engine performance. As of August 2002, the National Transportation Safety Board (NTSB, United States) requires the records of at least 88 flight parameters for a transport category airplane.[3] These data include values for all the degrees of freedom of the aircraft, velocities, acceleration rates, controls positions and engine state (temperatures, pressures, fuel flow, rotation speeds), and quantities as different as computer failure, autopilot engagement, icing, traffic alerts and collision-avoidance systems. The requirements for large helicopters include records for at least 26 different flight and state parameters. An additional set of 30 parameters is recommended though not required.

Modern FDR, based on solid-state memory, can record several hundred parameters for up to 24 hours. They can withstand accelerations of hundreds of *g*-s, temperatures in excess of 1,000°, and survive in ocean depths of 6,000 m, whilst emitting a locator beacon for about 30 days. If ditched into the ocean, they can be located by satellite.

One of the most challenging recoveries was that of an FDR from an Airbus A330 that disappeared off the coast of Brazil on 1 June 2009. The FDR was eventually found intact, nearly two years later, 4,000 metres below sea level, after searching 10,000 km$^2$ of sea floor.

International regulations mandate that commercial operators of aircraft heavier than 27,000 kg have their FDR data monitored. In practice, it is required to analyse the flight parameters, particularly in cases when problems of any nature may occur (too steep descent rate, hard landing, flight through clear sky turbulence, and so on). The scope of this analysis is at least twofold: 1) to guarantee safety, and 2) to understand whether the aircraft is operated efficiently.

Many performance indexes cannot be simply expressed by a single value; they are presented with charts. Some performance data are readily available from the manufacturer, other data can be inferred by appropriate analysis, others are clouded by secrecy or confidentiality, and others are difficult to interpret because the conditions under which the aircraft performs are not provided. Among the most common data covered by secrecy are the drag data, the stability characteristics, the excess power diagrams and the engine performance. Other examples are 1) the aircraft range, when the payload is not supplied together with the range, 2) the altitude at which this range is achieved, and 3) the radius of action of a military interceptor. This radius, in fact, may lie in the field of enemy fire.

The maximum take-off weight (MTOW) and the operating empty weight (OEW) are available for most aircraft. However, these data are not sufficient to calculate the maximum payload because the difference between MTOW and OEW must include the mission fuel. A weight advantage compared to heavier rivals translates into significant revenue-earning advantages, which in a competitive market is the most important factor for choosing and operating an aircraft. It is not uncommon to find manufacturers unhappy that their performance data and charts are published in the public domain. Performance charts allow customers and competitors to look at various options, to select the most competitive aircraft and to discover the flaws of the competitors' technology.

The operator of an aircraft is concerned that the performance parameters quoted by the manufacturer match the actual performance, and therefore accuracy of performance prediction methods is essential. Performance data in the Flight Crew Operating Manual (FCOM) are not always accurate for a variety of reasons: 1) the FCOM data are often extrapolated from a limited set of flight test data, 2) the actual flight conditions are different from the conditions in the FCOM, and 3) ageing effects on the airframe and the propulsion system may lead to performance sensibly different from new aircraft. One aircraft type may have several FCOM because each airline may have its own version.

## 1.2 Flight Optimisation

*Flight optimisation* is at the heart of design and operation of all modern aircraft. From the operational point of view, commercial aviation is driven by fuel prices, and operations at minimum fuel consumption are of great relevance. Performance optimisation requires notions of optimal control theory, a subject unfamiliar to the aerospace engineers. In the past 30 years these optimal conditions have been increasingly challenged by environmental concerns, including noise emission, air quality near airports, global climate change and sustainability.

Computer solutions of aircraft performance are now routine jobs and have reached a phenomenal level of sophistication to include the coupling among flight mechanics, aerodynamics, structural dynamics, flight system control and differential game theory.

There are two key types of optimisation: Optimisation of the aircraft performance during the design phase, and optimisation of the operational performance for the given airplane. In the former case, one can investigate the alternative changes in configuration that improve one or more performance parameters. This is more appropriately the subject of aircraft design. We will consider some cases of operational optimisation. An excellent source for optimisation problems with aircraft applications is the classical book of Bryson and Ho on optimal control[4] and Ashley[5] for a variety of flight mechanics problems. Today there are programs that plan optimal trajectory routes to minimise direct operating costs (DOC), whilst complying with several airline constraints. These programs have several types of input data: weather conditions, route, aerodynamics, aircraft performance and flight-specific information, such as payload, fuel cost, and so on. On output they provide the amount of fuel for optimal cruise altitude, climb and descent points, optimal cruise speed and flight path.

The flight controls themselves have reached a phenomenal level of sophistication, with several on-board computers and substantial software; they now represent a key aspect of the aircraft system. The best airplane cannot fly without its embedded software that satisfies the most stringent requirements. According to recognised standards – for example, NASA[6] – software controlling vehicles for human flight must have the highest level of quality assurance and is defined as *Class A*. Software failure in this class may cause a loss of life, or *catastrophic mission failure*. Next comes *Class B* software, which is designed for non-human flight: unmanned vehicles, rockets and satellites. Any software failure can lead to a total or partial loss of the vehicle but no loss of life (*partial mission failure*).

Figure 1.1 shows the flight control panel of the Gulfstream G550, with its peculiar four-screen view. The monitor display at the bottom centre of the photograph allows rapid performance calculation, including take-off speeds from basic input parameters, that can be quickly programmed by the pilot.

## 1.3 Certificate of Airworthiness

The certificate of airworthiness is a document that grants authorisation to operate an aircraft. It specifies the type of operations; the limits of operation of the vehicle in terms of weights, take-off and landing requirements; and a number of other

Figure 1.1. Cockpit of the Gulfstream G550 with Honeywell DU-1310 visual displays (author's photograph).

parameters, such as maintenance records, service history and compliance with safety regulations.

The certificate proves that the aircraft conforms to the type specified and it is safe to fly. The certificate is valid as long as the aircraft meets the type specification (commercial, commuter, utility, and so on), it is operated safely and all the airworthiness directives are met. The aircraft may lose its certificate for a number of reasons, including modifications upgrades and new directives approved by the international organisations that make the aircraft obsolete, not just unsafe to operate. Other documents are generally required, such as the type certificate data sheet, the airworthiness limitations, the flight log book, the certificate of maintenance and a list of other papers. The airworthiness limitations contain specific data on the number of flight hours, years of service or number of cycles for critical components and systems. These data can be used as limiters in flight simulation programs.

The *type certificate* contains various technical data, limitations and cautions, as well as reference to the appropriate technical manuals (operating manuals, maintenance manuals, and so on). Separate type certificates are issued for aircraft, propellers and engines. These certificates are issued by various national and international aviation organisations and are sometimes a duplication of effort in different countries. These certificates, which can run to several pages, contain at least some of the following data: airworthiness category, engines, engine limits, fuel, limit speeds, centre of gravity range, maximum certified weights, auxiliary power units, equipment, seating capacity, all weather capabilities and so on.*

* For example: European Aviation Safety Agency, Type Certificate Data Sheet, Airbus A380. TCDS A.110, Issue 04, Feb. 2009.

Certificates of airworthiness and the type certificates are issued by the Federal Aviation Administration (FAA) in the United States, the European Aviation Safety Agency (EASA) in Europe, the Civil Aviation Authority (CAA) in the United Kingdom and other national and international bodies around the world. Certification is a complex legal and technical matter that falls beyond the scope of this book. Note that the certificates issued by different bodies may contain different types of information. Therefore, it is sometimes useful to obtain the different certificates, particularly when mining for engine performance data.

## 1.4 The Need for Upgrading

Most aircraft performance applications focus on performance for design, performance optimisation and aircraft sizing. It seems naive to forget that more than 18,000 commercial airliners fly around the world every day. The same airlines will be flying for the next 20 years, as businesses try to recover their capital investments.

How many aircraft are actually sized every year? Most aircraft are likely to be upgraded and modified to fit the changing market and technological advances. The technology that is fitted over the years can be phenomenally different from the first design. The service time of a single aircraft is on the order of 20 to 25 years, and the life of an aircraft family may exceed 50 years. A single airplane program consists of several versions, derivatives, design improvements, weight configurations, power plants, avionics and systems.

A life-time career can be devoted to a single airplane. In the early days of aviation, a new aircraft could roll out of the factory in a few months. In 1936, it took just one year for the German aircraft designer Kurt Tank[7] to get from concept to first flight of the Focke Wulf Condor Fw-190, the first long-range passenger (and later reconnaissance and bomber) aircraft to fly from Berlin to New York without en-route stop (in 1938). By the 1960s, commercial airplane design and testing required thousands of man-years. The Boeing B747-100, that first flew in 1969[8;9], required 15,000 hours of wind-tunnel testing and about 1,500 hours of flight testing with five aircraft[1].

The B747-400 incorporated major aerodynamic improvements, including a more slender wing with winglets to reduce drag. A weight savings of approximately 2,270 kg was achieved in the wing by using new aluminium alloys. The version B747-400ER has an increased take-off weight of 412,770 kg. This allows operators to fly about 410 nautical miles (~760 km) farther or to carry up to an additional 6,800 kg payload, for a range up to 14,200 km (~7,660 n-miles). A larger version of this airplane, the B747-800, weighing about 443 metric tons, is due to carry on the 747-flag for many years to come.

An even older airplane is the Lockheed Hercules C-130A. Its first model was delivered to the U.S. military in 1956. The design of this aircraft actually started several years earlier. By the early 1960s, a VSTOL variant was designed[10]. Since then, the aircraft has progressed through at least 60 different variants. The version C-130J is actually a new airplane. Compared to the earlier popular version C-130E, the maximum speed is increased by 21%, climb time is reduced by 50%, the cruising altitude is 40% higher, the range is 40% longer and its Rolls-Royce AE-2100DE engines generate 29% more thrust, while increasing fuel efficiency by 15%. With new engines and new propellers, the C130-J can climb to 9,100 metres (~28,000 feet) in 14 minutes.

Technological advances in aerodynamics, engines and structures can be applied to existing aircraft to improve their performance. Over time weights grow, power plants become more efficient and are replaced, aerodynamics is improved by optimisation, fuselages are stretched to accommodate more payload and additional fuel tanks are added. Within the first few years of the Airbus A380 rolling out of the factory for the first time, new landing gear brakes design saved the weight of one passenger. At the same time, the wing was made about 350 kg lighter.

## 1.5 Military Aircraft Requirements

Military aviation has been a key aspect of national security since the very beginning of airplane engineering and is the source of many technological innovations. The fighter aircraft has evolved from a reconnaissance airplane of the First World War to the most complex aircraft of the modern day. Von Kármán[11] reported that they first flew over the battle fields of Europe to spy on enemy lines. Then enemy aircraft wanted to prevent this from happening, and hence their pilots started shooting at enemy aircraft with a pistol.

From reconnaissance to air strikes the step was short. The first recorded air bombing is attributed to the Italian Air Force during its Turkish Wars in North Africa in 1911–1912. The Italians operated a number of aircraft for reconnaissance (Taube, Deperdussin, Bleriot) but then started shooting at enemy forces on the ground. Finally, on 1 November 1911, they decided to drop four grenades weighing about 1.5 kg each. Lieutenant Giulio Gavotti carried out the air strikes in a Taube airplane[†]. Because the activation of the grenade required two hands and one hand was needed to steer the unstable airplane, the weapons were activated by snatching the plug with the teeth; then they were thrown overboard. Although the first bombs did not kill anybody (the blast was absorbed by the desert sand), the news was sensational. The Hague Convention (1899) prohibited the launching of projectiles and explosives from balloons. The Italians argued that the Convention did not apply to the powered airplane[‡]. Thus, the event was heralded as the start of a new era in warfare.

General Giulio Douhet was the first to distinguish among indiscriminate bombing, carpet bombing and strategic bombing (1912). In due course, the Germans used Zeppelin airships for the indiscriminate bombing of London in 1914 (the first urban bombing). The Hague Convention did not apply to them either. In response, the newly established Royal Air Force (RAF) performed the first air bombing of land targets with airplanes transported by ship near the theatre of war (Cuxhaven raid).

The psychosis of aerial bombing was sparked in the United Kingdom by H.G. Wells, who in 1907 published a science fiction book, *The War in the Air*, in which a

[†] Some sources report that the grenades were used for the first time on 24 October 1911 and landed on the desert; the day after the first casualty was recorded.

[‡] The Hague Convention, signed on 29 July 1899, entered into force on 4 September 1900. Chapter IV, in fact, limited the use of explosives from the air for five years, starting in 1900, by *balloon or other new methods of similar nature*.

fleet of airships attacked and bombed New York City. The Americans were not impressed but years later, science fiction turned into reality, with great loss of life.

Following the operations in North Africa in 1911, the scope of military reconnaissance was considerably expanded: now it was possible to observe from above, to provide aerial photography and update and improve maps with unprecedented details. By the start of the First World War, reconnaissance aircraft had two seats, one for the pilot and the other for an observer – without which the pilot would have had difficulty returning to base. The observer's equipment included a map, a pistol, a watch, a pair of binoculars, a one-way radio system and a life vest. Typically, the useful load was around 400 kg, including pilot, observer and their equipment. If we allow for about 180 kg for all this weight, there was about 220 kg left, to be shared between fuel and ordnance. The fuel flow was of the order of 32 kg per flight hour. Without ordnance, an airplane would have been capable of an endurance of 6.5 hours. At an average speed of 100–120 km/h (∼50–60 kt), the radius of action would have been 350 km (∼190 n-miles) in the most optimistic scenario.

In 1915, during the early days of the war, the Dutchman Anton Fokker[12], working at the service of the German Army, invented a system that synchronised the shooting of a machine gun through the propeller (*interrupter gear*) – mounted on a single-seater monoplane. With the interrupter, pilots had hands free to manoeuvre and fight at the same time. Occasionally the interrupter malfunctioned and claimed the life of such accomplished pilots as Max Immelmann.

By the end of the Great War, the European powers had thousands of aircraft at their disposal. It has been estimated that the total production of such aircraft exceeded a staggering 75,000 units in four years, 32,000 of which were British; note that only 15 years had passed since the invention of the airplane.

To follow the history of the development of the military aircraft means following the development of key aeronautical technologies over the past 100 years. Two relevant books on the birth of military airplanes are Driver[13] (on the British aviation) and Opolycke[14] (on the French aviation). Stevens[15] and Weyl[16] report more generic historical details. Jackson[17] published a chronology of events recording aerial warfare from the very beginning to the present.

The requirements for fighter aircraft now include multi-purpose missions, aircraft with complex flight envelopes, several configurations (changeable in flight), supersonic flight, combat capabilities, delivery of a wide range of weapon systems, manoeuvrability, all-weather and night-and-day operations. The aircraft has become a platform system of phenomenal complexity and cost. Yet history deserves another revolutionary weapon system: the military aircraft of the future is unmanned. Nevertheless, there will be a few decades before this happens. In the meantime, there can be problems of excess capacity coupled with excessive costs, which in turn make the military aircraft difficult to sell, operate and upgrade.

There are dozens of different mission scenarios[18]. Typical missions are basic, assault, combat, retrieval, close support, transport, refuel and reconnaissance. For each of these missions there is a specific take-off weight, mission fuel, payload, range, maximum rate of climb and service ceiling. This field is now so advanced that engineers use differential game theory and artificial intelligence to study the effectiveness of a given aircraft and the tactical manoeuvrability to incoming threats[19].

## 1.6 Review of Comprehensive Performance Programs

This review is limited to computer models that are documented in the open literature[20]. Both Airbus and Boeing have their own flight performance programs that model their own aircraft by using a combination of first principles and extensive flight data. These programs are not in the public domain. Versions of these programs are provided to airline operators in order to facilitate their performance analysis and their flight planning.

Computerised flight planning goes back to the late 1960s. One of the first attempts was due to Simpson *et al.*[21], after it was recognised that the optimal route selection across the North Atlantic could lead to considerable fuel savings. The method required to take into account the actual state of the atmosphere (specifically, winds and temperatures). The analysis required some basic performance data for the aircraft, such as climb and descent programs, cruise altitude, fuel consumption and other parameters. Climate-optimal routing problems have remained topical to this day[22].

Roskam's Advanced Aircraft Analysis (AAA) modules are based on Roskam's books[23]. These performance modules focus on aircraft design, from weight sizing to aerodynamic prediction, control and stability analysis. Another code in this technology area is ACSYNT, whose origins go back to the 1970s. In recent years this code has undergone major development to adapt it to aircraft conceptual design. The code integrates various disciplines, including performance, design, costs, noise and engineering process[24;25]. A flight optimisation system called FLOPS was developed in the 1980s[26] to address detailed performance during preliminary design.

The program DATCOM[27] provides calculations on static stability, high-lift performance, aerodynamic derivatives and trim conditions of the aircraft at subsonic and supersonic speeds. The program has been used extensively for the rapid estimation of the static and dynamic characteristics of high-performance aircraft in the preliminary design stage. The approach followed by this method is accurate enough for several applications[28].

A comprehensive performance simulation that is used for the air-traffic management is the so-called BADA Model, developed at Eurocontrol. BADA uses the lumped-mass approximation, a total energy model for the centre of gravity and a basic performance model for the prediction of the aircraft trajectory[29;30]. Its main application is the prediction of flight trajectories in terminal-area manoeuvre and for a management of traffic at the current conditions and at forecast growth. The basic equations used by BADA are ordinary differential equations for the centre of gravity and the total energy equation (i.e., balance among kinetic energy, potential energy and work done by the engines). Additionally, the model uses integral values of the essential parameters of the aircraft, such as operational limits (from the flight manual or other sources).

The total energy concept has been used[31] to predict the fuel consumption of transport aircraft, without having to rely on statistical databases. This approach has already been used in the 1980s by the FAA to provide a better automatic flight planning for air traffic control. In principle, the method is quite powerful because it does not require many details of the aircraft. In fact, it relies on the fundamental energy balance, on the gross weight and on the path profile of the aircraft.

ESDU[32] provides a suite of programs to calculate the performance of fixed-wing aircraft, including flight performance, airfield performance and mission performance (block fuel for a transport aircraft and radius of action for a military aircraft). The program consists of an implementation of several derivations published as data units, some of which are briefly discussed and cited in this book.

A comprehensive performance program of industry standards is PIANO[33]. This program includes preliminary design options, a large database of aircraft models and a detailed mission-performance analysis module. The program does a wide range of performance calculations and it is shown to match closely the manufacturer's performance, although the technical details are not disclosed.

## 1.7 The Scope of This Book

The subject of this book is the operational performance of aircraft, with all of the issues surrounding technology evolution (aircraft and engine upgrades) as well as the environmental impact. In this context, operational performance falls between conceptual design of aircraft and airport operations, which rely on limited information on the aircraft.

We present relatively advanced cross-disciplinary topics. We aim to solve accurately problems such as the best cruise altitude, the best cruise Mach number, and the best climb and descent procedures. We deal with commercial aircraft powered by either high by-pass turbofan engines or propellers-turboprop. We also discuss high-performance aircraft (mostly for military use) that are capable of supersonic flight. These topics are mapped into a computational framework that consists of a number of comprehensive computer programs. One such program should be able to accomplish at least some of the following tasks:

- trajectory optimisation and route planning
- mission analysis and field performance
- environmental emissions and fuel costs
- airframe–engine integration
- thermo-physics and structural dynamics
- aircraft noise trajectories
- noise impact around airports
- systems analysis
- verification of performance data
- competition analysis
- trade-off and parametric studies

A flowchart of the computer program is shown in Figure 1.2. This chart shows the key aspects of the sub-models of the performance framework, as well as the major items and sub-items of engineering analysis. Each of these items is discussed in the following chapters.

The flowchart for a supersonic high-performance aircraft is similar to Figure 1.2. However, the noise and emissions are not simulated, and there is more

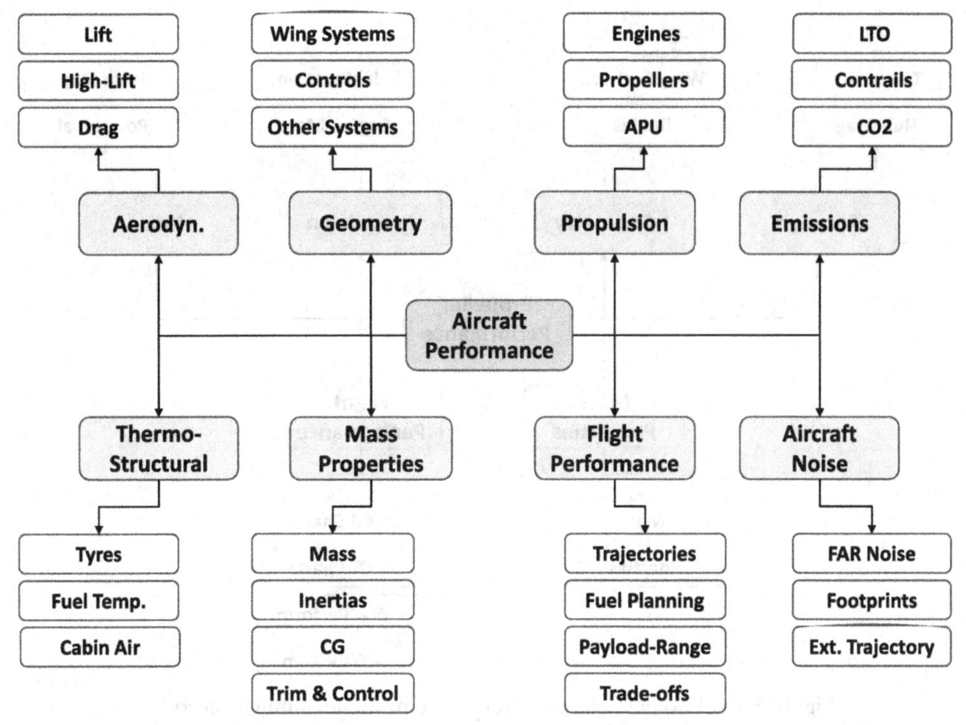


Figure 1.2. Flowchart of a comprehensive flight performance simulation code.

emphasis on the aerodynamics, which must contain models for transonics, supersonics and external stores. The thermo-structural models are limited to the aero-thermodynamic heating at high supersonic speeds. The propeller model does not apply to this type of airplane.

The propeller performance mind-map is shown in Figure 1.3. This model can be fully integrated into the flowchart of the turboprop aircraft. The optimal design block relies on the aerodynamics and geometry module and is not part of routine performance calculations.

The key sub-topics addressed in this book are as follows:

1. Development of an accurate models of airplanes, based on partial information available from the manufacturer. This model is validated against a number of different metrics, including areas, volumes, dimensions and reciprocal positions (Chapter 2).
2. Development of an accurate model for mass, inertias, centre of gravity of the airplane, along with location of the mass of each sub-system (Chapter 3).
3. Development of an accurate and rapid model for aerodynamic performance, including a variety of aerodynamic coefficients and derivatives at all flight conditions (Chapter 4).
4. Development of an accurate performance model for high by-pass turbofan engines, low by-pass turbojet engines for military use and turboprop engines. The models are designed to provide at least 20 different engine parameters over the full flight envelope (Chapter 5).

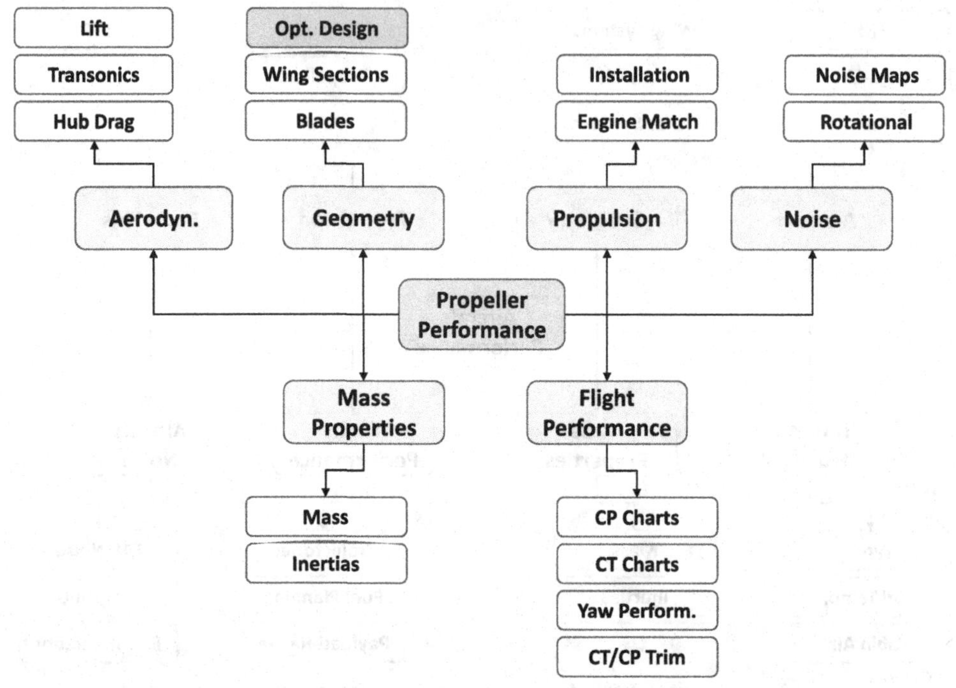


Figure 1.3. Flowchart of a comprehensive propeller simulation code.

5. Development of an accurate propeller model, based on partial information of its geometry, and on a specific aerodynamic model that simulates both axial and yawed flow conditions over the full flight envelope (Chapter 6).
6. Development of a static stability model for both longitudinal and lateral stability, including the determination of the optimal centre of gravity and the control required to establish equilibrium of forces and moments to the aircraft (Chapter 7). This chapter does not deal with unsteady response and short-time period of motion, which is a subject of flight dynamics.

All of the sub-models described previously are integrated into a flight mechanics framework that is capable of simulating all of the flight conditions of a commercial aircraft, as well as some selected conditions of a supersonic aircraft. Thus, there is a separate presentation of ground operations, with focus on take-off (Chapter 9); climb to initial cruise altitude (Chapter 10); descent, final approach and landing (Chapter 11); and optimal and sub-optimal cruise procedures (Chapter 12). Chapter 8 deals with flight envelopes, atmospheric models and some flight performance at supersonic Mach numbers, whilst Chapter 13 deals with turn and other manoeuvres in the horizontal and vertical plane, powered and unpowered, in still air and in downburst situations.

A number of special topics are presented in Chapter 14, which deals with thermo-structural performance. The items discussed include in-flight icing, fuel-temperature simulation, tyre heating and forcing at take-off and landing, as well as jet blast and jet plume diffusion.

All of these models are applied to the simulation of commercial missions in Chapter 15, where we deal with a number of diverse topics, such as fuel planning, optimal flight trajectories, payload-range analysis and other special cases. This chapter also deals with risk analysis in aircraft performance.

The final chapters of the book deal exclusively with environmental performance. Specifically, we have:

1. Development of sufficiently detailed sub-models for aircraft noise sources, for the propulsion system (including engine, propeller and auxiliary power unit [APU]) and the airframe (including landing gear) (Chapter 16).
2. Development of sub-models for noise interference, propagation, ground effect and other wall effects (Chapter 17).
3. Mapping of noise models into a noise-flight-mechanics framework for the prediction of noise signature on the ground for standard and arbitrary flight trajectories (Chapter 18).
4. Finally, discussion of methods for the prediction of the main environmental emissions from commercial aircraft, including carbon-dioxide, landing-and-take-off (LTO) emissions, contrail avoidance strategies and optimal free flights (Chapter 19).

Most computational procedures are described, although some of them require considerable effort to be translated into a computer program.

## 1.8 Comprehensive Programs in This Book

Multi-disciplinary flight code validation in the field of fixed-wing aircraft is not a well-developed discipline; the discipline, in fact, has not even been established. This gap is in sharp contrast with the comprehensive aero-mechanics codes widely used in rotorcraft engineering. Minimum validation requirements are required for configuration, aerodynamic models, propulsion systems, flight mechanics integration and aircraft noise. These major blocks are shown in Figure 1.4. At a deeper level, we may be interested in various aspects of thermo-structural performance, as also shown in Figure 1.2.

Within the configuration block, we address the geometry of the airplane, wetted areas (§ 2.4), surface matching through the analysis of the cross-sectional areas (§ 2.4.6), the mass distribution and the centre of gravity position (§ 3.8.2).

Within the aerodynamics block, we must verify the accuracy of the aerodynamic coefficients, the profile drag, the aerodynamic polars, the transonic effects, and the aerodynamic derivatives. With reference to the latter parameter we need some indirect types of tests, such as the calculation of the control speed in air (§ 7.2).

Within the propulsion block, we need to address at least basic parameters, such as fuel flows, turbine and combustor temperatures, rated thrust and power. If the engine is powered by propellers, a separate set of tests must be carried out on the propeller in isolation and with integrated mechanics (Chapter 6).

At the flight-mechanics level, we need to address the capability of the airplane and compare it with official data published in the FCOM, in the type certificate and in other documents. Typical examples of verification include the determination of the

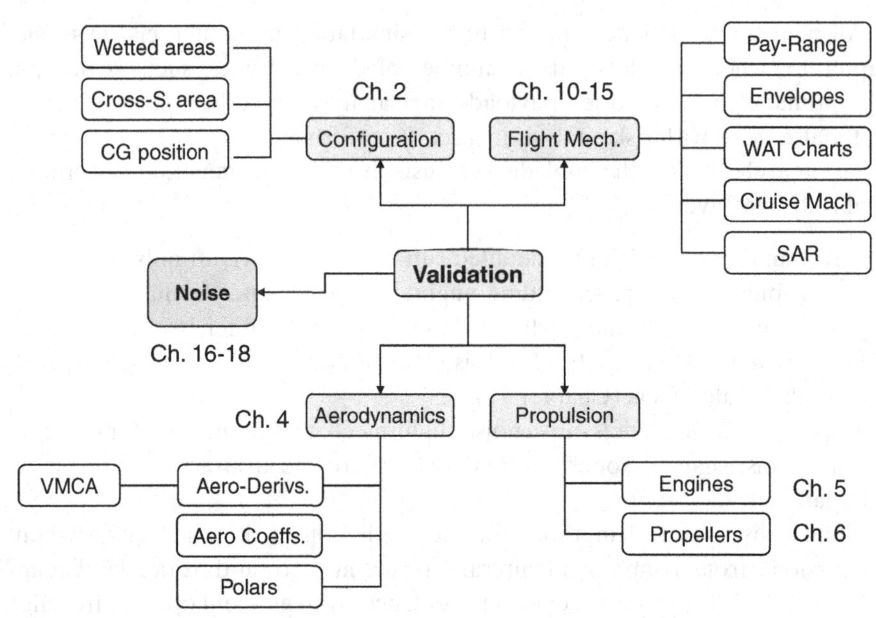


Figure 1.4. Flowchart of a verification and validation strategy for a comprehensive flight performance code.

payload-range chart (§ 15.2), the flight envelope in the Mach-altitude plane (§ 8.6), the weight-altitude-temperature charts for take-off performance, the optimal cruise Mach numbers (§ 12.3) and cruise altitude (§ 12.7), and the specific air range.

Finally, within the aircraft noise block, there are several tests that must be carried out at several levels: noise sources at the component level; noise sources at the integration level; and noise propagation, scattering, reflection, atmospheric effects and other effects arising from the operation of the airplane (Chapters 16–18).

## Bibliography

[1] The Boeing Corporation, 2004.
[2] CIA Factbook. Continuously updated (www.cia.gov).
[3] NTSB. *Flight Recorder Handbook for Aviation Accident Investigation*, Office of Research & Engineering, Washington DC, Dec. 2002.
[4] Bryson AE and Ho YC. *Applied Optimal Control*. Blaisdell, NY, 1969.
[5] Ashley H. *Engineering Analysis of Flight Vehicles*. Addison-Wesley, Reading, MA, 1974.
[6] NASA. Software assurance standard. NASA Technical Standard STD-8739.8, July 2004.
[7] Wagner W. *The History of German Aviation: Kurt Tank, Focke Wulf's Designer and Chief Pilot*. Schiffer Publ. Ltd, 1999.
[8] Sutter JF and Anderson CH. The Boeing model 747. *J. Aircraft*, 4(5):452–456, Sept. 1967.
[9] Lynn Olason M. Performance and economic design of the 747 family of airplanes. *J. Aircraft*, 6(6):520–524, 1969.
[10] Danby T, Garrand WC, Ryle DM, and Sullivan LJ. V/STOL development of the C-130 Hercules. *J. Aircraft*, 1(5):242–252, 1964.
[11] von Kármán T. *The Wind and Beyond: Theodore von Kármán Pioneer in Aviation and Pathfinder in Space*. Little Brown & Co., Boston, 1967.

[12] Dierikx M. *Fokker: A Transatlantic Biography*. Airlife Publishing Ltd, Shrewsbury, UK, 1997. Chapter 2.
[13] Driver H. *The Birth of Military Aviation: Britain 1903–1914*. The Royal Historical Society, 1997. ISBN 0 86193 234 X.
[14] Opolycke LE. *French Aeroplanes Before the Great War*. Schiffer Publishing Ltd, Agden, PA, 1999.
[15] Stevens JH. *The Shape of the Aeroplane*. Hutchinson & Co. Ltd, London, 1953.
[16] Weyl AR. *Fokker: The Creative Years*. Putnam & Sons, London, 1965 (reprint 1987).
[17] Jackson R. *The Guinness Book of Air Warfare*. Guinness, 1993.
[18] Gallagher GL, Higgins LB, Khinoo LA, and Pierce PW. *Naval Test Pilot School Flight Test Manual – Fixed Wing Performance*, volume USNTPS-FTM-No. 108. U.S. Navy Pilot School, Sept. 1992.
[19] Isaacs R. *Differential Games: A Mathematical Theory with Applications to Warfare, Pursuit, Control and Optimisation*. John Wiley & Sons, 1965.
[20] Filippone A. Comprehensive analysis of transport aircraft flight performance. *Progress Aero Sciences*, 44(3):185–197, April 2008.
[21] Simpson L, Bashioum DL, and Carr EE. Automated flight planning over the North Atlantic. *J. Aircraft*, 2(4):337–346, 1965.
[22] Irvine E, Hoskins B, Shine K, Lunnon R, and Froemming C. Characterizing North Atlantic weather patterns for climate-optimal aircraft routing. *Meteorological Applications*, 2012. DOI:10.1002/met1291.
[23] Roskam J. Design for minimum fuselage drag. *J. Aircraft*, 13(8):639–640, 1976.
[24] Jayaram S and Myklebust A. ACSYNT – A standards-based system for parametric computer aided conceptual design of aircraft. AIAA Paper 92-1268, Jan. 1992.
[25] Rivera F and Jayaram S. An object-oriented method for the definition of mission profiles for aircraft design. AIAA Paper 1994-867, Reno, NV, Jan. 1994.
[26] McCullers LA. Aircraft configuration optimization including optimized flight profiles. In *Experiences in Multidisciplinary Analysis and Optimization*, NASA CP-2327, pages 394–412. Jan 1984.
[27] Williams JE and Vukelich SP. The USAF stability and control digital DATCOM. Technical Report AFFDL-TR-79-3032, Vol. I, Air Force Flight Directorate Laboratory, April 1979.
[28] Sooy TJ and Schmidt RZ. Aerodynmic predictions, comparisons and validations using missile DATCOM (97) and Aeroprediction (AP98). *J. Spacecraft & Rockets*, 42(2):257–265, 2005.
[29] Nuic A. *User Manual for the Base of Aircraft Data (BADA) – Revision 3.6*. Eurocontrol Experimental Centre, Bretigny-sur-Orge, France, July 2004. Note 10/04.
[30] Nuic A, Chantal P, Iagaru MG, Gallo E, Navarro FA, and Querejeta C. Advanced aircraft performance modeling for ATM: Enhancements to the BADA model. In *Proceedings of the 24th Digital Avionics Systems Conference*, Washington, DC, 30 Oct. to 3 Nov. 2005.
[31] Collins BP. Estimation of aircraft fuel consumption. *J. Aircraft*, 19(11):969–975, Nov. 1982.
[32] ESDU. *Aircraft Performance Program. Part 1: Introduction to the Computer Programs for Aircraft Performance Evaluation*. 00031. ESDU International, London, Nov. 2006.
[33] Simos D. PIANO: A tool for preliminary design, competitor evaluation, performance analysis. In *ICAO Committee on Aviation Environmental Protection Working Group 2 Meeting*, Rome, Italy, May 2006.

# 2 Aircraft Models

## Overview

In this chapter, we present methods for the determination of the geometrical configuration of a modern transport aircraft (§ 2.1). We use the concept of control points and geometric wire-frames (§ 2.2). Thus, we establish a framework for the determination of linear dimensions, position of centroids, reference areas and volumes. The methods shows are valid for any other type of aircraft. We use a stochastic approach for more complex surfaces (§ 2.2.1). Section 2.3 deals with the geometry of the lifting surfaces. We then show a method for the determination of the wetted areas (§ 2.4), aircraft volumes (§ 2.5), including the wing tanks, and the mean aerodynamic chord (§ 2.6). We produce a few examples of verification of aircraft configurations (§ 2.7). Finally, we define some reference systems on the aircraft (§ 2.8).

**KEY CONCEPTS:** Transport Aircraft, Geometry Definition, Wetted Areas, Aircraft Volumes, Fuel Tanks, Mean Aerodynamic Chord, Reference Systems.

## 2.1 Model for Transport Aircraft

The aircraft geometry represents the basis of all of the aerodynamic calculations, fuel and payload capacity, centre of gravity position, stability and control requirements. The construction of an accurate aircraft model is done in a number of steps that require the choice of reference systems, the determination of the key sub-systems, and an assembly procedure to determine the geometric relationships between parts of the vehicle. Various reference systems are introduced to define geometrical relationships and operational conditions in flight. The procedures shown in this chapter are required to calculate reference areas, reference lengths and reference positions for a wide variety of calculations.

The aircraft is separated into clearly identified components, as shown in Figure 2.1. There are the fuselage and its sub-systems, the wing and its sub-systems, the tail surfaces and their sub-systems, the landing gear and its sub-systems, the propulsion system and the auxiliary units, the landing gear groups and the wheels. About two dozen different components are identified and then reconstructed. These sub-systems are mapped into the comprehensive flowchart in Figure 2.2: the aircraft

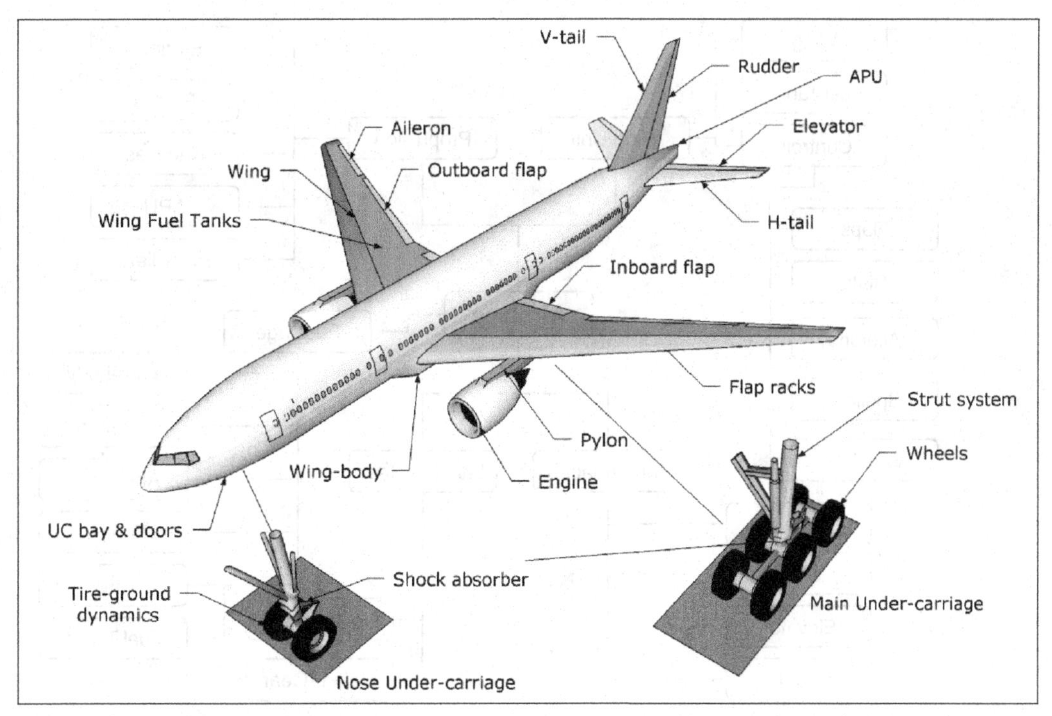


Figure 2.1. CAD model showing assembly of aircraft systems, from Ref.[1].

is decomposed into five major systems; each sub-system is decomposed into several major sub-systems.

Each component must be considered in order to have a sufficiently accurate representation of the aircraft. The components are assigned an identity card, a sequence of control parameters, and a set of rules to manipulate the control parameters. The reconstruction of the aircraft is done through the information read from the card and the reconstruction algorithms, as shown in Figure 2.3. The model is a function of the state parameters $\boldsymbol{x} = \{x_1, x_2, \cdots x_n\}$, some of which are time-dependent.

The state parameters are divided into separate categories:

- Geometry: control points, shape parameters, angle settings
- Assembly: sub-systems and complete model
- Systems: power-plant, fuel systems, APU, landing gears

**CONTROL POINTS.** The whole procedure is based on a database of control points. Each card corresponds to a component *and* a view (top, side and front). For example, in this model, there is a `wing-top`, `wing-side`, and `wing-front` card. This method can be generalised further to include the determination of the geometrical configuration via a recognised computer-aided design (CAD) standard, such as DXF and IGES. An example of a control-points file is reported in § A.0.1. The use of CAD formats, whilst useful in some instances for illustration purposes (for example, Figure 2.1), has limited value in this case.

Reliance on the few data published by the manufacturers (for example, wing area, wing span) is not sufficient. Most of the aircraft drawings in the open domain are for illustration purposes only; by themselves, they offer little accuracy. Further

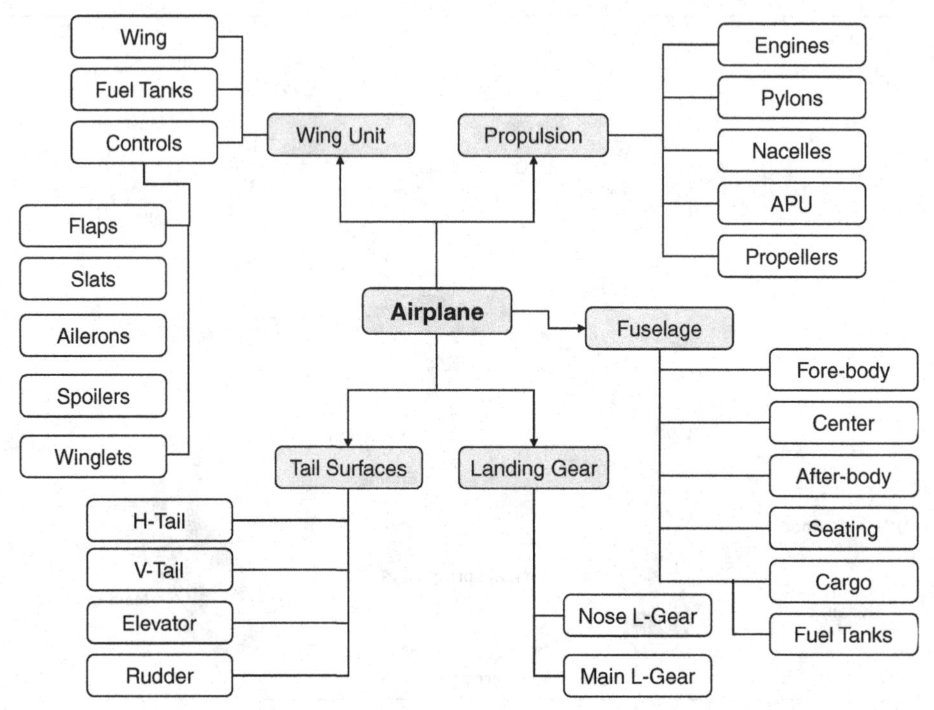


Figure 2.2. Flowchart of aircraft systems, mapped from Figure 2.1.

checks are required to validate all of the data extracted. For example, the longitudinal position of the control points is derived by using as a reference distance the fuselage length or the overall aircraft length. On this scale, connecting two points having a specified distance (for example, nose-to-engine-face) sometimes leads to a distance that does not match the value specified by the manufacturer.

**REFERENCE SYSTEM.** It is necessary to define one reference system to calculate the positions of each element. These reference systems will also be useful to calculate the mass distribution, the position of the centre of gravity and the moments of inertias (Chapter 3). All of the longitudinal positions ($x$) are calculated with respect

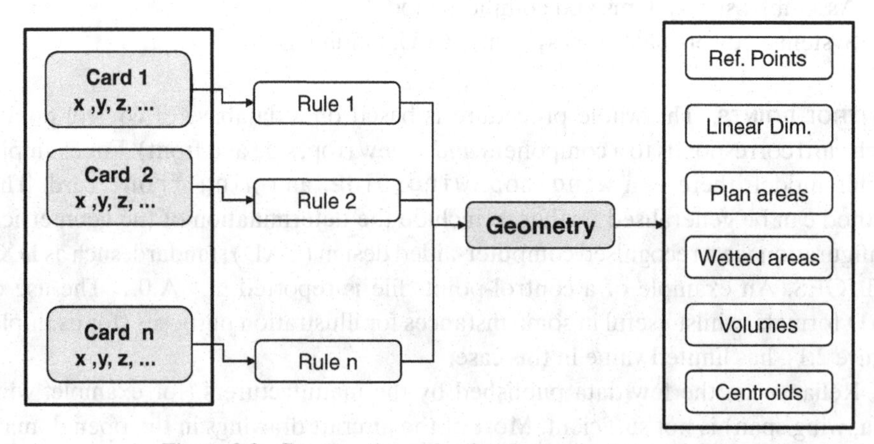


Figure 2.3. Construction rules for the geometry model.

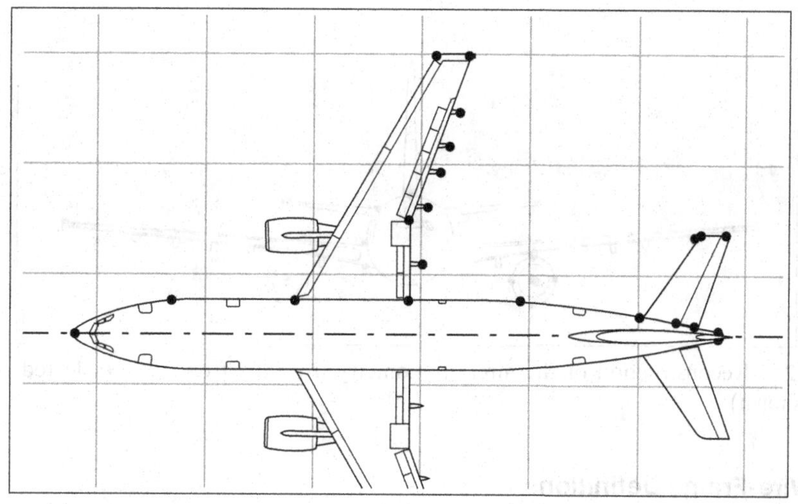

Figure 2.4. Reconstruction of the aircraft geometry from the top view (selected control points shown).

to the fuselage nose. All of the lateral positions ($y$) are calculated with respect to the vertical symmetry of the aircraft. The vertical positions ($z$) are calculated either with respect to the nose or to the ground, although the latter choice should account for the possibility there is a small pitch-down configuration when the aircraft is on the ground. The centre of this system has coordinates $O = \{0, 0, 0\}$; $O'$ is the nose point.

Data provided by the manufacturers often include the range of some key dimensions. Therefore, it is possible to find data that refer to an empty aircraft and an aircraft at maximum ramp weight; in the latter case, it is possible to verify that some quantities depend on the position of the centre of gravity.

Starting from the top view of the aircraft (Figure 2.4), each of the visible subsystems is defined by a sequence of points. The same is done from the side and front views, Figures 2.5 and 2.6. The accuracy can be improved by zooming in to the items of interest. The error increases as a component gets smaller, unless we are able to zoom in to the area of interest.

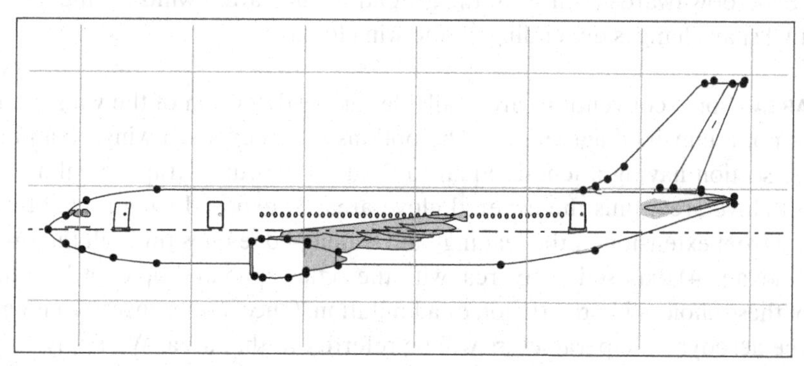

Figure 2.5. Reconstruction of the aircraft geometry from the side view (selected control points shown).

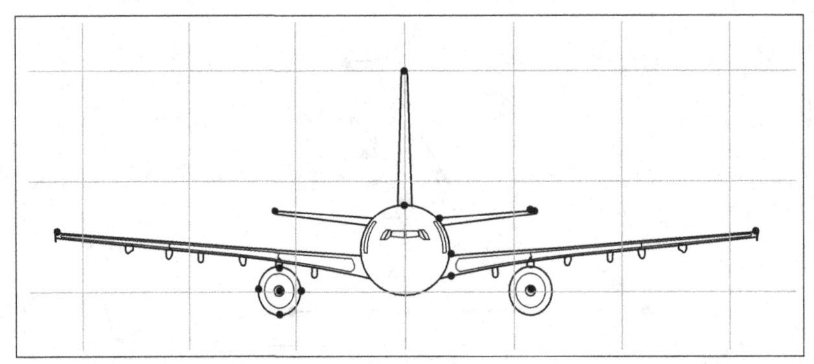

Figure 2.6. Reconstruction of the aircraft geometry from the front view (selected control points shown).

## 2.2 Wire-Frame Definitions

The determination of the wire-frames of each aircraft component is done through the rules shown in Figure 2.3, which allow the reconstruction of planforms and other well-recognised geometrical shapes.

**WING PLANFORM.** A top view of the wing is defined through a set of control points required to reconstruct the planform, the nominal span, the chord at the root and the tip, the chord distribution and the sweep angles. A frontal view is used to reconstruct the dihedral and the nominal height of the tip above the ground. A side view is not useful for deriving any wing parameter, due to the presence of the nacelles and the pylons.

The wing thickness and the twist distribution cannot be extracted from the drawings; these parameters represent a rough estimate. For the thickness distribution, we may have to refer to statistical data for known and *similar* aircraft. For the twist distribution, we can use aerodynamic analysis to derive an *optimum* planform geometry that yields minimum induced drag (Chapter 4).

Figure 2.7 shows the wire-frame of the wing of the Boeing B737-900W (with winglets). The dashed lines denote the approximate projection of the winglets onto the horizontal plane. The wing span is better defined as "over-the-winglet", that is, with the outermost point of the wing at nominal configuration. A loaded wing would bend downwards (whilst on the ground) or upwards (whilst in flight); hence the actual span changes depending on the wing loading.

**WING AREA.** Some conventions are available for the definition of the wing area, but there is not a universal agreement. The options are 1) exposed wing area plus the fuselage section having a length equal to the root chord; 2) trapezoidal area plus exposed glove area, plus the covered glove area; 3) exposed wing area, plus area resulting from extension of the leading- and trailing-edge lines through the fuselage centreline; and 4) exposed wing area, with the actual span and tip chord. Note that none of these choices is a restriction or a limitation. Once a reference area is agreed, all of the aerodynamic parameters will be referred to that area. We apply only the first definition.

The wing span of very large aircraft is not uniquely identified, especially if the wing is equipped with winglets. For example, at maximum fuel load, the wing span

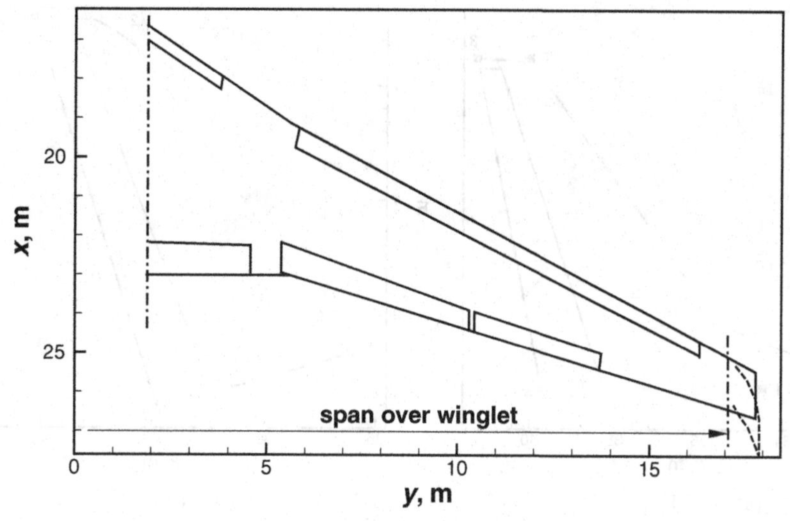


Figure 2.7. Wire-frame of the Boeing B737-900W wing.

of the Boeing B747-400 over the winglets is ~0.5 metres longer than the nominal value *jig position*. Therefore, a decision is needed to define uniquely the"wing span", something that is not possible to do by inspecting line drawings. The wing area is calculated with the methods explained in § 2.2.1 and is corrected for *dihedral effects* with the following formula:

$$A = \frac{A_{xy}}{\cos\varphi} \tag{2.1}$$

where $\varphi$ is the dihedral angle and $A_{xy}$ is the projection of the planform on the horizontal plane.

**TAIL SURFACES.** The tail surfaces (horizontal and vertical stabilisers) are determined with the same method as the wing. Figure 2.8a shows the wire-frame of the vertical tail of a model Boeing B737-900, including the rudder. The full geometry is given by the solid line. The dashed lines are the *simplified* wire-frame, with numbered control points. The wetted area and the centroid are calculated for the full geometry. The leading- and trailing-edge sweep angles, as well as the chord lengths and the mean aerodynamic chord, are calculated on the wire-frame.

Figure 2.8b shows the wire-frame of the horizontal tail of the Boeing B737-900, including the elevator. As in the case of the vertical tail, the rounding of some corners makes the definition of some key parameters (tip chord, leading edge sweep, mean aerodynamic chord) somewhat difficult. In both cases, the leading-edge sweep is calculated from the control points 1 and 2; the trailing-edge sweep is calculated from the control points 3 and 4.

### 2.2.1 Stochastic Method for Reference Areas

An essential aspect of our model is the determination of a general method for calculating surface areas, position of reference points (leading-edges, centroids), equivalent diameters and form factors. If the shape of the surface is trapezoidal, the calculation of the geometry is straightforward. If there are five or six control points, a

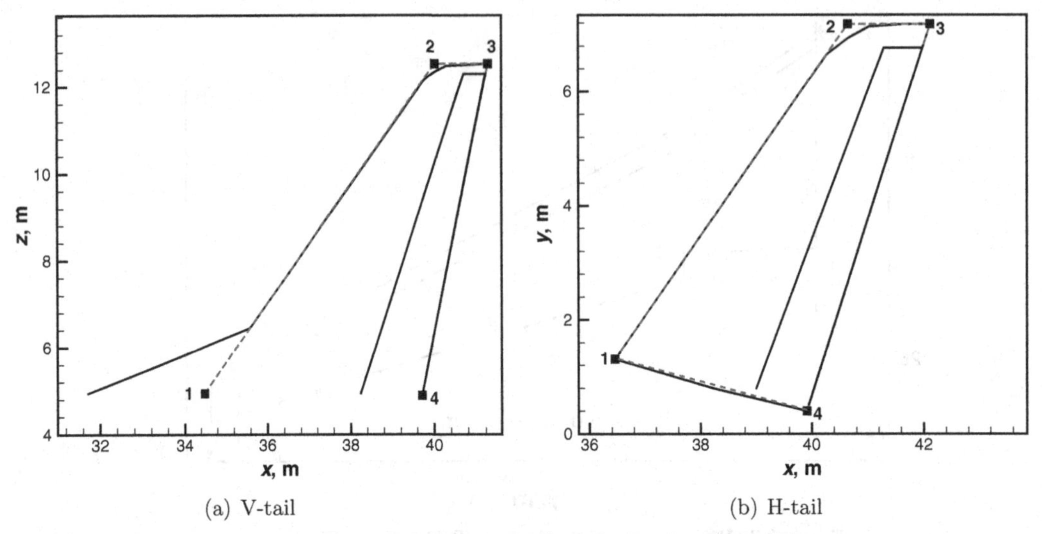


(a) V-tail (b) H-tail

Figure 2.8. Vertical tail of the Boeing B737-900.

convention is required to specify what these points denote. If the geometry is defined by several points, the calculation of the area is done with a stochastic method. The idea is explained in the following procedure:

1. Select the $n$ points $x$, $y$ that define the shape in a generic plane.
2. Define a closed curve with points $x$, $y$, with $x_1 = x_n$, $y_1 = y_n$.
3. Calculate the bounding box $[dx, dy]$ and the bounding-box area $A_{bbox}$.
4. Generate a random point $x_r$, $y_r$ within the bounding box.
5. Check whether the random point $x_r$, $y_r$ is inside or outside the curve.
6. Update the counter $k_i$ for points inside the curve.
7. After $k$ shots, the current area $A_k$ is defined by:

$$A_k = \frac{k_i}{k} A_{bbox}. \tag{2.2}$$

8. Check the convergence by using the criterion:

$$\frac{A_k - A_{k-1}}{A_{k-1}} < \epsilon, \tag{2.3}$$

   where $\epsilon$ is a tolerance that defines the relative change in area, typically $\epsilon = 10^{-4}$.
9. If the convergence is not satisfied, continue from point 4.

The centroids are calculated from the average of the coordinates of internal points. This method is particularly useful for the calculation of the fuselage nose area, cross-sectional area of non-circular fuselages and the area of a wing section. However, it is also useful for more complex wing shapes (including the filleting at the leading and trailing edge). A typical result of this numerical model is shown in Figure 2.9. This example is trivial because the wing is defined by only five points, and the method would not be used in this case. A few thousand shots are required to get the area within the specified tolerance, but the procedure will always converge. The key aspect of this procedure is the algorithm that decides whether a point falls inside or outside the contour.

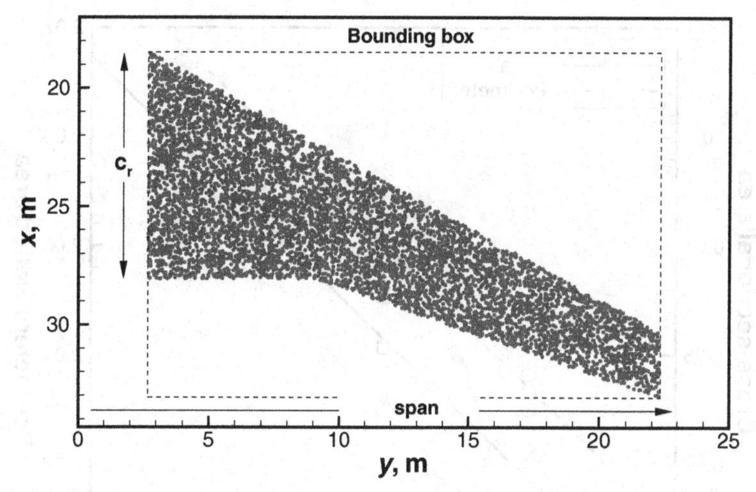


Figure 2.9. Stochastic calculation of wing area, aircraft shown in Figure 2.4.

## 2.3 Wing Sections

The exact geometry of the wing section of a modern transport aircraft is rarely known. However, we know that in most cases these sections are supercritical airfoils. If the spanwise thickness distribution $t/c$ is available, then it is possible to produce a model that is sufficiently accurate for performance calculations. Table 2.1 is a summary of geometrical characteristics of selected supercritical wing sections. The airfoils "SC" are NASA's second-generation profiles[2]. The perimeter and the cross-sectional areas are plotted in Figure 2.10. A linear fit of the tabulated data is the following:

$$A_c = 4.128 \cdot 10^{-3} + 6.452 \cdot 10^{-1}\,(t/c), \qquad p_c = 1.971 + 0.659\,(t/c) \qquad (2.4)$$

where $A_c$ and $p_c$ are the cross-sectional area and the perimeter for a unit chord, respectively. Note that these equations are only valid in the thickness range reported in Table 2.1.

Table 2.1. *Cross-sectional areas of selected supercritical wing sections*

| Airfoil | $t/c$ | Area | Perimeter | Notes |
|---|---|---|---|---|
| SC(2)-010 | 0.10 | 0.071047 | 2.03498 | symmetric |
| SC(2)-012 | 0.12 | 0.085258 | 2.04624 | symmetric |
| SC(2)-410 | 0.10 | 0.067538 | 2.03661 | |
| SC(2)-414 | 0.14 | 0.094633 | 2.06187 | |
| SC(2)-610 | 0.10 | 0.067510 | 2.03819 | |
| SC(2)-612 | 0.12 | 0.080860 | 2.05056 | |
| SC(2)-614 | 0.14 | 0.093931 | 2.06373 | |
| SC(2)-710 | 0.10 | 0.067424 | 2.03937 | |
| SC(2)-712 | 0.12 | 0.080638 | 2.05171 | |
| SC(2)-714 | 0.14 | 0.093730 | 2.06561 | |
| SC-1094-R8 | 0.094 | 0.066044 | | rotorcraft airfoil |
| SC-1095 | 0.095 | 0.065683 | | rotorcraft airfoil |

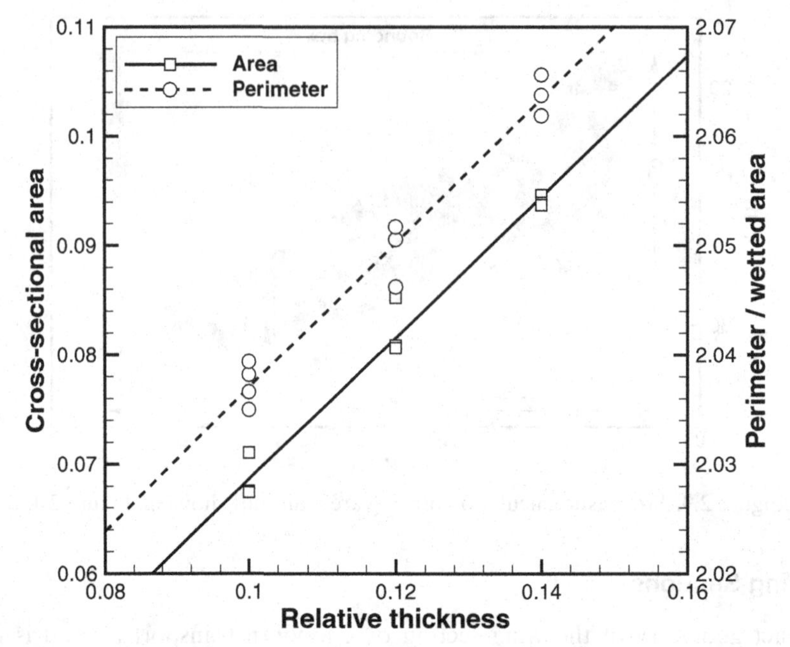


Figure 2.10. Geometric characteristics of some NASA supercritical wing sections.

Because the section characteristics reported in Table 2.1 refer to a unit chord, they must be rescaled to the actual chord:

$$A_c = A_{c_1} c^2, \tag{2.5}$$

where $A_{c_1}$ is the cross-sectional area of a wing section having unit chord. The analytical regressions show that the dominant parameter is the wing thickness. Therefore, in absence of detailed data, we can use Equation 2.4 to perform various geometry calculations, including wetted areas and volumes. Symmetric wing sections are to model the thickness horizontal and vertical stabilisers.

## 2.4 Wetted Areas

The wetted areas of the aircraft and its components are required for a wide range of aerodynamic and performance calculations. For the aircraft at landing or take-off configuration we need to account for the high-lift surfaces and parts of the landing gear. We consider the main elements separately.

### 2.4.1 Lifting Surfaces

The calculation of the wetted areas of all lifting surfaces require the planform (as previously calculated) and some additional information, which can have the form of average wing thickness, dihedral or anhedral angle. An alternative method consists in using the thickness distribution $t/c$ (for example, Equation 2.4), along with the spanwise chord distribution $c(y)$. The wetted area of a half-wing is then

$$A_{wet} = \int_o^{b/2} p_c(y) dy. \tag{2.6}$$

where $p_c(y)$ is the perimeter at spanwise position $y$.

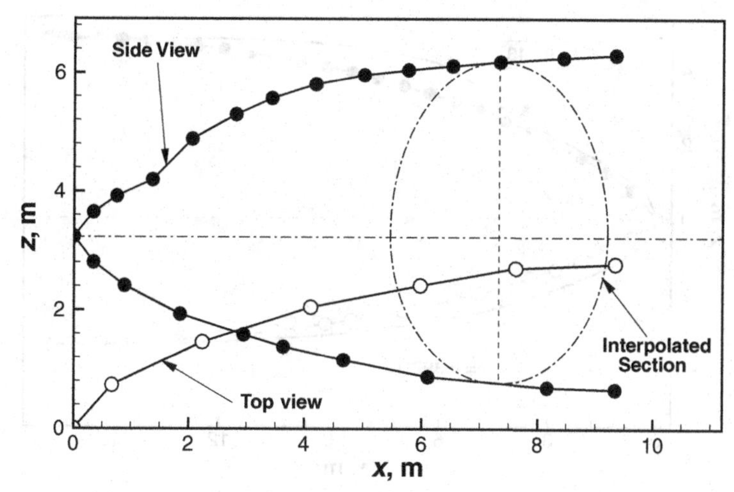


Figure 2.11. Control points for reconstruction of the fuselage forebody of the Airbus A300-600.

This equation is solved numerically, after discretisation of the wing in longitudinal strips. The area is then corrected for the effect of dihedral or anhedral (Equation 2.1).

### 2.4.2 Fuselage

For convenience, the fuselage is sectioned in three parts: the forebody (*nose*), the central section and the aft-body (*tail*), as illustrated in Figure 2.5. The central section is cylindrical. However, we must also account for the wing-body junction. The wetted area of the fuselage is then calculated from

$$A_{wet} = A_{wet_{nose}} + A_{wet_{centre}} + A_{wet_{tail}} + A_{wet_{blend}} - A_{wet_{wb}}. \tag{2.7}$$

We analyse separately the contributions in Equation 2.7.

**FUSELAGE FOREBODY.** The forebody is defined by a set of control points on the vertical symmetry plane and by a set of control points in the horizontal plane through the nose point. The procedure for interpolation consists in calculating the height $dz$ and the width of the airframe $dy$ at a given longitudinal position $x$. The cross-sectional shape of the airframe can be approximated by two half-ellipses (top and bottom half), as shown in Figure 2.11. This operation requires finding the centre of the ellipse and the vertical semi-axes $b_1$ and $b_2$, with $b_1 + b_2 = dz$. If the cross-section is not circular, it is possible to rescale the maximum cross-section along the longitudinal axis.

In any case, unless the cross-section of the fuselage is circular, this type of interpolation is not very accurate; in fact, there is the possibility that the forebody and the central sections show a discontinuity in the cross-sectional area distribution, as discussed further in § 2.4.6.

**SEMI-ANALYTICAL METHODS.** There is a simpler approach to the calculation of the wetted areas – one that does not require numerical integrations. The cross-sectional shape in the vertical plane is not too dissimilar from the fore section of a Sears-Haack

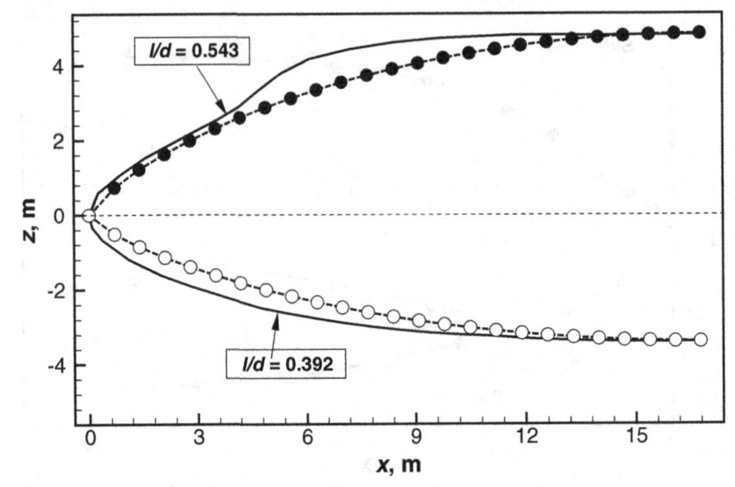


Figure 2.12. Forebody of the B747-400 fuselage, contour approximation through Sears-Haack bodies.

body. From the aerodynamic theory, we know that this is a body of revolution of minimum wave drag for given volume and length. The body is pointed at both ends, and therefore the comparison must be done between a fuselage forebody of length $l_{nose}$ and a Sears-Haack body of length $l = 2l_{nose}$. It can be shown that the wetted area of the fore section of the Sears-Haack body is

$$A_{wet} = 0.715\pi l_{nose} d. \tag{2.8}$$

We will assume that the actual wetted area is

$$A_{wet_{nose}} \simeq 0.72\pi \, l_{nose} \, d. \tag{2.9}$$

The numerical method shown in Figure 2.11 for the Airbus A300-600 yields $A =$ 123.5 m$^2$, whilst the Sears-Haack formula, Equation 2.9, yields $A = 119.1$ m$^2$.

The use of the Sears-Haack formula is not accurate for the Boeing B747-400, as it is shown in Figure 2.12. Two Sears-Haack bodies have been used from the nose point $O'$; the bottom one has a slenderness $l/d = 0.392$ and the top one has a slenderness $l/d = 0.543$.

**FUSELAGE AFT-BODY.** The construction of the aft-body (*tail*) is done with the same principle used for the forebody. In this case there is no clear waterline to follow. It will suffice to do the interpolation with two semi-ellipses whose axes are the local width and height of the airframe, as shown in Figure 2.13. Sometimes the cross-section cannot be approximated by elliptical shapes. For example, if the aircraft has a rear cargo door, it is likely that the shape of the tail cone is flat at the bottom. Thus, more information must be extracted from the drawings to create suitable cross-sectional shapes that make up the wire-frame of the rear fuselage.

**FUSELAGE CENTRAL SECTION.** In most cases the fuselage cross-section is cylindrical. Thus, the wet area is calculated from

$$A_{wet_{centre}} = \pi \, l_{centre} \, \overline{d}, \tag{2.10}$$

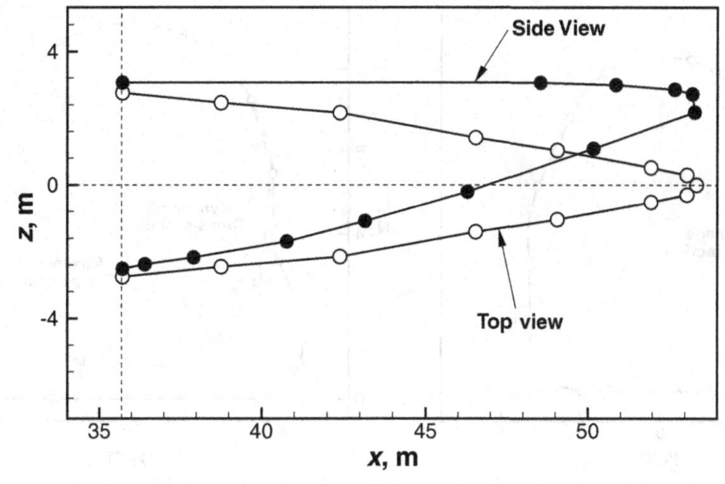


Figure 2.13. Control points for reconstruction of the fuselage after-body.

where $\overline{d}$ is an average diameter. Additional control points may be required to derive the complete shape, the perimeter, the cross-sectional area and the equivalent diameter. In some cases, it is sufficient to take as an average diameter the quantity $\overline{d}$ = height/width. We have verified that by using the average diameter on the Boeing B747-400, the error on the circumference is less than 0.4%. In fact, the double-decker section 5.50 m aft of the nose has a circumference of 19.54 m; by using the average diameter, the circumference is 19.60 m.

**WING-BODY BLENDING.** The major difficulty is in the calculation of the fuselage-wing intersection. We need to select sufficient control points from the drawings to define interpolating functions. For smaller aircraft, the intersection is defined by the wing section at the root. For aircraft such as the Airbus A320-200, the intersection can be calculated from the usual drawings and numerical interpolation. The area enclosed by the intersection will have to be mapped onto the central portion of the fuselage, which is cylindrical. For many aircraft the wing box crosses through the lower fuselage (notably, the B747 and the A380) and requires complex three-dimensional filleting.

We define some blending functions between the fuselage cross-section and the maximum cross-section. Call $l_p$ the cross-sectional perimeter and $l_{p_{max}}$ the maximum cross-sectional perimeter. If $l_{wb}$ is the longitudinal extension of the wing-body intersection, the perimeter function within this length is defined by

$$p(x) = (l_{p_{max}} - l_p) \sin\left(\frac{\pi x}{l_{wb}}\right), \tag{2.11}$$

where the coordinate $x$ is calculated from the most forward position of the surface blending. The sinusoidal function is arbitrary, but it serves well for this purpose. The additional wetted area due to the wing-body blending is calculated by a numerical solution of the integral

$$A_{wet_{blend}} = (l_{p_{max}} - l_p) \int_o^{l_{wb}} \sin\left(\frac{\pi x}{l_{wb}}\right) dx. \tag{2.12}$$

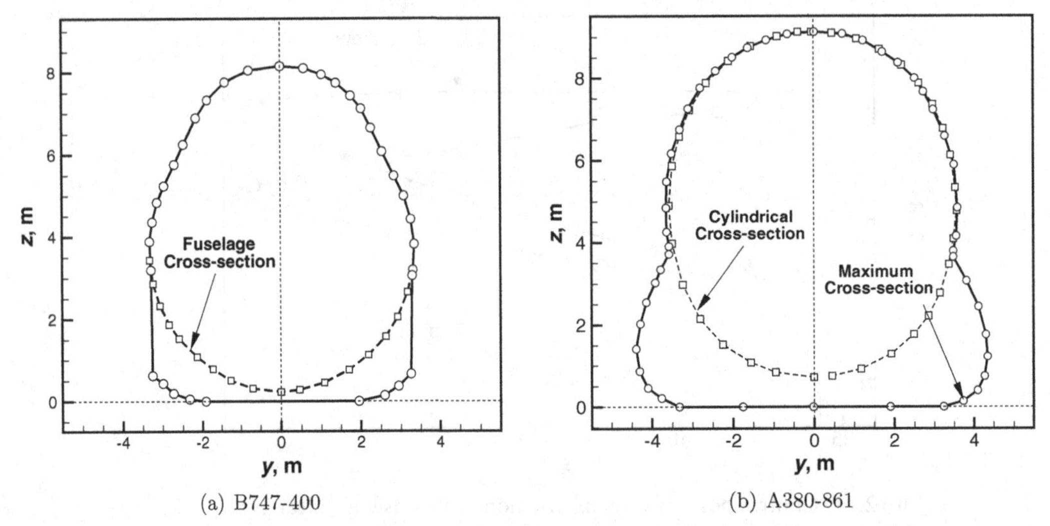


(a) B747-400 (b) A380-861

Figure 2.14. Wing-body intersection of very large aircraft (same scale).

For the B747, a further sectioning between centre-forward and centre-aft sections is required, along with a blending function for the two sections.

**WING-BODY INTERSECTION.** Finally, we need to subtract from the fuselage wetted area the contribution due to the wing-fuselage junction; see, for example, Figure 2.14. A first-order approximation is derived by subtracting the cross-sectional area of the wing root, $A_{wet_{wb}} = A_{c_{root}}$. To be more accurate, we need to map the wing root onto the fuselage. This is done numerically, by using the information from the control points.

### 2.4.3 Nacelles and Pylons

The engines are mounted on the aircraft with pylons and nacelles. The pylons are the structural connections between the nacelles and the wing. The front diameter, the maximum diameter, the aft diameter and the length of the nacelles can be derived either from line drawings or from manufacturer's data. With these data it is possible to calculate the approximate surface area of a body of revolution. Torenbeek[3] provides formulas for a more accurate calculation, requiring about 10 geometric parameters. A better accuracy is required because the wetted area of the engines can exceed 5%. Our approach is to use a set of control points from three-view drawings as shown in Figure 2.15, which has been elaborated from Airbus[†].

If the nacelle can be approximated by a body of revolution, the wetted area is estimated from

$$A_{wet_{nac}} = \pi A_{wet_{side}} - A_{inlet} - A_{outlet}. \tag{2.13}$$

[†] Airbus: Aircraft A300-600 Characteristics for Airport Planning, Blagnac, France, 1990.

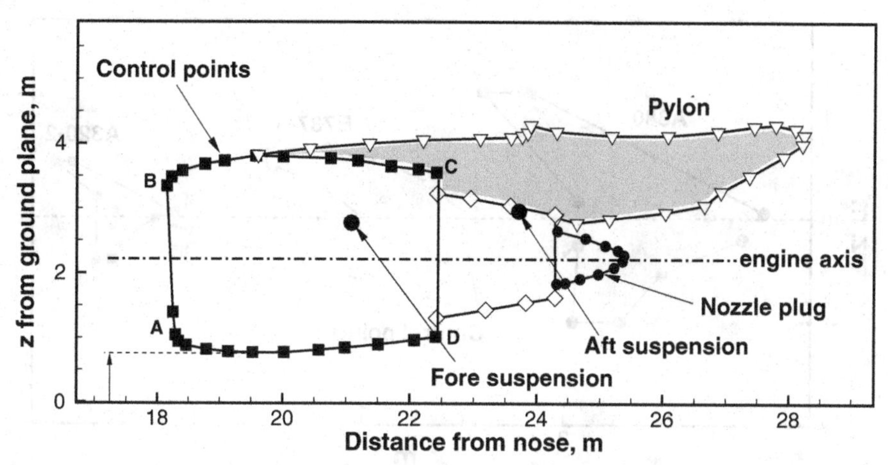


Figure 2.15. Control points for the engine–pylon assembly for the Airbus A300-600 aircraft with CF6 engines.

If the nacelle is not a body of revolution, as often is the case, from the side-view wetted area, we define an average diameter

$$\overline{d} = \frac{A_{wet_{side}}}{l_{nac}}. \tag{2.14}$$

Then the first term of Equation 2.13 becomes $\pi \overline{d}\, l_{nac}$. The wetted area of the pylon is calculated by multiplying the wetted area from a side view by ~2.1. The inlet and outlet areas are calculated from the respective diameters, which are derived from points A–B and C–D in Figure 2.15.

### 2.4.4 Winglets

A conventional winglet is mounted at the tip of the wing pointing upward. In some cases there can be a lower winglet. The essential parameters are the height from the base, the base chord, the tip chord and the cant angle. Simple geometric formulas are used to calculate the reference areas and the wetted areas. One detail worthy of note is that the canted winglet increases the wing span. For cases where a winglet does exist, we can define two wing spans: a nominal span to the root of the winglet and an effective span "over-winglet", that is, reaching the tip of the winglet. The wing loading causes the upper winglet to rotate outward, thus increasing the effective span. Figure 2.16 shows the side view of winglets from Airbus and Boeing aircraft. The graph shows the control points used for reconstructing the geometry. In the case of the B737 we also need control points from a frontal view.

Until recently, the development of the winglets was done as a retrofit technology (Boeing B737-300, B737-900, B757-500, B767-300ER). The modern trend is toward blended winglets (Gulfstream II; Boeing B787), which are better modelled by CAD systems. Manufacturers claim that the improved aerodynamics of these winglets can reduce fuel consumption at cruise by as much as 5%.

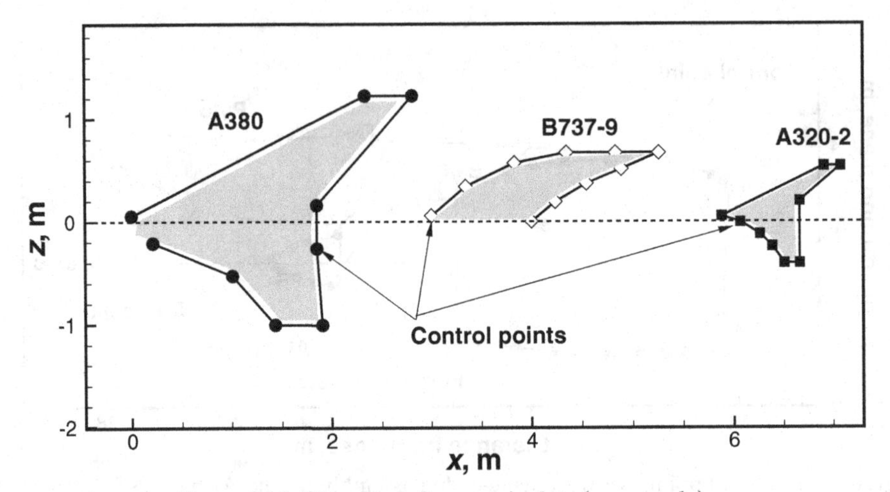


Figure 2.16. Side view of some winglets (same scale).

### 2.4.5 Flaps, Slats and Other Control Surfaces

The shape, planform area and wetted area of these devices are required when the aircraft is at take-off and landing. The exact shape of these surfaces cannot be inferred from the line drawings because the aircraft is shown with its control surfaces retracted. However, it is possible to calculate the span of each device (inboard and outboard flaps, slats, ailerons, spoilers, elevators). For the remaining quantities, we need to make assumptions or extract information from photographs. For example, the chord of a Kruger slat cannot be determined unless the slat is deployed; likewise, the chord distribution of the flaps cannot be determined from sources other than digital photographs of aircraft at landing or take-off.

**FLAP RACKS.** The flap racks contain the mechanical linkages that operate the flaps. Figure 2.5 shows five racks per side; the racks may not have the same dimension, although they are shaped like long and narrow keels partially exposed at the trailing edge. The wetted area of each of these racks is found by calculating the projected area on the side view multiplied by the perimeter of the maximum cross-sectional area. If some geometrical elements are missing, some further approximations are required. For example, if the shape of the side view cannot be extracted, then the wetted area of each rack is assumed to be

$$A_{rack} \simeq \frac{2}{3} p \, l_{rack}, \tag{2.15}$$

where $p$ is the perimeter of the maximum cross-sectional area and $l_{rack}$ is the overall length. The junction between the top of the racks and the bottom of the wing is subtracted from the total wetted area.

### 2.4.6 Model Verification: Cross-Sectional Area

The distribution of cross-sectional area is an indication of whether some of these interpolations have been done correctly. As mentioned, only the maximum cross-sectional area can be extracted, and the cross-sectional shapes at the nose and tail are

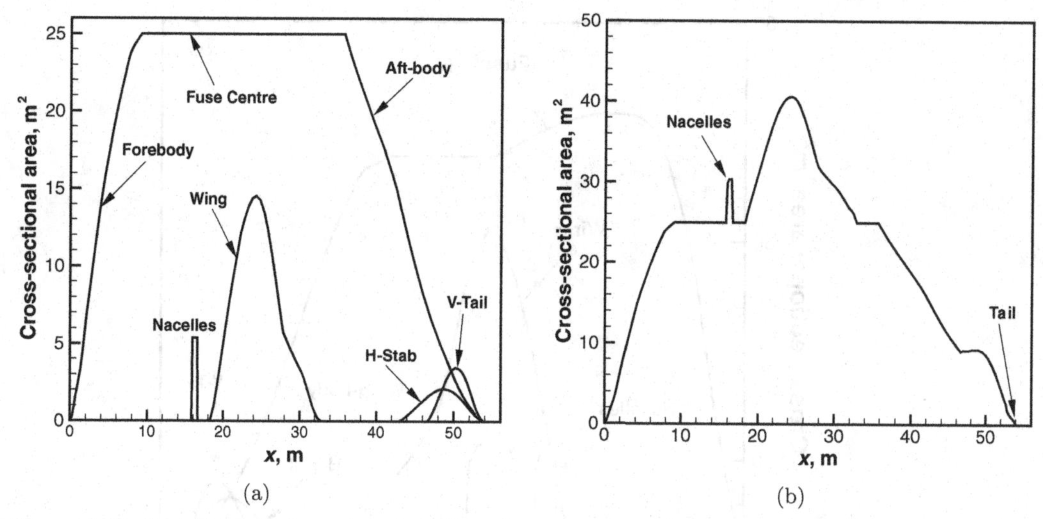


Figure 2.17. Distribution of cross-sectional area by component (left) and cumulative (right) for the Airbus A300-600.

approximated by ellipses. The main consequence of this approximation is that the cross-sectional area distribution shows a discontinuity at two points on the fuselage. To avoid this inaccuracy, the cross-sectional areas at the nose and the tail are adjusted iteratively. If $A_{cs}$ is the cross-sectional area at the end of the nose and $A_{cs_o}$ is the cross-sectional area of the fuselage, the correction to be introduced is

$$A_{cs} = A_{cs_o} k_1, \tag{2.16}$$

where $k_1$ is the ratio between the two areas. Blending functions can be used to scale the cross-sectional areas of the nose and the tail. The purpose of these blending functions is to generate smooth variations in cross-sectional areas from the wire-frame data given on the vertical and horizontal planes (data such as those shown in Figure 2.13).

Figure 2.17 shows the cross-sectional area of the fuselage and other major components of the Airbus A300-600 model. The fuselage area distribution is continuous, as specified (graph 2.17a); the sum of all cross-sectional areas is shown in graph 2.17b. Upon convergence, also the wetted areas of the forebody and the after-body are corrected.

A similar analysis was carried out for the Boeing B747-400 aircraft model, which has a more complex fuselage shape, Figure 2.18. This is due to the double-deck architecture of the forebody, which is then blended with a single-deck aft fuselage.

## 2.5 Aircraft Volumes

The calculation of the aircraft's volume or the volume of some of its systems is required for operational reasons (cargo and fuel capability), or for the estimation of the wave drag at supersonic speeds. Cargo volumes are generally available from the manufacturer and require no special treatment. To calculate the total volume of the

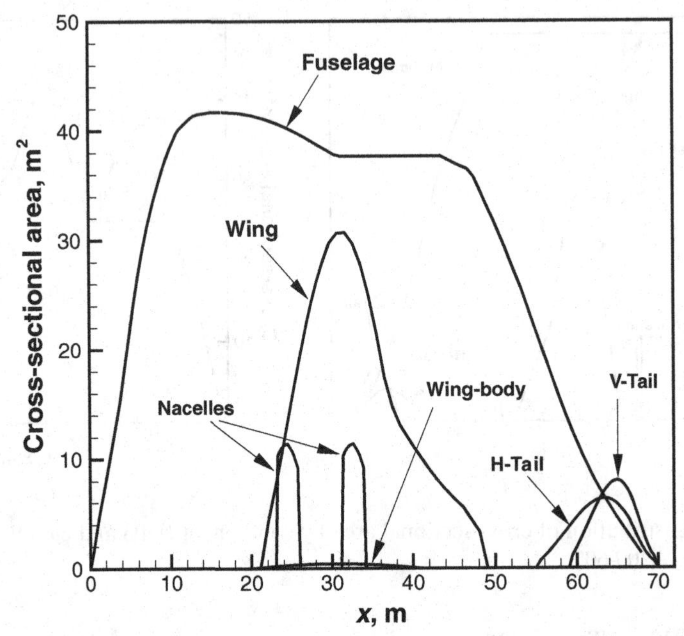


Figure 2.18. Distribution of cross-sectional area for the Boeing B747-400.

aircraft, we use the method of components:

$$\mathcal{V} = \mathcal{V}_{fuse} + \mathcal{V}_{wing} + \mathcal{V}_{ht} + \mathcal{V}_{vt} + \mathcal{V}_{eng} + \cdots \ldots \tag{2.17}$$

The items not listed in Equation 2.17 are minor contributions. The fuselage is split into nose, centre and tail section, as previously done in the wetted-area analysis (§ 2.4).

Volumes of cylindrical, cubic and tetrahedral bodies can be calculated by standard formulas. More difficult is to estimate the volume of complex geometries, such as a wing and a fillet in wing–body combination. For these cases, the use of CAD systems is inevitable. The volume of a wing is

$$\mathcal{V} = 2\int_{i}^{b/2} A_c dy, \tag{2.18}$$

where $A_c$ is the cross-sectional area of the wing. The integration is extended from the wing root to the tip. An example of volume estimates is shown in Table 2.2 for some commercial aircraft.

### 2.5.1 Case Study: Do Aircraft Sink or Float on Water?

A commercial aircraft lands "safely" on a body of calm water*, without any structural damage and with the landing gear retracted. Will it sink or will it float?

The first question to ask is whether the aircraft is water-tight. If this is the case, then the answer is: yes, for some time. If we consider the average aircraft density, $\overline{\rho}/\mathcal{V}$, with the volume taken from Table 2.2, then we find $\overline{\rho} \simeq 70$ to 130 kg/m$^3$. Therefore, from the law of buoyancy, the aircraft would float. To be more precise,

* Landing on water is defined as "ditching".

Table 2.2. *Volume breakdown of selected aircraft; all volumes in [m³]*

| Item | A300-600 | B777-300 | A380-861 | B747-400 |
|---|---|---|---|---|
| Fuselage nose | 129.3 | 179.3 | 400.6 | 468.0 |
| Fuselage centre | 669.8 | 1,347.4 | 1,440.2 | 1,167.0 |
| Fuselage tail | 185.4 | 329.2 | 608.1 | 455.3 |
| Wing–body blend | 0.0 | 55.6 | 296.6 | 54.3 |
| Fuselage | 984.6 | 1,911.6 | 2,745.5 | 2,144.5 |
| Wings | 99.6 | 237.2 | 715.7 | 337.0 |
| H-stabiliser | 11.4 | 27.3 | 90.3 | 43.6 |
| V-stabiliser | 22.6 | 29.2 | 75.5 | 55.4 |
| Nacelles | 39.3 | 74.9 | 134.0 | 45.7 |
| Total | 1,157.5 | 2,280.1 | 3,759.1 | 2,826.2 |

the engine/nacelles volumes should not be taken into account because the engines will be flooded with water and would drag the aircraft down. The aircraft is not completely water-tight. A minimum float time must be guaranteed to evacuate the aircraft from the doors above the wings.

**AIRCRAFT SCALING.** We make one further consideration on scaling down or up an aircraft. If we simply rescale the aircraft's dimension, the volume can be fitted by an equation such as

$$\left(\frac{\mathcal{V}}{\mathcal{V}}\right)_{ref} = \exp\left[a_1 \log\left(\frac{b}{b}\right)_{ref} + b_1\right]. \tag{2.19}$$

For the Airbus A380 the parameters $a_1$ and $b_1$ are $a_1 = 2.9975$, $b_1 = -3.8389$.

### 2.5.2 Wing Fuel Tanks

The calculation of the theoretical fuel capacity of the aircraft is based on the volume available for allocating the fuel tanks. The most common tanks are inside the wings. Depending on the aircraft, there can be central tanks (within the fuselage) and tail tanks (in the tail plane).

The calculation of the wing-tanks geometry relies on some assumptions regarding the position of the spars. Usually there are two major spars, at the leading and the trailing edge. Also, some information regarding the span of the tank is required. The most important geometrical parameters are $x_{spar_1}$ (position of the leading-edge spar, as a fraction of the chord); $x_{spar_2}$ (position of the leading-edge spar, as a fraction of the chord); $s_t$ (skin thickness); and $b_{wt}$ (span of the wing tank). With this information and the shape of the wing section at relevant spanwise sections, it is possible to write a simulation model for geometry of the tanks. The numerical model for calculating the wing tanks relies on a function

$$\mathcal{V}_{tank} = f(x_{spar_1}, x_{spar_2}, b_{wt}, s_t, \texttt{Wing\_Geometry}). \tag{2.20}$$

An example is shown in Figure 2.19; this graph displays the tanks of the Boeing B777-300. The available volume is estimated at 63.37 m³ per side, giving a theoretical fuel

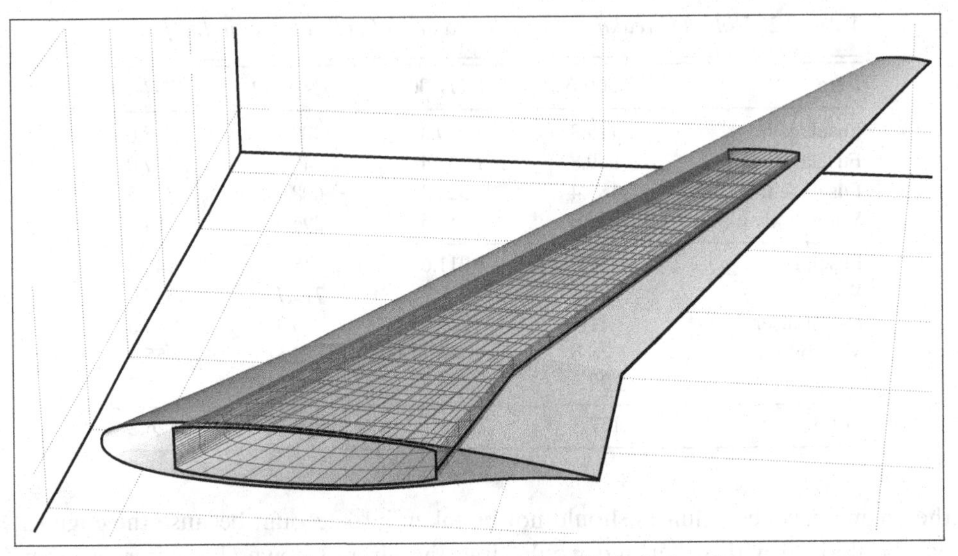

Figure 2.19. Reconstruction of the wing's fuel tanks.

capacity of 126,720 litres (~101,375 kg). Further discussion of the wing tanks is in § 3.7 and in § 14.3, where we present a numerical model that simulates the fuel temperature in the wing tanks.

## 2.6 Mean Aerodynamic Chord

There is one special definition of *wing chord* that is often used in aerodynamics and stability analysis. This is the *mean aerodynamic chord* (MAC, or $\bar{\bar{c}}$). This quantity is defined mathematically as the average of the squared chord distribution:

$$\text{MAC} = \bar{\bar{c}} = \frac{2}{A} \int_{o}^{b/2} c^2 dy. \tag{2.21}$$

This definition is derived from the strip theory of an untwisted blade with a constant wing section. Starting from the pitching-moment equation, the zero-lift pitching moment of a generic strip is

$$C_{mo}(y) = \frac{dM_o}{qc^2 dy}, \tag{2.22}$$

where $dM_o$ denotes the pitching moment of the wing strip and $q$ is the dynamic pressure. Integration of Equation 2.22 leads to

$$M_o = \int_{-b/2}^{b/2} C_{mo} q c^2 dy. \tag{2.23}$$

Next, assume that there is no change in the zero-lift pitching moment along the span. Thus, we can write

$$M_o = 2C_{Mo} q \int_{o}^{b/2} c^2 dy = C_{Mo} q \bar{\bar{c}} A, \tag{2.24}$$

from which we find the reference chord defined by Equation 2.21. For the simple case of a trapezoidal wing with taper ratio $\lambda$, wing span $b$, root chord $c_r$ and

Table 2.3. *Calculations of MAC for the Airbus A320-200 aircraft; graphs on the same scale*

| Wing | MAC | $y(MAC)$ | $x_{LE}(MAC)$ | Comments |
|---|---|---|---|---|
| 1 | 2.578 | 12.015 | 18.070 | |
| 2 | 4.309 | 7.975 | 15.891 | |
| 3 | 3.938 | 6.182 | 14.923 | Present choice |
| 4 | 5.009 | 6.629 | 15.164 | |
| Airbus | 4.120 | – | 16.310 | *Reference chord* |

gross area $A$, the MAC is

$$\bar{\bar{c}} = \frac{2}{3} c_r \left( \frac{1 + \lambda + \lambda^2}{1 + \lambda} \right). \tag{2.25}$$

The position of the CG is often given as a percent of the MAC. The position and the dimension of the MAC are calculated from the wing design. Once this is done, also the position of the point located at one quarter from the MAC's leading edge is known. The position of the CG with respect to the nose is called $x_{cg}$. The conversion of this datum to a percent of the MAC is

$$\% \, \bar{\bar{c}} = 100 \left( \frac{x_{cg} - x_{leMAC}}{\bar{\bar{c}}} \right), \tag{2.26}$$

where $x_{LE_{MAC}}$ is the position of the leading edge of the MAC with respect to the nose. The inverse formula, which allows for the calculation of the moment arm, is

$$x_{cg} = x_{le_{MAC}} + \frac{\% \, \bar{\bar{c}} \cdot \bar{\bar{c}}}{100}. \tag{2.27}$$

In Figure 2.20 we show two wing models, both having the root at the wing-body attachment. Wing No. 1 is the wire-frame of the Airbus A320-200 wing, projected onto the horizontal plane; Wing No. 2 is a wing obtained by joining the trailing edges of the root and the tip with a straight line and represents an "equivalent" trapezoidal wing. Note the difference in mean aerodynamic chord.

If we further extrapolate this wing to the symmetry section of the aircraft ($y = 0$), we have two additional wings: Wing No. 3 is representative of the true wing, extending through the wing box; Wing No. 4 has a straight trailing edge. Note again the difference in mean aerodynamic chord. Curiously, the MAC of Wing No. 3 occurs at the wing break point, although this is not always the case.

A summary of the MAC calculations is shown in Table 2.3 and compared with some data reported by the manufacturer. For this aircraft, airbus reports a "reference chord" that is relatively close to that of wing model No. 3; the difference is attributed to interpolation errors from technical drawings. In conclusion, for the calculation of the MAC, we consider a wing planform extending to the symmetry plane through the wing box.

## 2.7 Geometry Model Verification

The validation of the geometric model is done against a test case for which a detailed model exists. We have considered the wing-body combination F4 by DLR, which

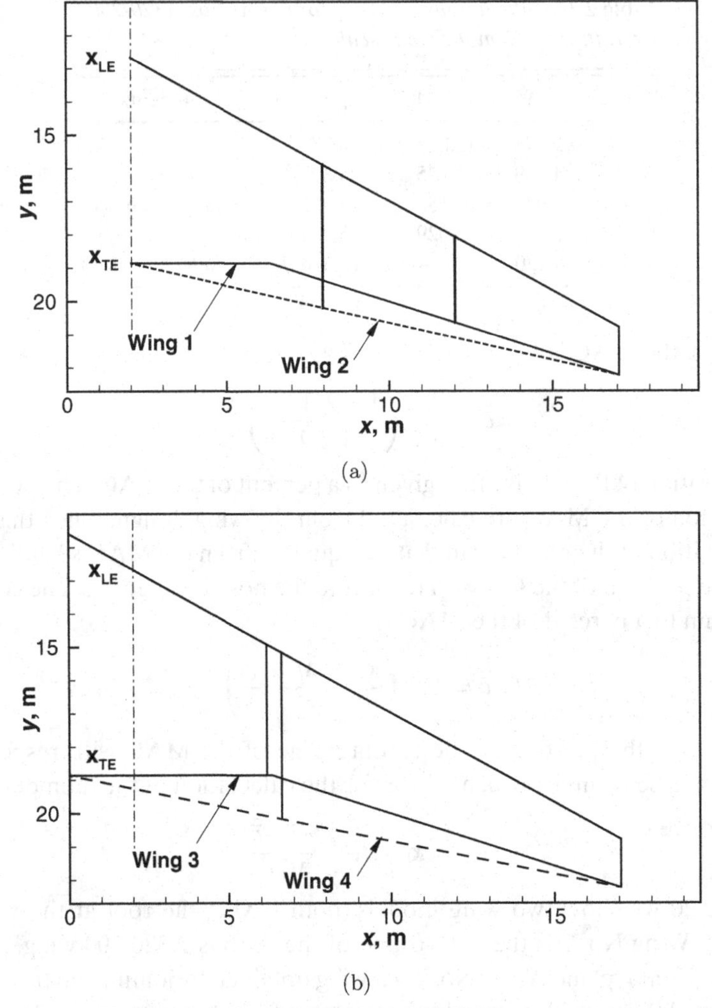


Figure 2.20. Four different ways of calculating the MAC of a transport wing; reference wing is that of the Airbus A320-200.

has been tested in a number of wind tunnels[4]. A finite-element version of the model has been produced for CFD validation and has been used for the validation of our model[5]. The aerodynamics is further discussed in Chapter 4. This geometry is a wing-body, without nacelles, pylons or tail surfaces. Table 2.4 is a summary of the areas and dimensions. The main difference is the estimation of the wing area and the aspect ratio, which do not correspond to the reference area. A correct calculation of this area relies on a clear definition of the wing planform.

### 2.7.1 Case Study: Wetted Areas of Transport Aircraft

We now apply the wetted-area method to the calculation of the wetted area to some model aircraft. The wetted-area breakdown is given in Table 2.5. The contributions are given in absolute value and in percentage of the total wetted area. The fuselage is split into fore, central and aft sections and wing-body blend. In all cases the most

Table 2.4. *Analysis of the geometry of the F4 aircraft model*

| Item | Calculated | Reference | Error, % |
|---|---|---|---|
| fuselage length | 1.1920 | 1.1920 | 0.0 |
| fuselage diameter | 0.1484 | 0.1484 | 0.0 |
| fuselage width | 0.1484 | 0.1484 | 0.0 |
| nose length | 0.2513 | 0.2600 | 3.3 |
| tail length | 0.5516 | 0.5600 | 1.5 |
| cross area | 0.4663 | 0.4663 | 0.0 |
| up-sweep angle | 12.4217 | n.a. | |
| area (exposed) | 0.1397 | n.a. | |
| area (reference) | 0.1772 | 0.1454 | 21.8 |
| area (apex) | 0.1789 | n.a. | |
| wing span | 1.1754 | 1.1754 | 0.0 |
| tip chord | 0.0606 | 0.0606 | 0.0 |
| root chord | 0.2401 | 0.2401 | 0.0 |
| taper ratio | 0.2526 | 0.2526 | 0.0 |
| aspect ratio | 8.7948 | 9.35 | 5.9 |
| MAC | 0.1261 | 0.1412 | 0.8 |
| sweep LE | 27.3233 | 27.10 | 0.8 |
| sweep QC | 23.1920 | 25.00 | 0.7 |
| dihedral | 4.8000 | 4.8 | 0.0 |
| average t/c | 0.1258 | n.a. | |

important contributions are the wing area and the central section of the fuselage. The wetted areas of some control surfaces are approximate.

The wetted areas calculated for a number of transport aircraft are shown in Figure 2.21 as a function of the manufacturer's empty weight (MEW). The solid line is a least-square fit of the data. In this case, the linear fit is a fairly good approximation of the data. The equation of the curve fit, based on a sample of eight aircraft, is

$$A_{wet}\,[m^2] = 2.14 \cdot 10^2 + 1.414W, \tag{2.28}$$

where $W$ denotes the MEW in metric tons.

## 2.8 Reference Systems

There are three reference systems to consider: the Earth system, the body system and the wind system. Local reference systems may be required for specific reasons (for example, wing aerodynamics). The aircraft is supposed to fly with respect to a Cartesian system fixed on the ground (*Earth Axes*), which for our purposes is considered flat. In fact, most of the performance calculations will be done for relatively short flight times and at relatively low altitudes. The Earth's curvature and rotation are important for inertial navigation systems and to take into account the Coriolis effects (accelerations) over a rotating Earth. The Coriolis acceleration is estimated at less $10^{-3}g$ in atmospheric flight mechanics. The gravitational field is characterised by acceleration of gravity, equal to the standard value of $g = 9.807$ m/s$^2$. The Earth system has the $x$ axis pointing North, the $z$ axis, normal to the ground and pointing downward; the $y$ axis points East, and makes a right-hand system with $x$ and $z$.

Table 2.5. *Wetted-area breakdown for the selected aircraft (calculated). All areas are in [m²]; ()* data are approximate*

| | A300-600 | | B777-300 | | B747-400 | | A380-861 | |
|---|---|---|---|---|---|---|---|---|
| *Item* | *A* | % | *A* | % | *A* | % | *A* | % |
| Fuselage | 772.4 | 50.85 | 1,255.8 | 55.3 | 1,325.8 | 46.49 | 1,496.5 | 37.70 |
| Nose | 123.5 | 8.13 | 156.3 | 6.64 | 332.8 | 11.67 | 286.5 | 7.22 |
| Centre | 459.6 | 30.25 | 815.5 | 34.62 | 609.6 | 21.37 | 700.1 | 17.66 |
| Tail | 189.2 | 12.47 | 265.1 | 11.23 | 356.9 | 12.52 | 403.9 | 10.17 |
| [Wing-Body] | | | 18.8 | 0.80 | 26.5 | 0.93 | 105.0 | 2.65 |
| Wing | 417.1 | 27.46 | 726.3 | 30.83 | 909.8 | 31.90 | 1,477.8 | 37.23 |
| Wing tips | 1.2 | | 0.43 | 0.02 | – | | – | |
| Winglets | – | | – | | 14.6 | 0.51 | 10.0 | 0.25 |
| Horizontal tail | 110.0 | 7.22 | 175.6 | 7.45 | 227.7 | 7.98 | 352.6 | 8.88 |
| Vertical tail | 94.5 | 6.22 | 35.9 | 1.52 | 160.0 | 5.61 | 237.8 | 5.99 |
| Nacelles | 109.7 | 5.42 | 111.1 | 4.72 | 117.2 | 4.11 | 236.2 | 5.95 |
| Pylons | 22.6 | 1.49 | 32.0 | 1.36 | 74.0 | 2.60 | 114.8 | 2.89 |
| Flap racks | 19.4 | 1.27 | 18.8 | 0.80 | 22.8 | 0.80 | 44.2 | 1.11 |
| *Total* | ~1,519 | ~100.0 | ~2,356 | ~100.0 | ~2,852 | ~100.0 | ~3,970 | ~100.0 |
| *Wing/Total area* | | 17.0 | | 18.6 | | 20.0 | | 22.2 |
| *Wet/Wing area* | 5.882 | | 5.367 | | 5.000 | | 4.505 | |
| All flaps* | 56.6 | 4.6 | 97.9 | 4.2 | 190.7 | 6.7 | 204.5 | 5.2 |
| All slats* | 41.0 | 2.8 | 79.9 | 3.4 | 110.0 | 3.9 | 137.2 | 3.5 |
| All spoilers* | 9.6 | 0.6 | 23.8 | 1.0 | 68.1 | 2.4 | 230.1 | 5.8 |
| All elevators | 17.4 | 1.1 | 21.9 | 0.9 | 29.7 | 1.0 | 49.4 | 1.2 |
| Rudder | 24.6 | 1.6 | 33.4 | 1.4 | 32.6 | 1.1 | 69.7 | 1.8 |

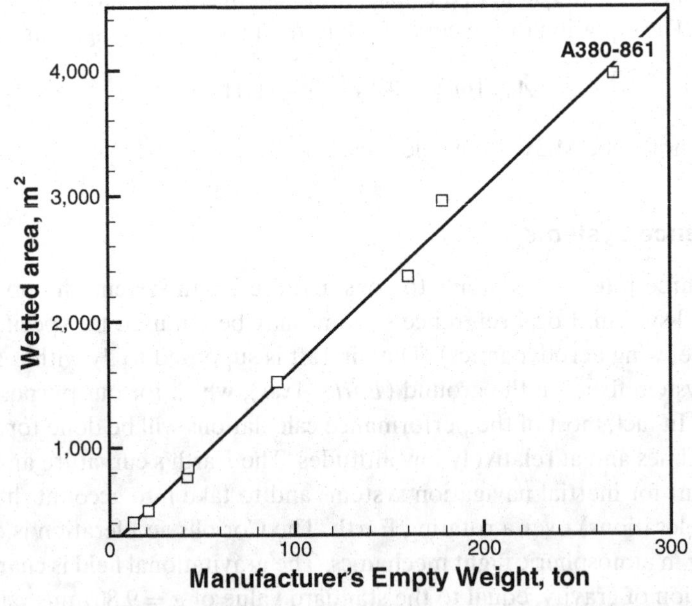


Figure 2.21. Distribution of aircraft wetted areas versus MEW (calculated).

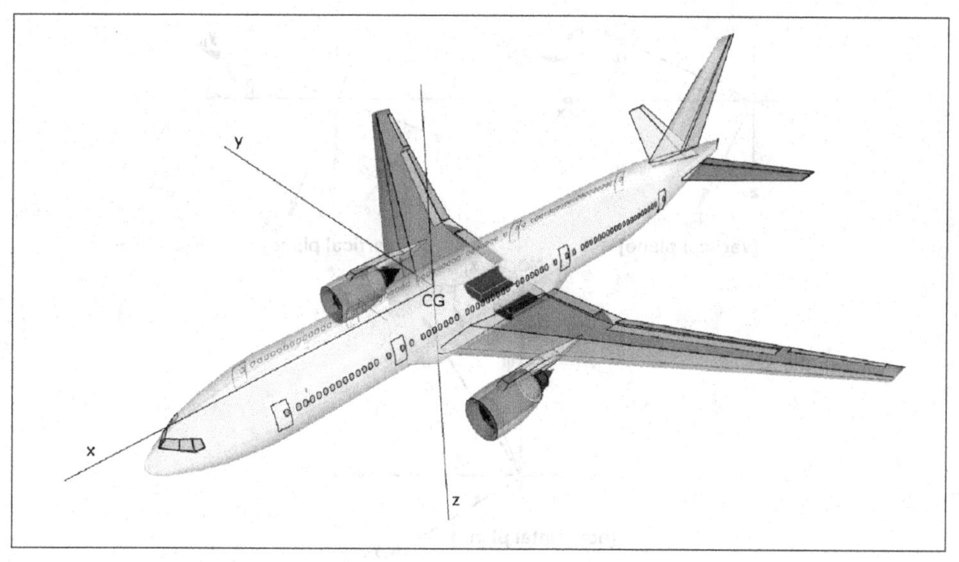


Figure 2.22. Body reference system and forces on the aircraft.

There are several ways to define reference axes on the aircraft; the choice will be limited by the fact that there is always one plane of symmetry. Thus, there will be a body-conformal orthogonal reference system, centred at the CG of the aircraft. The position of the CG is a known parameter (see Chapter 3). The reference system is shown in the three-dimensional view of Figure 2.22. The longitudinal axis $x$ is oriented in the direction of the forward speed (wind axis); the vertical axis $z$ is vertical (along the acceleration of gravity $\boldsymbol{g}$) and the $y$-axis makes a right-hand Cartesian system with $x$ and $z$. The positive $y$-axis is at the starboard side of the aircraft.

The correlation between body- and Earth-reference system is done by three attitude angles. The *pitch attitude* of the aircraft is the angle $\theta$ between the longitudinal axis and the horizontal plane (positive with the nose up). The *heading* is the angle $\psi$ between the aircraft's speed and the North-South direction. It is positive clockwise. The *bank attitude* $\phi$ is the angle between the aircraft spanwise axis $y_b$ and the horizontal plane.

The side force would not normally be present on the aircraft (and its occurrence should be avoided). It is mostly due to atmospheric effects (lateral gusts), asymmetric thrust and centre of mass off the symmetry line (due, for example, to a differential use of the fuel in the wing tanks). The presence of such forces may lead to a yawed flight condition. The yaw angle $\beta$ is the angle between the longitudinal axis and the true air speed vector.

The velocity (wind) axis reference system indicates the direction of the flight path with respect to the Earth system. At any given point on the trajectory the aircraft has a track and a gradient. The track is the angle on the horizontal plane between the flight direction and the North-South axis†. The gradient is the angle of

† The North-direction must be further specified: it can be magnetic North or geographical North.

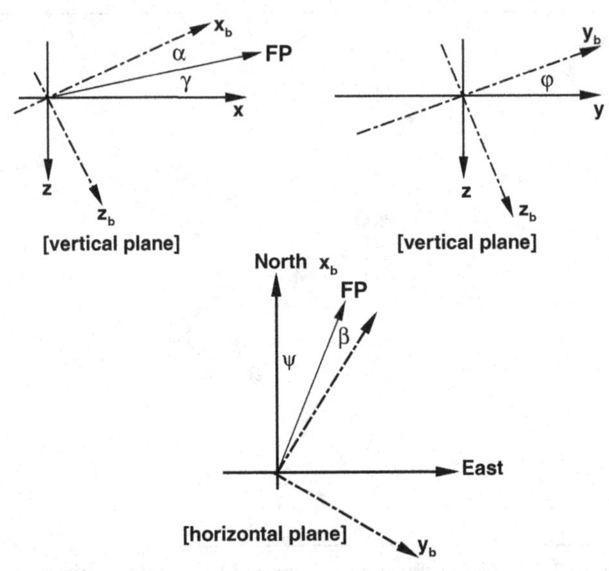


Figure 2.23. Relationships between angles and reference systems.

the velocity on the horizon. If $\boldsymbol{V}$ is the ground speed and $\boldsymbol{V}_w$ is the wind speed, the air speed is found from

$$\boldsymbol{V}_a = \boldsymbol{V} + \boldsymbol{V}_w. \tag{2.29}$$

The transfer of forces between one reference and the other is done through rotation matrices. The order of these rotations is important for the correct development of the flight-mechanics equations. For the full derivation of these equations we refer the reader to Yechout *et al.*[6].

### 2.8.1 Angular Relationships

We consider some simple flight cases to provide a correlation between angles in the different reference systems, Figure 2.23. First, consider the pitch angles. From the definitions, the pitch attitude $\theta$ is related to the angle of attack $\alpha$ and the flight-path gradient $\gamma$ by

$$\theta = \alpha + \gamma. \tag{2.30}$$

Next, consider a yaw. The heading $\psi$ is related to the track $\xi$ and the side slip by the equation

$$\psi = \xi - \beta. \tag{2.31}$$

If there is no side slip, the heading and the track are the same angle. Finally, consider a roll problem. The bank angle $\phi$ is the inclination of the spanwise body axis on the horizontal plane. The bank attitude is the same as the bank angle on a level flight path.

### 2.8.2 Definition of the Aircraft State

To conclude, we establish the set of parameters that fully describe the aircraft state in flight-performance calculations. The aircraft state is given by the vector:

$$\Psi = \{W, \mathbf{X}, \mathbf{V}, U, \mathbf{\Phi}, \mathbf{\Omega}, \alpha, \beta; \eta, \theta, \xi; I_{SF}, \mathsf{LG}\}. \tag{2.32}$$

The meaning of the symbols in Equation 2.32 is as follows:

- $W$ is the weight.
- $\mathbf{X} = \{x, y, z\}$ denotes the position of the CG with respect to a reference system on the ground. Often we call the geopotential altitude $h \equiv z$.
- $\mathbf{V} = \{u, v, w\}$ represents the velocity vector. In most cases we work with two-dimensional trajectories and assume $v = 0$. Hence, we call $V_g \equiv u$ and $v_c \equiv w$ the ground speed and the climb rate, respectively.
- $U$ denotes the true air speed. In level flight and in absence of winds, $U = V_g$.
- $\mathbf{\Phi} = \{\phi, \vartheta, \psi\}$ is the vector of aircraft angles with respect to the ground reference system; they are the bank angle, pitch attitude and heading angle, respectively. If we deal with two-dimensional trajectories, the heading angle is neglected.
- $\mathbf{\Omega} = \{\dot{\phi}, \dot{\vartheta}, \dot{\beta}\}$ is the vector of angular velocities. In most cases we work with quasi-steady trajectories; this vector is set to zero and is eliminated from the state equation.
- $\alpha, \beta$ are the effective mean inflow angle and the side-slip angle, respectively.
- The flight control systems are denoted by $\eta$ (elevator angle), $\theta$ (rudder angle), $\xi$ (aileron angle).
- $I_{SF}$ denotes the state of the high-lift system and is expressed by an integer number associated to a set of flap and slat deflection angles.
- LG is a binary flag for landing-gear position: retracted or deployed.

For a complete simulation, we need to add the engine state (Chapter 5) and the atmosphere.

## Summary

We have shown a strategy for building the configuration of a transport aircraft by careful use of technical documentation in the public domain. The strategy is based on the use of control points along three views and some rules that are used to reconstruct the aircraft geometry. By careful consideration of these rules, it is possible to elaborate distances, positions, planform areas, wetted areas and volumes. Not all of the geometrical characteristics are equally accurate; in fact, the procedure shown may fall short of expectations when considering the details of the control surfaces unless additional information is added to the database. In the absence of first-hand information from the manufacturer, this strategy is sufficiently accurate for aircraft performance. About two dozen different sub-systems have been identified and modelled. The geometry information will be used at several levels in later chapters. A verification example for a wing-body has been shown, as well as calculations for a number of aircraft.

## Bibliography


[1] Filippone A. Theoretical framework for the simulation of transport aircraft flight. *J. Aircraft*, 47(5):1679–1696, 2010.
[2] Harris CD. NASA supercritical airfoils: A matrix of family-related airfoils. Technical Report TP-2969, NASA, March 1990.
[3] Torenbeek E. *Synthesis of Subsonic Airplane Design*. Kluwer Academic Publ., 1985.
[4] AGARD. A selection of experimental test cases for the validation of CFD codes. Technical Report AR-303, Volume II, AGARD Advisory Group, Aug. 1994.
[5] Filippone A. Comprehensive analysis of transport aircraft flight performance. *Progress Aero Sciences*, 44(3):185–197, April 2008.
[6] Yechout TR, Morris SL, Bossert DE, and Hallgren WF. *Introduction to Aircraft Flight Mechanics: Performance, Static Stability, Dynamic Stability and Classical Feedback Control*. AIAA, 2003.


## Nomenclature for Chapter 2

| | | |
|---|---|---|
| $a_1$ $b_1$ | = | factors in Equation 2.19 |
| $A$ | = | area |
| $AR$ | = | wing aspect-ratio, $b^2/A$ |
| $A_{bbox}$ | = | area of bounding box |
| $A_{c_1}$ | = | cross-sectional area of wing with unit chord |
| $A_c$ | = | cross-sectional area of a wing |
| $A_{cs}$ | = | cross-sectional area |
| $A_{wet}$ | = | wetted area |
| $A_{xy}$ | = | projection of planform area on ground plane |
| $b$ | = | wing span |
| $b_w$ | = | wing span over winglet |
| $b_{wt}$ | = | overall span of wing tanks |
| $c$ | = | wing or airfoil chord |
| $c_b$ | = | chord at break-point |
| $c_r$ | = | root chord |
| $c_t$ | = | tip chord |
| $\bar{\bar{c}}$ | = | mean aerodynamic chord (also MAC) |
| $C_{mo}$ | = | sectional pitching moment |
| $C_{Mo}$ | = | pitching moment coefficient of a wing |
| $d$ | = | diameter |
| FP | = | flight path |
| $g$ | = | acceleration of gravity |
| $h$ | = | geopotential altitude |
| $k$ | = | iteration count, Equation 2.3 |
| $k_1$ | = | ratio between areas, Equation 2.16 |
| $I$ | = | moment of inertia |
| $I_{SF}$ | = | integer operator that defines flap/slat setting |
| $l$ | = | reference length |
| $l_{centre}$ | = | length of central fuselage section |
| $l_{wb}$ | = | length of wing–body intersection |

$l_{nose}$ = length of fore (nose) fuselage section
$l_p$ = perimeter of a fuselage cross-section
$l_{rack}$ = length of flap racks
$l_{tail}$ = length of aft (tail) fuselage section
LG = binary operator that defines state of landing gear
$m$ = mass
$M_o$ = pitching moment
$O$ = nose point
$p$ = perimeter, Equation 2.15
$p_c$ = perimeter, Equation 2.4
$q$ = dynamic pressure
$s_t$ = wing-skin thickness
$t/c$ = relative wing thickness
$t_{gap}$ = dimension of tail as shown in Figure 2.5
$\mathcal{V}$ = volume
$U$ = air speed or ground speed
$\boldsymbol{V}$ = vector velocity
$v_c$ = climb rate
$V_g$ = ground speed
$W$ = weight
$x, y, z$ = Cartesian coordinates
$x_r, y_r$ = coordinates of random point
$x_{cg}$ = longitudinal coordinate of the CG
$x_{le}$ = longitudinal coordinate of the leading edge
$x_{spar_1}$ = position of leading-edge wing spar
$x_{spar_2}$ = position of trailing-edge wing spar
$\boldsymbol{x}$ = aircraft state parameters
$\boldsymbol{X}$ = position vector

## Greek Symbols

$\alpha$ = angle of attack
$\beta$ = yaw or side-slip angle
$\gamma$ = flight-path angle
$\epsilon$ = tolerance quantity
$\eta$ = normalised spanwise coordinate
$\eta$ = elevator angle in Equation 2.32
$\lambda$ = wing taper ratio
$\xi$ = track angle; aileron deflection in Equation 2.32
$\theta$ = pitch attitude
$\rho$ = density
$\varphi$ = wing dihedral angle, Equation 2.1
$\phi$ = bank angle
$\boldsymbol{\Phi}$ = vector of aircraft angles in ground reference system
$\psi$ = heading angle
$\Psi$ = aircraft state vector, Equation 2.32
$\boldsymbol{\Omega}$ = vector of angular velocities

## Subscripts/Superscripts

| | | |
|---|---|---|
| $\overline{[\cdot]}$ | = | average value |
| $[\cdot]_a$ | = | air |
| $[\cdot]_b$ | = | body coordinates, see Figure 2.23 |
| $[\cdot]_{le}$ | = | leading edge |
| $[\cdot]_{nac}$ | = | nacelles |
| $[\cdot]_r$ | = | root, or random point |
| $[\cdot]_{ref}$ | = | reference conditions |
| $[\cdot]_t$ | = | tip, or tail |
| $[\cdot]_w$ | = | wing or *wind* |
| $[\cdot]_{wb}$ | = | wing-body |
| $[\cdot]_{wet}$ | = | wetted area |

# 3 Weight and Balance Performance

## Overview

The aircraft weight influences the performance of the airplane more than any other parameter. Weight has been of concern from the earliest days of aeronautics (§ 3.1). Therefore, we devote this chapter to the weight analysis and the relative aspects, such as operational and design weights (§ 3.2); weight management issues (§ 3.3); operational limits (§ 3.4); centre of gravity envelopes (§ 3.5); loading and certification issues, operational moments and corresponding errors (§ 3.6); the use of the wing tanks (§ 3.7); and, finally, mass properties, including determination of the CG and the moments of inertias of the empty airplane (§ 3.8).

**KEY CONCEPTS:** Aircraft Size, Design Weights, Operational Weights, Mass Distribution, Centre of Gravity Envelopes, Moments of Inertia, Fuel Tanks.

## 3.1 A Question of Size

By the start of the First World War, in the United Kingdom it was believed that the limiting weight of the airplane could not exceed 2,000 lb (~800 kg). Aircraft engineers of that time ignored that in 1913–1914 the great Igor Sikorsky designed, built and flew a four-engine aircraft known as *Ily'a Murometz*, or S-27, whose loaded weight was 5,400 kg. The total installed power was about 400 hp (~300 kW). The aircraft could carry as many as 16 passengers in its uncomfortable quarters[1;2].

Prior to this project, it was widely believed that a multi-engine aircraft would not be capable of flying. The main concern centred around the question of how to control the aircraft in the event of one engine failure – a likely occurrence in those days. With the *Ily'a Murometz*, Sikorsky proved that 1) multi-engined aircraft can be operated safely, even with two engines shut down; and 2) aircraft can get bigger for passenger transport and military operations. On 30 June 1914, Sikorsky and three crew members embarked on a record-breaking flight from St. Petersburg to Kiev and return, a total distance of about 2,600 km (~1,400 n-miles), covered in 26 hours.

Baumann[3] wrote a report in 1920, prompted by the construction of *giant airplanes* in Germany toward the end of the First World War – airplanes weighing as much as 15.5 tons and powered by 260 hp (~195 kW) engines.

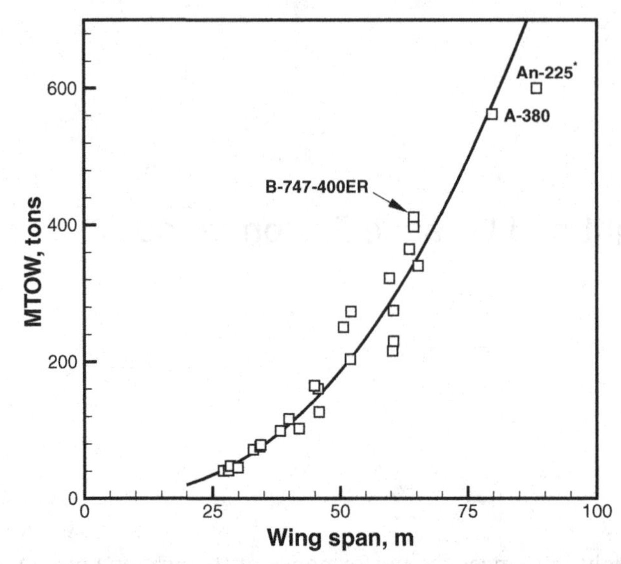


Figure 3.1. Wing span versus MTOW for large commercial aircraft; An-225 at design point.

Cleveland[4] predicted aircraft weights of 2 million pounds (800 tons) and nuclear power plants. The Lockheed Corporation persevered along the nuclear power plant option for some years and in 1976 Lange[5] proposed aircraft concepts in the 900 metric-ton class, including a 275 *MW* nuclear power plant. By contrast, the design office at Boeing proposed the span-loader concept[6] – a 1,270-ton aircraft that never went beyond the design office (project 759).

A detailed review focusing on configuration alternatives and economic viability of the big airplanes is available in Reference[7]. Cleveland's analysis contains a discussion of historical growth in size that leads to a *square-cube law*. The argument is that if technology had not improved, growth would have been halted by the fact that the load on airplane structures increases with the linear dimension, when the load is proportional to the weight. The load is proportional to the weight $W$; the weight is proportional to the cube of the linear dimension $\ell^3$; the cross-section of the beam is proportional to the square of the dimension $\ell^2$. Therefore, in first approximation, load/cross-section $\sim \ell$. At some point this increase in load reaches the structural limits of the material, and the beam may collapse under the effect of its own weight.

If the wing is scaled up while holding wing loading and structural stresses as constant, its weight will grow roughly as $W^{1.4}$. However, when one looks at the details of the components, they do not scale up with the same factor. Cleveland showed that by doubling the gross weight and the payload of the aircraft, the wing weight would have to increase by a factor ~2.69; the airframe would grow by a factor 1.84; the electrical systems would grow by a factor ~1.40.

Figure 3.1 presents a trend of aircraft maximum take-off weight (MTOW) and corresponding wing span $b$. The analysis shows that the wing span increases faster than the gross take-off weight, according to a power curve

$$W \simeq b^{2.417}, \tag{3.1}$$

Table 3.1. *Payload data for very large aircraft; X is the range at maximum payload*

| Aircraft | MTOW | PAY/MTOW | $X$ (km) | Notes |
|---|---|---|---|---|
| Airbus A380F | 560.0 | 0.268 | n.a. | |
| Airbus A-400M, *Grizzly* | 142.0 | 0.260 | | |
| Antonov AN-225, *Mrya* | 600.0 | 0.370 | 4,500 | design |
| Boeing C-17, *Globemaster* | 264.5 | 0.288 | 4,700 | |
| Boeing B747-400F | 396.8 | 0.284 | 8,240 | |
| Boeing B747-400ER F | 412.8 | 0.290 | 9,200 | |
| Boeing B747-800 F | 442.3 | 0.303 | 8,130 | |
| Lockheed C-5B, *Galaxy* | 381.0 | 0.311 | 5,526 | |
| Lockheed C-130J, *Hercules* | 79.4 | 0.245 | 5,400 | |
| Satic A300-600, *Beluga* | 155.0 | 0.305 | 1,666 | |

when the weight is given in [$N$] and the span is in [$m$]. From a productivity point of view, the most important factor is not the absolute weight and size of the aircraft but its useful payload. Historically, this has increased from about 10% to more than 30% of the current generation of airplanes; this ratio has also increased with the increased gross weight. The increase is driven by commercial requirements and by the need to move bulky equipment and machinery. Table 3.1 summarises the weight-payload data of the largest cargo airplanes currently in service; $X$ indicates the maximum range at maximum payload.

Freight is transported in standard containers called *pallets*. Loading and unloading of pallets is done quickly with ground-based vehicles with conveyors. The loading of pallets also rationalises the available space in the cargo hold. A suitable cargo performance parameter is

$$E = \frac{\text{PAY}}{\text{MTOW}} X. \tag{3.2}$$

This parameter is a measure of how much payload can be carried over a given distance, relative to the MTOW. It emphasises the fact that a certain payload can be flown over a longer or shorter distance. This parameter has been estimated for a number of aircraft and is shown in Figure 3.2.

The design of a new large airplane has often been met with external constraints of a different nature. For example, the Airbus A380 was constrained in wing span, overall length and overall height to fit the existing infrastructure as much as possible. Even then, considerable investment was required on ground infrastructure, such as increasing the width of runway, the gate capability, the ground handling, and the evacuation procedures. With regard to the latter points, the appearance of the Boeing B747 in the early 1970s caused a change in loading and unloading technology. At that time the lower deck was too large to be loaded manually, so new pallet sizes (or unit load devices [ULD]) were introduced.

## 3.2 Design and Operational Weights

There are several definitions of aircraft weights that are required for operations and certification. A comprehensive discussion of aircraft weights (including

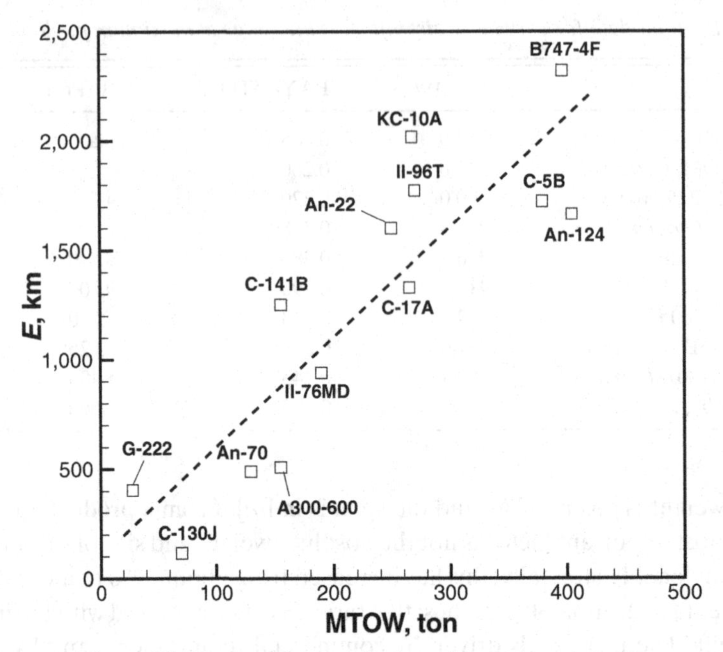


Figure 3.2. Payload factor of selected transport aircraft.

historical trends) is given by Staton *et al.*[8]. The most important definitions are given here.

**MANUFACTURER'S EMPTY WEIGHT (MEW).** This is the aircraft weight when the vehicle is delivered to the operator. This weight includes non-removable items, that is, all of the items that are an integral part of the manufacturer's configuration. There can be weight differences between airplanes in the same fleet, as discussed further in § 3.4.

**OPERATIONAL EMPTY WEIGHT (OEW).** This is the MEW plus a number of additional removable items carried for *operational* reasons, and inevitable causes, such as lubricant oils and the unusable fuel. The unusable fuel is the fuel that cannot be pumped into the engines under critical flight conditions, such as those specified by the aviation authorities. The operational items include the catering and entertaining equipment, flight and navigation manuals, life vests, passenger entrainment devices and emergency equipment.

**MAXIMUM ZERO-FUEL WEIGHT (MZFW).** This is the maximum weight of the airplane loaded only with bulk cargo. During flight, the wing fuel tanks generate a root bending moment that partially overcomes the stress created by the wing lift. When these tanks are empty and the aircraft is heavily loaded, there is a risk of exceeding the design limits for the wing root bending moment. The root loading is shown in Figure 3.3, points A and B. If the aircraft carries fuel only on the centre tanks, the resulting fuel weight contributes to the wing bending moments. Thus, it is advised to use first the centre tanks fuel and then the wing tanks fuel. For the same reason, the wing tanks are filled first (see further discussion in § 3.5.1).

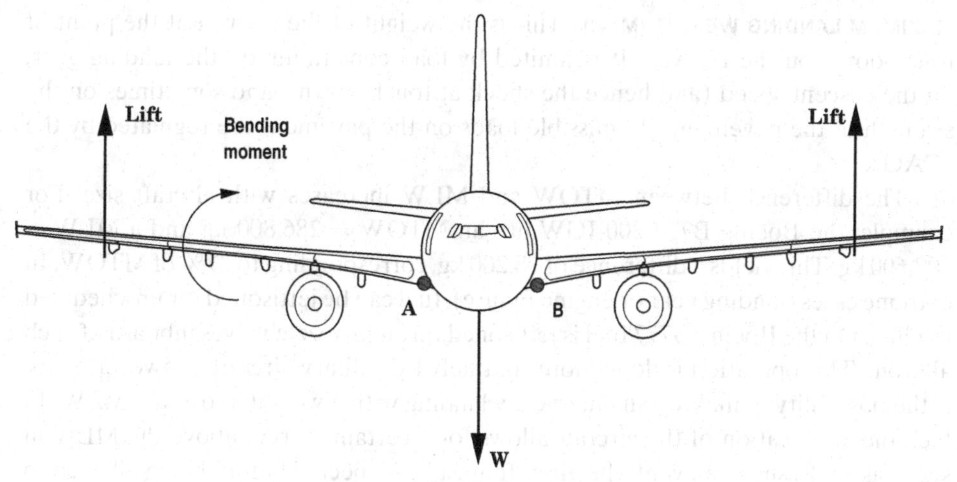

Figure 3.3. Wing root loads.

**MAXIMUM TAKE-OFF WEIGHT (MTOW).** This is the certified maximum weight at lift-off; that is when the front landing gears lose contact with the ground. The MTOW is reached by a combination of payload and fuel, as discussed in § 15.2.

**MAXIMUM RAMP WEIGHT (MRW).** This is the weight of the aircraft "ready to go", when the doors are closed, the systems are started and the vehicle starts taxiing. The difference between the maximum ramp weight and the maximum take-off weight corresponds to the amount of fuel burned between leaving the air terminal and the lift-off. This difference is only relevant for very large aircraft because it affects the operations on the runway. For example, the Boeing B747-400 has a MRW = 398,255 kg and a MTOW = 396,800 kg. The difference of 1,455 kg ($\sim$ 1,750 litres) is the fuel that can be burned from the moment the aircraft starts taxiing and the take-off point. A further analysis of taxi fuel is in § 9.10.

**MAXIMUM STRUCTURAL PAYLOAD (MSP).** This is the allowable weight that can be carried by the aircraft. The sum OEW + MSP is less than the MTOW and should give the MZFW. The payload can be virtually everything: passengers and their baggage, bulk cargo, military equipment and other hardware. The MSP is never reached for passenger aircraft. In general, there is a need to differentiate among MSP, volume available (space-limited payload, such as number of pallets) or capacity (due to seating limitations).

**MAXIMUM BRAKE-RELEASE WEIGHT (MBRW).** This is the maximum weight at the point where the aircraft starts its take-off run. Lower values of the MBRW, as they often occur, are called *brake-release gross weight* (BRGW). Some performance charts are drawn at constant BRGW.

**MAXIMUM TAXI WEIGHT (MTW).** This is the certified aircraft weight for taxiing on the runway and takes into account the loads on the landing gears and the CG envelope of the airplane.

**MAXIMUM LANDING WEIGHT (MLW).** This is the weight of the aircraft at the point of touch-down on the runway. It is limited by load constraints on the landing gear, on the descent speed (and hence the shock at touch-down), and sometimes on the strength of the pavement. Permissible loads on the pavement are regulated by the ICAO[9].

The difference between MTOW and MLW increases with aircraft size. For example, the Boeing B777-200-IGW has an MTOW = 286,800 kg and a MLW = 208,600 kg. This yields a difference of 78,200 kg, corresponding to 27% of MTOW. In extreme cases (landing gear or engine failure), fuel can be jettisoned for unscheduled landing. On the Boeing B777 fuel is jettisoned through nozzle valves inboard of each aileron. This operation is done more routinely by military aircraft. However, there is the possibility of making an emergency landing with a weight above the MLW. In fact, the certification of the aircraft allows for a certain margin above the MLW in such cases[†]. In such an event, the aircraft must be inspected before being allowed to restart its service.

**WEIGHT RELATIONSHIPS.** The relationship between the weights defined previously is the following:

$$MRW \geq MTW \geq MBRW \geq MTOW \geq MLW, \tag{3.3}$$

$$OEW + MSP + MFW > MTOW. \tag{3.4}$$

There are a number of other weight definitions in the aviation jargon. For example, the *all up weight* (AUW) generally refers to the weight of the aircraft in cruise conditions. There are, of course, the weights of the various aircraft components, such as the engines and the engine installation. For the engines, a *dry weight* is sometimes reported, which refers to the weight of the engine without lubricants and some ancillary systems.

The main weight limitations are reported in the *Flight Crew Operation Manual* (FCOM). The whole process of keeping track of the operational weights and managing the position of the CG is part of the *Weight and Balance Manual* (WBM) of the airplane. The manual is procured by the manufacturer and is given to the operator with all of the instructions needed to load the vehicle and to manage the position of the CG. Other important documents include:

- *Delivery Weight Report*, that establishes the OEW, the MEW and the corresponding CG positions.
- *Weighing Check List*, that contains a list of each item, its position and unit weight (item, weight, moment-arm, moment).

Other operational weight restrictions exist, such as the maximum weight for balanced field length, obstacle clearance, noise emission and available engine power. Methods for estimating the structural empty weight are given in a number of textbooks on aircraft design, for example Roskam[10], Torenbeek[11], Raymer[12], Staton[8] and more specialised publications[13]. Comparisons of weight breakdown for a number of aircraft were published by Beltramo *et al.*[14].

† The relevant regulations for fuel jettisoning are FAR/JAR $25.1001.

Most commercial airlines carry freight (cargo) as part of their business model. Some major airlines have a balanced split between passenger and cargo revenue. Such an important source of business has been made available by the standardisation of the ULD. There is a recognised international code for all ULD*. For each device, the information includes shapes, dimensions, volume and maximum weight. Loading can be manual or semi-automatic. There are specific loading rules, including weight limits, positions, tying-down, the effect of moving items and so on.

## 3.3 Weight Management

The structural weight of almost all aircraft grows over time during their service life, although there are exceptions. Weight grows due to a number of reasons, namely new performance specifications, re-engineering of the power-plant, exploitation of structural design margins, seating re-arrangement, and not least the correction of design flaws, that may come after several years of service. One, often overlooked, cause of weight increase is the collection of dust and dirt inside the airplane, along with the effects of humidity that increase the water carried by the airplane†.

Weight and balance logbooks are maintained to keep a check on all of the modifications done to the aircraft. The manufacturers also provide charts showing the basic weights and position of the CG. Loading of commercial aircraft is done according to the instructions provided. The airline flight management performs basic calculations of passengers and baggage. There are models that provide rapid solutions to the aircraft weight and balance as a function of passengers, baggage and fuel.

Fuel consumption must obey special priorities to maintain the balance of forces on the aircraft at different flight regimes. Fuel must be used from the inboard tanks first. An optimal distribution reduces the requirements on aircraft trim and therefore the drag associated with it, thereby maximising the profitability of the aircraft.

**AIRCRAFT WEIGHING.** Prior to weighing, the aircraft must be emptied of all its fuel. Three basic steps are required: 1) defuelling, 2) inventory of items, and 3) weighing. The first step consists of draining all of the fuel from the fuel tanks. The aircraft's nose is raised to the point where its pitch attitude is zero. Specific defuelling procedures are available in the WBM of the airplane. When the fuel level reads zero, all of the pumpable fuel has been drained. There remains the *unpumpable fuel*. Additional quantities can be eliminated by using a water drain. The remaining fuel after this operation is the *undrainable fuel*.

The second step is the inventory of all on-board items and equipment. This check list is to be compared with the list of fitted equipment from the manufacturer to determine the MEW as well as the OEW.

Finally, the weighing is performed, generally in a closed hangar, after measuring accurately the pitch attitude. The weighing is done by placing each wheel-set on

* IATA ULD Technical Manual.

† The water carried by the Airbus A380 is 1,700 $l$ (standard) or 2,500 $l$ (optional). The latter value corresponds to the mass of about 25 passengers and their luggage.

Table 3.2. *Standard passenger weights (rounded to full kg)*

| | Summer | Winter |
|---|---|---|
| Average adult | 86 | 88 |
| Average adult male | 91 | 93 |
| Average adult female | 81 | 83 |
| Child (2–13 years) | 37 | 39 |

a separate scale. The accuracy of the scale* must be ±20 kg for a weight range $2,000 < W < 20,000$ kg, or ±0.1% for a weight range $W > 20{,}000$ kg.

**CARGO LOADS.** Certified containers must be used. These containers have known values of the dimensions, volume, empty weight and CG. The containers must be weighted prior to loading in order to produce a loading chart. Loading of bulk payload causes localised stresses on the structure and the floor panels. Values for maximum distributed loads (kg/m$^2$ or other units) are provided by the manufacturer.

**PASSENGER AND PERSONNEL LOADS.** The balance of the aircraft is done by calculating as accurately as possible the distribution of the loaded weights; these are the flight crew, the cabin crew and the passengers. For each of these categories the moment arm is known through the cabin and seats layouts. In particular, each seat row is assigned a moment arm; the passenger weight is assumed to be acting at the centre of the seat.

Passenger weights for weight management have been standardised†. These weights are summarised in Table 3.2. The passengers' baggage must be weighted at check-in for a better estimation of the take-off weight.

## 3.4 Determination of Operational Limits

When a new aircraft is delivered by the manufacturer, it is weighted accurately in order to determine the manufacturer's empty weight and the corresponding position of the CG. In principle, each aircraft is unique, so there will be an individual value of the MEW and the corresponding CG. For example, there are at least 15 different weight versions of the Airbus A320-200, with MRW variable between 66,400 kg and 77,900 kg‡. The current regulations allow for the determination of the MEW from a set of aircraft (*a fleet*). This operation, in particular, is done by the operator, who determines the Dry Operating Weight (DOW) for a fleet in the same weight class[15]. This determination relies on a weight limit of ± 0.5% of the MLW around the DOW of the fleet. In mathematical terms,

$$\mathrm{DOW}_{min} = \mathrm{DOW} - 5 \cdot 10^{-3}\,\mathrm{MLW}, \qquad \mathrm{DOW}_{max} = \mathrm{DOW} + 5 \cdot 10^{-3}\,\mathrm{MLW}. \tag{3.5}$$

* JAR-OPS 1.605.

† Appendix 1 to JAR-OPS 1.620(g).

‡ Airbus: A320 Airplane Characteristics. Rev. 25, Sept. 2010, Airbus Customer Services, Blagnac, France.

An aircraft can be part of the same fleet if

$$\text{DOW}_{min} \leq \text{DOW} \leq \text{DOW}_{max}. \tag{3.6}$$

During the life of the aircraft, the DOW will change. However, the aviation regulations prescribe that the DOW of the aircraft be contained within 0.5% of the MLW for all purposes related to loading, balancing and trimming. For example, the Airbus A320-200 has a certified MLW = 64,500 kg and a DOW = 39,400 kg. The resulting variation of the DOW is 0.005 MLW, or 322.5 kg. Thus, the weight of the aircraft can vary between 39,077.5 kg and 39,332.5 kg. Therefore, the 15 weight configurations of the Airbus A320 cannot belong to the same fleet.

Similar considerations hold for the variation of the centre of gravity, which is allowed to oscillate by 0.5% MAC (or the reference chord) above and below the value indicated in the aircraft's documents. The corresponding quantity is called Dry Operating Centre of Gravity (DOCG). For the A320 the reference chord is 4.194 m. Therefore, the CG is allowed to shift by ± 0.021 m. If larger shifts occur, the loading and trimming instructions will generally not be valid.

The DOW and DOCG values are given for a configuration corresponding to landing gear deployed, high-lift surfaces in stowed position and potable water in the water tanks.

One of the first operators to perform this analysis understood that considerable weight savings could be achieved by controlling the use of drinkable water and catering to match the number of passengers on board[15].

The determination of the operational weight and CG is affected by uncertainty by weight contributions and their location (passengers, cargo, fuel and operational items). Let us call $E$ the moment error due only to the item location; the weight of the item is known accurately. This error can be written as

$$E = \mathcal{M}_{real} - \mathcal{M}^*, \tag{3.7}$$

or

$$E = W(x_{real} - x^*), \tag{3.8}$$

where $x$ is the moment arm and the asterisk denotes the "assumed" location. Next, assume that the weight has an inaccuracy of $\Delta W$, so that $W_{real} = W^* \pm \Delta W$. Such inaccuracy causes an error $E_w$ in both weight and CG position. The following cases may occur:

- If $\Delta W$ is forward of the CG, then a $+\Delta W$ moves the CG forward; $-\Delta W$ moves the CG aft.
- If $\Delta W$ is aft of the CG, then a $+\Delta W$ moves the CG aft; $-\Delta W$ moves the CG forward.

None of these events should move the CG beyond the certified limits.

## 3.5 Centre of Gravity Envelopes

The CG must be contained within specified limits, both longitudinal and lateral, to ensure that the aircraft is fully controllable and that the flight is safe. In this section we address the longitudinal movement of the CG and its dependence on both weight and loading. Thus, we must determine the CG envelopes of the airplane. A few definitions are required:

- The term *in-flight* denotes any flight condition other than take-off and landing.
- The term *aft* denotes rear part, or toward the tail.
- The term *go-around* denotes the end of final approach, which occurs when the aircraft is prevented from landing. In this case, the thrust is increased rapidly and the airplane is forced to pitch up. The pitch-up has to be compensated by the elevator. The elevator moment increases as the CG moves aft. There is the possibility that for some CG positions, the elevator is not capable of providing the necessary restoring moment. Go-around is further analysed in § 11.8.
- The term *α-floor protection* denotes a constraint in angle of attack and hence a reduction in the risk of airplane stall.

The movement of the CG is expressed by an *index* depending on the CG position in percent MAC. The index is defined by

$$\mathcal{I} = W(x_{CG\%} - x^*_{CG\%})\frac{\text{MAC}}{c} + k, \tag{3.9}$$

where $x_{CG\%}$ is the actual CG position in percent MAC; $x^*_{CG\%}$ is the reference CG position in percent MAC; $c$ and $k$ are two constant factors that are chosen to keep the index $\mathcal{I}$ within reasonable values. When $x_{CG\%} = x^*_{CG\%}$, the index is independent of the weight, $\mathcal{I} = k$. In all other cases, the index is proportional to both weight and MAC. Thus, on a graph $\mathcal{I}$-versus-$W$, each CG position is represented by straight lines; these lines have a negative slope if the CG is forward of the reference CG; they have a positive slope if the CG is aft of the reference CG. The iso-CG line is calculated from Equation 3.9. On this diagram it is possible to draw the operational limits of the aircraft at all flight conditions, as well as the zero-fuel limit. These limits generally represent a compromise between flight performance and aircraft loading.

### 3.5.1 CG Travel during Refuelling

Airplane refuelling is done according to procedures specified by the manufacturer. Refuelling is done without passengers, unless there are particular reasons to do so with passengers already on board. The operation is closely regulated to avoid incidents*.

The process of refuelling leads to a movement of the airplane's CG, which is sometimes called *fuel vector*, an example of which is shown in Figure 3.4. In this graph, segment 1 refers to refuelling of inner tanks 1–4 up to 3,000 kg; segment 2 corresponds to refuelling of the outer tanks up to 4,500 kg per tank; segment 3 corresponds to more refuelling of the inner tanks 2 and 3 until they reach 18,200 kg (the total fuel now being 81,200 kg). In segment 4 the inner tanks 2 and 3 are refuelled to 25,700 kg each, the centre tank is fuelled to 17,000 kg and the trim tank is filled to 2,400 kg. At this point, the total fuel load reaches 116,200 kg. In segment 5 the centre tank is filled to 40,000 kg and the trim tank reaches 5,900 kg (total fuel load 142,700 kg). For the A340-500 the rear centre tank is filled to 14,956 kg

* Relevant regulations are JAR-OPS 1.305 and FAR 121.750 (airplane evacuation).

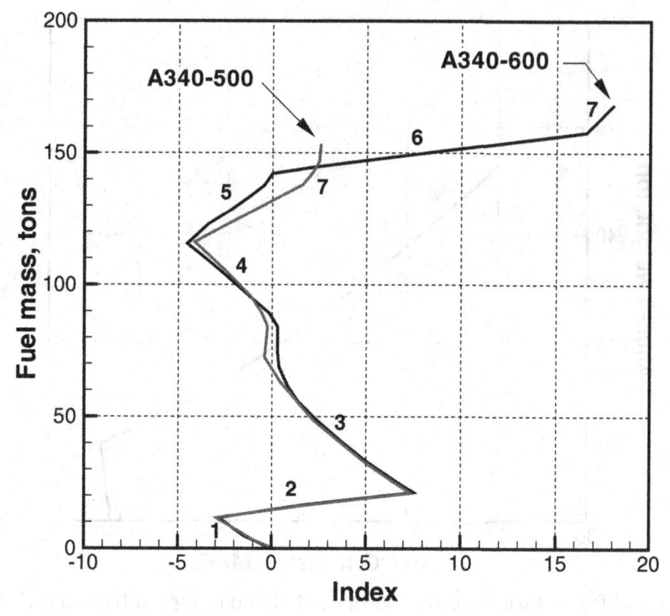


Figure 3.4. CG movement during refuelling of the Airbus A340-500 and A340-600 (adapted from the FCOM).

(segment 6). For both aircraft versions, the last segment of the fuel vector (segment 7) corresponds to all tanks full. The fuel vector depends on the fuel density and hence on the fuel temperature. The largest differences in CG position occur toward the high fuel loads. Note that the sequence indicated is common to other large aircraft and must be controlled by a fuel computer. In Figure 3.4, the index starts from zero when the fuel tanks are empty. The movement of the CG must be contained within the certified limits.

### 3.5.2 CG Travel in Flight

There are three reasons why the CG moves during the flight: 1) fuel burn, 2) change in configuration, and 3) passenger movement. Cargo movement should not occur.

**FUEL EFFECTS.** Due to the complex sequence of fuel pumping from the tanks, there is a CG movement, which if uncontrolled would cause an increase in fuel consumption (because of higher drag); in the worst case it may reach a point at which flight control becomes difficult. Due to the sophistication of modern airplanes, CG movement can only be controlled by the avionics systems, which are dedicated hardware and software systems (flight control system [FCS]). The FCS calculates the position of the CG and compares it to a target value. A control band is defined as a target CG and 0.5% MAC. The FCS commands fuel transfers in order to maintain the CG within the control band. Details of the operations done to maintain the CG position depend on the specific airplane. An example of a CG target is shown in Figure 3.5 for the Airbus A340. Figure 3.6 shows an example of the movement of the CG due to fuel depletion in flight.

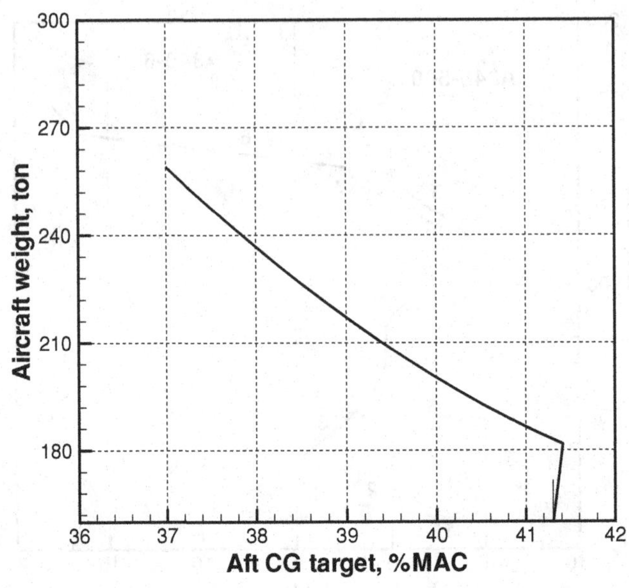


Figure 3.5. Aft CG target for the Airbus A340-300 (adapted from the FCOM).

**EFFECTS OF CONFIGURATION.** Such effects are due to landing-gear deployment and retraction, high-lift surfaces deployment and retraction, water movement, and movements in the cabin. The latter two examples are stochastic; therefore, a safety margin must be included in the estimation of the CG. The deployment of the flaps moves the CG position aft; the deployment of the slats moves the CG position forward. Landing-gear retraction moves the CG position forward. At landing the potable water will be less than at the start of the flight (some of it can be jettisoned) and part of it is in the waste tanks. The latter contribution cannot be discounted because for

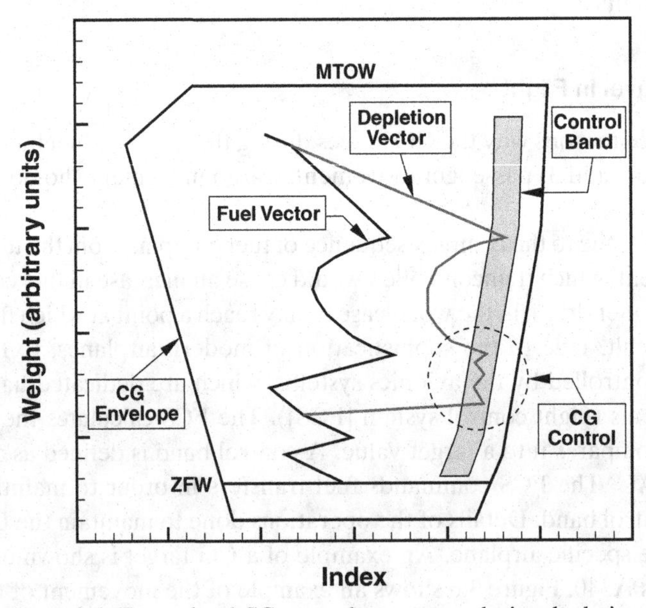


Figure 3.6. Example of CG control movement during depletion.

large aircraft, it can add several hundred kilograms to the operational weight (for the Airbus A330, the potable water capacity is 350 kg).

**PASSENGER MOVEMENT.** This is not an unusual situation when during a long-haul flight, passengers move away from their seats and cabin crew operate services up and down the aisle. The aircraft balancing is done with passengers at their seats, cabin crew at their position, and trolleys stowed. There are various movement scenarios used by the manufacturers (for design purposes) and the operators, including events such as passengers moving to the lavatories and back, passengers moving from back rows to front rows, trolleys moving a number of rows and so on. This is further analysed in § 3.6.

### 3.5.3 Design Limits on CG Position

Several requirements, including legal ones, have to be considered in the design of the CG envelopes and in the CG management during loading and in flight. The key aspects of the problem are listed as follows:

- **Take-off.** The forward CG limit is governed by the strength of nose landing gear, by the take-off rotation requirements, the manoeuvrability, and the deployment of the high-lift devices. The aft CG limit is governed by the strength of the main landing gear, by the nose gear adherence to the ground, the take-off rotation (*tail strike*) and the stability in level and manoeuvre flight. Some of these effects are further discussed in Chapter 9.
- **Landing.** The forward CG limit is set to the same value as the take-off limit; the aft limit is either superimposed to the take-off limit or aft of the take-off limit.
- **In-Flight.** The forward limits are constrained by fuel consumption, efficiency of the elevator and some aerodynamic requirements, including stall of the trimmable horizontal stabiliser. The aft limit requires stability in steady-state flight and during manoeuvres, *go-around* and $\alpha$-floor protection.

An example of CG limits at take-off is shown in Figure 3.7, where we have indicated the major elements of constraint. The top end of the scale is the maximum structural take-off weight, which is aft-limited by the structural integrity of the main landing gears (MLG) and is forward limited by the nose landing gear. At lower weights, the aft limit is labelled as *handling qualities*, which is a term that refers to aircraft rotation and pitch-up, deployment of high-lift systems and general flight stability. The forward limit is a compromise between performance requirements and aircraft loading. There is a grey area at the top left of the graph where the CG is unlikely to be placed. In fact, the compromise between performance and loading tends to prefer the former over the latter one. At the very low end of the weight, the aft CG limit is labelled as nose landing gear adherence. If the aircraft is improperly loaded, it will tend to lean back onto the main landing gear, which causes the tail strike.

Figure 3.8 shows the CG envelope for landing configurations. The aft and forward limits are similar to the take-off case, but the weight is considerably reduced for aircraft have MLW far below the MTOW.

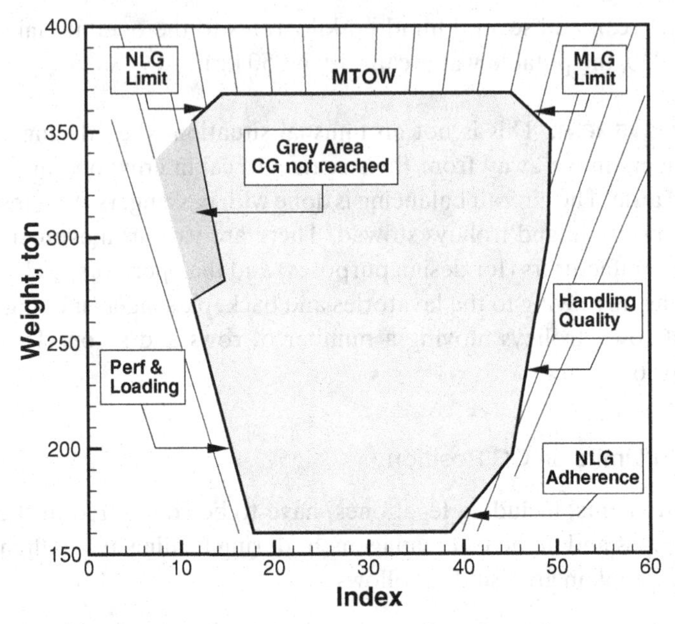


Figure 3.7. CG limits for take-off configuration.

A map of all of the operational limits is shown in Figure 3.9. The curve at the centre of the graph is the one shown in Figure 3.4. The solid lines represent the CG envelope at various flight conditions: take-off, flight and landing. The inclined lines represent loci of constant CG position. A forward CG limit is 17% MAC for a cruise condition and 18% MAC for take-off and landing; the aft CG limit is 43% MAC for cruise and 42% MAC for landing. Take-off at low gross weights is more restrictive.

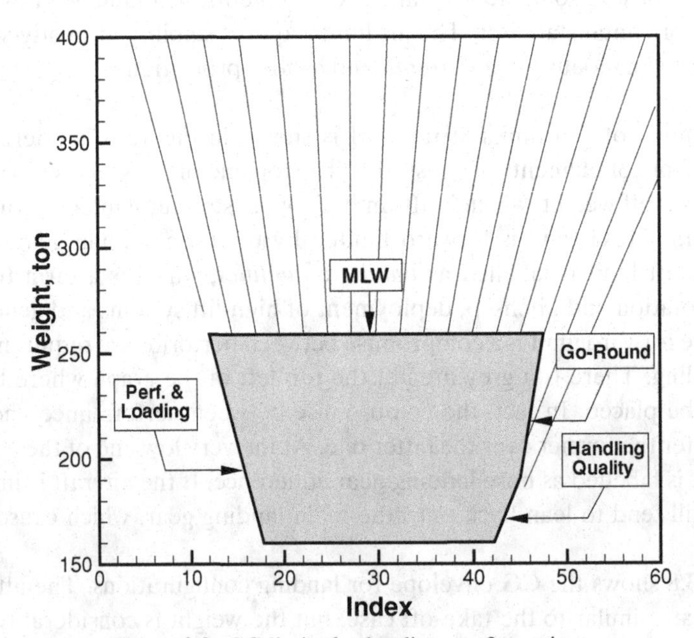


Figure 3.8. CG limits for landing configuration.

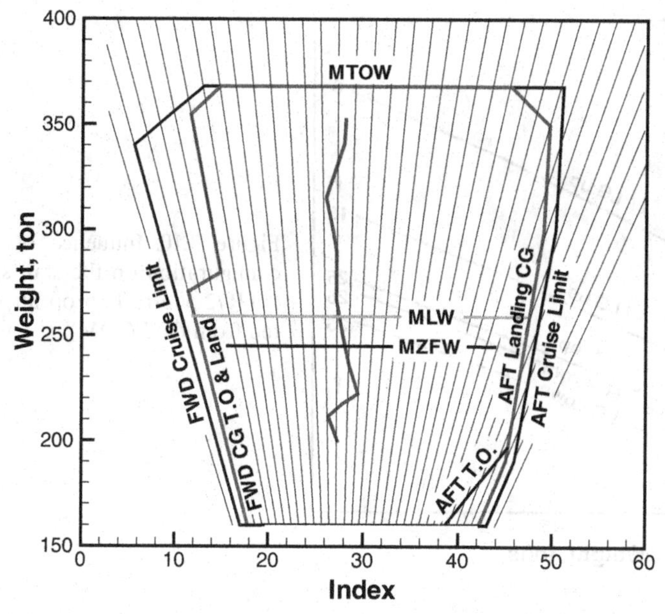


Figure 3.9. CG limits of the Airbus A340-600 at various flight conditions (adapted from the FCOM).

### 3.5.4 Determination of the Zero-Fuel CG Limit

The calculation of the zero-fuel CG limit is required to ensure that the aircraft can be controlled in all phases of flight, from take-off to landing. The reference axis for the calculation of the moments is a percentage of the MAC (generally 25%). If an item is loaded forward of the reference axis, it gives a negative (nose-down) pitching moment; if it is loaded aft of the reference axis, the pitching moment will be positive (nose up). The position of the zero-fuel CG is calculated from

$$\text{ZFCG} = \frac{\mathcal{M}_o + \mathcal{M}_{pax} + \mathcal{M}_p}{\text{ZFW}}, \tag{3.10}$$

where $\mathcal{M}_o$ denotes the pitching moment of the empty airplane, $\mathcal{M}_{pax}$ is the moment due to passengers and $\mathcal{M}_p$ is the payload moment. The take-off CG position (TOCG) is

$$\text{TOCG} = \frac{\mathcal{M}_o + \mathcal{M}_{pax} + \mathcal{M}_p + \mathcal{M}_f}{\text{GTOW}}, \tag{3.11}$$

where $\mathcal{M}_f$ is the moment arm of the fuel. The solution of Equations 3.10 and 3.11 requires the determination of the exact position of each operational item. Some of these components have a stochastic contribution. The operator can improve the balancing scenario by calculating the exact weight of the cargo items and their position on the airplane.

### 3.5.5 Influence of CG Position on Performance

This effect is best seen through a real-life example. Figure 3.10 shows the influence of the aircraft weight, as well as the configuration (landing gear up or down; flap deflection) on the stall speed of the ATR72-500 powered by Pratt-Whitney PW127M

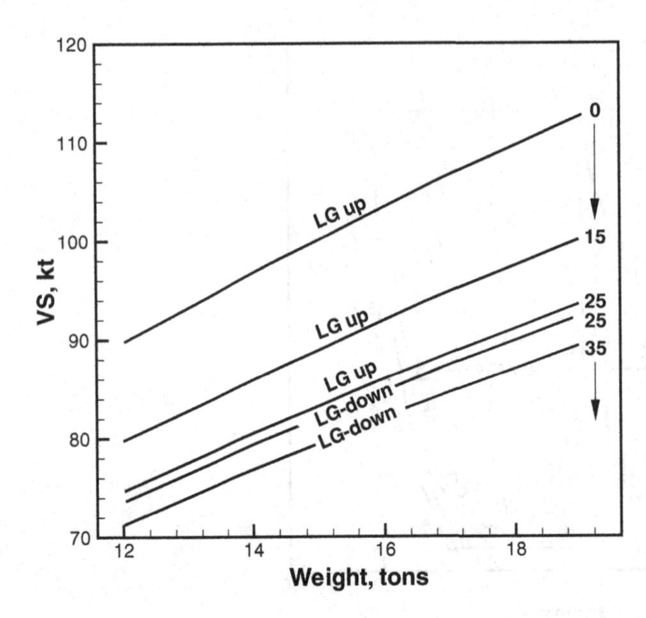


Figure 3.10. Influence of weight and configuration on the stall speed of the ATR72-500 turboprop airplane (adapted from the FCOM).

engines. The weight is always important, but the flap deflection has a strong effect on the stall speed as a result of the increased drag.

A forward CG increases the pitch-down moment, which commands the horizontal stabiliser to produce a restoring pitching moment. The converse is true for an aft CG. We conclude that the more forward the CG, the greater the stall speed.

## 3.6 Operational Moments

The calculation of the ZFW, the take-off weight (TOW) and the CG depend on 1) the weight and CG position before loading any item (fuel, cargo, passengers, service items and fuels); and 2) possible movements in the CG due to items moving during flight.

Some inaccuracies are inherent to this process, due to uncertainty in the weight components (for example, passenger weight distribution) and/or their position on the loaded aircraft. Also, uncertainties in the DOW and DOCG, as previously noted, require consideration of stochastic analysis.

By using the pitching moment, we define the *operational margin* as the difference between the actual aircraft moment and the assumed aircraft moment. The inaccuracy of an item location causes a moment

$$\mathcal{M}_i = W_i(x_i \pm \Delta x_i), \tag{3.12}$$

where $\Delta x_i$ is the uncertainty in the moment arm $x_i$ of item $i$ having a weight $W_i$. The moment error is

$$E_i = \pm W_i \Delta x_i \tag{3.13}$$

and increases with both the weight of the item and its uncertainty. The moment inaccuracy due to *both* weight and location is

$$\mathcal{M}_i = (W_i \pm \Delta W_i)(x_i \pm \Delta x_i), \tag{3.14}$$

which can be expanded into

$$E_i = W_i \Delta x_i + x_i \Delta W_i + \Delta W_i \Delta x_i. \tag{3.15}$$

The latter term is of second order and could be neglected.

**INACCURACY IN PASSENGER DISTRIBUTION.** The passenger distribution on the aircraft is one of the major sources of inaccuracy on both TOW and CG position. With reference to the latter point, each seat row has a known moment arm. If the allocation of seats is known exactly, then the balancing could be done rapidly on a spreadsheet. In practice, calculations are done by assuming an average arm for a cabin section and considering the worst-case scenario. This case occurs when all seats in front of the average moment arm are full and all seats behind the average arm are empty. However, this case is unlikely to occur. More realistically, most operators have various passenger-seat allocation procedures. Typically, window seats are filled first, aisle seats are filled next, and finally the middle seats (if any) are filled. The worst-case inaccuracy is determined, following the specific cabin layout.

With reference to the inaccuracy on the passenger weight, average weights are considered, as shown in Table 3.2, following the recognised regulations* but, in practice, there will be a distribution of weights that generates a *passenger weight inaccuracy*:

$$W_{pax} = \overline{W}_{pax} \pm \Delta W_{pax}. \tag{3.16}$$

Assuming a normal distribution around the average value $\overline{W}_{pax}$ and a standard deviation $\sigma$, we assume $\Delta W_{pax} = 3\sigma$. Thus, if $i = 1, \ldots, n$ denotes the cabin index, and $W_{pax_i}$ denotes the total passenger weight in cabin $i$, we have

$$W_{pax_i} = n_i \overline{W}_{pax} \pm \Delta W_{pax_i}. \tag{3.17}$$

The inaccuracy is the same for all cabins, so that

$$\Delta W_{pax} = \left( \sum_i^n \Delta W_{pax_i}^2 \right)^{1/2} = \sqrt{n(3\sigma)^2} = 3\sigma\sqrt{n}. \tag{3.18}$$

## 3.7 Use of Wing Tanks

The use of the fuel tanks is a complicated matter that depends on the specific aircraft. An example of wing fuel tanks is shown in Figure 3.11, that refers to the Airbus A340-600 airliner. Note that the inner tanks are further sectioned in two parts, one of which is shaded in the graph. The tank of other airplanes may have a considerably different layout.

Table 3.3 is a summary of fuel-tank capacities for some Airbus airplanes. Here is an example of how the fuel is used on the Airbus A340-600. Fuel is fed to the

* Federal Aviation Advisory Circular AC 120-27D, Nov. 2004.

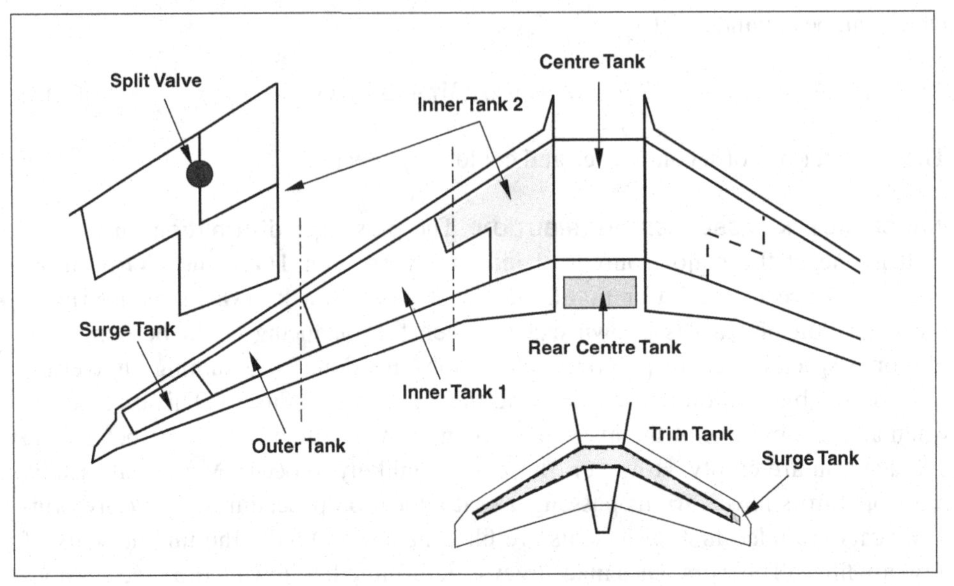


Figure 3.11. Fuel tanks of Airbus A340-600.

engines through the inner wing tanks only; therefore, there must be a supply system from all other tanks. The sequence is the following‡:

1. The ACT fuel (if applicable) is pumped to the centre tank.
2. The centre tank fuel (if applicable) is pumped into the inner wing tanks 1 and 4, to maintain full tanks, until all inner tanks are balanced.
3. The rear centre tank fuel is pumped to the centre tank when the latter one falls below 1,000 kg.
4. The centre tank fuel is pumped to the inner tanks, whilst maintaining the fuel level of the latter tanks between 17,200 and 18,200 kg, until the centre tank is empty.
5. The inner tanks are used down to 4,000 kg each.
6. The fuel from the trim tank is pumped to the inner tanks.
7. The inner tanks are used down to 2,000 kg each.
8. The outer tank's fuel is pumped to the inner tanks.
9. The inner tanks are used until empty.

The outer wing tanks are not used until the sequence has reached point 8 – a point relatively late in the cruise. The whole process is controlled by the flight computer.

## 3.8 Mass and Structural Properties

The determination of the structural mass distribution is generally a problem of aircraft design. However, because the position of the CG and the resulting moments of inertia are crucial to the aircraft performance, we need to estimate the distribution of these mass components on the airplane. The flowchart in Figure 3.12 indicates the

‡ Source: Airbus, Flight Operations Support Department.

Table 3.3. *Fuel tanks of some Airbus airplanes. ACT = Additional Central Tanks; Jet-A1 density at* 15 °C = 0.804 *kg/l*

| Airplane/Tank | Usable [kg] | Volume [l] | Notes |
|---|---|---|---|
| **A319/A320-200** | | | |
| Outer | 691 | 860 | |
| Centre | 6,476 | 8,095 | |
| Inner | 5,436 | 6,795 | |
| ACT-1 | 2,349 | 2,936 | optional |
| ACT-2 | 2,349 | 2,936 | optional |
| Vent tank | n.a. | | |
| **Total** | 18,730 | 23,412 | without ACT |
| **A330-200/300** | | | |
| Outer ×2 | 2,865 | 2,303 | |
| Centre | 32,625 | 26,231 | |
| Inner ×2 | 42,000 | 33,768 | |
| Vent tank | n.a. | | |
| Trim | 4,890 | 3,932 | |
| **Total** | 109,185 | 135,802 | |
| **A340-200/300** | | | |
| Outer ×2 | 2,865 | 2,303 | |
| Centre ×2 | 33,300 | 26,773 | |
| Inner | 33,578 | 26,997 | |
| ACT-1 | 5,652 | 4,544 | |
| ACT-2 | 5,652 | 4,544 | |
| Vent tank | n.a. | | |
| Trim | 4,890 | 3,932 | |
| **Total** | 111,076 | 89,305 | without ACT |
| **A340-600** | | | |
| Outer ×2 | 4,953 | 3,982 | |
| Centre ×2 | 43,279 | 34,796 | |
| Inner (1 & 4) | 19,402 | 15,599 | |
| Inner (2 & 3) | 27,322 | 21,967 | |
| Rear Centre | 15,600 | 12,542 | |
| Vent tank | n.a. | | |
| Trim | 6,269 | 5,040 | |
| **Total** | 168,503 | 135,476 | |

process required to establish the structural properties of the airplane. First, we need to build two separate databases: 1) the mass components, and 2) the position of the centre of each component. With these databases, we can proceed to the calculation of the airplane's CG and thence its moments of inertia with respect to the principal axes through the CG.

Therefore, we follow three steps to determine the relevant mass and structural properties: 1) mass distribution, 2) centre of gravity, and 3) moments of inertia.

The disassembly of the airplane components must be done with care and consideration of what data are available, what data can be inferred and what has to be left behind. In this framework, the critical components of a transport airplane are shown in Figure 3.13. Note that the exact distribution of the *systems* (hydraulics,

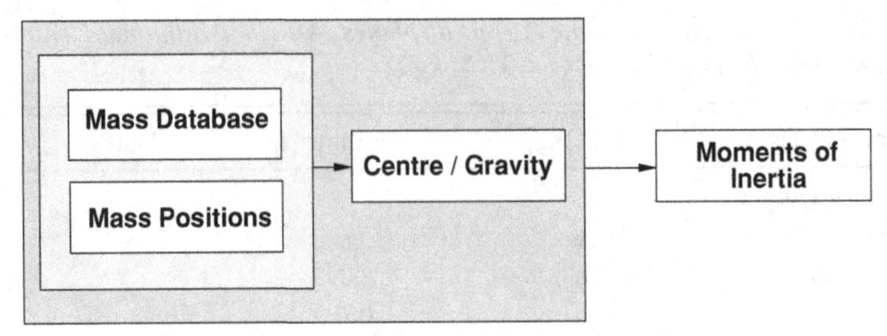


Figure 3.12. Flowchart for calculation of CG and moments of inertia from structural masses and their position.

pneumatics, air conditioning, navigation instruments, avionics, electrical systems, de-icing systems, load handling systems) is unknown. Thus, the systems mass is the difference between the manufacturer's empty mass and the allocated components, as shown in the flowchart in Figure 3.13.

### 3.8.1 Mass Distribution

The determination of the structural weight/mass distribution can be done with different methods, including empirical, statistical and semi-analytical methods. None of these methods is completely satisfying, and in absence of hard data on the airplane, we need to make a choice as to the most suitable candidate for this purpose. A method for the determination of fuselage and wing structural mass (or weight) has been proposed by Ardema *et al.*[16] and implemented in the aircraft design code

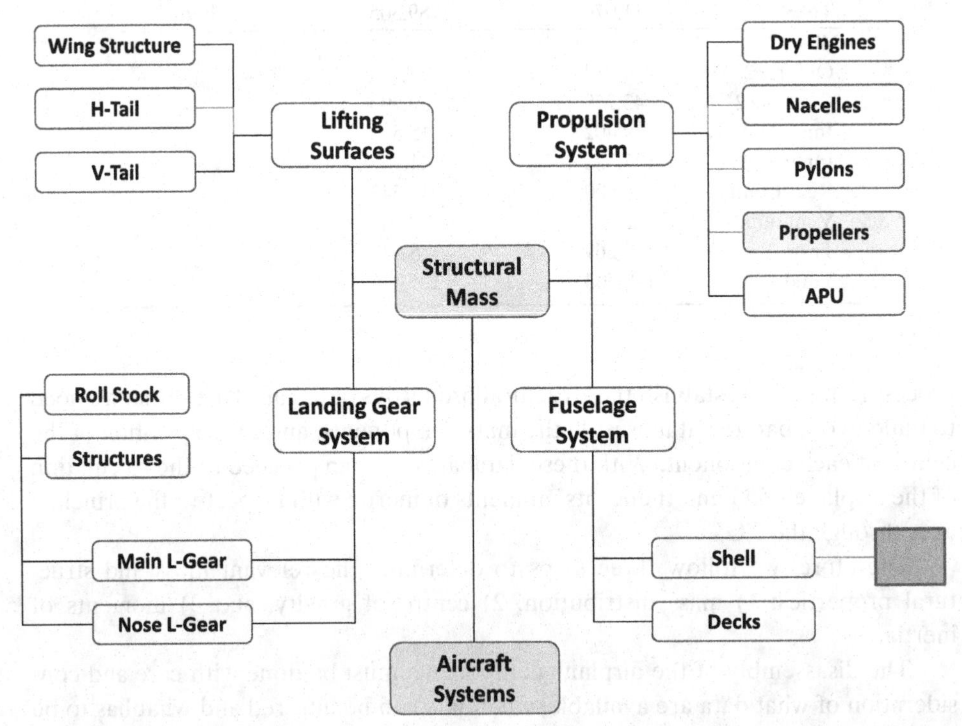


Figure 3.13. Flowchart of airplane mass distribution. The fuselage shell is further split into noise, centre and tail contributions.

ACSYNT. Various other methods exist for the estimation of these and other mass components, including Torenbeek[11], Raymer[12] and Roskam[10]. In this context, we take an intermediate approach, to take advantage of some systems modelling presented in Chapter 2.

**FUSELAGE MASS.** We split this contribution into four components: 1) nose section, 2) central section, 3) tail section, and 4) structural decks. We first calculate the structural mass of the shell fuselage (without the decks). For this purpose, we use Torenbeek's formula[11]:

$$m_{shell} = 0.23 S_g^{1.2} \sqrt{V_D \frac{L_t}{\overline{d}}}, \tag{3.19}$$

where $S_g$ is the wetted area of the shell $[m^2]$, defined as the wetted area of the fuselage, including the portions taken by the wing–body intersection, the empennage-intersection and the landing-gear bays. The tail arm $L_t$ [m] is defined as the distance between the root mid-chord of the wing and the root mid-chord of the horizontal tail-plane; $\overline{d}$ [m] is the average fuselage diameter*. These quantities can be calculated with the methods presented in Chapter 2.

The factor $V_D$ [kt] is the *design diving speed*, defined in terms of equivalent air speed (see also § 8.3). The value of the $V_D$ depends on the reference altitude. Torenbeek does not specify which altitude to consider. The airplane has a design dive Mach number; its corresponding dive speed requires the design dive altitude, which is not readily available.

A number of corrections have been proposed by Torenbeek himself, to include a variety of cases: If the engine is a turbofan, the mass given by Equation 3.19 increases by 8%; if it is a turboshaft, it increases by 5%. If the airplane is a cargo, the shell mass increases 10% and if the engines are mounted aft, it increases by 4%.

In this context, the shell fuselage mass is estimated and its breakdown into nose-central and tail contributions is done according to the relative wetted areas. Thus, we assume that the shell mass per unit area is uniform across the airframe. For example:

$$m_{fuse\_nose} = \left(\frac{A_{wet\_nose}}{A_{wet\_fuse}}\right) m_{shell}. \tag{3.20}$$

The mass estimation of the structural decks is more open to approximations. For example, the decks of a transport airplane have a mass range of $\rho_f$ = 5 to 10 kg/m$^2$; the largest values apply to heavy cargo vehicles. The deck area can be estimated from the airplane layout, and it should be straightforward for single-deck airplanes. Hence, the mass of the deck is

$$m_{deck} \simeq \rho_f A_{deck}. \tag{3.21}$$

* Notice that Equation 3.19 is dimensionally incorrect. If the dive speed is in [kt], the linear dimensions are in [m] and the areas are in [m$^2$], Equation 3.19 gives a mass in [kg] with the factor 0.23. A different factor must be used if the parameters are expressed in imperial units. Similar considerations hold for Equation 3.23.

**LIFTING SYSTEMS.** The lifting systems include the wing, the horizontal and the vertical tail plane. The control surfaces of each system are considered a part of the corresponding system and will not be split into separate contributions. The structural mass of the wing is calculated from a least-squares regression of existing wings in terms of the manufacturer empty mass, $m_e$ [kg]:

$$m_{wing} = -9.095 + 9.189 \cdot 10^{-2} m_e. \tag{3.22}$$

As for the fuselage case, Torenbeek makes use of equations that depend on the design diving speed. For example, the gross wing mass in [kg] is expressed by

$$m = \kappa A \left( 6.2 \cdot 10^{-2} \frac{V_D A^{0.2}}{(\Lambda_{QC} - 2.5)^{1/2}} \right), \tag{3.23}$$

where $A$ is the reference area of the surface in [m$^2$], $\Lambda_{QC}$ is the quarter-chord sweep angle in degrees, the design dive speed $V_D$ is in [kt] and $\kappa$ is a factor equal to unity, except for the trimmable horizontal stabiliser, in which case it is suggested to use $\kappa = 1.1$.

**PROPULSION SYSTEMS.** The propulsion system consists of the engines, the APU, the nacelles, the pylons and the propellers (if any). The dry engine, the APU and the propellers' mass can be extracted from the type certificate documents. The main unknown of this problem is the contribution of the structural masses (nacelles and pylons). We rely on statistical data and find that

$$W_{nac+pylon} \simeq 0.6 W_{eng}. \tag{3.24}$$

**LANDING-GEAR MASS.** The technical literature indicates that the average weight of the landing-gear systems is of the order of 4% MTOW. Following the method proposed by Torenbeek, if $m = \text{MTOW}/g$ is the maximum take-off mass expressed in kilograms, we calculate the nose landing-gear weight from:

$$m_{NLG} = A_1 + B_1 m^{3/4} + C_1 m + D_1 m^{3/2}, \tag{3.25}$$

with

$$A_1 = 9.1, \qquad B_1 = 0.082, \qquad C_1 = 0.0, \qquad D_1 = 2.97 \cdot 10^{-6}. \tag{3.26}$$

The landing-gear mass is split between rolling stock (tyres, wheels, brakes) and other masses (struts, actuators, shock absorbers, systems and controls). The mass of the tyres can be inferred from the tyre specifications, or calculated. The total mass of the rolling stock is variable, but for the nose landing gear, it is about double the mass of the tyres. Thus, the other masses can be calculated by direct difference. This split is important in the determination of the moments of inertia. Within the landing-gear group, the contribution of the brakes is relatively high. Although exact figures are not available, Boeing estimates that replacing steel brakes with carbon brakes on a Boeing 737 saves between 250 and 320 kg, depending on the aircraft version.

The main landing-gear mass is calculated in a similar fashion. The corresponding semi-empirical expression is obtained from Equation 3.25 with coefficients

$$A_1 = 18.1, \qquad B_1 = 0.131, \qquad C_1 = 0.019, \qquad D_1 = 2.23 \cdot 10^{-5}. \tag{3.27}$$

In this case, the mass of the rolling stock is about 3.0 to 3.1 times the mass of the tyres. The method is straightforward if the airplane has two main landing gears; it is more elaborate otherwise. For example, some airplanes have three main landing gears (Airbus A340) and others have four main landing gears (B747, A380). Not all of these systems have the same number of wheels. Thus, in the case of the Airbus A340, the single fuselage landing gear has two axles, whilst the wings' landing gears have three axles. In this case, the split between rolling stock is done according to the total number of wheels. If $m_{LG}$ is the total mass of the landing gear, then

$$m_{MLG} = m_{LG} - m_{NLG}. \tag{3.28}$$

The mass distribution of the main landing gears is

$$m^*_{MLG1} = m_{MLG} x_1, \qquad m^*_{MLG2} = m_{MLG} x_2, \tag{3.29}$$

with

$$x_1 \simeq \frac{(n_w n_{LG})_1}{(n_w n_{LG})_1 + (n_w n_{LG})_2}, \qquad x_2 \simeq \frac{(n_w n_{LG})_2}{(n_w n_{LG})_1 + (n_w n_{LG})_2}, \tag{3.30}$$

where $n_w$ denotes the number of wheels. Note that $x_1$ and $x_2$ are non-dimensional factors. To calculate the contribution of the structural masses, subtract the mass of the rolling stock, which is proportional to the number of wheels of each unit.

**FURNISHINGS.** The mass of the furnishings is roughly proportional to the number of seats. If $n_1$ is the number of economy seats and $n_2$ is the number of first-class seats, then

$$W_{furn} = n_1 w_1 + n_2 w_2, \tag{3.31}$$

where $w_1$ and $w_2$ are average industry values that take into account also the ancillary systems, such as overhead closets, entertainment systems and so on. A typical seat in economy class weighs about 20 to 30 kg, but there is considerable variation in the data. For a first-class passenger, the seat mass is at least double, and the new ultra-thin seat is about half of these values. No data are published to make this estimate more accurate. The furnishings for each passenger are estimated as 1.5 times the mass of their seat.

The mass of the furnishings is a key aspect in the overall financial viability of an airline. Note, in fact, that a reduction of 10% in the mass of the furnishings on an airplane having an average furnishings mass of 10% of the MEW would lead to a decrease of 1% in the OEW. For example, a Boeing B737-900 operating over a range of 2,000 km (1,079 n-miles) at full passenger load and 200 kg of cargo would save about 40 kg of fuel with a reduction of 1% in the OEW.

The methods presented are not always accurate enough, and variations of several percent around the correct values should be expected. Clearly, any substantial inaccuracy will be reflected in the determination of the CG and the moments of inertia. However, by using the method of components, there is scope for improvement at a later time if more data become available.

### 3.8.2 Centre of Gravity

To begin with, we need to establish some reference points and axes. The roll (longitudinal) axis of the airplane is through the centre of the central (cylindrical) section of the airplane. The pitch axis $y$ is along the span, pointing to the right, and the yaw axis $z$ makes a Cartesian coordinate system with $x$ and $y$. The intersection of these axes is initially indeterminate, and we assume that $x = 0$ and $y = 0$ correspond to the fuselage nose position. The coordinates of each airplane component $x_i$, $y_i$, $y_i$ are calculated with respect to this system. We then remove this indetermination by assuming that the pitch axis passes through the CG of the empty airplane. The centre of gravity is calculated on the basis of the distribution of the structural components:

$$x_{cg} = \frac{1}{m} \sum_i m_i x_i, \qquad y_{cg} = \frac{1}{m} \sum_i m_i y_i, \qquad z_{cg} = \frac{1}{m} \sum_i m_i z_i, \tag{3.32}$$

where $x_i$, $y_i$, $z_i$ are the distances between the centroids of the components and the reference zero point on the fuselage nose. For verification of the method used, we should convert the $x_{cg}$ into a percent MAC. If this value falls within the CG range of the airplane, the result is satisfactory.

### 3.8.3 Moments of Inertia

The determination of the airplane's moments of inertia with respect to its principal axes is important for several flight-mechanics calculations. The pitch moment of inertia determines the pitch-up response of the airplane at take-off, and therefore it is especially important. The determination of the moments of inertia is more elaborate than the estimation of the mass components because we need to determine their distribution on the airplane.

Let us start from the basics: the moment of inertia of a lumped mass $m$ from an axis is $I = md^2$, where $d$ is the distance between the mass and the axis. The moment of inertia is linear with respect to the mass, and therefore the moment of inertia of any number of masses can be added.

The empty airplane has different moments of inertia, depending on its configuration (for example, landing gear retracted or deployed). The accuracy of the method is only limited by the amount of information available. To start with, if the moments of inertia of the empty airplane are known, the moment of inertia of a loaded airplane can be estimated by adding the contribution of fuel and payload and any other item on board (assuming that their location is known). However, the main problem is the inertia of the empty airplane. If it is possible to build a finite-element version of the airplane, the problem of the determination of the moments of inertia could be solved with a fair approximation. However, unless the reader has access to commercial data, this is not quite possible. Therefore, we propose an alternative method. The key aspects of the calculations include the following steps:

- Calculate the position of the CG with respect to the reference point on the aircraft.
- Calculate the position of the centres of mass of each component.

- Estimate the mass of each component.
- Calculate the principal moments of inertia from the following equation:

$$I_{xx} = \sum_i m_i(y_i^2 + z_i^2), \qquad I_{yy} = \sum_i m_i(x_i^2 + z_i^2), \qquad I_{zz} = \sum_i m_i(x_i^2 + y_i^2), \tag{3.33}$$

where $x_i, y_i, z_i$ are the distances of the centroid of the sub-component from the centre of gravity, $x_{cg}, y_{cg}, z_{cg}$. Unless otherwise noted, the airplane is symmetric with respect to the vertical plane $x$-$z$, or $y_{cg} = 0$. For convenience, the centre of the axes is the fuselage nose (manufacturers may use an alternative reference). The corresponding radii of gyration are

$$r_x = \left(\frac{I_{xx}}{m}\right)^{1/2}, \qquad r_y = \left(\frac{I_{yy}}{m}\right)^{1/2}, \qquad r_z = \left(\frac{I_{zz}}{m}\right)^{1/2}, \tag{3.34}$$

which are sometimes written in normalised form:

$$\tilde{r}_x = \left(\frac{r_{xx}}{\ell}\right), \qquad \tilde{r}_y = \left(\frac{r_{yy}}{b}\right), \qquad \tilde{r}_z = \left(\frac{r_{zz}}{(b+\ell)/2}\right). \tag{3.35}$$

In the following analysis, we will consider the main items, with the only scope of identifying the largest contributors to the moments of inertia.

**FIXED MASS COMPONENTS.** The *fixed components* of the airplane are:

1. *Fuselage System*: This component is split between fuselage shell and structural floors (decks). The fuselage shell is split in nose, centre and tail sections. For each of these sections we calculate the position of the centroid $(x_c, y_c, z_c)$ with sufficient accuracy, except the central section, if there is a large wing–body blending.

   **Nose and Tail Section.** There are various levels of approximations that can be used. However, due to the uncertainty on other components (see following discussion), there is no need to be too clever. We calculate the position of the centroid of these sections and consider the nose and tail contributions as lumped masses.

   **Central Section.** If the central section can be considered cylindrical, the following equations are used for the component moment of inertia:

$$I_{xx} = \frac{1}{12}m\left[3(r_1^2 + r_2^2) + l^2)\right], \qquad I_{yy} = I_{zz} = \frac{1}{2}mr^2, \tag{3.36}$$

   where $r_1$ is the external diameter of the shell, $r_2$ is the internal diameter and $l$ is the length of the cylindrical body. The shell size can be inferred from cross-sections of the fuselage. In absence of data, the shell thickness is assumed to be $\sim 0.1$ m for a wide-body aircraft. Equation 3.36 defines the moments of inertia of the cylindrical body with respect to the cylinder's principal axes. A shift of the pitch axis is required to refer $I_{yy}$ to the airplane's pitch through the CG. The transformation is the following:

$$I_{yy} = I_{yy} + m\left(\frac{l}{2} - x_{cg}\right)^2. \tag{3.37}$$

**Deck Floors.** For a single-deck airplane, the deck has a known width $w_d$ and an area $A_d$. The equivalent length of a rectangular plate is $l_d = A_d/w_d$. The moments of inertia of this plate with respect to the principal axes are:

$$I_{xx} = \frac{1}{12} m l_d^2, \qquad I_{yy} = \frac{1}{12} m w_d^2, \qquad I_{zz} = \frac{1}{12} m (l_d^2 + w_d^2), \tag{3.38}$$

where in this case $m$ denotes the mass of the whole deck. As in the previous case, we need to transfer these estimates to the axes of the airplane. The transfer functions are:

$$I_{yy} \Rightarrow I_{yy} + m\left(\frac{l_d}{2} - x_{cg}\right)^2, \qquad I_{xx} \Rightarrow I_{xx} + m\left(z_d - z_{cg}\right)^2, \tag{3.39}$$

where $z_d$ is the elevation of the deck with respect to the bottom of the fuselage shell. The problem is more elaborate for double-deck aircraft. Because there are not many of these aircraft, the analysis is done on an *ad-hoc* basis.

2. *Wing system*: This system is modelled with a lumped mass. The mass itself is estimated with the methods discussed in § 3.8.1; the position of the centroid is calculated from the computerised geometry model; the data required are $(m, x_c, y_c, z_c)_{wing}$. This approximation is sufficiently accurate if the wing mass is uniformly distributed on the planform.
3. *Horizontal tail system*: It is treated like the wing; the data required are $(m, x_c, y_c, z_c)_{ht}$.
4. *Vertical tail system*: It is treated like the wing; data required are $(m, x_c, y_c, z_c)_{vt}$.
5. *Engines*: They are treated like full cylinders of known mass and dimensions, which are inferred from the type certificate document. The data required are $(m, x_c, y_c, z_c)_{eng}$. The formulas for the moments of inertia of the solid cylinder are

$$I_{xx} = \frac{1}{8} m d_{eng}^2, \qquad I_{yy} = I_{zz} = \frac{1}{12} m \left(\frac{3}{4} d_{eng}^2 + l_{eng}^2\right), \tag{3.40}$$

where $d_{eng}$ is the average diameter of the engine and $l_{eng}$ is its overall length. These moments are transferred to the principal axes of the airplane. Note also that $I_{xx}$ is proportional to the inertia of the rotating masses of the engine.

A comparison between the solid cylinder with the lumped-mass approximation indicates that there is a 1% difference in the moments of inertia.

6. *Propulsion system*: This contribution includes all of the propulsion mass except the engines. In practice, we split this contribution into pylons ($m_1$) and nacelles ($m_2$). The centroids of these components are known from the geometry model. The split between masses is done according to the relative wetted area (no other quantity is available). If $m$ denotes the total mass contribution of the pylons and the nacelles, then

$$m_1 = m\left(\frac{A_{wet\,pyl}}{A_{wet\,pyl} + A_{wet\,nac}}\right), \qquad m_2 = m\left(\frac{A_{wet\,nac}}{A_{wet\,pyl} + A_{wet\,nac}}\right). \tag{3.41}$$

The corrected centre of mass for the nacelle+pylon combination is

$$x = \frac{m_1 x_{c_{pyl}} + m_2 x_{c_{nac}}}{m}, \qquad z = \frac{m_1 z_{c_{pyl}} + m_2 z_{c_{nac}}}{m}, \tag{3.42}$$

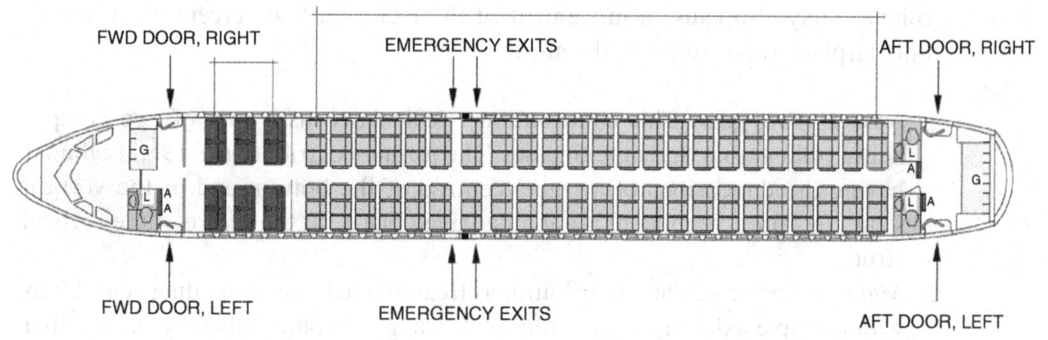


Figure 3.14. Cabin-floor arrangement for the Airbus A320-200 (adapted from Airbus).

which causes a shift of the CG with respect to the centroid of the nacelles:

$$x_{shift} = x - x_{C_{nac}}, \qquad z_{shift} = z - z_{C_{nac}}. \tag{3.43}$$

Note that for a conventional transport aircraft, this shift is backward and upward from the CG of the engine.

7. *Propellers*: The mass of the propeller is known from the type certificate document. The propeller location is in front of the engines; each propeller is treated as a point mass; the data required are $(m, x_c, y_c, z_c)_{prop}$.
8. *Auxiliary Power System*: The APU data are known from the type certificate. This component is generally located in the tail cone. Its position can be given with sufficient accuracy. Although the mass is relatively small, its pitch moment arm is large; the data required are $(m, x_c, y_c, z_c)_{APU}$.
9. *Furnishings*: There can be considerable indetermination on this contribution. In the first place, we split the furnishings between passenger seats and other masses. The passenger seats have known position from the cabin architecture, and therefore their moment arm is known. An example is shown in Figure 3.14 for the Airbus A320-200. However, to simplify the problem, we assume that the seat mass is locally distributed along the deck. For the moments of inertia we use Equation 3.38. A transformation of these contributions to the airplane's principal axes is required as in the previous cases.

   The contribution of the other furnishings masses is where the real problem is. We assume that this contribution is also uniformly distributed over the floor deck, but its vertical position is higher than the seats themselves. From Figure 3.14 we see the position of two galleys at the fore and aft end of the cabin*; these items have large moment arms and contribute to the $I_{xx}$.

**MOVABLE MASS COMPONENTS.** The *movable components* of the airplane are the landing-gear systems. Although the mass of the landing gear is relatively small in proportion to the structural mass, these components are somewhat dislocated and can yield very large moments of inertia with respect to the principal axes of the

* Adapted from Airbus: A320 Airplane Characteristics, Rev. Sept. 2010.

airplane. They also cause a movement of the CG by a few percent MAC on an empty airplane (depending on the airplane).

1. *Nose landing gear*: This contribution is calculated as a sum of two components: rolling stock and other masses. The former contribution is a solid cylinder having the dimensions of a wheel centred at the bottom end of the vertical strut; the latter component is a lumped mass placed at the centre of the vertical strut.
2. *Main landing gear*: This contribution is treated like the nose landing gear. Units with multiple axles are considered by dividing the rolling stock by the number of axles. Axles are displaced by about one wheel diameter.

When the contribution of the rolling stock is calculated with the cylinder approximation, it is important to determine the correct axes. When the landing gears are deployed and retracted, these axes rotate by about 90 degrees. Thus, it is important to determine if folding back into retraction is done longitudinally or laterally.

The control surfaces are associated to the wing system and the tail surfaces. They are not calculated separately and are not allocated among the *movable* components for practical reasons (there are too many unknown data). Any other contribution (systems, avionics, hydraulics, air conditioning, and so on) will be classified under *other items*. It may happen that the estimated structural mass is not the same as the manufacturer's empty mass. If the estimate is below the mark (as it is often the case), the remaining mass is associated to generic "systems", whose position is indeterminate. Therefore, it is suggested to assume this residual mass as *uniformly distributed on the airplane.*

The accuracy of the method proposed depends on the accuracy of its constituents. If the mass components and their position are inexact, one cannot expect great accuracy in the moments of inertia.

**OPERATIONAL MASSES.** When the airplane is loaded with fuel, cargo and passengers, the moments of inertia assume larger values. As in the previous cases, an accurate estimate of the position of these loads is necessary. When these data are unknown, some educated guess is required. For example, the passengers are assumed as uniformly distributed along the cabin (as it is done for the seats and other furnishings). This contribution is calculated with Equation 3.38, via the transformation to the principal axes of the airplane.

**FUEL TANKS.** The contribution of the fuel depends on which fuel tanks have been loaded. Figure 3.11 shows the wing tanks of the Airbus A340. Looking at the data in Table 3.3 for the Airbus A340-600, we note that about 90% of the fuel mass is allocated to the wing tanks. The size and position of the trim tanks are crucial in the overall flight management system because the corresponding fuel is pumped between tanks for trim conditions.

For reference, we report some mass components of the Airbus A320-100/200 in Table 3.4. The mass components are basically the same for the A318, A319-100 and A320-100/200, but the moment arms are different.

Table 3.4. *Weight breakdown of Airbus A320 airplanes; mass in [kg]*

| Wing System | $m$ | Propulsion | $m$ |
|---|---|---|---|
| All slats | 290 | Inlet cowl | 138 |
| Inner flaps | 228 | Fan cowl (avg) | 40 |
| Outer flaps | 244 | Thrust reverser (avg) | 202 |
| All spoilers | 132 | Dry engine, CFM56 | 2,778 |
| Wing tips & winglets | 52 | Complete engine | 3,501 |
| **Fuselage System** | | Pylon | 593 |
| Fwd pax door | 98 | **Landing Gear** | |
| Aft pax door | 96 | NLG system | 327 |
| Fwd cargo door | 121 | Wheel + tyre | 35 |
| Aft cargo door | 121 | MLG system | 1,010 |
| Tail-cone + APU | 460 | Wheel + tyre | 135 |
| | | Brake | 69 |
| **V-Stabiliser** | | **V-Stabiliser** | |
| Box | 365 | Box | 418 |
| Rudder | 88 | Elevator | 97 |
| Removable LE | 48 | Removable LE | 84 |

### 3.8.4 Case Study: Moments of Inertia

Following the methods presented in the previous sections, we have produced computer versions of a number of transport airplanes in configuration empty-landing-gear-out. We calculated the moments of inertia and the corresponding radii of gyration (Equation 3.34) with the FLIGHT code. Prior to these operations, we calculated the position of the centre of gravity, both in absolute coordinates and as a percent of the mean aerodynamic chord. These results are summarised in Table 3.5. Figure 3.15 shows the estimated moments of inertia plotted as a function of the MEW.

Table 3.5. *Airplane mass properties at take-off-empty (no fuel) configuration (calculated)*

| Airplane | mass [kg] | $I_{xx}$ [kgm$^2$] | $I_{yy}$ [kgm$^2$] | $I_{zz}$ [kgm$^2$] | $r_x$ [m] | $r_y$ [m] | $r_z$ [m] |
|---|---|---|---|---|---|---|---|
| ATR72-500 | 12,980 | $0.615 \cdot 10^6$ | $0.430 \cdot 10^6$ | $0.548 \cdot 10^6$ | 6.88 | 5.75 | 6.50 |
| A300-600 | 91,303 | $0.149 \cdot 10^8$ | $0.964 \cdot 10^7$ | $0.139 \cdot 10^8$ | 12.78 | 10.28 | 12.34 |
| A319-100 | 39,838 | $3.163 \cdot 10^6$ | $1.901 \cdot 10^6$ | $2.908 \cdot 10^6$ | 8.91 | 6.92 | 8.54 |
| A320-200 | 42,256 | $0.368 \cdot 10^7$ | $0.236 \cdot 10^7$ | $0.335 \cdot 10^7$ | 9.33 | 7.48 | 8.90 |
| A340-600 | 171,832 | $0.667 \cdot 10^8$ | $0.359 \cdot 10^8$ | $0.475 \cdot 10^8$ | 19.70 | 14.45 | 16.62 |
| A380-861 | 271,746 | $0.138 \cdot 10^9$ | $0.608 \cdot 10^8$ | $0.948 \cdot 10^8$ | 22.52 | 14.96 | 18.68 |
| B737-900W | 43,026 | $0.558 \cdot 10^7$ | $0.412 \cdot 10^7$ | $0.449 \cdot 10^7$ | 11.39 | 9.79 | 10.21 |
| B747-400 | 179,800 | $0.696 \cdot 10^8$ | $0.374 \cdot 10^8$ | $0.471 \cdot 10^8$ | 19.72 | 14.42 | 16.19 |
| B777-300 | 161,355 | $0.612 \cdot 10^8$ | $0.469 \cdot 10^8$ | $0.525 \cdot 10^8$ | 19.48 | 17.04 | 18.04 |
| Dash8-Q400 | 17,222 | $0.107 \cdot 10^7$ | $0.803 \cdot 10^6$ | $0.105 \cdot 10^7$ | 7.90 | 6.83 | 7.80 |
| G550 | 21,074 | $0.164 \cdot 10^7$ | $0.684 \cdot 10^6$ | $0.887 \cdot 10^6$ | 8.83 | 5.70 | 6.49 |

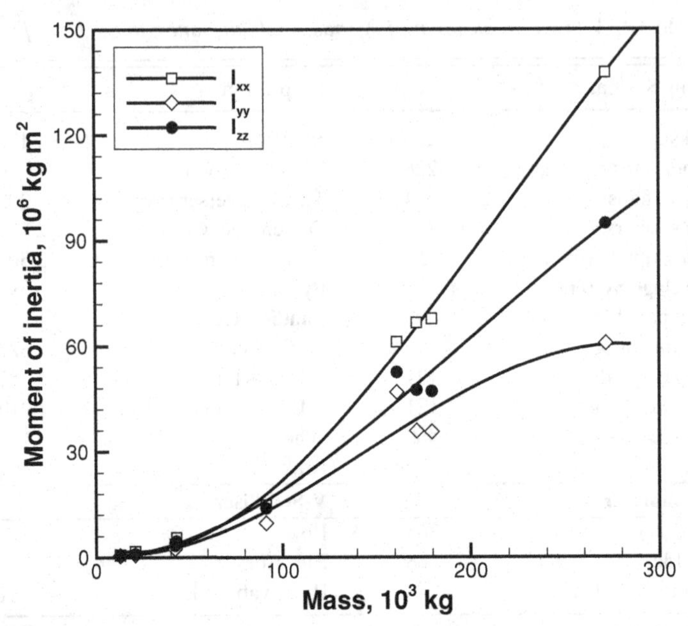


Figure 3.15. Roll, pitch and yaw moments of inertia of some commercial airplanes (calculated); empty airplanes, take-off configuration.

Based on our analysis, the following regression curve is proposed:

$$I = a + bm + cm^2 + dm^3, \tag{3.44}$$

where the mass $m$ is in metric tons and the moments of inertia are in $10^6$ kgm$^2$. The coefficients of this equation are given in Table 3.6. Equation 3.44 is valid for MEW > 10,000 kg. However, if the mass is restricted to MEW > 90,000 kg (large airplanes), the moments of inertia grow linearly with the mass. The corresponding least-square linear fit equation is:

$$I_{xx} = -49.597 + 0.681\,m, \qquad I_{yy} = -0.298 + 0.269\,m, \qquad I_{zz} = -26.11 + 0.441\,m, \tag{3.45}$$

where the empty mass $m$ is in tons ($10^3$ kg) and the resulting moments of inertia are $10^6$ kgm$^2$.

Table 3.5 shows also the calculated radii of gyration, $r_x, r_y, r_z$. A clear relationship with the wing span and the mass cannot be established.

Table 3.6. *Coefficients of Equation 3.44*

| | $a$ | $b$ | $c$ | $d$ |
|---|---|---|---|---|
| $I_{xx}$ | 3.2933 | $-1.6829 \cdot 10^{-1}$ | $4.1838 \cdot 10^{-3}$ | $-6.4162 \cdot 10^{-6}$ |
| $I_{yy}$ | 1.0628 | $-7.2817 \cdot 10^{-2}$ | $2.8159 \cdot 10^{-3}$ | $-6.4169 \cdot 10^{-6}$ |
| $I_{zz}$ | $1.5154 \cdot 10^{-1}$ | $-1.2659 \cdot 10^{-2}$ | $2.4202 \cdot 10^{-3}$ | $-4.0324 \cdot 10^{-6}$ |

## Summary

We have given a critical discussion of the aircraft size and its potential growth. We have also shown how the growth has been made possible by several advances in technology. Larger aircraft have higher cargo efficiencies, but consideration should be given to other factors, such as the actual value of the payload and the range required. We have given a definition of the key design and operational weights of an aircraft and explained how they are calculated. We developed a method for the estimation of the mass breakdown and for the CG; finally, we calculated the moments of inertia and the radii of gyration of a number of modern transport aircraft. We have emphasised the importance of estimating the position of the CG and the need to control and guarantee its movement, both on the ground (whilst loading the aircraft) and in flight. The position of the CG is critical to most flight conditions and will be discussed in the following chapters.

## Bibliography


[1] Finne KN. *Igor Sikorsky: The Russian Years*. Prentice Hall, 1988. ISBN 0874742749.

[2] Grant RG. *Flight: 100 Years of Aviation*. Dorling Kindersley Ltd, 2002.

[3] Baumann A. Progress made in the construction of giant airplanes in Germany during the war. Technical Report TN 29, NACA, 1920.

[4] Cleveland FA. Size effects in conventional aircraft design. *J. Aircraft*, 7(6):483–512, Nov. 1970.

[5] Lange RH. Design concepts for future cargo aircraft. *J. Aircraft*, 13(6):385–392, June 1976.

[6] Whitener PC. Distributed load aircraft concepts. *J. Aircraft*, 16(2):72–75, Feb. 1979.

[7] McMasters JH and Kroo I. Advanced configurations for very large transport airplanes. *Aircraft Design*, 1(4):217–242, 1998.

[8] Staton RN, editor. *Introduction to Aircraft Weight Engineering*. SAWE Inc., Los Angeles, 2003.

[9] Anon. Aerodrome design manual. Part 3. Pavements (Doc 9157P3), 2nd Edition, 1983 (reprinted 2003).

[10] Roskam J. *Airplane Design Part V: Component Weight Estimation*. Roskam Aviation & Engineering, 1985.

[11] Torenbeek E. *Synthesis of Subsonic Airplane Design*. Kluwer Academic Publ., 1985.

[12] Raymer D. *Aircraft Design: A Conceptual Approach*. AIAA Educational Series, 3rd edition, 1999.

[13] Udin SV and Anderson WJ. Wing mass formula for subsonic aircraft. *J. Aircraft*, 29(4):725–732, July 1992.

[14] Beltramo MN, Trapp DL, Kimoto BW, and Marsh DP. Parametric study of transport aircraft systems cost and weight. Technical Report CR 151970, NASA, April 1977.

[15] Galjaard ER. Real time related dry operating weight system. In *54th SAWE Annual Conference*, Paper 2278, Huntsville, AL, May 1995.

[16] Ardema MD, Chambers MC, and Patron AP. Analytical fuselage and wing weight estimation of transport aircraft. Technical Report TM-110392, NASA, May 1996.

## Nomenclature for Chapter 3

| | | |
|---|---|---|
| $a, b, c, d$ | = | coefficients of polynomial Equation 3.44 |
| $A$ | = | gross wing area |
| $A_1, B_1, C_1, D_1$ | = | coefficients in Equation 3.25 |
| $A_{wet}$ | = | wetted area |
| $b$ | = | wing span |
| $c$ | = | factor in Equation 3.23 |
| $d$ | = | diameter |
| $E$ | = | cargo performance index, Equation 3.2; error, Equation 3.13 |
| $k$ | = | factor in Equation 3.23 |
| $\mathcal{I}$ | = | index for CG position, Equation 3.9 |
| $I_{xx}$ | = | roll moment of inertia |
| $I_{yy}$ | = | pitch moment of inertia |
| $I_{zz}$ | = | yaw moment of inertia |
| $\ell$ | = | linear dimension; length; fuselage length |
| $L_t$ | = | horizontal tail arm |
| $m$ | = | mass |
| $m_e$ | = | manufacturer's empty mass |
| $M$ | = | flight Mach number |
| $\mathcal{M}$ | = | moment |
| $n$ | = | cabin index |
| $n_w$ | = | number of wheels |
| $n_1, n_2$ | = | number of seats, Equation 3.31 (economy, business) |
| $r$ | = | distance or radius of gyration, or diameter |
| $r_x, r_y, r_z$ | = | radius of gyration: roll, pitch, yaw, respectively |
| $S_g$ | = | wetted shell fuselage area |
| $V_D$ | = | design diving speed |
| $V_S$ | = | stall speed |
| $w$ | = | width |
| $w_1, w_2$ | = | seats weight (economy and business) |
| $W$ | = | weight |
| $x_1, x_2$ | = | non-dimensional factors in Equation 3.30 |
| $x_{cg}$ | = | position of the centre of gravity |
| $x^*_{cg}$ | = | position of the centre of gravity (reference) |
| $x_i$ | = | moment arm of $i$-th item |
| $x_c, y_c, z_c$ | = | position of centroid |
| $X$ | = | range |
| $z$ | = | vertical coordinate/distance |
| $z_d$ | = | deck's elevation with respect to the bottom of the fuselage |

### Greek Symbols

| | | |
|---|---|---|
| $\Lambda_{QC}$ | = | wing sweep at quarter chord |
| $\kappa$ | = | factor in tail mass Equation 3.23 |
| $\rho_f$ | = | specific mass (density) |
| $\sigma$ | = | standard deviation |

### Subscripts/Superscripts

$[.]^*$ = corrected value
$\overline{[.]}$ = average value
$\tilde{[.]}$ = normalised value
$[.]_c$ = centroid
$[.]_{cg}$ = centre of gravity
$[.]_d$ = deck
$[.]_e$ = empty
$[.]_{eng}$ = engine
$[.]_i$ = item index
$[.]_f$ = fuel
$[.]_{furn}$ = furnishings
$[.]_{LG}$ = landing gear
$[.]_{nac}$ = nacelles
$[.]_o$ = reference value
$[.]_p$ = payload
$[.]_{pax}$ = passenger
$[.]_{pyl}$ = pylons
$[.]_{ht}$ = horizontal tail
$[.]_{vt}$ = vertical tail
$[.]_{zf}$ = zero fuel

# 4 Aerodynamic Performance

## Overview

In this chapter, we present first-order methods, based on the principle of components, to determine the aerodynamic properties of the aircraft. One is tempted to develop methods of high accuracy, based on the physics. These methods are inevitably complex and computer intensive. The requirement for methods of general value inevitably limits the accuracy on any single aircraft. Aerodynamic data of real airplanes are closely guarded. We show how these low-order methods, with a compromise between physics and empiricism, yield results of acceptable accuracy.

We deal with aerodynamic lift in § 4.1. The wide subject of aerodynamic drag (§ 4.2) is split into several sub-sections that present practical methods. The analysis of the aerodynamic drag has a number of separate items, including a transonic model for wing sections (§ 4.3) to be used in the propeller model (Chapter 6), transonic and supersonic drag of wing–body combinations and bodies of revolution (§ 4.4), and buffet boundaries (§ 4.5). We present a short discussion on the aerodynamic derivatives in § 4.6. We expand the aerodynamic analysis to the estimation of the drag of a float-plane in § 4.7. We finish this chapter with a simple analysis of the vortex wakes and the aircraft separation distance (§ 4.8).

**KEY CONCEPTS:** Aircraft Lift, Aircraft Drag, Transonic Drag Rise, Supersonic Drag, Bodies of Revolution, Buffet Boundaries, Aerodynamic Derivatives, Float-Planes.

## 4.1 Aircraft Lift

The determination of the lift characteristics of the wing ultimately depends on the amount of information that is available. Often only the planform shape is known. The twist distribution and the camber of the wing sections, essential for a realistic calculation of the aerodynamics, are most likely unknown. We attempt to overcome this problem with the following simplified analysis. The lift coefficient at cruise configuration is given by

$$C_L = C_{L_\alpha}\overline{\alpha}_e + (C_{L_\alpha}\alpha)_{ht}, \tag{4.1}$$

where $\overline{\alpha}_e$ denotes the mean effective inflow angle across the span. Equation 4.1 shows the separate contribution of wing and tail plane. The lift must be provided with the fuselage nearly horizontal (zero attitude) to avoid passenger discomfort. The contribution of the tail plane must be very small, although it is important for longitudinal trim (Chapter 7). At cruise conditions we neglect the contribution from the horizontal stabiliser. For transport aircraft, values of the $C_L$ at cruise are about 0.4 to 0.6.

The effective mean incidence $\alpha_e$ is the sum of three contributions: 1) the mean incidence in nominal conditions $\alpha_o$, that is, the mean wing angle with respect to the aircraft's longitudinal axis (*a design parameter*); 2) the pitch attitude $\theta$; and 3) the effect of climb rate, $\gamma$:

$$\overline{\alpha}_e = \overline{\alpha}_o + \theta - \gamma. \tag{4.2}$$

At cruise conditions $\gamma = 0$. Therefore, the lift coefficient can be expressed by the following equation:

$$C_L = C_{L_\alpha}\overline{\alpha}_o + C_{L_\alpha}\theta = C_{Lo} + C_{L_\alpha}\theta. \tag{4.3}$$

The effect of wing thickness is of second order compared with the lifting properties of a cambered lifting surface. For the NACA 00XX series the lift increases linearly with the rate of $7 \cdot 10^3$ per 1% increase in relative thickness. Therefore, the angle of attack is far more important in a wing-planform analysis.

### 4.1.1 Calculation of Wing Lift

We use a lifting surface method for the calculation of the wing loading. This is the best method in terms of speed, flexibility and generality. It is flexible in terms of flight conditions that can be prescribed; it is general in the sense that most wing parameters can be added to the model, including winglets, controls surfaces and disjoint wings (such as a wing–tail combination). The major limitations of this approach are the linear aerodynamics underlying the models and the absence of viscous effects.

There are various numerical methods, but ultimately they are all equivalent from a theoretical point of view, although they differ in the computer implementation. A good summary of these methods is given by Katz and Plotkin[1]. Various lifting-surface programs exist in the open domain. Due to the nature and extent of the matter, we cannot devote the attention it deserves. In brief, the lifting-surface theory takes into account the actual shape of the planform, the basic angle effects (dihedral, angle of attack), the wing's mean camber and the winglets. This model is powerful enough to give results of engineering accuracy, including the aerodynamic coefficients, the moments and the aerodynamic derivatives. The calculation of the lift and lift distribution is closely related to the induced drag, discussed in § 4.2.1.

Results of the application of the method to the isolated wing and horizontal tail plane of the A300-600 are shown in Figure 4.1. The data shown are the spanwise distribution of lift coefficient and the product $c\, C_l$ (local chord × local lift coefficient).

### 4.1.2 Wing Lift during a Ground Run

During the initial stages of a ground run, the effective mean aerodynamic incidence $\overline{\alpha}$ changes due to two important factors: 1) the pitch-down attitude of the airplane, and 2) the ground effect. With reference to the former event, the pitch-down attitude

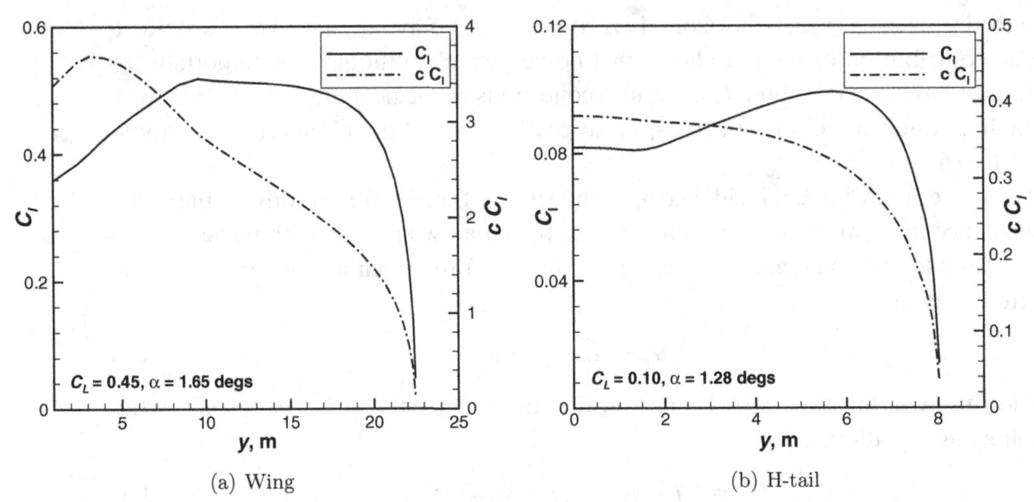


(a) Wing (b) H-tail

Figure 4.1. Spanwise lift distribution from the isolated wing and horizontal stabiliser of the Airbus A300-600 (calculated).

is of the order of 1 degree, although its precise value depends on the airplane. In this case the effective inflow consists of four contributions. In addition to the terms appearing in Equation 4.2, there is the ground effect $\Delta\alpha_g$, which is an apparent angle. Thus, we have:

$$\alpha_e = \overline{\alpha}_o + \theta - \gamma + \Delta\alpha_g. \tag{4.4}$$

Figure 4.2 shows all of the contributions except the ground effect. On further analysis, when the airplane is on the ground, then $\gamma = 0$; hence, the attitude is the result of two contributions: the nominal pitch-down attitude and the effect of the shock absorber of the nose landing gear. As the airplane gains speed, the nose landing gear is relieved of its normal load; the shock absorber relaxes to its unloaded position and contributes to the reduction of the nose-down pitch.

The ground effect is a more complex problem because it depends on the wing configuration, the ground clearance and the value of the $C_L$. In the lifting surface method, the proximity of the ground is accounted for with a *mirror* boundary condition that satisfies the constraint of no-through velocity at the ground.

Figure 4.3 shows examples of ground effect on some transport wings in their full configuration with control surfaces retracted. The winglets are included in the aerodynamic model. The ground effect is a linear function of the $C_L$. However,

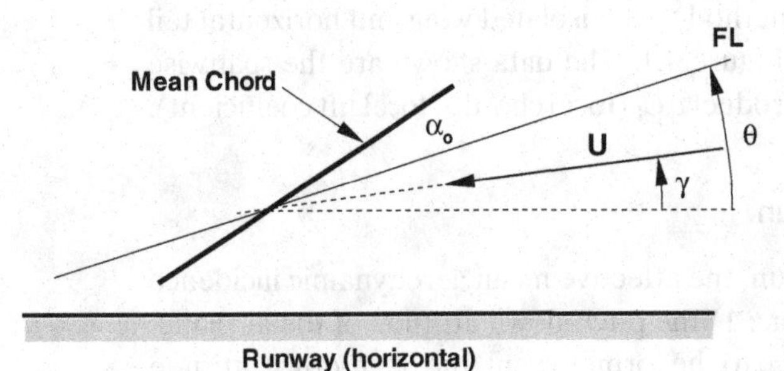


Figure 4.2. Calculation of effective angle of attack; FL = fuselage line.

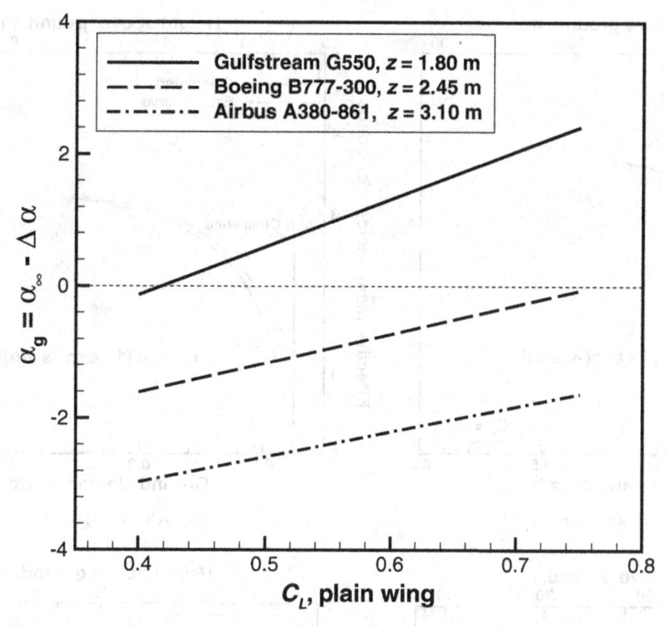


Figure 4.3. Change in mean incidence between OGE and IGE.

the general conclusion is that to achieve a specified $C_L$, the mean incidence has to decrease; in other words, for a given incidence, the wing provides a higher $C_L$. This difference vanishes as the aircraft becomes airborne.

The effect of ground clearance at fixed $C_L$ is shown in Figure 4.4 for four transport airplanes. From this limited analysis, we infer that an empirical expression for the ground effect on the effective mean inflow is

$$\alpha_{IGE} = a_1[1 + \exp(a_2 C_L)] + a_3, \tag{4.5}$$

where $a_1$, $a_2$, $a_3$ are constant factors. The change in effective incidence for a given configuration and $C_L$ is a function of the ground clearance $z/b$:

$$\Delta\alpha_g = \alpha_{IGE} - \alpha_{OGE} = f(z/b), \tag{4.6}$$

where the mean OGE inflow angle is taken from Equation 4.2. The change in mean incidence $\Delta\alpha_g$ is shown in Figure 4.3. For example, if a wing has $C_{Lo} \simeq 0.5$ out of ground effect, then the increase in mean incidence on the ground is estimated between 2.5 and 4 degrees. Assuming for the sake of discussion that $\Delta\alpha_g \simeq 3$ degrees, with $C_{L\alpha} \simeq 4.45$, then we have $\Delta C_L \simeq 0.233$, or $C_{Lg} \simeq 0.73$. This lift coefficient is relatively modest and certainly not high enough to ensure that an airplane will be able to take off.

### 4.1.3 Lift Augmentation

The term *augmentation* will be referred to all cases in which the zero-incidence lift is either increased or decreased by the use of suitable control surfaces, placed at the trailing edge or at the leading edge or both.

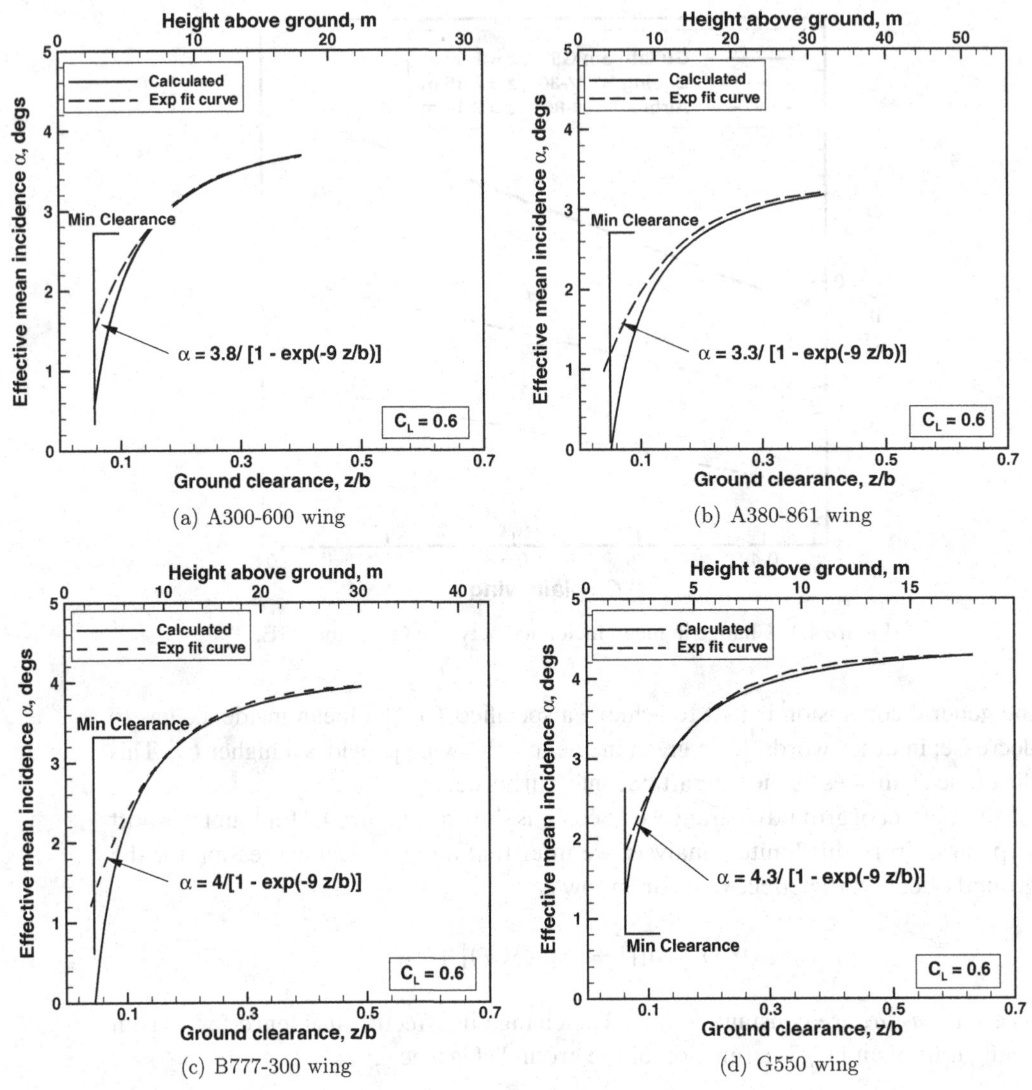


(a) A300-600 wing

(b) A380-861 wing

(c) B777-300 wing

(d) G550 wing

Figure 4.4. Effect of ground clearance on the effective mean incidence for fixed $C_L$ (calculated).

**TRAILING-EDGE FLAPS.** There are various types of flaps, identified by the number of independent lifting surfaces (single-, double- and triple-slotted) and by their mechanics (split-flap, Fowler flap, and so on).

Without going into such detail at this time, during take-off, climb out, final approach, landing and manoeuvring, the aircraft will have at least part of its control surfaces deployed. Most flight-operating manuals have limitations restricting the use of flap above a certain altitude and air speed. The manufacturers do not demonstrate, test or certify operations with flaps deployed at unconventional points in the flight envelope.

There are two classes of methods to consider in the present context: 1) semi-empirical methods, such as ESDU; and 2) first-order aerodynamic calculations based on the lifting-surface theory. Each class of methods has its advantages and drawbacks.

For example, the ESDU has no physics and relies on the engineering interpretation and interpolation of known wing performance. The essential aspect is the calculation of the drag increase at zero angle of attack of the main wing for a flap or slat deflection of a given angle. Each deflection causes a change in the effective wing chord (and hence the effective wing area). However, the coefficients must be referred to the original wing area.

From the ESDU method, the increment in lift at $\alpha = 0$, due to a trailing-edge flap deflection $\delta_f$ is calculated from

$$\Delta C_L = \left(\frac{c'}{c}\right) \Delta C'_L = \left(\frac{c'}{c}\right) C_{L_\delta} \delta_f, \tag{4.7}$$

where $\Delta C'_L$ is the increment in lift coefficient at $\alpha = 0$ based on the effective chord $c'$; $C_{L_\delta}$ is the lift-curve slope due to flap deflection $\delta_f$. The term $C_{L_\delta}$ is calculated from Glauert's thin-airfoil theory:

$$C_{L_\delta} = 2\left[\pi \cos^{-1} c_c + (1 - c_c^2)^{1/2}\right], \tag{4.8}$$

with

$$c_c = 2\left(\frac{c'}{c}\right) - 1. \tag{4.9}$$

For a three-dimensional wing we need to consider a number of corrections, essentially due to the finite span of the flap, its geometrical arrangement and the tip effects. A useful expression is provided by ESDU[2], on the basis of several published and unpublished data:

$$\Delta C_L(\text{wing}) = k_f \Delta C_L J_{p_o} (1 - k_{t_o}) \left(\frac{C_{L_\alpha}}{2\pi}\right) (\Phi_o - \Phi_i), \tag{4.10}$$

where $k_f$ is flap-type factor; $J_{p_o}$ is the efficiency factor of a plain flap on an unswept wing; $k_{t_o}$ is a correlation factor that depends on the geometry of the wing; $\Phi_i$ and $\Phi_o$ are part-span factors depending on the inboard and outboard position of the span and the wing's planform. The limits of validity of Equation 4.10 are given by ESDU[2].

The deployment of flaps, in addition to leading-edge slats, increases the lift to the highest level. If the spoilers are deployed, the lift decreases to an intermediate level. The combined effect of leading-edge slats, trailing-edge flaps and inlay spoilers is shown in Figure 4.5.

Along with a drastic change in lift characteristics, a spoiler deployment causes a large increase in drag. Practical methods for calculating the spoiler characteristics in the ground run are available from ESDU[3]. The lift is given from empirical correlations

$$\Delta C_L = -\left(c_1 \Delta C_{L_{sf}} + c_2 \Delta C_{L_{flap}}\right)_{OGE}, \tag{4.11}$$

where $\Delta C_{L_{sf}}$ is the increment in $C_L$ due to the deployment of split flaps having the same geometry as the spoilers; $\Delta C_{L_{flap}}$ is the increment in $C_L$ due to flap deflection only (spoiler retracted). The numerical factors $c_1$ and $c_2$ in Equation 4.11 are derived empirically. The aerodynamic coefficients on the right-hand side of Equation 4.11 are calculated *out of ground effect* at the angle of attack corresponding to the ground

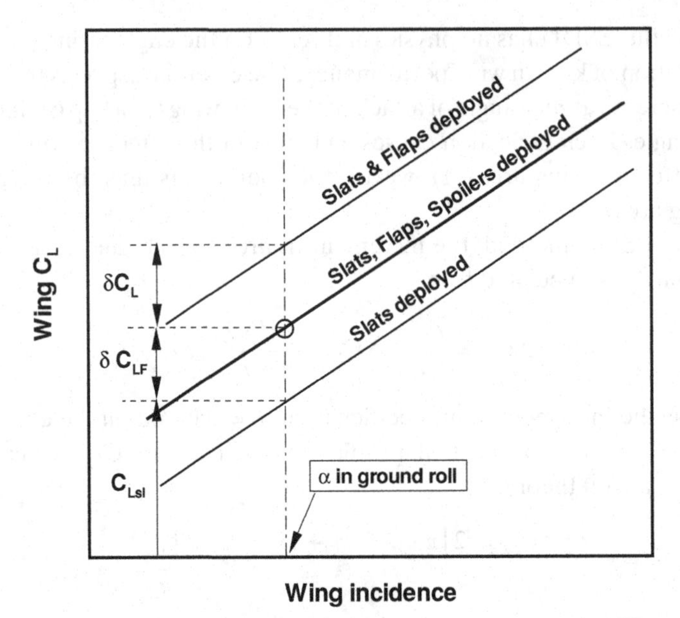


Figure 4.5. Effect of high-lift devices and spoilers on wing lift.

attitude of the aircraft. The change in $C_L$ due to the deployment of flaps and spoilers is

$$\Delta C_{L_r} = \left(\Delta C_{L_{flap}}\right)_{IGE} + \Delta C_L, \tag{4.12}$$

where, in this case, $\Delta C_{L_{flap}}$ denotes the change in $C_L$ due to flap deployment *in ground effect*. If these data are unknown, it is suggested to use the following equation:

$$\left(\Delta C_{L_{flap}}\right)_{IGE} \simeq 1.15 \left(\Delta C_{L_{flap}}\right)_{OGE}. \tag{4.13}$$

The net coefficient on the ground due to the deployment of all of the high-lift systems and the spoilers is

$$C_L = C_{L_{slat}} + \Delta C_{L_r}, \tag{4.14}$$

where $C_{L_{slat}}$ is the lift coefficient of the wing with leading-edge slats deployed and $\Delta C_{L_r}$ is taken from Equation 4.12.

Figure 4.6 shows the high-lift polar of a scaled Airbus A320 with flaps deployed in landing and take-off configuration. The wind-tunnel data[4] are compared with the calculated landing performance and show an acceptable agreement over a wide range of lift coefficients, although there are stronger non-linear effects in the wind-tunnel data at $C_L > 1.9$.

### 4.1.4 Maximum Lift Coefficient

The models presented so far do not include provisions for the determination of the maximum lift coefficient, $C_{L_{max}}$. This calculation is generally not possible with low-order methods, and even sophisticated aerodynamic methods may fail to provide acceptable answers, with the possible exception of two-dimensional airfoils at high Reynolds numbers. The theoretical value of the $C_L$ shown in Figure 4.6 would

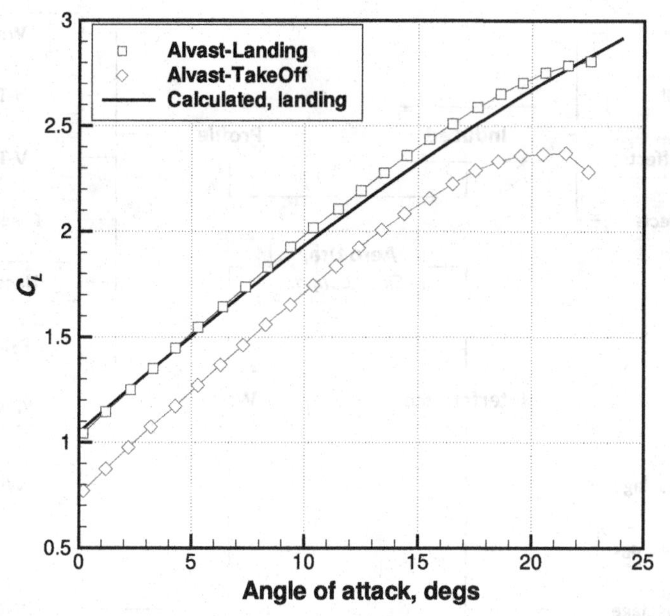


Figure 4.6. High-lift polar (landing and take-off) of a scale model of the Airbus A320.

continue to grow, unless a limiter is set. Again, we need to refer to semi-empirical methods, such as those by ESDU[5], which cannot be reviewed here.

## 4.2 Aircraft Drag

There are two distinct ways of calculating the drag of an aircraft, although both methods rely on the concept of components. These components are either the physical components of the aircraft or the physical components of the drag for a given system[6].

Figure 4.7 shows the logical connection between drag components in a clean configuration. In Figure 4.8 there is the combination of components at take-off and landing and manoeuvre. The remaining components are due to the atmospheric conditions and the state of the runway.

### 4.2.1 Lift-Induced Drag

Even within the constraints of our geometry model (Chapter 2), the lifting surface method described earlier is a better alternative for the calculation of the lift-induced drag. This is done on the Treffz plane by integrating the downwash velocity through the wing span. The lift-induced factor $k$ is calculated from

$$k = \frac{C_{Di}}{C_L^2}. \tag{4.15}$$

Equation 4.15 shows that the drag equation is parabolic. This equation includes all of the effects of twist, dihedral and winglets. When such a computational method is not available, we are forced to use some approximations from known closed-form solutions in low-speed aerodynamics. For example, the lift-induced drag of a

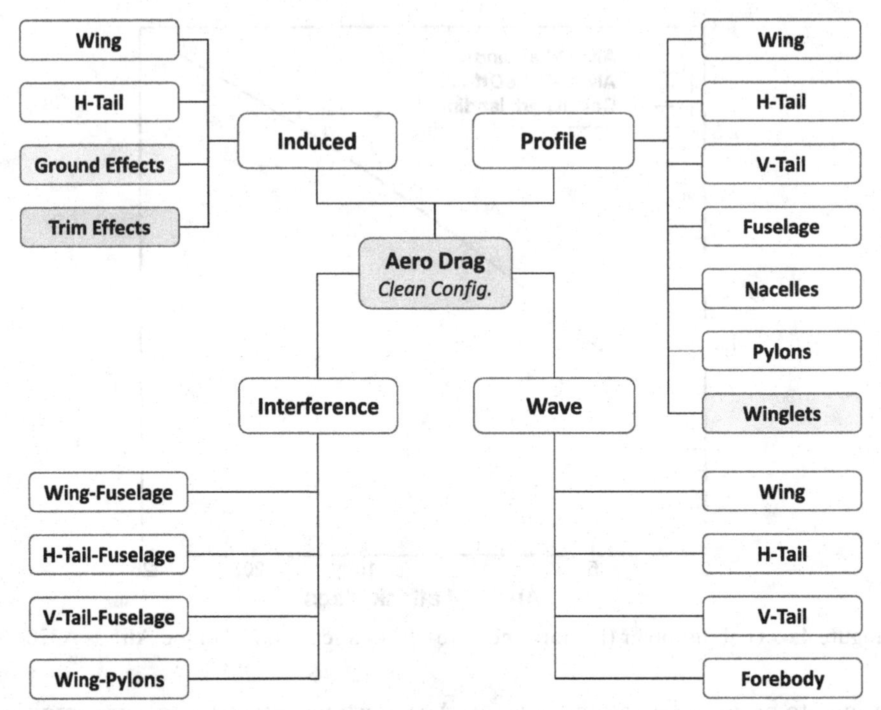


Figure 4.7. Model for the aerodynamic drag of a transport aircraft, clean configuration.

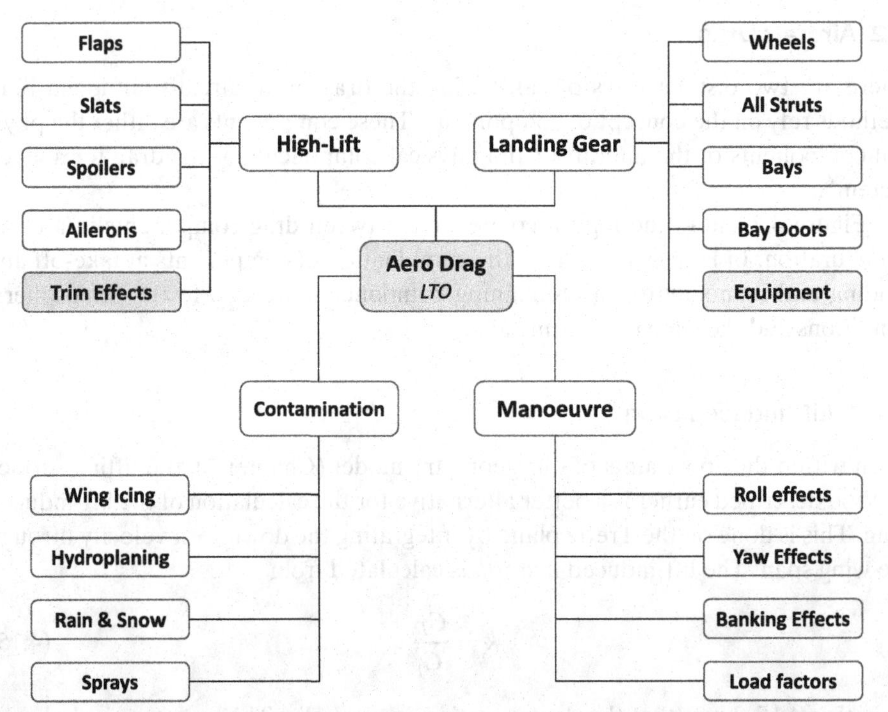


Figure 4.8. Model for the aerodynamic drag of a transport aircraft, take-off, landing and/or manoeuvre configurations.

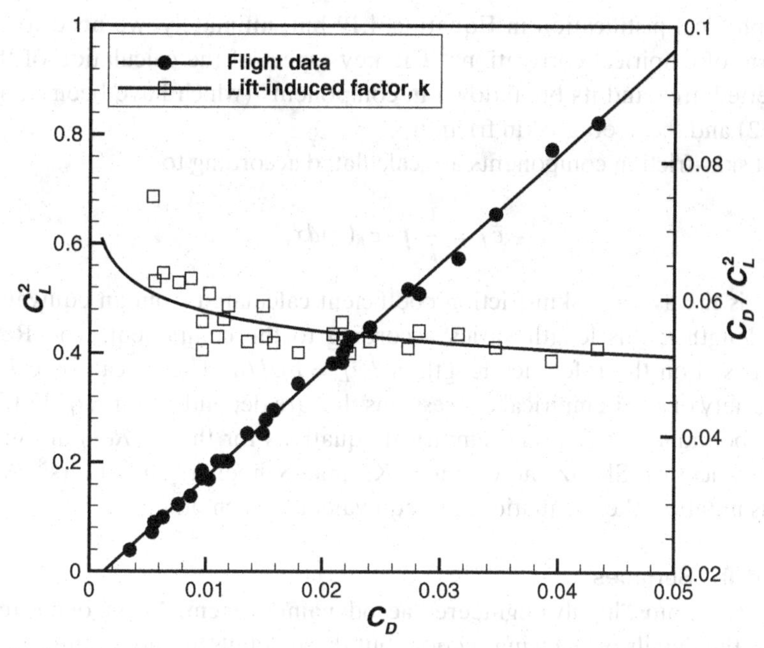


Figure 4.9. Aerodynamic polar of model Douglas DC-10 airplane. Adapted from Callaghan[7].

conventional wing system is

$$C_{D_i} = \frac{1}{e} \frac{C_L^2}{\pi AR}. \tag{4.16}$$

Suitable values of the factor $e$ are $e = 0.90$ at take-off, $e = 0.95$ at cruise. From Equation 4.16, the lift-induced drag factor $k$ becomes

$$k = \frac{1}{e} \frac{1}{\pi AR}. \tag{4.17}$$

Figure 4.9 shows the correlation between experimental data points of the Douglas DC-10 airplane[7] and the parabolic drag model in clean configuration

$$C_D = C_{D_o} + kC_L^2. \tag{4.18}$$

This equation is verified by the linearity between $C_D$ and $C_L^2$ (black points); the induced-drag factor is "nearly" constant, except at low values of the $C_D$ (white points).

### 4.2.2 Profile Drag

The profile drag of the aircraft is calculated by summing the contribution of all of the components. For a cruise configuration we have

$$C_{D_o} = C_{D_{fuse}} + C_{D_{wing}} + C_{D_{ht}} + C_{D_{vt}} + C_{D_{pylon}} + C_{D_{nac}}. \tag{4.19}$$

The engine's drag is accounted for in the calculation of the net thrust; therefore, it is not added to Equation 4.19. If the aircraft has a take-off or landing configuration, then we need to add the effects of the control surfaces and the landing gear. There is some physical justification in Equation 4.19 but, ultimately, we need to rely on some form of empirical correlations. The key aspect of the calculation of the $C_{D_o}$ is the wetted area and its breakdown in components (which have been derived in Chapter 2) and the average skin friction.

Most skin-friction components are calculated according to

$$\overline{c}_f = \frac{1}{l}\int_o^l c_f(x)dx, \tag{4.20}$$

where $\overline{c}_f$ is the average skin-friction coefficient calculated from an equivalent flat plate of length $l$; this length varies according to the component. The Reynolds number based on the reference length is $Re_L = \rho Ul/\mu$. The $\overline{c}_f$ can be calculated from a variety of semi-empirical expressions that are dependent on $Re_L$. For incompressible boundary layers, semi-empirical equations for the $c_f(Re_L)$ are given by Prandtl-Schlichting, Shultz and Grünow, Kármán-Schoenerr and others[8]. At flight Reynolds numbers these equations are equivalent to each other.

### Wing and Tail Surfaces

The wing is the most highly engineered aerodynamic system. Again, one is tempted to look at the details of the wing section, but these details are not readily available. Nevertheless, we can take a generic supercritical wing section of moderate thickness and apply it to all transport aircraft. We use a semi-empirical approach based on the *Eckert Reference Temperature* for the laminar flow, as shown by White[9] and Nielsen[10]; we use the *van Driest II theory*[11;12] for the turbulent flow. The calculation method will be based on 1) laminar skin friction, 2) turbulent skin friction, 3) laminar-turbulent transition point, and 4) correction due to form factors.

**1. LAMINAR SKIN FRICTION.** The first aspect to consider is the flat-plate approximation. If a suitable equivalent flat plate has been defined, the case is treated as a zero-gradient compressible flow at a free stream Mach number $M_\infty$ and flight altitude. The local skin-friction coefficient is based on the Blasius boundary layer, with an additional factor $C^*$ (Chapman-Rubesin constant)

$$c_f = \frac{0.664}{Re_x^{1/2}}\sqrt{C^*}. \tag{4.21}$$

The Chapman-Rubesin factor is related to the ratio between the reference temperature $T^*$ and the external temperature $T_e$, or the temperature at the edge of the boundary layer. Essentially, at the flight altitude this temperature is reduced to the air temperature. The value of $C^*$ resulting from the theory is

$$C^* = \left(\frac{T^*}{T^e}\right)^{1/2}\left(\frac{1+200T_e}{T^*/T_e + 200/T_e}\right), \tag{4.22}$$

with

$$\left(\frac{T^*}{T_e}\right) = 0.5 + 0.039M_\infty^2 + 0.5\left(\frac{T_w}{T_e}\right). \tag{4.23}$$

In Equation 4.23 $T_w$ is the wall temperature. Due to the heat transfer between flow and solid walls at compressible speeds, the wall temperature is different from the external temperature. Due to further heat transfer between the plate and the

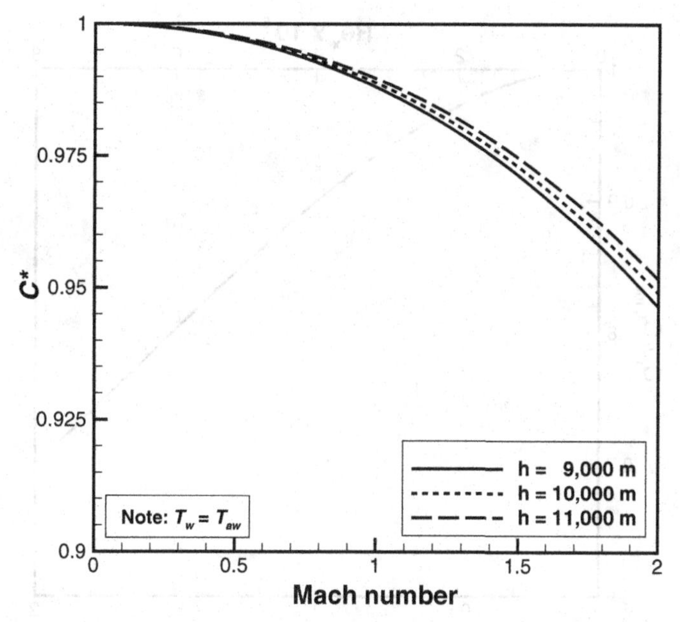


Figure 4.10. Chapman-Rubesin factor as a function of Mach number.

surrounding volume, the wall temperature can be different from the adiabatic temperature, $T_{aw}$. The latter quantity can be calculated from the compressible-flow relation:

$$\frac{T_{aw}}{T_e} = 1 + r\frac{\gamma - 1}{2} M_{\infty}^2, \tag{4.24}$$

where $r = \mathrm{Pr}^{1/2}$ is the recovery factor and $Pr$ is the Prandtl number. Equation 4.24 can be rewritten to show the ratio $T_w/T_{aw}$ or its inverse, so that we can make corrections whenever the wall temperature is known. In fact, multiply and divide Equation 4.24 by $T_w$ and re-arrange

$$\frac{T_{aw}}{T_e} = \frac{T_{aw}}{T_w}\frac{T_w}{T_e}, \tag{4.25}$$

$$\frac{T_w}{T_e} = \frac{T_w}{T_{aw}}\left[1 + r\frac{\gamma - 1}{2} M_{\infty}^2\right]. \tag{4.26}$$

For the calculation of the average skin friction (Equation 4.20), the extent of the laminar region must be estimated. In conclusion, the Chapman-Rubesin factor can be interpreted as the ratio between the turbulent skin friction at compressible and incompressible speeds. The computational procedure is the following:

1. Set the streamwise position and calculate $Re_x$.
2. Calculate the ratio $T_w/T_e$ from Equation 4.26, with $T_{aw}/T_w = 1$.
3. Calculate the ratio $T^*/T_e$ from Equation 4.23.
4. Calculate the Chapman-Rubesin factor $C^*$ from Equation 4.22.
5. Calculate $c_f$ from Equation 4.21.

Figure 4.10 shows the Chapman-Rubesin factor as a function of the Mach number. The flight altitude determines the external temperature and has some effect at high Mach numbers only.

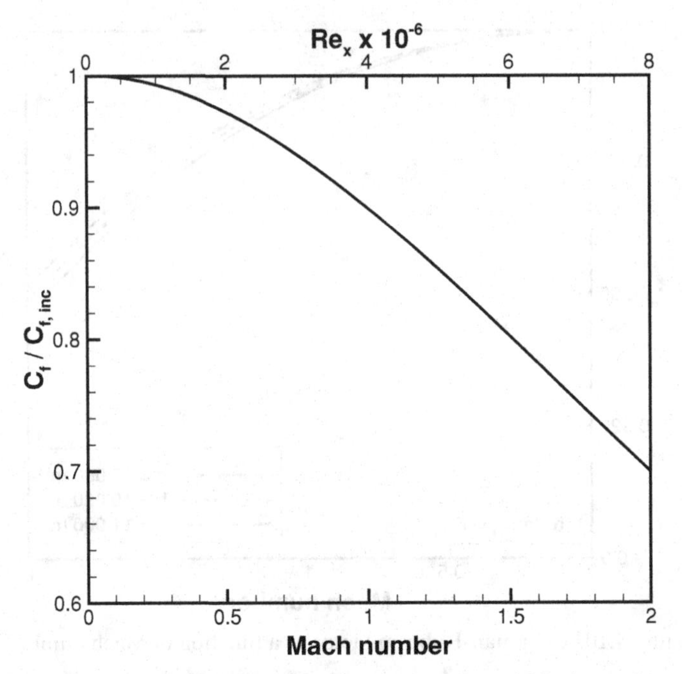


Figure 4.11. Turbulent skin friction as a function of Mach number.

Figure 4.11 shows the ratio between compressible and incompressible $c_f$ as a function of the Mach number. In this case, the flight altitude does not have an effect.

**2. TURBULENT SKIN FRICTION.** The turbulent skin friction is calculated from the van Driest theory, in the implementation shown by Hopkins and Inouye [13;14], to whom we refer for the numerical details. In brief, the incompressible skin-friction coefficient is calculated from the Kármán-Schoenerr formula

$$\frac{0.242}{\hat{c}_f} = \log_{10}(\hat{Re}_x \hat{c}_f), \tag{4.27}$$

where the hat quantities denote *incompressible conditions*. The relationship between incompressible and compressible Reynolds numbers is

$$\hat{Re}_x = \mathcal{F}_x Re_x. \tag{4.28}$$

Equation 4.27 is implicit in $\hat{c}_f$ and must be solved iteratively, for example, by using a Newton-Raphson method from a first-guess solution. The solution of Equation 4.27 is related to the *compressible* skin friction by

$$c_f = \frac{\hat{c}_f}{\mathcal{F}_c}, \tag{4.29}$$

where the factor $\mathcal{F}_c$ and the factor $\mathcal{F}_x$ are the key to the whole method; $\mathcal{F}_c$ is calculated from

$$\mathcal{F}_c = \begin{cases} m/(\sin^{-1}\alpha + \sin^{-1}\beta)^2, & M_e > 0.1 \\ 0.25\left(1+\sqrt{\vartheta}\right)^2, & M_e < 0.1 \end{cases}, \tag{4.30}$$

where $\vartheta$ is a relative temperature:

$$\vartheta = \frac{T_w}{T_e},$$

and

$$m = r\left(1 + \frac{\gamma - 1}{2} M_\infty^2\right),$$

$$\alpha = \frac{2A^2 - B}{(4A^2 + B^2)^{1/2}}, \qquad \beta = \frac{B}{(4A^2 + B^2)^{1/2}}, \tag{4.31}$$

$$A = \left(\frac{m}{\vartheta}\right)^{1/2}, \qquad B = \frac{1 + m - \vartheta}{\vartheta}. \tag{4.32}$$

The Reynolds factor $\mathcal{F}_x$ is a ratio between dynamic viscosity and is calculated from

$$\mathcal{F}_x = \frac{\mathcal{F}_\mu}{\mathcal{F}_c}, \qquad \mathcal{F}_\mu = \frac{\mu_w}{\mu_e}. \tag{4.33}$$

These viscosities are calculated with Sutherland's law (§ 8.1.1). In summary, the computational procedure is the following:

1. Set the streamwise position and calculate $Re_x$.
2. Calculate the factor $\mathcal{F}_c$ from Equation 4.30.
3. Calculate the factor $\mathcal{F}_x$ from Equation 4.33.
4. Calculate $\hat{Re}_x$ from Equation 4.28.
5. Solve Equation 4.27 using the Newton-Raphson method.
6. Calculate the compressible $c_f$ from Equation 4.29.

This procedure is valid at a given streamwise position $x$. Integration over the turbulent section of the wing yields the average value of the $c_f$ and hence the turbulent $C_{D_o}$. To make the procedure more accurate, we can divide the wing in streamwise strips and integrate both streamwise and spanwise.

**3. LAMINAR-TURBULENT TRANSITION.** Now we need an estimate of the laminar-turbulent transition point. This aspect can be subject to much debate, and it is in fact one of the most complex problems in aerodynamics. Again, we consider the flow past a flat plate and use the Blumer–van Driest semi-empirical correlation[15], which was verified by experimental data

$$Re_{trans}^{1/2} = 10^6 \frac{\sqrt{132500\, k_t + 1} - 1}{39.2\, k_t^2}, \tag{4.34}$$

where $k_t$ is an average level of free stream turbulence. For a value of $k_t = 0.2\%$, we find $Re_{trans} \simeq 0.165 \cdot 10^6$. This is a relatively high Reynolds number for turbulent transition.

**4. SHAPE EFFECTS.** The skin friction calculated with this method applies only to a flat plate without pressure gradients. To recover these gradients and the local curvature effects of the boundary layer, it is customary to correct the $c_f$ with a form factor. A suitable expression is the following:

$$f_f = 1 + 2.7(t/c) + 100\,(t/c)^4. \tag{4.35}$$

For example, if the average wing thickness is $t/c = 0.1$, then $f_f \simeq 1.28$. The calculation of the drag of the other lifting-surface units (horizontal and vertical tail) follows the same method as the main wing.

### Fuselage Drag

The fuselage contributes to the $C_{D_o}$ through the skin friction and the base drag. For convenience, the fuselage is divided into three main sections: forebody (or *nose*), centre and after-body (or *tail*), as in Chapter 2. A first-order method for the nose drag is the *turbulent nose cone* theory, derived first by van Driest and reported by White[9]. The theory leads to a correlation between the flat-plate drag and the drag of a nose cone at zero incidence. A necessary condition for this correlation is the same $Re_L$. For a turbulent flow

$$\frac{c_{f,cone}}{c_{f,plate}} \simeq 1.087 \div 1.176, \tag{4.36}$$

depending on the state of the boundary layer; the approximation is rather poor. The length of the plate is equal to the length of the nose section. If the flow is laminar, the ratio between skin-friction coefficients is $\sqrt{3}$. Clearly, this is not a good thing because the turbulent $c_f$ calculated from Equation 4.36 is only 9 to 17% above the reference value of the flat plate.

The drag of the central section can be calculated in at least two ways. First, by using White's semi-empirical expression for the drag of a very long cylinder

$$C_D = 0.0015 + \left[0.30 + 0.015\left(\frac{l}{d}\right)^{0.4}\right] \mathrm{Re}_l^{-1/3}. \tag{4.37}$$

Equation 4.37 is thought to be accurate to within 9%. The second option is to use the turbulent skin friction from van Driest's theory (as in the wing-drag analysis). In this case, the calculation is done with a Reynolds number growing from the value corresponding to the nose, $Re = \rho U l_{nose}/\mu$. The same method is applied for the calculation of the skin-friction drag of the tail section, with starting Reynolds number $Re = \rho U(l_{nose} + l_{centre})/\mu$.

**BASE DRAG.** A simple expression for the base drag of the fuselage is given by Hoerner's equation[16]:

$$\Delta C_D = \frac{0.029}{\sqrt{c_f}}\sqrt{\frac{A_{base}}{A_{wet}}}. \tag{4.38}$$

However, this equation lacks accuracy, and more detailed methods are needed to take into account effects such as the upsweep and the flow incidence. This is done, for example, with ESDU[17] that provide technical details regarding the effects of the upsweep and allow more precise calculations based on the actual geometry of the aft fuselage. The drag increment due to upsweep is calculated from

$$\Delta C_D = [G(\alpha, \beta_u, \epsilon) - G(\alpha, 0, \epsilon)]\overline{c}_d \frac{A_p}{A_{ref}}, \tag{4.39}$$

where $\alpha$ is the angle of attack of the fuselage; $\beta_u$ is the mean upsweep angle; $\epsilon$ is the downwash angle created by the wing half-way of the upswept section; $A_{ref}$ is the cross-sectional area of the fuselage; $A_p$ is the planform area of the upswept section of the fuselage; and, finally, $\overline{C}_D$ is the average drag coefficient of the rear fuselage (empirical value). The function $G$ is calculated from

$$G(\alpha, \beta_u, \epsilon) = \frac{\sin(\alpha - \beta_u)\sin^2(\alpha - \beta_u - \epsilon)}{\cos\beta_u}, \tag{4.40}$$

where $G$ is a tabulated function of its arguments.

### 4.2.3 Wave Drag

The wave drag is the result of a system of shocks, weak as well as strong ones, and shock-induced flow separation around the aircraft. In principle, all aircraft components are subject to these effects. However, we will consider the separate contribution from the lifting surfaces and from the fuselage forebody. Flight at higher Mach numbers, as is the case of high-performance aircraft, requires additional analysis from the wing–body junction.

**WAVE DRAG FROM LIFTING SURFACES.** From a practical point of view, an essential parameter is the divergence Mach number, defined by

$$M_{dd} = \kappa_A - \kappa C_L - \frac{t}{c}, \tag{4.41}$$

with $\kappa_A$ variable from 0.87 (NACA 6-series airfoils) to about 0.95 (modern supercritical wing sections) and $\kappa$ is another factor that represents the effects of the aerodynamic lift on the divergence Mach number

$$\frac{dM_{dd}}{dC_L} = -\kappa. \tag{4.42}$$

Average values for supercritical wing sections are $\kappa = 0.1 - 0.14$. Analysis of experimental data for low-speed airfoils, such as NACA 0012 and 23012, indicates that $\kappa \simeq 0.2$. A similar expression was reported by Malone and Mason[18]:

$$M_{dd} = \frac{\kappa_A}{\cos\Lambda_{LE}} - \kappa\frac{C_L}{\cos^2\Lambda_{LE}} - \frac{t/c}{\cos^3\Lambda_{LE}}. \tag{4.43}$$

Therefore, once the wing sweep is fixed and the $C_L$ is calculated at the flight conditions, the divergence Mach number is only a function of the "technology level".

There is a relationship between the wave drag and the critical Mach number. This relationship is found in Hilton[19]:

$$C_D = 20(M - M_c)^4, \quad M > M_c. \tag{4.44}$$

A relationship between $M_c$ and $M_{dd}$ can be calculated from Equation 4.44 and Equation 4.43. In fact, derive Equation 4.44 and recall the definition of divergence Mach number

$$\left(\frac{dC_D}{dM}\right)_{Mdd} = 80(M_{dd} - M_c)^3 = 0.1, \tag{4.45}$$

$$M_c = M_{dd} - 0.108. \tag{4.46}$$

The corresponding drag coefficient is now calculated from Equation 4.44. Note that $C_{D_w} = 0$ at $M \leq M_c$. This calculation must be repeated for the tail-plane and the fin.

**FUSELAGE FOREBODY.** An earlier example in the geometry analysis of the fuselage of some commercial aircraft (§ 2.4.2) has highlighted that the forebody is a complex shape that seldom can be associated to a recognised geometry. Thus, the determination of the wave drag should require higher-order methods for the solution of the flow field. Due to the computational efforts required to carry out these simulations, it is preferable in this context to outline criteria that allow the rapid estimation of the required coefficients, on the basis of data sets produced by higher-order methods. One such example is given in Ref.[20], to which we refer for further details. The forebody drag is associated to a limited number of shape parameters, namely the fineness $l/d$, the non-dimensional nose radius $2r/d$ and the free Mach stream number. For conventional shapes, the wave drag of the forebody is small, up to $M_\infty \simeq 0.85$.

### 4.2.4 Interference Drag

The interference drag is due to local flow disruption at the intersection between major components, in particular at the fuselage–wing, fuselage–tail–plane, fuselage–fin and fuselage–pylon intersections. To control separation, many airplanes have fine-tuned details such as aft-fuselage fences, flow diverters on nacelles, wing fences, vortex generators and blended wing-bodies. Many of these details are of second order and are not easy to model within the present context. However, the problem can be reduced to the determination of some interference factors, yielding a total interference drag

$$C_{D_{int}} = \sum_i C_{D_{int}}(i), \tag{4.47}$$

where the sum is extended to all of the relevant interference cases. A number of key cases are identified: intersection between a rectangular lifting wing and a wall; intersection between a swept lifting wing and a wall. A basic case is a wing intersecting a plane at 90 degrees. Hoerner[21] gives the following expression:

$$C_{D_{int1}} = \left[c_1(t/c)^3 + c_2\right]\frac{c^2}{A}. \tag{4.48}$$

A number of semi-empirical corrections are then introduced to take into account 1) the effects of wing sweep; 2) the effects of the intersection angle (for example, intersections at angles other than 90 degrees); and 3) the effects of wing lift. These effects are given by the following equations, respectively:

$$C_{D_{int2}} = \left(c_3\Lambda^2 + c_4\Lambda\right)\frac{c^2}{A}, \tag{4.49}$$

$$C_{D_{int3}} = \left(c_5\varphi^2 + c_6\varphi\right)\frac{c^2}{A}, \tag{4.50}$$

$$C_{D_{int4}} = c_7 C_L^2 \frac{c^2}{A}. \tag{4.51}$$

For a lifting body intersecting another lifting body at 90 degrees, there is a further term:

$$C_{D_{int5}} = \left[c_7(t/c)^4 + c_8(t/c)^2\right]\frac{c^2}{A}. \tag{4.52}$$

The coefficients $c_i$ are determined by using best-fit of experimental data. Some of these terms may be missing. For example, the fin produces no lift; hence Equation 4.51 yields zero. The effect of the dihedral $\varphi$ is to reduce the interference drag.

The equations provided were derived for cases without fillets and represent the worst possible scenario. The use of filleting at the junction can greatly reduce the interference drag to about one-tenth. Therefore, before proceeding with the final calculation of the interference drag, we need to check whether fillets are available. Note that all of these quantities do not depend on the flight conditions (altitude and Mach number), and therefore they are essentially constant values. More recent studies on interference drag are available[22–25] but, ultimately, the accuracy of these methods must be evaluated in the context of the accuracy achieved on the other components.

### 4.2.5 Drag of the Control Surfaces

The control surfaces discussed in this section are "flat-panel" components: spoilers, ailerons, elevators, rudders. Due to the limitation of the data that can be extracted from the geometry model, also the flaps are effectively considered as flat panels. The essential data of these components are the span, the average chord, the chordwise and spanwise position and the deflection (five parameters per component). We discuss in particular the aerodynamic effects caused by the spoilers, which can be used in conjunction with flaps and slats.

The term *spoiler* is used to denote a deployable panel mounted on the upper surface of an aircraft wing. Spoilers are used mostly for braking the aircraft on the ground upon landing. They disrupt the lift and create a down-force that adds weight to the wheels. Added weight increases the rolling resistance and facilitates the braking process. The effect on the drag is given by a combination of slats, flaps and spoilers. The spoiler terms give rise to a drag increment

$$\Delta C_D = \Delta C_{D_o} + \Delta C_{D_{out}} + \frac{1}{2}\left[K^2(\Delta C_{L_r})^2 + \frac{C_L^2}{\pi AR}\right], \tag{4.53}$$

where $\Delta C_{D_o}$ is the increase in profile drag due to the deployment of slats and flaps for those sections of the wing *also affected* by spoiler deflection; $\Delta C_{D_{out}}$ is the increment in profile drag due to the deflection of the flaps on those portions of the span *not affected* by spoiler deflection. The last term in Equation 4.53 is the lift-induced drag; in particular, this term denotes the lift-induced drag for the difference in lift coefficient. The term $K$ is a part-span factor. The complete calculation procedure, the accuracy and the limitations are discussed by ESDU[3].

A spoiler deflection above 80 degrees is equivalent to a flat plate in normal flow mounted on a wall (the wing's upper surface). For this case some drag data are available as a function of the aspect-ratio of the spoiler. For a ratio span/chord above 3, the normal force coefficient is basically constant, $C_N \simeq 1.2$, based on the spoiler's area. Hence, the contribution of one spoiler to the overall drag is

$$C_D \simeq C_N \left( \frac{A_{spoiler}}{A} \right). \tag{4.54}$$

The contribution to the lift is more difficult to estimate because the spoiler effectively blocks the flow on the upper side of the wing. In the context of the whole system, we can assume that the lift is reduced to zero along the span with the spoiler. Therefore, the residual lift is

$$C_L = C_{L_g} \left( \frac{b_{spoiler}}{b} \right). \tag{4.55}$$

The lift in ground effect is to be calculated by considering the flaps fully retracted. Like any other aerodynamic device, the spoilers are not effective at low speed. Therefore, during the braking process, they can be retracted when the speed decreases below a threshold. The remaining phase of the braking is done with the wheel brakes.

In the absence of more accurate data, the drag due to flap deflection can be estimated from the following equation[26]:

$$\Delta C_D \simeq k_{flap} \left( \frac{c_{flap}}{c} \right)^{1.38} \frac{A_{flap}}{A} \sin^2 \delta_f, \tag{4.56}$$

where $k_{flap} = 1.7$ for plain and split flaps; $k_{flap} = 0.9$ for slotted flaps. The calculation must rely on a reasonable value of the lift coefficient in ground effect, which provides the lift-induced contribution.

### 4.2.6 Landing-Gear Drag

The calculation of the landing-gear drag is an elaborate process, not without uncertainties, and relies exclusively on semi-empirical equations. The problem is split into manageable parts by considering isolated landing-gear units, as indicated in Figure 4.8.

The most comprehensive source of information on this subject is ESDU 79015[27], who provide various derivations and include interference factors. Roskam[28] provides semi-empirical relationships that are valid for smaller airplanes, based on earlier research, mostly at NACA. Yet, there is a limited amount of data to be used for validation and verification, despite considerable research over many years; most of the research focuses on aerodynamic noise and structural optimisation. A further analysis of the landing-gear assembly is in Chapter 16.

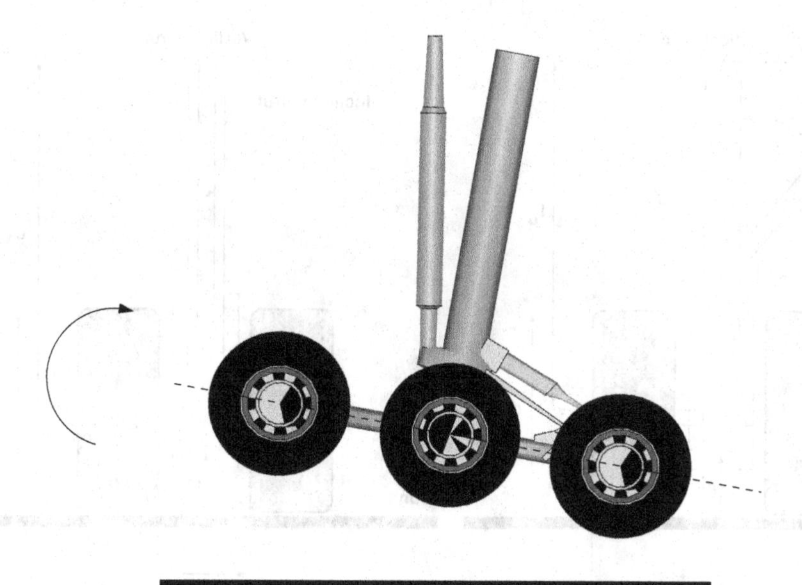

Figure 4.12. Multi-axle landing gear off the ground; wheels not aligned with incoming flow.

The present discussion is limited to retractable landing gears for transport aircraft. Each unit is made of a number of wheels, generally arranged in tandem; a main vertical strut; horizontal axles in numbers depending on the number of wheels; and various other struts, obliquely placed against the incoming flow (laterally or longitudinally), along with a series of complicating factors and surface roughness due to various systems. The whole unit must retract into a bay inside the fuselage or the wing. Therefore, there is a cavity bay of appropriate shape and bay doors. When the doors are open, they are roughly aligned with the incoming flow in order to offer least resistance to the air. In some modern airplanes, the doors are sectioned in several parts, some of which are closed to minimise the cavity drag.

For a landing gear placed under the wing, there are additional complications arising from the effects of flap deflection and wing thickness; both contribute non-linear effects to the drag. A systems approach to the determination of the drag is shown in the top right corner in Figure 4.8.

In the process of determination of drag coefficients, reference quantities must be assumed, which are inevitably dependent on the specific sub-system.

**1. WHEEL DRAG.** We must distinguish between rotating and non-rotating wheels and between wheel arrangements on one, two or three axles. When there is more than one axle and the wheels are above the ground, there is an angle between the line through the axles and the incoming flow. This arrangement is shown in Figure 4.12, which clearly indicates that the front wheels faces the free stream and the aft ones are partially shielded. In practice, the drag of the front-facing wheels is higher. The drag of a non-rotating multi-wheel combination is

$$\left(\frac{D}{qA}\right)_{bogie} = \left(\frac{C_D}{C_{Do}}\right) C_{Do} \left(\frac{b_u d_w - mn}{A}\right), \tag{4.57}$$

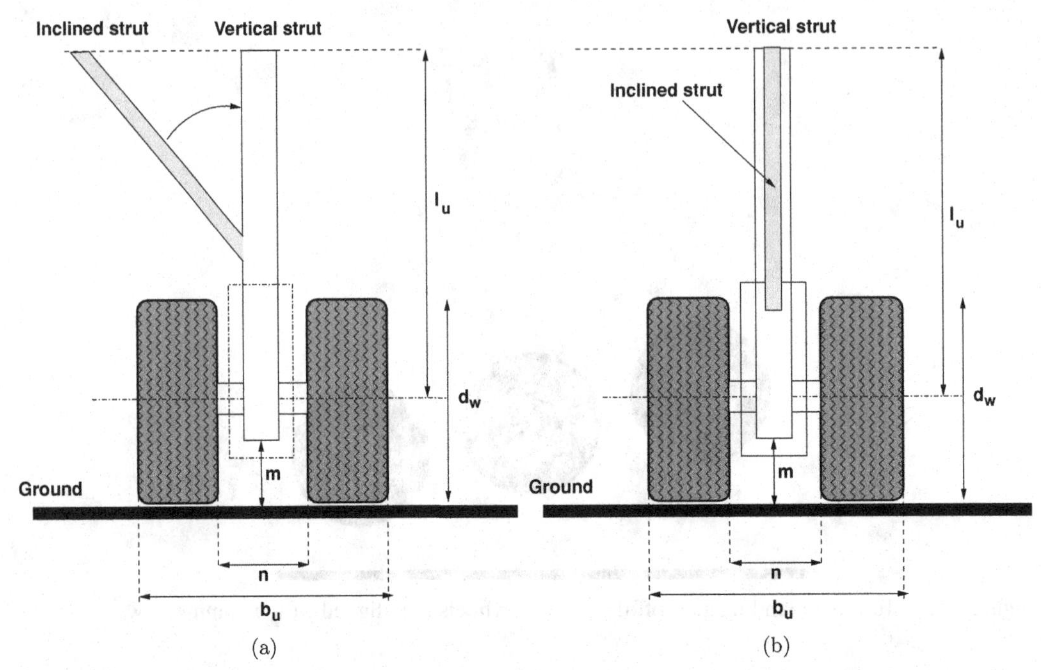


Figure 4.13. Typical landing gear of transport aircraft.

where $C_{Do} = 1.2$ for sub-critical Reynolds numbers ($Re_c < 5 \cdot 10^5$) and $C_{Do} = 0.65$ for supercritical Reynolds numbers. The Reynolds number is calculated by using the wheel's diameter and the free stream velocity. The definition of the other quantities is given in Figure 4.13. The parameter $b_u$ is always the maximum width of the wheel combination. If $s = d_w/b_u$, then the ratio $C_D/C_{Do}$ is calculated from

$$\frac{C_D}{C_{Do}} = \begin{cases} 0.642 - 0.2660s + 0.0846s^2 - 0.0081s^3, & 0.3 < s < 4.0, \quad Re < Re_c \\ 0.912 - 0.4850s + 0.1390s^2 - 0.0111s^3, & 0.3 < s < 5.0, \quad Re > Re_c \end{cases}. \tag{4.58}$$

Note that these equations are polynomial approximations of reference data and are only valid within the specified range. To take into account the presence of a plate wall near the wheels, Equation 4.57 is multiplied by an interference factor $c$, whose value is 2.5 when wheel and plane are in contact (that is, wheel on the ground) and equal to 1 when the diameter $d_w$ is equal to the distance from the plane.

**2. LANDING-GEAR STRUTS.** The drag of the vertical post is calculated as an isolated cylinder with a rough surface (to take into account pipes, brackets and other mechanical details). This is acceptable for cases in which the other main strut is lateral, as shown in Figure 4.13. For cases where there is a tandem condition, as in Figure 4.12, the spacing between struts, the diameter ratio and the slenderness of the struts must be taken into account. The Reynolds number in this case is based on the strut diameter. There is a critical Reynolds number around which there is a "jump" in drag. Methods based on semi-empirical analysis exist[29] to calculate the aerodynamics of various arrangements of finite-length cylinders.

**3. CAVITY BAY AND DOORS.** The dimensions of the bay can be inferred from the dimensions of the landing gear and from the retraction mechanics. In fact, for a main landing gear retracting sideways to its bay under the wing, the depth of the cavity must be of the order of the width $b_u$ to accommodate the wheels. The width must be about $d_w n_w/2$ ($n_w$ = number of wheels) and the spanwise extension must be of the order of the main strut.

The flow around and inside the bay is unsteady; any attempt to refer to steady-state of average characteristics is a challenge. The aerodynamic phenomena within the bay are complicated by the presence of the strut and other mechanical details. In some cases, the bays are occluded by partly closed doors, which may cause resonance phenomena and considerable acoustic emission. With these caveats in mind, the drag of the bay is written as

$$\frac{D}{qA} = C_{Db}\left(\frac{b_u l_u}{A}\right), \tag{4.59}$$

where the coefficients $C_{Db}$ is the drag coefficient of the bay at the relevant Reynolds number and given geometrical characteristics. Approximate empirical equations for the drag of the bay are the following:

$$C_{Db} \simeq \begin{cases} -0.0131 + 0.2363\left(\frac{d_w}{l_u}\right) - 0.4007\left(\frac{d_w}{l_u}\right)^2 + 0.1924\left(\frac{d_w}{l_u}\right)^3, & b_u/d_w = 1.0 \\ 0.0192 + 0.0586\left(\frac{d_w}{l_u}\right) - 0.1360\left(\frac{d_w}{l_u}\right)^2 + 0.0738\left(\frac{d_w}{l_u}\right)^3, & b_u/d_w = 2.0 \\ 0.0132 + 0.0537\left(\frac{d_w}{l_u}\right) - 0.1227\left(\frac{d_w}{l_u}\right)^2 + 0.0648\left(\frac{d_w}{l_u}\right)^3, & b_u/d_w = 4.0 \end{cases} \tag{4.60}$$

Equation 4.60 is only valid in the range $0.25 < d_w/l_u < 1$. If $d_w/l_u > 1$, assume $C_{Db} \simeq 0.015$. The bay doors will be considered aligned with the free stream, although this is not always the case. The drag is essentially a profile drag and can be calculated with the formula

$$C_D = \frac{0.455}{\log_{10} Re_l^{2.58}}. \tag{4.61}$$

In this case the Reynolds number is based on the length of the strut, $l_u$.

### Computational Procedure for Landing-Gear Drag

1. Set the basic geometrical data required for the landing gear.
2. Calculate drag of the isolated wheel unit, Equation 4.57.
3. Calculate drag of the vertical strut.
4. Calculate drag of the bay cavity, Equation 4.59.
5. Calculate drag of the bay doors.
6. Sum all drag components to find uninstalled drag of landing gear.

**4. INSTALLATION EFFECTS.** The installation drag for landing gear under the wing is of the order of 10 to 15%. The flap deflection is another factor that at take-off and landing is of the order of 0.6 to 0.7. The total landing-gear drag is found from summing the drag of all units. ESDU reports that this method may yield results that are off by 50% from the correct value, even when full consideration is given to each component. Therefore, the accuracy of the results must be evaluated with flight data. The wing-thickness effect is included with the following correction, extrapolated from ESDU data sheets[27]:

$$f_1 = 1 + (2.15 - 2.90833\, t_m + 1.91667\, t_m^2 - 0.64167\, t_m^3 + 0.08333\, t_m^4)(t/c), \quad (4.62)$$

where $t_m = 2\, l_{\text{v-strut}}/\text{MAC} - 0.2$ is the ratio between the length of the main vertical strut and the mean aerodynamic chord; $t/c$ is the average wing thickness. For a landing gear upstream of the flap, the effect of flap deflection is calculated from

$$f_2 = \left[1 - (0.0186 - 0.018t + 0.0053\, t^2)\, E\delta_f\right]^2, \quad (4.63)$$

with

$$E = 3.65833c_m - 5.373c_m^2 + 2.91667c_m^3, \quad (4.64)$$

and $c_m = c_{flap}/\text{MAC}$. Finally, we have

$$C_D = C_D f_1 f_2. \quad (4.65)$$

When the geometry of the landing gear is not known, we can use Torenbeek's semi-empirical expression (Torenbeek[30], Appendix G):

$$\frac{D}{qA} = k_u \frac{m^{0.785}}{A}, \quad (4.66)$$

where $k_u$ is a factor depending on the amount of flap deflection and $m$ is the aircraft's mass. It can be assumed that $k_u$ varies linearly with the flap angle $\delta_f$. The corresponding function is

$$k_u = \left(0.28 - 0.13\frac{\delta_f}{\delta_{f,\max}}\right) 10^{-3}, \quad (4.67)$$

with $\delta_f$ the flap deflection in degrees. The main difference between Equation 4.66 and the expanded ESDU method is that the latter is not dependent on the aircraft's weight. Hence, a comparison between the two approaches is necessarily contrived. Nevertheless, a comparison was made for the main landing gear of the reference transport airplane, and the result is shown in Figure 4.14, at the weights indicated. Equation 4.66 yields values of the $C_D$ within ±13%.

### 4.2.7 Environmental Effects

Environmental effects on the aircraft's drag are important at take-off and landing. These effects include the *precipitation drag*, a term that usually refers to the change in response due to a combination of displacement and spray. The displacement component is the drag produced when the aircraft rolls on a surface at least partially covered by standing water, snow, slush or ice. A practical method for the displacement drag consists of using the following equation:

$$D = \frac{1}{2}\rho^* S_{tyre} C_D K U^2. \quad (4.68)$$

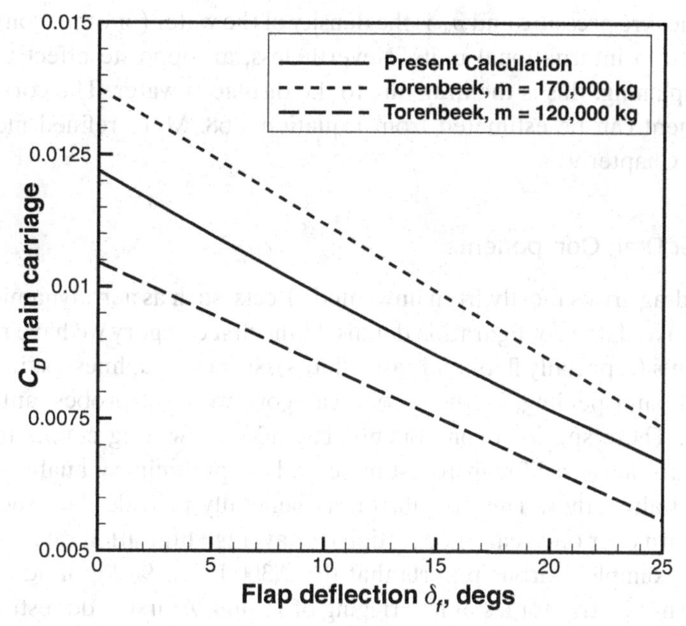


Figure 4.14. Main landing gear $C_D$ of reference transport airplane (calculated).

In Equation 4.68 $S_{tyre}$ is the frontal area of the tyre; $\rho^*$ is the density of the contaminant (water, snow, ice, and so on); $C_D \simeq 0.75$ and $K$ is a wheels coefficient, estimated at 1.5 to 1.6. When referring to the whole aircraft, it might be more practical to use the factor $0.75K$, in addition to the other drag components, in order to have the overall effect on the aircraft's speed on the ground. A correction is required to refer the coefficient to the wing area, so that the displacement $C_D$ is accounted for by

$$\Delta C_D = 0.75K\frac{S_{tyre}}{A}. \tag{4.69}$$

A more detailed method was developed by van Es[31] for rolling on a snow-covered runway. The method breaks down the displacement drag in components and grain-density functions.

The spray drag is created by the impingement of the spray from the tyres onto the lower parts of the aircraft (fuselage and wings). The spray drag can be estimated from

$$C_{D_{spray}} \simeq 24lC_{D_o}, \tag{4.70}$$

where $l$ is the length of the fuselage behind the point at which the spray plume hits the bottom of the fuselage; $C_{D_o}$ is the skin-friction drag of the fuselage.

**AQUAPLANING.** Aquaplaning (or *hydroplaning*) depends on the aircraft's speed and on the tyre conditions. The critical speed required for aquaplaning can be estimated from the following equation:

$$V = 17.5\left(\frac{p}{\rho_w}\right)^{1/2}, \tag{4.71}$$

where $p$ is the tyre pressure and $\rho_w$ is the density of the water (or other contaminant); $p$ and $\rho_w$ are in international units. Nevertheless, an opposite effect takes place during aquaplaning: this is the drag due to the displaced water. The corresponding drag coefficient can be estimated from Equation 4.68. More refined methods are discussed in Chapter 9.

### 4.2.8 Other Drag Components

Additional drag arises mostly from unwanted effects, such as aerodynamic deterioration and unavoidable configuration details. In the first category we have mis-rigging of components (especially flaps and cargo doors), surface roughness, skin dents, loss of seals and paint peeling. In the second category we have probes, antennas and design gaps. These spurious components may add a few drag counts to the total drag. They are always difficult to estimate and, in preliminary analysis, they can be neglected altogether. The manufacturers generally provide data such as additional fuel burn over one year of operation (for average utilisation rates) for specific details. For example, Airbus reports that the A300 burns 90 kg of additional fuel over a 2,000 n-mile trip for a slat mis-rigging of 15 mm. A first-order estimate of the corresponding drag is the following:

$$\Delta D = \Delta T = \frac{\Delta m_f}{f_j} = \frac{1}{2}\rho A \Delta C_D U^2, \tag{4.72}$$

with translates into

$$\Delta C_D = \frac{\Delta m_f}{f_j} \frac{1}{\rho A U^2} \simeq 0.06 \text{ drag counts.} \tag{4.73}$$

This value is considerably lower than the accuracy that can be achieved with the methods presented in this chapter. Equation 4.73 shows an example of coupling between aerodynamics and propulsion. Thus, the accuracy on the $\Delta C_D$ depends *also* on the accuracy of the specific fuel consumption used in the analysis.

### 4.2.9 Case Study: Aerodynamics of the F4 Wind-Tunnel Model

The geometrical model for this configuration was discussed in § 2.7. That geometry, which is just a wing–body combination, was the subject of extensive wind-tunnel testing[32;33] and is now used to verify the aerodynamic models shown in this chapter. Figure 4.15 shows the calculated aerodynamic polars and their comparison with selected wind-tunnel data. Figure 4.15b shows the drag polar plotted as $C_D$ versus $C_L^2$. If the induced-drag factor were constant, as also shown in Figure 4.9, this curve would be a straight line. Neither the wind-tunnel data nor the calculations show that this is the case. The correlation between the theoretical model and wind-tunnel data is completely acceptable, considering the simplification used and other inaccuracies which may arise in other parts of the comprehensive flight mechanics model. Furthermore, even sophisticated CFD models have difficulties in providing fail-proof results[34–36].

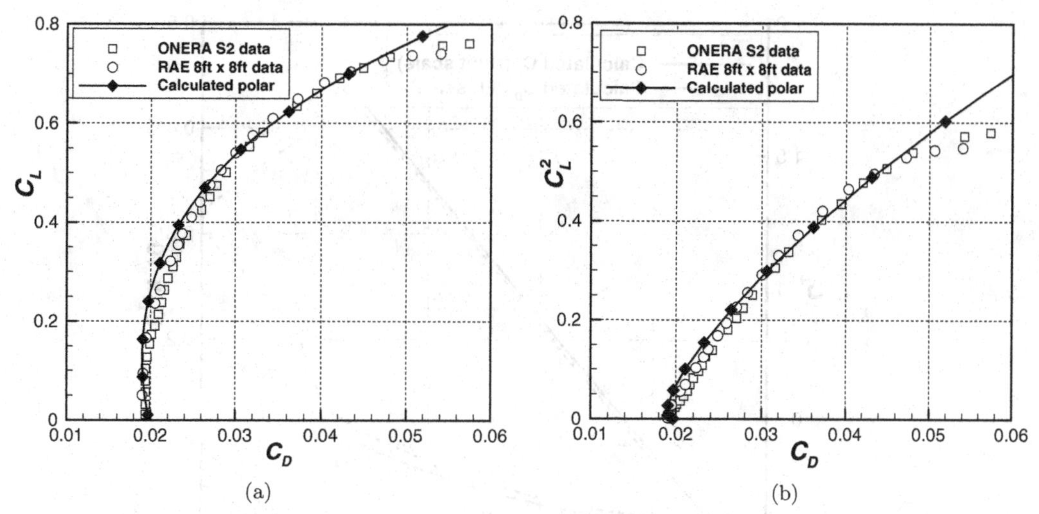


Figure 4.15. Aerodynamic analysis of the DLR F4 wing–body airplane model.

### 4.2.10 Case Study: Drag Analysis of Transport Aircraft

We now apply our method to the calculation of drag at cruise conditions to a typical aircraft. Examples for validation are scarce, but one useful set of data is the large report by Hanke and Nordwall[37] for the Boeing B747-100. This report includes most aerodynamic derivatives and parametric effects, not easily found for other airplanes.

Figure 4.16 shows the comparison between the present method and the data interpolated from Hanke and Nordwall for the Mach number effects on the $C_D$ at constant $C_L$. For this aircraft, MMO = 0.85; therefore, calculations at higher speeds

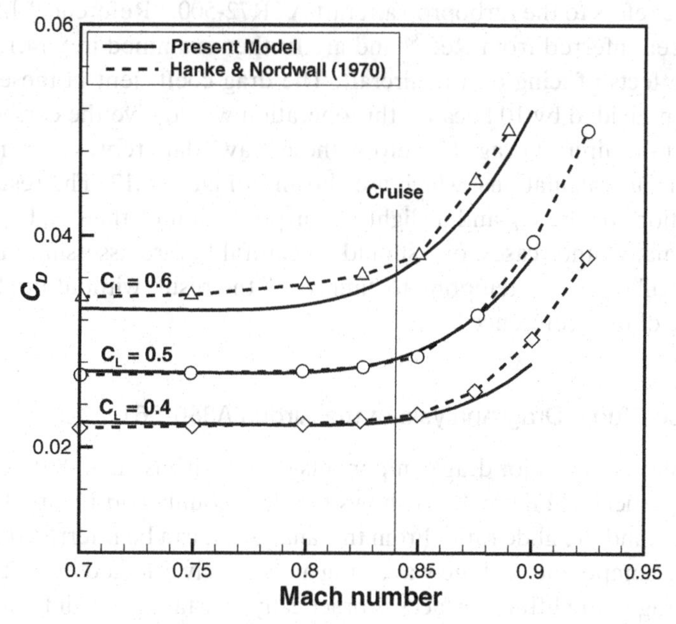


Figure 4.16. Mach number effects on the $C_D$ at constant $C_L$ for the Boeing B747-100.

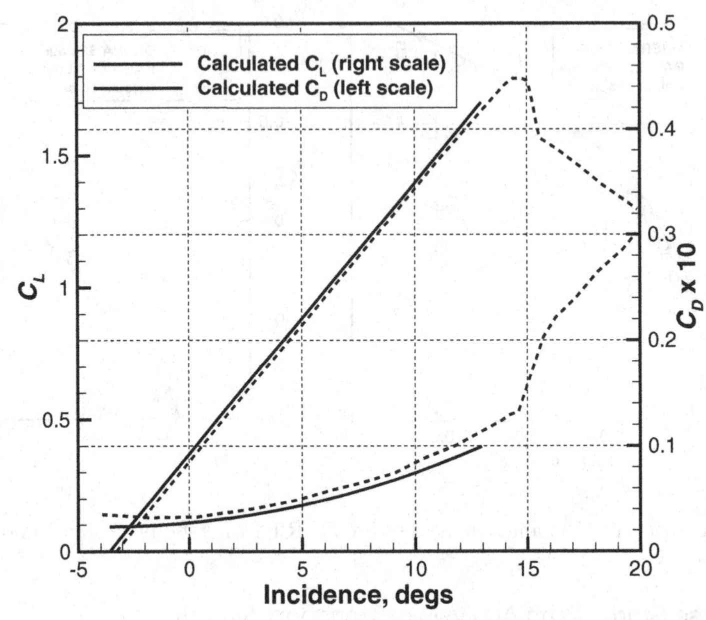


Figure 4.17. Lift and drag coefficients of the ATR72-500. Dashed lines are reference data, adapted from Ref.[38].

are not of practical interest. The typical cruise Mach number is shown in the graph for clarity.

### 4.2.11 Case Study: Drag Analysis of the ATR72-500

The next case refers to the turboprop aircraft ATR72-500. "Reference" lift and drag data have been inferred from Ref.[38] and are further examined in § 14.1, where we discuss the effects of icing on the aircraft. The drag coefficient in these reference data has been divided by 10 because this operation would give the correct order of magnitude of the aircraft drag. Therefore, these "raw" data represent some form of reference for our calculations, which are shown in Figure 4.17. The result shows a good prediction for the $C_L$ and a slight under-prediction of the $C_D$, by about 8 to 10%. As in many other cases, one should be careful before assessing this result in too much detail or even attempting to "improve" the result without full knowledge of the nature of the "reference" data.

### 4.2.12 Case Study: Drag Analysis of the Airbus A380-861

Figure 4.18 shows the major drag components of the Airbus A380-861 versus cruise altitude. In particular, Figure 4.18a shows the drag counts and Figure 4.18b shows the overall $C_D$ and the glide ratio. From this analysis, it can be inferred that both the induced-drag component and the wave drag are strongly affected by altitude, whilst the profile drag of the lifting surfaces is practically constant. The glide ratio reaches a maximum at a certain altitude, and then it decreases.

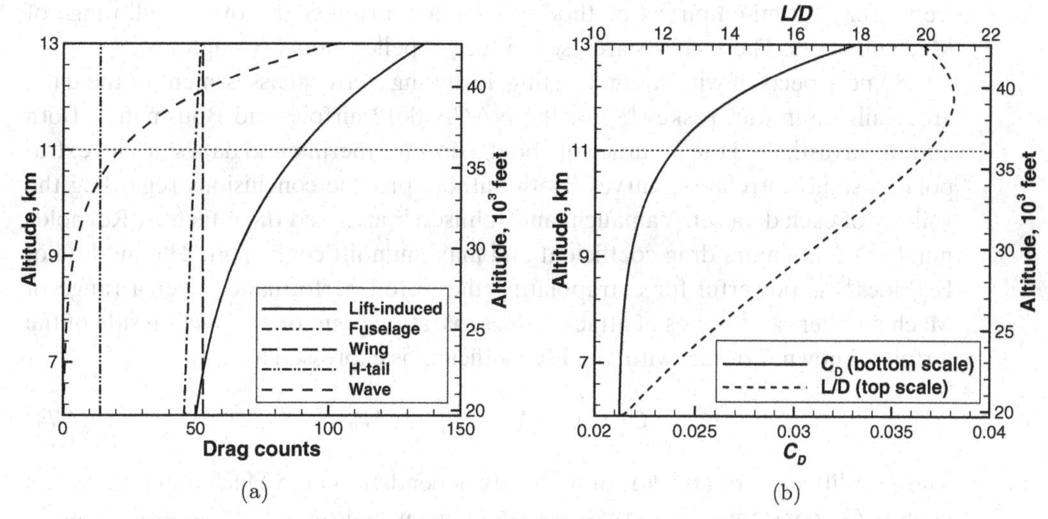


Figure 4.18. Drag analysis of the Airbus A380-861 at cruise conditions, $M = 0.850$.

**SENSITIVITY ANALYSIS.** We now carry out a sensitivity analysis to understand how the profile drag depends on inaccuracies in the estimation of the wetted-area (§ 2.4). We assume a 2% inaccuracy (in excess) on each of these components: fuselage, wing-body, wing system, horizontal tail, vertical tail, nacelles, pylons. The flight conditions are fixed; in this example we have $M = 0.80$ at 33,000 feet, $W = 420.0$ tons. The result is shown in Table 4.1 for a correction of 2% on each of the wetted-area components. In other words, if we selectively introduce a correction of 2% in each of the wetted areas, the $\Delta C_{Do}$ changes by less than one drag count at the specified flight conditions; $C_{Do}$ is the total profile drag coefficient. The result is dependent on the flight altitude.

## 4.3 Transonic Airfoil Model

The accurate calculation of high-speed rotors (propellers and fans) relies on airfoil data over a range of Mach numbers. Although there are plenty of aerodynamic data at low speeds, along with sophisticated methods that allow an accurate replication of the test data, the transonic effects are more difficult to address[39]. In this section

Table 4.1. *Profile drag sensitivity for the Airbus A380-861 resulting from $\Delta A_{wet} = 2\%$. All drag coefficients are given as drag counts*

| Configuration | $C_{Do}$ | $\Delta C_{Do}$ |
|---|---|---|
| Nominal | 189.73 | |
| Fuselage sensitivity | 190.55 | 0.82 |
| Wing sensitivity | 190.69 | 0.96 |
| H-Tail sensitivity | 189.84 | 0.11 |
| V-Tail sensitivity | 190.07 | 0.34 |
| Nacelles sensitivity | 189.98 | 0.25 |
| Pylons sensitivity | 189.95 | 0.22 |
| Nacelles sensitivity | 189.89 | 0.16 |

we present a semi-empirical method to generate airfoil charts over a full range of Mach numbers. These charts are used in the propeller model (Chapter 6).

Some aspects of wind-tunnel testing, including a critical assessment of the data, are available in McCroskey[40] (for the NACA 0012 airfoil) and Bousman[41]. Both authors investigated the accuracy of about a dozen experimental data sets and extrapolated some correlation curves. Both authors provide conclusions regarding the validity of each data set. Yamauchi and Johnson[42] analysed the effects of Reynolds number of minimum drag coefficient and maximum lift coefficient. The method of Beddoes[43] is powerful for extrapolating the airfoil performance over a range of Mach numbers and angles of attack if the separation point on the suction side of the airfoil is known. To start with, the lift coefficient is expressed as

$$C_L = C_{L_o}(M) + C_{L_\alpha}(\alpha, M)\alpha. \tag{4.74}$$

The zero-lift angle of attack is only weakly dependent on the Mach number; we can assume $C_{L_o}$ constant, or otherwise a useful approximation is

$$C_{L_o} = C_{L_o}(M^*) + \frac{\Delta C_{L_o}}{\Delta M}(M - M^*), \tag{4.75}$$

with $\Delta C_{L_o}/\Delta M \simeq 0.04$. The lift-curve slope depends on both the angle of attack and the Mach number. A suitable correction for the lift-curve slope is done by using the Kármán-Tzien second-order equation

$$C_{L_\alpha}(\alpha, M) = \frac{C_{L_\alpha}(\alpha, M^*)}{\beta_2}, \tag{4.76}$$

with

$$\beta_1 = \sqrt{1 - M^2}, \tag{4.77}$$

$$\beta_2 = \beta_1 + \frac{1}{2}\frac{M^2}{1 + \beta_1}. \tag{4.78}$$

This correction is valid at compressible Mach numbers lower than $M_{dd}$. If the angle of attack is fixed, then the $C_L$ produced at Mach numbers is higher than the value at the reference Mach $M^*$. If $M > M_{dd}$, the procedure is modified. In fact, the drag coefficient can be expressed as

$$C_D(M) = C_D(\alpha, M^*) + \Delta C_D(M). \tag{4.79}$$

Two corrections are required to Equation 4.79. First, we need a Reynolds number effect, due to an increased Mach number. Second, we need a Mach number correction to operate around the drag divergence point and beyond.

Assume that the airfoil polar is calculated at a Mach number $M^*$ (typically, $M^* = 0.2$). The Reynolds number effect is calculated from the definition of Mach number. In fact,

$$\mathrm{Re}_l = \frac{Ul}{\nu} = \frac{aMc}{\nu} \tag{4.80}$$

where $c$ is the chord; $a$ is the speed of sound. If the atmospheric conditions are fixed, then

$$Re(M) = Re^* + Re^*(M - M^*). \tag{4.81}$$

The profile drag coefficient scales with the Reynolds number as

$$C_D \propto \frac{1}{(\log_{10} Re)^{2.548}}. \tag{4.82}$$

Hence, the Mach number effect on the profile drag coefficient becomes

$$\frac{C_D}{C_D^*} = \frac{log_{10} Re^*}{\log_{10} Re}. \tag{4.83}$$

The $C_D$ can be calculated with reasonable accuracy over a range of angles of attack at low speeds. Note that it is not possible to make a correction on the drag component dependent on the pressure. It will be assumed that there is no appreciable change in pressure distribution with the increasing Mach number, although this cannot be correct at transonic speeds. The next step is to calculate the divergence Mach number of the airfoil. This is done through the use of Equation 4.43 with $\Lambda_{LE} = 0$, with $\kappa_A$ variable from 0.87 (NACA 6-series airfoils) to about 0.95 (modern supercritical wing sections). The $C_L$ to be used in Equation 4.43 must be the *corrected* value, obtained by a combination of Equations 4.76 and 4.74 (recall that the effective angle of attack is fixed). The relationship between the drag rise and the critical Mach number is calculated again from Equation 4.46. The corresponding drag coefficient is now calculated from Equation 4.44. The additional data required include relative thickness $t/c$, reference Mach number $M^*$, and reference Reynolds number $Re^*$. The factor $\kappa_A$ is a free parameter that must be chosen carefully because the transonic effects are strongly dependent on it. This problem can be overcome if the $M_{dd}$ is known at one value of the $C_L$.

From the analysis of several experiments, McCroskey[40] concluded that the best curve fit for the lift-curve slope of the NACA 0012 is

$$\beta_1 C_{L_\alpha} = 0.1025 + 0.00485 \log\left(\frac{Re}{10^6}\right), \tag{4.84}$$

with a maximum error of 0.0029. Equation 4.84 uses the Prandtl-Glauert compressibility correction (Equation 4.77) rather than the Kármán-Tzien correction. A result of this theory is shown in Figure 4.19. The comparison is done with selected wind-tunnel data[41]. Observe that the Reynolds number effects are captured correctly. The calculation was based on "reference data" at $Re^* = 2 \cdot 10^6$, $M^* = 0.3$, $\alpha = 0$ degrees, and $\kappa_A = 0.87$.

It is possible to verify that the choice of the factor $\kappa_A$ is correct. First, we need to find a best fit of the wind-tunnel data; then, we calculate the $M_{dd}$; and finally, we calculate $\kappa_A$ by solving Equation 4.43. There is some arbitrariness in fitting the reference data, but we find that the best correlation is a polynomial of order 3. This gives $M_{dd} = 0.818$. Solution of Equation 4.43 yields $\kappa_A \simeq 0.91$, a value higher than the one used.

The application of this method to the generation of tabulated data is shown in Figure 4.20 for the rotorcraft airfoil SC-1095, which has a 10% thickness ratio. We show the effects of Mach number on the lift-curve slope and the $C_{L_{max}}$ (Figure 4.20a), the drag polar at Mach numbers up to $M = 0.80$ (Figure 4.20b), the effects of Mach

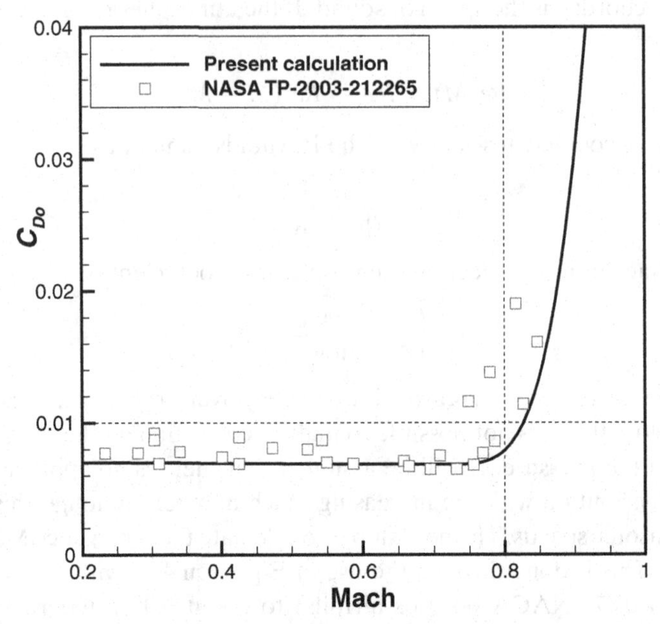


Figure 4.19. Transonic drag rise of the airfoil SC-1095 and comparison with wind-tunnel data.

number at increasing lift coefficients (Figure 4.20c) and the pitching-moment divergence at transonic Mach numbers (Figure 4.20d).

The comparison with wind-tunnel data show that 1) the Reynolds number effects can be captured with the same order of accuracy as the wind-tunnel data; 2) the lift-curve slope can be calculated with engineering accuracy before the lift stall; and 3) the divergence Mach number can be calculated accurately at zero angle of attack. For angles of attack other than zero it was not possible to make a verification of the method.

The method proposed allows to generate tabulated data of airfoil aerodynamics ($C_L$, $C_{L_\alpha}$, $C_D$) over a wide range of angles of attack, Reynolds numbers and Mach numbers. The data required include the airfoil thickness, the reference Reynolds and Mach numbers, and the corresponding aerodynamic polar. Rotor calculations based on combined momentum and blade-element theory with extrapolated airfoil data at transonic Mach numbers provide the correct estimation of the main rotor parameters, including the Mach number effects.

## 4.4 Aircraft Drag at Transonic and Supersonic Speeds

The determination of the drag of aircraft flying at transonic and supersonic speeds is a far more complicated matter. The complications arise from the wide range of Mach numbers at which the aircraft operate, the complexity of the wing system (even in absence of external stores), the wide range of angles of attack at which a high-performance aircraft is designed to operate and the substantial non-linearities that occur at transonic Mach numbers.

There are some classical methods available in the field of supersonic aerodynamics that are not possible to review here. The principle of components is still a useful

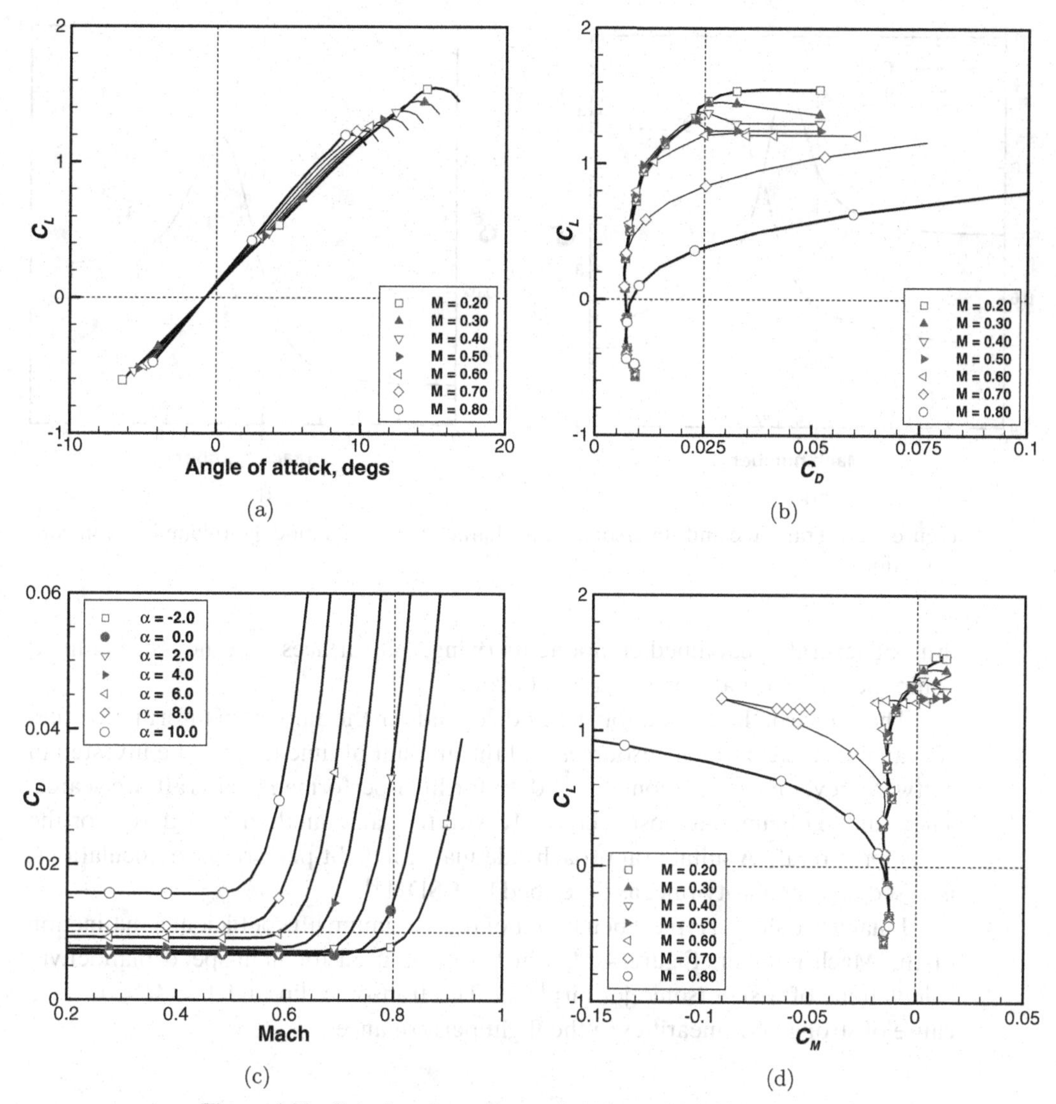


Figure 4.20. Calculated aerodynamic polars of SC-1095.

method, but the interference factors may have a stronger influence at high speed; if these factors are not properly addressed, the sum will not lead to an acceptable result. General aerodynamic methods are beyond the scope of this book. For the transonic drag we will assume the following equation:

$$C_D = C_{D_o} + \eta C_{L_\alpha}(\alpha - \alpha_o)^2, \tag{4.85}$$

where all of the coefficients are a function of the Mach number. A typical behaviour of the drag coefficients is displayed in Figure 4.21. The data in this figure, along with Equation 4.85, will be used for rapid calculations of transonic manoeuvres, including specific excess power.

The principle of drag decomposition can be applied (with some care) at supersonic Mach numbers. In the latter case, we need methods for the calculation of the

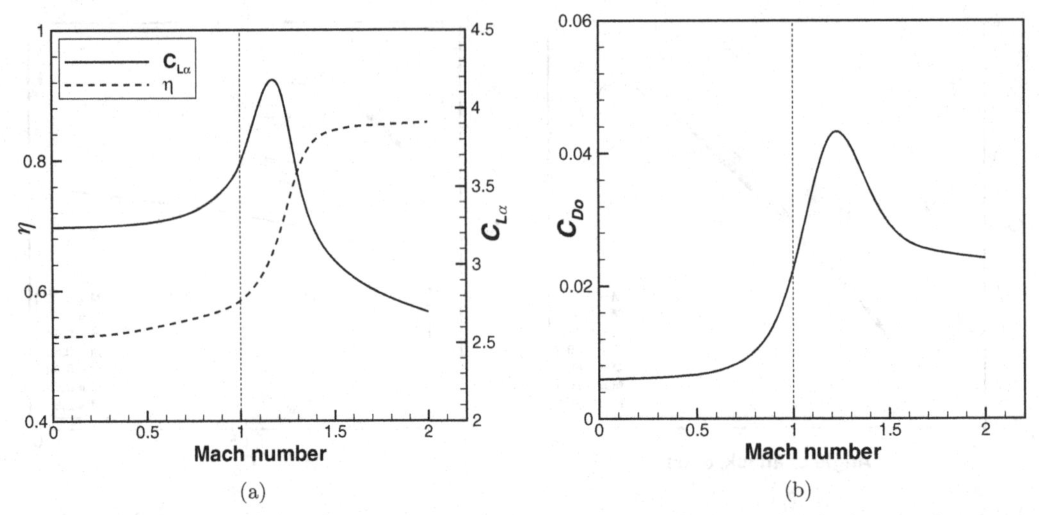


Figure 4.21. Transonic and supersonic drag characteristics of a model airplane in clean configuration.

drag of several streamlined components (wing, tail surfaces) and non-streamlined components (external stores and fuel tanks).

The choice of the calculation method depends on the amount of information that is available, the computer resources and the amount of time that can be invested in software development. Geometrical data for high-performance aircraft are scarce. They can be obtained almost exclusively with the same method considered for the transport aircraft. A suitable approach, adequate for flight-performance calculations, is based on a similarity rule, as described by ESDU[44].

Finally, we show a typical behaviour of the maximum lift coefficient as a function of the Mach number, Figure 4.22, which will be used for high-speed manoeuvre calculations of a supersonic jet airplane. The transonic dip at $1.1 < M < 1.3$ is a cause of strong non-linearities in the flight performance.

### 4.4.1 Drag of Bodies of Revolution

Most external stores and equipment on high-speed aircraft are bodies of revolution with pointed forebodies and various after-body shapes, including boat tails and blunt bodies. There is a variety of cases that is not possible to review in this context. We limit our discussion to a representative case, which is the body shown in Figure 4.23. This is a body of revolution having a forebody with "minimum wave drag" (to be established), a central cylindrical body, a boat tail and a blunt base. The body is characterised by its cross-sectional area $A$ (or diameter $d$) and length $l$. The forebody can have a number of shapes, but the most favourable from the point of view of low wave drag include the spherically blunted tangent ogive, the spherically blunted paraboloid and a modified ellipsoid. The boat tail can be conical (as shown) or circular arc or parabolic. The main characteristics of this body are overall length $l$ and maximum diameter $d$; forebody fineness ratio $l/d$; radius of curvature of the spherical nose $r$ (and nose-blunting ratio $2r/d$); boat-tail angle $\beta$ and base diameter $d_b$. The length of the forebody is fixed; the length of the boat tail can be derived

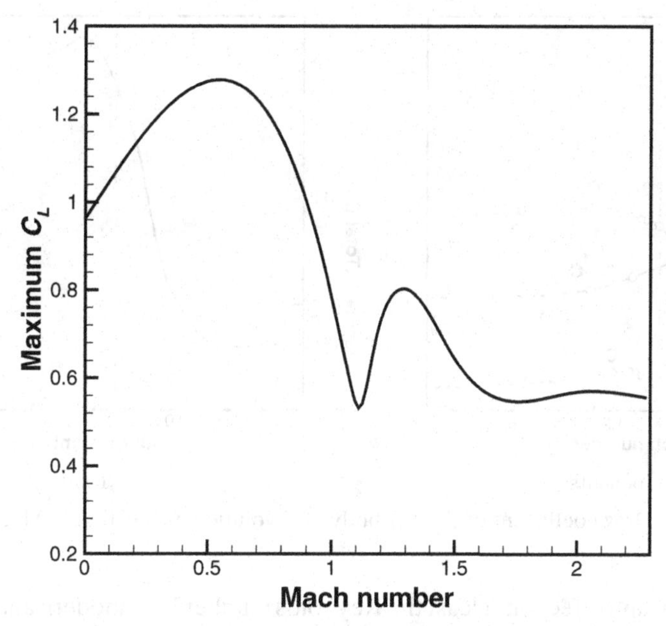


Figure 4.22. Transonic and supersonic behaviour of the maximum $C_L$ of a model aircraft.

from the boat-tail angle $\beta$. The length of the cylindrical body is arbitrary, but it must be at least $\sim 3d$ to be able to consider the various sections independently.

Our problem is the determination of the aerodynamic drag over the full range of Mach numbers, up to supersonic speeds, as in the previous case (see Figure 4.21). ESDU has a number of practical methods on both wave drag and after-body drag calculations[20;45–47]. The computational method available in *Missile DATCOM*[48] can estimate the drag of a variety of shapes, angles of attack and Mach numbers. The analytical methods of high-speed aerodynamics are limited to fully developed supersonic flow and to a few optimal shapes, such as the von Kármán ogive and the Sears-Haack body[49]. Then there is a wide body of experimental work covering a number of ogive-like geometries over the full transonic regime[50;51]. In general, no extrapolations are possible due to the non-linear nature of the transonic flow and

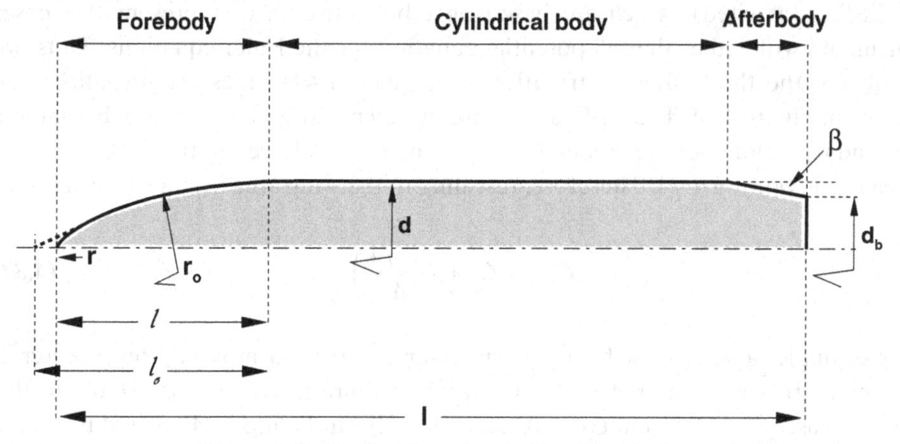


Figure 4.23. Typical low-drag body of revolution.

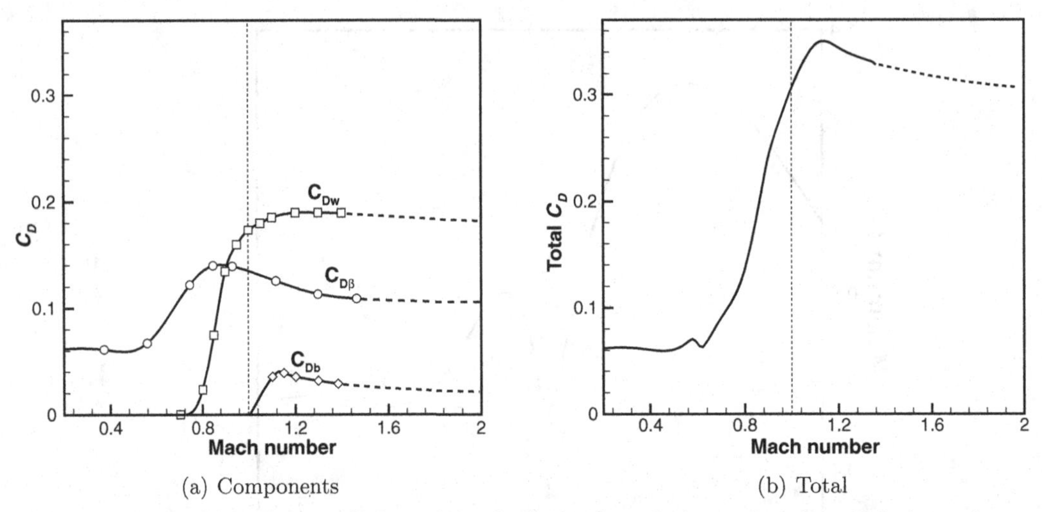


(a) Components (b) Total

Figure 4.24. Drag coefficient of slender body of revolution as function of Mach number.

various important effects, not least the Reynolds number[52]. A modern analysis would be based on CFD computations, that would be used to generate tabulated data of aerodynamic coefficients over the full range of Mach numbers for any fixed geometry. Even in this case, the results must be thoroughly validated with experimental data.

At a basic level, the drag components of the body shown in Figure 4.23 are 1) forebody wave drag $C_{Dw}$; 2) skin-friction drag $C_{D_o}$; 3) boat-tail drag $C_{D\beta}$; and 4) base drag $C_{Db}$. The second contribution is considerably smaller than the other contributions; in first analysis, it can be neglected (or calculated with the method shown in § 4.2.2). Thus, we consider the drag build-up with the following sum:

$$C_D \simeq C_{Dw} + C_{Db} + C_{D\beta}. \tag{4.86}$$

**CASE STUDY.** We consider a spherically blunted secant ogive forebody with a conical boat tail and a blunted base. The main characteristics of the body are forebody fineness ratio $l/d = 2$ and nose-blunting ratio $2r/d = 0.2$. We have used the methods of ESDU (as cited), which are based on tabulated data of representative cases calculated with either the full potential equations or the Euler equations. Thus, we estimated the three drag contributions in Equation 4.86; these contributions are shown in Figure 4.24. The dashed lines are our own extrapolation at Mach numbers beyond the validity of the model. Note that the $C_D$ is referred to the cross-sectional area of the body, $\pi d^2/4$. Therefore, rescaling to the wing area is done according to

$$C_D \to C_D \left( \frac{\pi d^2/4}{A} \right). \tag{4.87}$$

For example, a wing tank having a diameter $d = 0.51$ m mounted on an aircraft having a gross wing area $A = 28.9$ m$^2$ would require a scale factor $\sim 0.01$. In this specific case, the maximum $C_D \simeq 0.34$ at $M = 1.18$ in Figure 4.24b would be scaled to $C_D \sim 0.0034$, which is equivalent to 34 drag counts.

## 4.5 Buffet Boundaries

*Buffet* is the structural excitation caused by separated air flow. This excitation is represented by an amplitude, a frequency and a spectrum. The phenomenon is due a number of different events, such as upstream flow separation (tail-plane in the wake of the wing) and downstream separation (flow separation in the upsweep section of the fuselage). The spectrum is characterised by a broadband energy content, although occasionally there can be discrete frequencies due to coherent vortex shedding. *Buffeting* is the aeroelastic response to the buffet excitation. Buffet onset is determined from vibration measurements and is conventionally given as a vibration measuring at least $0.2g$ at the pilot's seat[53]. For a wing, the buffet excitation parameter is

$$\sqrt{\mathrm{St}\, G(\mathrm{St})} = \frac{2}{\sqrt{\pi}} \frac{m\ddot{z}}{qA} \sqrt{\zeta}, \tag{4.88}$$

where $St$ is the Strouhal number, $\ddot{z}$ is the RMS acceleration of the wing tip, and $\zeta$ is the damping ratio, or the ratio between the aeroelastic damping and the critical damping.

A considerable amount of research has been done on flow-induced separation and its effects on the structural response of aero structures, although most of it relies on wind-tunnel experiments and flight testing. Numerical simulation methods have lagged behind.

The discussion in this section is limited to subsonic transport aircraft. For this case, it is generally accepted that the aircraft must be able to perform at least a 1.3g manoeuvre in cruise configuration in response to buffeting. In other words, a 0.3g manoeuvre margin is imposed for any given flight altitude and true Mach number when buffeting occurs.

From a performance point of view, the best indicator of wing buffet is the behaviour of the $C_{L_{max}}$ at increasing Mach numbers. Here the variation can be quite striking, depending on the wing section, the wing planform and other geometrical factors. No generalisation is possible. However, there exists a Mach number above which the lift drops (*transonic dip*), before eventually recovering in the low supersonic regions.

A typical buffet boundary for a transport aircraft looks like the one in Figure 4.25, which represents the performance of the Douglas DC9-80 series of airplanes. At very low Mach numbers, the usable $C_L$ has relatively high values (point A). This point occurs around $1.2V_S$. At high subsonic speeds, the usable lift decreases dramatically and upon reaching the Mach number at point B, there is virtually no lift available. Most buffet curves can be defined through points A and B.

The buffet speed and Mach number are given by the intersection between the buffet boundary and the curve defined by the manoeuvre equation

$$V_B = a M_B = \sqrt{\frac{2nW}{\rho A C_{L_{max-buffet}}}}. \tag{4.89}$$

For a fixed weight and flight altitude, $V_B = f(n, C_{L_{max-buffet}})$. The normal load factor is set to $n = 1.3$ to account for the manoeuvre margin just described. In general, there will be two intersections, giving low and high buffet speeds.

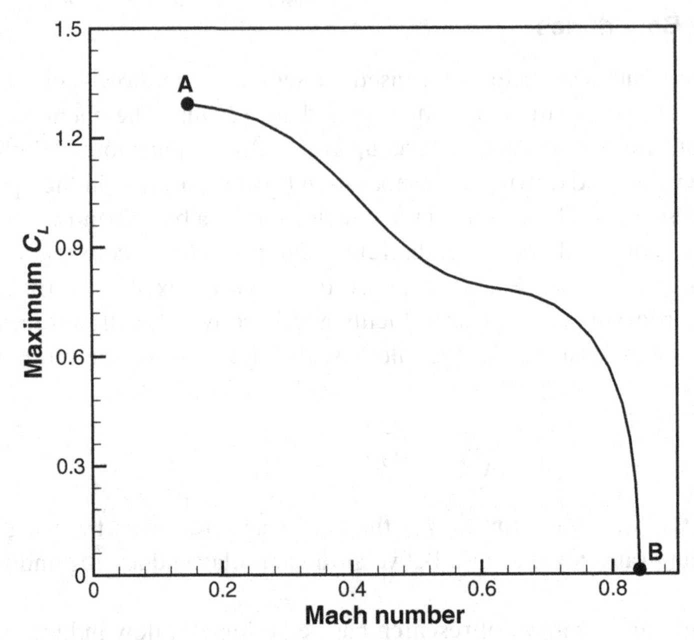


Figure 4.25. Buffet boundary of the Douglas DC9-80 series (elaborated from the FCOM).

## 4.6 Aerodynamic Derivatives

We limit the discussion to the main control surfaces of a modern transport airplane because they can be realistically modelled with the methods presented in this book. We assume that wings are equipped with *ailerons* and *spoilers*; horizontal stabilisers are equipped with *elevators* and vertical stabilisers have a *rudder*. In many cases these controls are sectioned in a number of parts. For example, the rudder and the elevator can have two or more sections. Elevators can also have small deflection tabs. The aerodynamic derivatives will be used for steady-state aircraft trim, stability and control. For dynamic effects, the reader is referred to books that specifically address flight dynamics, such as Stengel[54].

If $C_a$ denotes a generic aerodynamic coefficient that can be expanded in series around an initial value with respect to a control angle $\alpha$, then a linear approximation is the following:

$$C_a = C_{ao} + \left(\frac{\partial C_a}{\partial \alpha}\right) \alpha, \tag{4.90}$$

where $C_{ao}$ denotes the coefficient $C_a$ evaluated at $\alpha = \alpha_o$. The derivative in Equation 4.90 is an *aerodynamic derivative* and can be used to analyse the aerodynamic response of the system to sufficiently small perturbations of the state parameter $\alpha$ around its initial value. If the aerodynamics is linear, Equation 4.90 is in fact quite useful; if it is not, there are limitations to the extent of the perturbations that can be introduced.

For a simple wing, we can define perturbations with respect to the angle of attack $\alpha$ and the side-slip $\beta$. There are six aerodynamic coefficients: three force coefficients and three moment coefficients. These quantities can be referred either to body-conformal axes $\{x_b, y_b, z_b\}$ or wind axes $\{x, y, z\}$. In the latter case, we have

$C_D, C_Y, C_L$ for the force coefficients and $C_m, C_n, C_l$ for the moment coefficients. The aerodynamic derivatives are

$$\{C_D, C_Y, C_L, C_m, C_n, C_l\}_\alpha \qquad \{C_D, C_Y, C_L, C_m, C_n, C_l\}_\beta. \tag{4.91}$$

If the wing has a dynamic motion such a pitch rate p, roll rate q or yaw rate r, then we have the derivatives with respect to these quantities. If the wing has an aileron, there are changes in the coefficients due to aileron deflection $\delta$; the corresponding derivatives are:

$$\{C_D, C_Y, C_L, C_m, C_n, C_l\}_\delta. \tag{4.92}$$

Therefore, there are a large number of derivatives to deal with and, if the process is extended to a fully configured airplane, there can be in excess of 100 derivatives to account for, some of which are not always useful.

As long as the aerodynamics is linear (a reasonable assumption for lifting surfaces at moderate angles of attack and low subsonic Mach numbers), the derivatives can be calculated with a lifting-surface method. If the lifting surfaces are calculated separately, they need a reference area; if this is not the same for all of the components, the coefficients will have to be rescaled to the same reference area.

## 4.7 Float-Plane's Hull Resistance in Water

Because the drag coefficient will have to be rescaled with the wing area, it is preferable to calculate the ratio $D/q = C_D A$.

$$C_{D_{hull}} = C_{D_o} + C_{D_w} + k_w C_{L_{hull}}^2, \tag{4.93}$$

where $C_{D_w}$ is the wave drag, $k_w$ is a lift-induced drag factor and $C_{L_{hull}}$ is the lift produced by the immersed hull. The profile drag coefficient is calculated from the wetted area and an estimated value of average skin-friction coefficient as shown in § 4.2.2.

**GEOMETRY CALCULATIONS.** The wetted area to consider is the immersed area $A^*_{wet}$. The volume of the immersed floats, $V^*_{float}$, can be calculated from the buoyancy law at MTOW. If $W$ is this weight, then

$$V^*_{float} = \frac{W}{2g\rho_w}. \tag{4.94}$$

The actual volume of the float would have to be larger than this figure, for at least two reasons: first, the float can serve as a stepping point for boarding the aircraft; second, a temporary excess weight would sink the floats. Thus, we will assume (somewhat arbitrarily) that the actual volume of the floats is $V_{float} \simeq 1.5 V^*_{float}$. We can now define an equivalent cross-sectional area $A_{eq}$ for a float of length $l$ and an equivalent radius $r_{eq}$

$$A_{eq} = \frac{V_{float}}{l_{float}}, \qquad r_{eq} = \left(\frac{A_{eq}}{\pi}\right)^{1/2}. \tag{4.95}$$

The total wetted area is $A_{wet} \simeq 2\pi r_{eq} l_{float}$. The wetted area of the immersed portion of the floats is

$$A^*_{wet} \simeq \left(\frac{V^*_{float}}{\pi l_{float}}\right)^{1/2} l_{float}. \tag{4.96}$$

**WAVE DRAG.** Calculation of the wave drag of a ship hull is generally a fairly complicated problem. In this context, it will suffice to use a simplified expression of a theory of thin-ship resistance (Michell-Havelock), which is given by an improper integral†

$$C_{D_w} = \frac{4}{\pi C_p^2} \int_{x_o}^{\infty} (1 - \cos x) e^{-ax^2} \frac{x^2}{\sqrt{x^2 - x_o^2}} dx \tag{4.97}$$

with

$$C_p = \frac{V}{A_{eq} l}, \quad x_o = \frac{C_p}{\mathrm{Fr}^2}, \quad \mathrm{Fr} = \frac{U}{\sqrt{gl}}, \quad a = \frac{2\delta}{l}\left(\frac{\mathrm{Fr}}{C_p}\right)^2. \tag{4.98}$$

In these equations, the parameter "Fr" denotes the Froude number; $V$ is the submerged volume; $l$ is the length of the submerged hull; $A_{eq}$ is an equivalent (or maximum) cross-sectional area; and $\delta$ is the depth of the centroid of the maximum cross section $A_{eq}$. Equation 4.97 contains three important terms: 1) an oscillating factor $(1 - \cos x)$; 2) a damping factor $\exp(-ax^2)$; and 3) the improper factor $x^2/\sqrt{}$ that becomes singular at $x = x_o$. The integral in Equation 4.97 is more specifically called *improper Riemann integral* of the second kind and causes some difficulties. However, the solution of Equation 4.97 shows that above a critical value of the speed, the drag decreases uniformly.

In some cases, there can be interference between the internal leading-edge waves because they intersect each other and travel to the opposite float. This problem is similar to the *Busemann biplane*, in which mutual interference between the systems of shocks created by two slender bodies can lead to a decrease in total wave drag, if the system is properly designed (see, for example, Liepmann & Roshko[56]). A further analysis of the floats is shown in § 15.11, where we examine their effect on the mission range of a propeller airplane.

## 4.8 Vortex Wakes

An important side effect of lift generation by an aircraft wing is the wake downwash and the tip vortices. The downwash is the vertical component of the air speed behind the airplane. The tip vortices are generated by the pressure differences between the suction and the pressure side of the wing; they consist of a pair of counter-rotating swirling vortices and their strength increases with the wing loading.

From low-speed aerodynamic theory we can derive a first-order estimate of the downwash. The ideal condition, resulting from *elliptic loading*, yields the following expression of the average normalised downwash, or induced inflow angle (see also § 7.1):

$$\alpha_i = \frac{\overline{w}}{U} = \frac{2L}{\pi \rho b^2 U^2} \simeq \frac{2W}{\pi \rho b^2 U^2}. \tag{4.99}$$

† Alternative expressions are available for this equation; see Newman[55].

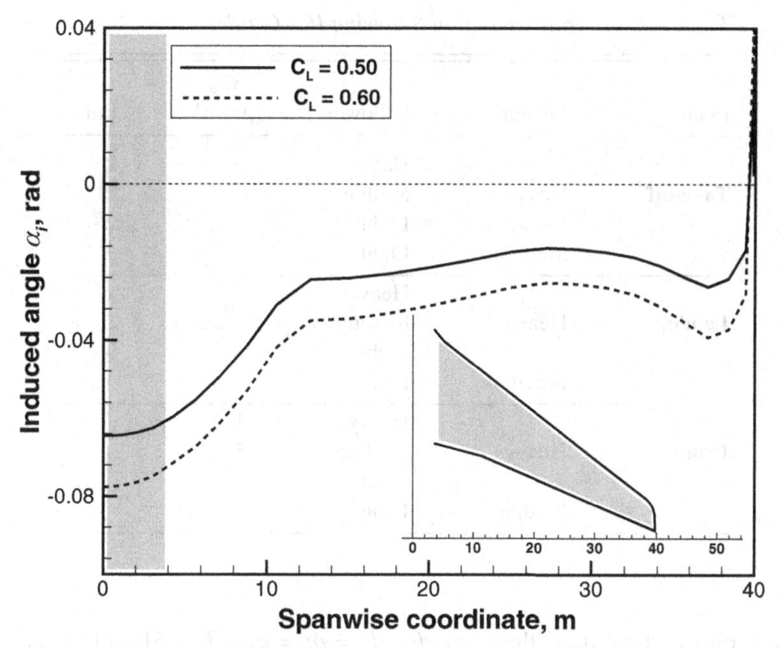


Figure 4.26. Calculated downwash of the Airbus A380-861 wing.

The induced inflow angle is associated with the local circulation. Therefore, in the general case, it is possible to calculate a local induced angle and a local downwash. This was done for the Airbus A380-861 and is shown in Figure 4.26 for two values of the $C_L$ by using a lifting-surface method. The average downwash $\overline{w}$ can be calculated from integrating the local downwash $w(y)$. A rapid variation of the downwash occurs at the tip, where the air flow is prevented from rolling up too quickly by the presence of the winglet.

The key operational parameters in the determination of the strength of the downwash is the air speed, the gross weight and the flight altitude, Equation 4.99. If the downwash is too strong, it may cause a hazard to aircraft following behind or crossing the flight path at a lower altitude. For this reason, there are regulations in place that specify safe separation distances, according to the data given in Table 4.2. Currently, there are only three weight categories: *heavy* refers to a weight $> 136{,}000$ kg; *medium* refers to a weight $7{,}000 \text{ kg} < W < 136{,}000$ kg; *light* refers to $W < 7{,}000$ kg. A special category exists for the Airbus A380 (*super heavy*). For the cases not listed in this table, which refer mostly to light aircraft, the minimum radar separation is enforced. The different weight classes and separation times are under review and may change in the future. However, the scientific basis for the vortex wakes is well established[57–59].Considerable effort has gone into the analysis of vortex decay, which appears to depend on several atmospheric parameters, including the background turbulence, the aircraft-induced turbulence, the ambient stratification and the wind shear. Vortex wakes are known to interact with the surrounding environment, including the ground and bodies of water.

To have an idea about the effect behind the aircraft, consider the case of the Airbus A380 on a climb-out at 3,000 feet at 220 KTAS and a gross weight of 380 tons. The average induced angle would be $\overline{w}/U \sim 0.025$, which corresponds to a

Table 4.2. *Aircraft separation following ICAO rules*

| Flight | Leader | Follower | $X_{sep}$ [n-m] | $t_{sep}$ [min] |
|---|---|---|---|---|
| | | Heavy | | 2 |
| **Take-off** | Heavy | Medium | | 2 |
| | | Light | | 2 |
| | Medium | Light | | 3 |
| | | Heavy | | 2 |
| **Landing** | Heavy | Medium | | 2 |
| | | Light | | 3 |
| | Medium | Light | | 3 |
| | | Heavy | 4 | |
| **Cruise** | Heavy | Medium | 5 | |
| | | Light | 6 | |
| | Medium | Light | 5 | |

$\overline{w} \sim 3$ m/s, or a vertical mass flow rate $dm/dt = \dot{m} = \rho \overline{w} U b \sim 31{,}100$ kg/s, or 27,000 $m^3$/s. If this downwash is incurred by a following aircraft at a similar speed, a loss in $C_L \sim -0.1$ should be expected.

## Summary

We presented comprehensive methods to be used for the prediction of the aerodynamic coefficients and their derivatives of a conventional airplane configuration. The lift has been elaborated from cruise conditions, as well as from the (approximate) effect of augmentation caused by flaps, slats and other devices. We have included the effects of the ground on the effective inflow angle. The drag analysis is somewhat more complex. It is calculated with the method of components. A separate model is developed for the transonic drag rise. The model is used both for the lifting surfaces and the blade sections of a propeller. A number of verification cases have been shown; these cases demonstrate the suitability of the models for aircraft performance work. We have shown relatively simple models for bodies of revolution (to be used in the drag estimation of the external tanks and other stores) and for a float plane in water. Finally, we have elaborated on the role of the vortex wakes from the point of view of the hazard created by trailing airplanes. We have explained that a minimum separation is required to limit the changes in aerodynamic inflow conditions that could cause the trailing aircraft to stall or lose control.

## Bibliography

[1] Katz J and Plotkin A. *Low Speed Aerodynamics*. McGraw-Hill, 1992.
[2] ESDU. *Wing Lift Coefficient Increment at Zero Angle of Attack due to Deployment of Plain Trailing-Edge Flaps at Low Speeds*. Data Item 97011. ESDU International, London, Nov. 2003.
[3] ESDU. *Lift and Drag due to Spoiler Operation in the Ground Run*. Data Item 76026. ESDU International, May 1977.

[4] Kiock R. The ALVAST model of DLR:. Technical Report IB 129 96/22, DLR, Lilienthal Platz, 7, D-38018 Braunschweig, Germany, 1996.
[5] ESDU. *Increments in Aerofoil Lift Coefficient at Zero Angle of Attack and in Maximum Lift Coefficient Due to Deployment of a Double-Slotted or Triple-Slotted Trailing-Edge Flap, with or without a Leading-Edge High-Lift Device, at Low Speeds*. Data Item 94031. ESDU International, London, Dec. 1994.
[6] ESDU. *Estimation of Airframe Drag by Summation of Components: Principles and Examples*. Data Item 97016. ESDU International, London, Dec. 1996.
[7] Callaghan JG. Aerodynamic prediction methods for aircraft at low speeds with mechanical high lift devices. In *Prediction Methods for Aircraft Aerodynamic Characteristics*, volume AGARD LS-67, pages 2.1–2.52, May 1974.
[8] White F. *Viscous Fluid Flow*. McGraw-Hill, 1974.
[9] White F. *Viscous Fluid Flow*. McGraw-Hill, 1974. Chapter 7.
[10] Nielsen J. *Missile Aerodynamics*. McGraw-Hill, 1960. Chapter 9.
[11] van Driest ER. The problem of aerodynamic heating. *Aeronaut. Eng. Rev.*, 15:26–41, 1956.
[12] van Driest ER. On turbulent flow near a wall. *J. Aero. Sci.*, 23(11):1007–1011, 1956.
[13] Hopkins EJ and Inouye M. An evaluation of theories for predicting turbulent skin friction and heat transfer on flat plates at supersonic and hypersonic Mach numbers. *AIAA J.*, 9(6):993–1003, June 1971.
[14] Hopkins EJ. Charts for predicting turbulent skin friction from the van Driest method II. Technical Report TN-D-6945, NASA, Oct. 1972.
[15] Blumer CB and van Driest ER. Boundary layer transition – Freestream turbulence and pressure effects. *AIAA J.*, 1(6):1303–1306, 1963.
[16] Hoerner SF. *Fluid Dynamic Drag*. Published by the Author, Bricktown, NJ, 1965.
[17] ESDU. *Drag Increment due to Fuselage Upsweep*. Data Item 80006. ESDU International, London, Feb. 1988.
[18] Malone B and Mason WH. Multidisciplinary optimization in aircraft design using analytic technology models. *J. Aircraft*, 32(2):431–437, March 1995.
[19] Hilton WF. *High Speed Aerodynamics*. Longmans & Co, London, 1952.
[20] ESDU. *Forebodies of Fineness Ratio 1.0, 1.5 and 2.0, Having Low Values of Wave Drag Coefficient at Transonic Speeds*. Data Item 79004. ESDU International, London, June 1979.
[21] Hoerner SF. *Fluid Dynamic Drag*. Published by the Author, Bricktown, NJ, 1965. Chapter 8.
[22] Tétrault PA, Schetz JA, and Grossman B. Numerical prediction of interference drag of strut-surface intersection in transonic flow. *AIAA J.*, 39(5):857–864, May 2001.
[23] Kubendran L, McMahon H, and Hubbard J. Interference drag in a simulated wing-fuselage junction. Technical Report CR-3811, NASA, 1984.
[24] Barber TJ. An investigation of wall-strut intersection losses. *J. Aircraft*, 15(10):676–681, Oct. 1978.
[25] Sakellaridis A and Lazaridis A. Experimental study of interference drag for multi-element objects. *Exp. Thermal & Fluid Science*, 26:313–317, 2002.
[26] McCormick BW. *Aerodynamics, Aeronautics and Flight Mechanics*. John Wiley, 2*nd* edition, 1995.
[27] ESDU. *Undercarriage Drag Prediction Methods*. Data Item 79015. ESDU International, London, March 1987.
[28] Roskam J. *Airplane Design, Part VI, Chapter 4*. DARCorporation, 2000 (paperback edition).
[29] ESDU. *Mean forces, pressures and moments for circular cylindrical structures: finite-length cylinders in uniform flow*. Data Item 81017, Amend. A. ESDU International, London, May 1987.

[30] Torenbeek E. *Synthesis of Subsonic Airplane Design*. Kluwer Academic Publ., 1985.
[31] van Es GH. Method for predicting the rolling resistance of aircraft tires in dry snow. *J. Aircraft*, 36(4):762–768, Oct. 1999.
[32] Redeker G. DLR-F4 wing-body configuration. In *A Selection of Experimental Test Cases for the Validation of CFD Codes*, AGARD AR-303, Volume II, pages B4–B21. Aug. 1994.
[33] Redeker G, Mueller R, Ashill PR, Elsenaar A, and Schmitt V. Experiments on the DFVLR F4 wing body configuration in several European wind tunnels. In *Aerodynamic Data Accuracy and Quality: Requirements and Capabilities in Wind Tunnel Testing*, volume AGARD CP-429, July 1988.
[34] Langtry RB, Kuntz M, and Menter FR. Drag prediction of engine-airframe interference effects with CFX-5. *J. Aircraft*, 42(6):1523–1529, Nov. 2005.
[35] O. Brodersen and A. Stürmer. Drag prediction of engine-airframe interference effects using unstructured Navier-Stokes calculations. In *19th Applied Aerodynamics Conference*, AIAA Paper 2001-2414. Anaheim, CA, June 2001.
[36] Wurtzler KE and Morton SA. Accurate drag prediction using Cobalt. *J. Aircraft*, 43(1):10–16, 2006.
[37] Hanke CR and Nordwall DR. The simulation of a large jet transport aircraft. Vol. II: Modeling data. Technical Report D6-30643, N73-10027 Boeing Doc, Sept. 1970.
[38] Caldarelli G. ATR-72 accident in Taiwan. In *SAE Aircraft & Engine Icing International Conference*, ICE 13, Sevilla, Spain, Oct. 2007.
[39] Filippone A. Rapid estimation of airfoil aerodynamics for helicopter rotor calculations. *J. Aircraft*, 45(4):1468–1472, July 2008.
[40] McCroskey J. A critical assessment of wind tunnel results for the NACA 0012 airfoil. Technical Report TM-100019, NASA, Oct. 1987.
[41] Bousman WG. Aerodynamic characteristics of SC1095 and SC1094-R8 airfoils. Technical Report TP-2003-212265, NASA, Dec. 2003.
[42] Yamauchi GK and Johnson W. Trends of Reynolds number effects on two-dimensional airfoil characteristics for helicopter rotor analyses. Technical Report TM-84363, NASA, April 1983.
[43] Beddoes TS. Representation of airfoil behavior. *Vertica*, 7(2):183–197, 1983.
[44] ESDU. *Similarity Rules for Application in Aircraft Performance Work*. Data item 97025. ESDU International, London, Sept. 2008.
[45] ESDU. *The Wave Drag Coefficient of Spherically Blunted Secant Ogive Forebodies of Fineness Ratio 1.0, 1.5 and 2.0 at Zero Incidence in Transonic Flow*. Data Item 89017. ESDU International, London, 1983.
[46] ESDU. *Pressure Drag and Lift Contributions for Blunted Forebodies of Fineness Ratio 2.0 for Transonic Flow ($M_\infty \leq 1.4$)*. Data Item 89033. ESDU International, London, 1989.
[47] ESDU. *Subsonic and transonic base and boat-tail pressure drag of cylindrical bodies with circular-arc boat-tails*. Data Item 96012. ESDU International, London, 1996.
[48] Blake WB. Missile DATCOM: User's manual – 1997 Fortran 90 revision. Technical report, US Air Force Research Lab, Air Vehicles Directorate, Wright-Patterson Air Force Base, OH, 1998.
[49] Ashley H and Landahl M. *Aerodynamics of Wings and Bodies*. Addison-Wesley, 1965.
[50] Harris RV and Landrum EJ. Drag characteristics of a series of low-drag bodies of revolution at Mach numbers from 0.6 to 4.0. Technical Report TN-D-3163, NASA, Dec. 1965.
[51] Wallskog HA and Hart RG. Investigation of the drag of blunt-noised bodies of revolution in free flight at Mach numbers 0.6 to 2.3. Technical Report RM L5314a, NACA, 1953.

[52] Bromm AF and Goodwind JM. Investigation at supersonic speeds of the variation with Reynolds number and Mach number of the total, base and skin-friction drag of seven boattail bodies of revolution designed for minimum wave drag. Technical Report TN 3708, NACA, 1956.
[53] ESDU. *An Introduction to Aircraft Buffet and Buffetting*. Data Item 87012. ESDU International, London, 1987.
[54] Stengel R. *Flight Dynamics*. Princeton Univ. Press, 2004.
[55] Newman JN. *Marine Hydrodynamics*. The MIT Press, 1977.
[56] Liepmann H and Roshko A. *Elements of Gas Dynamics*. J. Wiley & Sons, 1983.
[57] Rossow V. Lift-generated vortex wakes of subsonic transport aircraft. *Progr. Aerospace Sciences*, 35:507–660, Aug. 1999.
[58] Gerz T, Holzäpfel F, and Darracq D. Commercial aircraft wake vortices. *Progr. Aerospace Sciences*, 38:181–208, 2002.
[59] Holzäpfel F. Probabilistic two-phase wake vortex decay and transport model. *J. Aircraft*, 40, Mar. 2003.

## Nomenclature for Chapter 4

| | | |
|---|---|---|
| $a$ | = | speed of sound; term defined in Equation 4.98 |
| $a_1, a_2, a_3$ | = | factors in Equation 4.5 |
| $A$ | = | area; parameter defined in Equation 4.32 |
| $A_{c_1}$ | = | cross-sectional area of wing having unit chord |
| $A_{eq}$ | = | equivalent area |
| $A_p$ | = | planform area of the upswept section of the fuselage |
| $A_{ref}$ | = | maximum cross-sectional area of the fuselage |
| $A_{spoiler}$ | = | planform area of a spoiler |
| $A_{wet}$ | = | wetted area |
| $A^*_{wet}$ | = | wetted area of immersed float |
| $\mathcal{AR}$ | = | wing aspect ratio, $b^2/A$ |
| $b$ | = | nominal wing span |
| $b_{spoiler}$ | = | lateral dimension of the spoiler |
| $b_u$ | = | width of retractable multi-wheel landing gear |
| $b_w$ | = | wing span over winglet |
| $B$ | = | parameter defined in Equation 4.32 |
| $c$ | = | wing/airfoil chord |
| $c_c$ | = | corrected chord, Equation 4.9 |
| $c'$ | = | effective chord |
| $c_i$ | = | constant coefficients, $i = 1, 2, \cdots$ |
| $c_m$ | = | $c_{flap}$/MAC, Equation 4.64 |
| $C_a$ | = | generic aerodynamic coefficient |
| $C_D$ | = | drag coefficient |
| $C_{Db}$ | = | drag coefficient of a landing-gear bay; base drag coefficient, Equation 4.86 |
| $C_{D_{hull}}$ | = | hull-drag coefficient |
| $C_{Di}$ | = | induced-drag coefficient |
| $C_{D_{int}}$ | = | interference-drag coefficient |
| $C_{Do}$ | = | profile-drag coefficient |
| $C_{Dw}$ | = | wave drag; hull drag, Equation 4.97 |

| | | |
|---|---|---|
| $C_{D_\beta}$ | = | drag coefficient of boat-tail |
| $c_f$ | = | skin-friction coefficient |
| $C_L$ | = | lift coefficient |
| $C_{Lo}$ | = | lift coefficient corresponding to $\alpha = 0$ |
| $C_{L_g}$ | = | lift coefficient in ground effect |
| $C_{L_{hull}}$ | = | lift produced by immersed hull |
| $C_{L_{max}}$ | = | maximum lift coefficient |
| $C_{L_\alpha}$ | = | lift-curve slope |
| $\Delta C_{L_{flap}}$ | = | change in $C_L$ due to flap deployment in ground effect |
| $\Delta C_{L_{sf}}$ | = | increment in $C_L$ due to the deployment of split flaps |
| $\Delta C_{L_r}$ | = | change in $C_L$ due to flap and spoiler deflection |
| $C_N$ | = | normal force coefficient |
| $c_m$ | = | parameter in Equation 4.64, $c_m = c_{flap}/\text{MAC}$ |
| $C_m, C_n, C_l$ | = | roll/pitch/yaw moment coefficient |
| $C_p$ | = | term defined in Equation 4.98 |
| $C_Y$ | = | side-slip force coefficient |
| $C^*$ | = | Chapman-Rubesin factor, Equation 4.22 |
| $d$ | = | diameter |
| $d_b$ | = | base diameter, Figure 4.23 |
| $d_w$ | = | wheels diameter |
| $D$ | = | drag force |
| $e$ | = | Oswald factor, Equation 4.16 |
| $E$ | = | parameter defined in Equation 4.64 |
| $f_1$ | = | installation effects, Equation 4.62 |
| $f_2$ | = | installation effects (effects of flap deflection), Equation 4.63 |
| $f_f$ | = | form factor, Equation 4.35 |
| $f_j$ | = | thrust-specific fuel consumption |
| $F_c$ | = | parameter defined in Equation 4.30 |
| $F_x$ | = | parameter defined in Equation 4.33 |
| $F_\mu$ | = | non-dimensional dynamic viscosity, Equation 4.33 |
| Fr | = | Froude number, $\text{Fr} = V/\sqrt{gl}$ |
| $g$ | = | acceleration of gravity |
| $G$ | = | empirical function for fuselage upsweep drag, Equation 4.40 |
| $G$ | = | function of Strouhal number in buffet, Equation 4.88 |
| $J_{p_o}$ | = | efficiency factor of plain flap, Equation 4.10 |
| $k$ | = | induced-drag factor |
| $k_f$ | = | flat-type factor in Equation 4.10 |
| $k_{flap}$ | = | empirical parameter in Equation 4.56 |
| $k_t$ | = | free stream turbulence level, Equation 4.34 |
| $k_{t_o}$ | = | correlation factor in Equation 4.10 |
| $k_u$ | = | factor in the flap deflection, Equation 4.67 |
| $k_w$ | = | lift-induced-drag factor in hull-drag equation |
| $K$ | = | part-span factor (Equation 4.53); wheels coefficient (Equation 4.68) |
| $l_u$ | = | length of retractable multi-wheel landing gear on ground |
| $l$ | = | reference length |

| | | |
|---|---|---|
| $l_o$ | = | virtual forebody length, Figure 4.23 |
| $m$ | = | ground clearance of vertical strut, Figure 4.13 |
| $m_f$ | = | fuel mass |
| $M$ | = | Mach number |
| $M_B$ | = | buffet design Mach number |
| $M_c$ | = | critical Mach number, Equation 4.44 |
| $M_{dd}$ | = | divergence Mach number |
| $n$ | = | distance between wheels (Figure 4.13); normal load factor in Equation 4.89 |
| $n_w$ | = | number of wheels |
| $p$ | = | pressure |
| p, q, r | = | pitch/roll/yaw rates, respectively |
| Pr | = | Prandtl number, $\mathrm{Pr} = c_p\mu/k$ |
| $q$ | = | dynamic pressure, $q = \rho U^2/2$ |
| $r$ | = | recovery factor, Equation 4.24; radius of curvature in § 4.4.1 |
| $r_o$ | = | radius of secant ogive, Figure 4.23 |
| $r_{eq}$ | = | equivalent radius |
| $Re$ | = | Reynolds number, $\mathrm{Re} = \rho Ul/\mu$ |
| $\mathrm{Re}_c$ | = | critical Reynolds number |
| $\mathrm{Re}_l$ | = | Reynolds number based on length $l$ |
| $\mathrm{Re}_{trans}$ | = | Reynolds number for turbulent transition, Equation 4.34 |
| $\mathrm{Re}_x$ | = | local Reynolds number |
| $s$ | = | ratio between wheel's diameter and landing-gear width, Equation 4.58 |
| $S_{tyre}$ | = | tyre's frontal surface |
| St | = | Strouhal number |
| $t/o$ | = | relative thickness |
| $t_m$ | = | geometric parameter in Equation 4.62 |
| $T$ | = | net thrust |
| $\mathcal{T}$ | = | temperature |
| $\mathcal{T}_{aw}$ | = | adiabatic wall temperature |
| $\mathcal{T}_e$ | = | external temperature |
| $\mathcal{T}_w$ | = | wall temperature |
| $\mathcal{T}^*$ | = | reference temperature |
| $U$ | = | air speed |
| $V$ | = | volume; critical speed for aquaplaning, Equation 4.71 |
| $V_B$ | = | buffet design speed |
| $V_{float}$ | = | float volume |
| $V^*_{float}$ | = | float volume, immersed |
| $V_S$ | = | stall speed |
| $w$ | = | downwash velocity |
| $W$ | = | weight |
| $x_o$ | = | factor defined in Equation 4.98 |
| $x,y,z$ | = | Cartesian coordinates |
| $x_b$, $y_b$, $z_b$ | = | body-conformal reference system |
| $X$ | = | distance |

## Greek Symbols

| | | |
|---|---|---|
| $\alpha$ | = | angle of attack; parameter defined in Equation 4.31 |
| $\alpha_e$ | = | effective angle of attack |
| $\alpha_g$ | = | wing angle of attack in ground effect |
| $\alpha_i$ | = | induced angle of attack |
| $\alpha_o$ | = | zero-lift angle of attack |
| $\beta$ | = | side-slip angle; also $\beta = (M^2 - 1)^{1/2}$ |
| $\beta$ | = | parameter defined in Equation 4.31 |
| $\beta$ | = | boat-tail angle, Figure 4.23 |
| $\beta_1$ | = | Prandtl-Glauert factor, Equation 4.77 |
| $\beta_2$ | = | factor in Kármán-Tzien equation, Equation 4.78 |
| $\beta_u$ | = | average upsweep angle of rear fuselage |
| $\gamma$ | = | flight-path angle, § 4.1 |
| $\gamma$ | = | ratio between specific heats, $\gamma = 1.4$ |
| $\delta$ | = | depth of the centroid of the maximum float cross-section, Equation 4.98 |
| $\delta_f$ | = | flap angle |
| $\eta$ | = | normalised spanwise coordinate; induced-drag coefficient, Equation 4.85 |
| $\theta$ | = | aircraft attitude |
| $\vartheta$ | = | relative temperature in Equation 4.30, $\vartheta = \mathcal{T}_w/\mathcal{T}_e$ |
| $\kappa$ | = | lift factor in Equation 4.41 |
| $\kappa_a$ | = | airfoil technology factor, Equation 4.41 |
| $\Lambda_{LE}$ | = | leading-edge sweep angle |
| $\mu$ | = | dynamic viscosity |
| $\nu$ | = | kinematic viscosity |
| $\rho$ | = | air density |
| $\rho^*$ | = | density of contaminant |
| $\rho_w$ | = | water density |
| $\tau$ | = | wing's volume parameter |
| $\varphi$ | = | wing dihedral angle |
| $\Phi_i, \Phi_o$ | = | part-span wing factors, Equation 4.10 |
| $\zeta$ | = | damping ratio in buffet, Equation 4.88 |

## Subscripts/Superscripts

| | | |
|---|---|---|
| $\infty$ | = | free-stream conditions |
| $\overline{[.]}$ | = | average value |
| $\hat{[.]}$ | = | incompressible flow conditions |
| $[.]_e$ | = | external conditions |
| $[.]_{ht}$ | = | horizontal tail |
| $[.]_{IGE}$ | = | in ground effect |
| $[.]_{OGE}$ | = | out of ground effect |
| $[.]_{flap}$ | = | relative to flap |
| $[.]_{max}$ | = | maximum value |
| $[.]_{sf}$ | = | split-flap |

$[\cdot]_{sep}$ = separation
$[\cdot]_{trans}$ = turbulent transition
$[\cdot]_{vt}$ = vertical tail
$[\cdot]_{w}$ = wing
$[\cdot]_{\alpha}$ = derivative with respect to angle of attack
$[\cdot]_{\beta}$ = derivative with respect to yaw angle
$[\cdot]_{\delta}$ = derivative with respect to flap deflection
$(\cdot)^{*}$ = reference conditions

# 5 Engine Performance

## Overview

In this chapter we present a basic analysis for three key gas-turbine engine architectures: high by-pass turbofan, turboshaft and low by-pass turbojet with thrust augmentation. We consider general gas turbine engine architecture (§ 5.1), the thrust and power ratings (§ 5.2), the turbofan model (§ 5.3), the turboprop engine (§ 5.4) and the low by-pass engine model (§ 5.5). We briefly mention the methods of generalised engine performance (§ 5.6). We finally discuss the role of the auxiliary power unit (§ 5.7). An important aspect of the presentation is the strategy for the determination of the design point of the engine in absence of reliable data. In all cases we discuss the calculation of the engine state as a function of the main operational parameters.

**KEY CONCEPTS:** Gas Turbine Engines, Thrust/Power Ratings, Engine Derating, Turbofan Engines, Engine Design Point, Engine Simulation, Rubber Engines, Effects of Contamination, Turboprop Engines, Turbojet Engines, Auxiliary Power Units.

## 5.1 Gas Turbine Engines

The term *gas turbine* is associated with a jet engine consisting of a compressor, a combustion chamber, a turbine and an exhaust nozzle, although the name refers to both jet-thrust engines and shaft-power engines. The main types of gas turbine engines are the turbojet, the turbofan and the turboprop. The gas turbine is the core of the engine. However, there are other parts whose function is essential (inlet, fuel lines, fuel nozzles, sensors, collectors, thrust reverser).

The turbojet belongs to the first generation of gas turbine engines. It consists of a single gas flow. The operation of the engine requires a number of aero-thermodynamic stages: 1) compression of the inlet flow via a number of axial compressor stages; 2) the transfer of the compressed air into the combustion chamber, where it is mixed with fuel; 3) combustion in radially spaced combustion chambers; 4) discharge into a multi-stage turbine, rotating on the same shaft with the compressor; and 5) ejection of all of the exhaust gases as a high-speed hot jet through the nozzle. The air is captured by an inlet, whose other function is to provide pre-compression by an aero-thermodynamic mechanism called *ram compression*. This is an adiabatic compression in the engine inlet due to flow deceleration.

The main function of the turbine downstream of the combustion chamber is to operate the compressor. A considerable amount of power is generally required by the compressor. The rest of the thermal and kinetic energy associated to the mass flow is transformed into a high-speed, high-noise jet released from the nozzle.

The very first turbojets had centrifugal compressors, but as the engines became better understood, they were replaced by more efficient axial compressors. As the thrust requirements increased, the compressor's architecture became more complicated, with low- and high-pressure units, each with several rotor stages. The exhaust gas leaving the combustion chamber has a high temperature (about 1,000°C). When the compressor and the turbine are connected to the same shaft, their rotational speed is the same. This coupling is referred to as *spool*.

The gas turbine can have an additional combustion stage (re-heat, or after-burning). Fuel is injected after the primary combustion for the purpose of increasing the engine thrust. After-burning uses the excess air that does not support the primary combustion.

Gas turbines with this capability operate without re-heat most of the time because the increase in thrust is derived at the expense of a considerable fuel consumption. Their application is limited to some military jet aircraft.

A turbofan is a derivative of the turbojet. In this engine the excess air that does not support the combustion is channelled through an external annulus and by-passes the combustor. The *by-pass ratio* (BPR) is the ratio between the by-pass flow rate and the core flow. This ratio has been increasing over the years, from about 1.1 to values above 5 in modern engines. The General Electric GE-90 has a BPR = 8.4, and the P&W GP-7000 series has a BPR = 8.

The other difference, compared to the basic engine, is a large-diameter fan placed in front of a multi-stage axial compressor. The function of the fan is to increase the capture area of the inlet and to channel the by-pass flow through the annulus of the engine. The fan is powered by the engine itself, either on the same shaft as the compressor or on a separate shaft (dual-compressor engine). The advantages of this engine are that the exit flow has lower speed and lower average temperature and produces far less noise.

A *turboprop* is an aero engine consisting of a gas turbine unit coupled with a propeller. The thrust can be derived both from the jet engine and the propeller, although in practice, most of the useful thrust is imparted by the propeller. Due to the different speeds between the gas turbine and the propeller, these engines have a reduction gear. A gas turbine rotates at speeds of the order of 10,000 rpm, whilst the propeller's speed is less than one third of this, as limited by the tip Mach number.

Jet engines for helicopter applications are a variant of the turboprop and consist of one or two gas turbines, a reduction gear and a rotor shaft. The reduction in rotational speed is higher than the turboprop because helicopter rotor speeds do not exceed 300 rpm.

Modern engines come with an FADEC (Full Authority Digital Engine Control), which is the brain of the engine. A typical FADEC consists of a hardware unit and a software unit. The hardware unit includes sensors to determine in real time essential engine parameters and a digital computer (*electronic engine controller*) to elaborate on the input data and generate a control response. One of the operations that are possible with the electronic control is to balance the power from different engines. For example, if there is a loss of power from one engine, the FADEC reacts quickly to augment the power of the other engine. Other monitoring operations done by the FADEC include cycle counting, cold start, engine stop and general engine health.

The subject of aircraft engines is a specialised one and goes under the field of aerospace propulsion or gas turbines. For a more in-depth and specialised presentation, the reader is invited to consult the specialised literature, for example, Mattingly[1], Oates[2], Archer and Saarlas[3], and the citations thereof.

Modern gas turbine engines must be defined by dozens of geometrical and functional parameters. However, most critical data are hard to obtain because they represent proprietary information by the engine manufacturers. Some of the aero-engine programs are still considered a matter of national security. This aspect of flight performance is a bottleneck in most simulation procedures.

A number of programs of industry standard are available to evaluate the essential static and transient performance of typical engines. However, both simulation programs and engine manufacturers' data cannot predict exactly the flight performance at altitude and the installation losses. These losses depend on the integration between the engine and the airframe. The same airplane version can have different engines, with performance slightly different from each other. This is a choice of the airplane manufacturer, to reduce the dependency on any single engine manufacturer and to drive down prices.

Initial testing of the engine is done on a test bed in experimental facilities that allow the simulation of pressure altitude. Engine performance at different atmospheric conditions is done by specific correlation rules. Therefore, there is a widespread practice to work with normalised data from a mathematical model of the engine.

## 5.2 Thrust and Power Ratings

Aircraft engines have various thrust and power ratings, based on operation time and maximum allowable temperatures at selected sections of the engine. For a turbofan engine, typical thrust ratings include maximum take-off, maximum climb and maximum continuous thrust. For most ratings a time limit is associated to the thrust output. The atmospheric conditions, the altitude and the speed must be specified. The practice is to provide data for a standard day, although this is not always the case. One example is shown in Table 5.1. The engine ratings* are for temperatures below the values indicated in the table. Therefore, our interpretation is that these engines should provide the quoted power rating at any "reasonable" atmospheric temperature below the maximum temperature. Additional power ratings for this type of engine include a maximum contingency power (2.5 min) and an intermediate contingency power.

* Adapted from EASA: Type Certificate Data Sheet IM E041, PW100 Series, June 2008.

Table 5.1. *Power ratings for PW127 turboprop engine variants, sea level; maximum temperatures as indicated*

| Engine | MTOP[kW] (5 min) | MTOP[kW] (normal) | $T$[°C] (max) | MCP [kW] | $T$[°C] (max) |
|---|---|---|---|---|---|
| Basic | 2,051 | 1,846 | 32 | 1,864 | 41 |
| B | 2,051 | 1,846 | 30 | 1,864 | 41 |
| D | 2,051 | 1,846 | 33 | 2,051 | 33 |
| E | 1,790 | 1,611 | 45 | 1,790 | 45 |
| F | 2,051 | 1,846 | 35 | 1,864 | 44 |
| G | 2,178 | 1,973 | 35 | 2,178 | 35 |
| M | 2,051 | 1,846 | 39 | 1,864 | 48 |

In ground operations, the manufacturers often refer to a *breakaway thrust* (or power). This is the minimum thrust required to move the airplane. This thrust depends on the brake-release gross weight.

### 5.2.1 Engine Derating

Often the thrust required is lower than the thrust available. This event occurs at take-off weights lower than MTOW. Thus, to improve engine life and reduce maintenance and fuel costs, there is the possibility of "derating" an engine. Alternatively, excessive thrust may cause excessive structural loads, and derating is one way to prevent this event from occurring. Finally, a derated engine finds applications on different aircraft versions.

The term *derating* or *flexible* refers to the possibility of using the thrust appropriate to the take-off weight and atmospheric temperature. There are two important differences between derated and flexible thrust:

- An aircraft operating with a derated thrust cannot revert to full thrust if the speed is below the flap-retraction speed; an aircraft operating with flexible thrust can revert to full thrust if conditions require it.
- The aviation regulations allow the use of derated thrust but not the use of flexible thrust with contaminated runways.

An example of derated take-off thrust is shown in Figure 5.1. First, the engine is flat-rated and provides a net thrust that is not dependent on the outside air temperature (OAT) at temperatures below the flat rating. Above the derated temperature, the net thrust is limited by the exhaust gas temperature (EGT) and decreases as the OAT increases. If the actual TOW < MTOW, the thrust required is less than the available thrust: $T_{req} < T_{av}$. Flexible thrust can be used only if the actual temperature $T_{ref}$ is lower than the flexible temperature at the specified flexible thrust. In practice, there are a number of discrete derated levels, given in percent of the maximum thrust; each derated level must be certified. For example, D20 indicates a 20% derating, or 80% of the maximum thrust.

Opposite to derating is the engine "bump". This jargon refers to a temporary increase (5 to 10%) in the available thrust, to a value above the maximum take-off thrust.

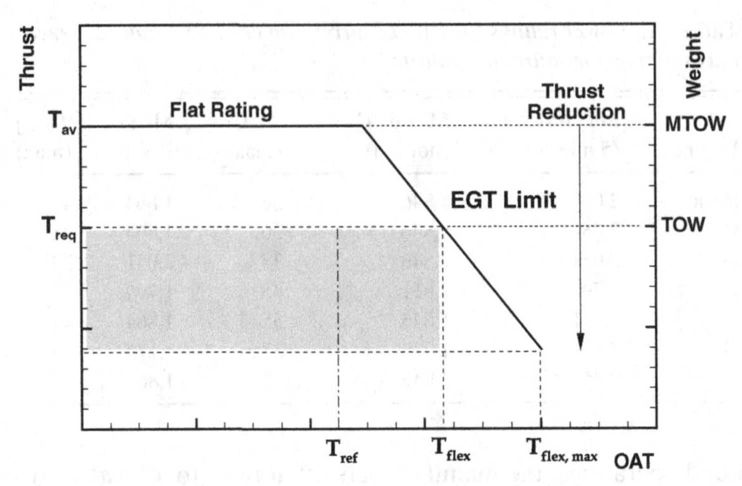


Figure 5.1. Engine thrust derate at take-off.

There is also a derated climb. As in the case of derated take-off, a lower climb thrust improves the engine life but, in general, causes an increase in fuel consumption and an increase in the time to climb. However, from the point of view of the operators, what counts most is the overall cost of operation, including items other than fuel. A derated climb is considered independent from a derated thrust.

Finally, an important advantage of the derated thrust is represented by the extended life of the engines. However, this gain must be evaluated against the increase in take-off fuel, the longer take-off distance and the lower initial climb rate.

### 5.2.2 Transient Response

The response of the engine to an increase in fuel flow (throttle) is not instantaneous. Times when sharp increases in thrust are required (take-off and go-around) are critical. Figure 5.2 shows a typical turbofan-engine response curve. The transient response depends on the specific engine. The engine must be protected against compressor and turbine stall as well as flame-out. Specific regulations* require a maximum of 5 seconds to accelerate from 15 to 95% of the go-around thrust. During this time, it is likely that the airplane loses altitude (see § 11.8); on recovering the go-around thrust, the engine must be able to ensure a minimum climb gradient (again, as specified by the relevant regulations). Turbofan and turboprop engines respond differently. The turboprop operates at nearly constant rpm and responds to an increase in shaft power; the turbofan must respond with an increase in rpm in order to increase its net thrust.

## 5.3 Turbofan Engine Model

The starting point of an engine model is the collection of essential data from the type certificate and other documents. Photographs of the engine will help as well. These data are indicated in the flowchart in Figure 5.3; they are split into two categories: configuration and design limitations. The grey-shaded boxes indicate data that are

* Federal Aviation Regulations, FAR 33.73: Airworthiness Standards, Aircraft Engines.

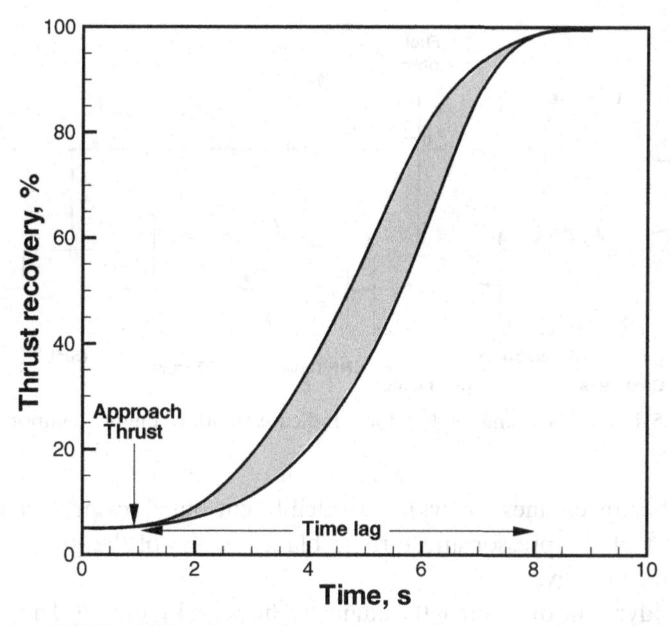


Figure 5.2. Typical turbofan thrust recovery in go-around manoeuvre.

generally unavailable; these include the number of blades and the diameters of all compressor and turbine stages (LPC, HPC, LPT, HPC), as well as the rotor-stator separation (RSS), the guide vanes and other smaller details, which may be needed for either the aero-thermodynamics or the determination of the acoustic sources (Chapter 16).

To study the transient performance, the inertias of the rotating parts are required to define the rotational acceleration. The unavailable data are estimated from a

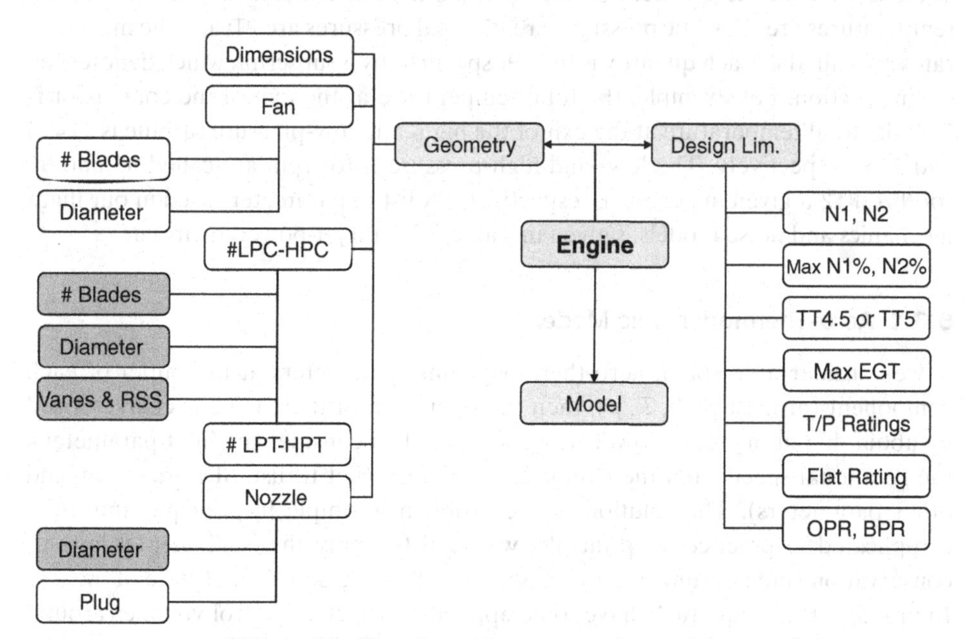


Figure 5.3. Flowchart of an aero-engine model.

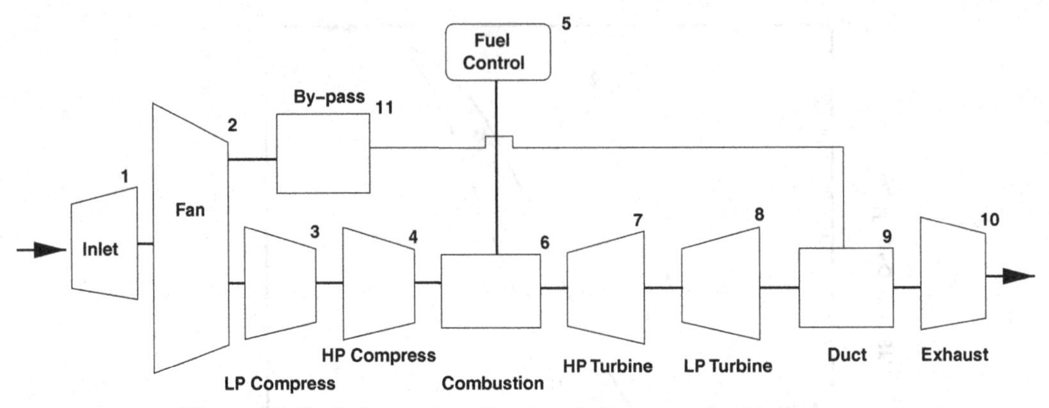


Figure 5.4. Turbofan engine. Numbers indicate standard engine components.

limited number of engines for which a reliable cut-out drawing is available; the actual number of compressor and turbine blades is one of the great mysteries of aero-engine technology.

A thermodynamic diagram of the engine is shown in Figure 5.4. The graph shows the logical connection between systems. Note that the inlet flow is split downstream of the fan. The core flow goes through the axial compressor. The by-pass annular flow is channelled to the duct and then mixed with the combustion gas. The numbers beside each component represent standard nomenclature. The flow through the engine is assumed as one-dimensional. This type of engine is typical of modern commercial aircraft, characterised by very large diameters – in some cases, limited only by the ground clearance. The BPR of these engines varies from about 4 to 6, sometimes higher.

There is a standard nomenclature to define the aero-thermodynamic parameters at all sections of the engine. For example, static temperatures are called `TS`; total temperatures are `TT`; static pressures are `PS`; total pressures are `PT`; and the mass flow rates are called `W`. Each quantity is further specified by a subscript, which denotes the engine section. For example, the total temperature at the exit of the combustor is `TT4`; the total temperature at the exit of the high- and low-pressure turbine is `TT4.5` and `TT5`, respectively. The low- and high-pressure rotor rpm are called `N1` and `N2` (or `N%1`, `N%2` if given in percent), respectively. A list of parameters used in our flight mechanics and noise models is given in Table 5.2 for a jet-powered aircraft.

### 5.3.1 Aero-Thermodynamic Model

If we consider a vector of aero-thermodynamic parameters at the outlet of each component (at least $\rho, \dot{m}, T, p$), then the operation of the engine is characterised by about 40 parameters. However, we also need to consider the inlet parameters, the rotational speeds and the overall performance (net thrust, SFC, fuel flow and other parameters). The solution of the problem is simple in principle and quite complicated in practice. In principle, we need to apply the fundamental laws of conservation (momentum, energy, mass), as well as the equation of state of the gas. In practice, these equations have to be applied to selected control volumes around each of the engine sub-components (as shown in Figure 5.4). The model can be

Table 5.2. *Turbofan-engine parameters used for flight and aircraft-noise calculations*

| Item | Symbol | Description | Text Symbol |
|---|---|---|---|
| 1 | W1 | Mass flow rate | $\dot{m}$ |
| 2 | WC2.5 | Core mass flow rate | |
| 3 | Wf6 | Fuel flow rate | $\dot{m}_f$ |
| 4 | N%1 | Low-pressure rotor rpm (%) | $N_1$ |
| 5 | N%2 | High-pressure rotor rpm (%) | $N_2$ |
| 6 | TT2.1 | Exit-fan temperature (core side) | |
| 7 | TT2.2 | Exit-fan temperature (by-pass side) | |
| 8 | TT3 | Combustor-inlet temperature | |
| 9 | TT4 | Exit-combustor temperature | |
| 10 | TT5 | Power-turbine temperature | |
| 11 | TS9 | Total static nozzle temperature | |
| 12 | TT14 | By-pass flow temperature, exit | |
| 13 | FN | Net thrust | $T$ |
| 14 | TSFC | Specific fuel consumption | $f_j$ |
| 15 | M9 | Nozzle Mach number (core side) | |
| 16 | TT2.5 | LP compressor-exit temperature | |
| 17 | PT2.5 | LP compressor-exit pressure | |
| 18 | PT3 | HP compressor pressure, exit | |
| 19 | PS9 | Total static nozzle pressure | |
| 20 | PT14 | By-pass flow total pressure, exit | |

improved by further considering secondary effects, such as heat flux between the gas and the external environment, bleed air at specified compressor sections, and the use of secondary power for oil and fuel pressure.

The equations are then assembled together with matching conditions between the outlet of one component and the inlet of the downstream component. The science behind this thermodynamics is better described in books on gas turbine engines (as cited). A number of computer programs of industry standard exist to help with this demanding task. One of those is the GSP program, developed at NLR[4], that we have used for our engineering analyses.

### 5.3.2 Determination of Design Point

The flowchart in Figure 5.3 does not show one of the most important parameters: the fuel flow (or the TSFC) at the design point. This datum is not available. Occasionally, there are data on TSFC published in the literature; these data are not reliable because the operation point is not fully specified. Notably, the ICAO databank does contain this datum*. However, it can be used as a guideline. A simulation method by itself is not capable of defining the design point. On the basis of the thermodynamic model, a number of parametric runs can be carried out so as to minimise the fuel flow at the nominal engine speed (100% rpm).

* ICAO Engine Data Bank. Updated regularly. Available in electronic form from the ICAO and other aviation authorities.

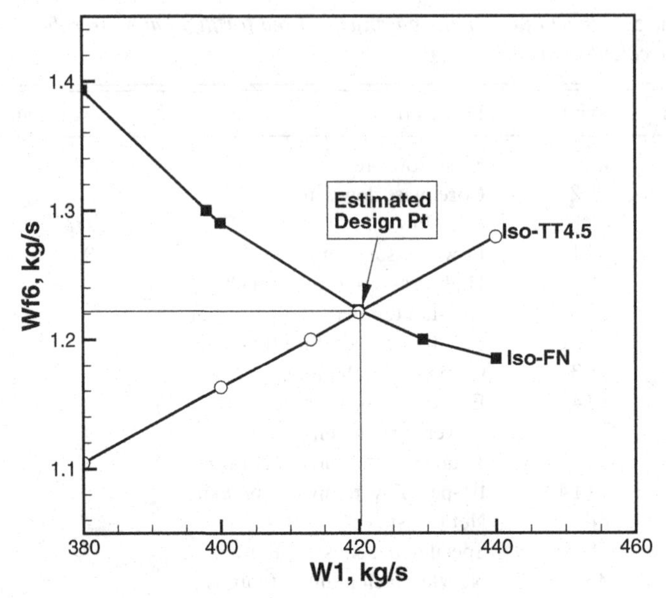


Figure 5.5. Example of determination of design point at `N%1` = 100, static conditions, sea level, standard day.

The design point is defined as the engine state corresponding to a design net thrust at sea level, static conditions, standard day, with the turbine temperature `TT5` (or `TT4.5`) equal to (or slightly below) the value of the limit temperature in the type certificate, with the engine running at 100% rpm. Now the procedure is as follows:

- Generate a table of performance data for variable mass flow `W1` (at fixed fuel flow `Wf6`) and variable `Wf6` (at fixed `W1`), around the estimated design point.
- The design point is determined as the intersection between the limit `TT4.5`-curve and the limit `FN`-curve. This point gives a unique value for `W1` and `Wf6`.

The determination of the design point is shown in graphic form in Figure 5.5. The iso-`FN` and the iso-`TT4.5` curve refer to the certified net thrust at the specified condition, that is, at sea level, static conditions, on a standard day.

There are some drawbacks in this procedure. In fact, on a hot day an engine operating at `N%1` = 100 (normal speed) would overheat; thus, the reference turbine temperature could exceed the certified limit temperature, although it should be able to operate normally. Alternatively, we can investigate the virtual design point corresponding to a limit engine speed (for example, `N%1` = 104) in standard atmosphere.

### 5.3.3 Case Study: General Electric CF6-80C2

We consider the General Electric CF6-80C2, an engine that powers aircraft such as the Airbus A300. Over the years, General Electric has developed at least 23 versions of this engine, each with a different pressure ratio and thrust rating. The fan has 38 blades, a 4-stage booster and 80 composite exit guide vanes. The low-pressure (LP) compressor consists of four stages and vanes mounted orthogonally. The high-pressure (HP) compressor consists of 14 stages, with inlet guide vanes; the first 5 stator rows have variable incidence. The combustion chamber is annular, with rolled

Table 5.3. *Selected engine data for the CF6-80C2A3; data with an asterisk * are estimated*

| | |
|---|---|
| **Thrust Ratings** | |
| Maximum continuous (S/L, static) | 24.853 kN |
| Take-off (5 min, S/L, static) | 26.739 kN |
| Flat-rating temperature (continuous) | 25°C |
| Flat-rating temperature (take-off) | 30°C |
| **Speeds** | |
| Design low-pressure rotor (N1) | 3,280 rpm |
| Design high-pressure rotor (N2) | 9,827 rpm |
| Low-pressure rotor (N%1), maximum | 117.0% |
| High-pressure rotor (N%2), maximum | 112.5% |
| **Maximum Permissible Temperatures** | |
| Turbine exhaust gas (TS9) | |
| Take-off (5 min) | 1,233 K |
| Maximum continuous | 1,198 K |
| Starting (40 s) | 1,143 K |
| **Bleeds** | |
| Compressor, stage 8, airflow (max) | 8.8% |
| Compressor, stage 11, airflow (max) | 1.5% |
| **Gas Flows** | |
| Design fuel flow, take-off* | 2.457 ks/s |
| Design mass flow | not available |

ring construction, aft-mounted, with film cooling. The high-pressure turbine has two stages and the low-pressure turbine has five stages. The control system is FADEC. The certification of the engine consists of a document that contains information regarding 1) limitations, and 2) permitted fuels, oils and spare parts. The limitations part is important in performance analysis because it contains data on rotor-speed limits, temperature limits of some components and rated thrust (Table 5.3).

The solution of the engine problem across the full range of fuel flows, Mach numbers and flight altitudes is shown in Figure 5.6. The graphs show the trends of selected parameters: net thrust, mass flow rate, TSFC, LPT total temperature on exit, nozzle Mach number and engine speed in percent. The graphs show the engine limits, which are set as constraints from the type certificate of this engine. The analysis is carried out at standard conditions, but it can be extended to include other atmospheric models, including cold and hot temperatures.

At moderate to high fuel flow, the relationship between the fuel flow and net thrust is roughly linear, which indicates that the TSFC is nearly constant. However, at low fuel-flow rates and idle conditions, the TSFC increases rapidly. An engine running idle on the ground would have an infinite TSFC because the engine generates no useful thrust. *The use of a constant TSFC in aircraft performance analysis is incorrect.* For an engine in idle mode, the fuel flow is a better indicator of performance.

The effect of fuel temperature is generally very small and certainly negligible when compared with other operational aspects. In practice, a colder fuel causes a slight increase in TSFC and a negligible decrease in net thrust.

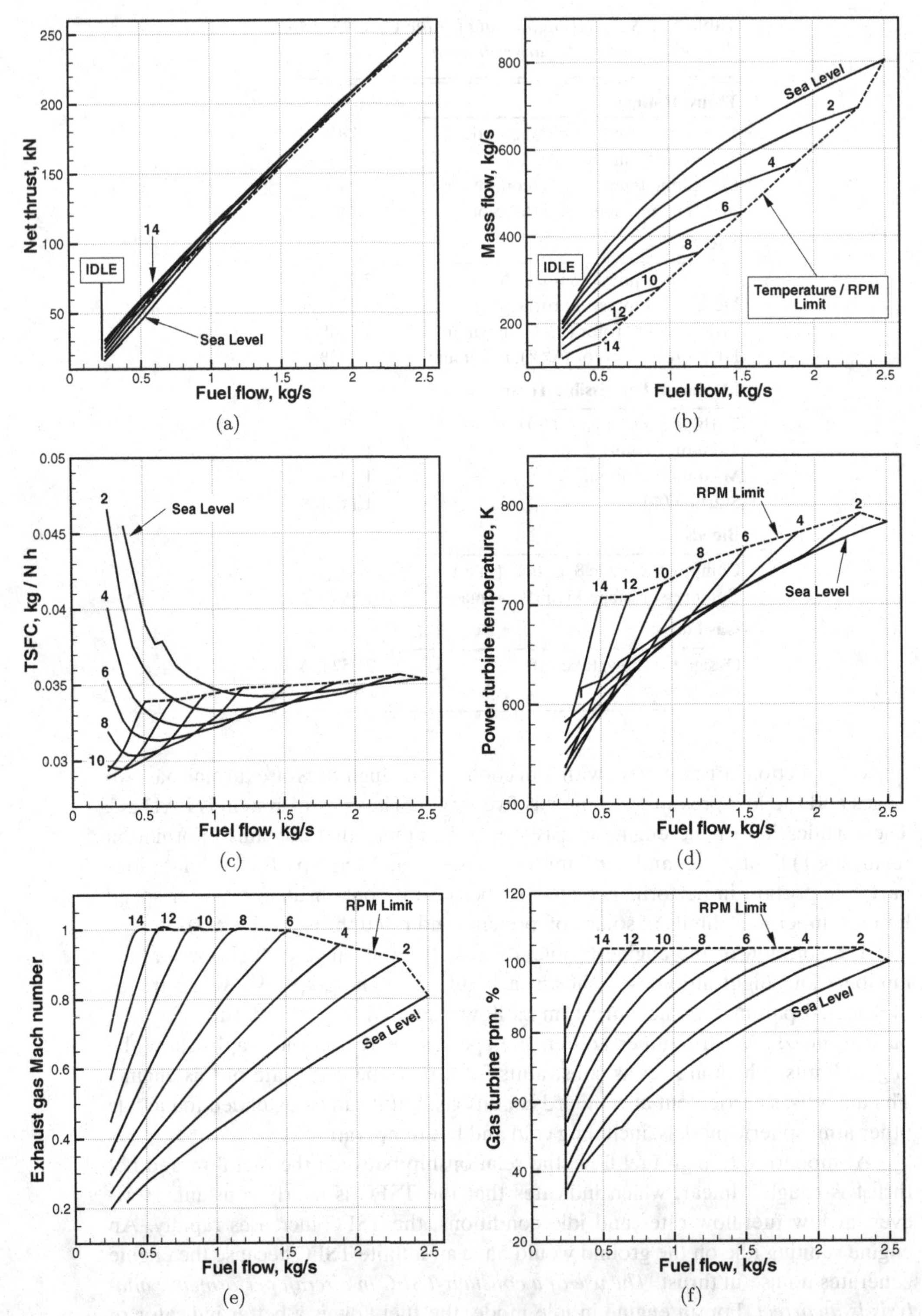


Figure 5.6. Simulation of General Electric CF6-80C2A3 turbofan engine; altitudes at 2,000 m intervals (indicated as 2, 4, ...); limit engine speed reduced to 105%; standard day.

**ENGINE INSTALLATION EFFECTS.** The thrust required for a flight condition (cruise or manoeuvre) is called $T_{req}$. The thrust delivered by the power plant depends on a number of factors. If $T$ is the uninstalled thrust, then

$$T_{req} = (1 + \kappa)T, \tag{5.1}$$

where the factor $\kappa$ takes into account the installation losses. The term *installation loss* refers broadly to all of the aero-thermodynamic effects that cause the engine to function differently when it is installed. Tests are carried out by the engine manufacturer *on the bench* or uninstalled conditions. Installation losses include reduction in mass flow rate (spillage drag), flow distortions, tail-pipe pressure drop, nozzle drag, incomplete combustion, loss due to associated systems, and so on. The precise estimation of each of these contributions is difficult because it depends on the flight conditions (Mach, altitude, thrust required) and must be done with very detailed engine models. An example of analysis is shown by Roth[5] on a turbojet engine.

### 5.3.4 Rubber Engines

A *rubber engine* is an engine that can be scaled to a lower or higher thrust. The scaling applies to engines in the same family, which differ for relatively small details from the point of view of comprehensive aircraft performance. For example, the engine architecture of the CFM56-5C2 (basic engine) is similar to at least 30 engines in the same family†. Differences may include redesigned fan blades, redesigned compressor stages, reduced thrust, different combustors, compression rates, by-pass ratios, and so on.

It is assumed that rubber engines perform identically, except for the thrust or power rating. Therefore, they should be able to provide increased or decreased levels of thrust and power with the same engine temperatures and specific fuel consumption, although the overall compression rates may be different. With reference to the SFC, at a given operation point we assume that a scaled turbofan engine provides a scaled thrust $T$ with a scaled fuel flow $\dot{m}_f = f_j T$, with TSFC equal to the "basic" engine. The corresponding mass flow rate is scaled by assuming that the fuel–air ratio is the same as the basic engine.

Because the compression rates of the rubber engine may be different, the pressures in the relevant sections are updated. These data are to be used in some noise sub-models, as described in Chapter 16.

There are some justifications for this type of operation: this level of detail of one engine and its similar counterpart is either unknown, or difficult to model, or unverifiable, or inconsequential from the point of view of flight simulation. The previous examples of redesigned fan blades and redesigned compressor stages are of little interest because they are not associated to a known fan and compressor efficiency that can be used in a one-dimensional aero-thermodynamic model. Likewise, a redesigned combustor that has an unknown limit temperature cannot be used to update the thermodynamic model.

† Federal Aviation Administration, Type Certificate Data Sheet E38NE: CFM International (CFM56 engines), Sept. 1996.

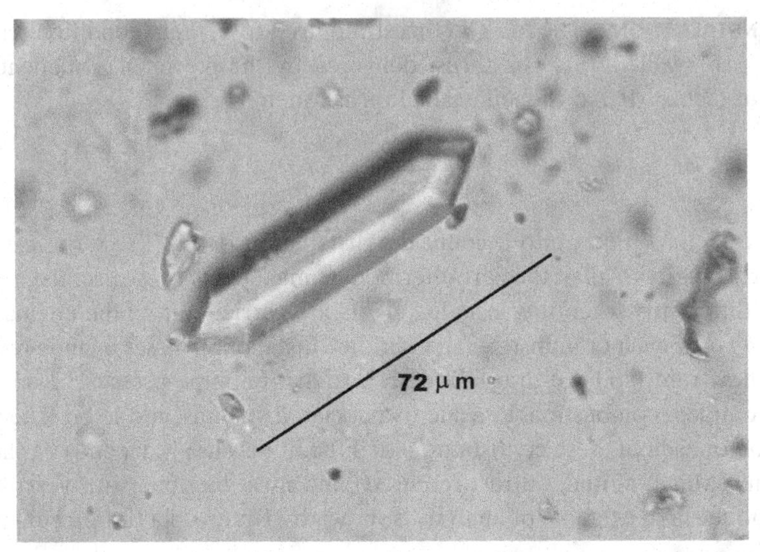


Figure 5.7. Volcanic ash particle from the Iceland eruption, March 2010, taken from a car in South Yorkshire, England [photo courtesy of John Chandler].

### 5.3.5 Effects of Contamination

There are various environmental effects that can cause degradation of engine performance and, in some serious cases, also engine shut-down within a few minutes. One of the most serious problems centres around ash clouds. Although damage can occur on all parts of the aircraft (airframe, windshields, and so on), this section deals exclusively with engine-related problems. About 80 aircraft have been reported to have been flown through volcanic ash between 1980 and 2000, with various degrees of severity. Critical regions are around the Pacific Ocean, although events in Northern and Southern Europe routinely affect flight operations. These clouds can move quickly around FL-300 (30,000 feet). They are made of very fine particulate ($\sim 1\mu$m) and may carry electrostatic charges.

Ingestion of volcanic ashes causes, at various degrees, erosion of the fan blades and compressor blades. If they manage to travel as far as the combustor, the ash particles are melted; then, they either stick to the combustor walls and glaze or travel further to the turbine stages (where they cause further structural damage), or remain within the vanes, thus disturbing the high-pressure out-flow. Side effects on the aircraft systems include damage to the Pitot tubes (essential for monitoring the air speed) and the air filters (essential for maintaining cabin pressure). Melting of volcanic ash is quite variable, from about ~1,250 K to ~1,600 K[6], whilst combustor temperatures can, in some cases, exceed this value.

Figure 5.7 shows an example of volcanic ash particle from the Eyjafjallajökull eruption in the spring of 2010. The exagonal particle is most likely quartz.

For reference, consider the combustor temperatures of the engine CF6-80C2A3, whose simulation is shown in Figure 5.6. The corresponding combustor temperature `TT4` is shown in Figure 5.8. The shaded area indicates the temperature `TT4` at flight altitudes between 8,000 and 12,000 m (26,250 and 39,270 feet, respectively). Therefore, it is very possible that melting occurs on the combustor, unless the engine

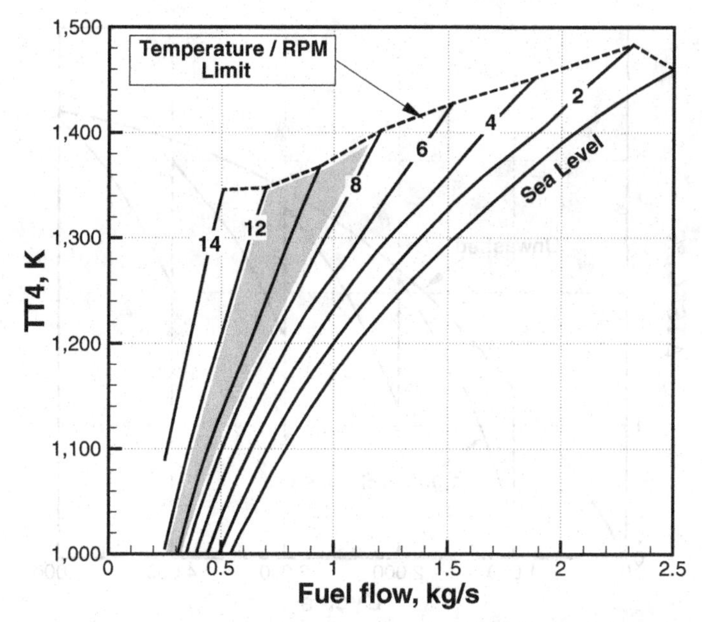


Figure 5.8. Combustor temperature of simulated engine CF6-80C2A3 (see also Figure 5.6).

operates at low thrust (low temperature) and the melting temperature of the ash is high.

Monitoring of volcanic ashes is a complex matter beyond the realm of a single aircraft performance‡. Actions required in the presence of such an event include a minimum clearance of 20 n-miles, possibly upwind of the affected area. If the cloud cannot be avoided, the aircraft is to be commanded to make a U-turn or decrease the thrust setting. This option limits the combustor temperatures and thus the risk of particles melting in the combustor.

Another form of engine contamination is due to dust and ice intake on the ground, either with the engines switched off or running. If the engines are to be switched off for any length of time, the intakes must be closed-off to contamination agents that would accumulate over time; the engine integrity could be compromised.

### 5.3.6 Performance Deterioration

Aircraft engines are subject to performance penalties that depend on the number of work hours or cycles. The loss in performance is quantified by the increase in TSFC. Reasons for such a deterioration include dirt accumulation in the first stages of the compressor, increased tip clearances due to differential heating, blade erosion and seal leakages. Most of these losses can be reduced by servicing the engine at specified intervals; the service guarantees the best performance over a large number of cycles. Typically, an improvement in excess of 1% in TSFC can be achieved. An example is shown in Figure 5.9. This graph was drawn using data extracted from Pratt & Whitney and refers to a typical high by-pass turbofan engine. The servicing is done at 1,000-cycle intervals. The data clearly point toward a loss of more than

‡ There are nine Volcanic Ash Advisory Centres around the world that deal with information to the civil aviation.

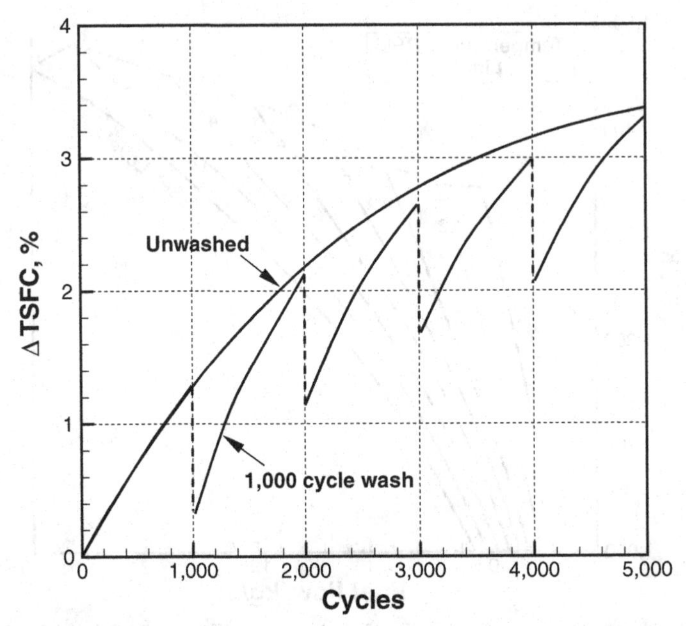


Figure 5.9. Loss of efficiency of turbofan engine with and without cycle wash (data adapted from Pratt & Whitney).

1% in TSFC within the service schedule, which for an engine in this class means a considerable amount of fuel burned to no benefit. Even with the service, after 4,000 cycles, there appears to be a permanent loss of TSFC of the order of 2%.

In the absence of detailed engine data, it will suffice to assume that the loss in TSFC can be expressed as

$$\Delta f_j\,(\%) = 2.55 \cdot 10^{-2} + 1.28 \cdot 10^{-3} n + p, \tag{5.2}$$

where $n$ is the number of cycles and $p$ is a permanent loss, which may depend on the number of services already performed.

### 5.3.7 Data Handling

Because our engines have to be integrated into an airframe and thence with a flight mechanics model, we need to take some further steps. Instead of solving the engine problem at every step, we generate a complete flight envelope at selected atmospheric temperatures. There are two reasons for this choice: 1) the solution of the engine problem at each time step is a slow process; and 2) the solution of the engine must be carried out in both direct and inverse mode. If the net thrust is a constraint, then we need to calculate the fuel flow and the engine state corresponding to the specified thrust (*inverse mode*). The functional relationship is

$$\texttt{EngineState}(T) = f(T, M, h, \Delta T). \tag{5.3}$$

If the fuel flow is specified, the engine is solved in *direct mode* to produce a thrust $T$. The functional relationship is

$$\texttt{EngineState}(\dot{m}_f) = f(\dot{m}_f, M, h, \Delta T). \tag{5.4}$$

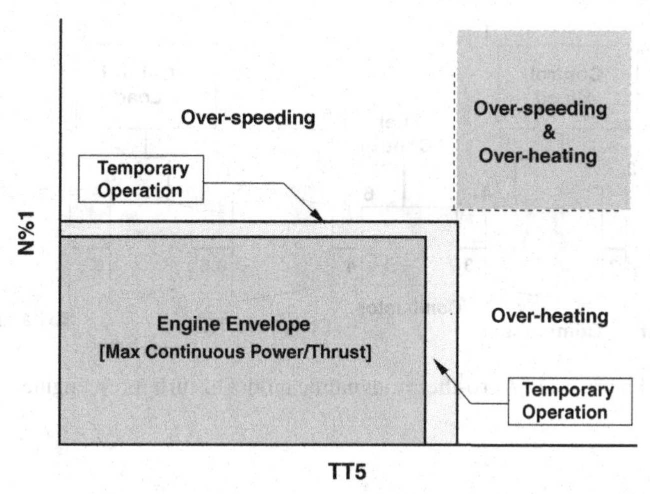


Figure 5.10. Engine limits in the space TT5, N%1 (gas turbine inlet temperature and rpm).

The simulation of the engine's envelope leads to a large matrix of data. Each envelope is a multi-dimensional matrix in the space defined by the vector $\{M, h, \Delta\mathcal{T}, \dot{m}_f\}$ for each engine state parameter. There are several dozen state parameters, which arise from the number of independent aero-thermodynamic parameters at all relevant sections of the engine. For the purpose of the analysis carried out in this and following chapters, we consider only a sub-set of these parameters, as listed in Table 5.2.

The envelope must be interpolated in either direct or inverse mode. For example, in direct mode, if $\mathcal{E}$ is a generic engine parameter, then its functional dependence on the operational conditions can be written as

$$\mathcal{E} = f(d\mathcal{T}, U, h, \dot{m}_f). \tag{5.5}$$

We use a higher-order multi-dimensional interpolation of the matrix at the operational point defined by the vector $\{d\mathcal{T}, U, h\}$. Thus, we construct a functional

$$\mathcal{E}_1 = f_1(\Pi). \tag{5.6}$$

In the engine analysis, the state parameter can be greater than one, which corresponds to an operational point that exceeds the design limits, either due to a too high gas generator temperature TT5 or a too high turbine rpm N1%, or both. To avoid engine operation beyond the design point, as shown in Figure 5.10, the state parameter $\Pi$ is calculated to match the worst possible condition. If the rpm is within the design limits, Equation 5.6 is limited by the power turbine temperature to avoid over-heating. If the engine is over-speeding (in general, about 104% the design rpm), the power turbine temperature is the one corresponding to the limit rpm.

## 5.4 Turboprop Engines

A turboprop is a gas turbine engine coupled to a propeller. It is designed to deliver shaft power instead of thrust force. For this reason, there are substantial differences in the engine architecture, although there is a degree of commonality between the two engines; the turboprop and the turbofan are both continuous mass flow engines with compressor and turbine units.

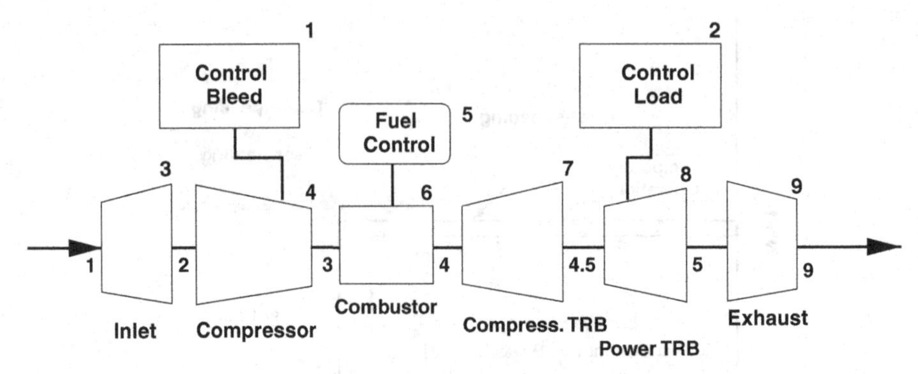


Figure 5.11. Aero-thermodynamic model of turboprop engine.

The turboprop is made of the engine itself, the cowlings, the engine mounts, the firewalls and the drain lines. In its basic form, the engine is composed of an axial compressor, a centrifugal compressor, an annular combustor, a centrifugal fuel injection system, a multi-stage gas turbine and a single-stage free power turbine. The gear shaft is considered part of the transmission system. The gas-generator turbine and the power turbine operate at different speeds and are mounted on two co-axial shafts (dual-spool engine). An aero-thermodynamic diagram of the engine is shown in Figure 5.11, with the standard nomenclature for the components and the engine sections. Only one compressor unit is modelled; the pressure ratio is provided by this component.

Air is drawn in through the inlet; it is passed to a multi-stage compressor and hence to a combustor. Fuel is added to the combustor through the fuel lines, with the aid of fuel pumps. The hot gas from the combustor expands into a gas-generator turbine, mounted on the same shaft as the compressor. The hot gas expands further in the power turbine, that is coupled to the power shaft, and hence to the rotor shaft, via a gear box. There are some conventions on the nomenclature that it is important to clarify. The nominal gas-generator rotational speed is indicated by $N_1$ and the power turbine speed is called $N_2$. These parameters are found in the certification documents of the engine.

A number of controls are indicated in the diagram. One control is the air bleed from the compressor that can be used to run auxiliary systems. The air bleed decreases the shaft power. Another control is the fuel flow into the combustor. Finally, there is a control on the shaft power or torque that is used for transient analysis of the engine.

The design point is determined in a similar fashion as explained in § 5.3.2. However, in this case we need to match the design shaft power instead of a thrust; there are no reference data to compare with because the ICAO databank is limited to turbofan engines. The data handling follows the same strategy as shown in § 5.3.7. The set of state parameters used is different. The net thrust is replaced by the shaft power; the nozzle parameters include the residual thrust $F_g$; the by-pass parameters do not apply to this case.

The engine model must be developed in a way such that both the direct and inverse solutions are possible, as in the case of the turbofan. In the direct mode the

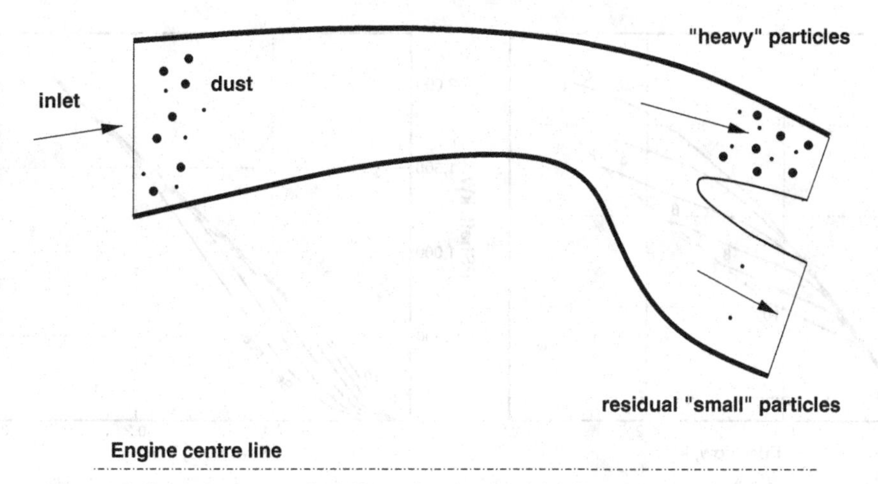


Figure 5.12. Axi-symmetric, bifurcating intake with inertial particle separator.

fuel flow is the independent parameter; in the inverse mode, the shaft power is the independent parameter.

### 5.4.1 Case Study: Turboprop PW127M

This turboprop engine is part of a family (see also Table 5.1). The engine has a two-spool gas generator driving the six-bladed Hamilton propeller F568-1 (§ 6.2.4). The main engine components are a centrifugal low-pressure compressor; a centrifugal high-pressure compressor (HPC), diffuser pipes, an annular perforated combustion chamber with 14 fuel nozzles; a high-pressure axial turbine (that drives the HPC); a low-pressure axial turbine (that drives the LPC); and a two-stage free-axial turbine that drives the reduction gearbox. The latter element is connected via a coupling drive-shaft and is mounted offset of the main axis of the engine. On the gearbox, there are a number of sub-systems, including the propeller brake, the over-speed governor and a number of pumps (oil, propeller feather).

The intake is offset and has an "S" shape to provide a uniform intake into the radial compressor. This intake also functions as an inertial particle separator that filters out any large foreign objects, as shown schematically in Figure 5.12. The intake air is also split between compressor and oil cooling.

A simulation of the Pratt & Whitney PW127M turboprop engine is shown in Figure 5.13. The graphs show the mass flow rate, the shaft power, the turbine temperature `TT5` and the engine speed `N%1` as a function of the fuel flow at the full range of altitudes. The calculations have been carried out with a static engine in standard atmosphere and can be extended to any flight Mach number and atmospheric temperature.

## 5.5 Turbojet with After-Burning

The turbojet is the basic gas turbine engine, Figure 5.14. All (or most) of the gas flow is sent to the combustor. Because there is excess air that has not been combined with the fuel, after-burning can be used to further augment the thrust.

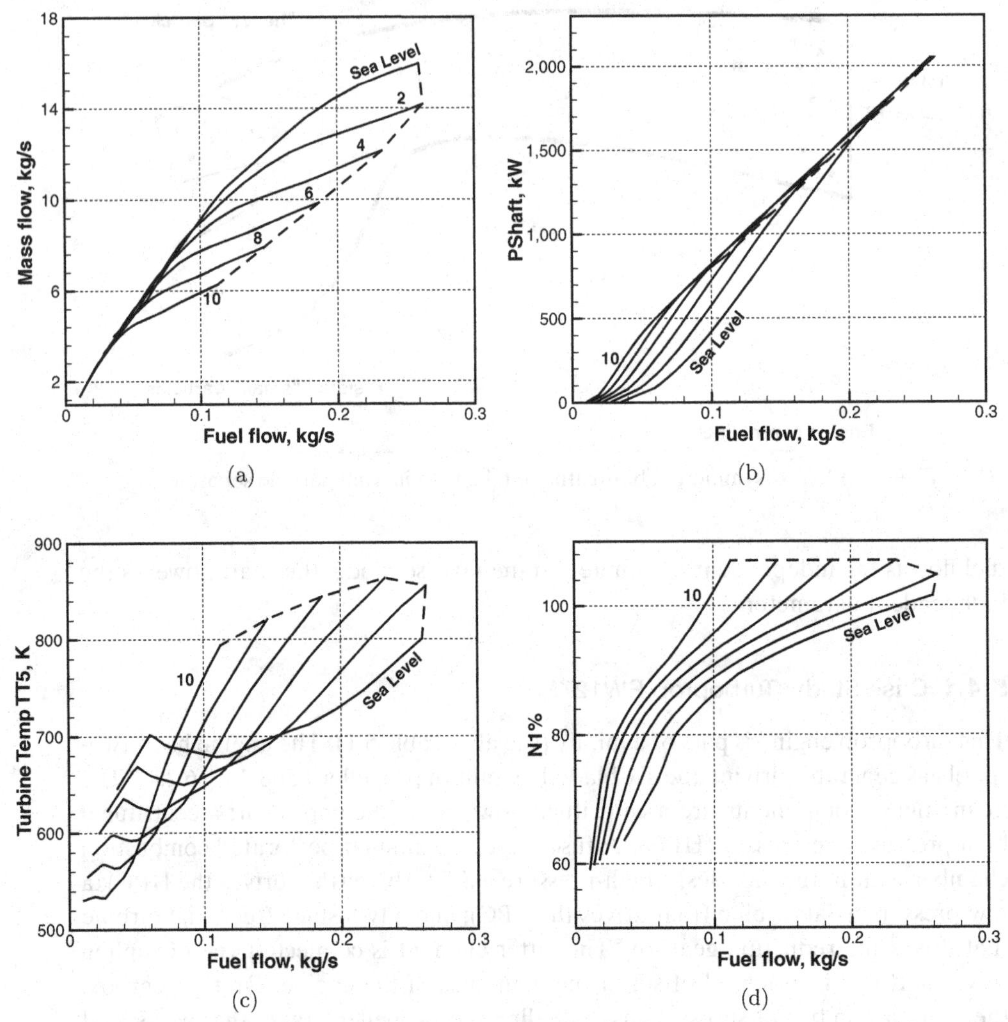


Figure 5.13. Simulation of PW127M turboprop engine; altitudes at 2,000 m intervals (indicated as 2, 4, ···); static conditions; standard day.

The model for this jet engine is the Pratt & Whitney PW F-100. This is a two-shaft augmented turbofan with small by-pass ratio, BPR = 0.7. The fan consists of three stages, with rotor blades in titanium. Its rotational speed is 10,400 rpm. The compressor consists of 10 stages with variable stators over the first three stages. Its nominal speed is 13,450 rpm, with OPR = 24.5. The combustion chamber is annular with film cooling and large-diameter air-blast fuel nozzles. Both the high-pressure and low-pressure turbine are made of two stages. The maximum inlet turbine temperature is about 1,340 °C. The after-burner is made of five concentric spray rings with downstream flame holders. Thrust vectoring is possible in later versions of the engine, with up to 20 degrees' deflection in all directions.

If there is no re-heat, the simulation of the engine is done in the same way as the turbofan engine (§ 5.3). Figure 5.15 shows selected engine parameters at a flight Mach number $M = 0.8$. If after-burning is used, we need to add a virtual combustor downstream the low-pressure turbine.

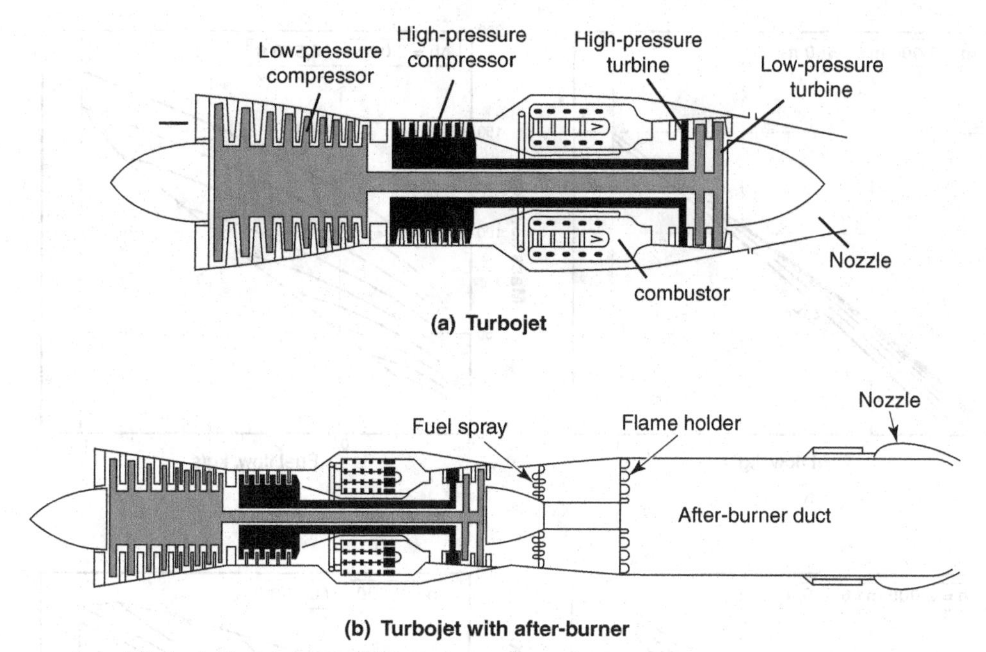


Figure 5.14. Diagram of a turbojet engine, with and without after-burner.

## 5.6 Generalised Engine Performance

There is a practical approach to engine performance which is based on generalised parameters. These parameters are identified through a dimensional analysis, but their determination requires some form of engine data in order to calculate best-fit curves. To begin with, the main propulsion parameters are:

- thrust $T$
- fuel flow $\dot{m}_f$
- rotational speed rpm
- air speed $U$ (or Mach number $M$)
- air temperature $\mathcal{T}$
- air pressure $p$
- engine diameter $d$

Because this is a relatively large number of parameters, the engine performance is described by a reduced number of non-dimensional or engineering quantities. The essential parameters are the air speed, the rpm, the net thrust and the mass flow rate $\dot{m}_f$. A normalisation of these parameters is performed by using the remaining quantities in the previous list.

First, the speed $U$ can be replaced by the Mach number with the definition $M = U/a$. The speed of sound is derived from the flight altitude. Second, the rotational speed of the engine is normalised with $a/d$, to yield the parameter rpm $d/a$. Third, the thrust $T$ is replaced by the parameter $T/pd^2$ (pressure $\times$ length$^2$ = force). Finally, the mass flow rate $\dot{m}_f$ is replaced by the dimensionless parameter

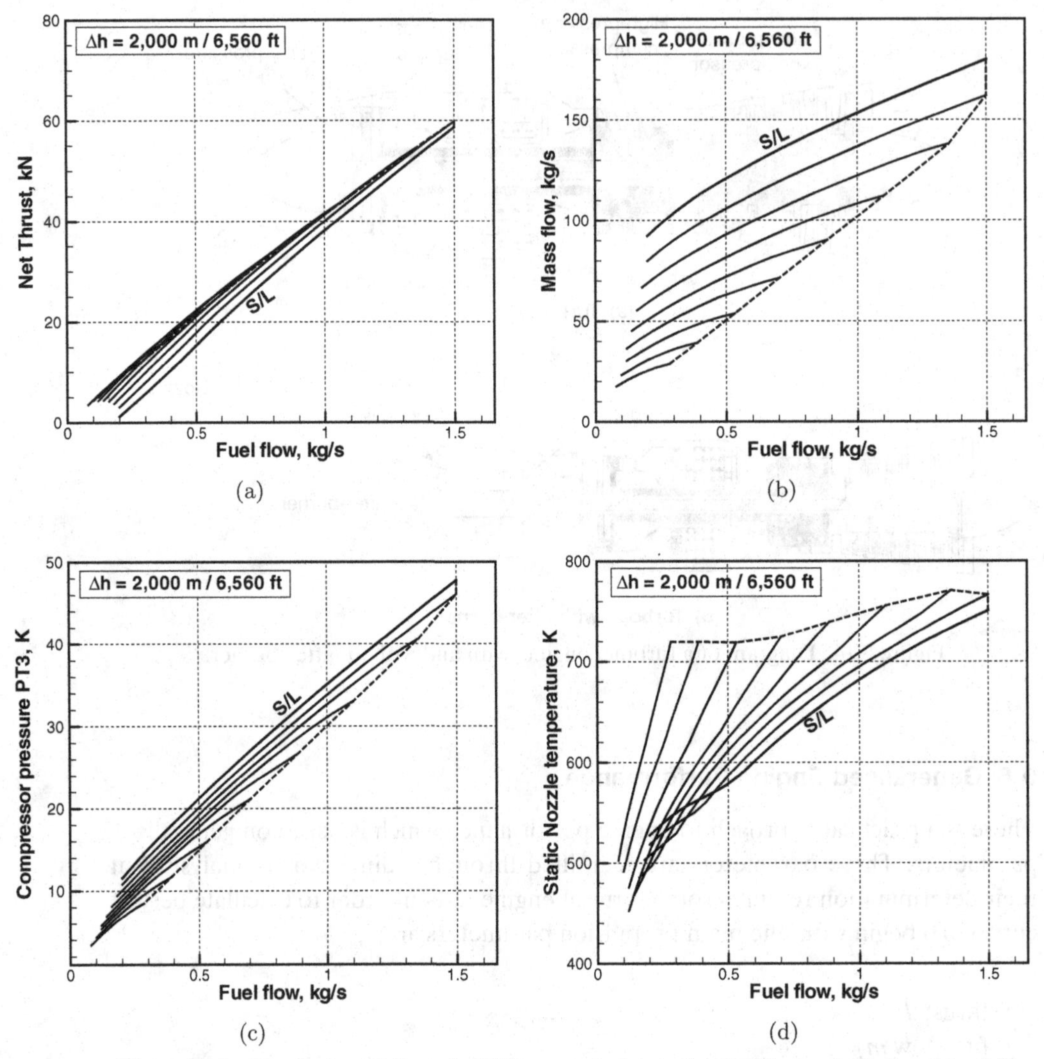


Figure 5.15. Simulation of Pratt & Whitney turbojet engine F100, standard day, $M = 0.8$.

$\dot{m}_f/(pd^2/a) = \dot{m}_f\, a/pd^2$. A further step is required for some parameters. By using the relative pressure $\delta$ we have

$$\frac{T}{pd^2} = \frac{T}{\delta}\frac{1}{p_o d^2}, \tag{5.7}$$

where $p_o$ is the standard atmospheric pressure at sea level. With the definition of speed of sound, $a = \sqrt{\gamma \mathcal{R} T}$, we have

$$\frac{\text{rpm}\, d}{a} = \frac{\text{rpm}}{\sqrt{T}}\frac{d}{\sqrt{\gamma \mathcal{R}}} = \frac{\text{rpm}}{\sqrt{\theta}}\frac{d}{\sqrt{\gamma \mathcal{R} T_o}}, \tag{5.8}$$

$$\frac{\dot{m}_f a}{pd^2} = \frac{\dot{m}_f \sqrt{\gamma \mathcal{R} T}}{pd^2} = \frac{\dot{m}_f \sqrt{T}}{\delta}\frac{\sqrt{\gamma \mathcal{R}}}{p_o d^2} = \frac{\dot{m}_f \sqrt{\theta}}{\delta}\frac{\sqrt{\gamma \mathcal{R} T_o}}{p_o d^2}. \tag{5.9}$$

Instead of using dimensionless parameters, it is a practice in engine performance to use the corrected thrust, rotational speed, and mass flow rate

$$\frac{T}{\delta}, \quad \frac{\text{rpm}}{\sqrt{\mathcal{T}}}, \quad \frac{\dot{m}_f\sqrt{\mathcal{T}}}{\delta}, \tag{5.10}$$

along with the Mach number $M$. The remaining factors in Equations 5.7 to 5.9 are all constant. The parameters in Equation 5.10 have dimensions. Thrust and mass flow rate are the parameters most widely used in aircraft performance simulation. The general dependence of the corrected parameters from the Mach number and the rpm is

$$\frac{T}{\delta} = f_1\left(M, \frac{\text{rpm}}{\sqrt{\theta}}\right), \qquad \frac{\dot{m}_f\sqrt{\theta}}{\delta} = f_2\left(M, \frac{\text{rpm}}{\sqrt{\theta}}\right). \tag{5.11}$$

The technical literature uses various equations to approximate Equation 5.11, including polynomials. The coefficients of these equations are extracted from databases. For example, the first of Equation 5.11 is written as a function of the air speed, the flight altitude and the rpm:

$$\frac{T}{\delta} = c_1 + c_2 M + c_3 h + c_4 h^2 + c_5 \mathcal{T} + c_5\left(\frac{N_1}{\sqrt{\theta}}\right) + c_6\left(\frac{N_1}{\sqrt{\theta}}\right)^2, \tag{5.12}$$

where the $c_i$ are coefficients determined empirically from a database. Notice that Equation 5.12 is not unique and that alternative forms can be found.

## 5.7 Auxiliary Power Unit

The APU is a gas turbine engine used to provide essential services to an aircraft in flight and on the ground, such as electricity, compressed air, and air conditioning, and to start the main engines. The first gas turbine APU was installed on the Boeing 727 (1963) and now is the standard technology on passenger-transport airplanes.

Modern airplanes operate at atmospheric conditions in which human bodies cannot survive. Thus, the airplane must provide suitable cabin pressure, temperature, humidity and air circulation. This is done through an Environmental Conditioning System (ECS) powered by an APU.

The APU is generally mounted on the back of the fuselage. The inlet is through the vertical stabiliser and the nozzle is through the tail-end of the fuselage. Typical APU system architecture includes the following: a single-stage centrifugal impeller with air-bleed valves, radial combustor with a variable number of fuel nozzles, a two-stage axial flow turbine that drives the gearbox, an exhaust and a health monitoring system. The gearbox is considered part of the system; it is used to reduce the rpm from the high speed of the shaft power to the rpm required to run an electric generator. Power must also be extracted for the fuel control units and the cooling system. Some of these units are required to self-start at all altitudes up to the service ceiling of the aircraft and provide critical service for ETOPS with one engine out.

An APU is required to be flexible in terms of services to be supplied. The output power depends on specific requirements. There are several operation modes. The term *ECS* refers to air bleed supplied to the air conditioning system, for typical aircraft gate operation; then there is the maximum ECS, which denotes the maximum aircraft environmental condition and may include some electric load. Other load

Table 5.4. *Typical APU fuel flow [kg/s], depending on load type and atmospheric conditions; RTL = Ready to Load; ECS = Environmental Conditioning System; EL = Electrical Load. All data are at sea level, standard day, given as kg/h*

| APU | RTL | RTL Max EL | Min ECS Max EL | Max ECS Max EL |
|---|---|---|---|---|
| 36-300 | 70 | 85 | 105 | 125 |
| 131-9A | 75 | 95 | 115 | 125 |
| 331-350 | 120 | 140 | 175 | 210 |
| 331-600 | 160 | 180 | 225 | 290 |

conditions include 1) no load (the same as idle); 2) maximum shaft load (electric shaft load only; no air bleed); 3) maximum bleed load (air extraction only; no electric power); 4) maximum combined load (electric and air bleed load); and 5) main engine start (air bleed supplied to engines turbines; some electric load possible). There are various other engine ratings but no agreement on APU rating standards. Note that the APU generate electrical power that is measured in kVA (kilo Volt-Ampere) and mechanical power measured in kW (kilo Watt). The conversion between the two units is 1 kW = 1.25 kVA.

Table 5.4 is a summary of basic APU performance and representative fuel flows. At a reduced load, the APU burns considerably less fuel. For example, the APU 331-600 uses 160 kg/h in ready-to-load condition and 225 kg/h at a maximum ECS condition. Other important data include the output power and the electrical power.

When the fuel consumption is unavailable, a rough estimate is still possible if the maximum combined power is available. Take the example of the PW980A APU used on the Airbus A380-800. This APU is rated for a maximum 1,800 hp (1,341 kW) power. Then take the best SFC available in the industry (about 0.35 kg/kW/h) and multiply the two numbers. The result is a maximum fuel flow ~470 kg/h.

Table 5.5 is a summary of estimated emissions from APU engines. To limit the emissions, running costs and noise whilst on the ground, the use of APU is strictly

Table 5.5. *Estimated APU power and emission database (adapted from several sources)*

| *APU* | Aircraft | $\dot{m}_f$ [kg/hr] | *P* [kW] | *HC* [g/kg] | *CO* [g/kg] | $NO_x$ [g/kg] |
|---|---|---|---|---|---|---|
| 36-150 | BAE-146 | 68.0 | | 0.61 | 6.45 | 5.10 |
| 36-300 | A-320 | 100.0 | | 0.15 | 2.05 | 10.10 |
| 36-4A | Fokker F-28 | 61.0 | | 0.36 | 13.47 | 5.10 |
| 85-129 | B-737-100/300 | 106.5 | | 1.03 | 17.99 | 4.75 |
| 131-9 | A-320, MD-90 | 115.5 | 447 | 0.37 | 4.88 | 6.64 |
| 331-200ER | B-757 | 121.5 | | 0.43 | 4.13 | 9.51 |
| 331-350 | A-330/340 | 186.0 | | 0.23 | 1.86 | 9.90 |
| 331-500 | B-777 | 243.0 | 750 | 0.20 | 1.89 | 11.41 |
| 331-600 | A-340-600 | 290.0 | 894 | | | |
| PW901A | B-747-400 | 391.3 | | 1.50 | 16.78 | 3.15 |
| PW980A | A-380-861 | 470.0 | 1,340 | | | |

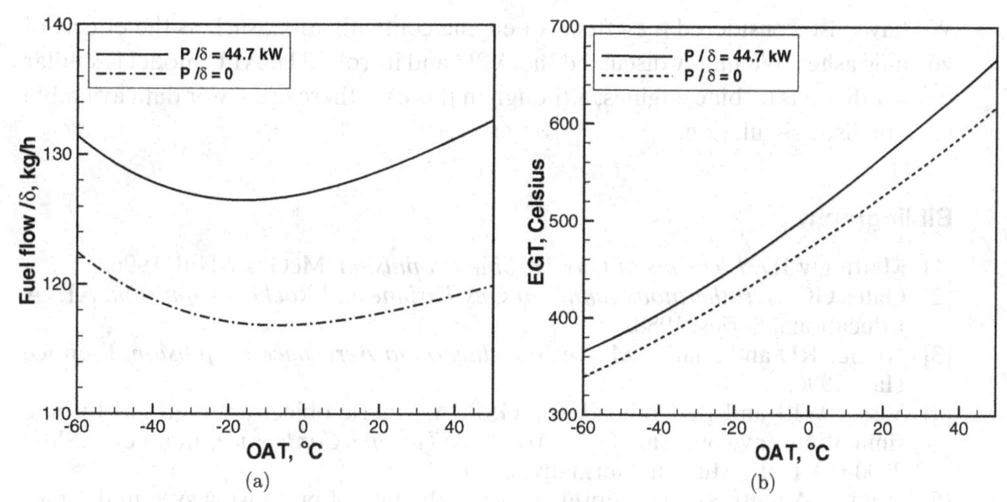


Figure 5.16. Estimated performance of the G550 APU system.

limited. The required services are provided via Ground Power Units (GPU), which are run from the electricity network of the airport.

The APU does not require a separate certification; therefore, only few data are released; compressor ratios are other key parameters that are not readily available. Therefore, the aero-thermodynamics of these engines cannot be simulated. Additional considerations are in § 16.5 that deals with APU noise. Landing and take-off emissions are discussed in § 19.3.

### 5.7.1 Case Study: Honeywell RE-220 APU

This APU powers the Gulfstream G550; this unit is capable of providing 40 KVA or 45 kW electrical power, and bleed air at 4.48 bar for starting the main engine and for environmental control. Its operating envelope extends to 45,000 feet, with self-start up to 43,000 feet. Basic data are as follows: $m = 108$ kg; inlet area = 0.677 m$^2$; exhaust area = 0.0613 m$^2$, rated EGT = 732 °C (maximum continuous); rotor speed = 45,585 rpm; maximum rotor speed is 106%. Figure 5.16 shows some graphs with the estimated performance of the unit. The ratio $P/\delta$ is the net shaft power divided by the relative pressure.

## Summary

We have given a brief description of the modern gas turbine engine. The problem at hand is the determination of engine models suitable for aircraft performance simulations; this is complicated by the lack of reliable data, especially the design point. Nevertheless, it is possible to define a fairly accurate one-dimensional thermodynamic model that provides most of the engine parameters required. The design point is defined as the intersection of constant-thrust and constant turbine temperature lines at the nominal engine rpm. The method is applied to high by-pass turbofans, turboprops and low by-pass military engines. Even these simplified models are always insufficient. Therefore, there is widespread practice use of generalised equations, which need the determination of some parameters, specific to the engine.

We have also considered the effects of engine contamination, such as the effects of volcanic ashes. We finally discussed the APU and its role. The APU model is similar to the other gas turbine engines, although in this case there are fewer data available for a realistic simulation.

## Bibliography


[1] Mattingly JD. *Elements of Gas Turbine Propulsion*. McGraw-Hill, 1996.
[2] Oates GC. *Aerothermodynamics of Gas Turbine and Rocket Propulsion*. AIAA Educational Series, 1988.
[3] Archer RD and Saarlas M. *An Introduction to Aerospace Propulsion*. Prentice Hall, 1996.
[4] Visser WPJ and Broomhead MJ. GSP: A generic object-oriented gas turbine simulation environment. In *ASME Gas Turbine Conference*, number ASME 2000-GT-0002, Munich, Germany, 2000.
[5] Roth B. A method for comprehensive evaluation of propulsion system thermodynamic performance and loss. AIAA Paper 2001-3301, 2001.
[6] Swanson SE and Beget JE. Melting properties of volcanic ash. In *First Int. Symp. on Volcanic Ash and Aviation Safety*, Bulletin 2047, Seattle, WA, July 1991. US Geological Survey.


## Nomenclature for Chapter 5

Additional symbols are given in Table 5.2.

$a$ = speed of sound
$c_i$ = constant coefficients
$d$ = diameter
$\mathcal{E}$ = generic engine parameter
$h$ = flight altitude
$h$ = altitude
$\dot{m}$ = mass flow rate (also called `W1`)
$\dot{m}_f$ = fuel flow rate (also called `Wf6`)
$M$ = Mach number
$n$ = number of wash cycles
$N_1$ = nominal gas-generator turbine rpm (also called `N1`)
$N_2$ = nominal power turbine rpm (also called `N2`)
$p$ = pressure; permanent loss of engine performance (Equation 5.2)
$P$ = engine (shaft) power
$\mathcal{R}$ = gas constant (dry air)
rpm = rotational speed
$T$ = thrust (also called `FN`)
$\mathcal{T}$ = temperature
$U$ = air speed
$W$ = weight

## Greek Symbols

$\gamma$ = ratio between specific heats
$\delta$ = relative pressure

$\theta$ = relative air temperature
$\kappa$ = installation losses, Equation 5.1
$\Pi$ = engine state; throttle setting
$\rho$ = air density

### Subscripts/Superscripts

$[.]_{av}$ = available
$[.]_{o}$ = standard conditions
$[.]_{ref}$ = reference
$[.]_{req}$ = required

# 6 Propeller Performance

## Overview

Aircraft propulsion by propeller is still the most widespread method of converting engine power into useful thrust. We seek the propeller parameters required to deliver specified thrust or power to the airplane (§ 6.1), depending on the flight condition. The simplest method for calculating the propulsive performance is the axial momentum theory (§ 6.2.1), which is useful when detailed data of the propeller itself are unknown. When detailed data are available, the combined momentum and blade element theory (§ 6.2.2), along with ancillary models for transonic flow, offers a powerful and accurate method for propeller analysis. The integration of the propeller with the flight mechanics is discussed in § 6.3. We show that when we trim to a specified thrust or power, the propeller generally does not operate at its most efficient point.

**KEY CONCEPTS:** Propeller Parameters, Propulsion Models, Momentum Theory, Blade Element Method, Flight Mechanics Integration, Propeller Installation.

## 6.1 Propeller Definitions

The basic parameters used in the analysis of the propeller are the forward speed $U$, the rotational speed rpm and the tip Mach number $M_{tip}$. In addition, there are several geometrical quantities: the number of blades, the diameter $d$, the pitch $\vartheta$, the type of blade section, the chord distribution, the tip geometry and the hub geometry.

The pitch is a measure of the orientation of the propeller on a plane normal to the axis of rotation, as shown in Figure 6.1. The reference line for the calculation of the pitch is the chord. Figure 6.1 shows how the local *inflow angle* is calculated. If the blade section is at a radial position $y$, the total inflow velocity is $\sqrt{U^2 + (\Omega y)^2}$. The direction of the resulting vector velocity is inclined by an angle $\alpha$ on the chord line: this is the local inflow angle, or angle of attack of the blade section.

The net thrust is the resulting aerodynamic force in the direction of the forward flight. The drag is parallel to the inflow speed; the lift is at a right angle with the drag of the blade element. The resulting thrust is parallel to the axis of rotation.

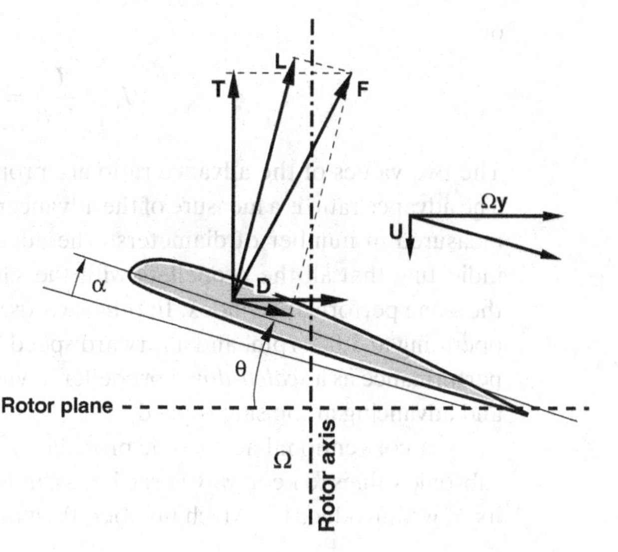


Figure 6.1. Nomenclature of forces on blade section.

The *solidity* $\sigma$ is the ratio between the blade area (projected on the rotor disk) and the rotor disk*:

$$\sigma = 2\frac{N\overline{c}}{\pi d}, \tag{6.1}$$

where $N$ is the number of blades and $\overline{c}$ is the mean chord. The solidity is an important design parameter; in fact, the propeller coefficients can be normalised by $\sigma$, to express the effective disk loading. To make the data more useful from an engineering point of view, we need a measure of performance. This is the efficiency, or the ratio between propulsive power and power at the shaft

$$\eta = \frac{TU}{P}. \tag{6.2}$$

The propulsive efficiency expresses the ability to convert power from the engine (or shaft power) into useful power to fly at a speed $U$. The dependence of the efficiency from the other parameters is expressed as

$$\eta = f(U, \text{rpm}, d, \vartheta, \cdots). \tag{6.3}$$

If the aircraft is stationary on the ground, the conversion efficiency is zero, and all the shaft power generated from burning fuel is lost. The energy $E$ is dissipated at the blades and transferred to the slipstream, that has an axial and rotational velocity. In general terms, the shaft power is

$$P = TU + E. \tag{6.4}$$

By using the dimensional analysis, the forward speed, the rotational speed and the diameter are replaced by another dimensionless group, the *advance ratio*

$$J = \frac{U}{\text{rpm}\, d}, \tag{6.5}$$

* Note that $\sigma$ is the same symbol as the relative density of the air. We maintain the convention currently used in propeller and helicopter analysis and use $\sigma$ for the solidity. Therefore, to avoid confusion, only $\rho$ will be used for the air density.

or

$$J_1 = \frac{U}{\Omega R} = \frac{U}{U_{tip}}. \tag{6.6}$$

The two values of the advance ratio are proportional to each other: $J_1 = J(60/\pi)$. The advance ratio is a measure of the advancement of the propeller in one revolution, measured in number of diameters. The advance ratio is also a scaling parameter, indicating that all the propellers with the same $J$, and *geometrically similar*, have the same performance index. In other words, a propeller with diameter of 4 metres, operating at 1,000 rpm, and a forward speed 70 m/s ($J = 0.33$; 136 kt) has the same performance as a *scaled-down* propeller having a 2 m diameter rotating at 2,000 rpm and advancing at the same speed.

For conventional aeronautic propellers, the tip speed $U_{tip} = \Omega R$ is restricted to subsonic values to keep wave drag losses, noise and vibrations under acceptable limits. If we introduce the Mach number, then the tip Mach number at flight conditions is

$$M_{tip} = \frac{U_{tip}}{a} = \frac{1}{a}\sqrt{U^2 + (\Omega R)^2}. \tag{6.7}$$

Propeller charts, where the parameter is the pitch angle of the blades, are to be used in the calculation of the range and the endurance of propeller-driven airplanes.

A number of other dimensionless coefficients are defined, and propeller charts can be found with these parameters, that are the power, thrust and torque coefficients:

$$C_T = \frac{T}{\rho A(\Omega R)^2}, \qquad C_P = \frac{P}{\rho A(\Omega R)^3}, \qquad C_Q = \frac{Q}{\rho A(\Omega R)^2 R}. \tag{6.8}$$

A useful relationship between the efficiency and the advance ratio involves some of the parameters in Equation 6.8:

$$\eta = J\frac{C_T}{C_P}. \tag{6.9}$$

Although of great importance for analysing the propeller performance, the coefficients given by Equation 6.8 are not useful for determining the engine power required to fly at a given air speed altitude and gross weight. In fact, propeller operation is analysed either at a constant rotational speed or a constant pitch. Operation with a mix of rotational velocities and pitch angles is also possible. For this purpose, it is necessary to have propeller data that do not depend on the rotational speed. If the pitch setting is unique, then the efficiency is a single curve that can only be changed with the advance ratio.

The essential propeller theory is available in Glauert[1], Theodorsen[2], and von Mises[3], among others. Data and propeller charts, for the purpose of basic performance calculations, can be found in some old NACA reports, such as Hartman and Biermann[4;5] and Theodorsen *et al.*[6]. More advanced concepts are available in AGARD CP-366[7]. Other important aspects of propeller performance are the interaction with the wing and the fuselage[8;9], compressibility effects[10;11] and the propeller noise, discussed as a separate problem in Chapter 16.

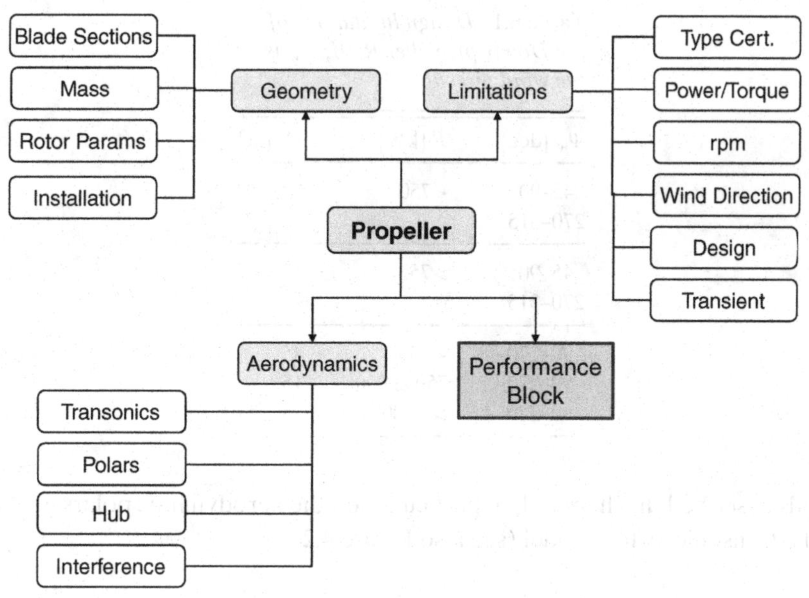


Figure 6.2. Flowchart of a propeller simulation model.

The establishment of a computational model for an aircraft propeller relies on the flowchart shown in Figure 6.2. There are three categories of data that must be assembled:

- **Limitations.** This category includes data from the type certificate (variable or constant pitch, variable or constant rpm, feathering and reversing modes, pitch selection, braking and other quantitative and qualitative parameters). Then we have power and torque limitations (both steady-state and transient), along with the design power, wind direction and relative speeds*. With specific reference to the power ratings, there are data on maximum continuous power (and torque), maximum take-off power (and torque) and maximum reverse power.
- **Geometry.** This category includes the geometry of the blade sections at selected radial stations, the rotor geometry (diameter, number of blades, twist, and so on), the mass distribution and any other relevant parameter. With very few exceptions, the wing-section details are unknown.
- **Aerodynamics.** This category includes the aerodynamic polars of the blade sections over a wide range of angles of attack and Mach numbers. A transonic aerodynamic model may be required if the data do not extend to high Mach numbers. For completeness, we need to account for the hub drag and the interference effects due to installation.

Some critical data are often missing: the blade-section geometry, the blade twist and the three-dimensional characteristics, such as the axial and radial twist (*scimitar* geometry). However, it is possible to derive some of these parameters on the basis of optimum propeller design. The "Aerodynamics" block in Figure 6.2 relies on the

* There can be additional data, such as software requirements, airplane integration, and de-icing equipment. See, for example, CAA: Type Certificate Data Sheet, Dowty R391 Propeller, No. 116, Issue 7, Feb. 2003.

Table 6.1. *Design limitations of the Dowty propeller R391; $\Psi_w$ is the wind direction*

| $\Psi_w$ [deg] | $P$ [kW] | $V_w$ [kt] |
|---|---|---|
| 45–90<br>270–315 | <750 | 60 |
| 45–90<br>270–315 | >750 | 15 |
| 90–270 | <750 | 45 |
| 90–270 | 750–3,000 | 15 |
| 90–270 | >3,000 | 5 |

methods described in Chapter 4, in particular, on the aerodynamic polars generated with the transonic airfoil model (see also Figure 4.20).

### 6.1.1 Propeller Limitations

The certification documents usually contain data such as wind speed and power-output limits. An example is given in Table 6.1 for the Dowty propeller R391 (Lockheed 382J and C-130J, Alenia C-27J). This is a six-bladed propeller having a diameter of 4.115 m, made of composite glass, carbon reinforced structure and nickel leading-edge sheath for erosion protection. The propeller has a variable pitch, a constant rpm, feathering, reversing and full hydro-mechanical control. Manual pitch selection is available for ground manoeuvring and aircraft braking. The type certificate documents report that the electronic control software meets the most stringent standards. Table 6.2 is a summary of data for some propellers and their application†.

Those propellers that operate at constant rpm may have different rpm for different flight conditions. One example is the Dowty R408 propeller, installed on the Dash8-Q400 turboprop airplane. The propeller has speed ratings of 1,020 rpm (take-off, as in Table 6.2), 900 rpm (maximum climb), and 850 rpm (cruise). These speeds, as well as the collective pitch, are governed by the propeller electronic control (called PEC in the technical jargon). The PEC governs the propeller parameters in response to the flight conditions. At a given shaft power, a decrease in tip speed causes an increase in $C_T$ and $C_P$.

## 6.2 Propulsion Models

The aerodynamic analysis is carried out in two steps of increasing complexity and accuracy by using the following theories: 1) axial momentum theory with constant or variable inflow (§ 6.2.1); and 2) blade momentum theory (§ 6.2.2), which in its complete extension includes elements of the momentum theory and allows the detailed calculation of propeller performance also in non-axial flight conditions.

† Data compiled from type certificates of propellers and/or their aircraft, published by the CAA, EASA and FAA. Note that there can be several propeller versions for each aircraft. Where several versions exist, data have been averaged. Consult the appropriate documents for details.

Table 6.2. *Some notable propellers and their applications; P is the engine's maximum take-off power; W is the propeller weight, including spinner; *depending on aircraft version; †estimated*

| Propeller | Airplane | $N$ | $d$ [m] | rpm | $P$ [kW] | $W$ [kg] |
|---|---|---|---|---|---|---|
| Hamilton 14SF | ATR42-300, CL-415 | 4 | 3.96 | 1,200 | 2,050 | 156 |
| Hamilton F568 | ATR72-500 | 6 | 3.93 | 1,200 | 2,050 | 164 |
| Hamilton 247F | ATR72-211 | 4 | 3.96 | 1,200 | 2,048 | 156 |
| Dowty R408 | Dash8-Q400 | 6 | 4.12 | 1,020 | 3,782 | 252 |
| Dowty R391 | C-27J, C-130J | 6 | 4.12 | 1,020 | 3,505 | 326 |
| Dowty R334 | CASA 212-300 | 4 | 2.79 | 1,591 | 689 | 87 |
| Dowty R175 | Fokker F27* | 4 | 3.66 | | 1,170 | |
| Dowty R193 | Fokker F27* | 4 | 3.51 | | <1,672 | 203 |
| Dowty R352 | Fokker F50 (F27-050)* | 6 | 3.65 | 1,200 | 1,624 | 172 |
| Dowty R410 | Fokker F50 (F27-050)* | 6 | 3.65 | 1,200 | 1,624 | 172 |
| Dowty R381 | Saab-2000 | 6 | 3.81 | 1,100 | 2,786 | 227 |
| Ratier-Figeac FH 386 | Airbus A400M | 8 | 5.31 | 850 | 8,195 | †360 |
| Hartzell HC-E5N | Piaggio Avanti | 5 | 2.16 | 2,000 | 634 | 81 |
| Hartzell HC-B3TN-3D | DeHavilland DHC6 | 3 | 2.59 | 2,110 | 462 | 50 |

### 6.2.1 Axial Momentum Theory

This theory is useful for the estimation of propeller performance with only a few data: diameter, rpm and number of blades. The theory is based on the most elementary theoretical basis for converting the rotating power of a propeller into useful thrust (Rankine-Froude momentum theory). This theory is conceptually simple and practically important because it provides a first-order estimate of the power and thrust of an open airscrew.

According to the basic momentum theory, the propeller is reduced to a rotating disk that imparts axial momentum to the air passing through it. For the purpose of this discussion, the air is incompressible. For reference, consider the sections 0 (far upstream), 1 (just upstream the disk), 2 (just downstream the disk) and 3 (far downstream), as shown in Figure 6.3. The subscripts will refer to quantities at these sections.

The theory assumes that there is no flux along the limiting streamlines and that the velocity is continuous through the propeller disk. The free stream velocity is equal to the propeller's speed $U = U_o$. The continuity equation written for the incompressible mass flow rate within the stream tube is

$$Au = \text{const.} \tag{6.10}$$

The propeller's thrust is equal to the rate of change of axial momentum at the disk,

$$T = A(p_2 - p_1). \tag{6.11}$$

Now we can apply the Bernoulli equation between sections 0–1 upstream and 2–3 downstream to the propeller (Figure 6.3),

$$p_o + \frac{1}{2}\rho U^2 = p_1 + \frac{1}{2}\rho U_1^2, \qquad p_o + \frac{1}{2}\rho u_3^2 = p_2 + \frac{1}{2}\rho U_2^2. \tag{6.12}$$

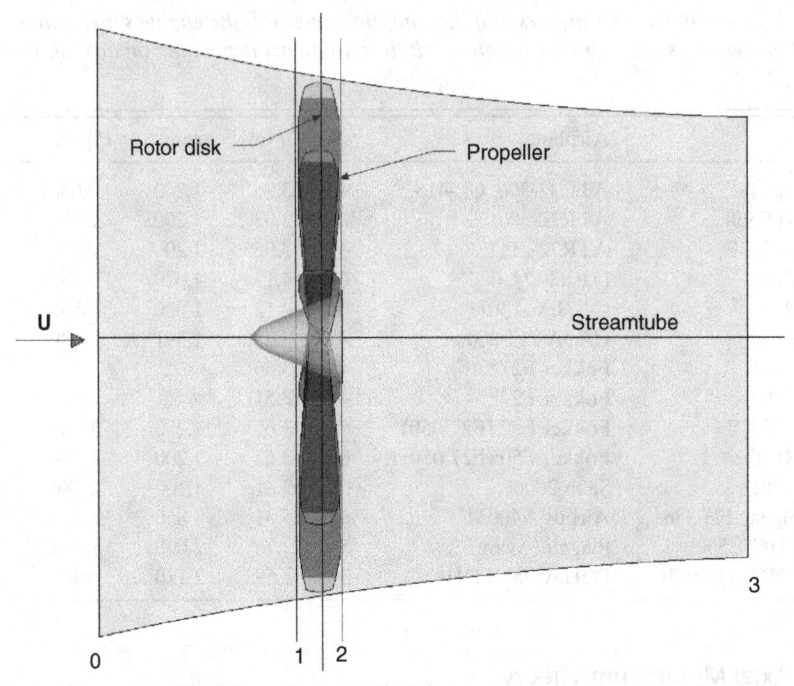


Figure 6.3. One-dimensional axial flow model of airplane propeller.

The difference between the two equations yields the pressure jump through the propeller

$$p_2 - p_1 = \frac{1}{2}\rho(u_3^2 - U^2), \tag{6.13}$$

because from the continuity equation the velocity is continuous at the rotor disk, or $u_1 = u_2$. From Equation 6.11 the thrust generated by the propeller becomes

$$T = \frac{1}{2}\rho A(u_3^2 - U^2) = \rho A_3 u_3(u_3 - U). \tag{6.14}$$

The total power is

$$P = T(u_1 + U) = \frac{1}{2}\dot{m}(u_3 + U)^2 - \frac{1}{2}\dot{m}U^2 = \frac{1}{2}\dot{m}u_3(u_3 + 2U). \tag{6.15}$$

The energy imparted to the slipstream is

$$E = \frac{1}{2}A_3\rho u_3(u_3^2 - U^2). \tag{6.16}$$

This energy is minimal when the slipstream velocity is equal to the propeller velocity. The total power is found from summing Equation 6.16 and Equation 6.15. The slipstream velocity far downstream, $u_3$, is related to the velocity at the disk $u_1$ and the free stream velocity $U$. The thrust absorbed by the propeller is also written as

$$T = \dot{m}(u_3 + U) - \dot{m}U = \dot{m}u_3. \tag{6.17}$$

By combination of Equation 6.15 and Equation 6.17, we have

$$u_1 = \frac{1}{2}u_3. \tag{6.18}$$

In conclusion, the air speed at the rotor disk is the average between the propeller's speed and the speed far downstream. Also, the downstream velocity is twice the induced velocity at the disk. For a static propeller this relationship provides the value of the *induced velocity* at the rotor disk:

$$u_1 = v_i = \frac{1}{2}u_3. \tag{6.19}$$

The corresponding induced power is $P_i = Tv_i$. The thrust becomes

$$T = 2\rho A(U + v_i)v_i, \tag{6.20}$$

where $\rho A(U + v_i)$ is the mass flow rate through the disk and $2v_i$ is the total increase in velocity. The corresponding power is the product between the thrust and the velocity through the disk

$$P = T(U + v_i) = 2\rho A(U + v_i)^2 v_i. \tag{6.21}$$

This power expression contains two terms: 1) a useful power $TU$, and 2) an induced power $Tv_i$ that is a loss due to the kinetic energy imparted to the flow. The propulsion efficiency is

$$\eta = \frac{TU}{T(U+v_i)} = \frac{U}{U+v_i}. \tag{6.22}$$

Therefore, the propulsion efficiency decreases as the induced velocity increases. If $U = 0$, the efficiency is zero, as anticipated. The induced velocity in terms of the thrust is found from Equation 6.20, that is quadratic in $v_i$. The only physical solution of Equation 6.20 is a positive $v_i$:

$$v_i = \frac{1}{2}\left[-U + \sqrt{U^2 + \frac{2T}{\rho A}}\right]. \tag{6.23}$$

Under static conditions, $U = 0$, we have

$$v_i = \sqrt{\frac{T}{2\rho A}}, \qquad P_i = \sqrt{\frac{T^3}{2\rho A}}. \tag{6.24}$$

An important aspect of this elementary theory is the behaviour of the pressure. For points upstream or downstream of the propeller, there is a conservation of total pressure (sum of the static pressure and dynamic pressure, $q = \rho U^2/2$). The total pressure is given by the Bernoulli equation, Equation 6.12. The engine power is used to increase the kinetic energy of the air passing through the disk. Therefore, we conclude that the theory involves a sudden increase of pressure through the propeller disk, whilst the velocity of the air in the stream tube is continuous. The flow undergoes an acceleration, and therefore the slipstream must contract, according to the continuity Equation 6.10.

**VARIABLE INFLOW CONDITIONS.** Various advancements and refinements can be added. With respect to the non-uniform axial inflow, one needs to integrate the change of axial momentum over the propeller disk, and the result is

$$T = \int_A \rho u_3(u_3 - U)\, dA. \tag{6.25}$$

An essential hypothesis is that of axi-symmetric flow through the disk, that is quite reasonable for a propeller in axial flight. This hypothesis leads to a slightly modified theory. Consider an annulus of width $dy$, corresponding to an area $dA = 2\pi y dy$. The element of thrust generated by the mass flow is

$$\dot{m} = \rho(U + v_i)dA, \tag{6.26}$$

through this annulus is

$$dT = 2\rho(U + v_i)v_i dA = 4\pi\rho(U + v_i)v_i y dy. \tag{6.27}$$

We define the induced velocity ratio

$$\lambda = \frac{U + v_i}{\Omega R} = \frac{U + v_i}{\Omega y}\frac{\Omega y}{\Omega R} = \left(\frac{U_n}{U_t}\right) r = \tan\phi\, r, \tag{6.28}$$

where $U_n$ and $U_t$ are the velocity components normal and tangential to the rotor plane, and $\phi$ is the inflow angle. In general, $U_n << U_t$; therefore we can make the approximation

$$\tan\phi \simeq \phi = \frac{U_n}{U_t} = \frac{\lambda}{r}. \tag{6.29}$$

If we introduce the inflow velocity ratio, Equation 6.28, we find

$$dT = 4\pi\rho\left(\frac{U + v_i}{\Omega R}\right)\left(\frac{v_i}{\Omega R}\right)(\Omega R)^2 y dy = 4\pi\rho\lambda\lambda_i(\Omega R)^2 y dy, \tag{6.30}$$

with $\lambda_i = v_i/\Omega R$. Note that $\lambda_i$ is the induced velocity ratio in absence of axial flight velocity. It can also be written as

$$\lambda_i = \frac{v_i}{\Omega R} = \frac{v_i}{\Omega R} + \frac{U}{\Omega R} - \frac{U}{\Omega R} = \lambda - \frac{U}{\Omega R} = \lambda - \lambda_c, \tag{6.31}$$

with $\lambda_c = U/\Omega R$. The resulting thrust is found from integration of Equation 6.30,

$$T = 4\pi\rho(\Omega R)^2 \int_o^R \lambda\lambda_i\, y dy. \tag{6.32}$$

The element of power is $dP = dTv_i$, and therefore

$$P = 4\pi\rho(\Omega R)^3 \int_o^R \lambda\lambda_i^2\, y dy. \tag{6.33}$$

These integrations can only be done if the radial distribution of induced velocity is known. A solution can be found by combining the results of the blade momentum theory (§ 6.2.2). Various further advancements are available; it is now possible to model the case of a propeller/rotor disk inclined by any angle on the free stream, with almost any load distribution on the rotor[12].

### 6.2.2 The Blade Element Method

The axial momentum theory provides integral quantities, such as thrust and power. These characteristics do not seem to depend on the propeller's geometry, which is clearly a shortcoming. With the blade element it is possible to take into account all of the geometrical information of the propeller. The resulting theory is more elaborate;

if efficiently programmed, it is a robust tool for engineering analysis in most flight conditions.

In this framework, the blade sections operate like two-dimensional sections, with the local inflow conditions derived by appropriate means in the rotating environment (see Figure 6.1). Although the definition of this inflow is not obvious, and the interference between elements on the same blade and between blades is not taken into account, the method is otherwise extremely powerful. It allows the calculation of basic performance from the geometrical details and the two-dimensional blade section aerodynamics.

The relationship among inflow angle, pitch angle and angle of attack is

$$\alpha = \vartheta - \phi. \tag{6.34}$$

The pitch is a geometrical setting, whilst the angle of attack of the blade section and the inflow velocity are operational free parameters. The lift and drag forces on this section are

$$dL = \frac{1}{2}\rho c C_L U^2 dy, \tag{6.35}$$

$$dD = \frac{1}{2}\rho c C_D U^2 dy. \tag{6.36}$$

These forces, resolved along the direction normal and parallel to the rotor disk, give the contributions to the thrust, torque and power for the single blade

$$dT = dL\cos\phi - dD\sin\phi, \tag{6.37}$$

$$dQ = (dL\sin\phi + dD\cos\phi)y, \tag{6.38}$$

$$dP = (dL\sin\phi + dD\cos\phi)\Omega y. \tag{6.39}$$

These elements can be written in non-dimensional form, using the definitions of thrust, torque and power coefficients. The total thrust, torque and power for $N$ blades will require integration of these expressions from the inboard cut-off point to the tip.

The integrals are, in fact, not solved directly but rather numerically. If we divide the blades into a number $n$ of elements, each having a radial width $dy_j$, then

$$T = N\sum_{j=1}^{n}\left(dL_j\cos\phi_j - dD\sin\phi_j\right), \tag{6.40}$$

$$Q = N\sum_{j=1}^{n}\left(dL_j\sin\phi_j + dD\cos\phi_j\right)y_j dy_j, \tag{6.41}$$

$$P = N\sum_{j=1}^{n}\left(dL_j\sin\phi_j + dD\cos\phi_j\right)\Omega y_j, dy_j, \tag{6.42}$$

with the forces evaluated from Equation 6.35 and Equation 6.36. All of the quantities appearing in the aerodynamic forces change with the radial position, including the chord, the air density (for high-speed flows), the $C_l$ and the $C_d$. The aerodynamic coefficients depend on the effective inflow angle $\alpha$, Reynolds number $Re$ and Mach number $M$. The key problem is to find the inflow angle $\phi$, the resultant inflow $\alpha$ and the actual inflow velocity $U$ at each blade section.

The aerodynamic coefficients of the blade section $C_l$ and $C_d$ are transformed into propeller coefficients $C_x$, $C_y$ by a rotation $\phi$:

$$C_x = C_l \sin\phi + C_d \cos\phi, \tag{6.43}$$

$$C_y = C_l \cos\phi - C_d \sin\phi. \tag{6.44}$$

The next step is to calculate the interference factors. According to Glauert[1], these factors are:

$$a = \sigma \frac{k}{F - \sigma k}, \qquad a' = \sigma \frac{k'}{F + \sigma k'} \tag{6.45}$$

with

$$k = \frac{C_y}{4\sin^2\phi}, \qquad k' = \frac{C_x}{4\sin\phi\cos\phi}. \tag{6.46}$$

The inflow angle is obtained from

$$\tan\phi = \frac{U}{\Omega y}\frac{1+a}{1-a'}. \tag{6.47}$$

The corrective factor $F$ in Equation 6.45 is calculated from

$$F = \left(\frac{2}{\pi}\right)\cos^{-1} e^{-f(r)}, \tag{6.48}$$

with

$$f(r) = \frac{N}{2}\frac{1-r}{r\phi}. \tag{6.49}$$

The inflow angle is unknown, and some iterations are required. A correction is usually applied to this procedure because the aerodynamic flow around the blade section does not follow that of the airfoil. Both the tip and the root are affected by three-dimensional effects, which lead to loss of lift and hence loss of thrust.

The corrected thrust will be $dTF(r)$. A plot of $F(r)$ shows that it is equal to one for most of the span, but it tends rapidly to zero near the tip, with a rate depending on the number of blades and the inflow angle.

The iterative procedure for the inflow angle shows poor convergence properties, unless a corrective action is taken. First, it is possible that the interference factors $a$ and $a'$ assume too large values. Following Adkins and Liebeck[13], we set a limiter $a \leq 0.7, a' \leq 0.7$. Second, and most important, unless some form of under-relaxation is applied, the procedure does not converge on a highly loaded propeller. If $i$ is the iteration counter at a generic blade section, the under-relaxation that we propose to use is

$$\tilde{\phi}_{i+1} = \frac{1}{2}\left(\phi_i + \phi_{i+1}\right), \tag{6.50}$$

where $\tilde{\phi}$ denotes the *corrected* inflow angle at iteration $i+1$ and $\phi_{i+1}$ is the *predicted inflow* angle. The different behaviour in convergence properties is shown in Figure 6.4. When we use Equation 6.50 the convergence is uniform and rapid; in absence of under-relaxation, the solution is hopeless.

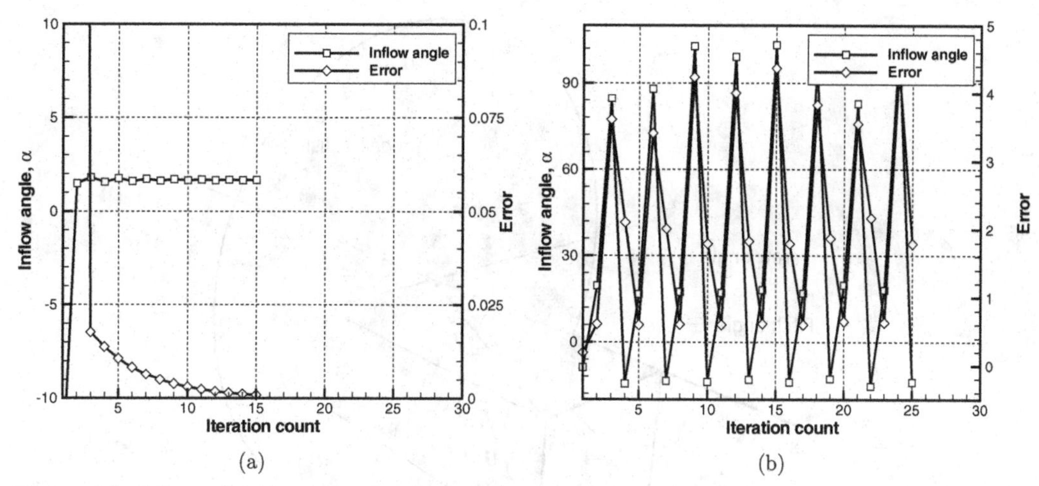


Figure 6.4. Effect of under-relaxation on inflow convergence of a highly loaded propeller, $y/R = 0.96$.

The procedure described neglects a number of practical difficulties. These are related to the format in which the aerodynamic data are available. Ideally, one would have data like $C_d = C_d(\alpha, M)$, $C_l = C_l(\alpha, M)$ in a matrix form.

**PROPELLER TRIM.** The propeller can be trimmed to provide a required thrust, power or torque. One way of doing this is to specify the flight conditions ($h$, TAS, rpm) and to determine the collective pitch corresponding to the required parameter. A suitable numerical method is the bisection. This method consists in bracketing the solution (i.e., the collective pitch) with a very low and a very large value, $\vartheta_1$ and $\vartheta_2$, respectively.

### 6.2.3 Propeller in Non-Axial Flight

Asymmetric flight occurs when the propeller axis is not aligned with the flight speed. This situation generates variable inflow across the actuator disk and hence some degree of unsteady flow and mechanical vibrations. The latter problem is typical of several turboprop aircraft. From a propulsion point of view, there will be changes in shaft power and net thrust. The resulting changes in in- and out-of-plane forces and moments are important in the analysis of flight control procedures.

To understand how the asymmetric inflow can be generated, consider the aerodynamic effects of the fuselage, which change the effective angle of attack at the disk, and the effects of aircraft attitude, especially at climb-out conditions.

To begin the analysis, let us consider a reference system based on the propeller. The convention is that the $x$-axis coincides with the rotational axis of the propeller; the $y$-axis is on the horizontal plane pointing to the right of the flight, and $z$ makes a right-hand Cartesian reference system with $x$ and $y$; $z$ points downward. A rotation of the propeller disk around $x$, $y$ and $z$ is called roll, pitch and yaw, respectively. The order in which these rotations take place is very important. In the following analysis, we will be limited to combinations of pitch and yaw because roll effects are generally

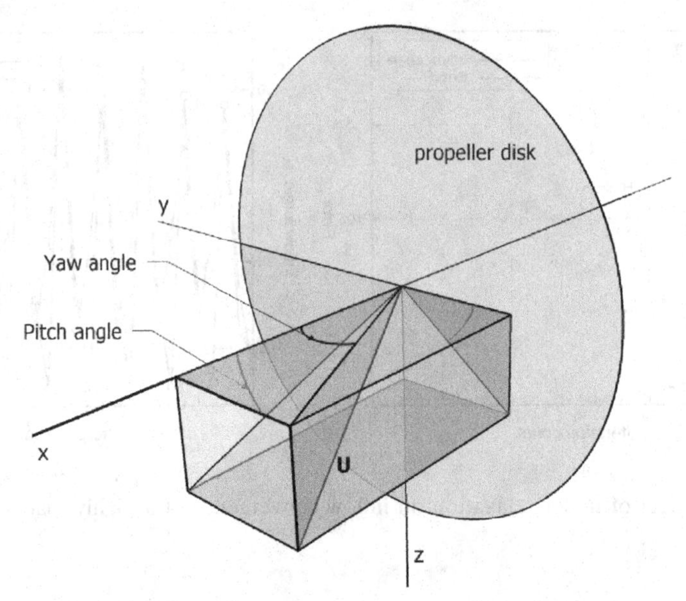


Figure 6.5. Propeller reference system and rotation angles.

small unless the aircraft is performing a rapid manoeuvre. A sketch indicating pitch and yaw angles is shown in Figure 6.5.

Figure 6.6 shows a field plot of the element of thrust coefficient $dC_T$ when the propeller advances through the azimuth. The case refers to a yaw angle of 5 degrees, a pitch angle of 10 degrees and typical cruise conditions. Although the blade sections would be affected by time-dependent inflow conditions, this case has been calculated in quasi-steady mode, whereby the blade sections move from one steady state to the next.

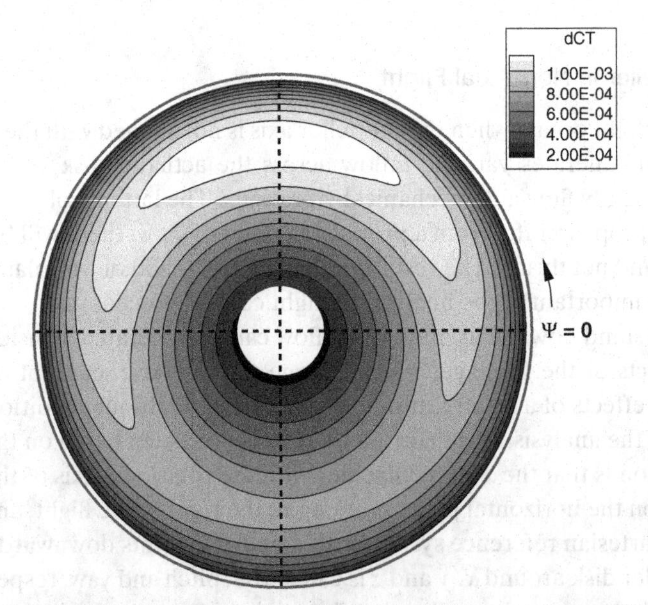


Figure 6.6. Loads on disk for asymmetric flight of the F568 propeller; pitch = 10 degrees, yaw = 5 degrees, $U = 250$ kt, $\vartheta = 40$ degrees.

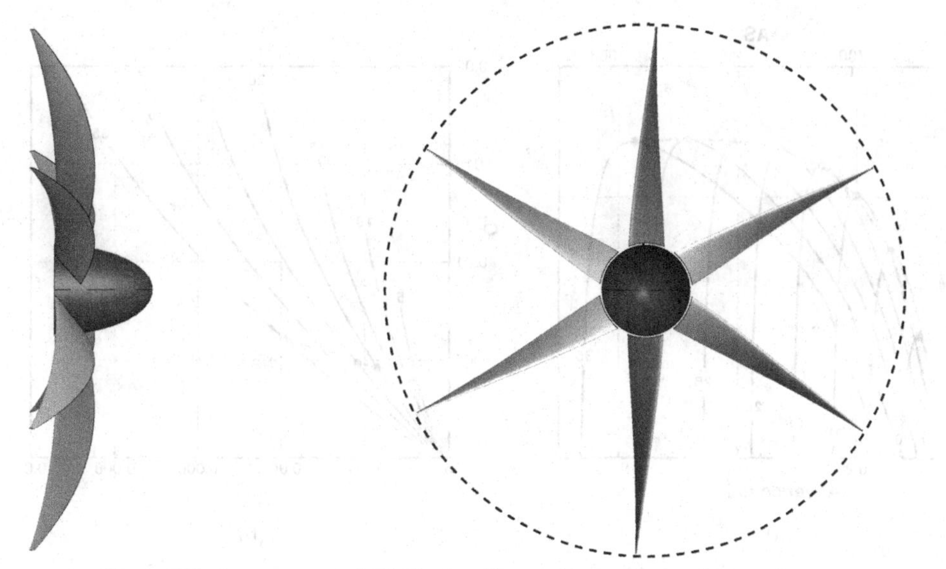

Figure 6.7. Reconstructed F568 propeller and hub, side and front views.

### 6.2.4 Case Study: Hamilton-Sundstrand F568 Propeller

The F568 is a modern propeller for turboprop-driven airplanes developed by Hamilton-Sundstrand† for the ATR42, ATR72, CASA C295 and Ilyushin-114-100. The propeller is six-bladed and has a nominal diameter of 3.93 m and a non-linear twist with swept-back tips. The blades are made of a Kevlar shell with a graphite spar. A titanium strip is applied at the leading edge to prevent blade erosion. The hub is made of steel. The propeller system is protected against low pitch angle in flight, over-speed and hydraulic pressure loss. The pitch is limited between −14 degrees (reverse) and 78.5 degrees (feather).

The nominal rotational speed is 1,200 rpm with a design shaft power of 2,244 kW; rotation is clockwise, looking forward. Electrical de-icing of the blades is available. Some blades are interchangeable with each other. The geometry of the reconstructed propeller is shown in Figure 6.7.

The propeller‡ is driven by a power turbine via a reduction gearbox. The collective pitch is controlled by a hydro-mechanic unit (propeller valve module, installed in each engine nacelle). The pitch control is electronic with hydro-mechanical backup. The propeller valve module is then controlled by an electronic control on each engine; a propeller interface unit transfers control to the flight deck. The propeller valve module has several important functions, including rpm control, over-speed control, reversing, feathering, low-pitch protection and *synchrophasing*. The latter term refers to the synchronisation of the phase between propellers, which may be important in controlling interference noise between propellers (see propeller noise in § 16.7).

† The manufacturer identifies the propeller with the numbers 568: 5 is the design number; 6 is the number of blades; 8 is the blade shank size; F denotes the flange mounting. Additionally, there can be another identifier, such as 568-x, where *x* denotes the airplane application.

‡ This information is adapted from the ATR72-500 Flight Crew Operating Manual.

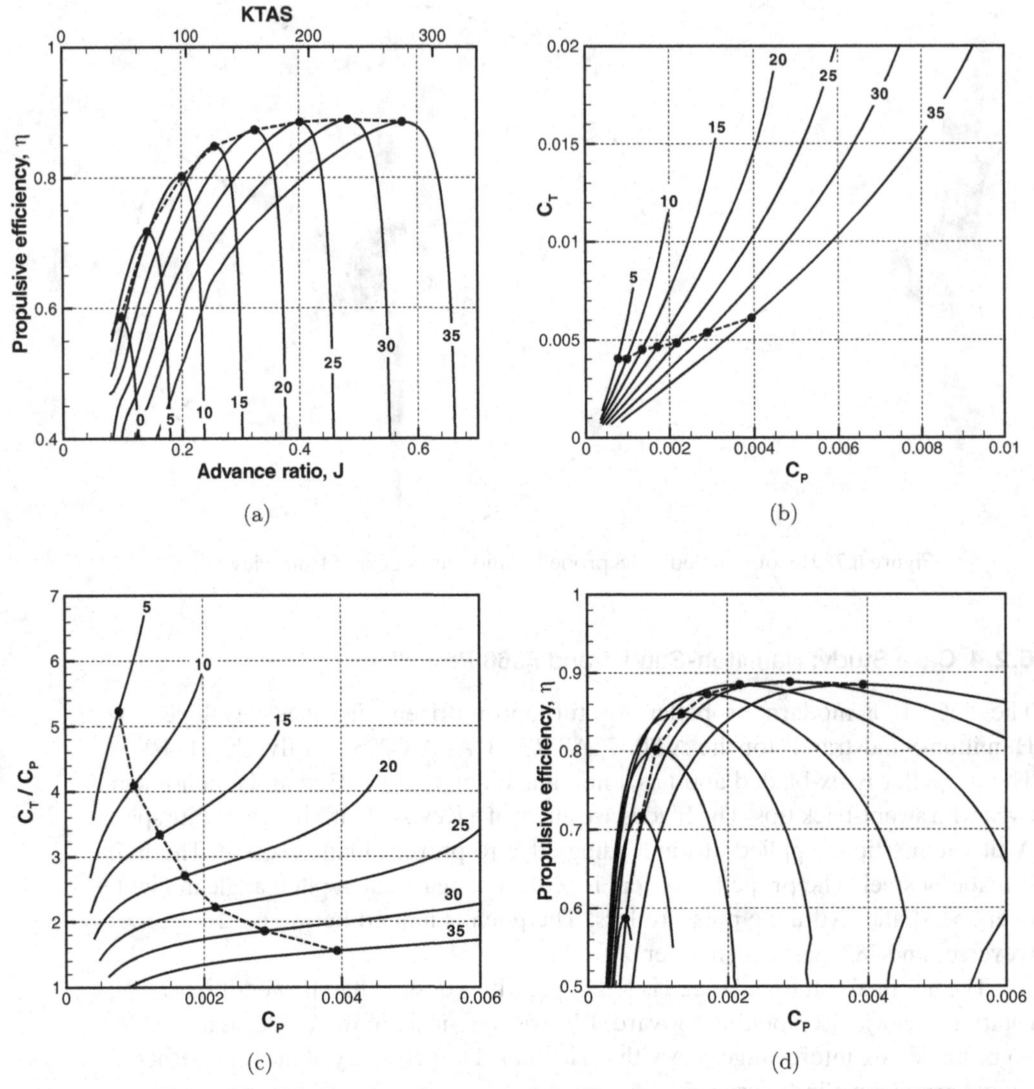


Figure 6.8. Calculated performance of the Hamilton-Sundstrand propeller F568-1.

The propeller electronic control is a hardware–software box that controls the pitch via a closed-loop feedback. The control input is the torque $Q = P/\Omega$. At low fuel flow, the control unit may be unable to adjust the pitch to match the torque; therefore, it reduces the rpm. In fact, the rpm can take a number of values, which are 82% (984 rpm) at climb-out and cruise, 100% at take-off and rejected take-off. This is an important fact: the propeller can assume different rpm, depending on the flight segment, but then it is maintained constant with an electronic pitch control.

Figure 6.8 shows the propeller performance (efficiency, shaft power, net thrust and maximum sectional $L/D$) as a function of the advance ratio (and KTAS) for selected pitch settings, as calculated with the FLIGHT code. Figure 6.8a shows that by appropriate choice of the collective pitch, it is possible to ensure nearly constant propulsive efficiency with increasing flight speed.

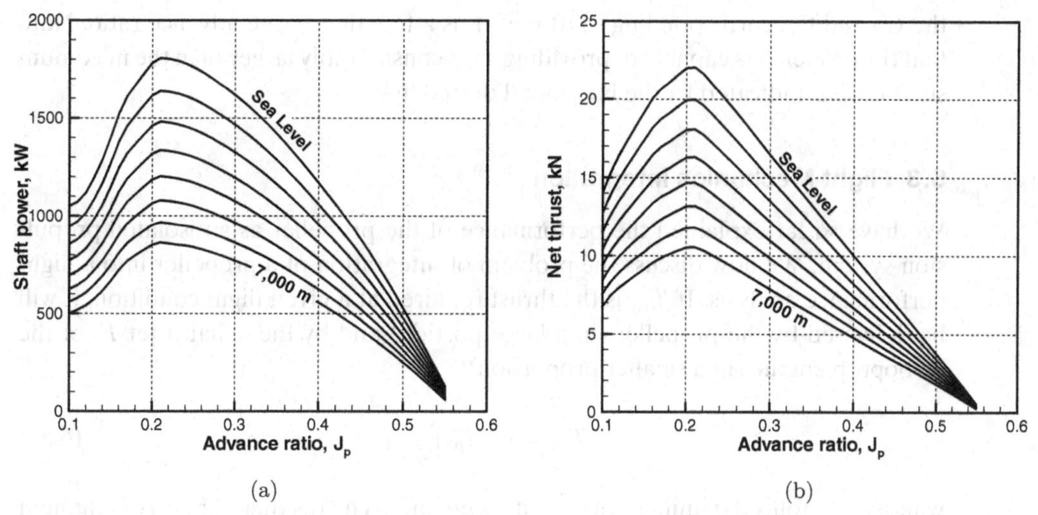


Figure 6.9. Calculated altitude performance of the Hamilton-Sundstrand propeller F568-1.

The flight altitude has negligible effect on the propulsive efficiency. However, the performance of the propeller (shaft power and net thrust) itself depends greatly on the flight altitude, as shown in Figure 6.9.

The aerodynamic analysis presented earlier referred to the relatively simple case in which the inflow is axial. However, it is not unusual that the propeller is required to work in yawed conditions, for which there will inevitably be some limitations. Another limitation is related to the matching between the propeller itself and the engine. An example is shown in the propeller chart in Figure 6.10. This figure displays

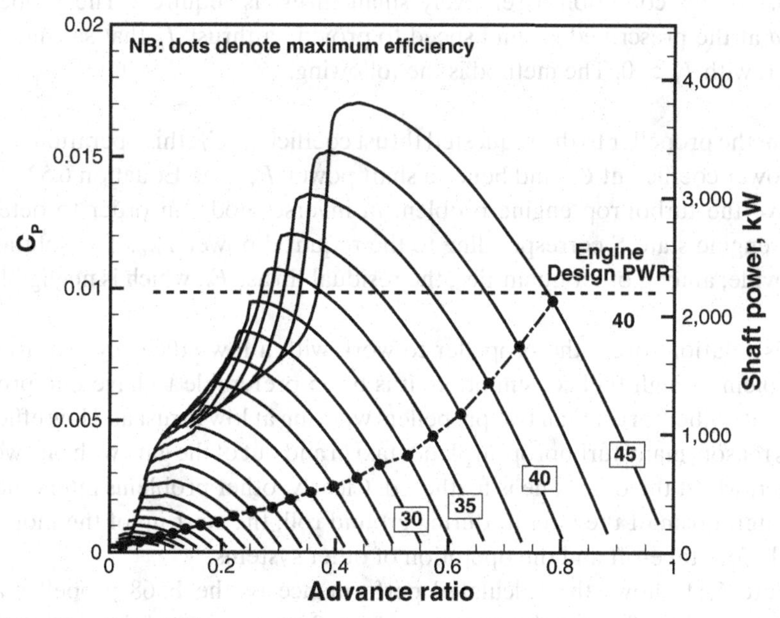


Figure 6.10. Simulated F568 propeller; power coefficient and shaft power; lines of constant pitch shown.

the $C_P$ and the corresponding shaft power as a function of the advance ratio. Note that the propeller is capable of providing a $C_P$ considerably larger than the maximum shaft power, indicated by the horizontal dotted line.

## 6.3 Flight Mechanics Integration

We have so far explained the performance of the propeller as an isolated propulsion system. We now discuss the problem of integration of a propeller into a flight performance analysis. If $T_{req}$ is the thrust required at a given flight condition, it will be delivered by the propeller (in a large portion) and by the exhaust jet $F_g$ of the turboprop engine (in a smaller proportion):

$$T_{req} = n_e(T_p + F_g), \tag{6.51}$$

where $n_e$ denotes the number of operating engines. The residual thrust $F_g$ is thought to be always in the direction of the propeller thrust $T_p$. The propeller thrust and the corresponding shaft power are given by Equation 6.8. Each flight condition is now discussed separately. In any case, once the propeller thrust $T_p$ and the corresponding $C_T$ have been calculated, we need to calculate the propeller $C_P$ and the propeller power $P_p$. The power required by the power-plant will be

$$P_{req} = \frac{1}{n_e} \frac{P_p}{\eta_g \eta_m \eta_i}, \tag{6.52}$$

which accounts for losses in the gearbox, for the mechanical efficiency of the engine and for the installation losses.

**TAXIING.** At this condition a relatively small thrust is required. The propeller is *trimmed* at the prescribed ground speed to provide a thrust $T_p$ that satisfies Equation 6.51, with $F_g \simeq 0$. The method is the following:

1. Trim the propeller to the requested thrust coefficient $C_T$; this operation provides a power coefficient $C_P$ and hence a shaft power $P_{req}$, via Equation 6.52.
2. Solve the turboprop engine problem in inverse mode, in order to determine the engine state $\mathcal{E}$ corresponding to the required power $P_{req}$. The solution will provide, among other quantities, the residual thrust $F_g$, which is negligible.

This solution forces the propeller to work with a low efficiency. To overcome the problem of high fuel consumption, it is often preferable to have one propeller working at higher thrust than two propellers working at low thrust and low efficiency. For this reason, many turboprop airplanes taxi in and out of the gate with one working propeller, whilst the other one is feathered. Clearly, other problems intervene, such as the lateral trim of the aircraft during ground roll, the start-up of the inoperative engine before take-off and the operation of other systems.

Figure 6.11 shows the calculated performance of the F568 propeller at taxi conditions for specified net thrust ranging from 2 kN to 8 kN. If the total net thrust required for rolling on the taxi-way is 4 kN (2 kN × 2 propellers), the propulsive

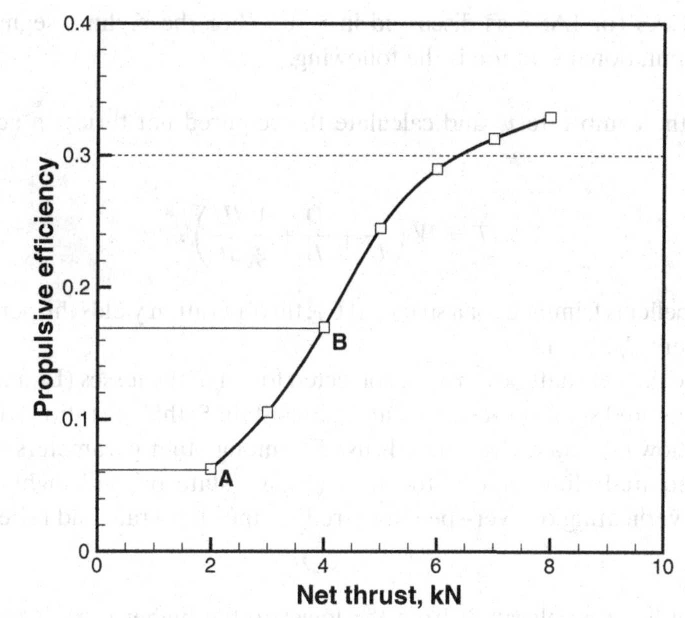


Figure 6.11. Propeller efficiency at taxi conditions; ground speed = 3.9 kt, altitude = 100 m.

efficiency is about 6% when two propellers are running and 17% when one propeller is running.

**TAKE-OFF.** At take-off it is required to operate the propeller at maximum thrust:

$$\max(T) = f(h, J_1, \vartheta_o). \tag{6.53}$$

This condition does not necessarily occur at maximum shaft power. Alternatively, the propeller can be trimmed to maximum power or a fraction of the maximum power to provide a net thrust. Again, the propulsive efficiency can be relatively low. The numerical procedure is the following:

1. Set $F_g = 0$.
2. Trim the propeller to the thrust or power condition; this operation provides the value of the collective $\vartheta$ and the corresponding shaft power or net thrust.
3. Solve the turboprop engine problem to determine the engine state $\mathcal{E}$ corresponding to the required power $P_{req}$, corrected for all of the conversion losses (Equation 6.52). The solution will provide the residual thrust $F_g$.
4. Calculate the corrected propeller thrust $T_p = T - F_g$ and trim the propeller to the net thrust $T_p$. Reiterate from point 3.

Sometimes there is the possibility of increasing slightly the rpm from its nominal value, which is quite useful in improving the thrust capabilities.

**EN-ROUTE CLIMB.** The climb problem does not have a unique solution, particularly because climb is performed in phases. There are at least two distinct climb segments at constant CAS (or IAS), as discussed in § 10.3. For these climb segments, one suitable computational solution is the following:

1. Specify the climb rate $v_c$ and calculate the required net thrust in accelerated flight:

$$T = W\left(\frac{v_c}{U} + \frac{D}{L} + \frac{1}{g}\frac{dU}{dt}\right). \tag{6.54}$$

2. The propeller is trimmed for a specified $C_T$; this operation yields the performance parameters $C_P$, $\vartheta$ and $\eta$.
3. Calculate the net shaft power $P_{req}$, corrected for all of the losses (Equation 6.52).
4. At the required shaft power, calculate engine state $\mathcal{E}$; this operation will provide the fuel flow $\dot{m}_f$ and the residual thrust $F_g$, among other parameters.
5. If the required climb rate is too high (engine state beyond flight envelope; engine overheating or over-speeding), reduce the climb rate and reiterate from point 1.

The level-flight acceleration from the lower to the higher CAS is treated in a similar manner, except that an acceleration value is specified instead of a climb rate.

**CRUISE.** Equation 6.51 is to be solved at a fixed altitude and Mach number (or TAS) because the ($h$,TAS) condition is determined in the computer simulation model as an optimal operation point. Thus, the net thrust is equal to the total aerodynamic drag:

1. Set $F_g = 0$.
2. Calculate the aerodynamic drag $D = T$ at the current flight condition.
3. Trim the propeller to the required $C_T$; calculate the propeller's parameters $C_P$ $\vartheta$, $\eta$; calculate the net shaft power, corrected for installation losses.
4. Calculate the engine state $\mathcal{E}$ corresponding to the required power; this operation yields $\dot{m}_f$, $F_g$ and other engine parameters.
5. Correct the net propeller thrust according to Equation 6.51.
6. Trim the propeller as at point 2.
7. Iterate until convergence, that is, when there is negligible change in the net propeller thrust (convergence is achieved in three or four iterations.)

**EN-ROUTE DESCENT.** This case is treated as the cruise problem, with the propeller required to deliver the required thrust.

Figure 6.12 shows the flight envelope of the turboprop engine PW-127M (ATR-72). The graph displays the behaviour of the residual thrust $F_g$ as a function of the shaft power over the full range of flight altitudes. At low power settings, this thrust is clearly negligible, but it grows almost linearly. The propulsive efficiency is not used in any of the flight conditions discussed. However, it is of some interest to calculate the net propulsive efficiency of the propeller-gas-turbine engine system, which is

$$\eta = \frac{(T_p + F_g)}{P_{shaft}} U. \tag{6.55}$$

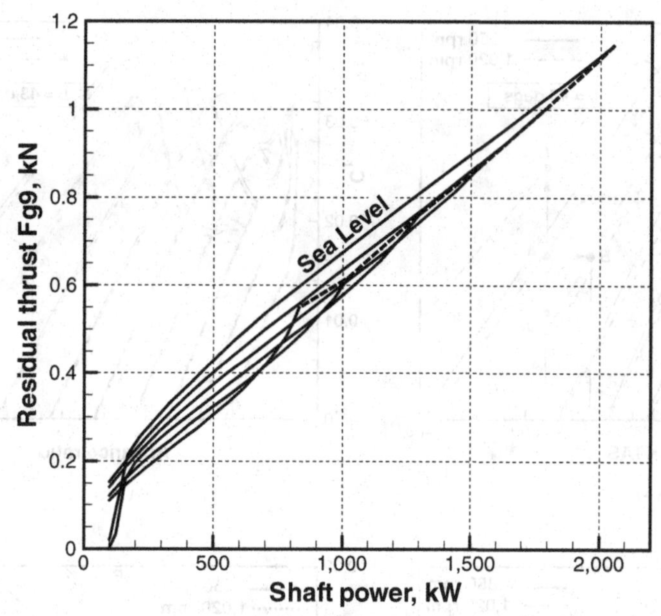


Figure 6.12. Estimated residual thrust $F_g$ versus shaft power for the turboprop engine PW127M; data at 1,000 m intervals, standard day.

Equation 6.55 would have to be corrected to take into account the gearbox efficiency $\eta_g$ (see § 6.4.1) and the shaft mechanical efficiency $\eta_m$. Thus, we have

$$\eta = \eta_g \eta_m \frac{(T_p + F_g)}{P_{shaft}} U. \tag{6.56}$$

The foregoing analysis has shown that the propulsive efficiency is not a key parameter in the airframe-propeller integration, as long as the propeller is trimmed to a specified thrust or power. There are instances when the propeller efficiency is low and other cases when it is high. The changes required in propeller pitch even within one flight condition (for example, en-route climb) make it impossible to manage without an appropriate flight control system. For this reason, modern variable-pitch propellers are controlled by software and actuators (FADEC), like most other airplane subsystems.

### 6.3.1 Propeller's Rotational Speed

Assume that the aircraft is required to fly with an air speed $U$; this condition also requires $T = D$, that is, a fixed value of the required thrust. If the propeller rpm is decreased, the following condition holds:

$$C_T(\Omega R)^2 = \frac{T}{\rho A} = \text{constant}. \tag{6.57}$$

A decrease in rotational speed $\Omega$ allows (in principle) the aircraft to fly faster by virtue of an offset in the effective tip Mach number (Equation 6.7). Consider the example shown in Figure 6.13, which refers to the Dowty R408 propeller that powers the Bombardier Dash8-Q400 (see Table 6.2). This propeller has a take-off speed

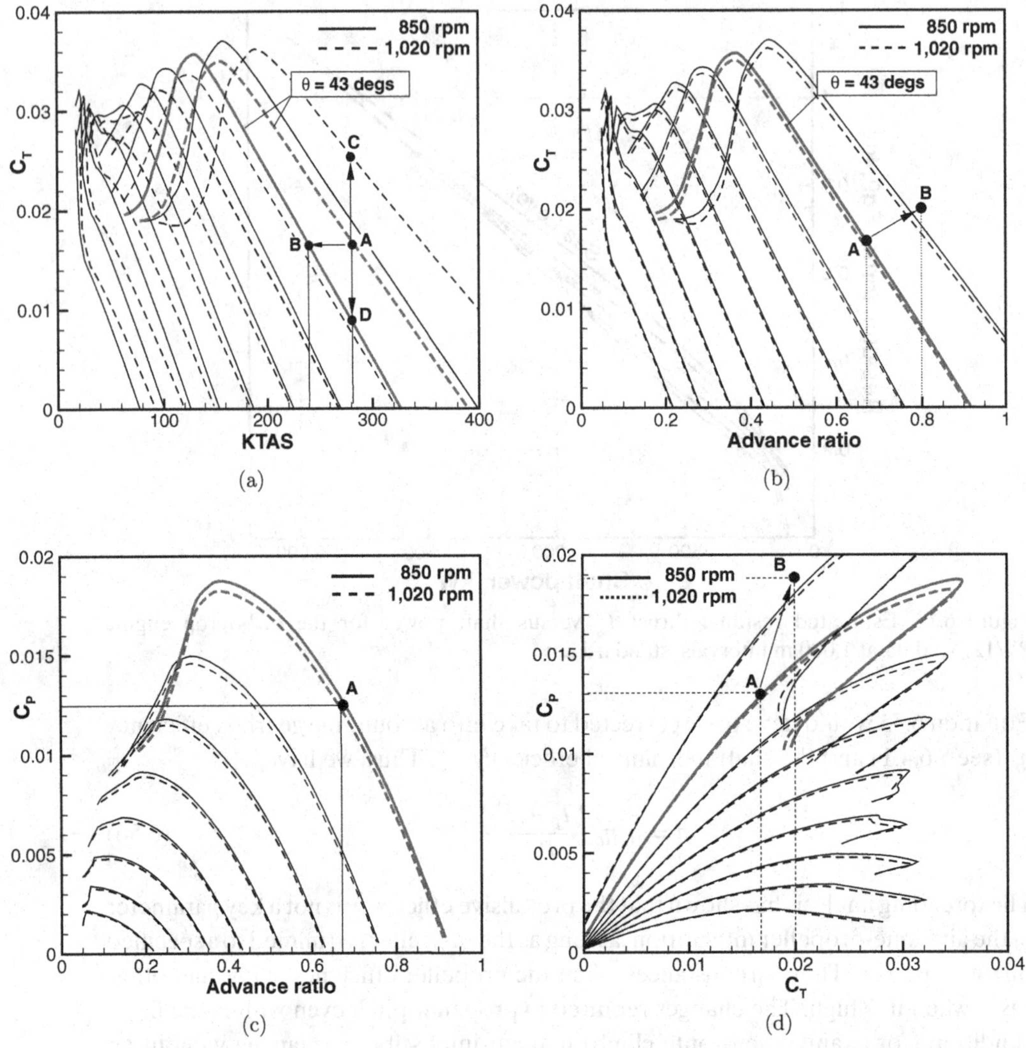


Figure 6.13. Effect of rotational speed on the $C_T$ of the Dowty R408 propeller (calculated). The curves are plotted at constant pitch values with a 2.5-degree interval.

equal to 1,020 rpm and a landing speed of 850 rpm. Assume that the aircraft cruises at 280 KTAS. There are three trim options:

1. With a propeller speed 1,020 rpm, the aircraft is trimmed with a pitch $\vartheta \simeq 43$ degrees and a $C_T \simeq 0.0167$ (point A in Figure 6.13a). To maintain the air speed whilst reducing the rpm, the $C_T$ must increase; this change can be obtained by increasing the collective pitch (point A to C).
2. If the pitch is unchanged, reducing the rpm causes the aircraft to slow down (point A to B in Figure 6.13a).
3. The final option shows the propeller moving to point D in Figure 6.13a as the propeller rpm is reduced. This point corresponds to a lower $C_T$. The only way Equation 6.57 can be satisfied is by increasing the density altitude: in other words, the aircraft must descend.

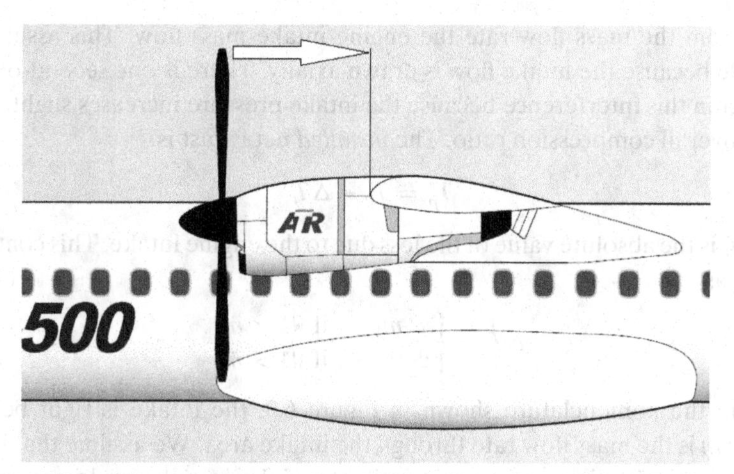


Figure 6.14. Side view of the propeller-nacelle-wing installation on the ATR72-500 (adapted from the manufacturer's documents).

Figures 6.13b and 6.13c show that when the $C_T$ and $C_P$ curves are plotted against the non-dimensional speed $J_1$, they are basically overlapping. This is a useful result. In fact, we calculate only one flight envelope, for example, at the take-off rpm rating (1,020 rpm). To perform the analysis at landing conditions (850 rpm), we use the chart in Figure 6.13b. For the case of a propeller trimmed to a required thrust, we have the state vector $\mathcal{S} = \{h, U, T, \Omega\}$. The corresponding $C_T$ is calculated from Equation 6.57. For example, in Figure 6.13b, point A corresponds to a $J_A \simeq 0.67$ and $C_{T_A} \sim 0.0167$ at 1,020 rpm. With the lower rpm, we have $C_{T_B} \simeq (1{,}020/850) \times 0.0167 \simeq 0.02$. The advance ratio is increased to the new value $J_B = U/\Omega R \simeq 0.80$. The corresponding operation point moves to B, which requires an increase in collective pitch from $\vartheta_A \simeq 43$ degrees to $\vartheta_B \simeq 46$ degrees. The power coefficient $C_P$ is calculated at $(J_B, \vartheta_B)$, as shown in Figure 6.13d.

## 6.4 Propeller Installation Effects

The foregoing analysis has presented performance issues of an isolated propeller in axial flight. Three key aspects must now be considered: one is the effect of installation, such as the combined effect of propeller hub and downstream or upstream elements (such as engine mounting and wing); the second aspect deals with the coupling of the propeller to an engine shaft via a gearbox, which causes some losses in propulsive efficiency; and the third is the issue of non-axisymmetric flight (yaw and angle of attack).

**PROPELLER-WING INTERFERENCE.** The propeller is mounted, via a nacelle, onto a wing. The most common case is that of a *tractor* propeller, whereby the wing is placed downstream of the rotor disk. Due to the combination of swirl effects and increased air speed in the slipstream, the net effect on the wing is an increase in drag. If the wing is placed at least one propeller radius downstream (Figure 6.14) and the air-flow over the wing remains attached, the effect is minimal.

**PROPELLER-INTAKE INTERFERENCE.** The engine intake is generally within the working section of the propeller; thus, it collects some of the airflow and prevents the development of the full slip-stream. One way of accounting for this effect is to subtract from the mass flow rate the engine intake mass flow. This assumption is reasonable because the intake flow is drawn axially. There is one second-order positive effect in this interference because the intake pressure increases slightly, and so does the overall compression ratio. The *installed* net thrust is

$$T_p^* = T_p - \Delta T_i \tag{6.58}$$

where $\Delta T_i$ is the absolute value of the loss due to the engine intake. This contribution is written as

$$\Delta T_i = \begin{cases} \simeq \dot{m} v_i & \text{if } \texttt{W1} < \dot{m} \\ \simeq 0 & \text{if } \texttt{W1} > \dot{m} \end{cases} \tag{6.59}$$

because in the nomenclature shown in Figure 6.3, the intake is right behind the propeller; $\dot{m}$ is the mass flow rate through the intake area. We assume that if W1 $< \dot{m}$ then Equation 6.59 gives the correct estimate of the installation losses on the net thrust; if W1 $> \dot{m}$, we assume that $\Delta T_i \sim 0$ because the engine is drawing more mass flow than it would normally be passing through the intake area if the engine were not there.

The induced velocity must be calculated locally. The use of Equation 6.23 gives a too high engine inlet mass flow. A more accurate estimation for the induced velocity is $v_i \simeq dP/dT$ at the centre of the inlet. Thus, the average mass flow rate passing through the inlet area $A_i$ is $\dot{m} \sim \rho A_i v_i$. In summary, when the first instance in Equation 6.59 is true, the thrust correction is

$$\Delta T_i = \rho A_i v_i^2. \tag{6.60}$$

For the purpose of this analysis, the induced velocity is taken at the centre of the inlet and is given by $v_i = U(1 + a)$, where $a$ is the local axial interference factor calculated with the blade-element method (see § 6.2.2). It is not recommended to use the basic momentum theory (Equation 6.51) because it grossly over-estimates the value of the intake mass flow. In any case, the installation effects are introduced iteratively, starting from Equation 6.51, according to the following procedure:

1. Specify the total thrust $\overline{T}$ and set $F_g = 0$; this condition specifies the $C_T$ (without engine).
2. Trim the propeller to the $C_T$; calculate the $C_P$ and the uninstalled power $P$.
3. Solve the turboprop problem in the inverse mode to deliver the power $P$; the solution also yields $F_g$ and W1.
4. Calculate the installation thrust difference from Equation 6.60.
5. Assemble the total thrust from $T = T_p + F_g - \Delta T_i$.
6. If $T > \overline{T}$, then reduce the fuel flow and restart from point 3; otherwise increase the fuel flow and restart from point 3.
7. Iterate until the difference in $T_p$ between iterations is negligible.

The increase or decrease in fuel flow is indeterminate. A more rational approach is to build the computations method around a bisection method.

There are cases when the propeller is mounted aft, such as in the *Piaggio Avanti*. In this case, we must account for the interference between the exhaust gas and the propeller blades, which creates a complicated case of blade interaction and further noise emission.

### 6.4.1 Gearbox Effects

The propeller requires a power train for reducing the large rpm of the gas turbine engines to the more moderate rpm required to maintain the blade tips within a moderate transonic Mach number. This is done through a gearbox that bears all the torque transmitted by the engine to the propeller and reduces the rpm by a factor of 5 to 10. There are various technological aspects related to the gearbox, but in this instance we limit our discussion to performance problems, namely the power losses. The loss of power can be monitored indirectly by the heat generation. As a first-order analysis one can assume that this quantity is proportional to the square of the pressure between gear wheels or to the square of the transmission torque. A semi-empirical expression for gearbox efficiency is the following:

$$\eta_g = 1 - g_r \left( \frac{Q}{Q^*} \right)^2, \tag{6.61}$$

where $Q$ is the actual torque, $Q^*$ is the design torque, $g_r$ is the reduction ratio, or

$$g_r = \kappa \frac{\Omega}{N_1}, \tag{6.62}$$

where $N_1$ denotes the actual rotational speed of the power-turbine shaft and $\kappa$ is an empirical factor. The heat generated within the gearbox is estimated from

$$\dot{Q} = \frac{dQ}{dt} = (1 - \eta_g) P_{shaft}. \tag{6.63}$$

For example, at a nominal $P_{shaft} \sim 10^3$ kW with $\eta_g = 0.98$, we have $\dot{Q} \simeq 20$ kJ/s, which is a considerable amount of heat, especially considering that lubricating oils have heat capacities of the order of 20 kJ/kg K.

## Summary

We have defined the key performance parameters of an aircraft propeller. We then proposed a method for modelling a realistic propeller based on a mix of manufacturers' data, certified data, photographs and educated guesses. The propeller performance is then computed by using a combined blade element and momentum theory. The model is extended to include flight in yaw conditions. Propeller charts are shown over the entire range of advance ratios, pitch settings and flight altitude. The propeller model is fully integrated into the flight mechanics program by a combination of the propeller thrust/power, the engine power and the residual jet thrust. Various trim conditions are possible, including thrust and power. It is shown how the efficiency of the propeller is greatly dependent on the flight conditions; in particular, propellers operating at very low speed (taxi and ground manoeuvre) are inefficient. We also demonstrated that the residual thrust from the power-plant can contribute to the total thrust. Finally, we considered secondary effects, such as engine/propeller installation and gearbox losses.

## Bibliography


[1] Glauert H. *Airplane Propellers*, volume 4 of *Aerodynamic Theory*. Dover ed., 1943.
[2] Theodorsen T. *Theory of Propellers*. McGraw-Hill, 1948.
[3] von Mises R. *Theory of Flight*. Dover Publications, 1959.
[4] Biermann D and Hartman EP. Wind-tunnel tests of four- and six-blade single- and dual-rotating tractor propellers. Technical Report 747, NACA, 1942.
[5] Hartman EP and Biermann D. The aerodynamic characteristics of full-scale propellers having 2, 3, and 4 blades of Clark Y and R.A.F. 6 airfoil sections. Technical Report R-640, NACA, 1938.
[6] Theodorsen T, Stickle GW, and Brevoort MJ. Characteristics of six propellers including the high-speed range. Technical Report R-594, NACA, 1937.
[7] *Aerodynamics and Acoustics of Propellers*, AGARD-CP-366, Feb. 1985.
[8] Wieselberger C. Contribution to the mutual interference between wing and propeller. Technical Report TM-754, NACA, 1934.
[9] McHugh J and Eldridge H. The effect of nacelle-propeller diameter ratio on body interference and on propeller and cooling characteristics. Technical Report R-680, NACA, 1939.
[10] Delano JB. Investigation of the NACA 4-(5)(08)-03 and NACA 4-(10)(08)-03 two-blade propellers at forward Mach numbers to 0.725 to determine the effects of camber and compressibility on performance. Technical Report R-1012, NACA, 1951.
[11] Stack J, Delano E, and Feldman J. Investigation of the NACA 4-(3)(8)-045 two-blade propellers at forward Mach numbers to 0.725 to determine the effects of compressibility and solidity on performance. Technical Report R-999, NACA, 1950.
[12] Conway JT. Exact actuator disk solutions for non-uniform heavy loading and slipstream contraction. *J. Fluid Mech.*, 365:235–267, 1998.
[13] Adkins CN and Liebeck RH. Design of optimum propellers. *J. Propulsion & Power*, 10(5):676–682, Sept. 1994.


## Nomenclature for Chapter 6

$A$ = disk area
$A_i$ = engine intake area
$a$ = speed of sound
$a$ = axial-interference factor, Equation 6.45
$a'$ = radial-interference factor, Equation 6.45
$c$ = chord of a blade section
$\overline{c}$ = mean chord
$C_d$ = sectional-drag coefficient
$C_D$ = drag coefficient
$C_l$ = sectional-lift coefficient
$C_L$ = lift coefficient
$C_P$ = power coefficient
$C_Q$ = torque coefficient
$C_T$ = thrust coefficient
$C_x$ = axial-force coefficient
$C_y$ = normal-force coefficient
$d$ = diameter

$D$ = aerodynamic drag
$E$ = energy
$\mathcal{E}$ = engine state
$f$ = tip/root correction factor
$F$ = Prandtl factor, Equation 6.48; total aerodynamic force (Figure 6.1)
$F_g$ = residual thrust from turboprop engine (also `Fg9`)
$g$ = acceleration of gravity
$g_r$ = gear-reduction ratio
$h$ = flight altitude
$i$ = blade-element counter; iteration counter
$j$ = blade counter
$J, J_1$ = advance ratios
$k, k'$ = parameters defined by Equation 6.46
$L$ = aerodynamic lift
$\dot{m}$ = mass flow rate (also `W1`)
$\dot{m}_f$ = fuel flow (also `Wf6`)
$M$ = Mach number
$M_{tip}$ = tip Mach number
$N$ = number of blades
$N_1$ = nominal gas generator turbine rpm (also called `N1`)
$n_e$ = number of operating engines
$p$ = pressure
$P$ = shaft power
$P_i$ = induced power
$P_p$ = propeller power
$P_{req}$ = required power
$q$ = dynamic pressure, $\rho U^2/2$
$Q$ = shaft torque
$Q^*$ = design shaft torque
$\mathcal{Q}$ = thermal energy, Equation 6.63
$r$ = radial position, $y/R$
$R$ = rotor radius
$t$ = time
$T$ = net thrust
$T_p$ = propeller thrust
$T_p^*$ = installed net thrust
$T_{req}$ = required thrust
$u$ = air speed (local; reference stations 1, 2, 3)
$U$ = air speed
$U_n, U_t$ = normal and tangential velocity components
$U_{tip}$ = tip speed, $\Omega R$
$v_c$ = climb rate
$v_i$ = induced velocity at propeller disk
$y$ = radial station
$v_c$ = climb rate
$v_i$ = induced velocity at propeller disk

| | | |
|---|---|---|
| $V_w$ | = | wind speed |
| $W$ | = | weight |
| $x,y,z$ | = | Cartesian coordinate system |

## Greek Symbols

| | | |
|---|---|---|
| $\alpha$ | = | effective angle of attack |
| $\eta$ | = | propulsive efficiency |
| $\eta_g$ | = | gearbox efficiency |
| $\eta_m$ | = | mechanical efficiency |
| $\eta_i$ | = | effect of installation on propulsive efficiency |
| $\lambda$ | = | tip-speed ratio (advancing) |
| $\lambda_c$ | = | tip-speed ratio (static) |
| $\kappa$ | = | empirical factor in Equation 6.62 |
| $\rho$ | = | air density |
| $\phi$ | = | inflow angle |
| $\tilde{\phi}$ | = | corrected-inflow angle |
| $\vartheta$ | = | pitch angle |
| $\vartheta_o$ | = | collective-pitch angle |
| $\sigma$ | = | rotor solidity |
| $\Psi_w$ | = | wind-direction angle |
| $\Omega$ | = | rotational speed |

# 7 Airplane Trim

## Overview

The problem of airplane trim involves the determination of the control requirements to maintain a stable flight or to perform specified manoeuvres. The position of the centre of gravity (CG) is essential, and some consideration is given to this effect on the cruise flight. In this chapter we consider the problem of static longitudinal trim (§ 7.1), lateral trim and airplane control under asymmetric thrust (§ 7.2). We consider only steady-state conditions. Transient conditions are the subject of flight dynamics and therefore are not considered in this chapter.

**KEY CONCEPTS:** Longitudinal Trim, Trim Drag, Stick-Free Trim, Thrust Asymmetry, Lateral Control.

## 7.1 Longitudinal Trim at Cruise Conditions

The airplane in free flight is subject to a number of forces that must be balanced to ensure steady-state flight. In the following analysis, *cruise condition* is a term that is extended to the airplane climbing and descending, subject to the airplane being in clean configuration. We consider the role of the tail-plane and the elevator in providing longitudinal control (longitudinal trim). Because the wing lift and the CG are not at the same point, the airplane will have a nose-down or nose-up pitching moment.

Although the CG is generally on the vertical symmetry plane, neither the propulsive forces nor aerodynamic forces are on that plane. To begin with, we assume that the contributions from engines and wings are symmetric, so that it is possible to reduce the problem to a balance of forces and moments in one plane. For longitudinal equilibrium, two equations must be satisfied: the equation for the pitching moment and the momentum equation in the vertical direction.

The reference distances can be calculated in a number of ways. For example, we can assign global longitudinal coordinates with respect to the nose and call $x$ the distances from this point. Note, however, that in most cases the manufacturers have their own method of assigning longitudinal sections from a reference point that is not at the nose.

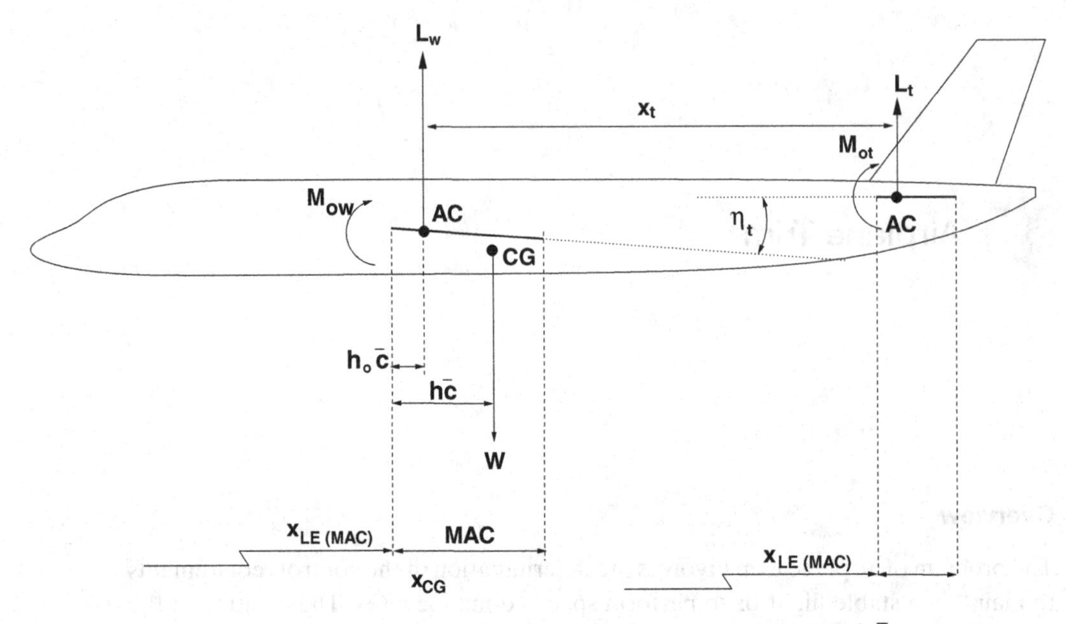


Figure 7.1. Nomenclature for longitudinal trim, adapted from Ref.[1]; $\bar{\bar{c}}$ = MAC.

In stability calculations, we make reference to the mean aerodynamic chord of the wing (MAC; see § 2.6); the MAC is itself referred to the coordinate system centred at the nose. Another possible reference is the CG. However, this point moves in flight, due to fuel burn, and with the airplane loading (see Chapter 3).

The nomenclature we use is the following: $x_{LE}$ is the root leading edge of the wing, $x_{LEt}$ the root leading edge of the stabiliser; and $x_{MAC}$ and $x_{MACt}$ are the leading edges of the MAC of wing and stabiliser, respectively. The distance between the CG and $x_{LE}(MAC)$ is a fraction of the MAC and is called $h$; this quantity is not to be confused with the flight altitude*. The distance between the aerodynamic centre AC and $x_{LE}(MAC)$ and is fraction of the MAC and is called $h_o$. These and other relevant symbols are shown in Figure 7.1.

**EQUATION FOR VERTICAL EQUILIBRIUM.** This equation is:

$$L_w + L_t - W = 0. \tag{7.1}$$

If we divide Equation 7.1 by the product $qA$, we obtain a normalised form, that is, an equation containing only lift coefficients:

$$C_{Lw} + C_{Lt} - C_L = 0. \tag{7.2}$$

Note that all of the lift coefficients in Equation 7.2 are calculated with respect to the *wing area*. This detail is quite important because in most cases the aerodynamic coefficients of a wing are calculated with respect to the planform area of the wing itself. The lift coefficients can be written as

$$\begin{aligned} C_{Lw} &= C_{L\alpha}(\alpha - \alpha_o), \\ C_{Lt} &= C_{L\alpha t}\alpha_t. \end{aligned} \tag{7.3}$$

* The symbol $x$ is a dimensional quantity (a longitudinal coordinate), whilst $h$ is non-dimensional.

The effective inflow angle of the horizontal stabiliser depends on the configuration and the downwash created by the wing. With reference to the latter point, if $\eta_t$ is the angle between the neutral lines of the horizontal tail and the wing, and $\Delta\alpha$ denotes the change in angle of attack due to the wing's downwash, then the inflow angle of the horizontal tail is

$$\alpha_t = \alpha + \eta_t - \Delta\alpha. \tag{7.4}$$

The angle $\eta_t$ depends on the configuration of the airplane; $\Delta\alpha$ depends on the operational conditions, including weight, air speed and flight altitude. Next, assume that the lift distribution on the wing is elliptical. This distribution generates a uniform downwash $\overline{w}$, such that

$$\frac{\overline{w}}{U} = \frac{2L_w}{\pi\rho b^2 U^2}. \tag{7.5}$$

A span efficiency factor can be introduced to take into account the departure from elliptic loading; otherwise the downwash can be calculated from the effective wing-loading distribution. The change in tail plane incidence as a result of this downwash is

$$\Delta\alpha \simeq \frac{\overline{w}}{U} \simeq \frac{C_L}{\pi \mathcal{AR}}, \tag{7.6}$$

where $\mathcal{AR} = b^2/A$ denotes the wing's aspect-ratio. If a numerical solution of the lifting surface is available, the equivalence in Equation 7.5 is not required: the normalised downwash is calculated numerically (see also § 4.8). Finally, solve Equation 7.2 in terms of the wing's angle of attack with the inflow angle for the horizontal tail (Equation 7.4 and Equation 7.6). The result is the following:

$$C_{L\alpha}(\alpha - \alpha_o) + C_{L\alpha t}\left(\alpha - \frac{\overline{w}}{U} + \eta_t\right) - C_L = 0, \tag{7.7}$$

which solved in terms of the inflow angle yields

$$\alpha = \frac{1}{C_{L\alpha} + C_{L\alpha t}}\left[C_L - C_{L\alpha}\alpha_o + C_{L\alpha t}\left(\frac{\overline{w}}{U} + \eta_t\right)\right]. \tag{7.8}$$

The additional condition for longitudinal trim is derived from the balance of the pitching moments, which is derived next.

**PITCHING-MOMENT EQUATION.** With reference to Figure 7.1, the pitching-moment equation is:

$$\mathcal{M} = \mathcal{M}_{ow} + \mathcal{M}_{ot} + (h - h_o)\overline{c}\, W - L_t x_t. \tag{7.9}$$

The moment contributions in Equation 7.9 are (from left to right): moment of the wing $\mathcal{M}_{ow}$ and tail-plane $\mathcal{M}_{ot}$; moment due to the CG offset $(h - h_o)$; and moment of the tail-plane lift. If we divide Equation 7.9 by the quantity $qA\overline{\overline{c}}$, we obtain a non-dimensional form of the equation:

$$C_M = C_{Mow} + C_{Mot} + (h - h_o)C_L - C_{Lt}V_t, \tag{7.10}$$

where the factor $V_t$ denotes the *tail volume coefficient*

$$V_t = \frac{A_t}{A}\frac{x_t}{\bar{\bar{c}}}. \tag{7.11}$$

The term $C_{Mo} = C_{Mow} + C_{Mot}$ is the pitching moment of the airplane, interpreted as the sum between the contribution of the isolated wing and horizontal stabiliser. The pitching moment coefficients are defined by

$$C_{Mow} = \frac{\mathcal{M}_{ow}}{qA\bar{\bar{c}}}, \qquad C_{Mot} = \frac{\mathcal{M}_{ot}}{qA\bar{\bar{c}}}, \qquad C_{Mo} = \frac{\mathcal{M}_{ow} + \mathcal{M}_{ot}}{qA\bar{\bar{c}}}. \tag{7.12}$$

With the inflow angle of the horizontal tail given by Equation 7.4, the pitching-moment equation becomes:

$$C_M = C_{Mo} + (h - h_o)C_L - V_t\left[C_{L_{\alpha t}}\left(\alpha - \frac{\overline{w}}{U}\right) + C_{L_{\alpha t}}\eta_t\right]. \tag{7.13}$$

The trim of the airplane can be done by using elevator and/or tab deflection. If elevator and tab are present, the tail lift is then written in linearised form (thanks to the small angles):

$$C_{Lt} = C_{L_{\alpha t}}\alpha_t + C_{L_\delta}\delta + C_{L_\beta}\beta, \tag{7.14}$$

where $\delta$ and $\beta$ denote the elevator and tab deflection, respectively. By including these terms in Equation 7.13, the non-dimensional pitching-moment equation becomes:

$$C_M = C_{Mo} + (h - h_o)C_L - V_t\left[C_{L_{\alpha t}}\left(\alpha - \frac{\overline{w}}{W}\right) + C_{L_{\alpha t}}\eta_t + C_{L_\delta}\delta + C_{L_\beta}\beta\right]. \tag{7.15}$$

At least one control deflection is required to trim the airplane. The angle of attack of the wing is taken from Equation 7.8. In the general case, Equation 7.15 has infinite solutions. The best solution is the one that yields *minimum trim drag* (§ 7.1.1).

**TRIM WITH LANDING GEAR DEPLOYED.** The solution of the problem is similar when we consider the airplane with landing gears deployed. The landing gear adds parasite drag and has no effect on the induced drag. However, because of the location of the landing gear, there can be an additional contribution to the pitching moment of the airplane. The pitching moment Equation 7.15 will contain the additional term $C_{Mlg}$ for the landing gear. This term arises from the bluff-body drag of the landing gear, which tends to cause a nose-down contribution. An estimate of this moment contribution is as follows:

$$\mathcal{M}_{lg} \simeq D_{Nlg}z_N + D_{Mlg}z_M, \tag{7.16}$$

where $z_N$ and $z_M$ are the vertical coordinates (with respect to the CG) that define the line of action of the nose- and main-landing-gear drag, respectively. To be more precise, the drag contributions from the landing-gear system are split between the following contributions: 1) the wheels drag acts at the centre of the wheel; 2) the struts drag acts at the centre of the strut; and 3) the drag of the bay doors acts approximately at the hinge points. The drag components are calculated in § 4.2.6. The lines of action of these drag components with respect to the CG are calculated with the methods shown in § 3.8.2 and the geometry model (see Chapter 2).

The corresponding pitching-moment coefficient $C_{Mlg}$ is calculated by dividing Equation 7.16 by $qA\bar{c}$. Therefore $C_{Mlg}$ is *added* to the right-hand side of Equation 7.15 to yield a larger absolute value of the pitching moment.

### 7.1.1 Trim Drag

One aspect of the trim is that the deflection of the tail surfaces creates the *trim drag*. The trim drag is calculated from

$$C_{D_{trim}} = kC_L^2 - \left[kC_{L_w}^2 + k_t C_{L_t}^2 \left(\frac{A_t}{A}\right)^2\right]. \tag{7.17}$$

Equation 7.17 expresses the difference in induced drag between a condition of zero tail-plane lift (no tail-plane trim required; all lift generated by the wing) and the induced drag in the trimmed condition (lift split between wing and tail-plane). The tail-plane term is multiplied by a corrective factor $A_t/A$ only if the $C_{Lt}$ is referred to the tail-plane area. Unless we perform this rescaling, there is an inconsistent mix of coefficients. In fact,

$$\frac{L_t}{qA} = C_{L_t}\left(\frac{A_t}{A}\right) = C_{L_t}^A, \tag{7.18}$$

where the right-hand term indicates that the $C_L$ is referred to the wing area $A$. Note that there are cases in which the trim drag is *negative*.

### 7.1.2 Solution of the Static Longitudinal Trim

From Equation 7.13 we wish to find the elevator angle $\delta$ that trims the airplane. In the absence of a trim tab, this angle is

$$-\delta_{trim} = \frac{C_{Mo}}{V_t C_{L_\delta}} + \frac{(h - h_o)C_L}{V_t C_{L_\delta}} - \frac{1}{C_{L_\delta}}\left[C_{L_{\alpha t}}\left(\alpha - \frac{\overline{w}}{U}\right) + C_{L_{\alpha t}}\eta_t,\right]. \tag{7.19}$$

For a given weight and air speed, $\delta_{trim}$ depends *linearly* from the position of the CG. Thus, the solution will have the form

$$\delta_{trim} = \delta_o + C_L f(h), \tag{7.20}$$

with

$$\delta_o = \frac{C_{Mo}}{V_t C_{L_\delta}} - \frac{1}{C_{L_\delta}}\left[C_{L_{\alpha t}}\left(\alpha - \frac{\overline{w}}{U}\right) + C_{L_{\alpha t}}\eta_t\right], \tag{7.21}$$

$$f(h) = \frac{h - h_o}{V_t C_{L_\delta}}. \tag{7.22}$$

The value $\delta = \delta_o$ is the elevator angle when the longitudinal position of the CG is coincident with the aerodynamic centre of the wing. When $h - h_o > 0$ (as shown in Figure 7.1), the function $f(h)$ is positive and the trim angle increases as the CG moves aft. If the landing gears are deployed, the increase in pitching moment causes an increase in control deflection $\delta_{trim}$.

The trim moment must be provided by the control system. If the control is manual, the pilot would have to maintain the stick *fixed* at the position required for trim. Alternatively, the elevator deflection can be set to zero if the tail is *trimmable*: the horizontal tail can be rotated so that the angle $\eta_t$ satisfies Equation 7.20 with $\delta_{trim} = 0$:

$$\eta_t = \frac{1}{C_{L_{\alpha t}}} \left[ \frac{C_{Mo}}{V_t} + \frac{(h - h_o)C_L}{V_t} - C_{L_{\alpha t}} \left( \alpha - \frac{\overline{w}}{U} \right) \right]. \tag{7.23}$$

The longitudinal trim can also be obtained with a fixed stabiliser and a $\delta = 0$, provided that the CG position satisfies the following condition from Equation 7.19:

$$h = h_o - C_{Mo} + \frac{C_{L_{\alpha t}}}{C_L} \left( \alpha - \frac{\overline{w}}{U} + \eta_t \right). \tag{7.24}$$

### 7.1.3 Stick-Free Longitudinal Trim

Because the condition $C_M = 0$ alone generates a hinge moment due to elevator deflection, a more restrictive trim condition arises from forcing also the hinge moment to be zero, that is: $C_H = 0$. In this case, the trim is stick-free (*hands-off*). The hinge moment is a linear combination of the inflow angle of the tail-plane, the elevator and the trim tab angles:

$$C_H = \left( \frac{dC_H}{d\alpha_t} \right) \alpha_t + \left( \frac{dC_H}{d\delta} \right) \delta + \left( \frac{dC_H}{d\beta} \right) \beta. \tag{7.25}$$

To simplify the equations, the derivatives in Equation 7.25 are replaced by the following symbols:

$$b_1 = \frac{dC_H}{d\alpha_t}, \qquad b_2 = \frac{dC_H}{d\delta}, \qquad b_3 = \frac{dC_H}{d\beta}. \tag{7.26}$$

Thus, the hinge moment can be rewritten as

$$C_H = b_1 \left( \alpha - \frac{\overline{w}}{U} \right) + b_1\eta_t + b_2\delta + b_3\beta. \tag{7.27}$$

The elevator deflection required to produce a zero hinge moment is:

$$-\delta = \left( \frac{b_1}{b_2} \right) \left( \alpha - \frac{\overline{w}}{U} \right) + \left( \frac{b_1}{b_2} \right) \eta_t + \left( \frac{b_3}{b_2} \right) \beta. \tag{7.28}$$

Equation 7.28 shows that the elevator is *geared to the tab* via the gear ratio $-b_3/b_2$. If we insert the elevator angle from Equation 7.28 into Equation 7.10, we have a new pitching moment

$$C_M = C_{Mo} + (h - h_o)C_L - V_t\mathcal{H}, \tag{7.29}$$

with

$$\mathcal{H} = \left[ C_{L_{\alpha t}} - C_{L_\delta} \left( \frac{b_1}{b_2} \right) \right] \left( \alpha - \frac{\overline{w}}{U} \right) + \left[ C_{L_{\alpha t}} - C_{L_\delta} \left( \frac{b_1}{b_2} \right) \right] \eta_t + \left[ C_{L_\beta} - C_{L_\delta} \left( \frac{b_3}{b_2} \right) \right] \beta. \tag{7.30}$$

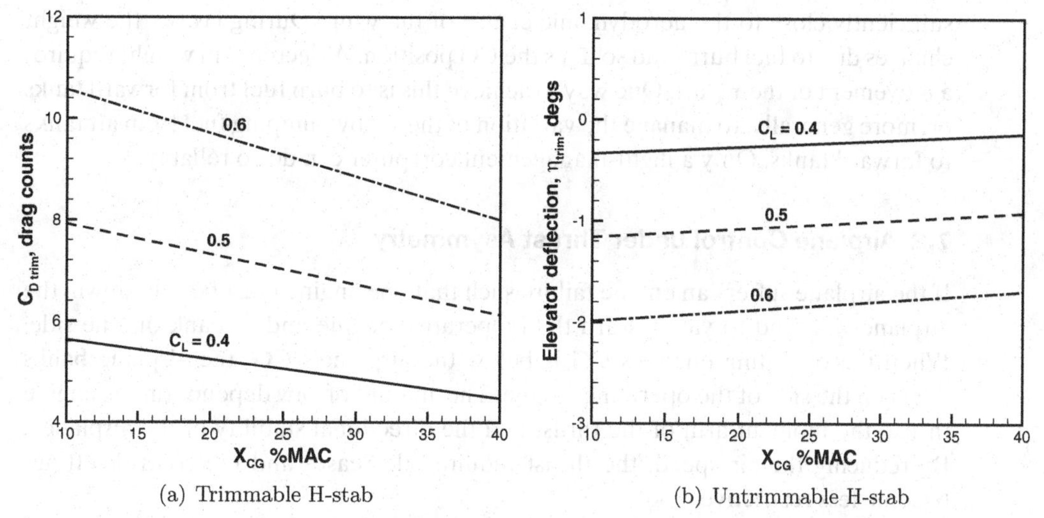


(a) Trimmable H-stab (b) Untrimmable H-stab

Figure 7.2. Longitudinal trim of a model Airbus A320-200; trim drag and elevator deflection at cruise configuration, $M = 0.78$, $h = 10{,}000$ m (~32,800 feet); standard day.

To simplify this equation, assume

$$a_1 = \left[C_{L_{\alpha t}} - C_{L_\delta}\left(\frac{b_1}{b_2}\right)\right], \qquad a_2 = \left[C_{L_\beta} - C_{L_\delta}\left(\frac{b_3}{b_2}\right)\right]. \tag{7.31}$$

Russell[2] proves that these coefficients are the lift-curve slopes of the tail-plane. Equation 7.29 is thus reduced to

$$C_M = C_{Mo} + (h - h_o)C_L - V_t\left[a_1\left(\alpha - \frac{\overline{w}}{U}\right) + a_1\eta_t + a_2\beta\right] = 0. \tag{7.32}$$

Equation 7.32 is the *stick-free pitching-moment* equation. The hinge moment is zero, but to trim the airplane, a tab angle must be found. This is done by solving the equation in terms of the tab deflection $\beta$:

$$\beta = -\frac{C_{Mo}}{a_2} - (h - h_o)\frac{C_L}{a_2} + V_t\left[\frac{a_1}{a_2}\left(\alpha - \frac{\overline{w}}{U}\right) + \frac{a_1}{a_2}\eta_t\right]. \tag{7.33}$$

The solution depends on the lift coefficient and on the position of the CG. It can be verified that, in general, there is only a restricted range of parameters that ensures longitudinal trim.

**NUMERICAL RESULTS FOR THE LONGITUDINAL TRIM.** We consider again the case of the model Airbus A320-200. The calculated trim drag is shown in Figure 7.2a. The condition of trimmable stabiliser has been used, which leads to $\delta_{trim}$ identically zero. For a fixed value of the $C_L$ (or weight), the trim drag decreases as the CG moves aft; for a given position of the CG, the trim drag increases with the $C_L$ (or weight).

Figure 7.2b shows the elevator's angle required for longitudinal trim at cruise conditions. In this case it was assumed that the horizontal stabiliser is not trimmable (which is not the case for this type of airplane). The calculation is done iteratively because the downwash from the wing (Equation 7.5) depends on the wing lift; the effective angle of attack of the tail-plane depends on the split between $L_w$ and $L_t$. To avoid using relatively large elevator angles, it is necessary to maintain the CG sufficiently close to the aerodynamic centre of the wing. During cruise, the weight changes due to fuel burn, and so does the CG position. A decrease in weight requires a movement of the $x_{cg}$ aft. One way to achieve this is to burn fuel from forward tanks or, more generally, to manage the variation of the $x_{cg}$ by pumping fuel from aft tanks to forward tanks. Only a flight-management computer can do so reliably.

## 7.2 Airplane Control under Thrust Asymmetry

If the airplane suffers an engine failure such that one engine must be shut down, the airplane will tend to yaw toward the inoperative engine and to bank on one side. When the operating engine's CG is below the airplane's CG, the airplane banks down on the side of the operating engine. The amount of yaw depends on the engine thrust, the moment arm of the thrust and the directional stability of the airplane[3]. By reducing the air speed, the thrust required decreases and the control settings become less restrictive.

An asymmetry situation may also arise in cases when the CG is off the vertical symmetry of the airplane, with all engines functioning. This situation is not unusual and includes cargo displacements, passenger location and fuel asymmetry. In fact, most commercial airplanes are certified with lateral CG limits or with a maximum load asymmetry. For example, the A300-600 was cleared to operate with a maximum fuel asymmetry of 2,000 kg.

If the airplane is a turboprop, the drag created by the inoperative propeller will contribute further to the yaw. With the airplane yawing toward the inoperative engine, the asymmetry of the inflow increases the drag of the "advancing" wing; at the same time, the lift decreases. The opposite effect takes place on the "retreating" wing. As a result, the airplane rolls and the advancing wing banks down. This effect can be compounded by a CG offset and by the drag of the propeller. Airplanes must be able to overcome this potential loss of control by combined use of directional stability, aileron and rudder deflections. In summary, a conventional airplane is subject to the following events:

- destabilising effect of the working engine
- destabilising effect of the CG offset (due to banking or otherwise)
- stabilising effect of the side force on the fuselage and vertical tail
- stabilising effect of the side force on the rudder

For a turboprop airplane, other factors that may have to be included are 1) the propeller's torque, which tends to roll the airplane in the opposite direction of the rotation; 2) the drag of the inoperative propeller (minimised by feathering the blades); 3) the asymmetric blade effect, arising when the axis of rotation of the propeller disk is not aligned with the inflow; and 4) the effect of the propeller's slipstream[4]. The first case may not be a concern if the torque-induced roll tends to lift the wing with the inoperative engine.

### 7.2.1 Dihedral Effect

A yaw/side-slip generates a rolling moment. Consider a wing with dihedral $\varphi$ operating with a small side-slip $\beta$. The wing has a higher angle of attack on the forward

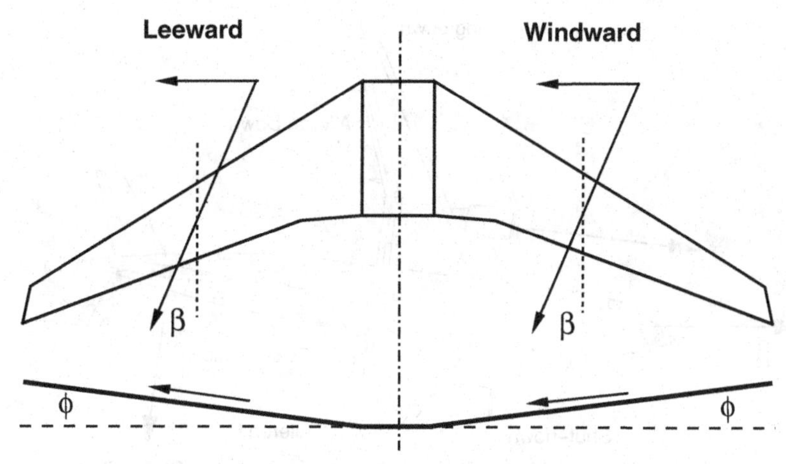


Figure 7.3. Sketch of wing flow with side-slip.

side than on the rearward side. If the local chord is $c$, the fluid particles travel inward or outward by an amount $c\varphi$. These fluid particles travel toward a lower point on the forward side and toward a higher point on the rearward side. The amount by which the air is displaced depends on the geometry of the wing, as shown in the sketch in Figure 7.3.

If the trailing edge is a straight line, then the spanwise offset does not depend on the side of the airplane and is equal to $c\varphi\beta$. Because the fluid particles have fluctuated by an amount $c\varphi\beta$ over the chord $c$, the change in inflow conditions is

$$\Delta\alpha = \pm\frac{c\varphi\beta}{c} = \pm\varphi\beta. \tag{7.34}$$

This condition tends to generate different values of the lift on the forward and rearward half-wing; hence a rolling moment. Furthermore, there will be a change in wing drag that would tend to reduce the yaw. (This situation arises, for example, when the airplane enters a downburst, as described in § 13.7.) The additional restoring moments would be

$$\Delta C_l = \Delta C_L\left(\frac{b_d}{b}\right) = -C_{L_\alpha}\Delta\alpha\left(\frac{b_d}{b}\right) = -2C_{L_\alpha}\phi\beta\left(\frac{b_d}{b}\right), \tag{7.35}$$

$$\Delta C_N = \Delta C_D\left(\frac{b_d}{b}\right) = -4k(C_{L_\alpha}\phi\beta)^2\left(\frac{b_d}{b}\right), \tag{7.36}$$

where $b_d$ is the moment arm of the wing drag (an unknown quantity). If these effects are not already included in the aerodynamic derivatives $C_{l_\beta}$ and $C_{N_\beta}$, they should be added to the equations for the rolling moment and the yaw moment, respectively.

**PROBLEM FORMULATION.** The following analysis deals with jet-powered airplanes. For reference, consider a twin-engine airplane. The control of the airplane is achieved by a combined rudder and aileron deflection, Figure 7.4. The fuselage in yaw has an effect on the derivative of the pitching moment with respect to the lift: $dC_N/dC_L$. There is little or no effect on the $C_L$ and the $C_{L_\alpha}$, as demonstrated in Ref.[5], and an increase in drag, due to flow separation on the lee side. Assume that the response of the airplane is linear; in other words, it can be calculated as a linear combination

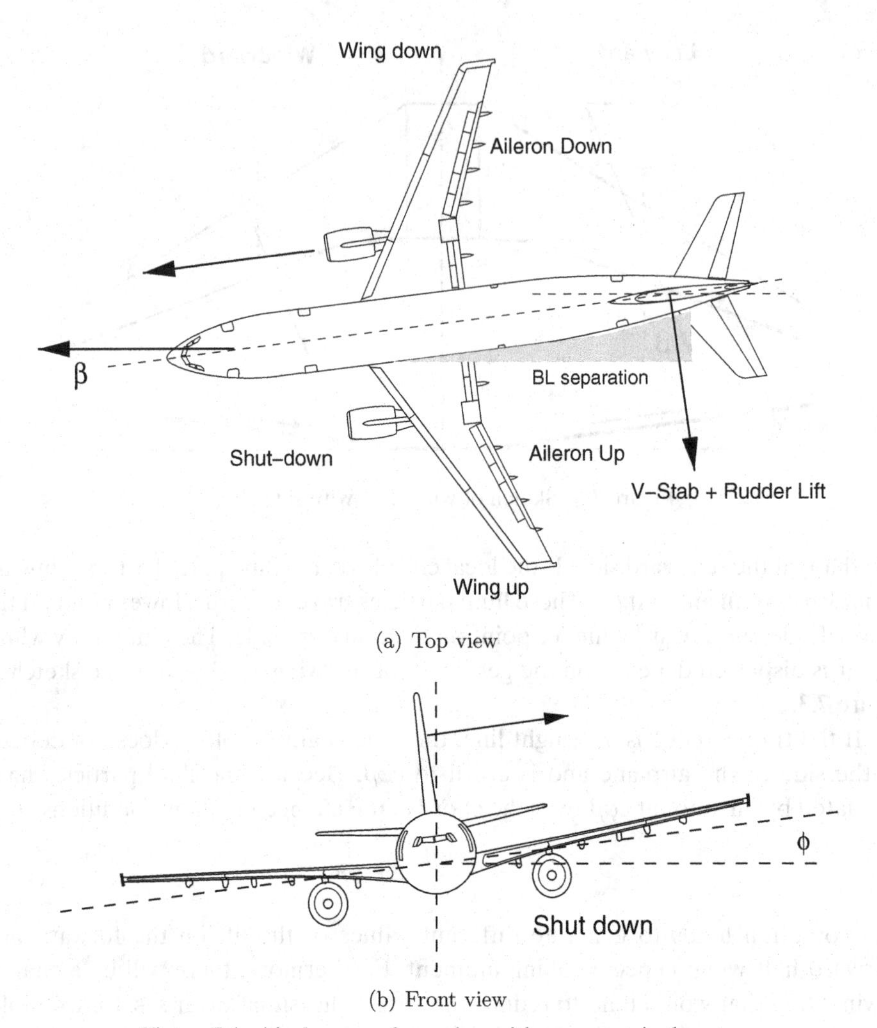


(a) Top view

(b) Front view

Figure 7.4. Airplane configuration with asymmetric thrust.

of the rudder and aileron deflections, for any reasonable value of the side-slip and bank angle. With these hypotheses, the static side-force coefficient is

$$C_Y = C_{Y_\beta}\beta + C_{Y_\xi}\xi + C_{Y_\theta}\theta + \frac{W \sin\phi}{\rho A U^2/2} = 0, \tag{7.37}$$

where

$$C_Y = \frac{Y}{\rho A U^2/2} \tag{7.38}$$

is the side-force coefficient, $\beta$ is the side-slip angle, $\phi$ is the roll angle, $\xi$ is the aileron deflection (assumed symmetrical) and $\theta$ is the rudder-deflection angle; $C_{Y_\beta}$, $C_{Y_\xi}$ and $C_{Y_\theta}$ are the derivatives of the side-force coefficient with respect to the yaw angle $\beta$, aileron angle $\xi$ and rudder angle $\theta$, respectively. The last term in Equation 7.37 is the side-force coefficient created by the offset of the CG. It can also be written as $C_L \sin\phi$. The roll-moment coefficient required to stabilise the flight is

$$C_l = C_{l_\beta}\beta + C_{l_\xi}\xi + C_{l_\theta}\theta = 0. \tag{7.39}$$

The linearised static yawing moment coefficient is

$$C_N = C_{N_\beta}\beta + C_{N_\xi}\xi + C_{N_\theta}\theta + \frac{\Delta T b_t}{\rho A U^2 b/2} + C_{D_e}\left(\frac{b_t}{b}\right) = 0, \tag{7.40}$$

where $\Delta T$ is the thrust asymmetry, $b_t$ is the moment arm of the thrust asymmetry and $C_{D_e}$ is the idle engine's drag coefficient. In the OEI case, assume that $\Delta T$ is the thrust of the functioning engine. Because Equations 7.37 to 7.40 reflect static conditions for the airplane, they represent a limit case.

Because the thrust depends on the air speed, the problem has to be solved iteratively. Furthermore, if the airplane has to maintain this air speed, the component of the thrust along the vector velocity must balance the resulting drag. The latter quantity cannot be derived from cruise conditions because the combination of yaw and deployment of control surfaces contributes to an increase in the aerodynamic resistance.

If we use the strip theory with the wing aerodynamics, the yaw does not cause any difference in profile drag. There can be differences in induced drag due to the difference in lift force. As far as the fuselage is concerned, there will be a roughly parabolic behaviour of the drag with respect to yaw angles for $-10 < \beta < 10$ degrees[6].

From Equation 7.37, the side-slip angle can be written in terms of the remaining parameters as follows:

$$\beta = -\frac{1}{C_{Y_\beta}}\left[C_{Y_\xi}\xi + C_{Y_\theta}\theta + \frac{2W\sin\phi}{\rho A U^2}\right]. \tag{7.41}$$

The side-slip angle increases with the weight and decreases with the increasing speed. However, the value of the aerodynamic coefficients is important in limiting the yaw effects. If the inoperative engine is at the starboard side, the airplane will veer to the starboard. The rudder will have to deflect in the opposite direction to create a restoring moment. The convention is that $\beta < 0$ and $\theta > 0$. The bank angle $\phi$ depends on the position of the thrust with respect to the centre of gravity of the airplane.

**NUMERICAL SOLUTION.** Given the airplane data (weight, wing area, flight altitude) and the stability coefficients, we need to calculate the minimum speed that guarantees full control (longitudinal, lateral and directional). This is the *minimum control air speed*, VMCA.

Calculation of the control derivatives is quite elaborate, but there are computer programs designed for this purpose, for example, the U.S. Air Force Stability and Control Digital Datcom[7], ESDU[8] and a number of more recent vortex lattice aerodynamic programs. These coefficients are sensitive to the airplane configuration and may change considerably from one airplane to another. Some stability derivatives for the Boeing B747-100, Lockheed C-5A, Grumman F-104A and other representative airplanes are available in Heffley and Jewell[9].

For a commercial jet liner, FAR §25.149 requires that control is assured in the worst possible scenario, for example, with one engine at the take-off thrust, the airplane at MTOW, the CG in the aft position and the flaps at take-off position and the landing gear retracted. A maximum bank angle of 5 degrees is allowed. The

Table 7.1. *Stability derivatives for calculation of airplane response to asymmetric thrust; model Boeing B747-100*

| | | | | | |
|---|---|---|---|---|---|
| $C_{l_\beta}$ | −0.2210 | $C_{N_\beta}$ | 0.1500 | $C_{Y_\beta}$ | −0.9600 |
| $C_{l_\xi}$ | 0.0460 | $C_{N_\xi}$ | 0.0064 | $C_{Y_\xi}$ | 0.0000 |
| $C_{l_\theta}$ | −0.0070 | $C_{N_\theta}$ | 0.1090 | $C_{Y_\theta}$ | −0.1750 |

unknown parameters in Equations 7.37–7.40 are the minimum control speed $U =$ VMCA, the side-slip angle $\beta$, the aileron deflection $\xi$ and the rudder deflection $\theta$. For a fixed rudder deflection $\theta = \theta_{max}$, the solution system is written as

$$\begin{bmatrix} C_{l_\beta} & C_{l_\xi} & 0 \\ C_{N_\beta} & C_{N_\xi} & 2\Delta T b_t/\rho A b \\ C_{Y_\beta} & C_{Y_\xi} & 2W\sin\phi/\rho A \end{bmatrix} \begin{bmatrix} \beta \\ \xi \\ 1/U^2 \end{bmatrix} = - \begin{bmatrix} C_{l_\theta}\theta \\ C_{N_\theta}\theta + C_{D_e} b_t/b \\ C_{Y_\theta}\theta \end{bmatrix}, \tag{7.42}$$

where the unknowns are $\beta$, $\xi$, $1/U^2$. For a fixed aileron deflection $\xi = \xi_{max}$, the solution system is

$$\begin{bmatrix} C_{l_\beta} & C_{l_\theta} & 0 \\ C_{N_\beta} & C_{N_\theta} & 2\Delta T b_t/\rho A b \\ C_{Y_\beta} & C_{Y_\theta} & 2W\sin\phi/\rho A \end{bmatrix} \begin{bmatrix} \beta \\ \theta \\ 1/U^2 \end{bmatrix} = - \begin{bmatrix} C_{l_\xi}\xi \\ C_{N_\xi}\xi + C_{D_e} b_t/b \\ C_{Y_\xi}\xi \end{bmatrix}. \tag{7.43}$$

The value of $C_{D_e}$ is calculated with the method described by Torenbeek[10]:

$$C_{D_e} = 0.0785 d^2 + \frac{2}{1 + 0.16 M^2} A_j \frac{U}{U_j}\left(1 - \frac{U}{U_j}\right), \tag{7.44}$$

where $U_j$ is the velocity of the jet at the nozzle, $d$ is the fan diameter and $A_j$ is the area of the nozzle. Typical values of $U/U_j$ are 0.92 (high by-pass turbofan), 0.42 (low by-pass jet engine) and 0.25 (straight turbojet and turboprop). Equation 7.44 shows that the $C_{D_e}$ depends on the Mach number.

To maintain the linearity of the system, we can calculate the $C_{D_e}$ a posteriori, or iteratively. For example, guess the VMCA, calculate the Mach number, calculate the $C_{D_e}$ from Equation 7.44 and solve the system again. The solution of the system requires the inversion of the matrix on the left-hand side. For certain values of the aerodynamic derivatives, the solving system has no solution; hence it would not be possible to control the airplane by any combination of aileron–rudder deflections. This situation must clearly be avoided. From a computational point of view, the non-existence of a solution (when a solution is known to exist) is an indication that the derivatives are most likely incorrect.

A calculation is shown for the Boeing B747-100, whose aerodynamic coefficients are known from Nelson[11] and summarised in Table 7.1. Figure 7.5 shows the side-slip angle of the airplane for varying aileron deflection at a fixed rudder deflection, at a fixed weight and air speed, as indicated in the graph. The side-slip is linear and weakly affected by the rudder deflection.

Figure 7.6 shows the calculated VMC as a function of the airplane's AUW. It shows the VMC limited by the maximum rudder and aileron deflection. Both speeds are above the stall speed, calculated with a $C_L = 1.8$. With an AUW of 180 tons or less, it is not possible to ensure full control of the airplane. In some cases the rudder

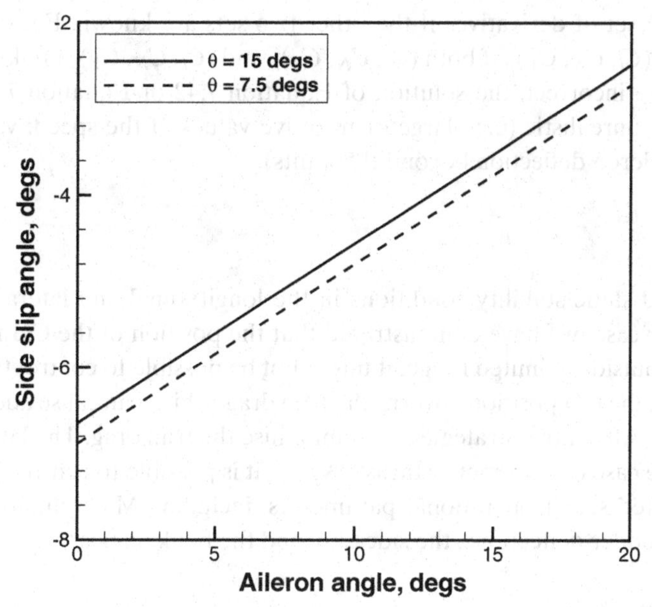


Figure 7.5. Side-slip angle versus aileron deflection at maximum rudder for the Boeing B747-100. The flight conditions are sea-level, standard day, MTOW, OEI, $M = 0.53$, $\phi = -5$ degs.

deflection has an opposite trend to the one shown, and a branched solution can be found.

**CALCULATION OF AERODYNAMIC DERIVATIVES.** The calculation of these quantities is not trivial. If we have FDR data, with the airplane parameters $U$, $W$, $\phi$, $T$, $\beta$, $\xi$, $\theta$ known quantities, then the system Equation 7.42 could be used to determine

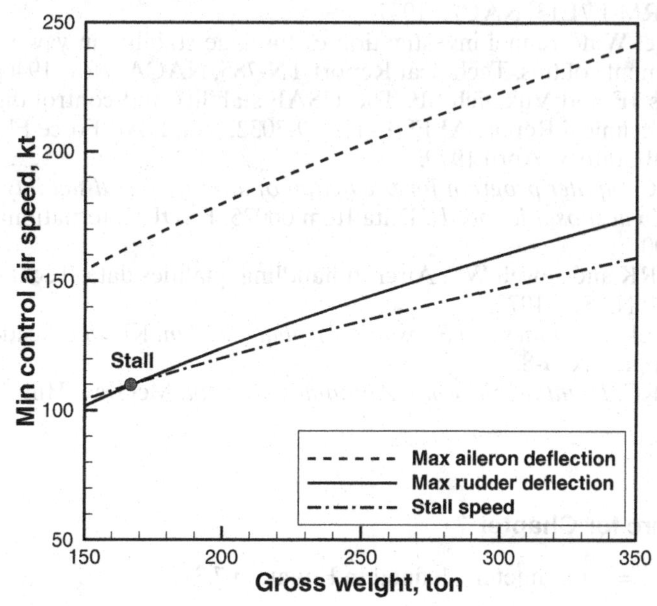


Figure 7.6. Calculated VMC versus gross weight for the Boeing B747-100. Bank angle $\phi = -5$ degrees.

indirectly one set of derivatives if the other two sets are known. For example, we can calculate $(C_l, C_N, C_Y)_\beta$ if both $(C_l, C_N, C_Y)_\theta$ and $(C_l, C_N, C_Y)_\xi$ are known. If the derivatives are incorrect, the solution of Equation 7.42 or Equation 7.43 is either impossible or unrealistic (e.g., large or negative values of the speed; values of the rudder and aileron deflection beyond the limits).

## Summary

We presented static stability conditions in the longitudinal and lateral directions. In the former case, we have demonstrated that the position of the CG is crucial. If the CG falls outside a limited range, it might not be possible to control the airplane. Furthermore, the CG position governs the trim drag, which can cause additional fuel consumption. All control strategies must minimise the trim drag. The lateral control deals with the case of asymmetric thrust. Again, it is possible to control the airplane within a limited set of operational parameters, including Mach number, altitude, aileron and rudder deflections, the side-slip, and the bank angles.

## Bibliography


[1] Filippone A. *Encyclopaedia of Aerospace Engineering*, volume 5, chapter 252. John Wiley, 2010.
[2] Russell JB. *Performance and Stability of Aircraft.* Butterworth-Heinemann, 2003.
[3] ESDU. *Loading on a Rigid Aeroplane in Steady Lateral Manoeuvres.* Data Item 01010. ESDU International, London, Oct. 2001.
[4] ESDU. *The Influence of Propeller Slipstream on Aircraft Rolling Moment due to Sideslip.* Data Item 06012. ESDU International, London, Aug. 2006.
[5] Salmi R and Conner W. Effects of a fuselage on the aerodynamic characteristics of a 42-degrees sweptback wing at Reynolds numbers up to 8,000,000. Technical Report RM-L7E13, NACA, 1947.
[6] HH Page. Wind tunnel investigation of fuselage stability in yaw with various arrangements of fins. Technical Report TN-785, NACA, Nov. 1940.
[7] Williams JE and Vukelich SP. The USAF stability and control digital DATCOM. Technical Report AFFDL-TR-79-3032, Vol. I, Air Force Flight Directorate Laboratory, April 1979.
[8] ESDU. *Computer program for prediction of aircraft lateral stability derivatives in sideslip at subsonic speeds.* Data Item 00025. ESDU International, London, Oct. 2000.
[9] Heffley RK and Jewell WF. Aircraft handling qualities data. Technical Report CR-2144, NASA, 1972.
[10] Torenbeek E. *Synthesis of Subsonic Airplane Design.* Kluwer Academic Publ., 1985. Appendix G-8.
[11] Nelson RC. *Flight Stability and Automatic Control.* McGraw-Hill, 2nd edition, 1998.


## Nomenclature for Chapter 7

| | | |
|---|---|---|
| $a_1, a_2$ | = | parameters defined in Equation 7.31 |
| $A$ | = | wing area |
| $A_t$ | = | tail-plane area |
| $A_j$ | = | nozzle area |

| | | |
|---|---|---|
| $\mathcal{A}R$ | = | wing-aspect ratio |
| $b$ | = | nominal wing span |
| $b_i$ | = | derivatives of the hinge moment, $i = 1, 2, 3$ |
| $b_d$ | = | moment arm of wing drag force |
| $b_t$ | = | thrust arm |
| $c$ | = | local wing chord |
| $\bar{\bar{c}}$ | = | mean aerodynamic chord |
| $C_l, C_N, C_Y$ | = | roll/pitch/yaw moment coefficients |
| $C_{l_\beta}, C_{N_\beta}, C_{Y_\beta}$ | = | roll/pitch/yaw moment derivatives with respect to side-slip |
| $C_{l_\theta}, C_{N_\theta}, C_{Y_\theta}$ | = | roll/pitch/yaw moment derivatives with respect to rudder deflection |
| $C_{l_\xi}, C_{N_\xi}, C_{Y_\xi}$ | = | roll/pitch/yaw moment derivatives with respect to aileron deflection |
| $C_{D_e}$ | = | drag of inoperative engine |
| $C_{D_{trim}}$ | = | trim-drag coefficient |
| $C_H$ | = | hinge moment, Equation 7.25 |
| $C_L$ | = | wing or total-lift coefficient |
| $C_{Lt}$ | = | horizontal-tail lift coefficient referred to h-tail area |
| $C_{Lt}^A$ | = | horizontal-tail lift coefficient referred to wing area |
| $C_{L_\alpha}$ | = | lift-curve slope |
| $C_{L_{\alpha t}}$ | = | lift-curve slope of the horizontal tail-plane |
| $C_{L_\eta}$ | = | elevator's lift-curve slope |
| $C_{L_\delta}$ | = | lift-curve slope of the elevator |
| $C_M$ | = | pitching-moment coefficient |
| $C_{M_o}$ | = | pitching-moment coefficient (reference, or zero-lift) |
| $C_{M_{lg}}$ | = | pitching-moment coefficient, landing gear |
| $C_{M_{ot}}$ | = | pitching-moment coefficient, tail-plane |
| $C_{M_{ow}}$ | = | pitching-moment coefficient, wing |
| $C_N$ | = | static yawing-moment coefficient, Equation 7.40 |
| $d$ | = | fan diameter; diameter |
| $D$ | = | aerodynamic drag |
| $f(h)$ | = | function of the CG position, defined in Equation 7.22 |
| $L$ | = | total lift |
| $\mathcal{H}$ | = | function defined in Equation 7.30 |
| $h$ | = | position of the aerodynamic centre, fraction of the MAC (non-dimensional) |
| $h_o$ | = | position of the centre of gravity, fraction of the MAC (non-dimensional) |
| $k$ | = | wing's induced-drag coefficient |
| $k_t$ | = | tail-plane's induced-drag coefficient |
| $L_t$ | = | tail-plane lift |
| $L_w$ | = | wing lift |
| $M$ | = | Mach number |
| $\mathcal{M}$ | = | pitching moment |
| $\mathcal{M}_{ot}$ | = | pitching moment of the horizontal tail |
| $\mathcal{M}_o$ | = | pitching moment of the wing |
| $T$ | = | net thrust |
| $q$ | = | dynamic pressure, $q = \rho U^2/2$ |

$U$ = air speed
$U_j$ = jet velocity
$V_t$ = tail-volume coefficient, Equation 7.11
$\overline{w}$ = average downwash
$W$ = airplane weight
$x_{ac}$ = distance of wing's aerodynamic centre from leading edge (longitudinal)
$x_w$ = distance of wing's aerodynamic centre from nose (longitudinal)
$x_t$ = distance between tail-plane's aerodynamic centre and nose (longitudinal)
$x_{wt}$ = distance between wing's and tail's aerodynamic centre (longitudinal)
$x_{cg}$ = distance of centre of gravity from nose (longitudinal)
$Y$ = side force
$z$ = vertical coordinate

### Greek Symbols

$\alpha$ = angle of attack
$\alpha_e$ = effective angle of attack
$\alpha_t$ = angle of attack, tail surface
$\beta$ = tab deflection angle; side-slip angle
$\delta$ = elevator angle
$\delta_o$ = elevator angle when CG coincides with AC, Equation 7.21
$\delta_{trim}$ = elevator angle that trims the airplane
$\eta$ = elevator's angle
$\eta_t$ = angle between the neutral lines of tail and wing
$\theta$ = rudder deflection
$\phi$ = bank angle
$\varphi$ = dihedral angle
$\rho$ = air density
$\xi$ = aileron deflection

### Subscripts/Superscripts

$[.]_{cg}$ = centre of gravity
$[.]_j$ = jat quantity
$[.]_o$ = reference conditions
$[.]_{max}$ = maximum value
$[.]_M$ = main landing gear
$[.]_N$ = nose landing gear
$[.]_{lg}$ = landing gear
$[.]_t$ = tail
$[.]_w$ = wing
$[.]_\beta$ = derivative with respect to side-slip angle
$[.]_\theta$ = derivative with respect to rudder deflection
$[.]_\xi$ = derivative with respect to aileron deflection

# 8 Flight Envelopes

## Overview

This chapter deals with flight envelopes in the speed-altitude space. We present various atmospheric models (§ 8.1), standard as well as non-standard. We give several operating air speed definitions (§ 8.2), as well as design speeds (§ 8.3) and the techniques required to measure them. For the steady-state level flight we derive two optimal conditions: minimum drag and minimum power (§ 8.4). We discuss the flight corridors at constant altitude and the ceiling performance airplanes (§ 8.5). We discuss the flight envelopes of subsonic transport aircraft and their limitations (§ 8.6), including the effects of the cabin pressure. We conclude the chapter with flight envelopes at supersonic Mach number (§ 8.7), including dashspeed, supersonic accelerations and propulsion limitations.

**KEY CONCEPTS:** International Standard Atmosphere, Atmosphere Models, Operating Speeds (EAS, CAS, TAS), Transition Altitude, Design Speeds, Optimum Level Speeds, Ceiling Performance, Cabin Pressure, Flight Envelopes, Supersonic Dash, Supersonic Acceleration.

## 8.1 The Atmosphere

Most performance calculations are done with a conventional atmosphere that has been declared standard (§ 8.1.1), although the almost totality of flights take place in far more complex atmospheres, with variation in the thermodynamic properties in both the vertical and horizontal directions. To overcome the limitations of the standard model, alternative atmosphere models are available to take into account extreme temperatures (very hot and very cold), as explained in § 8.1.2. The effect of air humidity is generally neglected in aerodynamics and engine performance, although it is important for other aspects, such as noise propagation.

### 8.1.1 International Standard Atmosphere

Nearly all of the basic calculations of aircraft performance are done in International Standard Atmosphere (ISA) conditions (*standard day*), whose parameters at sea level are given in Table 8.1.

Table 8.1. *Sea-level data of the International Standard Atmosphere*

| Parameter | Symbol | Sea-Level Value |
|---|---|---|
| Temperature | $\mathcal{T}_o$ | 15.15 °C |
| Pressure | $p_o$ | $1.01325 \cdot 10^5$ Pa |
| Density | $\rho_o$ | 1.225 kg/m$^3$ |
| Viscosity | $\mu_o$ | $1.7894 \cdot 10^{-5}$ N s/m$^2$ |
| Humidity | $\mathcal{H}$ | 0% |

Observations on the state of the atmosphere at sea level go back hundreds of years, but they have become systematic in the last century with aviation, rocket and satellite data and perfect gas theory. A number of *standard* versions exist: NACA's atmosphere[1], the ARDC[2], the U.S. standard[3] (1962, amended in 1976) and the ICAO standard[4]. These tables are basically equivalent to each other up to about 20 km (~65,600 feet), which is the altitude of interest for most airplanes.

The atmosphere is divided into a number of layers. The atmosphere below 11,000 m (36,089 feet) is called *Troposphere.* It is characterised by a decreasing temperature from sea level and reaches a standard value of −56.2 °C. The altitude of 11,000 m is called *tropopause.* The level above is called *Lower Stratosphere* and covers an altitude up to 20,000 m (65,627 feet), in which the temperature remains constant. The air density keeps decreasing with the increasing altitude. The upper limit of this layer includes most of the atmospheric flight vehicles powered by air-breathing engines. The *Middle Stratosphere* reaches up to an altitude $h = 32{,}000$ m (104,987 feet). In this layer the atmospheric temperature increases almost linearly from the value of −56.2 °C. The edge of space is generally considered to be at an altitude ~100 km. This altitude is somewhat arbitrary. At the edge of space, a vehicle flying on aerodynamic lift alone would have to maintain a velocity larger than the orbinal velocity. Therefore, the problem is to find the altitude at which these velocities are the same.

A number of functions are used to approximate the ICAO data. In the troposphere a linear expression is used:

$$\mathcal{T} = \mathcal{T}_o + \lambda h, \tag{8.1}$$

where $h$ is the altitude in metres and $\lambda = -0.0065$ K/m is the temperature lapse rate. Interestingly, such a temperature gradient causes the atmosphere to be stable to vertical movements. The limit gradient for vertical stability, corresponding to an adiabatic atmosphere, is about −9 K/km, as demonstrated by Prandtl and Tietjens[5]. The atmosphere is well described by the equation of the ideal gas:

$$\frac{p}{\rho} = \mathcal{R}\mathcal{T}, \tag{8.2}$$

where $\mathcal{R} = 287$ J/kg K is the gas constant (air). Equation 8.2 written at two different states leads to the equivalence

$$\frac{p}{p_o} = \frac{\rho}{\rho_o}\frac{\mathcal{T}}{\mathcal{T}_o}. \tag{8.3}$$

The relative density is called $\sigma$, the relative pressure is $\delta$ and the relative temperature is $\theta$. We call the altitude corresponding to a given air density *density altitude*. If, instead, we relate the altitude to the local air pressure, then we have a *pressure altitude*. To find the pressure-altitude and the density-altitude relationships, we use the buoyancy law for still air along with Equation 8.1, to find the rate of change of the pressure with altitude. The buoyancy law is

$$\frac{\partial p}{\partial h} = -\rho g. \tag{8.4}$$

If we insert the differential form of Equation 8.1, the buoyancy law becomes

$$\frac{\partial p}{\partial T} = -\frac{\rho g}{\lambda}. \tag{8.5}$$

Finally, use Equation 8.2 to eliminate the density, rearrange the equation and integrate from the sea-level altitude. The result is

$$\ln\left(\frac{p}{p_o}\right) = \frac{g}{\lambda \mathcal{R}} \ln\left(\frac{T}{T_o}\right), \tag{8.6}$$

$$\delta = \theta^{(g/\lambda \mathcal{R})}. \tag{8.7}$$

The value of the power coefficient is $g/\lambda\mathcal{R} = 5.25864$. If we insert Equation 8.1 in Equation 8.7, we have a *pressure-altitude* correlation:

$$\delta = \left(1 - \frac{0.0065}{T_o} h\right)^{5.25864} = (1 - 2.2558 \cdot 10^{-5} h)^{5.25864}, \tag{8.8}$$

with $h$ expressed in [m]. Equation 8.8 is in good agreement with the ICAO data. A more accurate expression is

$$\delta = (1 - 2.2558 \cdot 10^{-5} h)^{5.25588}. \tag{8.9}$$

In Equation 8.9 the altitude is related to the pressure ratio, therefore it can be read directly from the altimeter that is calibrated with the ISA reference value of $p_o$. To find a *density-altitude* correlation in the lower atmosphere, we use the equation of state; the relative pressure is taken from Equation 8.9. The result is

$$\sigma = \frac{(1 - 2.25577 \cdot 10^{-5} h)^{5.25588}}{1 - 0.0065\, h/T_o}. \tag{8.10}$$

Figure 8.1 shows the ratios of density, pressure, temperature and speed of sound from sea level to the altitude of 25,000 m. If $\gamma$ denotes the ratio between specific heats, the speed of sound in the atmosphere is calculated from

$$a = \sqrt{\gamma \mathcal{R} T}. \tag{8.11}$$

The non-dimensional form of Equation 8.3 is $\delta = \sigma\,\theta$. The quantities can be approximated by exponential functions in the troposphere. For a standard day, these functions are

$$\sigma \simeq \theta^{4.25}, \qquad \delta \simeq \theta^{5.25}, \qquad \delta\sqrt{\theta} \simeq \theta^{4.75}. \tag{8.12}$$

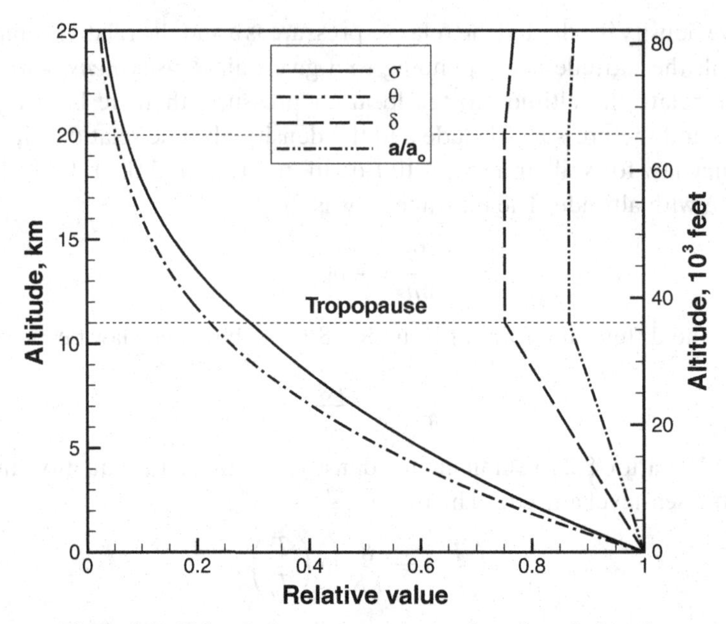

Figure 8.1. ISA relative parameters as a function of altitude.

The relative temperature is readily calculated from Equation 8.1. The inverse problem (calculation of the altitude $h$ corresponding to relative density $\sigma$) is more elaborate because it requires to solve a non-linear equation in implicit form. The solution can be found with a bisection method. For the bisection method to work, one has to choose two points at which the function has opposite values. It is safe to choose $\sigma_1 = 0.01$ and $\sigma_2 = 1$, to make sure that the method converges to a solution.

Finally, the value of the air viscosity from the air temperature can be found from Sutherland's law

$$\mu = \mu_o \left(\frac{T}{T_o}\right)^{3/2} \left(\frac{T_o + C}{T + C}\right), \tag{8.13}$$

where $C = 111\ K$ is the Sutherland constant for air; $\mu_o$ is the viscosity at the reference temperature (273.15 $K$). Note that the dynamic viscosity can be expressed as $kg/ms$ or N s/m$^2$. The two units are equivalent.

Other physical quantities that are sometimes used in performance calculations include the thermal conductivity. The conductivity of dry air is

$$\kappa_d \simeq 0.023807 + 7.1128 \cdot 10^{-5}(T - T_o) \quad \text{[W/mK]}. \tag{8.14}$$

The conductivity of moist air is

$$\kappa_a \simeq \kappa_d \left[1 - \left(1.17 - 1.02\frac{\kappa_v}{\kappa_d}\right)\right] \frac{n_v}{n_v + n_d} \quad \text{[W/mK]}, \tag{8.15}$$

where $n_v$ and $n_d$ are the number of moles in the vapour and dry phases, respectively.

### 8.1.2 Other Atmosphere Models

The ISA model is useful to compare aircraft performance over all the range of atmospheric altitudes. This is an idealised case that does not occur anywhere. Figure 8.2

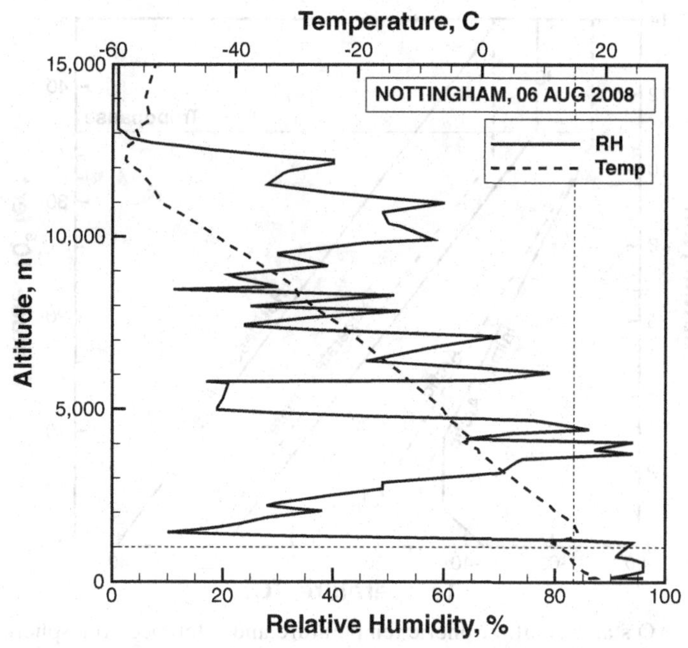


Figure 8.2. Temperature and relatively humidity versus altitude detected by radiosonde above central England.

shows a stratified atmosphere above central England in the dog-days of summer, as reported by a radiosonde probe. There is a temperature inversion just above 1,000 m. The distribution of humidity does not follow a recognised pattern. Clearly, these data cannot be associated to a standard atmosphere. Therefore, some alternative models are needed. The relative humidity is not used for flight-mechanics calculations but it important in the propagation of acoustic sources (§ 17.2).

We recognise that a realistic analysis of aircraft performance requires consideration of large deviations from the standard values to deal with extreme environmental conditions: winters in the northern hemisphere, very hot weather on the ground. In addition, airport altitude, humidity and precipitations, atmospheric winds, lateral gusts and global air circulation have strong influence on flight and safety. Rain and snow can be so heavy that take-off may have to be aborted. To simplify these matters, the U.S. Department of Defense defines four non-standard atmospheres, referred to as hot, cold, tropical and arctic (MIL-STD-210A). These profiles are shown in Figure 8.3.

Three important classes of weather-related flight problems are icing (§ 14.1), downbursts (§ 13.7) and atmospheric turbulence. Turbulence is a more familiar weather pattern to the frequent flyer. It includes cases of free air and convective air turbulence, atmospheric boundary layers and mountain ridge waves. No airplane is immune to the powerful gusts of the atmosphere. The aviation industry refers more often to *wind shear* rather than gust. Wind shear can occur in any direction, although it is classified as vertical and horizontal wind shear. A vertical wind shear causes turbulence that affects climb and descent; the amplitude of the wind speed variation can reach 30 kt per 1,000 feet (305 m). A horizontal wind shear can cause a shift from a head- to a tail-wind, and vice versa, with an amplitude up to 100 kt per

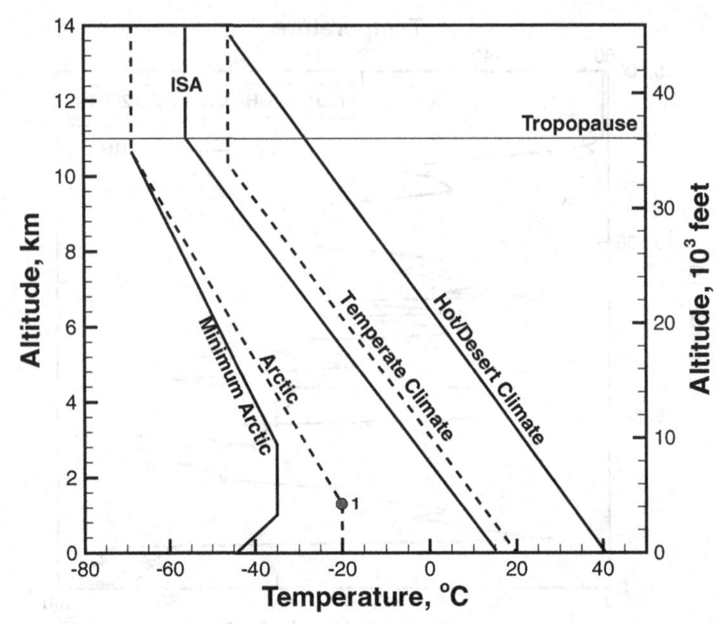


Figure 8.3. ICAO standard atmospheric temperature and reference atmospheres for flight-mechanics calculations.

n-mile. Such a situation can arise in a downburst. A detailed discussion of turbulence and flight is available in Etkin[6] and Houbolt[7]. An introduction to weather processes and climatic conditions was published by Barry and Chorley[8]. A comprehensive reference on aviation meteorology is published by HMSO[9].

Among the global circulation effects there is the jet stream. This is an atmospheric wind with an East-West prevalence, strongest around the troposphere, that affects trans-Atlantic flights between Europe and North America. These streams can be thousands of kilometers wide and affect the weather on a continental scale. Two typical cases occur. First, in some regions of the jet, air *converges* to the stream. As the jet accumulates air it becomes heavier and increases the air pressure at lower altitudes down to sea level. Another effect of compression is due to downward movement of the air, which in turns causes an increase in temperature and prevents the formation of clouds. This case is characterised by high pressure and clear skies.

The second case is when air *diverges* from the jet. The events are generally opposite than the case described herein. There is a loss of atmospheric pressure as the atmosphere above the jet becomes lighter. The reduction in pressure extends to sea level and causes updrafts and faster evaporation. Rising air cools down, and the humidity contained in the air freezes to form clouds. This event is also on a very large scale (thousands of kilometers wide).

Figure 8.4 shows a snapshot of the North Atlantic jet stream. The shaded areas show wind speeds in excess of 50 kt in the N-E direction. Particularly strong are the winds above the British Islands, that eventually turn South over France. The map refers to an altitude corresponding to a pressure of 300 mbar, or about 9,750 m (~32,000 feet). This is generally the altitude at which the winds are strongest. GFS

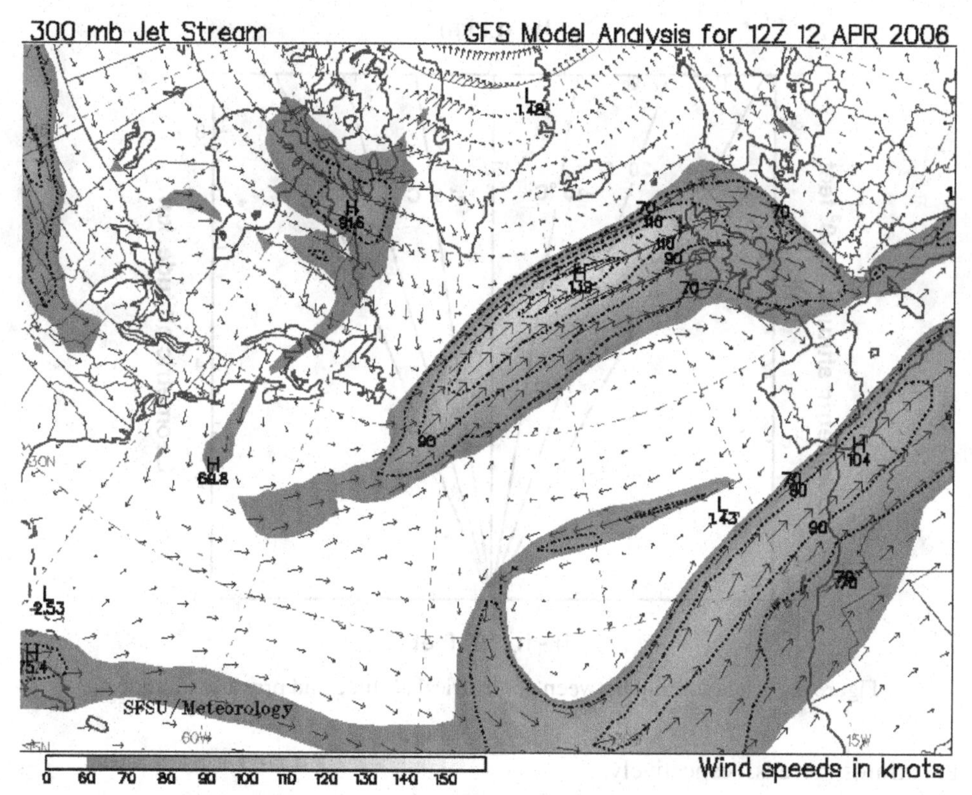


Figure 8.4. Jet stream over the North Atlantic [adapted from the U.S. National Center for Environmental Protection].

is the *Global Forecast System*; this is a computer simulation of the weather made available to the public over the Internet and other channels.

Overall, temperature variations in the atmosphere can be of the order of 80 degrees (−40 °C to +40 °C). It is not uncommon to encounter temperature inversions, that is, cooler air at the ground level and warmer air at low altitudes, contrary to the standard model. At the tropopause the temperature can fall below −70 °C in winter time at moderate to high latitudes.

Because the temperature does not appear explicitly in any of the performance equations, a useful relationship with the pressure and density is required. If the temperature has a constant deviation from the standard value, say a constant $\pm\Delta T$, the method of § 8.1.1 can still be used because the temperature gradient is the same (Equation 8.1). The only difference is that the symbol $T_o$ denotes the sea-level temperature, whatever that may be. A simple model for non-standard atmosphere consists in adding or subtracting a constant value from the standard temperature profile. Thus, the temperature distribution Equation 8.1 is replaced by

$$T = T_{ISA} + dT, \tag{8.16}$$

where $dT$ is the temperature shift; a cold day is obtained with $dT < 0$ and a hot day is $dT > 0$. The atmospheric pressure in the troposphere, at the tropopause and in

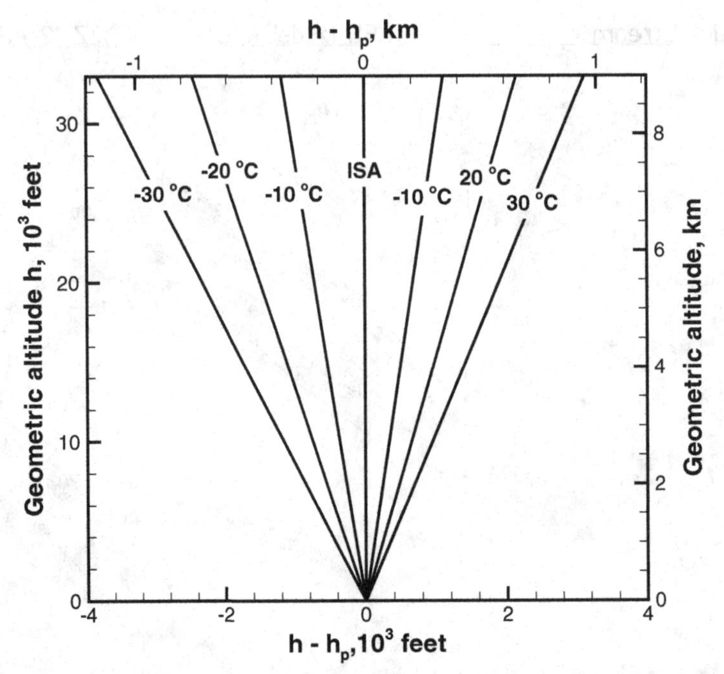


Figure 8.5. Relationship between geometrical altitude and pressure altitude.

the stratosphere is, respectively:

$$p = p_o \left( \frac{\mathcal{T} - \Delta\mathcal{T}}{\mathcal{T}_o} \right)^{-g/\lambda\mathcal{R}}, \tag{8.17}$$

$$p_t = p_o \left( \frac{\mathcal{T}_t - \Delta\mathcal{T}}{\mathcal{T}_o} \right)^{-g/\lambda\mathcal{R}}, \tag{8.18}$$

$$p = p_t \exp \left[ -\frac{g}{\mathcal{R}T_{ISA,t}} (h - h_t) \right]. \tag{8.19}$$

In Equations 8.18 and 8.19 the subscript "t" denotes conditions at the tropopause. These equations lead to a definition of two important concepts: the *pressure altitude* and the *density altitude*. The flight altitude is not measured directly; the pressure and the temperature are measured; thus, only a pressure altitude would be available from Equations 8.17 through 8.19.

In this model, the temperature depends only on the altitude. Therefore, it is possible to solve the pressure equation for a known pressure and unknown altitude. If the temperature is below the standard value, the pressure altitude is lower than the geopotential altitude; vice versa, if the temperature is above the standard value, the pressure altitude is higher than the geopotential altitude.

The correlation between pressure altitude and geopotential altitude is shown in Figure 8.5. This figure shows the difference between the geometrical altitude and the pressure altitude at increasing geometrical altitude for specified temperature shifts.

## 8.2 Operating Speeds

The *ground speed* $V_g$ is the aircraft speed measured with respect to a fixed point on the ground. The *air speed* $U$ is the aircraft speed relative to the air and accounts for the presence of atmospheric winds $\pm V_w$. If an atmospheric wind is aligned with the aircraft speed, the air speed is defined as

$$U = V_g \pm V_w, \tag{8.20}$$

where $V_w$ is the wind speed (negative for a *tail wind*, positive for a *head wind*). Measuring the air speed is of considerable importance. For example, an error in defect may cause the aircraft to run into a stall condition. At incompressible speeds, the air speed can be evaluated with the Bernoulli equation,

$$p + \frac{1}{2}\rho U^2 = p^*, \tag{8.21}$$

that gives the *true air speed* (TAS)

$$U = \text{TAS} = \sqrt{\frac{p^* - p}{2\rho}} = \sqrt{\frac{\Delta p}{2\rho_o}} \frac{1}{\sqrt{\sigma}}. \tag{8.22}$$

In Equation 8.22 $p$ is the free-stream atmospheric pressure and $p^*$ is the stagnation pressure and $\rho$ is the density; the subscript "o" still denotes sea-level standard conditions. An instrument that measures a difference in pressure between the free-stream conditions and the static conditions is the Pitot probe. With the Bernoulli equation the probe converts a pressure difference into air speed, as long as the air density is known. Therefore, the instrument would not work without additional readings. By taking a temperature reading and a static pressure reading at the same time, we can find the value of the density $\rho$ from the equation of ideal gases:

$$\rho = \frac{1}{\mathcal{R}} \left(\frac{p}{T}\right)_{static}. \tag{8.23}$$

If we refer to atmospheric conditions to sea level, we have an *equivalent air speed*:

$$\text{EAS} = \sqrt{\frac{\Delta p}{2\rho_o}}. \tag{8.24}$$

The relationship between TAS and EAS is derived from Equation 8.22:

$$\text{EAS} = \text{TAS}\sqrt{\sigma}. \tag{8.25}$$

An Air Speed Indicator (ASI) based on this principle is of no use in high-speed flight, where compressibility is important. From compressible aerodynamic theory we know that the ratio between the stagnation pressure $p^*$ and the static atmospheric pressure $p$ corresponding to an isentropic deceleration from a Mach number $M$ is

$$\left(\frac{p^*}{p}\right)^{\gamma-1/\gamma} = 1 + \frac{\gamma - 1}{2} M^2, \tag{8.26}$$

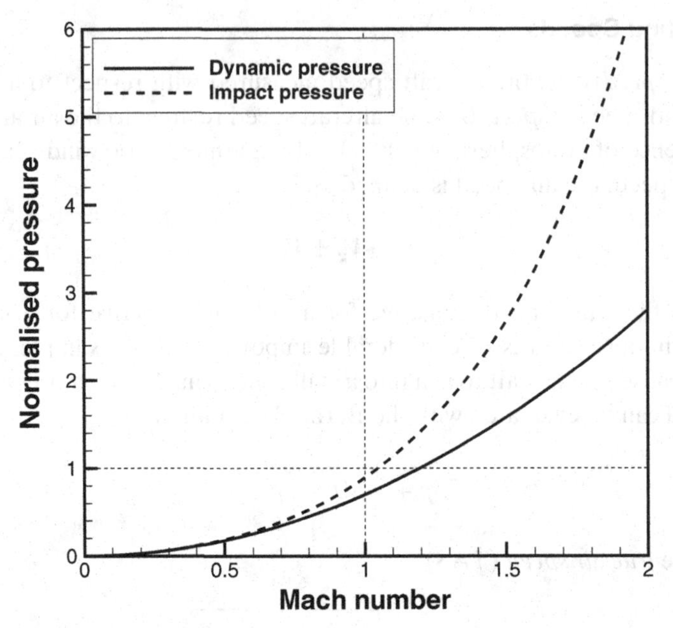


Figure 8.6. Impact pressure and dynamic pressure versus Mach number.

where $\gamma$ is the ratio between specific heats at constant pressure and constant volume. Solving for the Mach number, we find

$$M^2 = \frac{2}{\gamma - 1}\left[\left(\frac{p^*}{p}\right)^{\gamma-1/\gamma} - 1\right]. \tag{8.27}$$

The true air speed will be found from Equation 8.27, the definition of speed of sound (Equation 8.11) and the equation of ideal gases (Equation 8.2):

$$\text{TAS} = \sqrt{\frac{2\gamma}{\gamma - 1}\left(\frac{p}{\rho}\right)\left[\left(\frac{p^* - p}{p} + 1\right)^{\gamma-1/\gamma} - 1\right]}. \tag{8.28}$$

Solving Equation 8.26 for the stagnation pressure $p^*$, we find

$$p^* = p\left(1 + \frac{\gamma - 1}{2}M^2\right)^{\gamma/\gamma-1}, \tag{8.29}$$

$$p^* - p = p\left[\left(1 + \frac{\gamma - 1}{2}M^2\right)^{\gamma/\gamma-1} - 1\right]. \tag{8.30}$$

The pressure difference $p^* - p$ is the so-called *impact pressure*. At low Mach numbers, when the flow is practically incompressible, the impact pressure is equal to the dynamic pressure, as shown in Figure 8.6. An instrument that measures the impact pressure, the local static pressure and the local temperature provides the TAS (or KTAS, if given in knots).

Sometimes it is more useful to calibrate the speed to sea-level conditions. The local pressure $p$ and density $\rho$ are replaced by the sea-level values. The resulting speed is called *Calibrated Air Speed*; it is abbreviated CAS or KCAS (CAS in knots):

$$\mathrm{CAS} = \sqrt{\frac{2\gamma}{\gamma - 1}\left(\frac{p_o}{\rho_o}\right)\left[\left(\frac{p^* - p_o}{p_o} + 1\right)^{\gamma-1/\gamma} - 1\right]}. \tag{8.31}$$

It can be verified that the relationship between TAS and CAS is

$$\mathrm{CAS} \simeq \mathrm{TAS}\sqrt{\frac{T_o}{T}} = \frac{\mathrm{TAS}}{\sqrt{\theta}} = \mathrm{TAS}\sqrt{\frac{\delta}{\sigma}}. \tag{8.32}$$

In other words, the CAS is equal to the true air speed at standard sea level. If the local static pressure can be measured and the local density is replaced with the sea level $\rho_o$, then we find again the equivalent air speed. EAS is the TAS corrected for changes in atmospheric density,

$$\mathrm{EAS} = \sqrt{\frac{2\gamma}{\gamma - 1}\left(\frac{p}{\rho_o}\right)\left[\left(\frac{p^* - p}{p} + 1\right)^{\gamma-1/\gamma} - 1\right]}, \tag{8.33}$$

a result equivalent to the case of incompressible flow. For a given TAS, the EAS increases with the increasing flight altitude. The *Indicated Air Speed* (IAS) is the aircraft speed indicated by the instrument that can be affected by errors (position, time and pressure lag): IAS = CAS + *error*. Here, we assume that the error is negligible, so that we can call CAS = IAS.

When the aircraft flies at supersonic speed, the instrument is unable to *sense* the actual free-stream conditions. A normal shock establishes ahead of the instrument. Two events must be taken into account: 1) a normal shock wave ahead of the probe that produces a subsonic Mach number, and 2) an isentropic deceleration from a subsonic Mach number to stagnation pressure $p^*$ in the probe. This pressure is related to the static pressure $p$ by the Raleigh equation that is found in most textbooks dealing with high-speed aerodynamics[10]:

$$\frac{p^*}{p} = \left[\frac{(\gamma + 1)^2 M^2}{4\gamma M^2 - 2(\gamma - 1)}\right]^{\gamma/\gamma-1}\left[\frac{1 - \gamma + 2\gamma M^2}{\gamma + 1}\right]. \tag{8.34}$$

To find the Mach number, this equation must be solved for $M$.

Figure 8.7 shows the combination of CAS-TAS-altitude and flight Mach number. An operation point is uniquely determined by two parameters, such as Mach-TAS combination, Mach-altitude, or CAS-altitude and so on (for example, point A). An aircraft climbing at constant CAS would have an increasing TAS if also the Mach number increases; if the Mach number is kept constant, then the TAS decreases as the aircraft climbs – a result that can be inferred also from Equation 8.32.

**TRANSITION ALTITUDE.** Finally, we introduce the transition (or cross-over) altitude, which is defined as the geopotential altitude at which the CAS and Mach number represent the same TAS. This altitude is defined by the equation:

$$h_{trans} = \frac{10^3}{\lambda}\left[(T_o + dT)(1 - \theta_{trans})\right]. \tag{8.35}$$

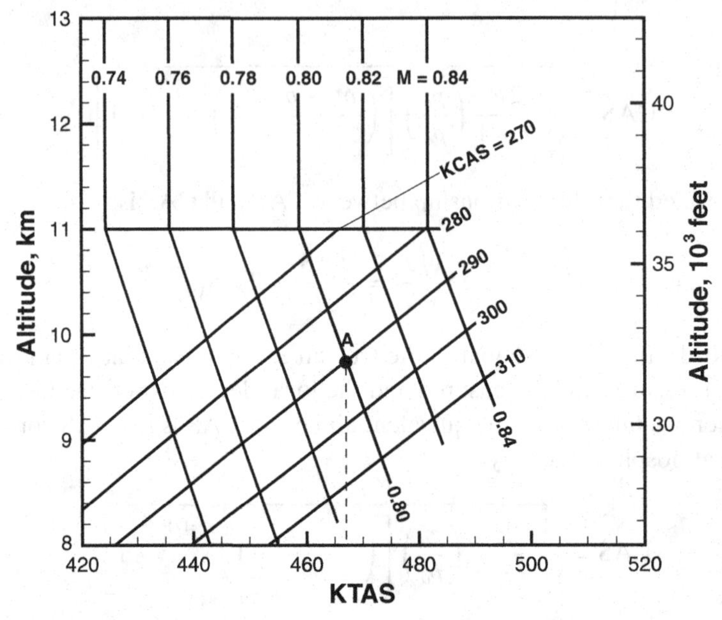


Figure 8.7. CAS-TAS-Mach-Altitude chart.

In Equation 8.35 the transition altitude is in [m]; $\theta_{trans}$ is the relative temperature at the transition altitude

$$\theta_{trans} = \delta_{trans}^{-\lambda\mathcal{R}/g}. \tag{8.36}$$

The relative pressure at the transition altitude is

$$\delta_{trans} = \frac{[1 + \kappa M_{CAS}^2]^{\gamma/\gamma-1}}{[1 + \kappa M^2]^{\gamma/\gamma-1}} \tag{8.37}$$

with

$$\kappa = \frac{\gamma - 1}{2}, \qquad M_{CAS} = \frac{V_{CAS}}{a_o}. \tag{8.38}$$

A plot of the transition altitude over a range of cruise Mach numbers is shown in Figure 8.8. The transition altitude varies greatly with the air speed. For example, at 250 KCAS and $M = 0.85$ the transition altitude is ~11,400 m (~37,400 feet).

## 8.3 Design Speeds

Aircraft speeds have been increasing over the years. These speeds reached a point of diminishing returns around 1970, when most of the commercial long-range airplanes were powered by gas-turbine engines. Cruise speeds converged toward an average $M = 0.78$ to 0.82. Some advances in aerodynamics have allowed a slight increase in the cruise speed for the latest generation of commercial jet aircraft, which is $M \simeq 0.85$. A number of other design speeds and Mach numbers are required, as listed here:

1. *Structural design Mach number*, $M_C$: This Mach number is specified by the aircraft manufacturer. For operational flexibility it is assumed that $M_C$ is equal

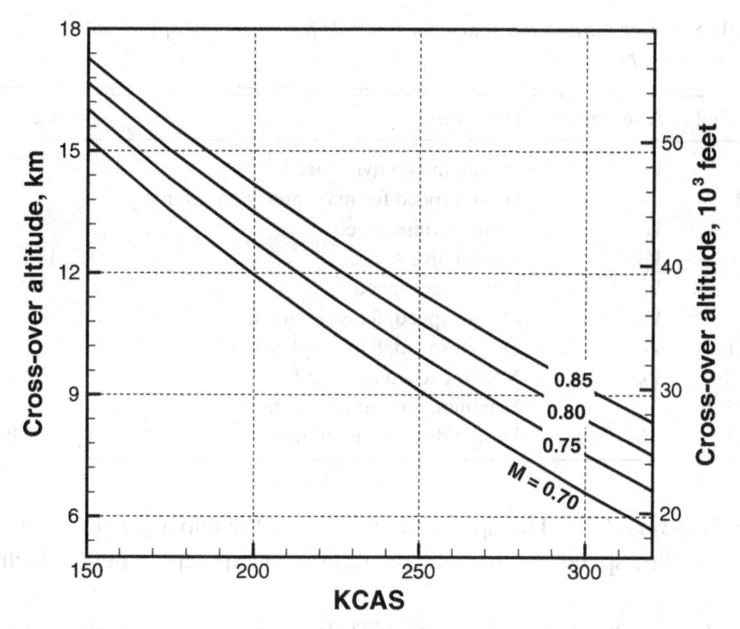


Figure 8.8. Transition altitude versus KCAS at selected Mach numbers.

to the maximum operating Mach number, MMO. For a transonic transport airplane, MMO $\simeq M_{cruise} + 0.04$. The corresponding speed is called *structural design speed*, $V_C$.

2. *Design dive Mach number*, $M_D$: This Mach number is larger than the maximum operating Mach number, with $M_D \simeq 1.07$ MMO, or $M_D \simeq 1.25 M_C$. Specific conditions are set forth in FAR § 25.335, which includes clauses for verification and demonstration. The corresponding speed (*design dive speed*) is $V_D \simeq 1.15 V_C$, based on a 20-second dive on a 7.5-degree glide angle. The value of $M_D$ must be selected such that design cruise Mach number does not exceed $0.8 M_D$.
3. The *Never-to-exceed speed* VNE is determined by the structural limits of the aircraft. For a given aircraft and given gross weight, this speed depends on the flight altitude.
4. *Design manoeuvre speed*, $V_A$: This is the minimum equivalent air speed at which the airplane can produce lift with flaps retracted, at the design load factor and design weight. If $n = 1$, we have the stall speed; otherwise, $L = nW$. According to FAR § 25.335, $V_A \geq V_{S1}\sqrt{n}$ ($V_{S1}$ is the stall speed with flaps retracted).
5. *Design speed for maximum gust intensity*, $V_B$: This speed must satisfy the following condition:

$$V_B \geq V_{S1} \left[ 1 + \frac{a_1 K_g U_{ref} V_C}{W_L} \right]^{1/2}, \tag{8.39}$$

where $U_{ref}$ is the reference equivalent air speed of the gust; $W_L = W/A$ is the average wing loading, $a_1$ is the normal-force curve slope and $K_g$ is a factor, as defined by FAR § 25.335. The parameters in Equation 8.39 must be given in coherent units (the FAR documents give imperial units).

Table 8.2. *Recognised international symbols for design air speeds and Mach numbers*

| Symbol | Alternative | Definition | FAR 25 |
|---|---|---|---|
| VA | $V_A$ | Design manoeuvre speed | |
| VB | $V_B$ | Design speed for maximum gust intensity | |
| VC | $V_C$ | Design cruise speed | |
| VD | $V_D$ | Design dive speed | $1.15V_C$ |
| VF | $V_F$ | Design flap speed | |
| VSo | $V_{So}$ | Design speed, flaps retracted | |
| VS1 | $V_{S1}$ | Design speed, flaps deployed | |
| VNE | $V_{NE}$ | Not-to-exceed air speed | |
| MMO | $M_C$ | Maximum operating Mach number | |
| MD | $M_D$ | Design dive Mach number | 1.07 MMO |

6. *Design flap speed*, $V_F$: This speed is a function of the flap angle, the flight altitude and the stalling speed. It is in fact a restriction on flap deployment, which is given as:
   $V_F > 1.6V_{S1}$ with take-off flaps and MTOW.
   $V_F > 1.8V_{S1}$ with approach flaps and MLW.
   $V_F > 1.8V_{S0}$ with landing flaps and MLW, where $V_S$ indicates the flaps-up condition.

A summary of these speeds is given in Table 8.2. Other operational and design speeds are defined in Chapters 9 and 13.

## 8.4 Optimum Level Flight Speeds

We derive some closed-form solutions for steady-state flight that correspond to minimum drag and minimum power. It is possible to demonstrate that a level speed can be maintained by the aircraft if

$$\frac{T}{W} \geq 2\sqrt{C_{D_o}k}. \tag{8.40}$$

The limiting condition occurs at the absolute ceiling of the aircraft. The minimum speed coincides with the maximum speed. At the absolute ceiling there is only one possible speed

$$U = \frac{T}{\rho A C_{D_o}}. \tag{8.41}$$

The danger of this situation is that the aircraft cannot accelerate (because of insufficient thrust), it must not decelerate (because it could enter a stall condition) and it must not perform a turn (because it could stall, as a consequence of increasing drag). Theoretically, the only way out of this situation is a descent at constant air speed.

We now calculate the speeds relative to minimum drag and minimum power of a generic aircraft, jet- or propeller-driven. Consider an aircraft whose drag equation is parabolic. The drag force on the airplane is

$$D = \frac{1}{2}\rho A U^2 (C_{D_o} + kC_L^2). \tag{8.42}$$

The speed corresponding to minimum drag is found from the condition that

$$D = \left(\frac{D}{L}\right) L = \left(\frac{D}{L}\right) W \tag{8.43}$$

is at a minimum. This implies that the glide ratio $C_L/C_D$ is at a maximum. The derivative of Equation 8.43 is to be calculated with respect to the $C_L$. The condition $\partial(C_L/C_D)/\partial C_L = 0$ yields

$$C_L = \sqrt{\frac{C_{D_o}}{k}}. \tag{8.44}$$

This is the $C_L$ corresponding to minimum drag. The corresponding speed will be

$$U_{md} = \sqrt{\frac{2}{\rho}\frac{W}{A}}\left(\frac{k}{C_{D_o}}\right)^{1/4}. \tag{8.45}$$

An alternative solution is to consider $C_L \sim U^2$, that is, a minimum with respect to $C_L$ will be a minimum with respect to the aircraft speed (and vice versa). The result will be the same, although more elaborate. The conclusion of Equation 8.45 is that at a given altitude, the speed of minimum drag increases with the wing loading and with the profile-drag coefficient; it decreases with the increasing lift-induced factor. All other parameters being constant, $U_{md}$ increases with the flight altitude. The speed corresponding to minimum engine power for the same aircraft is

$$P = DU = \frac{C_D}{C_L^{3/2}}\sqrt{\frac{2W^3}{\rho A}}. \tag{8.46}$$

At a given altitude the terms under square root are constant; therefore the *speed of minimum engine power is the speed that minimises the factor* $C_D/C_L^{3/2}$. The condition of minimum power is found from

$$\frac{\partial}{\partial U}\left(\frac{C_{D_o} + kC_L^2}{C_L^{3/2}}\right) = 0. \tag{8.47}$$

After working out the derivative in Equation 8.47, we find

$$3C_{D_o}c_1^{-3/2}U^4 - kc_1^{1/2} = 0, \tag{8.48}$$

with $c_1 = 2W/\rho A$. The solution of the latter equation is the speed of minimum power

$$U_{mp} = \sqrt{\frac{2}{\rho}\frac{W}{A}}\left(\frac{k}{3C_{D_o}}\right)^{1/4}. \tag{8.49}$$

This speed corresponds to a lift coefficient

$$C_L = \sqrt{3\frac{C_{D_o}}{k}}. \tag{8.50}$$

The relationship between the $C_L$ of minimum drag and minimum power is simply $\sqrt{3}$, whilst the ratio between corresponding speeds is

$$\frac{U_{md}}{U_{mp}} = \sqrt[4]{3} \qquad U_{md} \sim 1.32U_{mp}. \tag{8.51}$$

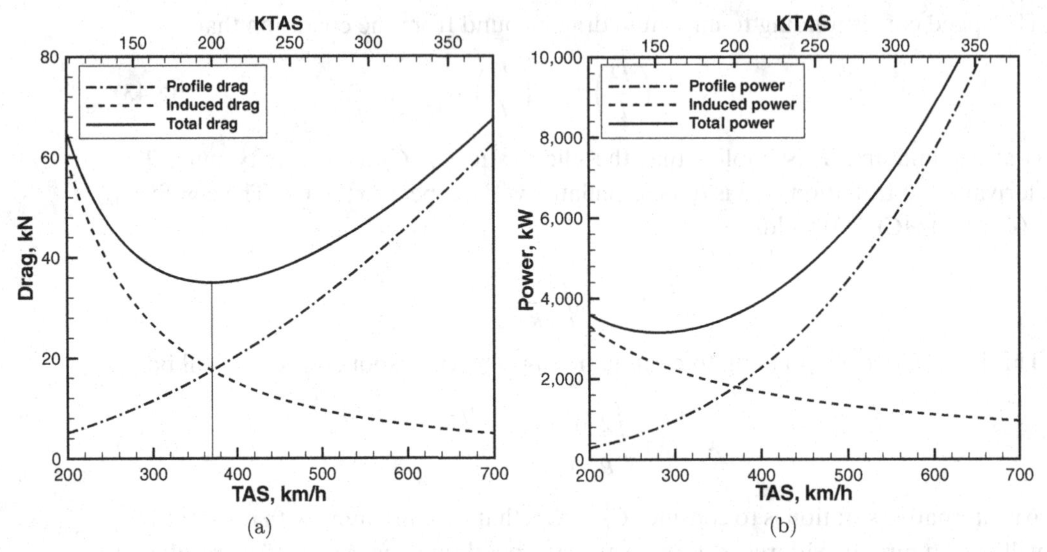


Figure 8.9. Drag and power characteristics of generic subsonic airplane.

This proves that the speed of minimum drag is about 32% higher than the speed of minimum engine power. Both optimal velocities change with altitude like $1/\sqrt{\sigma}$.

The variation of drag and power with the aircraft speed is shown in Figure 8.9 for a generic subsonic jet aircraft. The induced component decreases with the speed; the profile drag grows as $U^2$; the corresponding power grows as $U^3$. The sum beween the two contributions has a minimum at an intermediate value of the air speed.

## 8.5 Ceiling Performance

The *absolute ceiling* is the maximum altitude at which an aircraft can keep a steady level flight. Above this altitude, the engine power or thrust is not enough to overcome the aircraft drag. The aircraft can zoom past the absolute ceiling by exchanging part of its kinetic energy for potential energy (i.e., altitude). This can be done with an inertial climb, even in absence of sufficient engine power.

The ceiling altitude depends on the type of aircraft. It is about 6,000 to 7,000 m (~19,700 to 23,000 feet) for turboprop aircraft; 10,000 to 12,000 m (~32,800 to 39,400 feet) for commercial jet aviation; and it increases to 18 or 19 km (above 60,000 feet) for high-performance military aircraft. It is seldom above this altitude. The Lockheed SR-71 had a cruise altitude of 27,000 m (~88,500 feet) and an estimated ceiling of 30,700 m (~100,000 feet).

### 8.5.1 Pressure Effects on Human Body

The flight altitudes described could not be achieved without cabin pressurisation. In fact, pressurisation is required at altitudes above 3,000 m (~9,840 feet). The first airliner to be fully equipped with a pressurised cabin was the Lockheed Constellation in the 1940s.

At about 10,000 feet (~3,050 m), pilots without an artificial oxygen supply begin to suffer from hypoxia: their brains begin to show poor response and they speak gibberish. Up to about 34,000 feet (~10,360 m), the sea-level equivalent amount of oxygen can be supplied by increasing the percentage of oxygen in the air. At about 40,000 feet (~12,200 m), all of the oxygen must be supplied artificially. At about 50,000 feet (~15,200 m), gases trapped in the body expand, swelling intestines and rupturing lung tissues. In 1959, U.S. pilot William Rankin had to bail out of his Vought F8U Crusader aircraft at 50,000 feet (~15,200 m) and survived the fall thanks to his oxygen mask. However, his abdomen stretched and bloated from the expanding gas, air exploded through his ears, and he bled from his eyes, nose and mouth.

### 8.5.2 Cabin Pressurisation

The cabin pressure is monitored by an automatic control system. The system allows air to escape from the cabin by controlling an outflow valve. As the airplane changes altitude, the valve is repositioned so that pressure changes are within the comfort limits. Maintaining a sea-level pressure is not possible or even economical. A typical cabin pressure is equivalent to about 2,000 m (~6,560 feet). The relevant regulations for cabin-pressure altitude are given by FAR, Part 25, § 841. A loss of cabin pressure causes a rapid loss of consciousness. For this reason, oxygen masks are fitted as part of the essential safety kit on passenger airplanes. Even then, the airplane is required to descend rapidly.

The Gulfstream 550 (Appendix A) has an automatic *emergency descent procedure* that, after a 90-degree left turn, can take the aircraft from 40,000 feet (~12,190 m) down to 15,000 feet (~4,570 m) at the maximum descent rate (about 5,000 feet/min; 25.4 m/s). In some cases, for flight above 41,000 feet (~12,500 m), the FAA regulations also require that one pilot wear an oxygen mask to ensure that a loss of cabin pressure does not hinder their physical capabilities. A typical cabin pressure as a function of the flight altitude is shown in Figure 8.10.

Airplanes such as the Boeing B737 have two air conditioning systems. If one system does not work, the aircraft can still fly but it has to maintain a cruise altitude below 25,000 feet (7,620 m, $\delta = 0.3711$). At this altitude, it is possible to breathe only with difficulty. In case of failure of the air conditioning system, the aircraft must descend to 14,000 feet (~4,270 m, $\delta = 0.5875$), for which about 4 minutes are required. In these emergency situations, oxygen masks provide air supply to the passengers and the crew.

## 8.6 Flight Envelopes

The set of all flight corridors from sea level to the absolute ceiling defines the *flight envelope*. In other words, the flight envelope is the closed area in the $M - h$ diagram that includes all operating conditions *for a particular aircraft configuration at a given weight*. The flight envelope depends on a large number of factors, including weight, aerodynamics, propulsion system, structural dynamics, cabin-pressure limits and atmospheric conditions.

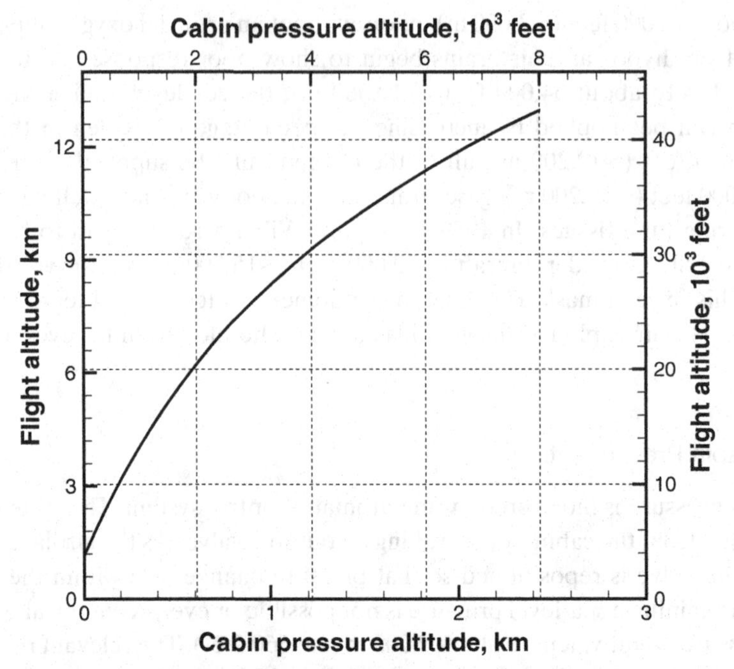


Figure 8.10. Typical cabin pressure on a modern airliner.

Figure 8.11 shows a generic flight envelope for a commercial subsonic jet. The graph shows lines of constant CAS and constant Mach numbers. The flight is limited at high speed by the maximum operating speed (VMO) or the maximum operating Mach number, $M_{mo}$. In general, the aircraft will cruise at a speed slightly lower than this, at the recommended cruise speed (indicated by a thick dashed line).

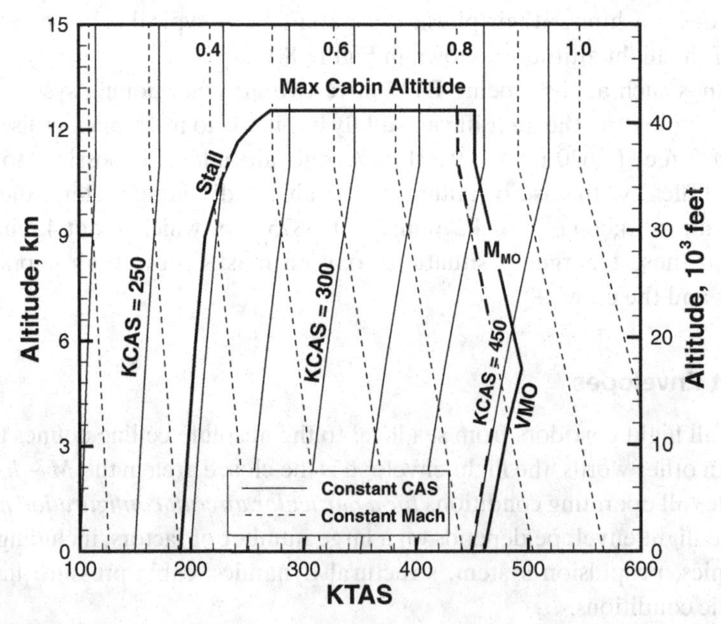


Figure 8.11. Generic flight envelope of commercial subsonic jet aircraft.

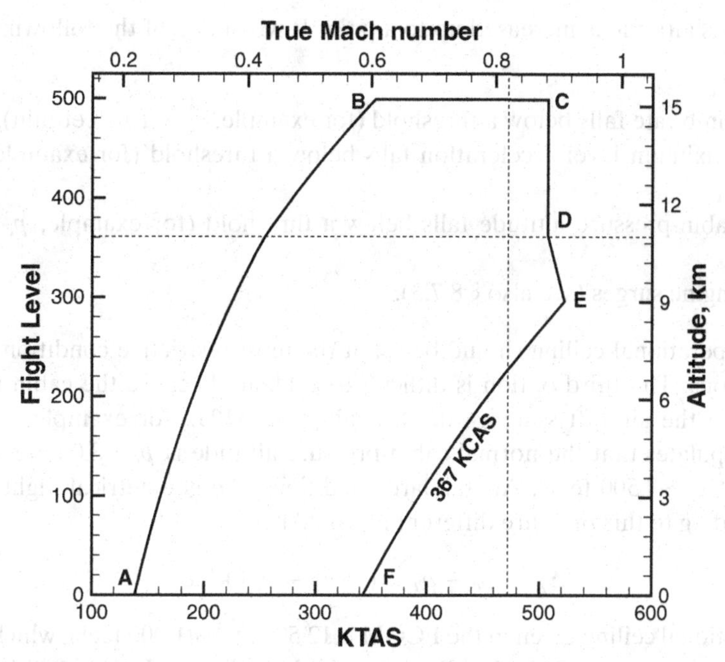


Figure 8.12. Flight envelope of the Airbus A320-200-CFM; standard day, no wind; $m = 58{,}800$ kg (calculated).

### 8.6.1 Calculation of Flight Envelopes

The subject of this section is the determination of the flight envelope of a commercial or transport aircraft. This envelope consists of three main segments:

1. *Stall Line*: This is the low-speed segment A-B from sea level (A) to the operational ceiling B (to be determined), in clean configuration. At low altitudes, there can be an extension of the envelope thanks to the deployment of the high-lift surfaces (Figure 8.12).
2. *Operational Ceiling*: This is the corridor B-C between the stall speed (or Mach number) B and the maximum speed (or MMO).
3. *High-Speed Line*: This is the segment C-F from the operational ceiling (C) to sea level (F), in clean configuration. This segment can be split into two sub-segments if the operational ceiling is above the tropopause.
   - The high-altitude segment C-D-E is a constant-Mach line; point D is at the tropopause.
   - The low-altitude segment E-F is constant-CAS line.

For the calculation of the stall line, the reference stall speed is

$$V_S = K_s \left( \frac{2W}{\rho A C_{L_{max}}} \right)^{1/2}, \tag{8.52}$$

where $K_s$ is the stall margin and $C_{L_{max}}$ is the maximum (or usable) lift coefficient of the airplane in clean configuration. The stall speed depends on the flight altitude; this altitude is increased up to a point B when one of the following events occurs:

1. The climb rate falls below a threshold (for example, $v_c < 100$ feet/min).
2. The maximum level acceleration falls below a threshold (for example, $a/g < 0.2$).
3. The cabin-pressure altitude falls below a threshold (for example, $p_c < 8{,}000$ feet).
4. The engine surges (see also § 8.7.3).

The operational ceiling should be set at the most restrictive conditions among these options. The third option is difficult to evaluate because the cabin pressure depends on the aircraft systems, in particular the APU. For example, the A320 FCOM stipulates that the normal cabin-pressure altitude is $p_c \sim 8{,}000$ feet, with a warning at $p_c \sim 9{,}500$ feet*. At standard conditions, the geometrical flight altitude corresponding to this pressure differential would be

$$\Delta p = p_c - p_a \simeq (35.65 - p_a) \text{ kPa}. \tag{8.53}$$

The operational ceiling given in the FCOM is 12,500 m (~41,000 feet), which corresponds to a pressure $p_a \simeq 17.934$ kPa at standard conditions. If the 8,000-feet cabin pressure is maintained, the cabin-to-exterior difference would be $\Delta p \simeq 17.71$ kPa, which is about one third of the limit pressure differential for the fuselage shell.

The fourth option requires a full engine simulation, with an analysis of the compressor map. The first and second options are manoeuvre requirements (excess power and excess thrust, § 10.2.1). For the purpose of the analysis carried out in this context, we use the first two options. Therefore, we define the operational ceiling as

$$z_{op} = \min \left\{ h(v_c^*), h(a^*) \right\}, \tag{8.54}$$

where $h(v_c^*)$ denotes the flight altitude corresponding to a $v_c < v_c^*$; $h(a^*)$ denotes the flight altitude corresponding to an acceleration $a/g < 0.2$. At the operational ceiling $z_{op}$, we need to establish the maximum speed of the airplane, C. We define this point as the most restrictive among the following options:

1. The maximum level acceleration falls below a threshold (for example, $a/g < 0.2$).
2. The engine intake suffers a buzz (see also § 8.7.3).

The first option is to allow for some excess power to manoeuvre away from the operational ceiling. Once this point is determined, we calculate the constant-Mach limit of the envelope at high altitude, C-E. The limit flight Mach number is the MMO. The lowest altitude E can be determined again as a minimum manoeuvre limit. At the point E, we calculate the KCAS. At altitudes below E (segment E-F), the limit of the envelope is determined by a constant KCAS, as previously calculated. The segment E-F is sometimes split into two sub-segments, corresponding to $\text{KCAS}_1$ at

* For this airplane, the maximum differential pressure (cabin-to-outside) is $\Delta p = 8.6$ psi (~59.3 kPa); this limit is never reached.

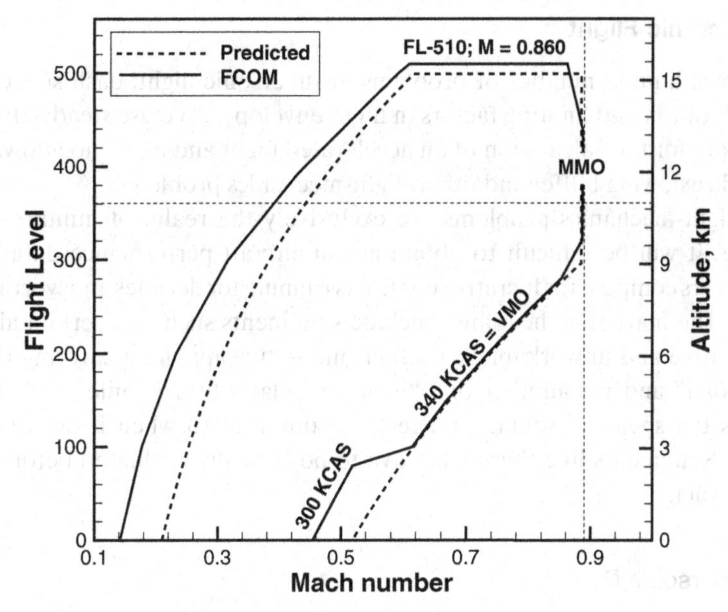


Figure 8.13. Flight envelope of the Gulfstream G550; standard day, no wind; $m = 35{,}500$ kg.

altitudes above 3,000 feet and $KCAS_2$ at altitudes below 3,000 feet. Because the manoeuvre limits can be arbitrary, further information may have to be gathered.

## 8.6.2 Case Study: Flight Envelopes of the A320 and G550

First, we consider the Airbus A320-200 with CFM engines. Figure 8.12 shows the flight envelope of this airplane. The operational points A through F are indicated in the graph. Without a limit on the cabin pressure, the operational ceiling is estimated at 13,900 m (~41,600 feet), well above the limit of 12,500 m given in the FCOM. Once it has reached the point C, the airplane moves along the MMO line down to a flight altitude 26,000 feet (E). Below this point we estimated a limit line at 338 KCAS down to sea level (F).

The second case refers to the Gulfstream G550. The present calculations are compared with data available in the FCOM (Figure 8.13). We predict an operational ceiling at 15,290 m (49,900 feet), just short of the value of 51,000 feet quoted by the manufacturer. The stall line is not well predicted. However, if the FCOM is correct, it would mean that the aircraft is capable of manoeuvring at low altitude with ~100 KTAS, which seems extraordinary. We estimate that we need at least 135 KTAS to fly at sea level.

On the high-speed segment, the upper portion of the FCOM indicates that the aircraft must not operate above $M = 0.86$ at altitudes above 43,500 feet (see also Figure A.1 in Appendix A). No such limit is found on the basis of excess thrust alone. Therefore, we must conclude that this slightly reduced Mach number is related to intake buzz and other engine issues.

The limit speed 340 KCAS is well predicted as a result of a good prediction of the limit point E. However, the lower portion of the high-speed segment in the FCOM indicates that the airplane must not exceed 300 KCAS below 9,000 feet.

## 8.7 Supersonic Flight

We present a limited number of problems in supersonic flight: dash speed, supersonic acceleration and limiting factors in flight envelopes. We use steady-state aerodynamic data for the calculation of an accelerated flight and make no allowance for transient thrust, wing buffet and other flight-mechanics problems.

The flight-mechanics problems are exclusively the realm of military aircraft. In practice, it will be difficult to obtain actual aircraft performance data because manufacturers compete for lucrative contracts running for decades and worth billions of dollars. The marketing headlines include statements such as a certain aircraft is "uncompetitive and unworkable", another one is "uncapable of carrying the same weapons load" and yet another one "would regularly fly 700 miles at 40,000 feet at just-less the speed of sound, and feels no different to when it does not carry anything". Statements like these will have to be critically evaluated before signing such a contract.

### 8.7.1 Supersonic Dash

We begin with an aircraft whose engine thrust and transonic drag rise are given in tabulated format from flight testing. We show how the solution of this problem leads to a root finding of non-linear algebraic equation. The problem may lead to non-physical and non-unique solutions. The conditions that must be satisfied by the aircraft in level flight are:

$$T = D, \qquad L = W. \tag{8.55}$$

The solution is given by the angle of attack that provides the maximum Mach number compatible with Equation 8.55. From the equilibrium in the vertical direction we find the angle of attack and the lift coefficient, respectively:

$$\alpha = \alpha_o + \frac{2W}{\rho A a^2} \frac{1}{C_{L_\alpha} M^2}, \tag{8.56}$$

$$C_L = C_{L_\alpha}(\alpha - \alpha_o). \tag{8.57}$$

We will consider $\alpha_o \simeq 0$, but the solution procedure is unaffected by $\alpha_o$. The equilibrium in the horizontal direction is

$$T = \frac{1}{2}\rho A a^2 (C_{D_o} + \eta C_{L_\alpha}\alpha^2) M^2. \tag{8.58}$$

By further simplification, we have

$$T = c_1 C_{D_o} M^2 + c_2 \frac{\eta}{C_{L_\alpha}} \frac{1}{M^2} \tag{8.59}$$

with

$$c_1 = \frac{1}{2}\rho A a^2, \qquad c_2 = \frac{2W^2}{\rho A a^2}. \tag{8.60}$$

In Equation 8.59 the aerodynamic coefficients are also a function of the Mach number. The solutions are

$$M^2 = \frac{T^2}{\rho A a^2} \pm \frac{1}{\rho A a^2}\sqrt{T^2 - 4W^2\frac{\eta C_{D_o}}{C_{L_\alpha}}}. \tag{8.61}$$

Both solutions are positive. Although both transonic and supersonic speed are possible, Equation 8.61 does not give an indication as to how the aircraft can reach these speeds. It is possible to demonstrate that Equation 8.61 has either no physical solutions or a single solution or multiple solutions, depending on flight altitude and gross weight.

### 8.7.2 Supersonic Acceleration

The next problem is to calculate the acceleration of the supersonic jet fighter from a cruise Mach number (say, $M = 0.8$) to supersonic speed. There are different ways to achieve this. One is an acceleration at constant altitude; another one is an acceleration at constant angle of attack and constant attitude. Finally, the aircraft can accelerate to phenomenal speeds by doing a zoom-dive. Several flight-mechanics studies exist in the technical literature to address acceleration problems[11;12], including transient and steady-state solutions. In this context, we will solve only the relatively simple case of acceleration at constant altitude. Any other profile requires a zoom-climb or a zoom-dive and is discussed in Chapter 10.

#### Acceleration at Constant Altitude

We now proceed to the calculation of such accelerations. The equation of motion in the flight direction is

$$m\frac{\partial U}{\partial t} = T - D. \tag{8.62}$$

If we use the definition of speed of sound and rearrange the equation, we find

$$\frac{\partial M}{\partial t} = \frac{1}{a}\frac{T}{m} - \frac{1}{2}\frac{\rho a A}{m}\left(C_{D_o} + \eta C_{L_\alpha}\alpha^2\right)M^2. \tag{8.63}$$

The angle of attack is established from the equilibrium condition in the vertical direction, $L = W$, Equation 8.56. As the aircraft accelerates in the horizontal direction, it must decrease its lift coefficient in order to keep level flight. Solution of the problem requires integration of the ordinary differential equation. A typical simulation is shown in Figure 8.14. All other conditions being the same, the weight is important in setting the limits on maximum acceleration and supersonic dash. At $m = 10{,}000$ kg, the drag nearly equals the available thrust; the aircraft is capable of going past the $M = 1.25$ limit to reach an $M \simeq 2.1$. With a higher weight, the aircraft would not be capable of performing this acceleration. A look at the function $(\partial U/\partial t)/g$ shows relatively low accelerations – compared with accelerations in high-speed turns and pull-ups. The acceleration is about $0.35g$ at the start and goes down to nearly zero at the transonic point T. It then grows again, thanks to the reduction in aerodynamic drag; thus, the airplane can accelerate to its maximum speed (for this configuration and altitude), as indicated by point M.

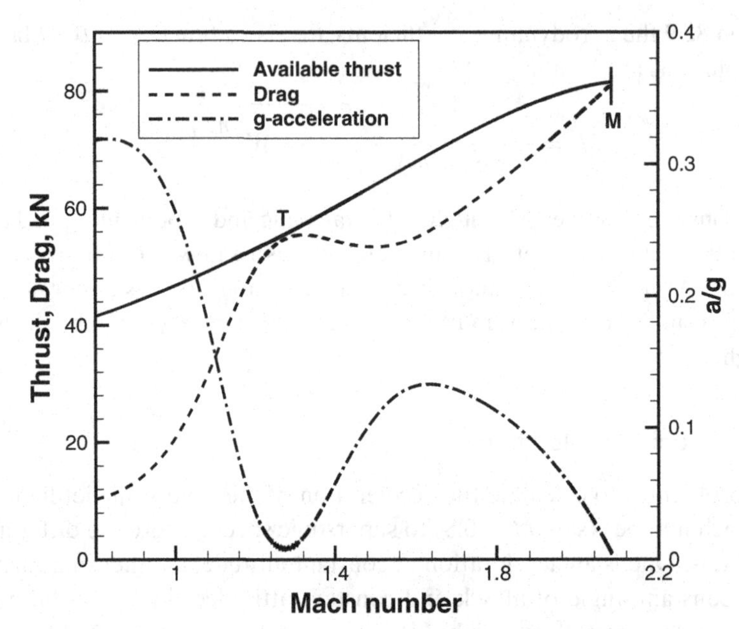


Figure 8.14. Acceleration of supersonic aircraft model, constant flight altitude $h = 8{,}000$ m (~26,250 feet); $m = 10{,}000$ kg; standard day, no wind, clean configuration.

By comparison, Concorde's best acceleration took the aircraft from $M = 1$ to $M = 2$ in about 7 minutes. After-burning was used in the range $M = 0.96$ to 1.70. In the military arena, the Vought F8U-3 with after-burning thrust was capable of accelerating from $M = 0.98$ to $M = 2.2$ in 3 minutes and 54 s at 11,500 m (~37,700 feet). The F-15 Strike Eagle is capable of accelerating from $M = 1.1$ to $M = 1.8$ in 56 s at about 32,000 feet (~9,750 m).

### 8.7.3 Supersonic Flight Envelopes

Figure 8.15, adapted from Abercrombie[13], shows the flight envelope of a high-performance supersonic jet aircraft. The outer envelope is the limit of all operation points from a statistical analysis. The graph shows a series of points that represent operational flight conditions, including acceleration, deceleration, climb, descent and cruise. The straight lines join extreme points of the acceleration or deceleration envelope. The shaded area indicates the normal envelope. The maximum speed is supersonic at all altitudes.

The envelopes discussed take into account only the basic aerodynamic and propulsion characteristics but not other factors, such as structural limits of the aircraft, aero-thermodynamic heating and other forcing at the extreme flight conditions. Some of the limitations are described in this section and schematically shown in Figure 8.16. There are at least four phenomena that need consideration: intake buzz, wing flutter, skin temperature limits and blade stall.

The *intake buzz* is the interaction between the oblique shock wave at the ramp of the engine inlet and the ramp boundary layer (Seddon & Goldsmith[14]). It creates oscillating conditions inside the intake that assume the shape of forced vibrations.

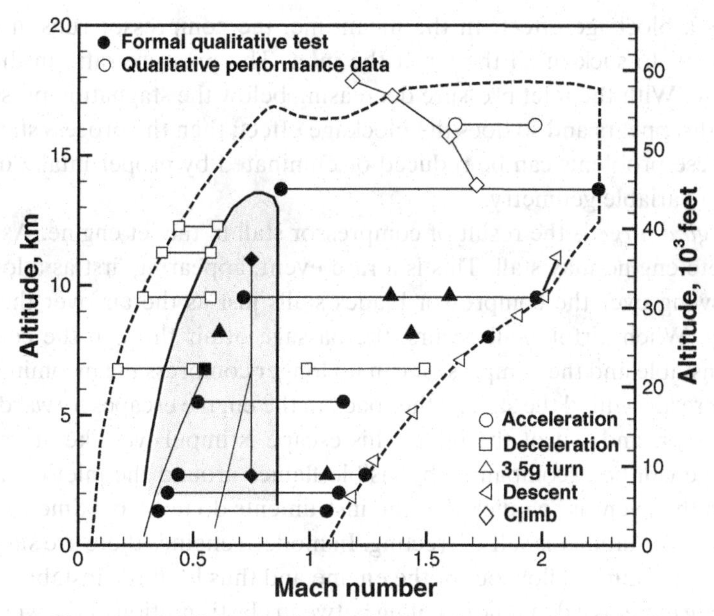


Figure 8.15. Flight envelope of high-performance jet aircraft.

A typical frequency is about 10 $Hz$. If a shock occurs, it will be formed by two waves: an oblique shock from the external compression flow and a normal shock. These two shocks meet at a point and expand outward, as a λ-shock. When intake buzz starts, the shock intersection point is inside the cowl lip and a shear flow is established.

For a given Mach number, if this shear flow is internal to the diffuser, it creates flow separation that reduces the mass captured by the inlet because the separated

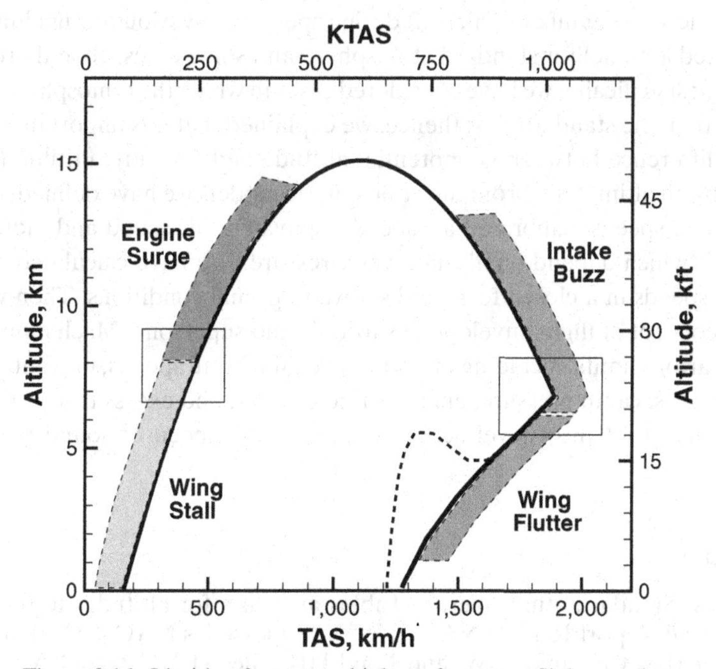


Figure 8.16. Limiting factors on high-performance flight envelope.

flow causes a blockage effect. In the meantime, the compressor runs at constant *rpm* and tends to suck in all the air at the inlet. The pressure ratio in the engine will decrease. With the inlet pressure decreasing below the stagnation pressure, the shear flow disappears and so does the blockage effect; then the process starts again. Some of these problems can be reduced or eliminated by proper intake design, in particular, a variable geometry.

The *engine surge* is the result of compressor stall in the jet engine. As a result, the complete engine may stall. This is a rare event, appearing first as a loud bang. The air flowing over the compressor blades stalls just as the air over the wing of an airplane. When airfoil stall occurs, the passage of air through the compressor becomes unstable and the compressor can no longer compress the incoming air. The high-pressure air behind the stall farther back in the engine escapes forward through the compressor, and out of the inlet. This escape is impulsive, like an explosion. Engine surge can be accompanied by visible flames around the inlet or in the tail pipe. Often the event is so quick that the instruments do not have time to respond. Generally, the instability is self-correcting. In modern engines there are surge valves that pump the disturbed flow out of the engine and thus limit the instability.

The *wing flutter* is a dynamic coupling between elastic motion of the wing and the unsteady aerodynamic loading. This dynamic response is at first stable but increases with the Mach number. It depends on the geometry of the wing (aspect-ratio, sweep angle), its stiffness, its moment of inertia and other parameters. This response (frequency and damping) may depend on the acceleration rate of the aircraft. The response can be quantified by the flutter number.

## Summary

We have reviewed a number of aircraft design speeds and various design limitations. We presented a model for standard atmosphere and some cases when there is deviation from it; specifically, we have considered cases in which the atmosphere is colder or warmer than the standard day; thence, we explained that it is important to understand the difference between geopotential altitude and pressure altitude (which is measured by an altimeter). From the atmospheric models we have defined a number of operational speeds: calibrated air speed, equivalent air speed and indicated air speed, all of which depend on altitude and pressure. We have calculated optimum level flight speeds in a closed form and showed optimal conditions. Then we introduced the concept of flight envelope at subsonic and supersonic Mach numbers and discussed various limits, including operational ceiling, stall speeds, maximum speeds and their causes: cabin pressure, engine surge or buzz, and excess thrust. We finally addressed the problem of level acceleration past the speed of sound (supersonic dash.)

## Bibliography

[1] NACA. Standard atmosphere – Tables and data for altitudes to 65,800 feet. Technical Report R-1235, NACA, 1955. (Supersedes NACA TN-3182).
[2] Minzer RA, Champion SW, and Pond HL. The ARDC Model Atmosphere. Technical Report 115, Air Force Surveys in Geophysics, 1959.

[3] Anon. U.S. Standard Atmosphere. Technical report, U.S. Government Printing Office, Washington, DC, 1962.
[4] Anon. Manual of the ICAO Standard Atmosphere, extended to 80 kilometres (262,500 feet). Technical report, ICAO, 1993, 3rd edition.
[5] Prandtl L and Tietjens OG. *Fundamentals of Hydro- and Aeromechanics*. Dover, 1957. Chapter 2.
[6] Etkin BE. Turbulent wind and its effect on flight. *J. Aircraft*, 18(5):327–345, May 1981.
[7] Houbolt JC. Atmospheric turbulence. *AIAA J.*, 11(4):421–437, April 1973.
[8] Barry RG and Chorley RJ. *Atmosphere, Weather and Climate*. Routledge, London, 8th edition, 2003.
[9] Anonymous. *Handbook of Aviation Meteorology*. HMSO, London, 3rd edition, 1994.
[10] Kuethe AM and Chow CY. *Foundations of Aerodynamics*. John Wiley, 5th edition, 1997.
[11] Miele A. *Flight Mechanics. Vol. I: Theory of Flight Paths*. Addison-Wesley, 1962.
[12] Bilimoria KD and Cliff EM. Singular trajectories in airplane cruise-dash optimization. *J. Guidance, Control and Dynamics*, 12(3):303–310, May 1989.
[13] Abercrombie JM. Flight test verification of F-15 performance predictions. In *Performance Prediction Methods*, CP-242. AGARD, 1978.
[14] Seddon J and Goldsmith EL. *Intake Aerodynamics*. Blackwell Science, 1999.

## Nomenclature for Chapter 8

Symbols defined in Table 8.2 are not repeated here.

| | | |
|---|---|---|
| $a$ | = | speed of sound; acceleration |
| $a_1$ | = | normal-force curve slope, Equation 8.39 |
| $A$ | = | wing area |
| $c_i$ | = | constant factors, $i = 1, 2 \cdots$ |
| $C$ | = | constant in Sutherland's law, Equation 8.13 |
| $C_{D_o}$ | = | profile-drag coefficient |
| $C_L$ | = | lift coefficient |
| $C_{L_\alpha}$ | = | lift-curve slope |
| $C_{L_{max}}$ | = | maximum-lift coefficient |
| $D$ | = | aerodynamic drag |
| $g$ | = | acceleration of gravity |
| $h$ | = | altitude |
| $h_p$ | = | geopotential altitude |
| $h_{trans}$ | = | transition altitude |
| $\mathcal{H}$ | = | relative humidity |
| $k$ | = | induced-drag factor |
| $K_g$ | = | factor in Equation 8.39 |
| $K_s$ | = | stall margin |
| $L$ | = | aerodynamic lift |
| $m$ | = | mass |
| $M$ | = | Mach number |
| $n$ | = | number of moles, Equation 8.15 |

$p$ = pressure
$p_c$ = cabin pressure
$p_o$ = sea-level standard pressure
$p^*$ = stagnation pressure
$P$ = power
$q$ = dynamic pressure
$\mathcal{R}$ = gas constant
$t$ = time
$T$ = net thrust
$\mathcal{T}$ = temperature
$U$ = air speed
$U_{md}$ = speed of minimum drag
$U_{mp}$ = speed of minimum power
$U_{ref}$ = reference (equivalent) gust speed, Equation 8.39
$v_c$ = climb rate
$V_g$ = ground speed
$V_w$ = wind speed
$W$ = weight
$W_L$ = wing loading, $W/L$
$z_{op}$ = operational ceiling

## Greek Symbols

$\alpha$ = angle of attack
$\alpha_o$ = zero-lift $\alpha$
$\gamma$ = ratio between specific heats
$\delta$ = relative air pressure
$\delta_{trans}$ = relative pressure at transition altitude
$\eta$ = propeller efficiency; lift-induced drag coefficient
$\theta$ = relative air temperature
$\theta_{trans}$ = relative temperature at transition altitude
$\lambda$ = temperature-lapse rate
$\kappa$ = parameter defined by Equation 8.38
$\kappa_a$ = thermal conductivity of moist air
$\kappa_d$ = thermal conductivity of dry air
$\kappa_v$ = thermal conductivity of water vapour
$\mu$ = dynamic viscosity
$\rho$ = air density
$\rho_o$ = air density at sea level, standard conditions
$\sigma$ = relative air density

## Subscripts/Superscripts

$[.]_o$ = standard conditions
$[.]_a$ = air
$[.]_d$ = dry air

| | | |
|---|---|---|
| $[.]_g$ | = | ground |
| $[.]_t$ | = | troposphere |
| $[.]_v$ | = | water vapour |
| $[.]_w$ | = | wind |
| $[.]^*$ | = | reference conditions |

# 9 Take-Off and Field Performance

## Overview

This chapter deals with a realistic model of the aircraft at take-off conditions. After an introduction to general take-off procedures, we deal with transport-type airplanes (§ 9.1) and write the general take-off equations (§ 9.2). We carry out numerical solutions with all-engines operating §9.3 and consider the case of one-engine inoperative, both accelerate-and-go and abort-and-stop (§ 9.4). The takeoff of a turboprop aircraft is solved in §9.5. Lateral control issues in both cases are dealt with in the analysis of the minimum control speed (§ 9.6). We discuss the important cases of aircraft braking (§ 9.7) and performance on contaminated runways (§ 9.8), including extreme cases such as hydroplaning and other forms of contamination. For the sake of completeness, we present more rapid closed-form solutions (§ 9.9). The chapter ends with considerations on field performance, including taxiing and turning (§ 9.10) and the effects of bird strike (§ 9.10.2).

**KEY CONCEPTS:** Take-Off Equations, Normal Take-Off, OEI Take-Off, Balanced Field Length, Decelerate-Stop, Accelerate-Go, Minimum Control Speed, Contaminated Runways, Ground Manoeuvring, Bird Strike.

## 9.1 Take-Off of Transport-Type Airplane

The take-off phase is dependent on a vast array of aircraft parameters and external circumstances – including the pilot's skills. Take-off lasts little time and burns a considerable amount of fuel.

There are several different types of aircraft take-off and landing. We will refer exclusively to Conventional Take-Off and Landing (CTOL). This class includes most civil, commercial and military vehicles, that is, subsonic jet transport, turboprop airplanes, and cargo airplanes that take off and land horizontally. These airplanes operate from normal airfields with a paved runway; they have moderate thrust ratios or power loading.

We consider a take-off operation on a vertical plane. Due to the level of approximations that it is possible to achieve, in the calculation of the take-off of a

Table 9.1. *International symbols for take-off of a transport airplane*

| Symbol | Math | Definition | FAR/JAR 25.107 |
|---|---|---|---|
| VFE | $V_{FE}$ | maximum speed with flaps extended | |
| VEF | $V_{EF}$ | engine failure speed | $V_{EF} \leq V_1$ |
| VS | $V_S$ | stall speed | |
| VMCG | $V_{MCG}$ | minimum control speed on ground | $< V_{LO}$ |
| VMC | $V_{MC}$ | minimum control speed in air | |
| VMU | $V_{MU}$ | minimum unstick speed | $\geq V_S$ |
| VR | $V_R$ | rotation speed | $> 1.05V_s$ |
| V1 | $V_1$ | OEI decision speed | $\geq V_{MCG}$ |
| VLO | $V_{LO}$ | lift-off speed | $> 1.1V_{MU}$; $> 1.05V_{ME}$(OEI) |
| V2 | $V_2$ | speed over obstacle (take-off) | $> 1.2V_S$ |
| VMBE | $V_{MBE}$ | maximum brake energy speed | $\geq V_1$ |
| VTIRE | $V_{TIRE}$ | maximum tyre speed | $V_{TIRE} \geq V_1$ |

conventional airplane, we will assume that the runway reference system and the ground reference system are equivalent. The sequence of airplane speeds during take-off are given in Table 9.1 and further explained graphically in Figure 9.1.

With reference to Figure 9.2a, take-off is performed with all engines operating. The airplane accelerates, the nose landing gear lifts off and the airplane rotates with a speed $V_R$; the airplane lifts off the ground with a speed $V_{LO}$ and follows a rectilinear flight path to clear a conventional screen at 35 feet from the ground. In Figure 9.2b an engine failure occurs when the airplane has reached a speed $V_{EF}$. The pilot decides to continue with the remaining engine thrust (after stabilising the airplane) to take off over a longer distance. In Figure 9.2c, following the engine failure, the pilot decides to abort the flight and stop the airplane on the runway. The sequence of events in Figure 9.2 is explained in further detail in the following sections. According to the airworthiness requirements, there are four key distances on the ground:

**A.** All-engines-operating take-off distance from brake release to clearing of screen.
**B.** All-engines-operating accelerate-stop distance.

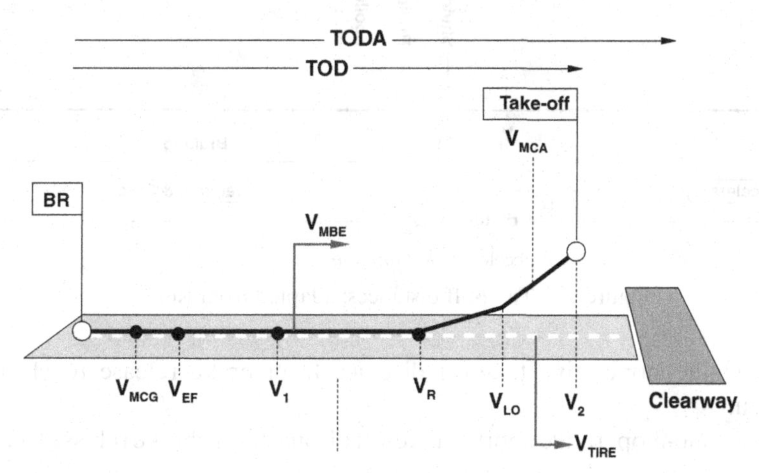


Figure 9.1. Sequence of reference speeds in a conventional take-off; TOD = take-off distance.

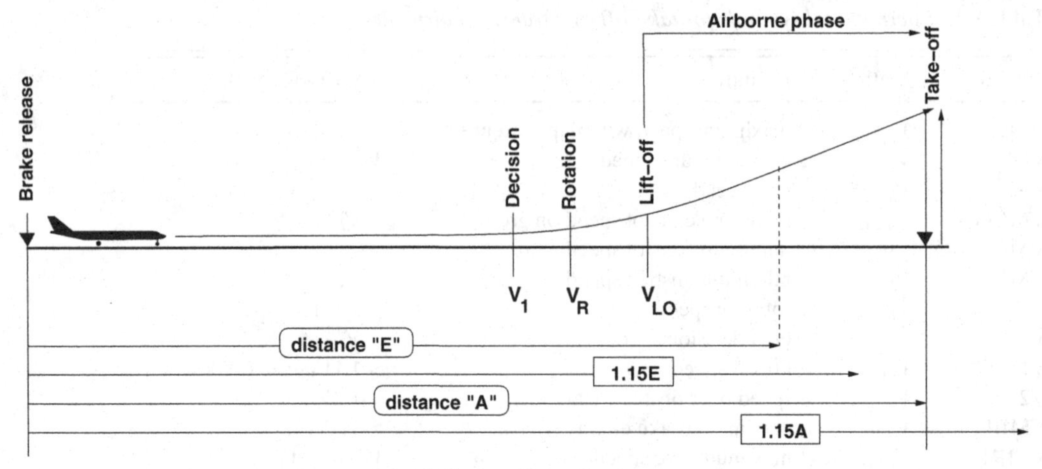


(a) All Engines Operating

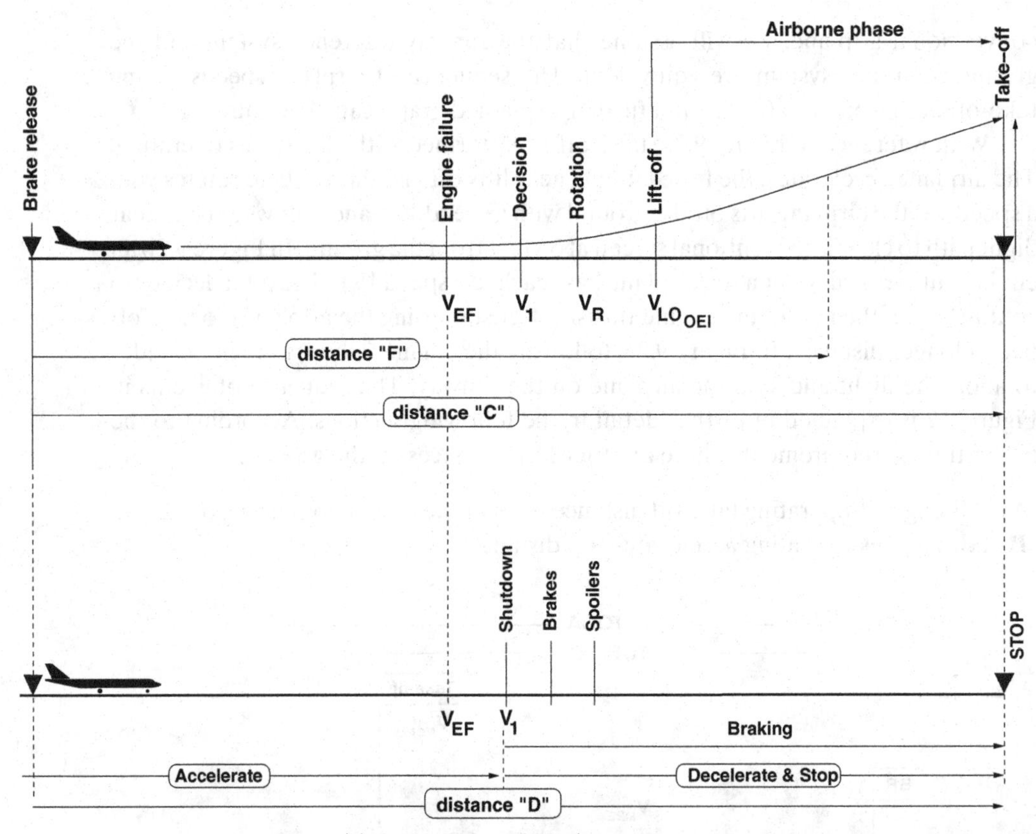


(b) OEI accelerate & Go/Stop

Figure 9.2. Take-off distances; adapted from Ref.[1]

**C.** One-engine-inoperative take-off distance from brake release to clearing of screen.

**D.** One-engine-inoperative aborted take-off distance from brake release to airplane halt.

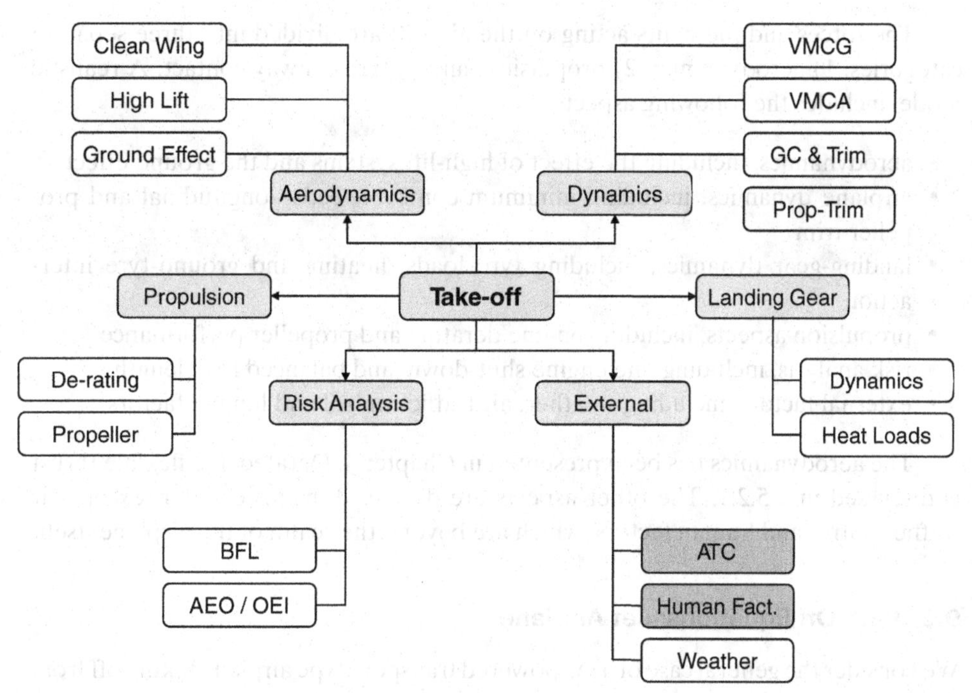


Figure 9.3. Flowchart of take-off problem.

A number of critical distances are defined. For the all-engines-operating case we have:

$$E = \frac{1}{2}(x_{lo} + x_{to})_{AEO}. \tag{9.1}$$

This is the AEO distance from brake-release to the halfway point between lift-off and take-off. Similarly, for OEI we have:

$$F = \frac{1}{2}(x_{lo} + x_{to})_{OEI}. \tag{9.2}$$

The Take-Off Distance Required (TODR) is defined by

$$\text{TODR} = \max\{1.15\text{A}, \text{C}\}. \tag{9.3}$$

The take-off distance available (TODA) must be longer than the TODR. The Take-Off Run Required (TORR) is defined by

$$\text{TORR} = \max\{1.15\text{E}, \text{F}\}. \tag{9.4}$$

The Accelerate-Stop Distance Required (BFL) is defined as

$$\text{BFL} = \max\{\text{B}, \text{D}\}. \tag{9.5}$$

Once the airplane becomes airborne and flies past the conventional screen, it starts its initial climb. This flight segment is discussed separately in Chapter 10. The main aspects of the take-off problem are illustrated in the flowchart in Figure 9.3, as they are implemented in the `FLIGHT` program.

The forces and moments acting on the aircraft are divided into three separate categories: 1) aerodynamics, 2) propulsion, and 3) tyre-runway contact. A realistic model includes the following aspects:

- aerodynamics, including the effect of high-lift systems and the ground effect
- airplane dynamics, including minimum control speeds, longitudinal and propeller trim
- landing-gear dynamics, including tyre loads, heating and ground-tyre interaction
- propulsion aspects, including engine derating and propeller performance
- risk analysis, including one engine shut-down, and balanced field length
- external factors, including weather, air traffic control and human factors

The aerodynamics has been presented in Chapter 4. Derated and flexible thrust is discussed in § 5.2.1. The other aspects are discussed in this chapter, except air traffic control and human factors, which are beyond the realm of the airplane itself.

## 9.2 Take-Off Equations: Jet Airplane

We consider the general case of a jet-powered transport-type airplane taking off from a prepared runway. We write the equations for the centre of gravity of the airplane moving on a vertical plane, that is, on a straight line whilst on the ground, and on a straight climb. The computational model is built as a set of ordinary differential equations that are solved from the brake-release point to the point of clearing a conventional screen.

A detailed solution method for the take-off of transport airplanes is given by ESDU[1-3]. Powers[4] published a method for the estimation of the balanced field length. An algorithm shown by Krenkel and Salzman[5] used such an integration scheme to calculate the take-off performance and the balanced field length.

**BASIC NOMENCLATURE.** We need to specify the nomenclature before getting started: $V$ denotes the total speed with respect to a reference on the ground. If $\gamma$ is the flight-path angle, then the velocity components are $u = V\cos\gamma$ and $v = V\sin\gamma$. In the ground run, with all wheels on the runway, $u = V$ and $v = 0$. The air speed is called $U$. If there is no wind, then $U = V$. If the wind has only a horizontal component, the relationship between air speed and ground speed is

$$U = \left[(V\cos\gamma \pm U_w)^2 + V^2\sin^2\gamma\right]^{1/2}. \tag{9.6}$$

The air speed $U$ is also called TAS (or KTAS, if expressed in knots).

The main forces and moments are shown in Figure 9.4. The weight is split between main and nose landing wheels. The wheels are subject to the reaction of the roadway, which has two components: normal (which operates against the weight) and parallel to the ground (which is effectively the rolling resistance). The forces on the aircraft are reduced to weight (operating at the CG), aerodynamic lift from the wing and the horizontal tail (operating on the aerodynamic centre of the wing), and the corresponding aerodynamic moments. Finally, there is the propulsive thrust, which is symmetric with respect to the vertical plane and therefore moved to the

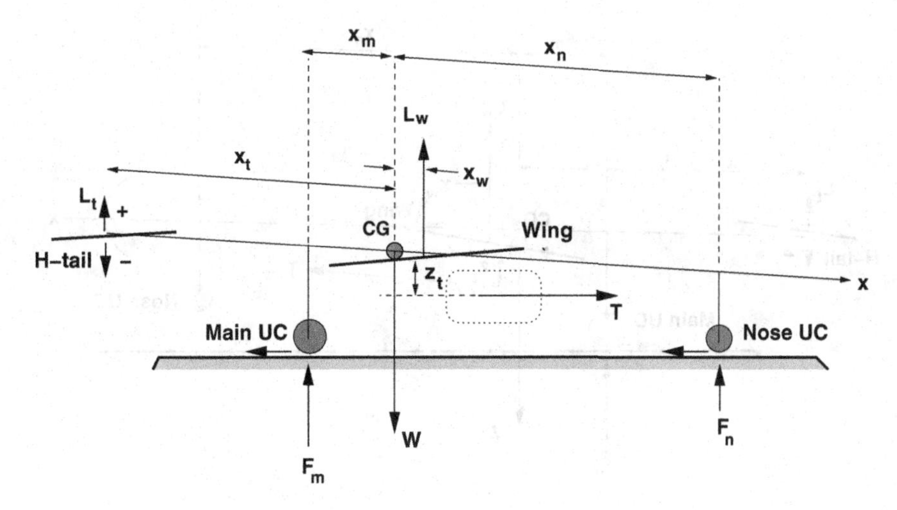


Figure 9.4. Aircraft forces on the ground.

symmetry plane. In Figure 9.4, we show a thrust vector that is horizontal and below the wing.

Because the wing has a complex planform, it is useful to replace it with the equivalent mean aerodynamic chord (MAC), which is calculated as shown in § 2.6. The wings shown in Figure 9.4 must be interpreted as MAC.

### 9.2.1 Ground Run

To allow the aircraft to pitch up (nose under-carriage lift-off first), the aerodynamic lift on the wing must be in the forward position with respect to the CG. The more aft the CG, the higher the pitching moment. The dynamic equation that provides the rotation of the airplane around the CG is

$$\mathcal{M}_y = I_y \mathsf{q}, \tag{9.7}$$

where $\mathcal{M}_y$ is the moment of the aerodynamic forces, $I_y$ is the polar moment of inertia around the spanwise axis $y$ (calculated in § 3.8.3) and $\mathsf{q}$ is the pitch rate, $\mathsf{q} = \dot{\omega}_y$. With reference to Figure 9.4, the total rigid-body moment around the CG is

$$\mathcal{M}_y = \mathcal{M}_{ow} + \underbrace{(L_t x_t + L_w x_w)}_{lift} + \underbrace{(F_m x_m + F_n x_n)}_{reaction} + \underbrace{(\mu_{rm} F_m z_m + \mu_{rn} F_n z_n)}_{friction} + T z_t \tag{9.8}$$

where $\mathcal{M}_{ow}$ is the wing's pitching moment and the underscores "t" and "w" denote horizontal tail plane and wing, respectively*; $\mu_r$ is the rolling coefficient; the remaining quantities are indicated in Figure 9.4. Equation 9.8 does not contain the drag. Its application point is not determined, so we assume that it lies on the line of the

* This analysis assumes that the aerodynamic coefficients are calculated by normalising with their own wing area, $C_{L_t} = L_t/qA_t$, $C_L = L_w/qA$.

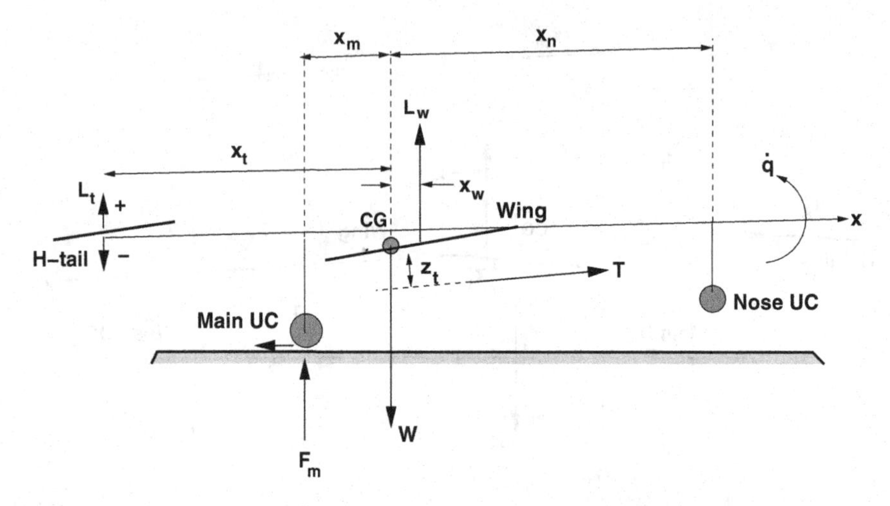


Figure 9.5. Aircraft forces on the ground during rotation.

longitudinal axis through the CG. If we solve in terms of the angular acceleration, we have

$$\frac{d^2\alpha}{dt^2} = \frac{1}{I_y}\left[\mathcal{M}_{ow} + (L_t x_t + L_w x_w) + (F_m x_m + F_n x_n) + (\mu_{rm} F_m z_m + \mu_{rn} F_n z_n) + T z_t\right], \tag{9.9}$$

$$\frac{d\alpha}{dt} = \omega_y. \tag{9.10}$$

During the ground run, the longitudinal axis of the airplane has a slight negative attitude. This contributes to the generation of a small downforce from the horizontal stabiliser, unless a corrective action is taken. A small downforce from the horizontal stabiliser improves the rotation of the airplane at lift-off. Due to the rotation of the airplane around the main landing gears (Figure 9.5), the CG has two additional rigid-body velocity components:

$$u_r = -\omega_y x_m \sin\alpha, \qquad v_r = \omega_y x_m \cos\alpha. \tag{9.11}$$

Hence, the air speed around the main wing is augmented by $\omega_y x_m$. The angle of attack, instead, is increased by

$$\alpha_1 = \tan^{-1}\left(\frac{v_r}{U - u_r}\right). \tag{9.12}$$

The next problem is to resolve the reaction force on the wheels before rotation. Because there is no rotation, the sum of all moments should be zero. The total resistance can be split between main and nose landing gear:

$$R = \mu_{rm} F_m + \mu_{rn} F_n. \tag{9.13}$$

With reference to Figure 9.4, the equilibrium of a rigid body with all of these vertical forces is

$$F_m + F_n = L_w + L_t - W, \tag{9.14}$$

$$(L_t x_t + L_w x_w) + (F_m x_m + F_n x_n) + (\mu_{rm} F_m z_m + \mu_{rn} F_n z_n) + T z_t + \mathcal{M}_{ow} = 0. \tag{9.15}$$

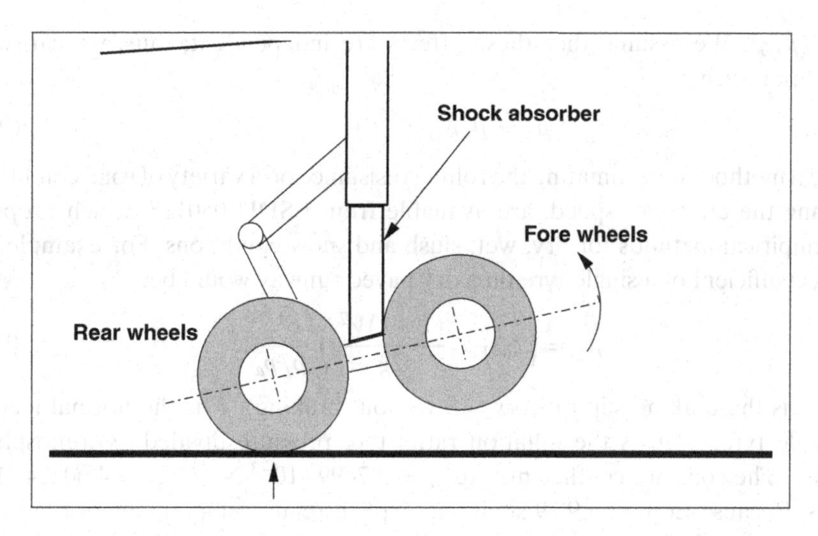


Figure 9.6. Rotation of main under-carriage at take-off and landing.

The latter equation can be rearranged in a more useful form that shows the unknown ground reaction forces $F_m$ and $F_n$:

$$(x_m + \mu_{rm} z_m) F_m + (x_n + \mu_{rn} z_n) F_n = -(L_t x_t + L_w x_w) - T z_t + \mathcal{M}_{ow}. \quad (9.16)$$

The system of Equation 9.14 and Equation 9.16 can be cast into matrix form as follows:

$$\begin{bmatrix} 1 & 1 \\ (x_m + \mu_{rm} z_m) & (x_n + \mu_{rn} z_n) \end{bmatrix} = \begin{bmatrix} F_m \\ F_n \end{bmatrix} = \begin{bmatrix} L_w + L_t - W \\ -(L_t x_t + L_w x_w) - T z_t + \mathcal{M}_{ow} \end{bmatrix}. \quad (9.17)$$

The solution of this system is straightforward, although for some load conditions there is no solution. A few points are worthy of note. First, the contribution of the wing pitching moment is normally “nose-down”. Therefore, an unusually large $\mathcal{M}_{ow}$ would prevent a load relief on the nose landing gear and thence would increase the ground run. Second, if the line of action of the engine thrust is below the CG, it helps with the rotation and thus contributes to relieving the load on the nose landing gear. When the airplane starts rotating, $F_n = 0$; the total rolling resistance is reduced to $\mu_{rm} F_m$.

A more accurate model should be used if the main under-carriage has more than one axle. In this case, at take-off the rear wheels are the last ones to lift off and the first to touch down, as also shown in Figure 9.6. In both cases the rear wheels sustain the largest loads.

### 9.2.2 Rolling Coefficients

The calculation of the rolling resistance in the take-off model requires the determination of the coefficients $\mu_{rm}$ and $\mu_{rn}$ (main and nose landing gear, respectively). In general, there are three contributions to consider: 1) the decelerating force created during the tyre roll, as a consequence of the mechanical deformation of the tyre ($\mu_{rR}$); 2) the effects of ground contamination ($\mu_{rF}$); and 3) the braking forces ($\mu_{rB}$). We assume that these effects are independent; thus, we can write an equation such as:

$$\mu_r = \mu_{rR} + \mu_{rF} + \mu_{rB}. \tag{9.18}$$

Practical methods for estimating the rolling resistance on a variety of road conditions, including the effects of speed, are available from ESDU 05011[6], which proposes semi-empirical methods for dry, wet, slush and snow conditions. For example, the rolling coefficient of a single tyre on a dry paved runway would be:

$$\mu_r = \left(\zeta_o + \zeta_1 \frac{(1-s)V^2}{2g}\right) \frac{F^{1/3}}{p/p_a}, \tag{9.19}$$

where $s$ is the braking slip ratio ($s = 0$ without braking), $F$ is the normal load on the single tyre, $p/p_a$ is the inflation ratio: tyre pressure divided by atmospheric pressure. The constant coefficients are $\zeta_o = 3.7699 \cdot 10^{-3}$ N$^{-1/3}$, $\zeta_1 = 4.60824 \cdot 10^{-5}$ N$^{-1/3}$/s. Because Equation 9.19 shows an explicit relationship among normal load, tyre pressure and rolling coefficient, the rolling resistance is a variable quantity that depends on the aircraft loading as well as the runway conditions. For example, if the main landing gears have $n_w$ wheels, then $F = F_m/n_w$ must be introduced in Equation 9.19. The total resistance of the main landing gear is then found from $\mu_{rm} F_m$ if the runway is dry or wet; if the runway is covered in snow, $F_m$ must account only for the leading tyres. Thus, if each under-carriage has four wheels in tandem, then the rolling resistance would be $\mu_{rm} F_m/2$. The second retarding effect, due to contamination, is discussed separately in § 9.8.

## 9.3 Solution of the Take-Off Equations

The solution of the take-off problem is achieved by integrating the system of ordinary differential equations (ODE). A suitable method of integration is a fourth-order Runge-Kutta procedure with a constant time step. There are three segments to consider: 1) ground run with all wheels on the runway to the point of nose wheels lift-off; 2) ground run to the point of lift-off (all wheels off the ground); and 3) airborne phase to the height of the screen. The system consists of eight differential equations:

$$\frac{du}{dt} = \frac{1}{m}\left[T\cos(\epsilon + \alpha + \gamma) - D\cos\gamma - L\sin\gamma\right], \tag{9.20}$$

$$\frac{dv}{dt} = \frac{1}{m}\left[T\sin(\epsilon + \alpha + \gamma) - D\sin\gamma + L\cos\gamma - W\right], \tag{9.21}$$

$$\frac{dx}{dt} = u, \tag{9.22}$$

$$\frac{dh}{dt} = v, \tag{9.23}$$

$$\frac{dm}{dt} = -\dot{m}_f, \tag{9.24}$$

$$I_y \frac{d\omega_y}{dt} = \left[\mathcal{M}_{ow} + (L_t x_t + L_w x_w) + (F_m x_m + F_n x_n) + (\mu_{rm} F_m z_m + \mu_{rn} F_n z_n) + T z_t\right], \tag{9.25}$$

$$\frac{d\alpha}{dt} = \omega_y, \tag{9.26}$$

with

$$\gamma = \tan^{-1}\left(\frac{v}{u}\right). \tag{9.27}$$

Before lift-off, $\gamma = 0$. This system of equations is closed with Equation 9.17. When the aircraft is airborne, the rolling resistance and the normal load terms disappear. The aircraft can still be treated like a rigid body in the vertical plane. This body has three degrees of freedom and is described by three differential equations similar to the climb phase. Three more aspects of the take-off problem must be considered:

**1. AERODYNAMIC FORCES.** The aerodynamic forces in the system of equations are lift and drag. These forces are dominated by the position of the control surfaces and the ground effect. Each flap/slat setting is associated to an index $I_{SF} = 0, 1, \ldots$. For each value of the setting $I_{SF}$, we have a well-defined flap and slat angle ($\delta_f$, $\delta_s$, respectively), which depend on the aircraft. The calculation procedure for the aerodynamic terms is the following:

- Set a tentative value of $I_{SF}$ and determine the angles $\delta_f$, $\delta_s$.
- Calculate the lift and drag effects of $\delta_f$, $\delta_s$ following the methods in § 4.1.3 and § 4.2.5.
- Calculate the ground effects with the methods shown in § 4.1.2.

The aerodynamic forces are calculated at the air speed $U$, which is different from the ground speed whenever there are atmospheric winds (Equation 9.6).

**2. NET THRUST.** The net thrust is delivered in equal amounts by a number of engines $n_e$. Cases when this does not occur include OEI (discussed separately) and small imbalances in the fuel flow. The flight control system must be able to correct any thrust imbalance; otherwise the consequences could be catastrophic. For the case of all engines operating (AEO), we use

$$T = F_N n_e. \tag{9.28}$$

At each time step, the net thrust $F_N$ is calculated numerically from the engine deck (see Chapter 5).

At the brake-release point the thrust is increased rapidly. The normal procedure is 1) brakes off, 2) stick forward about halfway, and 3) stick forward to full thrust once the engines have stabilised. The engine thrust increases rapidly. Most airplanes are not designed to operate with brakes on whilst the engines are spooling up. From a flight simulation point of view it is sufficient to have a ramp function such as

$$F_N = T_o \tanh t, \tag{9.29}$$

where $T_o$ is the rated take-off at static conditions and $t$ is the time, calculated from brake-release. The thrust stabilises to the rated thrust within 3 seconds.

**3. LONGITUDINAL TRIM.** A major problem is the aircraft trim. If this is not done, it is quite likely that the numerical solution causes the aircraft to pitch up too fast, and the aircraft would enter a dangerous, if not impossible, trajectory. This happens if the estimate of the pitching moment and/or the pitch moment of inertia is inaccurate. On modern airplanes there is a control system called $\alpha$-floor protection, which prevents

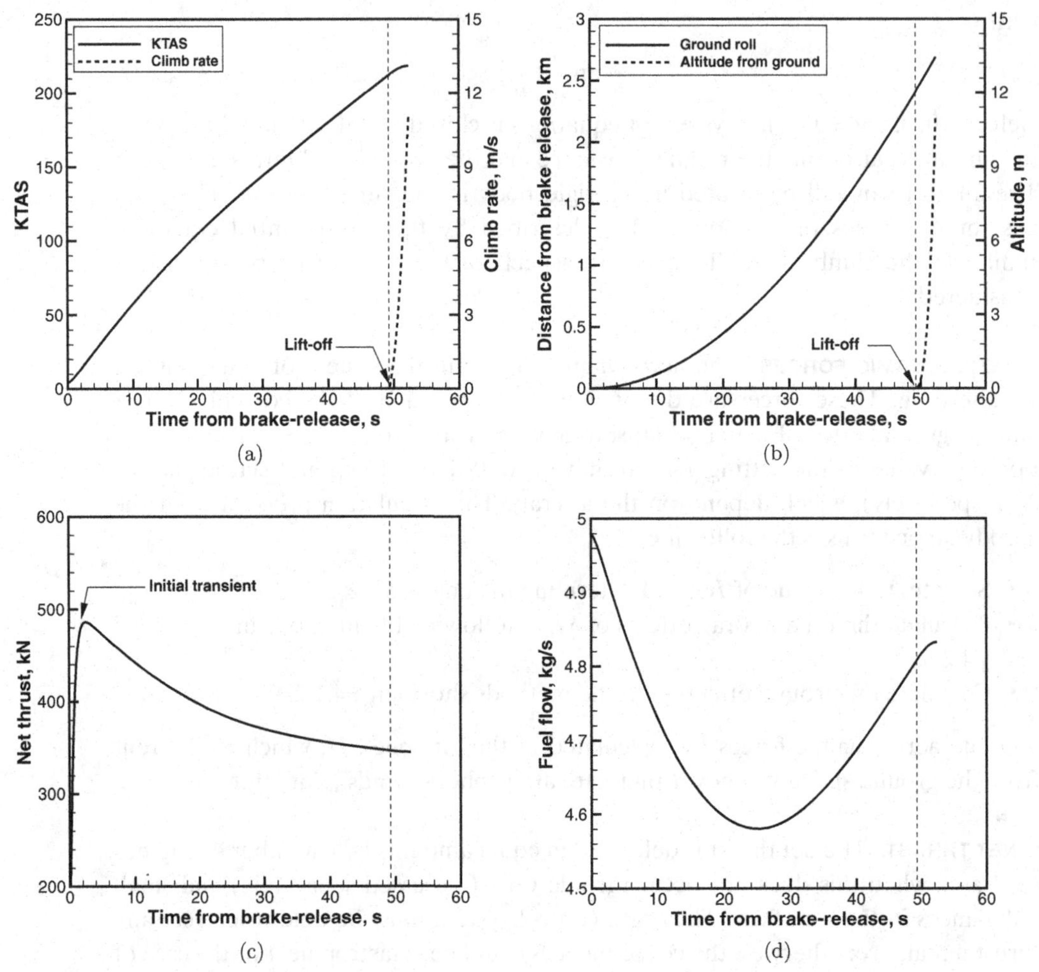


Figure 9.7. AEO take-off analysis; BRGW = 128,500 kg; headwind = −2 m/s; $h$ = 50 m; take-off thrust (no bleed); flap $\delta_f$ = 20 deg; $x_{CG}$ = 25% MAC; standard day (part I).

the aircraft from entering a flight trajectory with a too-high attitude. Many transport airplanes have a rotation rate $\dot{\alpha} = \omega_y$ = 2 to 4 degrees/s. One possible solution is to assume that the aircraft is trimmed at all times. Because

$$\ddot{\alpha} = \frac{d\dot{\alpha}}{dt} = \frac{\mathcal{M}_y}{I_y} \tag{9.30}$$

if $\ddot{\alpha}$ is above a limit value, the pitching moment is corrected. This is equivalent to adding a damping function to Equation 9.25. At the end of the rotation we force the airplane to proceed on a linear trajectory.

### 9.3.1 Case Study: Normal Take-Off of an Airbus A300-600 Model

In Figures 9.7 and 9.8 we show a calculation of the AEO take-off of a model Airbus A300-600 powered by CF6-80C2 turbofan engines. The data plotted include the distance from brake release, the ground speed, the true air speed (KTAS), the climb

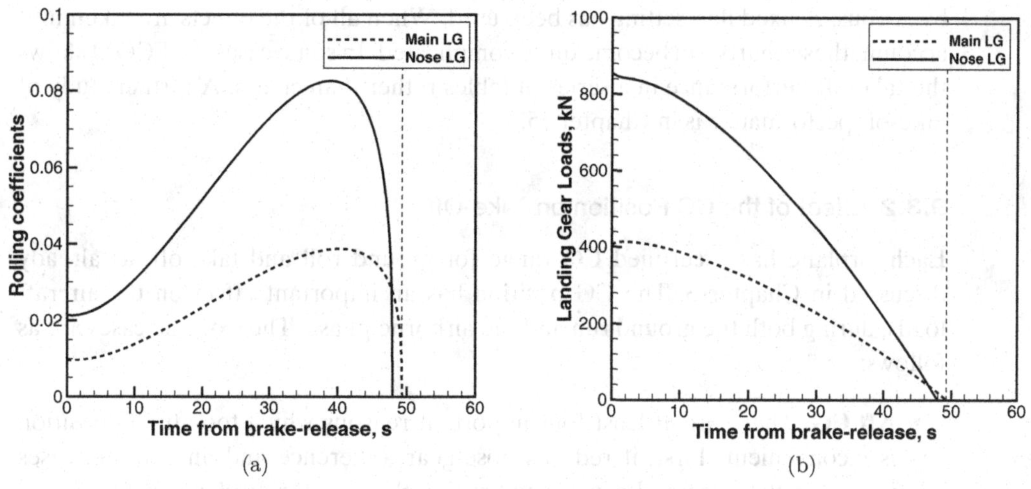


Figure 9.8. AEO take-off analysis (part II).

rate, the rotation speed, the loads on the landing gears, the net thrust, the total rolling resistance, the lift-over-weight ratio and the fuel burn. The calculation ends when the aircraft has passed the screen. The graphs show the aircraft trajectory $(x, h)$, its velocity components $(u, v_c)$, the net thrust, the fuel flow, the rolling coefficients and the rolling resistance. Note the short time between nose lift-off and airplane lift-off.

Several parametric studies can be carried out to examine the effects of all of the relevant parameters, including brake-release gross weight, airfield altitude, wind speed, atmospheric temperature, runway gradient and runway conditions (six independent parameters). This is in fact done in the W-A-T charts (weight-altitude-temperature) that are part of the FCOM. One such chart is shown in Figure 9.9. Calculations have been carried out over the full range of BRGW, at selected altitudes on a standard day, and are compared with some reference data published

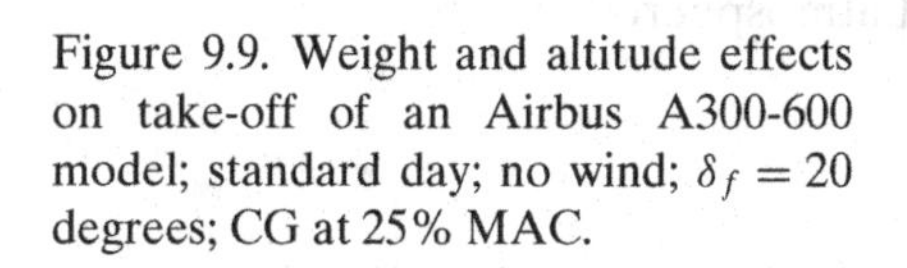

Figure 9.9. Weight and altitude effects on take-off of an Airbus A300-600 model; standard day; no wind; $\delta_f = 20$ degrees; CG at 25% MAC.

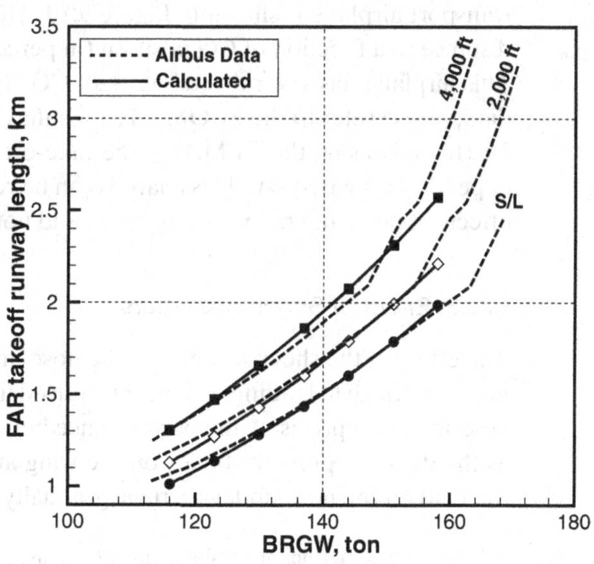

by Airbus. A fixed flap setting has been used. When all of the effects are taken into account, these charts can become quite complicated. In some cases the FCOM shows the take-off performance in a series of tables rather than charts. A further study of take-off performance is in Chapter 15.

### 9.3.2 Effect of the CG Position on Take-Off

Each airplane has a certified CG range for ground roll and take-off, as already discussed in Chapter 3. The CG position has an important effect on the aircraft loads, during both the ground run and the airborne phase. The extreme cases are as follows:

- **Aft CG.** There are at least four important reasons why a too-aft CG position is inconvenient. First, it reduces nose-gear adherence and, in extreme cases it can lift the nose and cause a tail strike. Second, for airplanes with engines mounted below the wings, there is a further pitch-up effect caused by the thrust force being located below the CG; the stalling speed is reduced and the take-off speed $V_2$ increases. Third, an aft CG reduces the ability to steer the aircraft whilst taxiing. Finally, lateral control at low speeds must be secured by the nose wheels because the vertical tail and rudder do not have the necessary dynamic pressure to generate lateral control forces (see discussion in § 9.6).
- **Forward CG.** A too-forward CG position makes it more difficult for the airplane to rotate (pitch up); thus, it increases the take-off run. Furthermore, this CG position causes higher structural loads on the nose tyres, increased heating and higher wear (as discussed in Chapter 14). When determining the aircraft take-off performance, the calculation is always performed at the most forward CG position.

An aircraft trimmed for take-off would have to take into account these effects. A parametric study of the effect of CG position on the take-off performance of a transport airplane is shown in Figure 9.10. This graph shows the change in take-off distance as a function of CG position (in percent MAC) for three different BRGW. This airplane has several versions with CG slightly different limits. For the correct data, please refer to the FCOM*. The results clearly indicate that by moving the CG aft (by increasing the % MAC), the take-off length decreases, by an amount that depends on the BRGW. This analysis can be repeated at different altitudes, with the effects of head and tail winds and cold and hot atmospheres.

### 9.3.3 Effect of Shock Absorbers

The effect of the shock absorbers (the nose under-carriage at take-off and the main under-carriage at landing) can be introduced in the dynamic model of the airplane to describe more precisely the performance before and after nose-gear lift-off. In fact, as the dynamic pressure builds on the wing and the aircraft is trimmed to take off, the load on the nose under-carriage gradually decreases, and the aircraft undergoes

* Airbus recommends an aft-shift of the CG by about 2% to improve the take-off performance, with a CG variable between 25% MAC and 30% MAC for many of its aircraft.

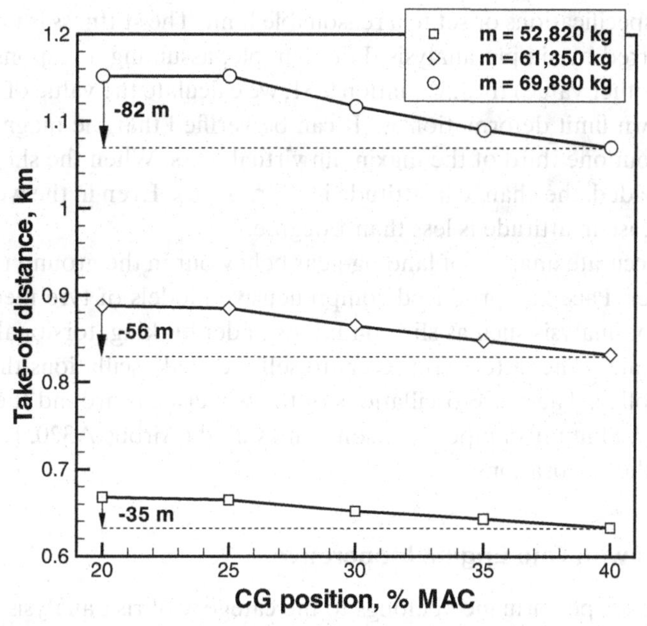


Figure 9.10. Calculated effect of CG position on the take-off run of an Airbus A320-200 model; standard day, no wind, sea level (do not use for flight planning).

a small upward rotation whilst on the ground. This increase in attitude contributes to the increase in wing lift. Nose-gear lift-off occurs when the shock absorber has no load. Thus, the main change in the model is the relaxation of the shock absorbers, as indicated in Figure 9.11. The model is equivalent to a spring-damper in parallel. The spring stiffness is $\kappa$ and the damping is $\mathcal{D}$. If $m = F/g$ is the virtual mass arising from the normal load $F$ on the landing gear, then the system would have a natural oscillation $\omega_o = \sqrt{\kappa/m}$ and a damping ratio $\zeta = \mathcal{D}/2\sqrt{m\kappa}$.

By using the critical damping $\mathcal{D}_{crit} = 2\sqrt{\kappa m}$, the damped response is a return to the undisturbed position in minimum time. The response of the system is

$$x(t) = x_o(1 + \omega_o t)\exp(-\omega_o t), \tag{9.31}$$

where $x_o$ is the initial displacement under load $F$ and $\dot{x}_o = 0$. The normal load is generally known, and the maximum deformation of the unit can be found from

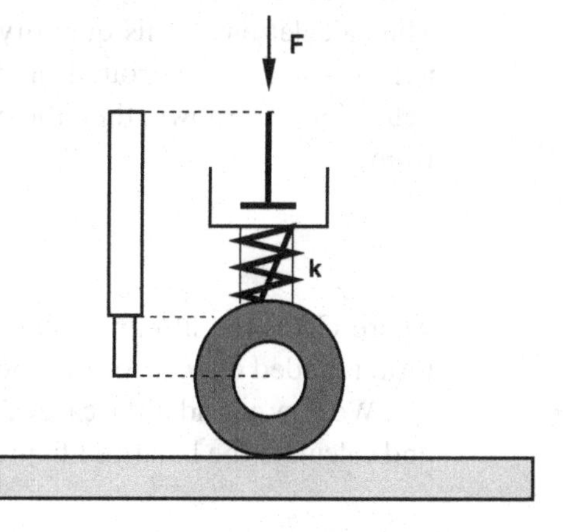


Figure 9.11. Shock-absorber model on nose landing gear.

landing-gear specifications or set to a reasonable limit. The stiffness is unknown, but it can be inferred by simple analysis. For example, assuming a response time of 1 second (or another value) from Equation 9.31, we calculate the value of the stiffness from the known limit deformation $x_o$. It can be verified that the magnitude of the stiffness is about one third of the maximum virtual mass. When the shock absorber has been unloaded, the change in attitude is $\Delta\alpha \simeq x_o/w_b$. Even in the worst possible case, the increase in attitude is less than 1 degree.

A more accurate analysis of landing-gear behaviour in the ground run is a complicated matter. Pacejka[7] produced comprehensive models of tyre mechanics that can be used for analysis such as slip conditions under braking, torsional effects and *shimmy* dynamics. The latter term refers to self-induced oscillations that resemble a shopping trolley. These are oscillations of the wheels axis around the axis of the main strut[8–10]. Shimmy dampers are sometimes used (Airbus A320, Boeing B737) to overcome these vibrations.

## 9.4 Take-Off with One Engine Inoperative

The OEI take-off performance belongs to the category of risk analysis indicated in Figure 9.3. Human factors are involved as well because a decision must be made to either accelerate and take off or stop and abort the flight. The task is to carry out an analysis for the case when one engine shuts down at some point during the ground run. The scenario is limited to the loss of thrust, but it can be expanded into the causes of the shut-down and what may happen in more extreme cases, such as fuel leak.

The governing equations are the same as the AEO case, except that the inoperative engine creates asymmetric loads. With reference to the latter point, the yawing moment must be restored by a combination of corrective actions. This is only possible if the failure has occurred and has been detected at speeds above the minimum control speed. The increase in aerodynamic drag is due to a combination of 1) inoperative engine, and 2) rudder deflection:

$$\Delta C_D = \Delta C_{D_e} + \Delta C_{D_{rudder}}. \tag{9.32}$$

The engine contribution in Equation 9.32 is calculated from Equation 7.44. The rudder deflection is required to stabilise the airplane when one engine shuts down. The calculation of this quantity is made difficult by the interaction between tyre performance on the ground[7] and the aircraft control system. However, if the rudder deflection $\xi$ is known, then the corresponding increase in induced drag is estimated from

$$\Delta C_{D_{rudder}} = (C_{L_\xi}\xi)\xi = C_{L_\xi}\xi^2, \tag{9.33}$$

where $C_{L_\xi}$ is the lift-curve slope of the vertical-tail-plus-rudder combination. This term is added only at speeds above the VMCG.

We now consider the case of decelerate-stop (abort) and accelerate-go (take-off) and calculate the balanced field length.

### 9.4.1 Decelerate-Stop

When take-off is to be aborted following one engine failure, a sequence of events takes place:

- Engine failure occurs at time $t_1$, when the aircraft has a speed $V_1$ and has run a distance $x_1$ from brake release. These are called *critical* quantities.
- There is a time lag, $t_{lag}$, between engine failure and the decision to abort the flight, due to the natural reaction of the pilot (*human factors*). This time is usually ~3 seconds. When the clock indicates a time $t_D = t + t_{lag}$, the aircraft has reached a speed $V_D$ at a position $x_D$ from brake-release point.
- At the decision point, the thrust is cut, the brakes are applied and the aircraft is prepared to a braking configuration by deployment of the inlay spoilers, which also contribute to dumping the lift. Alternative braking methods are possible.
- The aircraft decelerates; when it is halted, it reaches the so-called stopping distance.

A model is required for the transient thrust, both for the failed engine and the running engine. It is not possible to make general statements because the transient response depends on both engine and type of failure. For the running engine, we need a model for the transient thrust when the throttle is cut back to zero. For the running engine, we can define a cut function

$$F_{cut}(t) = e^{nt}, \tag{9.34}$$

where $n$ is an arbitrary power that defines the transient behaviour and $t$ is the time. Given the time $t_o$ at which the net thrust has decayed to zero, the exponent can be found from solving Equation 9.34 in terms of $n$. For modern high bypass turbofan engines, $n \simeq -1.5$.

When one engine shuts down, the aircraft has an asymmetric thrust for the time $t_{lag} + t_o$. The yawing moment due to the combination of inoperative engine and engine operating at full thrust is

$$\mathcal{M}_z = T y_b + D_e y_b, \tag{9.35}$$

where $y_b$ is the engine's arm, $T$ is the thrust of the operating engine and $D_e$ is the drag of the inoperative engine. This yawing moment may be balanced by the tyre-runway contacts and by the rudder. The total yawing moment is

$$\mathcal{M}_z = T(t) y_b + D_e y_b - \mu_{sm} F_m - \mu_{sn} F_n - C_{L_\xi} \xi, \tag{9.36}$$

where $\mu_s$ is the friction coefficient between tyre and runway for sideways forcing (when the tyres are in fact forced to skid laterally). It cannot be guaranteed that during the transient the resulting moment is zero.

There remains the question of whether the pilot hits the brakes at the same time as he or she cuts the thrust. A further time delay can be included in the model to take into account this synchronisation. ESDU[1;2] considers the effect of the time lag on brakes, spoilers and engine response. A summary of average time delays is given in Table 9.2.

Table 9.2. *Delay in response time after activation for selected systems*

| System | Delay[s] | Comment |
|---|---|---|
| Pilot response | ~3 | |
| Throttle cut | 3–8 | |
| Fuel cut | 2–4 | |
| Prop pitch change | 2–4 | |
| Spoiler deployment | 0.5–1.0 | |
| Wheel brake | 0.5 | Manual |
| | 0.1–0.2 | Automatic |

### 9.4.2 Accelerate-Stop

The alternative possibility is to continue the acceleration with the remaining functioning engine, provided that the aircraft has been trimmed and stabilised both longitudinally and laterally.

It is possible that the specified flap setting is not capable of securing enough aerodynamic lift. In this case, there is a need to increase the flap deflection. Two cases are possible: 1) the lift is not enough for take-off, and 2) the lift is not enough to climb with the minimum gradient specified by the international regulations. This event is dependent on the gross take-off weight.

#### Computational Procedure for Accelerate-Stop

- Estimate the rotation speed, $V_R = 1.1V_S$.
- Bracket the interval $V_1 < V_{crit} < V_2$ with $V_1 = 0.25V_{crit}$, $V_2 = V_{crit}$.
- Guess the critical speed from $V_{crit} = (V_1 + V_2)/2$.
- Calculate the AEO take-off up to $V_{crit}$.
- When the aircraft has reached $V_{crit}$, do the following:

  1. Continue acceleration with engine failure till the aircraft has cleared reference screen. The take-off distance is $x_{to_1}$.
  2. Decide to stop. Cut down thrust, hit the brakes, deploy the spoilers, dump lift and calculate the stopping trajectory $x_{to_2}$ (these operations are done with time delays).

- Compare the accelerate-stop distances. If the stopping distance $x_{to_2}$ is greater than the acceleration distance $x_{to_1}$, then set $V_1 = V_{crit}$; or else set $V_2 = V_{crit}$.
- Repeat the iterations with new $V_{crit}$ till a satisfactory convergence is achieved.

The result of this iterative procedure is shown in Figure 9.12. The right scale shows the variation of the critical speed $V_{crit}$. The convergence criterion is a difference of 5 metres between OEI take-off and OEI abort. A smaller tolerance serves only to increase the number of iterations and does not guarantee a better result. No reverse thrust is used in the calculation.

Figure 9.13 shows the simulation for an Airbus A300-600 model. The data in the graph are the ground speed and true air speed from the brake-release point to the stopping point on the runway. The graph shows the AEO and OEI runs, the

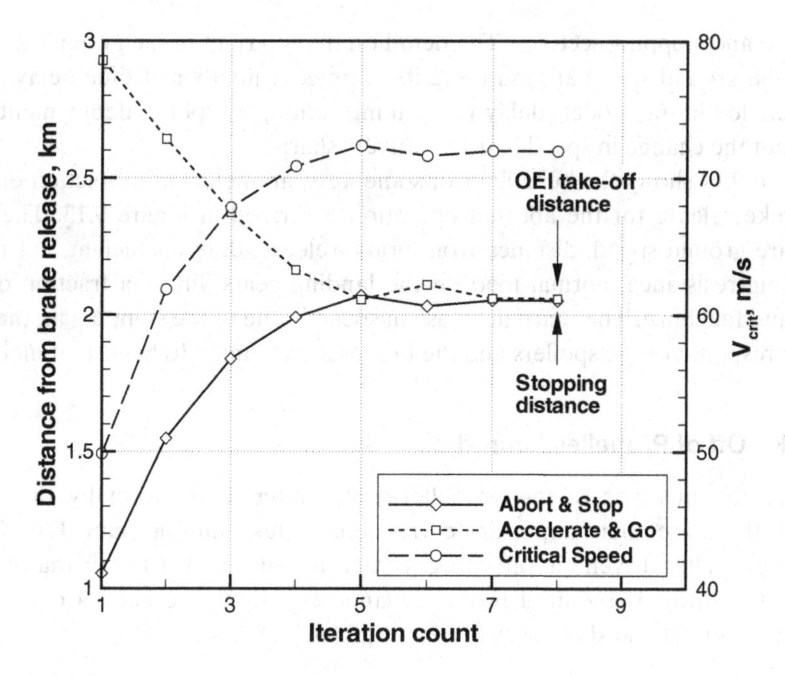


Figure 9.12. Accelerate-stop analysis of ground speed for an Airbus A300-600 model; sea level, standard day, no wind.

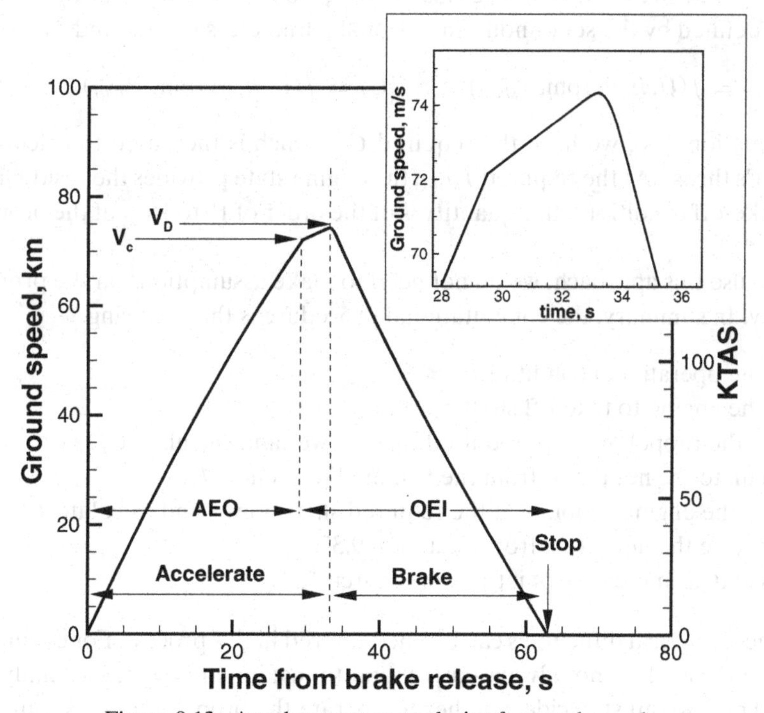


Figure 9.13. Accelerate-stop analysis of ground speed.

accelerate and stopping sections. The detail on the top right of the graph is a closer look at the ground speed around the critical time. A number of time delays have been included in the model (delay for braking action, for spoiler deployment, and so on), but the change in speed is still relatively sharp.

Figure 9.14 shows the behaviour of some key parameters as a function of time from brake release for the aborted operation described in Figure 9.13. The data shown are ground speed, distance from brake-release, drag coefficient, net thrust and rolling resistance, normal load on the landing gears, lift as a fraction of the weight and fuel burn. The sharp decrease in speed is due to the stopping method. In fact, the response of the spoilers and the brakes are assumed to be instantaneous.

## 9.5 Take-Off of Propeller Aircraft

The take-off equations of the propeller-driven aircraft are formally the same, although there are some important corrections to take into account. First is the fact that propeller-driven aircraft are governed by shaft power rather than thrust. Because the thrust is required in the dynamic equations, we need a correlation between net thrust and shaft power. We start from

$$T = T_p + F_g, \tag{9.37}$$

where $T_p$ is the net propeller thrust and $F_g$ is the residual thrust provided by the engine exhaust. The engine throttle is set to take-off power. The propeller is trimmed to the required shaft power. Once this is done, we have the state of the propeller, which is defined by the set of non-dimensional parameters $C_T$, $C_P$ and $J$:

$$C_P = f(U, h, \vartheta, \text{rpm}, P_{shaft}), \qquad C_T = f(U, h, \vartheta, \text{rpm}, P_{shaft}). \tag{9.38}$$

With Equation 9.38, we have the required $C_T$, which is then used to calculate the propeller's thrust. At the required $P_{shaft}$, the engine state provides the residual thrust $F_g$. At take-off conditions, this quantity is of the order of 10 to 15% of the propeller's thrust.

If we use this approach, we do not need to make assumptions on the propeller's efficiency. In summary, the computational procedure is the following:

- Set the operational conditions.
- Set the engine to take-off shaft power.
- Trim the propeller at the required shaft power and calculate $C_T$.
- Calculate the net thrust from the trimmed propeller, $T_p$.
- Solve the engine problem at the required shaft power and calculate $F_g$.
- Calculate the net thrust from Equation 9.37.
- Proceed as in the case of jet-powered aircraft.

Some numerical difficulties can be encountered in this process. For example, the maximum thrust does not always correspond to maximum power, especially at low speeds. Thus, we must decide whether to operate the propeller at maximum thrust or maximum power. In most cases, the rpm is fixed and the propeller is required to operate at its nominal speed. However, sometimes it is allowed to have a small

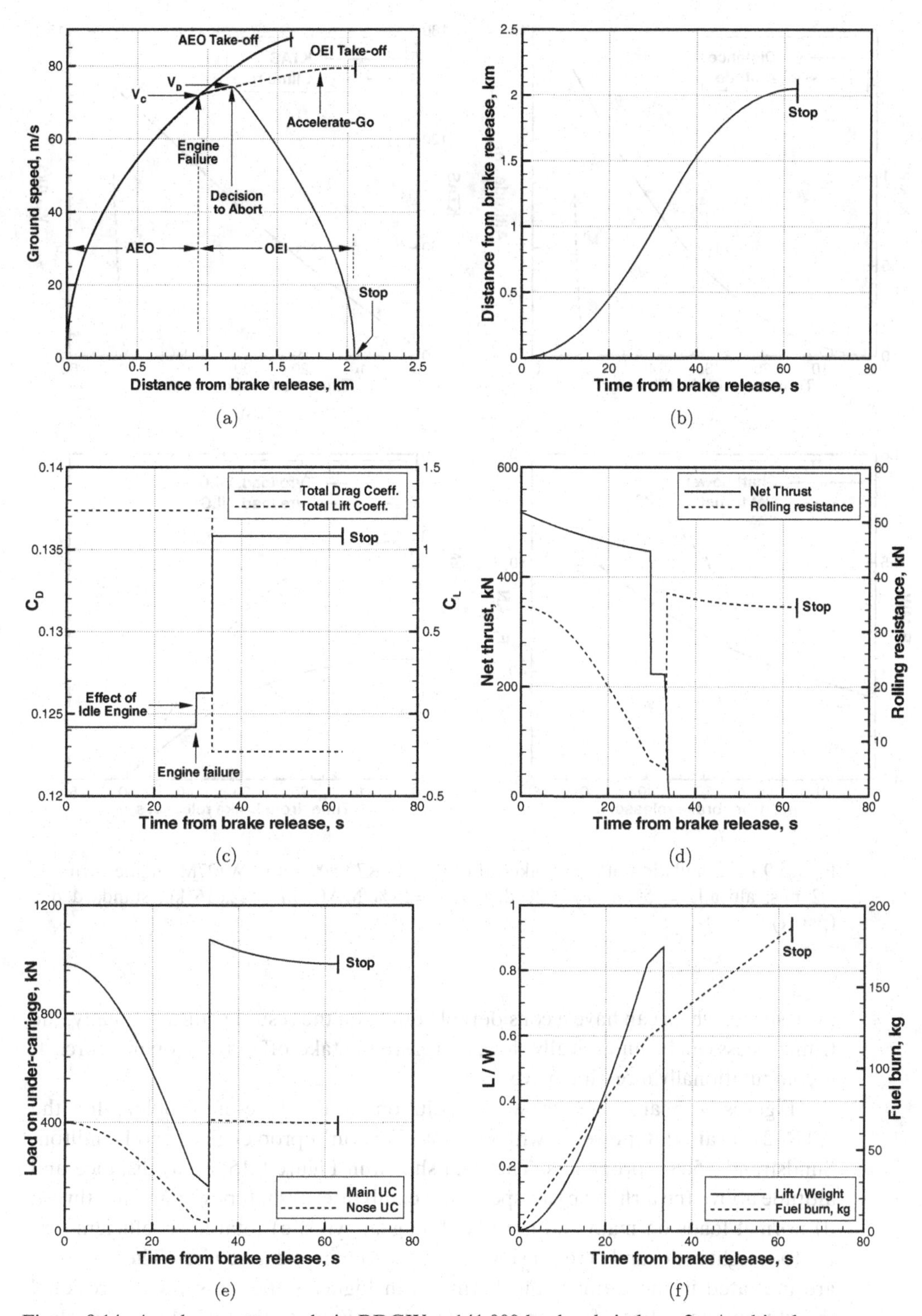


Figure 9.14. Accelerate-stop analysis; BRGW = 141,000 kg; headwind = −2 m/s; altitude = 50 m; take-off thrust (no bleed); flap $\delta_f = 20$ degrees; $x_{CG} = 25\%$ MAC; standard day.

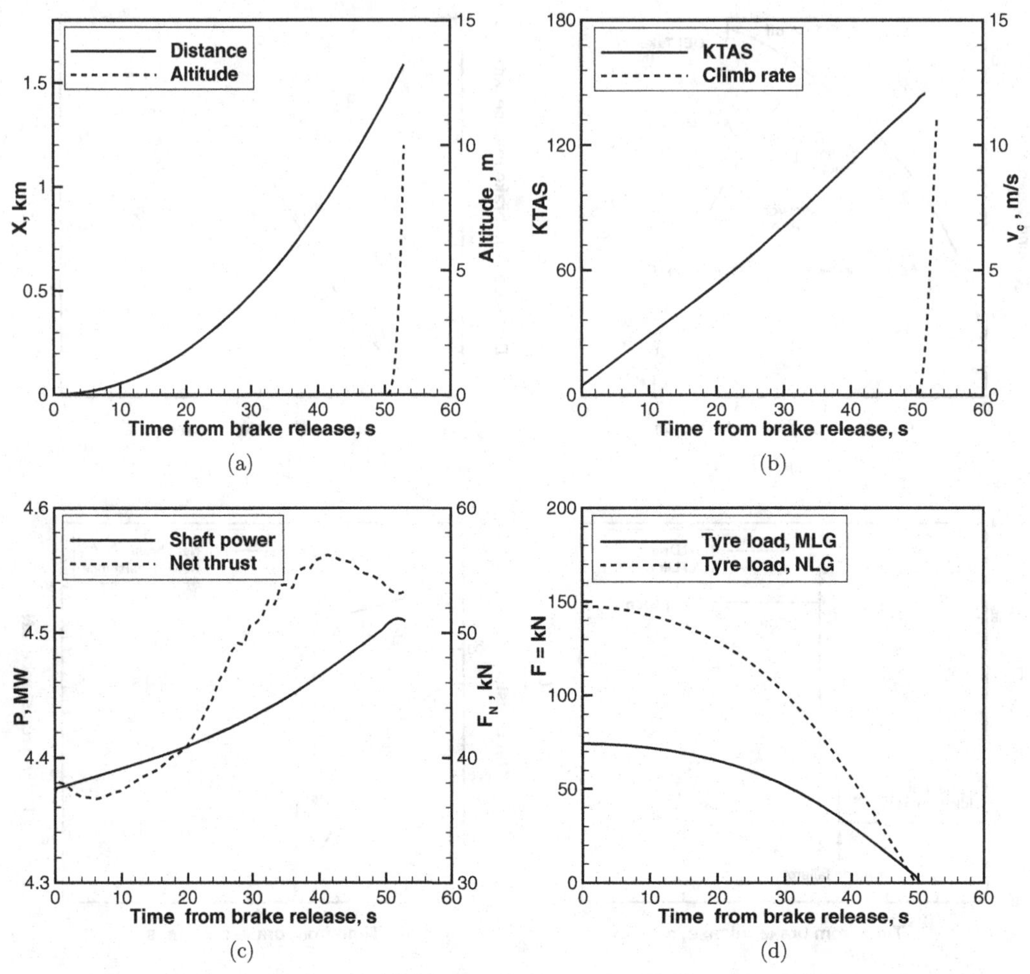


Figure 9.15. Simulation of AEO take-off of the ATR72-500 with PW127M engines: wind = −2 m/s; altitude = 50 m; $\delta_f = 20$ deg; $x_{CG}$ =25% MAC; $m =$ 22,616 kg; standard day (part I).

over-speed, which can have a considerable effect on the resulting thrust. Finally, the trim process can be numerically slow; therefore the take-off of the propeller aircraft is computationally more intensive.

Figures 9.15 and 9.16 show the solution of the take-off problem for the ATR72-500 aircraft, powered with two PW127M turboprop engines and Hamilton-Sundstrand F568-1 propellers. The data shown in Figure 9.15 are a) distance and altitude above the airfield; b) air speed and climb rate; c) shaft power and net thrust; d) normal loads on main and nose landing-gear tyres; e) total lift coefficient and lift-to-weight ratio; and f) rolling coefficients for main and nose tyres. The case data are indicated in the caption. Furthermore, in Figure 9.16d we show the residual thrust from the turboshaft engines and its value relative to the total thrust. The data clearly indicate that the residual thrust is about 10 to 15%; as anticipated, it is not a negligible portion of the propeller thrust and must be accounted for in a proper engineering analysis.

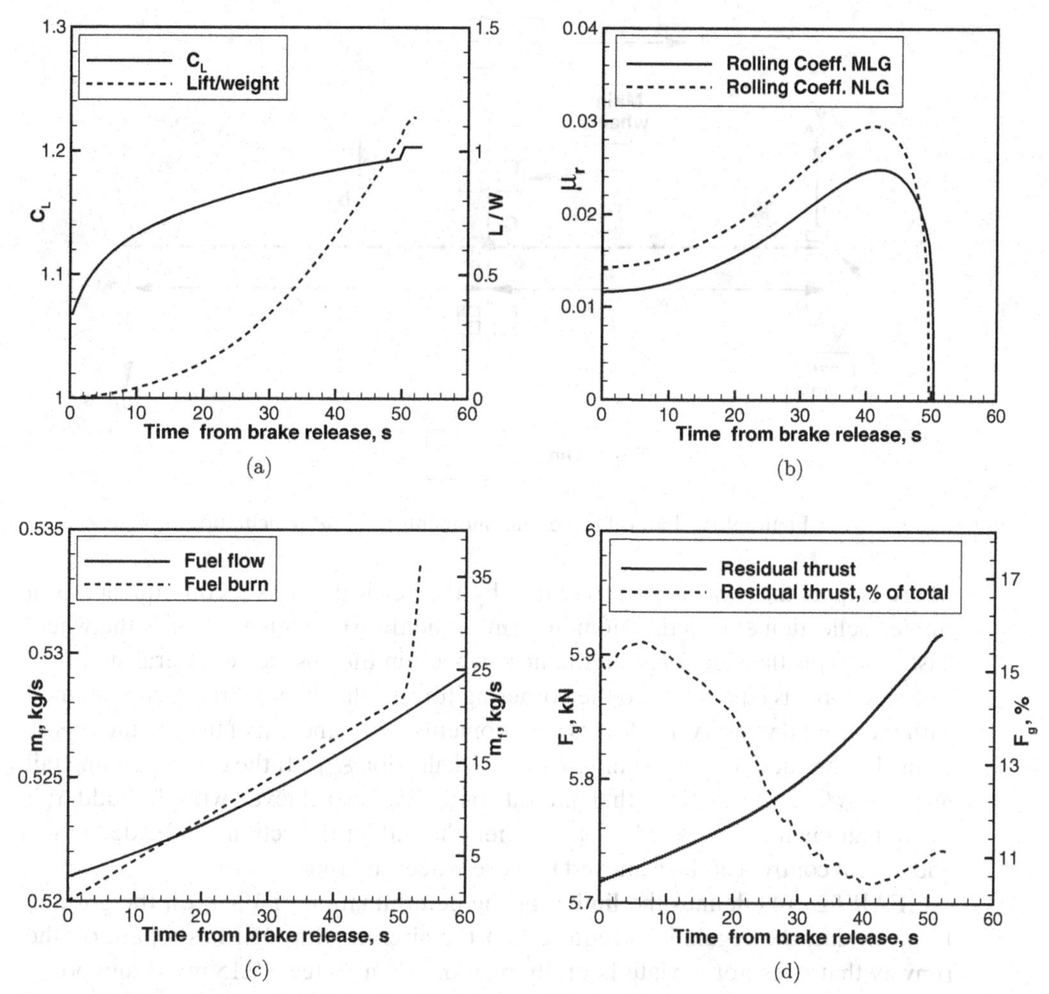


Figure 9.16. Simulation of AEO take-off of the ATR72-500 with PW127M engines. Data as in Fig 9.15 (part II).

**RESPONSE OF A FAILED PROPELLER.** The analysis carried out for a jet engine applies in general terms also to the turboprop aircraft. However, following an engine failure, the response of the aircraft is different. Unless an automatic corrective action is taken on the failed engine-propeller, the drag on that side of the aircraft would greatly increase. The procedure is to set the propeller to feathering position as quickly as possible. This operation can take a few seconds. The rate of pitch change is about 15 to 25 degrees/s.

## 9.6 Minimum Control Speed

With reference to the twin-engine aircraft shown in Figure 9.17, the yaw moment balance around the point C (centre of the main landing gear) is

$$\mathcal{M}_{vt} = Tb_t - Y_n w_b, \tag{9.39}$$

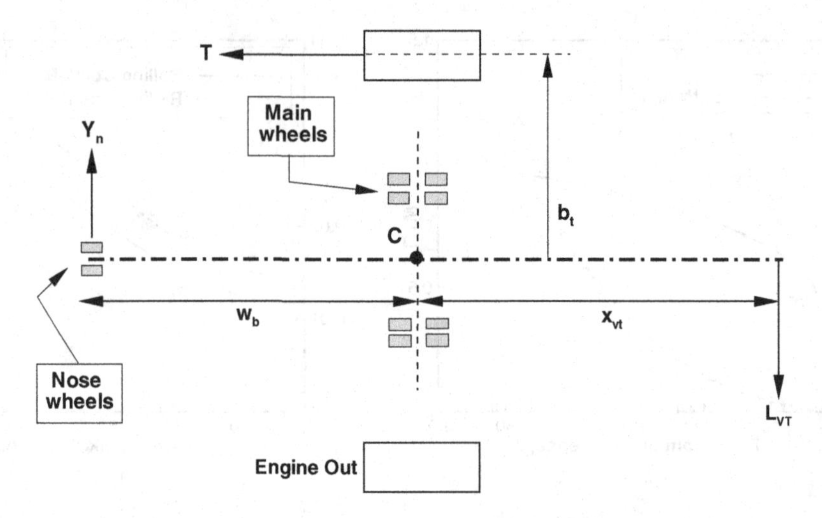


Figure 9.17. Lateral forces and moments for $V_{MCG}$ calculation.

where $\mathcal{M}_{vt}$ is the yaw moment created by the vertical tail as a consequence of a rudder deflection $\xi$; $b_t$ is the moment arm of the thrust asymmetry; $w_b$ is the wheel base; and $Y_n$ is the side force on the nose wheels in the absence of lateral skidding. The side force is equivalent to the cornering force if the plane of rotation is aligned with the ground velocity. All forces and moments are a function of the ground speed. Consider the rudder in its maximum design deflection $\xi_{max}$. If the corresponding tail moment $\mathcal{M}_{vt} > Tb_t - Y_n w_b$, then lateral control can be achieved with the rudder; if the tail moment is $\mathcal{M}_{vt} < Tb_t - Y_n w_b$, then the rudder deflection can be decreased and lateral control can be achieved by nose-wheel steering.

FAR §25.149 defines the limits for the determination of $V_{MCG}$ on the ground. In particular, the regulations require that the aircraft with OEI has a path on the runway that does not deviate laterally by more than 30 feet (9.15 m) at any point. The minimum control speed has to be lower than the lift-off speed. The $V_{MCG}$ must be determined with the aircraft at the most critical condition, the most unfavourable position of the CG and take-off weight. The moment generated by the vertical tail is

$$\mathcal{M}_{vt} = L_{vt} x_{vt} = \frac{1}{2}\rho U^2 A_{vt} C_{L_\xi} \xi x_{vt}, \tag{9.40}$$

where $A_{vt}$ is the vertical tail's area and $x_{vt}$ is the distance between the line of action of the rudder lift and the centre C of the main landing gear (Figure 9.17). Assuming $\xi = \xi_{max}$ and solving for the ground speed $U = V_{MCG}$, the yaw equation becomes

$$V_{MCG}^2 = \frac{2(Tb_t - Y_n w_b)}{\rho A_{vt} C_{L_\xi} \xi_{max}}. \tag{9.41}$$

The solution of Equation 9.41 depends on the limit side force on the nose tyres. A semi-empirical method for calculation of this force is given by ESDU[11], although it requires the yaw angle of the tyre[7]. We will consider the case of an unbraked wheel.

For a given value of the normal load on the tyre, $F_n$, we are interested in finding the limit side force (*cornering force*) on the tyre $Y_n$ before the tyre starts skidding sideways. In practice, before such condition occurs, the tyre will operate in yaw conditions. This tyre yaw introduces a major complication. In the first place, as

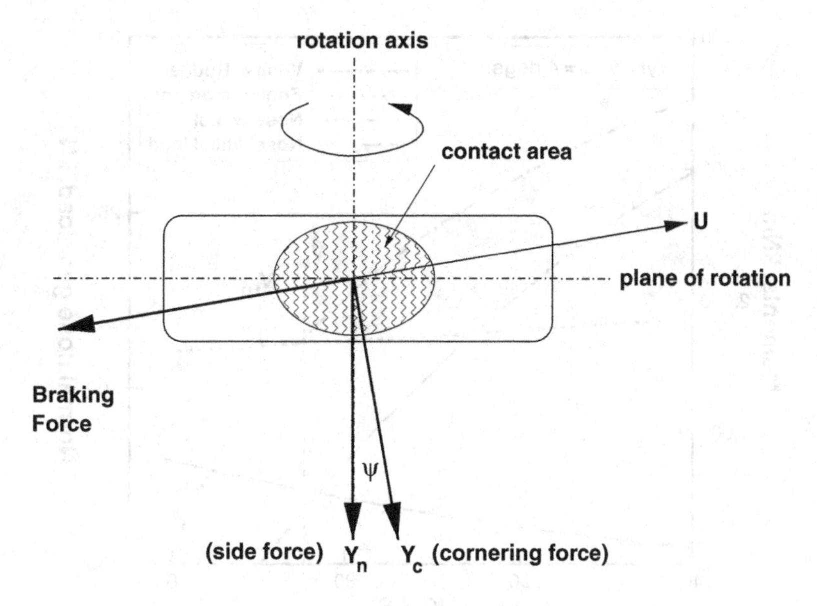


Figure 9.18. Cornering force on tyre.

the aircraft accelerates it gains lift and unloads the nose gear. As the nose gear is unloaded, its cornering force decreases. Thus, unless the dynamic lift grows quickly, the velocity $V_{MCG}$ will be inconveniently high. In fact, it is possible that there is a range of ground speeds at which it is not possible to control the aircraft, either by nose-gear steering or rudder deflection. It is necessary to decide whether the nose tyres are operating in yaw because the cornering force depends on this angle.

Now assume that the tyre's yaw angle $\psi$ is sufficiently small (Figure 9.18). The ratio between these two forces is the lateral friction coefficient

$$\mu_s = \frac{Y_n}{F_n}. \tag{9.42}$$

From the analysis of data published by Dreher and Tanner[12] on a number of tyres under dry, wet and flooded runways, the sideways-friction coefficient can be written as

$$\mu_s = \mu_s c_1(V, F_n)\psi + c_1(V, F_n)\psi^2. \tag{9.43}$$

A suitable computational procedure consists in iterating at increasing air speeds, as follows:

- Set the maximum rudder deflection $\xi = \xi_{max}$.
- Fix the true air speed, $U = V \pm U_w$.
- Calculate the yaw moment $Tb_t - Y_n w_b$.
- Calculate the rudder moment $\mathcal{M}_{vt}$.
- If $\mathcal{M}_{vt} < Tb_t - Y_n w_b$, increase the true air speed.
- If $\mathcal{M}_{vt} < Tb_t - Y_n w_b$, the corresponding air speed is the $V_{MCG}$; exit the loop.

An example of $V_{MCG}$ calculation is shown in Figure 9.19 for the conditions specified in the caption. The result indicates that $V_{MCG} < V_{lo}$, as expected. The

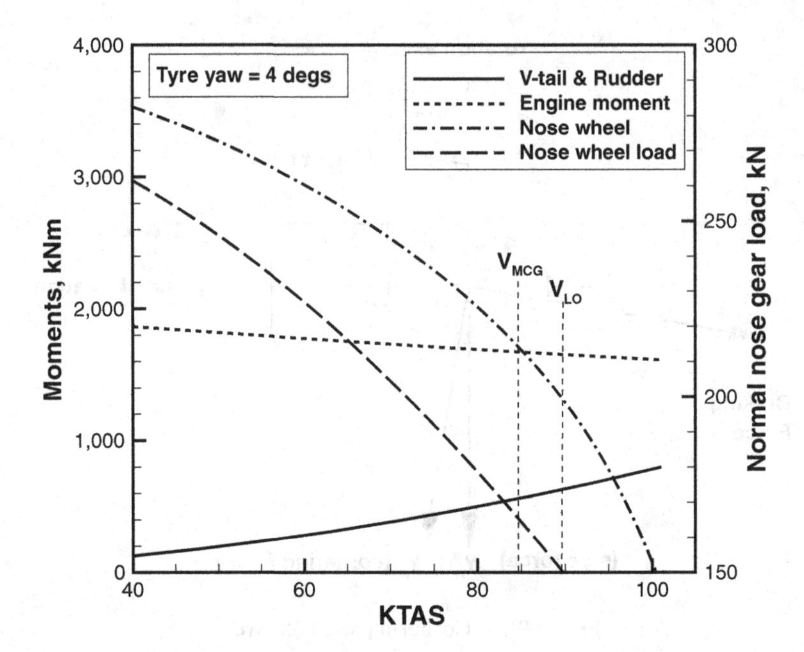


Figure 9.19. Calculated $V_{MCG}$ for a reference aircraft; $m = 145{,}000$ kg, standard day, no wind, dry runway.

calculation assumed that the maximum rudder deflection is 10 degrees and that the aircraft drifts laterally by 30 feet (9.15 m) over the distance of 1,000 m.

The result depends on several factors, including conditions of the runway, tyre tread (ribbed or smooth), wind speeds and direction (headwind or tailwind), position of the CG, tyre yaw angle, maximum rudder deflection and so on. The method described does not take into account roll effects.

For a four-engine aircraft, we consider the case of only one engine inoperative. The worst-case scenario is when one of the outer engines fails. The remaining steps of the analysis would be the same.

**EFFECTS OF DERATED THRUST.** A few additional clarifications are required with respect to the derated thrust. A derated thrust causes a reduction on the minimum control speed on the ground and a change in take-off speeds. A reduced take-off thrust must not result in loss of any functions and systems operations*. In the case of one engine inoperative (OEI), a derated thrust reduces the destabilising yawing moment. This moment has to be balanced by the vertical tail that is capable of providing control only at speeds above the VMCG. A lower yawing moment leads to a lower VMCG; a reduced VMCG leads to a reduced accelerate-stop distance. For each derated level, there must be a certified VMCG.

## 9.7 Aircraft Braking Concepts

Aircraft braking on the ground is achieved by a combination of systems. The primary system consists of wheel brakes; the secondary system is based on aerodynamics, in

* Federal Aviation Regulations: Reduced and Derated Takeoff Thrust (Power) Procedures, AC-25-13, April 1988.

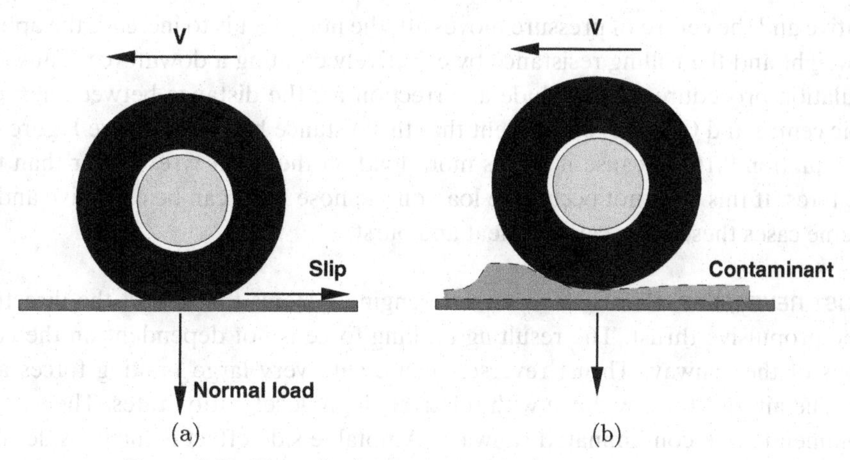


Figure 9.20. Tyre slip (normal operation) and aquaplaning.

particular spoilers. Finally, engine thrust reversers can be used. Each system can be characterised by a mathematical equation in terms of the relevant parameters (aircraft weight, position of the CG, ground speed, runway conditions).

**MECHANICAL BRAKES.** The aircraft can be brought to a halt by using appropriate wheel brakes. The overall resistance is the sum of the rolling resistance of the wheel–tyre combination. The larger the load on the wheel, the higher the braking force. However, an important factor is the tyre–runway speed (or skid). A free-rolling wheel has a rotational speed equal to the aircraft's speed. A fully locked wheel has no rolling speed. At intermediate condition the tyre skids on the runway. The ratio $\Omega r/V$ between the rotational speed of the tyre and the aircraft's speed is the so-called slip ratio.

The slip between the tyre and the roadway is shown in Figure 9.20. At normal operating conditions, the tyre slip is equal to the ground speed (100% slip-ratio, or free-rolling speed); when there is partial loss of contact, the tyre slip is lower and causes skidding. The limiting case is when the tyre slip is zero; that is, the tyre is locked (the slip-ratio is zero). The friction force depends on the slip-ratio: a free-rolling tyre has zero-resistance and hence zero friction coefficient and zero braking action. Clearly, this is not a good situation because it would be difficult to stop the aircraft with mechanical brakes. Conversely, a skidding tyre has a poor braking performance in addition to considerable risks of damage and explosion. Wheel-lock is prevented with anti-locking (or anti-skid) systems. These systems compare the speed of each wheel with the rolling speed of the aircraft. The maximum braking performance is at slip-ratios between 10 and 15%. FAR §23.109 recommends maximum braking coefficients for wet conditions in the form

$$\mu_b = c_1 V^3 + c_2 V^2 + c_3 V + c_4, \tag{9.44}$$

in which the coefficients are dependent on the tyre pressure $p$.

**AERODYNAMIC SPOILERS.** For a transport aircraft, braking is achieved by dumping lift and mechanical brakes on the wheels. Lift is dumped from the wing by a rapid deployment of the spoilers. When the spoilers are deployed, the wing lift becomes negative and the centre of pressure moves aft; the net effect is to increase the apparent weight and the rolling resistance by effectively creating a downforce. Thus, the calculation procedure must include a correction for the distance between aerodynamic centre and CG. It is convenient that this distance be reduced (see Figure 9.4 and Equation 9.16) because it forces more load on the main tyres rather than the nose tyres. If this does not occur, the load on the nose tyres can be excessive and in extreme cases these tyres can overheat and burst.

**THRUST REVERSERS.** Thrust reversers are engine systems that revert the direction of the propulsive thrust. The resulting braking force is not dependent on the conditions of the runway. Thrust reversers can create very large braking forces and cause the aircraft to slow down with relatively high deceleration rates. Their use is recommended on contaminated runways. A notable side effect is the considerable acoustic emissions they generate. Consequently, their use is regulated; the aircraft must demonstrate appropriate deceleration and stopping in the absence of thrust reversers.

## 9.8 Performance on Contaminated Runways

There is a variety of conditions for paved runways. Each condition affects the rolling resistance of the aircraft and the braking efficiency. The official classification in order of increasing severity includes damp, wet, standing water, slush, wet snow, dry snow, compacted snow and icy surface. Investigations on rolling-friction coefficients have been carried out over the years. See, for example, Wetmore[13], Harrin[14], Yager[15] and Agrawal[16].

The runway is *contaminated* when more than 25% of the runway surface area to be used is covered by water more than 3mm deep, or by slush or loose snow equivalent to at least 3mm of water. A damp runway contains small amounts of humidity. A wet surface has a thin layer of water that makes it shiny; however, this water is not deep enough to cause hydroplaning. Standing water is generally localised to areas where drainage is poor after a rain storm. Slush is water mixed with snow at temperatures above freezing. Wet snow is found after fresh precipitation; it is light-weight and can be easily shaped. A dry snow is present at colder temperatures and dry weather; compacted snow has been compressed by heavy machinery. In icy conditions, the rolling resistance is lowest ($\mu_r < 0.005$). Specifically, braking performance degrades with slippery roadways; take-off acceleration and lateral control decrease.

Under *wet* conditions, the friction coefficient between the tyre and the road is lower than normal because the contaminant cannot be pushed out by the tyre-road contact; consequently there is a reduced contact and a reduction of the traction force. The contact between tyre and contaminant does not generate a useful tractive force. To increase the friction coefficient, it is necessary to have tyres capable of squeezing the contaminant out of the contact area. In practice, this can be done at relatively low speeds, but as the speed increases, the time available for the contaminant to be squeezed out decreases and the roadway becomes *more slippery*, with potentially catastrophic consequences.

An extreme case is *hydroplaning* also shown in Figure 9.20. Hydroplaning is that condition in which the tyre is no longer in contact with the runway but travels over a deep layer of water. The water film between the tyre and the runway reduces the rolling resistance. Under these conditions, braking becomes ineffective and directional control is lost. Additional problems that are incurred in this event include the risk of spray ingestion into the engines, the flaps, the sensors and other delicate aircraft systems.

A rule-of-thumb that is used to determine the friction coefficient under wet conditions is to take half the value of the friction coefficients on dry runways. However, this rule does not satisfy the requirements for safety, and new methods have been developed that take into account the tyre inflation pressure, its wear state, the type of runway and the anti-skid systems.

### 9.8.1 Contamination Drag

In the presence of loose runway contamination (water, slush and snow), an additional form of drag appears. This drag is due to two factors: the contamination drag due to the effects of displacement of the contaminant by the tyres (§ 9.8.1) and the resulting spray impingement on the airframe (§ 9.8.2). The former contribution is always larger than the latter. The maximum value of the contamination drag increases with the roll speed up to a maximum, and then it decreases.

Figure 9.21 shows the effects of contamination on the airframe due to spray plumes lifting from the nose and main tyres. In particular, at point A we have the effect onto the fore fuselage due to the nose tyres. The same plume hits the lower wing, inboard of the engine (B). It would be more damaging if this plume expands over a wider angle to impinge on the engine intakes. The effects of the nose spray plume on the central fuselage are shown by point C. Finally, the effect of the main landing gear on the rear fuselage shows impingement of the spray on the horizontal stabiliser (D) and on the lower fuselage (D).

For a single tyre, ESDU[17] gives the following expression for the displacement drag:

$$\begin{cases} G_{F_{1,1}} = \frac{1}{2}\rho_w \varsigma V^2 C_{D_{1,1}} b_s d, & V/V_p \leq 1 \\ G_{F_{1,1}} = f_L G_{F_{1,1}}(V/V_p = 1), & V/V_p > 1 \end{cases} \tag{9.45}$$

where $b_s$ is the tyre width at the intersection with the contaminant, $d$ is the tyre diameter, $C_{D_{1,1}}$ is the drag coefficient due to displacement of the contaminant, and $\varsigma = \rho_c/\rho_w$ is the ratio between the density of the contaminant and that of the water ($\rho_w$). The factor $f_L$ accounts for the effects of ground speed $V$ relative to a reference speed $V_p$. The latter parameter is defined as

$$V_p = 34.28 \left(\frac{p}{\varsigma}\right)^{1/2}, \tag{9.46}$$

where the pressure is given in bar and the reference speed $V_p$ is in knots. The displacement drag of twin tyres, $G_{F_{2,1}}$, is given by

$$\begin{cases} G_{F_{2,1}} = f_{2,1} G_{F_{1,1}}, & V/V_p \leq 1 \\ G_{F_{1,1}} = f_L G_{F_{2,1}}(V/V_p = 1), & V/V_p > 1 \end{cases}. \tag{9.47}$$

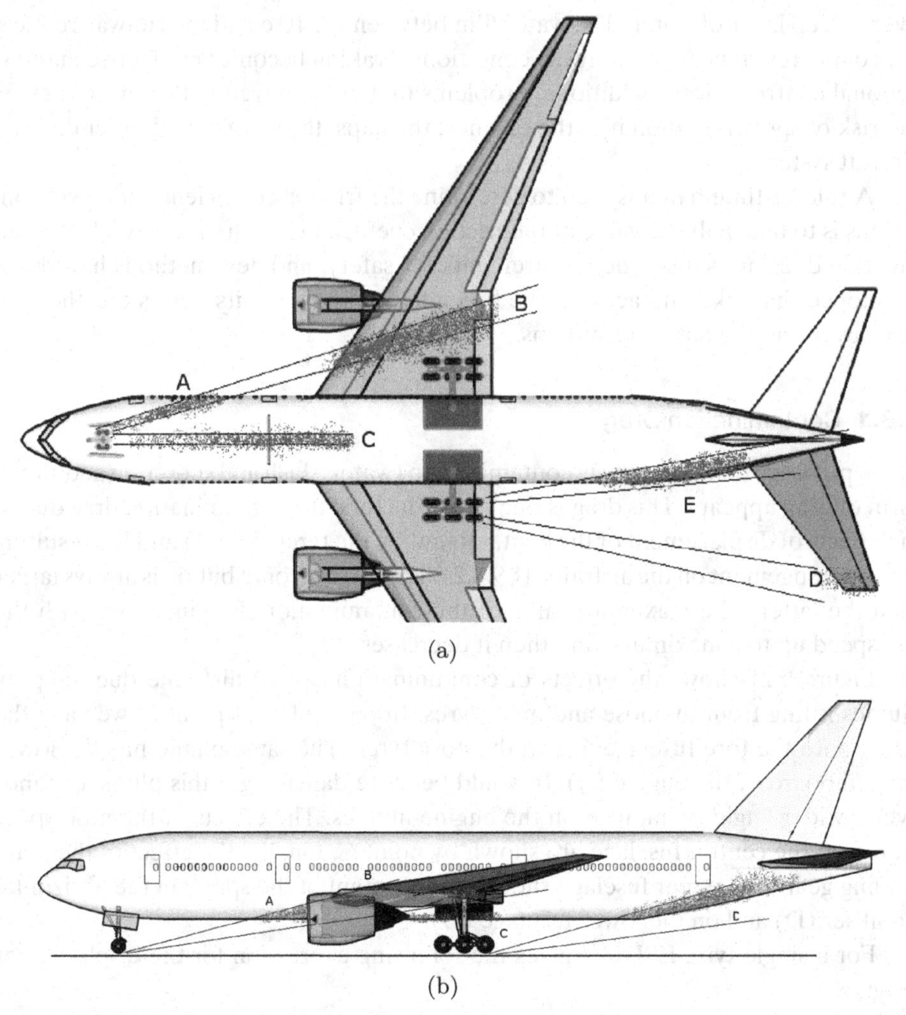


Figure 9.21. Effect of spray plumes from contaminated runways onto a transport aircraft.

The displacement drag of dual-tandem tyres $G_{F_{2,2}}$ (four tyres) is estimated from the sum of two components:

$$G_{F_{2,2}} = G_{F_{2,1}} + G_{F_i}, \tag{9.48}$$

where the first contribution is recognised as the displacement drag of twin tyres (the leading pair and associated mechanical systems). The second contribution is an impingement contribution on the trailing tyres, defined as

$$G_{F_i} = f_i G_{F_i}(V/V_p = 1), \tag{9.49}$$

with

$$G_{F_i}(V/V_p = 1) = \rho_w \varsigma V_p^2 b_s d. \tag{9.50}$$

Both factors $f_L$ and $f_i$ are given in Ref.[17] in graphical form. An example of calculation is shown in Figure 9.22, which shows the breakdown of the retarding forces due to standing slush having a constant depth of 1 cm (0.01 m) for an Airbus A320-200 model. The landing-gear contribution refers to the sum of two separate

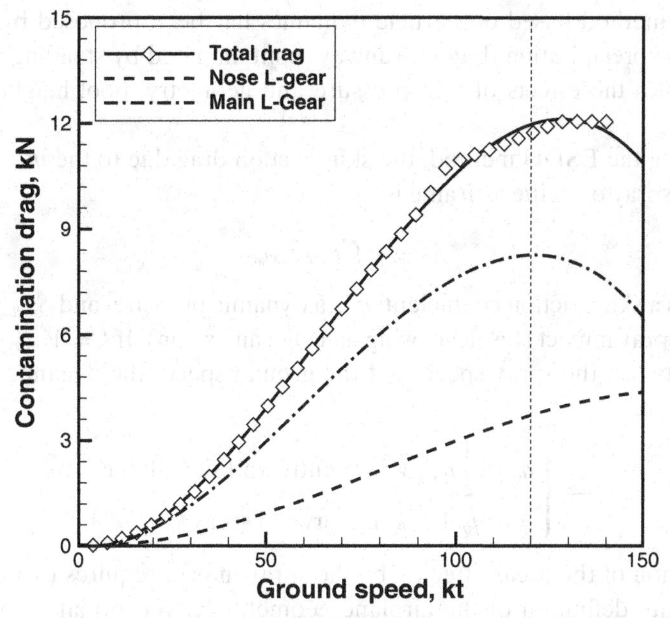


Figure 9.22. Contamination drag of a transport airplane (calculated).

under-carriages. The tyre deformation was assumed as 2.5% of the tyre diameter in both nose and main under-carriage (*in lieu* of a full aircraft trim), with tyre inflation pressures equal to 190 and 110 psi (*pounds-per-square inch*), respectively (~1.31 and ~0.76 MPa). The contamination drag increases to a maximum at about 120 kt, and then it decreases.

### 9.8.2 Impingement Drag

A first-order estimate proposed by the Joint Aviation Regulations (JAR) for the displacement drag is

$$D_{displ} = \frac{1}{2}\rho_c S_{tyre} V^2 K C_D, \tag{9.51}$$

where $\rho_c$ is the density of the contaminant (water, slush, snow); $S_{tyre}$ is the tyre's frontal; area within the contaminant; $K$ is the wheels coefficient (as discussed in § 4.2.7); and $C_D = 0.75$ for both water and slush.

The impingement drag depends on the complex interaction between the geometry of the airframe and the fluid dynamics of the displaced contaminant. The current practice is to perform tests on the runway with various levels of contamination for certification purposes. However, a body of research is available to guide in the estimation of this effect. ESDU[18] has basic methods for the estimation of the shape of the plumes generated by the contact between tyre and contaminant, the displacement drag[17], as well as the estimation of the impingement drag[19]. On the basis of a set of tests with a number of different aircraft with both standing water and snow, Giesberts[20] reported that Equation 9.51 is not very accurate. In particular, it is shown that the precipitation drag increases up to a ground speed, and then it gradually decreases. Equation 9.51 predicts a uniformly increasing drag. An engineering method based on particle dynamics has been proposed by Gooden[21] to predict the precipitation drag on runways contaminated by standing water. This model includes the effects of tyre pressure and geometry, pool height and other factors.

Following the ESDU method, the skin-friction drag due to the $m$–th impingement of the spray onto the airframe is

$$G_{F_m} = C_{F_m} q S_{wet_m}, \tag{9.52}$$

where $C_{F_m}$ is a skin-friction coefficient, $q$ is a dynamic pressure, and $S_{wet_m}$ is the area affected by spray impact (fuselage, wing, nacelle, and so on). If $k = V_{spray}/V$ denotes the ratio between the spray speed and the ground speed, the dynamic pressure is calculated from

$$q = \begin{cases} q_o = \frac{1}{2}\rho_w \varsigma V^2 & \text{centre and side plumes} \\ q = q_o(1 - k^2) & \text{forward spray} \end{cases}. \tag{9.53}$$

The calculation of the areas affected by the spray impact requires two ingredients: an appropriate definition of the airplane geometry (CAD format or other useful formats) and an estimate of the spray envelopes as they travel downstream. If these data are available, it is possible to describe a numerical algorithm that calculates the intersection between the airframe geometry and the spray envelope.

The last term required in Equation 9.52 is the average skin-friction coefficient in the area affected by spray impact. Because the air flow is most likely turbulent and the airframe is nearly flat relative to the size of the spray particles, we can use well-known expressions for the turbulent skin friction over a flat plate. One of these expressions is the equation of the skin-friction coefficient $C_F$ in incompressible turbulent flat plate (Kármán-Schoenerr):

$$\frac{1}{\sqrt{C_F}} = 4.3 \log_{10}(C_F Re_x). \tag{9.54}$$

In Equation 9.54 $Re_x$ denotes the Reynolds number based on the reference length $x$: $Re = \rho U x/\mu$, with $\mu$ and $\rho$ the dynamic viscosity and the density of *clean water*, respectively. The reference length is taken as the maximum length of the $m$–th area affected by spray impact. Simplifications can be applied at this point. For example, the use of the root chord $c_r$ (which is readily available) is not too different from the length of the spray impingement on the wing. Finally, note that the mean $C_F$ from Equation 9.54 is *added* to the skin-friction drag occurring in normal conditions, that is, due to air flow around the airplane.

## 9.9 Closed-Form Solutions for Take-Off

Closed-form solutions are obtained with a number of simplifications. They are useful for a first-order investigation of the field performance and represent methods that are useful in conceptual design. In fact, there is some value to be gained from approximations such as lumped mass, parabolic drag equation, constant thrust and a number of other assumptions. We consider the cases of the jet and propeller aircraft separately.

Table 9.3. *Average rolling coefficient for some runway conditions*

| Runway Condition | $\mu_r$ |
|---|---|
| Dry concrete/asphalt | 0.02 |
| Hard turf & gravel | 0.04 |
| Short & dry grass | 0.05 |
| Long grass | 0.10 |
| Soft ground | 0.10–0.30 |

### 9.9.1 Jet Aircraft

The ground run is the distance between brake-release and lift-off of the forward wheels. To calculate the ground run, we write the dynamics equations on the centre of gravity of the aircraft in the horizontal and vertical directions, for a take-off from a horizontal runway. The engine thrust is aligned with the vector velocity. Thus, we have:

$$m\frac{\partial u}{\partial t} = T - D - R, \tag{9.55}$$

$$R = \mu_r(L - W), \tag{9.56}$$

where $R$ is the ground resistance. Some average data for rolling-resistance coefficients are given in Table 9.3.

The acceleration of the aircraft can be written as

$$a = \frac{\partial u}{\partial t} = u\frac{\partial u}{\partial x} = \frac{\partial}{\partial x}\left(\frac{1}{2}V^2\right). \tag{9.57}$$

With this definition, Equation 9.55 becomes

$$\frac{1}{2}m\frac{\partial V^2}{\partial x} = T - D - R. \tag{9.58}$$

Integration of Equation 9.55, performed between brake-release ($x = 0$, $t = 0$) and the lift-off point, yields

$$\frac{1}{2}mV_{lo}^2 = \int_o^x (T - D - R)\, dx. \tag{9.59}$$

Integration of the right-hand side of Equation 9.59 requires additional information as to how the aircraft forces depend on the speed. For example, the front wheels raise from the ground before the rear wheels. When this occurs, there is a step change in the rolling resistance.

First, consider the lift-off speed. This speed can be estimated from the stall speed. It is allowed to have $V_{lo} \simeq 1.1\, V_S$, which implies that $C_{L_{lo}} \simeq 0.83 C_{L_{max}}$; this is a reasonably safe margin. FAR Part 25 (§25.103) defines the stall speed as

$$V_{stall} \geq \frac{V_{CLmax}}{\sqrt{n}}, \tag{9.60}$$

where $n$ is the load factor in the direction normal to the flight path; $V_{CLmax}$ is the calibrated airspeed obtained when the load-factor-corrected lift $nW/qA$, is a maximum. The regulations prescribe the methods that must be used to calculate the $V_{CLmax}$. A stall warning must be clear at all critical flight conditions. Therefore, the lift-off speed is

$$V_{lo}^2 = \frac{2}{\rho}\frac{W}{A}\frac{1}{C_{L_{lo}}}. \tag{9.61}$$

The take-off time is obtained by integration of the speed,

$$t_{lo} = \int_0^{x_{lo}} \frac{dx}{V}. \tag{9.62}$$

**SIMPLIFIED SOLUTIONS.** A first-order approximation is to take an average value of the thrust, drag and rolling resistance over the ground run. Under these circumstances, the result of the integration is

$$\frac{1}{2}mV_{lo}^2 = (\overline{T} - \overline{D} - \overline{R})\, x_{lo}. \tag{9.63}$$

Suitable values of the parameters in Equation 9.63 are:

$$\overline{T} = T_o, \qquad \overline{D} = \frac{1}{2}\rho A C_D \left(\frac{V_{lo}}{2}\right)^2, \qquad \overline{R} = \mu_r(W - \overline{L}) \simeq \frac{1}{2}\mu_r W. \tag{9.64}$$

By replacing the equivalences in Equation 9.63, we find

$$\frac{1}{2}mV_{lo}^2 = \left(T_o - \frac{1}{8}\rho A C_D V_{lo}^2 - \frac{1}{2}\mu_r W\right) x_{lo}, \tag{9.65}$$

that solved in terms of $x_{lo}$ yields

$$x_{lo} = \frac{4mV_{lo}^2}{8T_o - \rho A C_D V_{lo}^2 - 4\mu_r W}. \tag{9.66}$$

For a more refined solution, consider the thrust independent of the speed and equal to the static thrust. Express the drag force as

$$D = \frac{1}{2}\rho A (C_{D_o} + kC_L^2) V^2, \tag{9.67}$$

where the $C_L$ and $C_D$ are relative to the ground configuration of the aircraft. At this point we note that in the aircraft's ground configuration the control surfaces are extended.

Returning to the analysis of the momentum equation, the rolling resistance is

$$R = \mu_r \left(W - \frac{1}{2}\rho A C_L V^2\right), \tag{9.68}$$

with $W$ = constant. The solution is

$$\frac{1}{2}mV_{lo}^2 = (T_o - \mu_r W)x_{lo} - \frac{1}{2}\rho A \int_o^x (C_{D_o} + kC_L^2 + \mu_r C_L) V^2\, dx. \tag{9.69}$$

The $C_L$ to be introduced in Equation 9.69 is a function of the angle of attack of the aircraft (see § 4.1), but not the speed. A suitable approximation is that these surfaces are set to take-off positions, and so the lift coefficient in ground effect is constant.

Therefore, Equation 9.69 becomes

$$\frac{1}{2}mV_{lo}^2 = (T_o - \mu_r W)x_{lo} - \frac{1}{2}\rho A(C_{D_o} + kC_L^2 + \mu_r C_L)\int_o^x V^2\,dx. \tag{9.70}$$

The last step in solving the dynamics equation is the knowledge of the speed $V(x)$. If an acceleration $a(V) = a_o + c_1 V^2$ is assumed, then it can be demonstrated that if the total acceleration during the ground roll does not exceed 40% of its initial value at lift-off, the "exact" and "constant" acceleration solutions for the ground distance differ by about 2% – by all means an acceptable error from an engineering point of view. By using a constant acceleration, the relationship between the speed and time is quadratic. In fact,

$$a = \frac{\partial u}{\partial t} = \frac{\partial u}{\partial x}\frac{\partial x}{\partial t} = \frac{1}{2}\frac{\partial V^2}{\partial x} = \text{const.} \tag{9.71}$$

Integration of this leads to $V^2 \simeq c\sqrt{x}$, bar a constant of integration. To match the lift-off conditions $\{x_{lo}, V_{lo}\}$, the relationship between speed and ground run must be

$$V^2(x) = \frac{V_{lo}^2}{x_{lo}}x. \tag{9.72}$$

With this result, the solution of the integral term in Equation 9.70 is

$$x_{lo} = \frac{mV_{lo}^2/2}{(T_o - \mu_r W) - \rho A(C_{D_o} + kC_L^2 + \mu_r C_L)V_{lo}^2/4}. \tag{9.73}$$

The final method of solution of the take-off equation consists in using the actual thrust and drag. The engine thrust is expressed by polynomial functions, that is, functions such as

$$T(U) = T_o(1 + c_1 U + c_2 U^2), \tag{9.74}$$

with $c_i$ constant coefficients. Note that the thrust depends on the airspeed $U$, not the ground speed $V$. The take-off Equation 9.59 will be written as

$$\frac{\partial V}{\partial t} = \frac{1}{m}\int_o^x (T - D - \mu_r R)\,dx. \tag{9.75}$$

The best method of integration relies on a fourth-order Runge-Kutta method with variable time stepping, which is stable and accurate. The stopping criterion is the lift-off point that we assume occurs when $L = W$.

**ROTATION AND INITIAL CLIMB.** The lift-off point is reached with at least one landing gear on the ground. The rotation of the aircraft must be done with a small angle or else there is a risk of a tail strike on the runway. Tail strike is not unusual with a wide-body aircraft, and it has potentially serious consequences. Figure 9.23 shows that the tail strike depends on the geometrical configuration of the aircraft, in particular the landing-gear height $h_g$ and the distance between the centre of the wheels and the strike point, $l_1$. Pinsker[22] proposed the solution to a number of problems that

Figure 9.23. Tail strike at the take-off flare; the limit rotation angle depends on the distance between the point of contact with the ground and the main landing gear.

occur at lift-off, including conditions for tail strike and minimum and ground clearance required for simultaneous banking and pitching.

At the start of the lift-off there is an abrupt change in rolling resistance. The airborne phase starts with a further rotation of the aircraft. Then the aircraft climbs along a straight path in order to fly past the reference screen. During flare the aircraft has a centripetal acceleration

$$n = \frac{V^2}{\chi}, \tag{9.76}$$

where $\chi$ is the radius of curvature of the flight path. The distance on the ground run during this phase will be $x_3 \simeq \chi\gamma$, and the height reached will be

$$h_1 = \chi(1 - \cos\gamma) \simeq \chi\frac{\gamma^2}{2}, \tag{9.77}$$

where $\gamma$ is the climb angle. A first-order estimate for a commercial subsonic jet indicates that $h_1 \simeq 1$ m (3 feet) and the distance covered is of the order of 12 m (~40 feet), which is negligible in most cases. The initial climb is done with a constant climb angle, so that

$$x_a = \frac{h - h_1}{\tan\gamma} \simeq \frac{h}{\tan\gamma}. \tag{9.78}$$

The total take-off distance will be

$$x_{to} \simeq x_{lo} + x_a. \tag{9.79}$$

An alternative calculation of the airborne phase can be done with an energy method. The change in total energy from lift-off to the point of clearing the screen is

$$Wh + \frac{1}{2}m\left(V^2 - V_{lo}^2\right) = (\overline{T} - \overline{D})x_a, \tag{9.80}$$

where $u$ is the minimum airborne control speed. Therefore, the airborne distance is

$$x_a = \frac{Wh + m\left(V^2 - V_{lo}^2\right)/2}{\overline{T} - \overline{D}}. \tag{9.81}$$

There are a number of other cases of interest. Take-off under icing conditions has been studied extensively. One practical example is van Hengst[23], who also provides some lift curves with and without de-icing fluids for the Fokker 50 and Fokker 100.

The take-off performance with iced surfaces requires at least the knowledge of the correct aerodynamic coefficients.

The take-off of seaplanes and flying boats is complicated by the additional resistance of the water and the water waves. Essentially, we need to find a good estimate of the resistance of the aircraft. This is the sum of the aerodynamic and hydrodynamic drag (see also § 4.7). Relevant studies on this subject are those of Perelmuter[24] and Parkinson *et al.*[25] DeRemer[26] showed take-off measurements from shallow lakes of the Cessna 180.

### 9.9.2 Propeller Aircraft

The method of calculation follows closely that of the jet aircraft, although there are a number of complications due to replacement of the jet thrust with a propeller thrust, which depends on a number of operational parameters (see Chapter 6). A low-order solution is found with the balance of forces along the horizontal direction is

$$m\frac{\partial V}{\partial t} = \eta\frac{P}{V} - D - R, \tag{9.82}$$

or

$$\frac{\partial}{\partial x}\left(\frac{1}{2}V^2\right) = \frac{1}{m}\left(\eta\frac{P}{V} - D - R\right), \tag{9.83}$$

where both the engine power and the propeller efficiency depend on the aircraft speed. We make a number of simplifying assumptions. If the power term in the right-hand side of Equation 9.83 is much larger than the sum of the other two, then

$$\frac{\partial}{\partial x}\left(\frac{1}{2}V^2\right) \simeq \frac{\eta}{m}\frac{P}{V}. \tag{9.84}$$

In addition, if we take an average engine power and propeller efficiency, then we have a first approximated value for the ground run by integrating Equation 9.84,

$$d\left(\frac{1}{2}V^2\right) = \frac{\eta}{m}\frac{P}{V}dx, \tag{9.85}$$

and finally*

$$x_{lo} \simeq \frac{1}{4}\frac{m}{V_{lo}^2}\frac{1}{\overline{\eta P}}. \tag{9.86}$$

This equation can be improved by replacing the proper values of the aerodynamic drag and the rolling resistance. The result is

$$\frac{\partial V^2}{\partial x} = \frac{1}{m}\left[2\eta\frac{P}{V} - \frac{1}{4}\rho A C_D V_{lo}^2 - \mu_r W\right]. \tag{9.87}$$

Equation 9.87 is solved for the lift-off distance $x_{lo}$ and yields

$$x_{lo} = \frac{mV_{lo}^2}{2\eta P/V - \rho A C_D V_{lo}^2/4 - \mu_r W}. \tag{9.88}$$

* For the integration use the equivalence $V^2 = z$, $V = z^{1/2}$, which leads to $\int z^{1/2}dz = z^{-1/2}/2$.

We finally consider the case in which the drag and the rolling resistance change with the speed. The governing equation can be written

$$\frac{1}{2}\frac{\partial V^2}{\partial x} = \frac{1}{m}\left[\eta\frac{P}{V} + c_1 V^2 - \mu_r W\right], \tag{9.89}$$

where the coefficient $c_1$ is

$$c_1 = -\frac{1}{2}\rho A(C_{D_o} + kC_L^2 + \mu_r C_L). \tag{9.90}$$

We consider $C_L$ a constant parameter in the ground run. Integration of Equation 9.89 yields

$$V_{lo}^2 = \frac{2}{m}\int_o^{x_{lo}}\left(\frac{\eta P}{V} + c_1 V^2 - \mu_r W\right)dx. \tag{9.91}$$

This equation can only be solved if the propeller's efficiency and the engine power as a function of the speed are known. The problem is to determine if the aircraft is operated with a variable pitch/fixed rpm or vice versa.

## 9.10 Ground Operations

Ground operations include taxiing out of the gate and back to the gate once the aircraft has landed. In a typical taxi operation, the aircraft moves at a low speed along a paved ground, performs a turn and sits idle, whilst waiting for air traffic control to give a go-ahead. Taxi times are greatly dependent on the airfield and on airfield congestion. The fuel requirements for a taxi operation are calculated as follows:

$$m_f = m_{f_1} + m_{f_2} + m_{f_3}, \tag{9.92}$$

where the first term denotes the fuel burned during taxiing at a constant speed $u_{taxi}$; the second term denotes the fuel burned in idle mode; and the last term is the fuel required to accelerate the aircraft from rest to roll speed. If the aircraft is halted more than once, $m_{f_3}$ must account for the number of accelerations. Solution of Equation 9.92 requires the distance from the gate to the brake-release point and the total taxi time – data that depend on the airport, the position of the aircraft around the terminal building and other factors. However, if these data are known, the calculation is done from

$$m_{f_1} = \left[\dot{m}_f\left(\frac{x}{V}\right)\right]_{taxi}, \tag{9.93}$$

$$m_{f_2} = (\dot{m}_f t)_{idle} \tag{9.94}$$

with

$$\dot{m}_{f_{taxi}} = f_j T = f_j\left[\mu_r(W - L) - D\right], \tag{9.95}$$

$$t_{idle} = t_{taxi} - \left(\frac{x}{V}\right)_{taxi}. \tag{9.96}$$

Table 9.4. *Estimated fuel burn during a taxi-out: V = 5 m/s; X = 2.0 km from ramp to brake-release; all fuel burn in [kg]; times in minutes; sea-level airfield; standard day; 6,500 km (3,510 n-miles) mission.*

| | A380-861-GP | | B747-400-GE | |
|---|---|---|---|---|
| Parameter | 20′ | 10′ | 20′ | 10′ |
| Fuel rest-to-roll | 24.1 | 24.1 | 23.3 | 23.3 |
| Fuel in idle mode | 960.0 | 240.0 | 652.8 | 163.2 |
| Fuel in roll mode | 381.9 | 376.6 | 391.9 | 402.0 |
| Total taxi fuel | 1365.9 | 640.4 | 1,068.0 | 588.9 |
| Time in idle | 16.7 | 6.7 | 16.7 | 6.7 |
| Time in roll | 3.3 | 3.3 | 3.3 | 3.3 |

The acceleration fuel is calculated from

$$m_{f_3} = \dot{m}_f \left( \frac{mV}{R} \right), \tag{9.97}$$

where $R$ is the total resistance (rolling plus aerodynamic drag); the fuel flow $\dot{m}_f$ is calculated by solving the engine problem with $T \simeq R$ at the prescribed speed, airfield altitude and atmospheric conditions.

The method proposed is only approximate because it does not take into account effects such as one-engine-out taxi and asymmetric thrust during a turn. Some manufacturers occasionally recommend operating the aircraft with one engine out for part of the taxiing in order to conserve fuel. This recommendation comes with some caution in case of heavy aircraft, severe weather conditions and large turns on the taxiway. Important side effects of this procedure include 1) higher risk of loss of braking capability and nose-wheel steering; 2) engine start-up failure (which would require return to the gate); and 3) higher jet blast of the operating engine. The latter problem is discussed in Chapter 14.

A summary of taxi-out fuel burn for some large aircraft is in Table 9.4 for two different taxi times (10 and 20 minutes), for a fixed distance from the gate to the brake-release point. We have assumed that the aircraft does not stop at intermediate points, but the result can be expanded to include these effects, as well as the effects of turning (see following discussion). In any case, the fuel burn required is staggering. Therefore, *large aircraft should be given queue priority at departure.*

### 9.10.1 Ground Manoeuvring

Before reaching the point of brake-release and after completing the landing operation, transport aircraft must taxi out and back to the gate. In doing so, they must manoeuvre slowly with the engines running at low efficiency. Ground manoeuvres, including steering and turning, are the source of considerable fuel waste, as none of these operations is done efficiently. An example of turning is shown in Figure 9.24 for the A380, adapted from Airbus*.

Other ground operations include moving the airplane in reverse from the gate and moving the airplane to the gate. The operation is done by ground support

* Airbus A380 Airport Planning Manual, Oct. 2009.

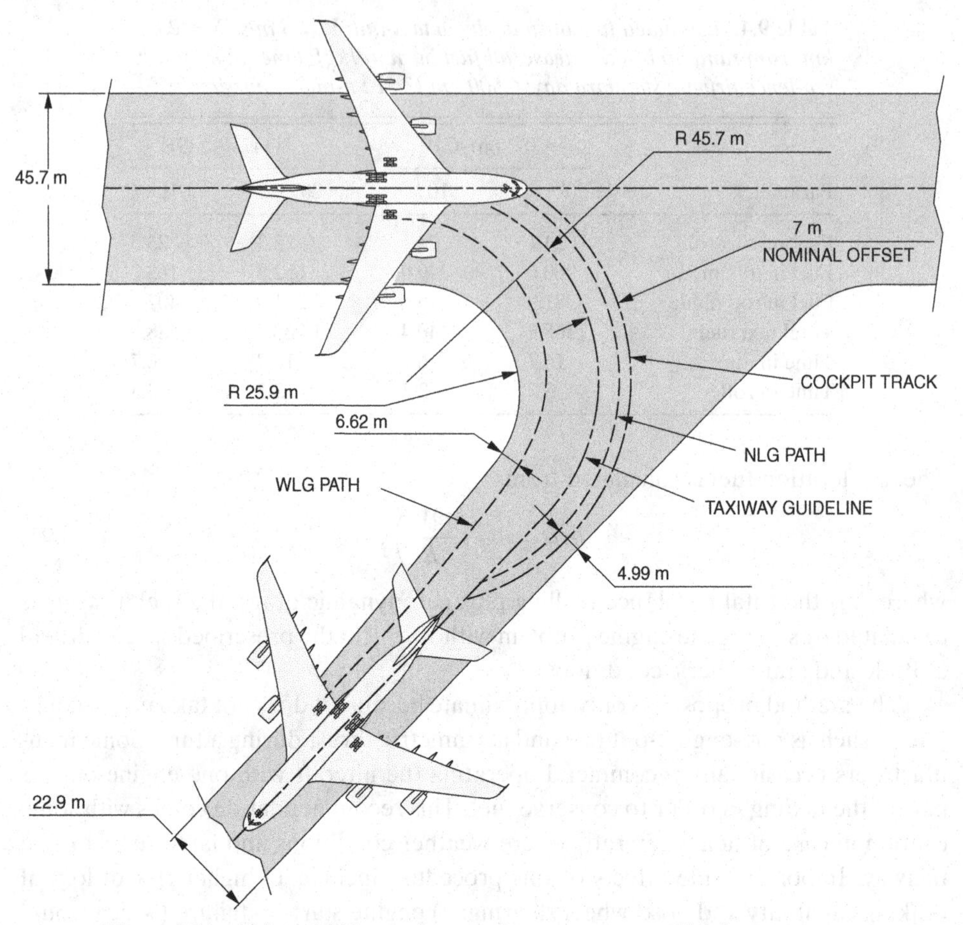


Figure 9.24. Airbus A380 turn from taxi-way to runway, 135 degrees turn.

equipment; the second one is often a manoeuvre done by the pilot. The manufacturers provide data and charts on ground manoeuvres, including minimum radius of turn and procedures to steer the aircraft, including nose steering and the use of asymmetric thrust.

### 9.10.2 Bird Strike

Birds and aircraft engines do not get along too well. It is not unusual that birds enter into a collision course with an aircraft, something that is commonly defined as a *bird strike*. Sometimes the problem is so serious that it leads to an accident or aborted flight. Data from the Flight Safety Foundation[27] indicate that there have been more than 52,000 bird strikes on commercial aircraft in the decade ending in 1998. Analysis of these data indicates that the most likely strike points are the engines (41%), the nose and the windshield (41%), the wing (7%) and the fuselage (7%). Only 15 to 20% of these strikes cause damage and require a turn-back. In the most severe cases, engine shut down is required at take-off. Approximately 50% of bird strikes into an engine cause damage, resulting at least in fan-blade damage, which causes an immediate increase in vibration and increase in exhaust-gas temperature. The damage increases with the engine thrust setting. There are two important parameters to consider in the collision: the mass of the bird and the relative speed $\Delta U$. The impact energy is $E = m\Delta U^2/2$.

Bird-strike frequency increases as the airplane is close to the ground (take-off, climb-out, final approach and landing). At least half of these strikes take place on the ground and about 80% at altitudes below 500 feet (~150 m) from the airfield. Removing the birds from the airport area is the responsibility of the airport‡.

Aside from the research on aircraft and engine response to a bird strike, considerable research has gone into understanding birds' behaviour in the proximity of the airfield. In fact, most of the strikes occur at take-off and landing (90% of all strikes), although strikes at higher altitudes are not uncommon. The introduction of higher by-pass turbofan engines may cause higher accident rates, due to the large mass flows required by these engines and their lower noise emission. As a result, the birds do not have much advance notice to avoid the advancing aircraft.

The most widespread practice is to make the airfields unfriendly to birds. An active protection method consists in landscaping the airfield with feeding areas (short grass and water bodies) and areas where birds are unlikely to loiter (tall grass). By contrast, sometimes lakes and ponds are covered with plastic balls that make birds' landing impossible; vegetation that produces seeds is removed from the area.

These days, more technological help is already at hand. Bird radars have been developed and tested at major airports and are due to be installed at control towers. These radars are programmed to peg the birds' flight parameters (altitude, direction, speed) and the birds' species. This is important because geese are recognised as more dangerous than other birds (they are relatively big and fly in large flocks). Gulls seem unable to avoid collisions, whilst red-tail hawks learn to stay away from airplanes' flight paths. The radars are also programmed to identify soaring flight and flapping wings. Birds that flap fast generally do so for a short time and then glide. To be accurate, these radars have to work on the triangulation principle; three radars are more effective than two. Bird-strike prevention strategies recommended by Airbus include:

- To keep aircraft lights on at altitudes below 10,000 feet to help the birds locate the aircraft.
- If bird strike is encountered on take-off and the speed is greater than the decision speed $V_1$, the flight must continue.
- If bird strike is encountered on take-off and the speed is less than $V_1$ but greater than 100 kt, the take-off must be aborted, to allow inspection of the engines.
- If bird strike is encountered on landing, the aircraft is to fly through the flock and land, whilst maintaining low thrust setting.
- If bird strike is encountered on landing, reverse thrust must not be used to stop the aircraft because this operation would cause further damage to the engine.

‡ Various national organisations operate wildlife control and provide advice and recommendations to the airports and crews. They can be found on the Internet with the keywords "bird strike". The first recorded bird strike is due to the Wright Brothers (1908).

## Summary

In this chapter, we presented problems of field operations of a transport aircraft powered by either turbofan or a turboprop engine. Specifically, we solved the problem of normal take-off and have identified the key parameters that affect the balanced field length: the position of the CG, the weight, the airfield altitude, the wind, the air temperature and other conditions. We have made a risk analysis with particular reference to the loss of one engine thrust/power and calculated the accelerate-go and decelerate-stop distances. In the latter case, we have further considered the problem of lateral control on the ground with asymmetric thrust. We demonstrated that control can be achieved with the rudder and vertical stabiliser, provided that the aircraft has reached a sufficient speed (minimum control speed). For some case studies we have shown the behaviour of the key flight-mechanics parameters. The atmospheric effects at the airfield are represented by the wind, by the runway contamination and not least by the risk of bird strike. For completeness, we have shown some approximate closed-form solutions.

## Bibliography

[1] ESDU. *Example of Take-off Field Length Calculations for a Civil Transport Aeroplane.* Data Item 87018. ESDU International, London, Oct. 1987.
[2] ESDU. *Calculation of Ground Performance in Take-off and Landing.* Data Item 85029. ESDU International, London, Mar. 2006.
[3] ESDU. *Force and Moment Components in Take-off and Landing Calculations.* Data Item 85030. ESDU International, London, Nov. 1985.
[4] Powers SA. Critical field length calculations for preliminary design. *J. Aircraft*, 18(2):103–107, 1981.
[5] Krenkel AR and Salzman A. Take-off performances of jet-propelled conventional and vectored-thrust STOL aircraft. *J. Aircraft*, 5(5):429–436, Sept. 1968.
[6] ESDU. *Comprehensive Method for Modelling Performance of Aircraft Type Tyres Rolling or Braking on Runways Contaminated with Water.* Data Item 05011. ESDU International, London, May 2005.
[7] Pacejka HB. *Tyre and Vehicle Dynamics.* Butterworth-Heinemann, 2002.
[8] Pritchard JI. An overview of landing gear dynamics. Technical Report TM-1999-209143, NASA, 1999.
[9] Fallah MS and Bhat R. Robust model predictive control of shimmy vibration in aircraft landing gears. *J. Aircraft*, 45(6):1872–1880, Nov. 2008.
[10] Gordon J. Perturbation analysis of non-linear wheel shimmy. *J. Aircraft*, 39(2):305–317, Mar. 2002.
[11] ESDU. *Frictional and Retarding Forces on Aircraft Tyres, Part IV: Estimation of Effects of Yaw.* Data Item 86016. ESDU International, London, Oct. 1992.
[12] Dreher RC and Tanner JA. Experimental investigation of the braking and cornering characteristics of 30 11.5 × 14.5 type VIII aircraft tires with different tread patterns. Technical Report TN D-7743, NASA, Oct. 1974.
[13] Wetmore JW. The rolling friction of several airplane wheels and tires and the effect of rolling friction on take-off. Technical Report R-583, NACA, 1937.
[14] Harrin EN. Low tire friction and cornering forces on a wet surface. Technical Report TN-4406, NACA, Sept. 1958.
[15] Yager TS. Factors influencing aircraft ground handling performance. Technical Report TM-85652, NASA, June 1983.
[16] Agrawal SK. Braking performance of aircraft tires. *Progress Aerospace Sciences*, 23(2):105–150, 1986.

[17] ESDU. *Frictional and Retarding Forces on Aircraft Tyres. Part V: Estimation of Fluid Forces*. Data Item 90035. ESDU International, London, 1990.
[18] ESDU. *Estimation of spray patterns generated from the sides of aircraft tyres running in water or slush*. Data Item 83042. ESDU International, London, 1983.
[19] ESDU. *Estimation of airframe skin-friction drag due to impingement of tyre spray*. Data Item 98001. ESDU International, London, 1998.
[20] Giesberts MKH. Test and evaluation of precipitation drag on an aircraft caused by snow and standing water on a runway. Technical Report NLR-TP-2001-490, NLR, Amsterdam, NL, Nov. 2001.
[21] Gooden JHM. CRspray – Impingement drag calculation of aircraft on water-contaminated runways. Technical Report NLR-TP-2001-204, NLR, Amsterdam, NL, Oct. 2001.
[22] Pinsker WJG. The dynamics of aircraft rotation and liftoff and its implication for tail clearance especially with large aircraft. Technical Report ARC R& M 3560, Aeronautical Research Council, 1967.
[23] Van Hengst J. Aerodynamic effects of ground de/anti-icing fluids on Fokker 50 and Fokker 100. *J. Aircraft*, 30(1):35–40, Jan. 1993.
[24] Perelmuter A. On the determination of the take-off characteristics of a seaplane. Technical Report TM-863, NACA, May 1938.
[25] Parkinson J, Olson R, and House R. Hydrodynamic and aerodynamic tests of a family of models of seaplane floats with varying angles of dead rise – NACA models 57-A, 57-B, and 57-C. Technical Report TN-716, NACA, 1939.
[26] De Remer D. Seaplane takeoff performance – Using delta ratio as a method of correlation. *J. Aircraft*, 25(8):765–766, Aug. 1987.
[27] Anon. Bird strikes found most common at low altitude in daylight. *Flight Safety Foundation Digest*, 19(2):14–16, 2000.

## Nomenclature for Chapter 9

Symbols listed in Table 9.1 are not repeated here.

| | | |
|---|---|---|
| $a$ | = | acceleration |
| $A$ | = | area or wing area |
| $b$ | = | width of tyre contact on roadway |
| $b_s$ | = | tyre width in Equation 9.45 |
| $b_t$ | = | moment arm of engine thrust |
| $c_i$ | = | constant factors |
| $c_r$ | = | root chord |
| $c_1$ | = | factor in sideways friction coefficient, Equation 9.43 |
| $C_D$ | = | drag coefficient |
| $C_{D_e}$ | = | engine-drag coefficient |
| $C_{D_g}$ | = | drag coefficient in ground effect |
| $C_{D_o}$ | = | profile-drag coefficient |
| $C_F$ | = | skin-friction coefficient |
| $C_L$ | = | lift coefficient |
| $C_{L_g}$ | = | lift coefficient in ground effect |
| $C_{L_{max}}$ | = | maximum lift coefficient |
| $C_{L_\xi}$ | = | lift-curve slope of vertical tail due to rudder deflection |
| $C_P$ | = | power coefficient of a propeller |
| $C_T$ | = | thrust coefficient of a propeller |
| $d$ | = | diameter |

$D$ = aerodynamic drag
$D_e$ = drag of inoperative engine
$\mathcal{D}$ = damping, Equation 9.31
$\mathcal{D}_{crit}$ = critical damping
$D_e$ = idle engine's drag
$E$ = energy
$f(..)$ = functional relationship
$f_i$ = factor in impingement drag of dual-tandem tyres
$f_j$ = thrust-specific fuel consumption
$f_L$ = factor in displacement drag of leading tyres
$F$ = normal load on individual tyre
$F_{cut}$ = thrust-cut function, Equation 9.34
$F_g$ = residual thrust in turboprop engine (also `Fg9`)
$F_m$ = normal load on main under-carriage
$F_n$ = normal load on nose under-carriage
$F_N$ = net thrust from a single engine (also `FN`)
$g$ = acceleration of gravity
$G_{F_i}$ = skin-friction drag due to i-th impingement on airframe
$G_{F_{1,1}}$ = displacement drag function, Equation 9.47
$G_{F_{2,1}}$ = displacement drag of twin tyres, Equation 9.47
$G_{F_{2,2}}$ = displacement drag of twin-tandem tyres, Equation 9.48
$I_{SF}$ = index denoting flap and slat setting
$I_y$ = pitch moment of inertia of airplane
$h$ = geometric altitude
$h_1$ = height at the end of the flare
$J$ = advance ratio of a propeller
$K$ = wheels coefficient, Equation 9.51
$k$ = induced-drag factor
$k$ = ratio between spray speed and ground speed, $V_p/V$
$k$ = ratio between spray velocity and airplane velocity, $V_{spray}/V$, Equation 9.53
$L$ = lift
$m$ = mass
$\dot{m}_f$ = fuel flow (also `Wf6`)
$M$ = Mach number
$\mathcal{M}_{ow}$ = pitching moment of a wing
$\mathcal{M}_y$ = pitching moment
$\mathcal{M}_z$ = yaw moment
$n$ = power factor in Equation 9.34
$n$ = normal-load factor in Equation 9.60
$n_e$ = number of engines
$n_w$ = number of wheels/tyres
$p$ = pressure
$\mathsf{q}$ = pitch rate, $\mathsf{q} = \dot{\omega}_y$
$q$ = dynamic pressure
$r$ = radius

$R$ = rolling resistance
Re = Reynolds number
$\mathrm{Re}_x$ = Reynolds number based on length $x$
$s$ = braking-slip ratio
$S_{wet}$ = wetted area (in spray-contamination analysis)
$S_{tyre}$ = frontal area of a tyre
$t$ = time
$t_D$ = decision time
$t_{lag}$ = lag time
$T$ = net thrust or total thrust
$T_o$ = rated take-off thrust, static conditions
$T_p$ = propeller thrust
$u, v$ = aircraft velocity components
$u_r, v_r$ = rotational velocity components, Equation 9.11
$U$ = air speed
$V_{C_{Lmax}}$ = airspeed corresponding to maximum $C_L$, Equation 9.60
$V_{crit}$ = critical speed
$V$ = ground speed
$V_{lo}$ = lift-off speed
$V_p$ = reference speed in spray analysis
$V_{spray}$ = speed of contaminant spray
$w$ = tyre width
$w_b$ = wheel base
$W$ = weight
$x$ = distance travelled
$x_a$ = airborne distance at take-off, Equation 9.81
$x_o$ = initial displacement of a spring
$x_{CG}$ = longitudinal position of the CG
$x_m$ = distance between main UC and CG on longitudinal axis
$x_n$ = distance between nose UC and CG on longitudinal axis
$x_o$ = spring displacement
$x_t$ = distance between aerodynamic centre of H-tail and CG
$x_w$ = distance between aerodynamic centre of wing and CG
$y$ = vertical distance
$y_b$ = engine arm
$Y_c$ = cornering force
$Y_n$ = side force
$z$ = dummy variable
$z_m$ = vertical distance between centre of main wheel and CG
$z_n$ = vertical distance between centre of nose wheel and CG
$z_t$ = vertical distance between line of thrust and CG

## Greek Symbols

$\alpha$ = attitude or angle of attack
$\alpha_1$ = change in inflow angle of attack due to rotation, Equation 9.12

$\delta_s$ = slat angle
$\gamma$ = flight-path angle
$\gamma_r$ = runway incline
$\delta_f$ = flap angle
$\epsilon$ = thrust vector
$\varsigma$ = ratio between density of contaminant and water
$\zeta$ = damping ratio
$\zeta_o, \zeta_1$ = coefficients in Equation 9.19
$\eta$ = propeller efficiency
$\vartheta$ = propeller pitch
$\kappa$ = spring stiffness
$\mu$ = dynamic viscosity
$\mu_r$ = rolling coefficient
$\mu_s$ = sideways rolling coefficient
$\mu_b$ = braking coefficient on runway
$\nu$ = kinematic viscosity
$\xi$ = rudder deflection
$\rho_c$ = contaminant density
$\rho$ = density
$\rho_w$ = water density
$\sigma$ = relative air density
$\chi$ = radius of curvature of a trajectory
$\psi$ = tyre's yaw angle
$\omega$ = rotational speed
$\omega_o$ = natural frequency of harmonic oscillator
$\omega_y$ = $d\alpha/dt$ = angular velocity

## Subscripts/Superscripts

$\overline{[.]}$ = average quantity
$[.]_a$ = atmosphere
$[.]_B$ = braking
$[.]_{crit}$ = critical quantity
$[.]_f$ = fuel quantity or flap
$[.]_F$ = effect of contaminant
$[.]_g$ = in-ground effect or gas quantity
$[.]_{lo}$ = lift-off
$[.]_m$ = main landing gear; spray impingement index
$[.]_n$ = nose landing gear
$[.]_o$ = reference quantity
$[.]_p$ = propeller or propulsive
$[.]_r$ = rolling
$[.]_t$ = tyre or tail
$[.]_{to}$ = take-off
$[.]_{vt}$ = vertical tail
$[.]_y$ = with respect to $y$-axis
$[.]_w$ = wing or wind

# 10 Climb Performance

## Overview

The aircraft climb includes a variety of flight problems in which the airplane usually (not always) gains altitude. We define the general governing equations for flight in the vertical plane and thence some closed-form solutions for the propeller and jet airplane (§ 10.2). We then present the general problem of climb by a transport aircraft and show numerical solutions for both types of airplanes (§ 10.3), including the case of one engine inoperative (§ 10.3.6). The case of the turboprop aircraft is discussed in § 10.4. A powerful method for dealing with aircraft climb, particularly at transonic and supersonic Mach numbers, is the total energy approach (§ 10.5). There is a wide range of climb problems addressed with the energy methods. We report a few of these cases in § 10.6.

Accelerated climb problems are exclusively the domain of numerical solutions. Some of these methods are mathematically involved and will not be discussed in sufficient detail. The assumption of *quasi steady flight* is valid for many conventional aircraft.

**KEY CONCEPTS:** Closed-Form Solutions, Climb to Initial Altitude, Energy Method, Specific Excess Power, Optimal Climb, Climb Trajectories.

## 10.1 Introduction

There are two methods for solving climb problems: by solution of the differential equations that govern the motion of the centre of gravity and by the use of energy methods. There is a difference in the climb characteristics of propeller- and jet-driven aircraft. Although modern flight programs routinely include a turn during climb-out and descent, we will restrict the discussion to a vertical plane. There is no unique solution to a climb problem because there can be several free parameters and different types of constraints.

The rate of climb is the aircraft's velocity normal to the ground. As the aircraft climbs, the power plant delivers less thrust. The aircraft will reach a point where it can no longer climb; when this is the case, the aircraft reaches the absolute ceiling. Rates of climb are often given in the technical literature as feet/minute. It is possible for the manufacturer to provide the maximum instantaneous climb rate in such a unit, although this represents a peak value that may not be maintained for a full minute.

In some operations, the aircraft can zoom past the absolute ceiling by trading its kinetic energy into potential energy (*zoom climb*). Past the absolute ceiling, the aircraft might not be able to sustain controllable flight. Zoom-climb maximisation of the F-4C and F-15 aircraft was carried out in the 1970s for stratospheric missions reaching 27,000 m (~88,600 feet). The maximum known rates of climb are around 18,000 m/min (300 m/s; ~984 feet/s).

## 10.2 Closed-Form Solutions

Closed-form solutions are obtained by algebraic manipulation of the state equations and are useful in a preliminary analysis. We introduce these methods before addressing the more realistic (and complicated) flight trajectories of a real-life airplane. We consider separate cases for the jet- and propeller-driven airplane.

### 10.2.1 Steady Climb of Jet Airplane

Assume that the thrust angle is zero. The governing equation for the centre of gravity of the aircraft in arbitrary flight is given by

$$T - D - W \sin \gamma = m \frac{\partial U}{\partial t}. \tag{10.1}$$

Next, multiply this equation by the aircraft speed $U$, to find

$$TU - DU - W v_c = mU \frac{\partial U}{\partial t}. \tag{10.2}$$

The climb velocity in general accelerated flight is

$$v_c = \frac{T - D}{W} U - \frac{U}{g} \frac{\partial U}{\partial t}. \tag{10.3}$$

We need to define three important concepts. First is the specific excess thrust SET $= (T - D)/W$, which represents how much thrust is available for acceleration relative to the weight of the aircraft. This quantity is a non-dimensional number and depends on a number of operational factors, such as weight, altitude and atmospheric conditions.

The second concept is the specific excess power (SEP), which represents how much power is available for vertical climb, relative to the weight of the aircraft. The SEP has the dimensions of a velocity and is equal to the climb rate:

$$\text{SEP} = v_c = \frac{T - D}{W} U. \tag{10.4}$$

The corresponding climb angle is

$$\sin \gamma = \frac{v_c}{U} = \frac{T - D}{W} \tag{10.5}$$

and is equivalent to the SET. Finally, we define the normal load factor as the ratio between the lift and the weight:

$$n = \frac{L}{W}. \tag{10.6}$$

If we use the normal load factor, the general expression of the climb angle is

$$\sin\gamma = \frac{T}{W} - \frac{n}{L/D}. \tag{10.7}$$

As a further simplification, if the thrust is not dependent on the flight speed, then the maximum angle of climb is obtained with the speed corresponding to maximum glide ratio, $(L/D)_{max}$.

We now seek the optimal solution of a steady-state climb at subsonic speeds, for an airplane whose drag equation is parabolic. We optimise the climb with respect to a single parameter, the $C_L$. Changes in $C_L$ can be achieved by change of configuration (i.e., by deployment of the high-lift systems) and by changes in the angle of attack. The fastest climb is ensured by maximum rate of climb. The mathematical condition is found from

$$\frac{\partial v_c}{\partial C_L} = 0 \tag{10.8}$$

or

$$\frac{\partial}{\partial C_L}\left[\left(\frac{T}{W} - \frac{C_D}{C_L}\right)\sqrt{\frac{2}{\rho}}\sqrt{\frac{W}{A}\frac{1}{\sqrt{C_L}}}\right] = 0. \tag{10.9}$$

If we replace the parabolic drag in Equation 10.9 and work out the derivatives, we find

$$3C_{D_o}C_L^{-2} - \frac{T}{W}C_L^{-1} - k = 0. \tag{10.10}$$

Equation 10.10 is a quadratic equation in the unknown $C_L^{-1}$. The positive solution is

$$C_L = \frac{6C_{D_o}}{T/W + \sqrt{(T/W)^2 + 12C_{D_o}k}}. \tag{10.11}$$

Equation 10.11 expresses the optimal $C_L$, which is a function of both aerodynamics and propulsion and structural factors.

### 10.2.2 Steady Climb of Propeller Airplane

Assume a propeller aircraft, whose longitudinal axis is aligned with the speed and the thrust generated by the propellers. Start from Equation 10.2, with zero acceleration. Recall that the effective power of the propeller aircraft is

$$TU = \eta P. \tag{10.12}$$

The rate of climb of the aircraft is found by rearranging Equation 10.2 with Equation 10.12. The result is

$$v_c = \frac{\eta P - DU}{W}. \tag{10.13}$$

From the definition of lift coefficient, replace the speed $U$ with

$$U = \sqrt{\frac{2W}{\rho A C_L} \cos\gamma}. \tag{10.14}$$

Equation 10.14 neglects the centrifugal acceleration but, in this case, it is an acceptable approximation. By replacing Equation 10.14 into Equation 10.13, we find

$$v_c = \frac{\eta P - DU}{W} = \frac{\eta P}{W} - \cos^{3/2}\gamma \frac{C_D}{C_L^{3/2}} \sqrt{\frac{2}{\rho} \frac{W}{A}}. \tag{10.15}$$

For a small climb angle ($\gamma < 10$ degrees), assume that $\cos^{3/2}\gamma \sim 1$. The approximation of a small climb angle is not required when solving the climb problem with numerical methods.

**FASTEST CLIMB OF PROPELLER AIRPLANE.** The fastest climb is a flight program requiring the climb rate to be at a maximum at all altitudes. At a *given altitude h* Equation 10.15 is only a function of the aircraft speed. The necessary optimal climb condition requires that the derivative of the climb rate $v_c$ with respect to the relevant flight parameters will be zero. Assuming that the throttle is at full position, for a propeller engine, the free parameters are the flight speed $U$ and altitude $h$, the advance ratio $J$ and the pitch setting $\theta$ of the propeller: $v_c = f(h, U, J, \theta)$. The climb rate can be optimised on a point-to-point basis; therefore the altitude can be taken out of the list. Because both the propulsive efficiency and the shaft power are dependent on the air speed, a closed-form solution of this problem cannot be found, unless we make major simplifying assumptions, such as $\eta P =$ constant. In the latter case, the propulsive effects are eliminated, and the condition of maximum climb rate is given by

$$\frac{\partial v_c}{\partial U} = 0, \quad \text{or} \quad \frac{\partial}{\partial U}\left(\frac{C_{D_o} + kc_L^2}{C_L^{3/2}}\right), \tag{10.16}$$

which can then be solved in closed form.

### 10.2.3 Climb at Maximum Angle of Climb

The angle of climb is given by Equation 10.5. The condition of maximum for this angle is

$$\frac{\partial}{\partial U}(\sin\gamma) = \frac{\partial}{\partial U}\left(\frac{\eta P - D}{W}\right) = 0. \tag{10.17}$$

If we assume again that the change in weight is negligible, then

$$\frac{\partial}{\partial U}\left(\frac{\eta P}{W} - \frac{D}{L}\right) = 0. \tag{10.18}$$

The optimisation problem proceeds in the same fashion as the preceding case. The optimal equation is

$$\frac{\partial}{\partial U}(\eta P) - W\frac{\partial}{\partial U}\left(\frac{D}{L}\right) = 0. \tag{10.19}$$

If we assign

$$c_1 = \frac{2}{\rho}\frac{W}{A}, \tag{10.20}$$

then

$$\frac{\partial}{\partial U}\left(\frac{D}{L}\right) = 2\left(\frac{C_{D_o}}{c_1}U - \frac{c_1 k}{U^3}\right), \tag{10.21}$$

and the optimal equation becomes

$$\frac{\partial}{\partial U}(\eta P) - 2W\left(\frac{C_{D_o}}{c_1}U - \frac{c_1 k}{U^3}\right) = 0 \tag{10.22}$$

or

$$\left(\frac{\partial \eta}{\partial J}\right)\frac{P}{\Omega R} + \left(\frac{\partial P}{\partial U}\right)\eta - 2W\left(\frac{C_{D_o}}{c_1}U - \frac{c_1 k}{U^3}\right) = 0. \tag{10.23}$$

Propeller performance charts are required to solve this equation. Only by neglecting the propulsive effects can a closed-form solution be found.

## 10.3 Climb to Altitude of a Commercial Airplane

The climb procedures of a modern commercial airplane are nowhere similar to the cases presented so far. These procedures are sometimes called "optimal", but often they are not because they involved a series of manoeuvres defined to go around obstacles, noise constraints and air-traffic management. We discuss the noise-related strategy separately (Chapter 16). The climb to initial cruise altitude (ICA) of a transport aircraft is done in several phases, some of which are considered "standard", as discussed herein.

### 10.3.1 Climb Profiles

We describe three typical profiles, named "Standard", ICAO A and ICAO B. These procedures apply below 10,000 feet (FL-100). The sub-steps are equivalent to each other upon reaching a speed of 250 KCAS above 3,000 feet. Whenever acceleration to 250 KCAS is required, it is assumed that the aircraft has not yet reached that speed (this is not always the case). The thrust cut-back must be such that the resulting climb gradient is not below the climb gradient obtainable with one engine inoperative.

#### Standard Procedure

- Take-off is done at maximum thrust (or power), with an initial climb to 1,000 feet.
- Cut back thrust (or power) and accelerate past the flap-retraction speed.
- Climb out to 3,000 feet at constant KCAS.
- Pitch over and accelerate to 250 KCAS.
- Upon reaching 250 KCAS, climb to 10,000 feet.

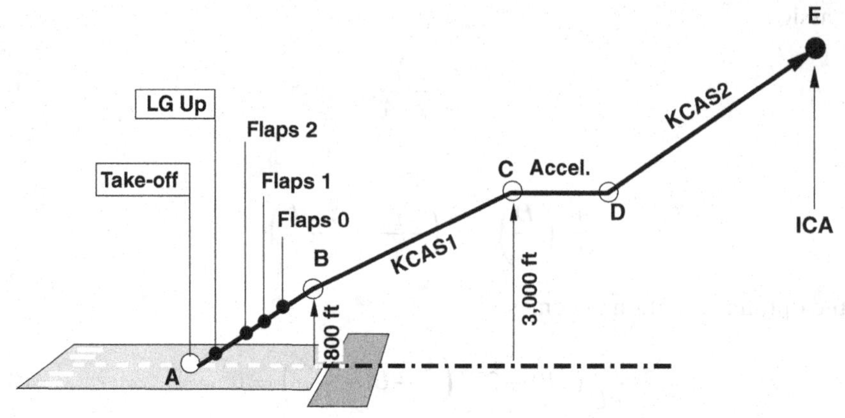


Figure 10.1. Typical climb schedule of transport aircraft.

**ICAO A Procedure**

- Take-off is done at maximum thrust (or power), with an initial climb to 1,500 feet.
- Cut back thrust (or power) at 1,500 feet.
- Climb at constant KCAS to 3,000 feet.
- Pitch over and accelerate, whilst retracting the flaps, to 250 KCAS.
- Upon reaching 250 KCAS, climb to 10,000 feet.

**ICAO B Procedure**

- Take-off maximum thrust (or power), with an initial climb to 1,000 feet.
- Accelerate at full power, whilst retracting the flaps/slats.
- At clean configuration cut back thrust (or power).
- Climb at constant KCAS to 3,000 feet.
- Pitch over and accelerate to 250 KCAS.
- Upon reaching 250 KCAS, climb to 10,000 feet.

The actual procedures followed by the airlines are not exactly as described. Furthermore, there are some variants. For example, in the ICAO B procedure the first target altitude can be 800 feet, 1,000 feet or 1,500 feet. The main operational parameters are the gross weight, the flap/slat angles and the thrust setting. In Chapter 18 we describe the NADP1 and NADP2 procedures, which are used for noise-abatement strategies. One example of climb profile is shown in Figure 10.1. The steps are as follows:

1. The aircraft starts at take-off (A) with state vector $\boldsymbol{V}_A$ (using maximum take-off thrust or appropriate derated thrust).
2. The aircraft performs an acceleration on a linear flight path. At least two options are possible:
   - To reach a target altitude (in this case, 800 feet) above the field with a take-off thrust (B).
   - To reach the target climb speed CAS1 with a take-off thrust. The airplane state vector at the end of this segment is $\boldsymbol{V}_B$.
3. The aircraft performs a climb at constant CAS1 with a reduced thrust (or power), to be determined, till it reaches a target altitude of 3,000 feet.

Table 10.1. *Approximate limit speeds for selected commercial aircraft*

| Airplane | VFE [kt] | Setting | VLO [kt] | Type |
|---|---|---|---|---|
| ATR72-500 | 185 | Flap 15 | 170 | retract |
| | 150 | Flap 30 | 160 | deploy |
| A320-200 | 215 | Flap 0 | | |
| | 200 | Flap 1+F | | |
| | 170 | Flap 2 | | |
| | 160 | Flap 3 | | |
| | 150 | Full | | |
| A330-300 | 240 | Flap 0 | | |
| | 215 | Flap 1+F | | |
| | 205 | Flap 2 | | |
| | 196 | Flap 3 | | |
| | 186 | Full | | |
| B737-800 | 250 | Flap 1 | | |
| | 250 | Flap 2 | | |
| | 215 | Flap 5 | | |
| | 205 | Flap 10 | | |
| | 190 | Full 15 | | |
| | 185 | Full 25 | | |
| | 165 | Full 30 | | |
| | 160 | Full 35 | | |

4. The aircraft performs a level acceleration at 3,000 feet to reach a target CAS2 (the thrust must increase).
5. The aircraft performs a climb with constant CAS2 to reach the target initial cruise altitude and cruise Mach number (the thrust may decrease in this phase).
6. If a further altitude gain is required, it is done at constant Mach number; if an increase in Mach number is required (although this is not normally the case), the aircraft performs a level acceleration.

The cross-over altitude is defined as the altitude at which the constant CAS meets the constant Mach number curve in the $M - h$ envelope. This altitude depends on the cruise Mach number and on aircraft parameters, such as gross weight and available thrust. However, when this phase is required, the TAS and the CAS decrease as the aircraft climbs to the tropopause.

Because the climb to the ICA shown in Figure 10.1 is done en-route, a longer climb means a lower cruise segment. Vice versa, a faster climb would require a longer cruise. The overall optimisation of the fuel burn has to be done in synchrony with the cruise (see § 12.3).

**CONFIGURATION CHANGES.** Several events take place in the initial stages of the climb, and these need further specification. In the first phase, from A to B, the aircraft changes state several times, with a landing-gear retraction, and then a retraction, in short sequence, of the high-lift devices. An example of flap and slat settings is shown in Table 10.1 for selected aircraft. Landing-gear retraction is done as soon as possible. When this event takes place, the drag decreases sharply by about 50%, and the thrust required to maintain the TAS also decreases sharply. This event is not associated to a change in lift. When a change in flap/slat setting takes place, there is change in both lift and drag, although nowhere as dramatic as in the case of the landing gear. Slat and flap operation takes place *after* landing-gear retraction, with a time lag sufficient for stabilising the aircraft. The flight-mechanics equations include switches such as

$$\texttt{LGearPosition} = \{\texttt{up}, \texttt{down}\} = f(h, U, W). \tag{10.24}$$

A condition required for landing-gear retraction is that the aircraft has reached a target speed. High-lift systems retraction is done *after* the retraction of the landing gear. For the slat/flap position, we propose the following.

### Computational Procedure

- The climb procedure starts with a $C_L = C_{L_{to}}$.
- The flap/slat setting is given by an index $I_{SF} = 0, 1, 2 \ldots$, each index corresponding to a pre-defined configuration of the flaps and slats (depending on the airplane).
- Within the climb-segment A-B determine the lift $C_L^*$ required to sustain a 1-g flight. This approximation is acceptable if the aircraft is on a linear trajectory; that is, normal load factor $n = 1$:

$$C_L^* = \frac{2W}{\rho A U^2}. \tag{10.25}$$

The lift is to be provided by the wing with the high-lift devices. The clean wing lift is $C_L = C_{L_o} + C_{L_\alpha}\alpha_e$, where $\alpha_e$ is the effective mean angle of attack of the wing (presently unknown). The difference

$$dC_L = C_L^* - C_L \tag{10.26}$$

is the difference in $C_L$ that must be obtained with the high-lift system. The possible cases are:

1. If $dC_L \simeq 0$, then $I_{SF} = 0$ (no flap deflection).
2. If $dC_L < 0$, then there is a need for a flap deflection.
3. Set tentatively $I_{SF} = 1$; calculate the $dC_{Lf}$ obtained with the flaps/slats and check the difference Equation 10.26.
4. If $dC_L \simeq 0$, then $I_{SF} = 1$ is the correct setting; otherwise increase the setting to $I_{SF} = 2$ and repeat the procedure until $dC_L \simeq 0$ is satisfied.

Each value of the index $I_{SL}$ corresponds to fixed angles $\delta_s$ and $\delta_f$ (slat and flap deflection, respectively). To each deflection there corresponds a change in wing lift and drag, which will be calculated with the aerodynamic model. With this procedure, we are able to calculate the impulsive change in aircraft configuration,* the corresponding change in aerodynamics and finally the change required on the engine settings.

There is the possibility that the best ICA cannot be reached, either for distance limitations or for weight limitations. The first instance occurs when the airplane is to perform a very short-range service: The en-route climb distance would not leave time for cruise. In the second case, the large take-off weight would cause the airplane

* In practice, the change in flap/slat setting is not impulsive; it may take three to five seconds, both on deployment and retraction.

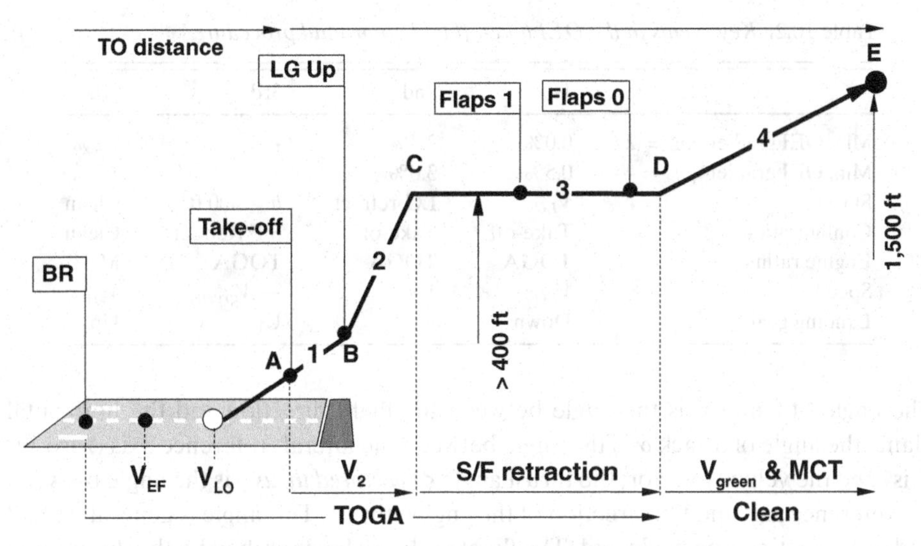


Figure 10.2. Flight path of OEI take-off and go-around.

to have a sluggish climb at intermediate altitudes and possibly insufficient climb rates at higher altitudes. In both cases the target ICA must be decreased to a sub-optimal value.

### 10.3.2 OEI Take-Off and Go-Around

Safety requirements determine the conditions that aircraft have to comply at take-off if one engine fails in the most critical part of the flight. In case of engine failure, following the procedures described in Chapter 9, if the option accelerate-go prevails, the aircraft climbs out with the remaining engine power. The key events of this procedure (also called TOGA, for take-off and go-around) are shown in Figure 10.2 and listed in Table 10.2. The graph shows various events from brake-release (BR) to the point when the aircraft reaches a target altitude of 1,500 feet above the airfield. Engine failure takes place with the aircraft on the ground with a speed $V_{EF}$. The go-around phase consists of a four-segments climb. Upon reaching the threshold altitude of 35 feet, the aircraft initiates the climb at point A. The first segment terminates at B, where the landing gears are retracted and the aircraft flies with the take-off speed $V_2$. The second segment terminates at C, where the aircraft has reached a minimum target altitude of 400 feet. The third segment terminates at D, where the aircraft has fully retracted its high-lift surfaces whilst accelerating to a target green-dot speed $V_{green}$. The final segment terminates at E, where the aircraft has reached a target altitude of 1,500 feet with the green-dot speed and maximum continuous thrust. The minimum climb gradients are given in Table 10.2.

### 10.3.3 Governing Equations

For a jet aircraft on an arbitrary flight path on the vertical plane, the equations of motions are

$$m\frac{\partial U}{\partial t} = T\cos(\alpha+\epsilon) - D - W\sin\gamma, \tag{10.27}$$

$$mU\frac{\partial \gamma}{\partial t} = T\sin(\alpha+\epsilon) + L - W\cos\gamma. \tag{10.28}$$

Table 10.2. *Key events in the OEI take-off and go-around procedure*

| | 1st | 2nd | 3rd | 4th |
|---|---|---|---|---|
| Min. OEI gradient, $n_e = 2$ | 0.0% | 2.4% | — | 1.2% |
| Min. OEI gradient, $n_e = 4$ | 0.5% | 3.0% | — | 1.7% |
| Start | $V_{LO}$ | LG-retract | $h \geq 400$ ft | Clean |
| Configuration | Take-off | Take-off | S/F Retract | Clean |
| Engine rating | TOGA | TOGA | TOGA | MCT |
| Speed | $V_{LO}$ | $V_2$ | $\rightarrow V_{green}$ | $V_{green}$ |
| Landing gear | Down | Up | Up | Up |

The angle of climb $\gamma$ is the angle between the flight direction and the horizontal plane; the angle of attack $\alpha$ is the angle between the aircraft reference axis (zero-lift axis) and the velocity vector; the thrust angle $\epsilon$ (*vectored thrust*) is the angle between the reference axis and the direction of the engine thrust. This angle is generally small and can sometimes be neglected. The flight path will be described by the differential equations

$$\frac{\partial x}{\partial t} = V_g \cos \gamma, \tag{10.29}$$

$$\frac{\partial h}{\partial t} = V_g \sin \gamma \tag{10.30}$$

where $V_g$ is the ground speed. The fuel flow is also part of the problem because it affects the aircraft's gross weight. The corresponding equation is

$$\frac{\partial m}{\partial t} = -\frac{\partial m_f}{\partial t} = -\dot{m}_f. \tag{10.31}$$

The problem is to be closed with a set on initial conditions:

$$t = 0, \quad U = U_o, \quad \gamma = \gamma_o, \quad x = x_o, \quad h = h_o, \quad m = m_o. \tag{10.32}$$

The aircraft can climb in an infinite number of ways, but a limited number of climb programs deserve special word. There are climb programs that include local optimal conditions and fixed starting conditions. These are typically initial-value problems. These programs are 1) fastest climb, 2) climb at maximum climb angle, and 3) climb at minimum fuel. The maximum-angle-of-climb problem is only important to clear an obstacle in emergency situations, but it is not a normal way of operating the aircraft. The minimum fuel to climb is also the most economical climb. There are also special climb programs that require final conditions, such as speed or Mach number at a given altitude. These problems are called two-value boundary problems.

### 10.3.4 Boundary-Value Problem

Once the configuration switches have been accounted for, the climb procedure still depends on a number of parameters. The aircraft state at point E, $\boldsymbol{V}_E = f(h, M, W)$ can be pre-calculated by optimisation of the cruise conditions with the methods described in Chapter 12. Thus, we are able to establish the final conditions of the en-route climb. The initial conditions $\boldsymbol{V}_A$ are also known from the take-off performance. We now have a two-value boundary problem, with known conditions at the start and

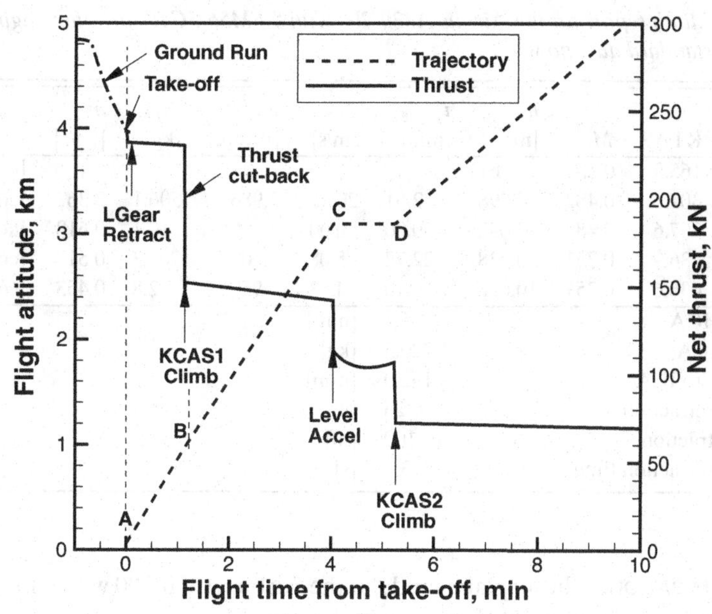


Figure 10.3. Simulated en-route climb for the Airbus A320-200-CFM; standard day, no wind.

at the end ($V_A$, $V_E$), and a number of functional constraints, such as specified thrust on the first segment and constant KCAS climb on the second and fourth segments. The free parameters are the thrust level of the first segment (A-B), the effective mean angle of attack of the airplane in the first segment (A-B) and the second-segment target climb speed CAS1.

The mean angle of attack will have to be selected in a way to minimise the flap setting and hence the fuel consumption. The net thrust following take-off can be maintained to the target altitude or CAS, but often there is no need for such a level of thrust.

**NUMERICAL SOLUTIONS.** A typical result for the Airbus A320-200 model with CFM engines is shown in Figure 10.3. The figure shows the trajectory, with the points A, B, C, D previously indicated in Figure 10.1. The thrust schedule shows the decreasing net thrust over the take-off run, the nearly constant thrust in the first segment of the climb, the thrust cut-back at the start of the second segment and all of the other events up to 5,000 m (~16,400 feet) above the airfield. The sharp decrease in thrust at point D is due to the fact that the airplane converts from acceleration to climb.

Another case, implementing the so-called NADP2 (Noise Abatement Departure Procedure 2) (see details in Chapter 17), is shown in Figure 10.4. In this case, we have a thrust cut-back at 3,000 feet with a minimum climb gradient of 4 degrees, up to a flight altitude of 4,000 feet, when the aircraft restores its normal climb rate.

Table 10.3 is a summary of computed results. The table has five reference points: 0) airplane above the screen at 35 feet; 1) airplane at the end of the initial climb, when it has reached the first target CAS; 2) airplane at the end of the first segment climb at constant CAS; 3) airplane at the end of the level acceleration; and 4) airplane at the top of climb (initial cruise altitude). At the bottom of the table there is a summary of climb data.

Table 10.3. *Climb report for the Airbus A320-200 with CFM56-5C4P turbofan engines and 331-9 APU; standard day, no wind*

| | KCAS | KTAS | $M$ | $h$ [m] | $t$ [min] | $v_c$ [m/s] | $v_c$ [ft/min] | $m_f$ [kg] | $\dot{m}_f$ [kg/s] | |
|---|---|---|---|---|---|---|---|---|---|---|
| 0 | 165.1 | 165.5 | 0.251 | 61 | | | | | | [over screen] |
| 1 | 266.0 | 307.6 | 0.482 | 3,098 | 2.51 | 20.25 | 3,986 | 506.1 | 3.362 | [const CAS] |
| 2 | 266.0 | 307.6 | 0.482 | 3,098 | 0.00 | 0.00 | 0 | 10.1 | 0.618 | [accelerate] |
| 3 | 266.0 | 436.9 | 0.751 | 10,038 | 22.77 | 5.08 | 1,000 | 701.3 | 0.513 | [const CAS] |
| 4 | 265.4 | 436.6 | 0.751 | 10,058 | 0.10 | 4.83 | 950 | 2.8 | 0.453 | [const M] |
| | Time to ICA | | | | 25.4 | [min] | | | | |
| | Fuel to ICA | | | | 1,220.3 | [kg] | | | | |
| | Distance to ICA | | | | 152.0 | [n-m] | | | | |
| | L-Gear retraction | | | | 26 | [m] | | | | |
| | Flaps retraction | | | | 100 | [m] | | | | |
| | L-Gear retraction time | | | | 3.5 | [s] | | | | |

**CLIMB OPTIMISATION.** The airplane model is the Airbus A320-200 with CFM engines. The assumptions of this study include the following:

- NADP2 procedure (this is a low-noise procedure fully described in § 18.2)
- Specified BRGW (as indicated)
- Climb target: specified by the best ICA-Mach combination. In this instance we have climb to FL-310 with a target $M = 0.752$, which was frozen for all cases.
- Free parameters are KCAS1 and first segment $\overline{v}_c$; the climb thrust is calculated to satisfy the KCAS1-$v_c$ combination.

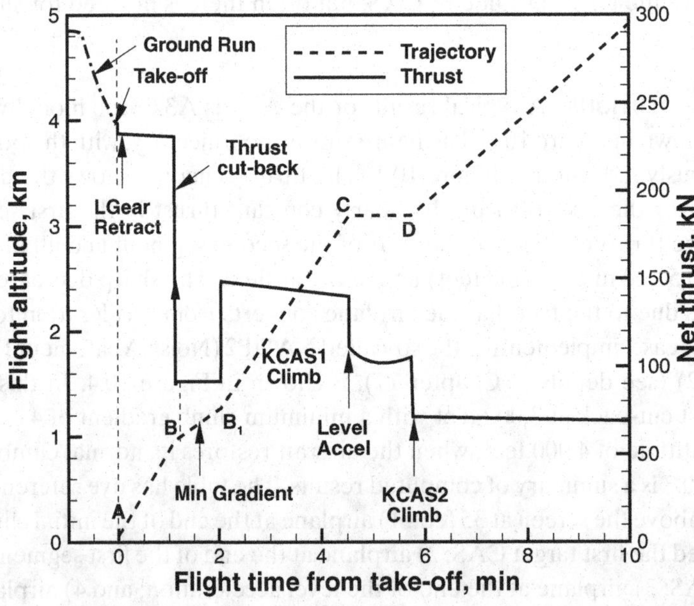


Figure 10.4. Simulated en-route climb for the Airbus A320-200-CFM, with NADP2 procedure; standard day, no wind.

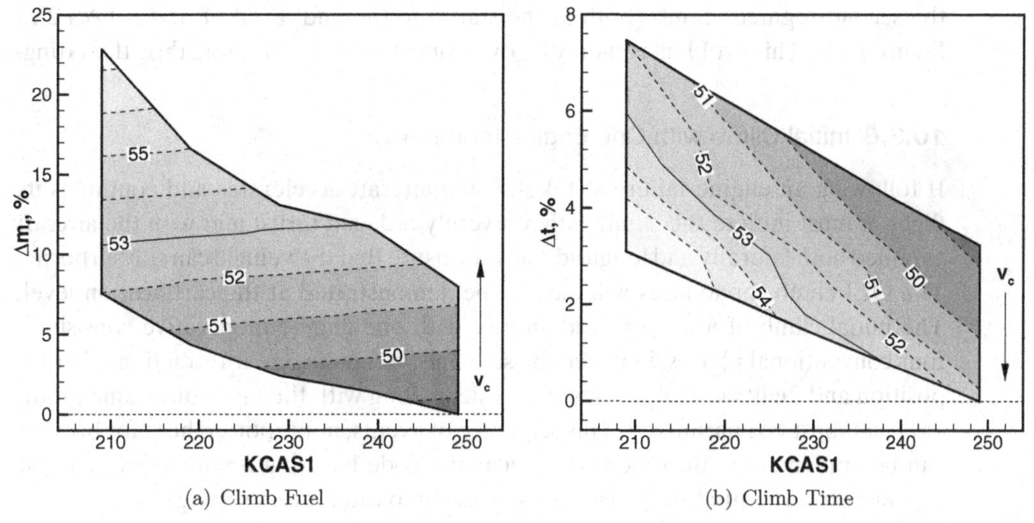

(a) Climb Fuel (b) Climb Time

Figure 10.5. Parametric study of climb to ICA of a model Airbus A320-200; $m = 55{,}200$ kg. The iso-levels denote a constant cost index.

- Climb distance constrained to $X_{target} = 170$ km (∼91 n-miles). If the target ICA-Mach is reached at a distance $x_c < X_{target}$, we add the cruise fuel to the target distance; if $x_c < X_{target}$, the solution is discarded.
- Standard day; headwinds $U_w = -4$ kt measured 2 m above the airfield.

The cost index is a weighted average of the cost of fuel and the cost of time. This index is discussed in § 12.9*. The result of our parametric analysis is shown in Figure 10.5. The arrows point toward increasing climb rates. A minimum fuel trajectory requires a high first-segment KCAS1 as well as a *low* climb rate; a minimum climb time requires a high first-segment KCAS1 as well as a *high* climb rate. The darker shades refer to lowest cost index, that is, most economical climb.

The second-segment KCAS2 is kept constant to the value corresponding to the best ICA-Mach combination (§ 12.3). However, it is possible to introduce a further parameter, the second-segment climb rate. Some manufacturers show climb charts for increasing KCAS2 and various cruise Mach numbers. These constraints are unlikely to satisfy the best cruise conditions.

### 10.3.5 Numerical Issues

From a computational point of view, when the landing gears are retracted there is an impulsive change in drag. This causes an impulsive change in aerodynamics, climb rates and flight path, unless a corrective action is applied. This action would require a more detailed analysis of the retraction process, a method for estimating the drag of a moving landing gear and the use of a much smaller time step than is required for ordinary flight-path integration.

Another problem, perhaps more evident, is that the application of the conventional climb strategy would lead to an aircraft suffering a jump in climb rate in

* This parameter is not the same as that in the FMS. Do not use for flight planning.

the second-segment climb (both at the start and the end; KTAS1 and KTAS 2 in Figure 10.1). This problem can only be overcome by using a variable throttle setting.

### 10.3.6 Initial Climb with One Engine Inoperative

If following an engine failure at take-off the aircraft accelerates and continues its flight, it must initiate the climb with a severely reduced thrust and with the aircraft trimmed both laterally and longitudinally to ensure that the vehicle can stay airborne. The OEI climb capabilities will have to be demonstrated at the certification level. The initial climb of a jet-powered aircraft with one engine inoperative consists of four conventional phases, in which the sequence of landing-gear retraction, high-lift position and flight control actions is specified, along with the minimum climb gradients that must be guaranteed. This sequence of events, not reported here for brevity, can be implemented into the flight-mechanics code by using impulsive switches at trigger points (flight altitude, true air speed, climb rate, thrust setting).

## 10.4 Climb of Commercial Propeller Aircraft

We now carry out a numerical analysis for a commercial flight of a propeller airplane. This analysis is similar to the case of the turbofan-powered airplane. The climb segments are as described earlier (§ 10.3). However, there are a number of important differences in the flight-mechanics model. The numerical procedure is described here:

- After take-off the airplane maintains full power up to a point when the landing gears are retracted. The engine-propeller matching is done at the limit engine power. In other words, at the given speed-altitude the turboprop is capable of providing a shaft power $P_{shaft}$. The corresponding power coefficient is $C_P = P_{shaft}/qU_{tip}^3$. The propeller is then trimmed to provide this value of the $C_P$. When this operation is done, the corresponding thrust coefficient $C_T$ is used to calculate the net propeller thrust: $T_p = qU_{tip}^2\, C_T$. Because there is a residual jet thrust $F_g$ from the engine, the total propulsive thrust with $\eta_e$ operating engines is

$$T = (T_p + F_g)\eta_e. \tag{10.33}$$

- When the landing gears are retracted (numerically, this is an impulsive event), the shaft power required to maintain speed or accelerate at the specified acceleration rate is considerably diminished. An acceleration and a climb rate are set. This condition leads to a required *thrust*

$$T_{req} = m\frac{dU}{dt} + D + W\sin\gamma = T_p + F_g. \tag{10.34}$$

  The required thrust is provided by a combination of propeller and engine thrust. Equation 10.34 is solved iteratively:

  1. Set $F_g = 0$, $T_p = T_{req}$.
  2. Trim the propeller to the required thrust $C_T = T_p/qU_{tip}^2$.
  3. Calculate the $C_P$ and $P_{shaft}$ corresponding to the trimmed condition.

4. Solve the engine model in the inverse mode in order to determine the engine state `EngineState` $= f($`W1, WF6, TT5, Fg9, ...`$)$. Set $F_g =$ `Fg9`.
5. Correct the required propeller thrust: $T_p = T_{req} - F_g$.
6. Reiterate from point 2 until the difference in residual thrust $F_g$ is negligible.

- At the first reference altitude (1,000 feet or otherwise) there is a power cut-back. Thus, the shaft power is reduced to a fraction determined by a minimum climb rate and a minimum acceleration. The actual power required is calculated iteratively; now the propeller is trimmed to a required power.
- Upon reaching the reference altitude with flaps and slats retracted, and with a target KCAS = KCAS1, the airplane starts a constant-CAS climb. At each altitude in the trajectory the TAS is specified by the $h$-CAS combination. The thrust required for climb is

$$T_{req} = W\left(\frac{v_c}{U}\right) + mv_c\left(\frac{dU}{dh}\right) + D. \tag{10.35}$$

In this equation, the acceleration term $dU/dt$ is converted to $dU/dt = v_c(dU/dh)$. The derivative $dU/dh$ depends only on the $h$-CAS combination. Therefore, for a given gross weight, the required thrust $T_{req}$ is a function of the rate of climb. This parameter is normally set to a value recommended by the manufacturer or the ATC.

Alternatively, a tentative climb rate is set, the propeller is trimmed to the required $T_{req}$ and the corresponding $P_{shaft}$ is calculated; if this power is excessive, the climb rate is reduced. In any case, Equation 10.35 is solved with the same algorithm as Equation 10.34.

- Upon reaching a target altitude, for example 3,000 feet above the airfield, the airplane performs a level acceleration to a target KCAS = KCAS2. The required thrust is

$$T_{req} = m\frac{dU}{dt} + D \tag{10.36}$$

and depends *critically* on the acceleration rate, which can be the dominant term. Once the acceleration is determined iteratively, Equation 10.36 is solved as Equation 10.34.

- The final segment of the climb to initial altitude is done at constant KCAS = KCAS2.

Figures 10.6 and 10.7 show the simulated climb trajectory from take-off of the ATR72-500 with PW-127M engines and F568 propellers.

These graphs show a number of impulsive events, as described in the captions. The first impulsive event (A) is when the landing gear is retracted: the drag decreases and the required thrust/power also decreases. The second impulsive event is at the power cut-back (after C), then at the first constant-CAS climb (E), the level-flight acceleration (F) and the start of the second-segment climb (G). Note that there is also an impulsive change in the efficiency of the propeller, which is maintained within the range $\eta = 0.7$ to 0.86. The fuel consumption increases rather uniformly.

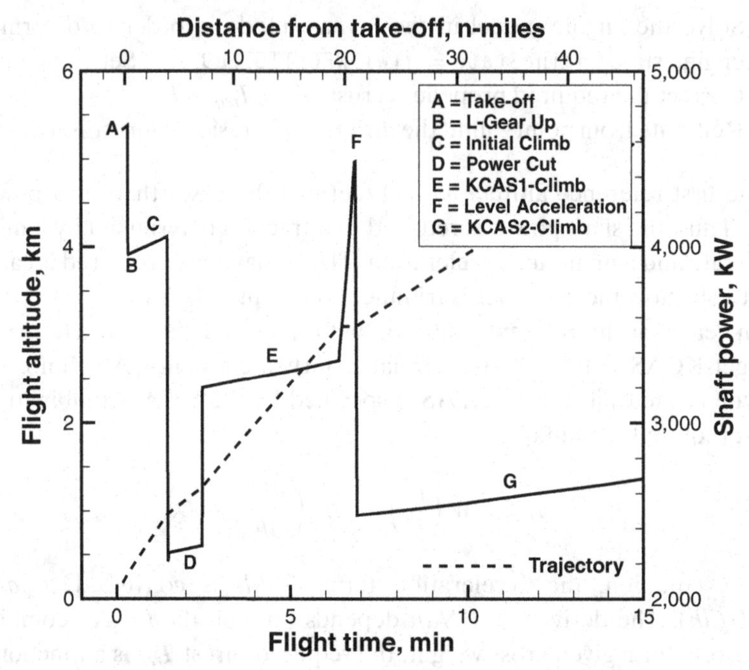


Figure 10.6. AEO climb to ICA of an ATR72-500 model with PW127M engines; standard day, no wind.

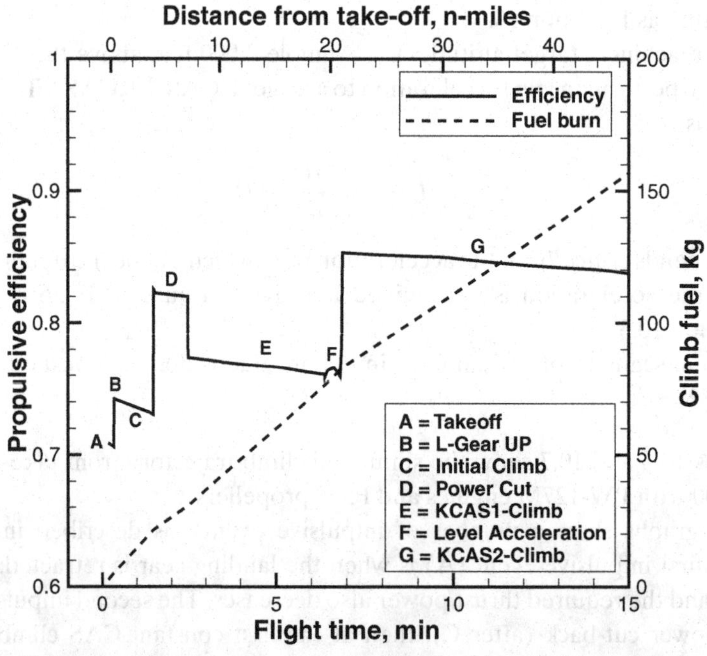


Figure 10.7. AEO climb to ICA of an ATR72-500 model; standard day, no wind.

Table 10.4. *Climb report for the case shown in Figure 10.6*

| | KCAS | KTAS | $M$ | $h$ [m] | $t$ [min] | $v_c$ [m/s] | $v_c$ [ft/min] | $m_f$ [kg] | $\dot{m}_f$ [kg/s] | |
|---|---|---|---|---|---|---|---|---|---|---|
| 0 | 161.9 | 162.3 | 0.246 | 61 | | | | | | [over screen] |
| 1 | 196.4 | 228.0 | 0.358 | 3,098 | 7.96 | 6.39 | 1,257 | 93.8 | 0.197 | [const CAS] |
| 2 | 196.4 | 228.0 | 0.358 | 3,098 | 0.00 | 0.00 | 0 | 0.0 | 0.186 | [accelerate] |
| 3 | 196.5 | 288.2 | 0.479 | 7,615 | 14.82 | 5.08 | 1,000 | 144.8 | 0.163 | [const CAS] |
| 4 | 196.4 | 288.2 | 0.479 | 7,620 | 0.07 | 2.54 | 500 | 0.5 | 0.126 | [const Mach] |
| | Time to ICA | | | | 22.8 | [min] | | | | |
| | Fuel to ICA | | | | 239.1 | [kg] | | | | |
| | Distance to ICA | | | | 92.0 | [n-m] | | | | |
| | L-Gear retraction | | | | 26.0 | [m] | | | | |
| | Flaps retraction | | | | 100.0 | [m] | | | | |
| | L-Gear retraction time | | | | 8.7 | [s] | | | | |

Table 10.4 is a summary of climb calculations, with the various segment climbs, the altitude climbed, the flight time, the fuel burned, the average fuel flow and other quantities.

## 10.5 Energy Methods

The typical approach to the climb problems is done by using steady-state models. However, this cannot be correct because as the climb rate and the optimal climb rate change with increasing altitude, the aircraft *must accelerate*. The difference between steady state and accelerated flight is particularly important for high-performance aircraft. Bryson and Denham[1] proved that an optimal accelerated climb to a target altitude requires about half the time of an optimum *quasi-steady* climb. Optimal problems have been published by Rutowski[2], Kelley[3], Schultz and Zagalsky[4], Calise[5] and others. Work on the subject includes optimum climb profiles of supersonic transport aircraft with noise minimisation[6], and optimal trajectories, including minimum fuel, minimum time or cost for fixed-range[7].

Instead of considering forces on the centre of gravity of the aircraft, it is sometimes useful to write balance equations for the total energy of the aircraft in its climbing flight. Methods of this nature are called *energy methods*. The first methods based on the concept of total aircraft energy go back to the 1940s. A widely acclaimed paper in this field is that of Rutowski[2], although the original idea seems to belong to the German engineer F. Kaiser, who developed the concept of "resultant height" for the optimal climb profile of the Messerschmidt Me 262, the first-ever jet fighter. Kaiser's concept is reviewed by Merritt *et al.*[8], and the climb technique is sometimes called the "Kaiser procedure". Kelley[9] contended that trading kinetic energy for potential energy is essential in maximising aircraft performance. These methods, although approximate, are extremely powerful, as evidenced by later research[10]. These methods have been expanded to be defined as a *total energy method*; they have become an industry standard for the air traffic control centres in Europe[11] and have widespread use in the professional field[12].

### 10.5.1 Total-Energy Model

The total-energy balance for the aircraft, reduced to a lumped mass, is the following:

$$(T - D)U = W\left(\frac{dh}{dt}\right) + mU\left(\frac{dU}{dt}\right)\left(\frac{dh}{dt}\right), \tag{10.37}$$

where $h$ is the flight altitude. Equation 10.37 contains separate terms for the vertical and horizontal speed components. Solving for the *vertical speed* $dh/dt$, we have

$$\frac{dh}{dt} = \frac{T - D}{W} U \left[1 + \left(\frac{U}{g}\right)\left(\frac{dU}{dh}\right)\right]^{-1}. \tag{10.38}$$

The *rate of climb* $v_c$ is the variation of the geopotential pressure altitude $h_p$ with time, $dh_p/dt$:

$$\frac{dh_p}{dt} = \frac{T - \Delta T}{T}\left(\frac{dh}{dt}\right), \tag{10.39}$$

where $\Delta T$ denotes the temperature differential with respect to the standard atmospheric model. On a standard day, the rate of climb is equivalent to the vertical speed; on a hot day ($\Delta T > 0$) the rate of climb is lower than the vertical speed.

The term within square brackets in Equation 10.38 is a function of the Mach number and is sometimes called *energy share factor*, $E(M)$. In the stratosphere, where the aircraft maintains a constant flight Mach number, we have

$$E(M) = 1 \tag{10.40}$$

because $dU/dh = d(aM)/dh = 0$ (the temperature is constant, and so is the speed of sound). In the lower atmosphere, where there is a temperature lapse rate $\lambda$, the energy-share factor is

$$E(M) = \left[1 + \frac{\gamma R \lambda}{2g} M^2 \left(\frac{T - \Delta T}{T}\right)\right]^{-1}. \tag{10.41}$$

An interpretation of this result is that a climb at constant Mach in the troposphere causes a *gain* in climb rate, or a conversion of kinetic energy into potential energy – thanks to the decreasing air temperature.

A third case is relative to a constant-CAS climb in the troposphere, a common practice for commercial airplanes. It can be demonstrated that the energy share factor can be written as

$$E(M) = \left\{1 + \frac{\gamma R \lambda}{2g} M^2 \left(\frac{T - \Delta T}{T}\right) + \left(\frac{\rho}{\rho_o}\right)\left[\left(\frac{p_o}{p}\right) - 1\right]^{-1}\right\}^{-1} \tag{10.42}$$

with

$$\frac{\rho}{\rho_o} = \left(1 + \frac{\gamma - 1}{2} M^2\right)^{1/1-\gamma}, \qquad \frac{p_o}{p} = \left(1 + \frac{\gamma - 1}{2} M^2\right)^{\gamma/\gamma-1}. \tag{10.43}$$

A sub-case of this equation is a constant-CAS climb in the stratosphere. In this case, the temperature is constant and the energy-share function, although formally the same, has a lower value. The difference of $E(M)$ in the troposphere and the stratosphere for a constant-CAS climb is shown in Figure 10.8. In both cases, the function is less than one. A constant-CAS climb requires an increase in true air

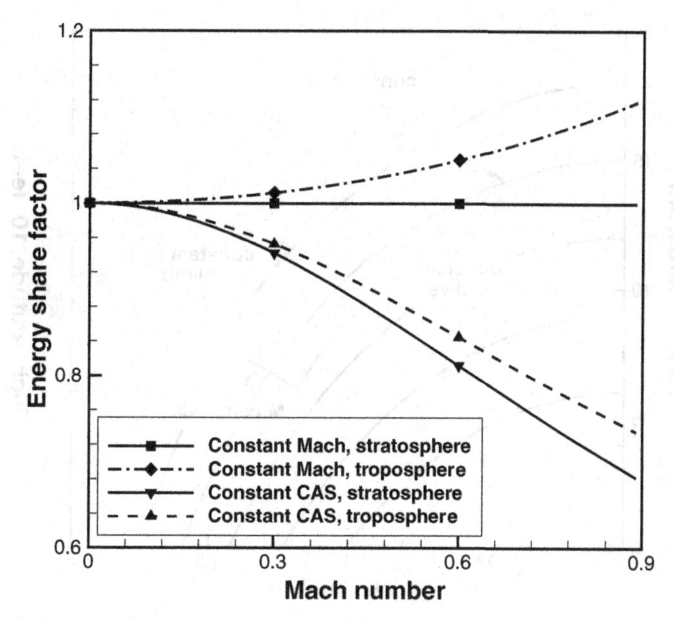


Figure 10.8. Energy-share factor for four different climb/descent rates; standard day.

speed, as a consequence of a decreasing air density; part of the climb energy must be spent for acceleration.

Consider now Equation 10.37. This equation can also be written as

$$(T - D)U = m\frac{\partial}{\partial t}\left(\frac{1}{2}U^2 + gh\right). \tag{10.44}$$

The term within parentheses is the total energy $E$; this quantity, divided by the acceleration of gravity $g$, has the dimensions of a distance and is called *energy height* $h_E$,

$$h_E = \frac{E}{g} = \frac{U^2}{2g} + h. \tag{10.45}$$

The energy $E$ is the sum of the kinetic and potential energy of the aircraft per unit of mass. The energy height Equation 10.45 represents the altitude at which the aircraft would climb if it were to convert all of its kinetic energy to potential energy. The time derivative of the total energy is the work done by the power plant

$$\frac{\partial E}{\partial t} = \frac{TU - DU}{m}. \tag{10.46}$$

The time derivative of the energy height is equal to the specific excess power

$$\frac{\partial h_E}{\partial t} = \frac{TU - DU}{W} = \text{SEP}. \tag{10.47}$$

Some methods used for flight-path optimisation use Equation 10.45, with the additional assumption that the aircraft can *instantaneously* exchange kinetic energy with potential energy, and vice versa. This approximation is a fairly good one if the short period of motion is neglected. However, it leads to sharp changes in direction in the flight path, which are unreasonable.

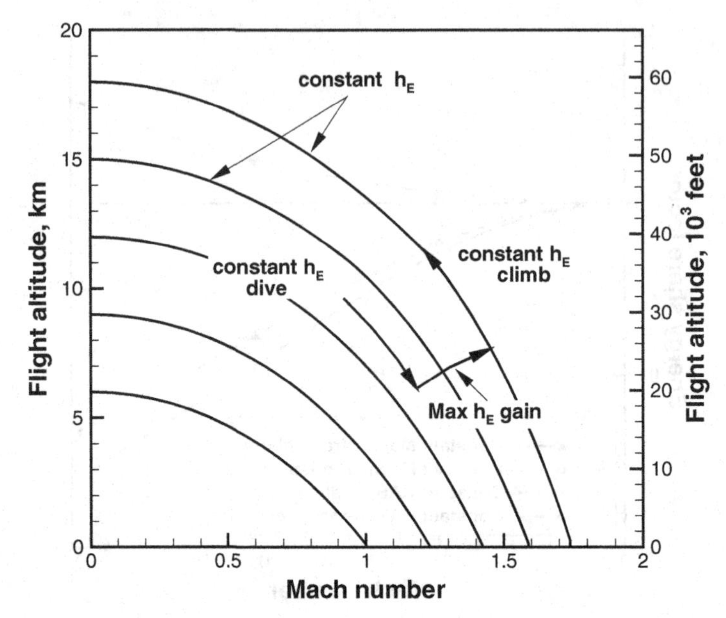


Figure 10.9. Constant energy levels on the plane $M - h$.

Figure 10.9 shows lines of constant energy height in the $M - h$ plane for a unit mass. These lines show a knee at a point located at altitude $h = 11{,}000$ m (36,089 feet). This is due to the change in atmospheric conditions at the tropopause. The lines of constant energy are found from

$$h = \frac{E - U^2/2}{g} = \frac{E - (aM)^2/2}{g}, \tag{10.48}$$

with $E$ the assigned energy level. When the $M = 0$, then $h = E/g$. This means that all the energy is potential energy. The geopotential height decreases with the increasing speed.

### 10.5.2 Specific Excess Power Charts

Diagrams of the SEP summarize the total performance of an aircraft in the altitude-speed plane and complement the flight envelope that is discussed in Chapter 8. Lines of constant specific excess power, SEP, are only valid for a fixed configuration, weight, load factor, angle of attack, engine throttle and atmospheric conditions. In practice, only a limited number of SEP curves are drawn. An example is shown in Figure 10.10. This figure shows the SEP distribution on the Mach-altitude plane. Selected levels of SEP are shown. The line SEP $= 0$ denotes the limits of operation stabilised level flight. There are two hills, at subsonic and supersonic speeds. At subsonic speeds, maximum values of SEP are found at sea level; the SEP decreases rapidly with the altitude. At supersonic speeds, maximum values of SEP are at intermediate altitudes, which indicates that a supersonic aircraft is most effective at altitude manoeuvring only within a limited altitude range. Outside this range, its acceleration capabilities are severely impaired by high drag and reduced thrust.

The next plot, Figure 10.11, refers to the same aircraft without after-burning thrust. This figure contains further information. For example, the excess power at

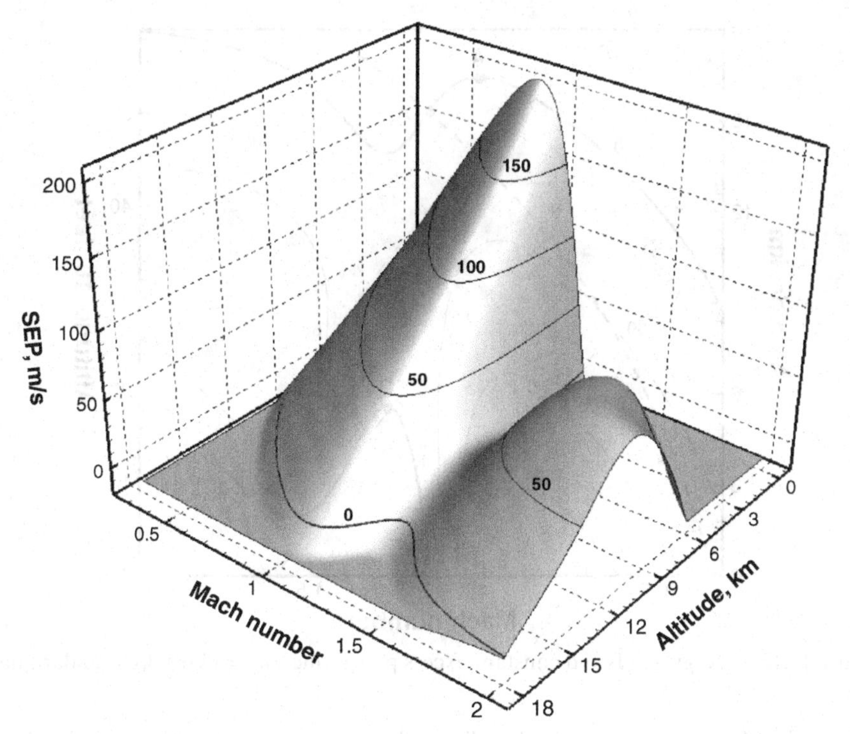


Figure 10.10. Specific excess power plot with after-burning thrust at supersonic speed; mass $m = 12{,}000$; standard day.

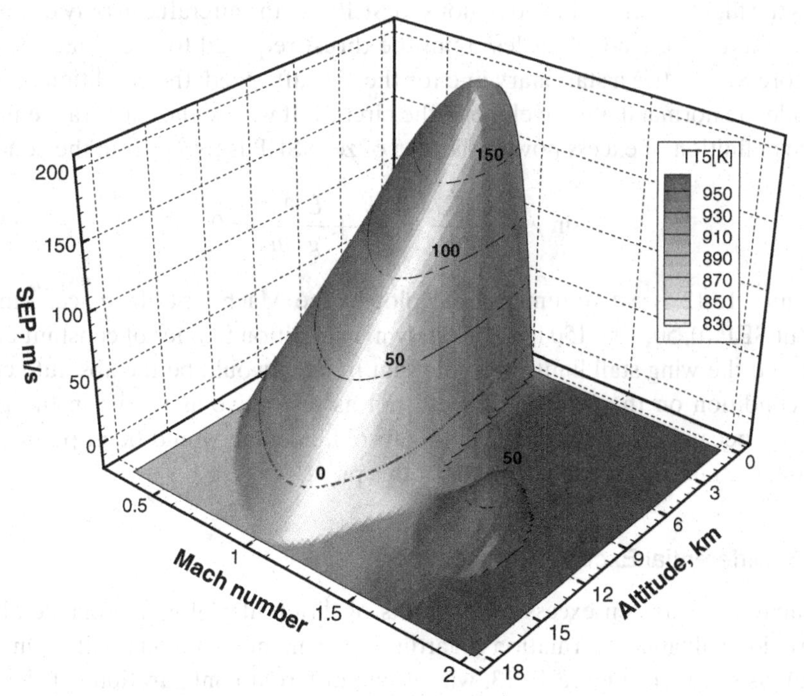


Figure 10.11. Specific excess power plot without after-burning thrust at supersonic speed; colour map based on turbine temperature TT5; standard day.

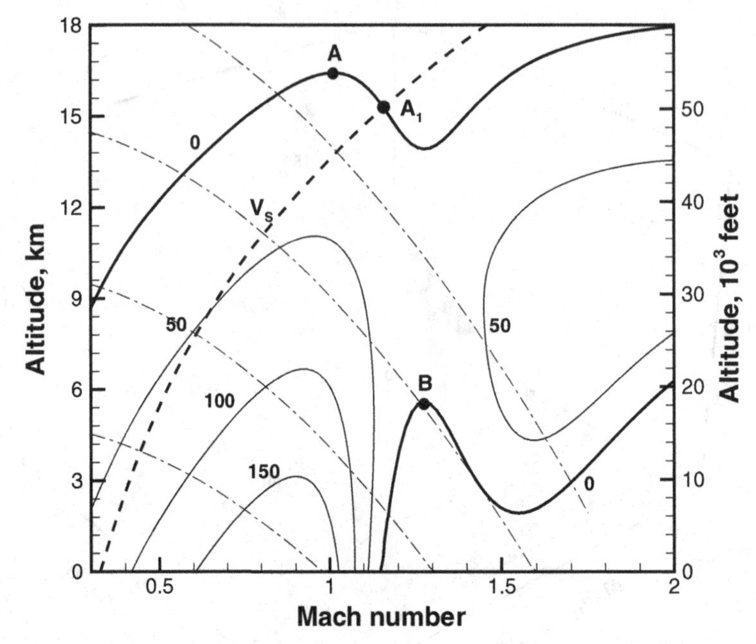


Figure 10.12. Energy levels and constant excess power lines; $m = 12{,}000$ kg; standard day.

supersonic Mach numbers is rather limited, and the aircraft would be unable to manoeuvre. Second, the plot shows the distribution of temperature TT5 at the exit of the turbine. The line corresponding to zero excess power is a limiting case because it divides the flight envelope in two regions. If SEP $< 0$, the aircraft can only decelerate because the thrust available is less than the thrust required to overcome the drag. Therefore SEP $= 0$ is a stationary line for the aircraft speed; the condition SEP $< 0$ is outside the normal flight envelope of the aircraft. If we use the climb rate equation and the definition of excess power, the line of zero SEP is found from the condition

$$U \sin\gamma + \frac{U}{g}\frac{\partial U}{\partial t} = \frac{\partial h}{\partial t} + \frac{U}{g}\frac{\partial U}{\partial t} = 0. \tag{10.49}$$

Figure 10.12 is a two-dimensional plot in the Mach-altitude space. Lines of constant SEP (0, 50, 100, 150 m/s) are shown, in addition to lines of constant energy height and the wing stall limit, $V_S$. Note that point $A$ would be the absolute ceiling if the condition on the stall speed were not as restrictive as shown in the graph. However, because of this restriction the absolute ceiling would be at point $A_1$. In any case, the aircraft should stay clear of this point.

### 10.5.3 Differential Excess Power Charts

A comparison between excess power plots of different high-performance aircraft can provide valuable information regarding their manoeuvre capability. One such example is shown in Figure 10.13, where two different configurations of the same aircraft have been selected. From the performance point of view, these configurations differ only in the transonic and supersonic drag characteristics; they have the same engine and the same weight. At any point in the plane $M - h$ we can define the

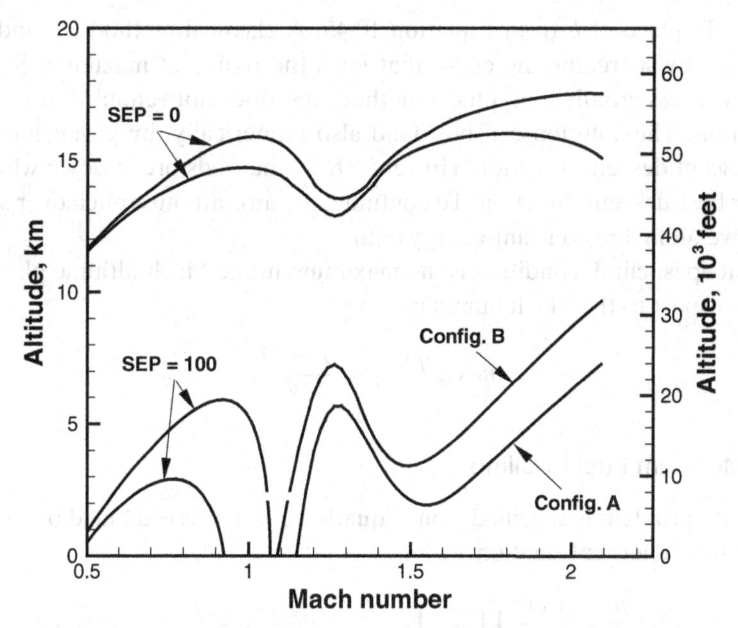


Figure 10.13. Differential specific excess power for two configurations.

difference in excess power as

$$\Delta\,\mathrm{SEP} = \mathrm{SEP}(B) - \mathrm{SEP}(A). \tag{10.50}$$

Charts like those in Figure 10.13 tend to get complicated because one aircraft can have a manoeuvring advantage in a region of the envelope and a handicap in another region. In the present case, we have plotted only two SEP lines. One line is obviously SEP = 0, that expresses the limits of the manoeuvre envelope. Another value is SEP = 100 m/s. This is close to the maximum SEP for both aircraft at the given gross weight.

These comparisons become less straightforward if the two aircraft are substantially different. The choice of weights is essential, and only comparisons at similar weights make sense. By reading this difference, the performance and design engineers can improve some operations; the pilots learn to avoid the flight condition in which they are likely to have a handicap.

## 10.6 Minimum Problems with the Energy Method

The energy method is most suitable for the analysis of the climb performance of supersonic aircraft that accelerate past the speed of sound. A number of flight programs are of interest: fastest climb (or minimum-time to climb), steepest climb and minimum-fuel climb.

### 10.6.1 Minimum Time to Climb and Steepest Climb

The minimum time to climb to a target altitude is specified from the condition that the gain in energy height is maximum with respect to the Mach number. This is equivalent to maximising the climb rate (see Equation 10.4). The energy height is introduced in place of $h$ from Equation 10.45. A classical method for finding this path is done by searching the curve that joins the points of maximum SEP at all altitudes. It is essentially a graphical method that does not require the solution of any equation. The solution can be found also numerically, by advancing along a steepest ascent/descent direction. However, both methods break down when there is no clear best-descent direction. To continue, the aircraft must pitch over and start a zoom-dive along the constant-energy path.

The steepest climb condition is the maximum in the Mach-altitude plane of the SEP with respect to the Mach number:

$$max_M \left(\frac{v_c}{U}\right) = \frac{T-D}{W}. \tag{10.51}$$

### 10.6.2 Minimum Fuel to Climb

A fuel-climb problem is specified from Equation 10.31 that is divided by the rate of change of total energy: Equation 10.46,

$$\frac{\dot{m}}{\dot{E}} = \left(\frac{\partial m}{\partial t}\right)\left(\frac{\partial t}{\partial E}\right) = \frac{\partial m}{\partial E} = -\dot{m}_f \frac{m}{(T-D)U}. \tag{10.52}$$

When we separate the differentials $dm$ and $dE$, we find

$$\frac{dm}{m} = -\dot{m}_f \frac{dE}{(T-D)U}. \tag{10.53}$$

The ratio $dm/m = dm_1$ is the *specific* change in aircraft mass, that is, the change in mass due to fuel flow divided by the aircraft mass. A *minimum-fuel-climb* problem is formulated by the condition that minimises $dm/m$ for a given energy level; that is,

$$\frac{dm_1}{dE} = -\frac{\dot{m}_f}{(T-D)\,U}. \tag{10.54}$$

Therefore, a minimum-fuel-climb flight path requires the minimisation of the right-hand side of Equation 10.54 or a maximisation of its inverse. We construct a function

$$f(h, M) = -\frac{(T-D)U}{\dot{m}_f}, \tag{10.55}$$

that is proportional to the excess thrust. Note that $f(h, M)$ is a negative function in the flight envelope of the aircraft because $T \geq D$. The minimum fuel to climb is the locus of the maximum energy increase per unit of fuel burned at a fixed energy $E$.

### 10.6.3 Polar Chart for the Climb Rate

The polar diagram shows the split between power required to keep the speed and the power required to climb, in steady state. Diagrams of this type are calculated at a constant mass, flight altitude and throttle setting. In general, only a few polars in this family are of interest: the polar at maximum thrust, the power-off polar, the polar at MTOW and the polar at absolute ceiling. The polar is obtained by joining two segments: the climb and descent segments.

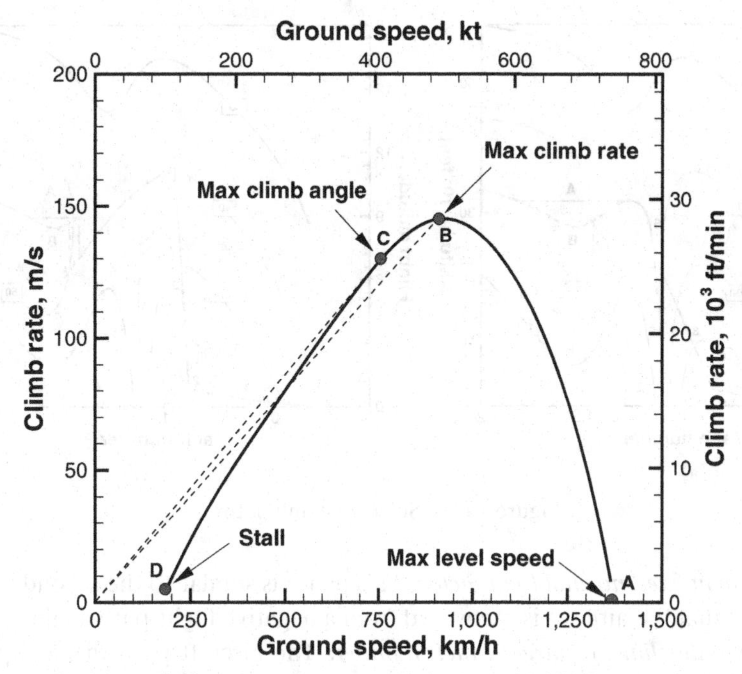


Figure 10.14. Positive climb polar of supersonic jet $m = 11{,}000$ kg, starting altitude $h =$ 2,000 m (~6,560 feet), steady flight, military thrust.

An example of calculation is shown in Figure 10.14. As the aircraft starts climbing, the horizontal speed must decrease because a portion of the power must be used to keep the air speed. The maximum $v_c$ is achieved at an intermediate forward speed, as indicated. At forward speeds below this point, the climb rate decreases. The climb at maximum climb angle is a yet lower speed $v_c/U$. The speed of maximum climb angle is lower than the speed of maximum climb rate; this corresponds to the tangent point of the polar with the line through the centre of the reference system.

### 10.6.4 Case Study: Climb to Specified Mach Number

We use the energy method to calculate some climb profiles of a supersonic aircraft model. The climb consists of the following segments:

1. *Climb at constant flight-path angle from take-off.* This segment starts from given conditions at take-off $(M, h, \gamma, W)$ and proceeds at full thrust on a given flight-path angle until the combination of Mach-altitude reaches a value corresponding to maximum SEP.
2. *Pull-up at fixed normal load factor.* The aircraft performs a pull-up at a fixed load factor to assume a climb rate corresponding to the maximum SEP.
3. *Subsonic climb at maximum SEP or maximum energy gain.* The aircraft climbs at maximum SEP until this parameter reaches a relatively low value (to be specified); alternatively, it climbs between levels of constant energy.
4. *Pull-out and zoom dive.* The aircraft pulls out and starts a dive at a fixed value of the flight path (to be specified) down to an altitude also to be specified.

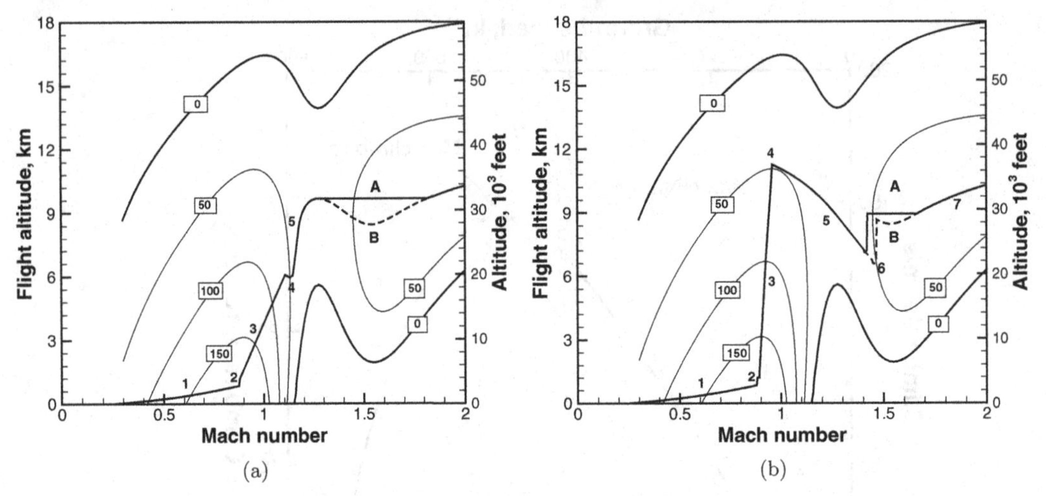


Figure 10.15. Selected climb paths.

5. *Pull-up at fixed normal load factor.* This phase is similar to the second segment, except that the aircraft is recovered from a negative flight-path angle.
6. *Supersonic climb to target Mach number.* The aircraft performs a supersonic climb at maximum SEP until it reaches the target Mach number. The final altitude is unspecified.

There are at least three free parameters: 1) the minimum-climb gradient before the aircraft pulls out and starts a zoom-dive; 2) the flight path gradient of the zoom-dive; and 3) the minimum altitude of the zoom-dive. We can specify secondary parameters, such as the normal load factors in the pull-out manoeuvres, the initial climb rate after take-off and other quantities. The constraints are the following:

- **Engine:** Operational limits set on `N%1` = 104, `TT5` = 970 K at maximum continuous power, to avoid over-speeding and over-heating at the turbine section of the engine.
- **Aerodynamic Heating:** Maximum skin temperature at stagnation $T_{stag} = 200$ (this temperature is calculated with an adiabatic model).
- **Flight Mechanics:** Speed sufficiently larger than the stall speed: $U > V_S$, and below the wing buffet limit, $U < V_{buffet}$.

The pull-up segments 2 and 4 are calculated by integration in the time domain of the equations defined in § 13.6.

Figure 10.15 shows four different sub-optimal climb paths from take-off ($M = 0.25$; $m = 12{,}000$ kg; $\gamma = 8$ degrees; after-burning for supersonic flight; sea-level, standard day) to a target $M = 2.0$ (without specified final altitude). In all cases, the sequence of climb segments is as described previously and denoted by the numbers $1, 2, \ldots$. Both graphs show selected levels of SEP and the stall speed, calculated with a stall margin $K_S = 1.15$.

In Figure 10.15a the aircraft performs a maximum energy-gain climb after a 4$g$ pull-up. Then it performs a supersonic maximum-SEP acceleration. For the case denoted by A, the aircraft is prevented from descending, whilst in case B it follows

Table 10.5. *Climb time and fuel for the flight paths shown in Figure 10.15*

| Flight path | *A* | *B* | *C* | *D* |
|---|---|---|---|---|
| Climb time [min] | 3.12 | 4.12 | 4.22 | 4.41 |
| Climb fuel [kg] | 160.0 | 173.2 | 244.1 | 249.8 |

the path of maximum supersonic SEP. In Figure 10.15b the aircraft performs a maximum-SEP climb after a 4*g* pull-up. It then performs a long zoom-dive to reach about $M = 1.4$. For the case denoted by C, it is prevented from descending (with the dive halted at a lower Mach), whilst in case D it follows the path of maximum SEP. A summary of climb times and climb fuel is given in Table 10.5. The results indicate clearly that the best climb (among those examined) is the flight path A, giving both minimum time and minimum fuel.

### 10.6.5 Minimum Flight Paths

The case discussed herein was based on quasi-steady flight-mechanics analysis, which in some cases is not of sufficient accuracy. Furthermore, there are cases when we need to specify *both* initial and terminal conditions, subject to a minimum cost function (fuel, time, ground distance). The case of Figure 10.15 belongs to this class of problems, known as *trajectory optimisation*, with or without terminal constraints. The total-energy method is a viable approach to minimum flight paths, although it assumes that the aircraft can convert instantaneously its kinetic energy into potential energy; therefore, it leads to flight paths with singularities (*sharp turns*) that are unrealistic (see Figure 10.15). More general methods for the solution of these problems are based on optimal control theory, a subject that falls beyond the scope of aircraft performance. The methods are mathematically involved, but they can be programmed on a variety of platforms.

The first relevant unsteady analysis of the aircraft climb is the one published by Miele[13;14]. An example of three-dimensional climb-turn analysis is that of Neuman and Kreindler[15;16], who derived climb-out and descent trajectories from and to runways, including optimisation for minimum fuel consumption. These results show that the velocity profiles for straight and turning flight are almost identical, except for the final horizontal accelerated or decelerated turn.

**CONTROL PROBLEM DEFINITION.** The problem is to take the aircraft from its initial state $\boldsymbol{x}_o = \{U, \gamma, h, m\}_o$ to its final state $\boldsymbol{x}_e = \{U, \gamma, h, m\}_e$. If the engine is always at full throttle, the problem is reduced to finding a control variable program $\alpha(t)$. However, this option is unrealistic for most flight conditions, and the throttle must be considered as a free parameter, to be found as a part of the solution.

The nomenclature of the problem is as follows: $\alpha(t)$ is the control variable program (the angle of attack); $\boldsymbol{x}(t) = \boldsymbol{x}(x_1(t), \ldots, x_n(t))$ is the vector of state variables programs (velocity, altitude, climb rate ...); $\psi = (\psi_1, \ldots, \psi_p)$ is the vector of terminal constraints functions ($\psi$ is a known function of the terminal point and the vector of state variables); and, finally, $\phi$ is the cost function that depends on the final point and the state variables $\boldsymbol{x}(t_e)$ at the final point. Therefore, the problem is to take the aircraft to the terminal state subject to the performance criterion being a minimum or maximum. The objective function depends on the final state of the aircraft and on *the course of events* between the start and the terminal point. The aircraft can be subject to some terminal constraints. For example, one may want to have a final Mach number, a final altitude or a final climb angle. The most important cases are the following: *minimum-time climb* and *minimum-fuel climb*. To solve the problem, we must use a form of the Lagrange multipliers to construct the so-called *adjoint* equations. The derivation of the conditions for the Lagrange multipliers can be elaborate. Bryson and Ho[17] and Ashley[18] showed the entire derivation procedure. The initial conditions are assigned. The terminal conditions can be assigned or left free. The problem so formulated is a boundary-value problem. The stopping condition is found from the altitude at the terminal point. The aircraft will reach that altitude with a time $t_e$, which is assured to be a minimum.

The adjoint equations contain derivatives of the lift, drag, thrust and fuel flow. These derivatives are calculated numerically. There are a variety of methods available to solve the optimal problem, including gradient methods, multiple shooting algorithm and dynamic programming. Bryson and Denham[1], in their solution of the problem for the F-4 interceptor, used the steepest-descent method. The method can be found in the technical literature (for example, Press *et al.*[19]). The method requires a first-guess of the solution; it then proceeds, by local linearisation around the current point, in the direction of the steepest ascent or descent. The *multiple shooting method* of Bulirsch and Stoer[20;21] has been used by Bŕuning and Hahn[22] for a variety of optimal-climb problems.

## Summary

This chapter has discussed aircraft climb in the vertical plane. We have shown some closed-form solutions, including fastest climb conditions. However, the climb to altitude by commercial aircraft is done in steps, which are not always optimal. In fact, there are constant-CAS segments with level acceleration. We have presented some standard ICAO climb profiles. One important solution strategy is based on the use of total energy (energy methods), which applies best to high-performance supersonic aircraft. These methods allow the rapid analysis of the specific excess power in graphical form; thus, they are suitable for graphic interpretation rather than analytical treatment. The total-energy methods have been applied to a number of case studies, including differential analysis between competing aircraft. We have shown that even with the simplifications required by the instantaneous energy transfer, it is possible to analyse problems such as air superiority and fastest climbs. However, there are sophisticated methods in the technical literature. These methods are based on control theory and allow the solution of climb problems with optimality conditions and initial and terminal constraints. These methods have only been annotated; the readers should consult the literature on control theory for theory and applications.

## Bibliography

[1] Bryson AE and Denham WF. A steepest-ascent method for solving optimum programming problems. *J. Applied Mechanics*, 29(2):247–257, 1962.

[2] Rutowski ES. Energy approach to the general aircraft performance problem. *J. Aero. Sci.*, 21(3):187–195, Mar. 1954.
[3] Kelley HJ and Edelbaum TN. Energy climbs, energy turns, and asymptotic expansions. *J. Aircraft*, 7(1):93–95, Jan. 1970.
[4] Schultz RL and Zagalsky NR. Aircraft performance optimization. *J. Aircraft*, 9(2):108–114, Feb. 1972.
[5] Calise AJ. Extended energy management method for flight performance optimization. *AIAA J.*, 15(3):314–321, Mar. 1977.
[6] Berton JJ. Optimum climb to cruise noise trajectories for the high speed civil transport. Technical Report TM-2003-212704, NASA, Nov. 2003.
[7] Ardema MD, Windhorst R, and Phillips J. Development of advanced methods of structural and trajectory analysis for transport aircraft. Technical Report CR-1998-207770, NASA, Mar. 1998.
[8] Merritt SR, Cliff EM, and Kelley HJ. Energy-modelled climb and climb-dash – the Kaiser technique. *Automatica*, 21(3):319–321, May 1985.
[9] Kelley HJ. An investigation of optimal zoom climb techniques. *J. Aero Sci*, 26:794–803, 1959.
[10] Kelley HJ, Cliff EM, and Weston AR. Energy state revisited for minimum-time aircraft climbs. Number AIAA Paper 83-2138. Aug. 1983.
[11] Anon. *User Manual for the Base of Aircraft Data (BADA), Rev. 3.8.* Eurocontrol Experimental Centre, April 2010. EEC Technical Rept. 2010-003.
[12] ESDU. *Energy Height Method for Flight Path Optimisation.* Data Item 90012. ESDU International, London, July 1990.
[13] Miele A. Optimum flight paths of a turbojet aircraft. Technical Report TM-1389, NACA, Sept. 1955.
[14] Miele A. General solutions of optimum problems of non-stationary flight. Technical Report TM-1388, NACA, Oct. 1955.
[15] Neuman F and Kreindler E. Optimal turning climb-out and descent of commercial jet aircraft. Number SAE Paper 821468, Oct. 1982.
[16] Neuman F and Kreindler E. Minimum-fuel turning climbout and descent guidance of transport jets. Technical Report TM-84289, NASA, Jan. 1983.
[17] Bryson AE and Ho YC. *Applied Optimal Control.* Blaisdell, New York, 1969.
[18] Ashley H. *Engineering Analysis of Flight Vehicles.* Addison-Wesley, 1974.
[19] Press WH, Teukolsky SA, Vetterling WT, and Flannery BP. *Numerical Recipes.* Cambridge University Press, 2nd edition, 1992.
[20] Bulirsch R and Stoer J. Numerical treatment of ordinary differential equations by extrapolation methods. *Numer. Math*, 8(1):1–13, Jan. 1966.
[21] Stoer J and Bulirsh R. *Introduction to Numerical Analysis.* Springer-Verlag, 2nd edition, 1993.
[22] Brüning G and Hahn P. The on-board calculation of optimal climbing paths. In *Performance Prediction Methods*, volume AGARD CP-242, pages 5.1–5.15, May 1978.

## Nomenclature for Chapter 10

$a$ = speed of sound
$A$ = wing area
$c_i$ = constant coefficients, $i = 1, 2 \ldots$
$C_D$ = drag coefficient
$C_{D_o}$ = profile-drag coefficient
$C_L$ = lift coefficient
$C_L^*$ = lift coefficient for 1-$g$ flight
$C_{L_{to}}$ = take-off lift coefficient
$C_P$ = power coefficient of a propeller

| | | |
|---|---|---|
| $C_T$ | = | thrust coefficient of a propeller |
| $dm_1$ | = | specific change in aircraft mass, $dm/m$ |
| $D$ | = | drag force |
| $E$ | = | energy level; energy-share factor |
| $F_g$ | = | residual jet thrust (also `Fg9`) |
| $g$ | = | acceleration of gravity |
| $h$ | = | geodetic altitude |
| $h_p$ | = | geopotential pressure altitude |
| $h_E$ | = | energy height |
| $k$ | = | induced-drag factor |
| $K_s$ | = | stall margin |
| $I_{SF}$ | = | index for slat/flap setting |
| $J$ | = | advance ratio of a propeller |
| $k$ | = | induced-drag factor |
| $L$ | = | lift force |
| $\mathcal{L}$ | = | function of the course of events |
| $m$ | = | mass |
| $m_f$ | = | fuel mass |
| $\dot{m}_f$ | = | fuel flow (also `Wf6`) |
| $M$ | = | Mach number |
| $n$ | = | normal-load factor |
| $\eta_e$ | = | number of engines |
| $p$ | = | pressure |
| $P$ | = | power |
| $R$ | = | propeller radius |
| $\mathcal{R}$ | = | gas constant |
| $q$ | = | dynamic pressure |
| $t$ | = | time |
| $T$ | = | net thrust |
| $T_p$ | = | propeller thrust |
| $T_{req}$ | = | required thrust |
| $\mathcal{T}$ | = | temperature |
| $\mathcal{T}_{stag}$ | = | stagnation temperature |
| $U$ | = | air speed |
| $U_w$ | = | wind speed |
| $U_{tip}$ | = | tip speed of a propeller |
| $v_c$ | = | climb rate |
| $V_g$ | = | ground speed |
| $V_2$ | = | take-off speed |
| $V_{EF}$ | = | engine-failure speed |
| $V_{LO}$ | = | lift-off-speed |
| $\boldsymbol{V}$ | = | airplane state vector in flight |
| $V_{buffet}$ | = | buffet speed |
| $V_{green}$ | = | green-dot speed |
| $V_S$ | = | stall speed |
| $x$ | = | flight distance |
| $\boldsymbol{x}$ | = | state vector in control analysis |

$X$ = required distance
$W$ = weight

### Greek Symbols

$\alpha$ = angle of attack
$\gamma$ = climb angle; ratio between specific heats in § 10.5.1
$\delta_f$ = flap angle
$\delta_s$ = slat angle
$\varepsilon$ = thrust angle
$\eta$ = propulsive efficiency
$\theta$ = propeller pitch
$\lambda$ = temperature lapse rate in the atmosphere, Equation 10.41
$\phi$ = cost function
$\psi$ = terminal constraint
$\rho$ = air density
$\Omega$ = rotational speed

### Subscripts/Superscripts

$[.]_e$ = end point
$[.]_f$ = fuel
$[.]_o$ = initial state
$[.]_\gamma$ = flight-path angle
$\dot{[.]}$ = time derivative
$\bar{[.]}$ = average value
$[.]_{req}$ = required quantity

# 11 Descent and Landing Performance

## Overview

The descent deals with that segment of the flight when the airplane decreases its flight altitude in a controlled mode. Landing still requires good pilot skills, which are best appreciated in bad weather conditions. The descent can be a large portion of the stage length, reaching in excess of 100 n-miles. Several distinct phases are identified. We will consider separately the phase of en-route descent down to 1,500 feet altitude above the airfield (§ 11.1) and the final approach down to ~50 feet above the airfield (§ 11.2). We also discuss two unconventional flight procedures: the continuous descent approach (§ 11.3), which has some advantages in terms of fuel consumption and noise emissions, and the steep-descent approach, which is a more complex manoeuvre (§ 11.4). We analyse the case of airplanes placed on holding stacks (§ 11.6) and optimal performance issues. Landing consists of an airborne phase and a landing run (§ 11.7). Effects of side gusts are considered, including crab landing and wing strike. This chapter ends with considerations of go-around trajectories, which are associated to aborted landing (§ 11.8).

**KEY CONCEPTS:** En-Route Descent, Continuous Descent Approach, Steep Descent, Unpowered Descent, Holding Procedure, Landing Performance, Crab Landing, Go-Around.

## 11.1 En-Route Descent

The aircraft starts descending from its cruise altitude well ahead of its destination. The flight computer will indicate the distance to the airfield and the estimated en-route descent, that is, the distance and time to landing, based on a number of factors, including speed, altitude and winds. The en-route descent from the final cruise altitude is done in stages. One such procedure is shown in Figure 11.1. First, the aircraft will descend at a constant Mach number to an altitude where it reaches a target CAS (or IAS). In the second stage, it maintains constant CAS (or IAS) to an altitude of 10,000 feet (FL-100, 3,048 m). In the third stage it decelerates at a constant altitude to reach a target CAS (below the previous one). The final stage is a descent at a constant CAS to 1,500 feet. Below this altitude, the aircraft is in the

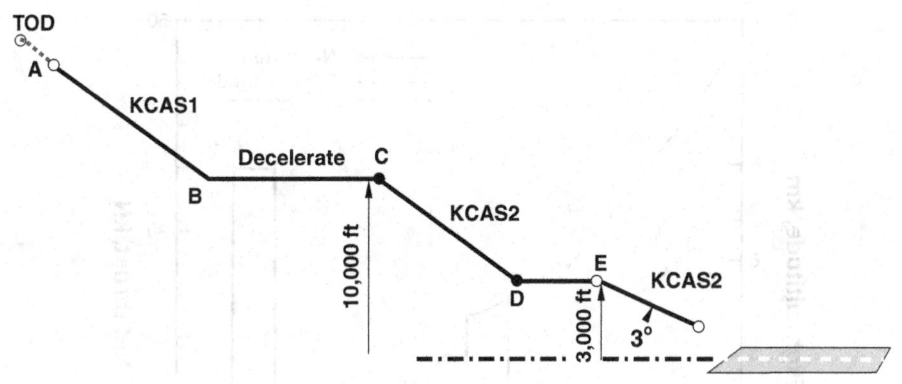


Figure 11.1. Multi-stage en-route descent (graph not on scale).

terminal manoeuvre area. This is by no means a unique descent trajectories. There are more complicated cases, involving more steps, constant-CAS specifications and fixed descent rates.

The trajectory is characterised by a number of parameters: the final cruise altitude (FCA) or initial descent altitude (IDA), the second-segment CAS, and the fourth-segment CAS. For some flight conditions, the first-stage, constant-Mach descent is absent because the aircraft is already within the specified CAS. An alternative to this procedure is to avoid operation at FL-100 ($KCAS_2$ in Figure 11.1) and start the descent later. This procedure is meant to keep the aircraft at the optimum altitude for a longer time.

If the engine is set to idle, the air speed is controlled by the flight altitude. A higher weight increases the descent distance because it decreases the glide angle. In fact, in stabilised flight we have

$$\gamma \simeq \frac{T - D}{W} \simeq -\frac{D}{W}. \tag{11.1}$$

A high weight decreases the flight-path angle and causes a longer, more shallow approach; the descent fuel increases because the engines have to run for a longer time (even in idle mode they burn a considerable amount of fuel); a longer descent shortens the cruise distance and hence the cruise fuel. A low weight combined with high drag will cause a steeper descent path. However, a too fast descent would cause a too rapid change in cabin pressure. To avoid the steep descent, power may be applied, which decreases the value of $\gamma$ in Equation 11.1. In any case, the flight time is involved, and one must evaluate whether it is preferable to start a descent from higher altitude and lower gradient or otherwise. Cruise and descent must be considered jointly in a flight optimisation procedure.

A number of optimal problems can be defined; these problems include minimum fuel for a given distance and initial altitude, minimum time for a given distance and initial altitude, minimum time or fuel with unconstrained distance and a constrained initial altitude, minimum cost index, and so on. We consider the fuel consumption versus the descent time for a fixed weight at the FCA = IDA = TOD (Top of Descent). The free parameter is the descent CAS (or IAS). The relationship between descent time and descent fuel is evaluated at a constant cost index. Because the pilot

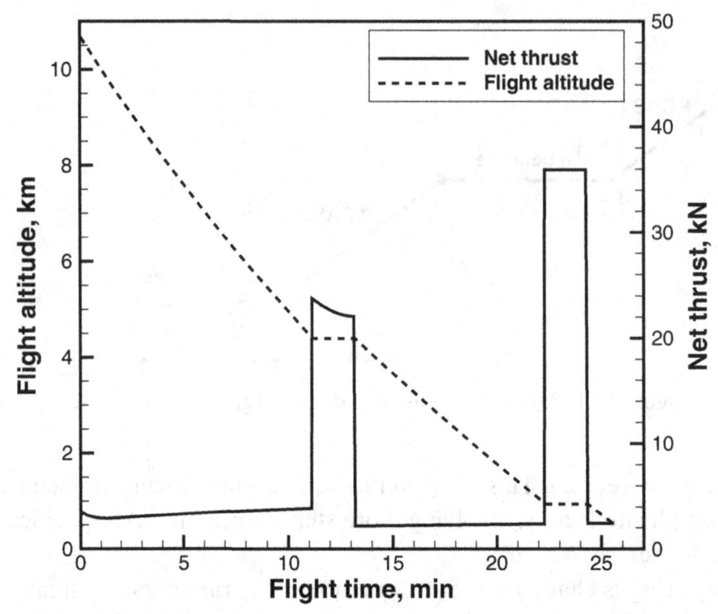


Figure 11.2. Simulated flight trajectory of the Airbus A320-200, standard day, no wind, from FL-350 (~10,670 m). Aircraft mass at the top of descent is 69,380 kg.

cannot be aware of all of these parameters, the Flight Management System (FMS) calculates the optimum flight procedures. The numerical method is the following:

- The airplane is at the end of the cruise at the top of descent.
- If the top of descent is $h > 11{,}000$ m (~36,000 feet), descend with the (constant) cruise Mach number to the tropopause (FL-360). The engines are set to idle mode, and the flight-path slope is calculated from Equation 11.1.
- Descend from FL-360 (or any other flight level) at constant $KCAS_1$. This speed is defined from the cruise Mach number and the flight level (see § 8.2 and Figure 8.7). The target descent altitude of this segment is 10,000 feet (FL-100). The flight-path angle is constrained to $-3$ degrees.
- Set the aircraft to a flight level FL-100 and decelerate to $KCAS_2$. This speed is calculated from the green dot speed $V_{green}$ (defined in § 11.2) at the end of the flight segment (1,500 feet above ground level). Assume that the third-segment CAS is $KCAS_2 = V_{green} + 5$ kt. Therefore, the calibrated air speed moves from $KCAS_1$ to $KCAS_2$. Again, we may choose to constrain the flight-path angle.
- Descend at constant $KCAS_2$ to 1,500 feet and enter the terminal manoeuvre area.

Occasionally, there is an intermediate step, with a level flight at 7,000 feet or 5,000 feet. Following this procedure, we calculated the en-route trajectory and the corresponding net thrust for the Airbus A320-200 airplane model. The result is shown in Figure 11.2. The engines are in idle mode during the descent segments but are forced to spool-up in order to maintain speed in level flight.

A problem of interest in descent performance is the minimum-fuel or minimum-time from the final cruise altitude, with or without a constraint on the en-route distance. We consider the case of a descent with a distance constrained to

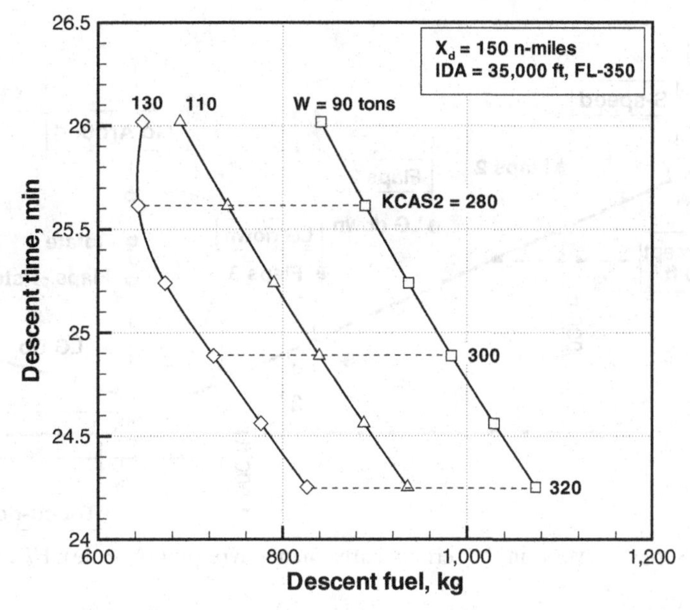


Figure 11.3. Parametric effects on descent performance for the Airbus A300-600 at a fixed descent distance (calculated).

150 n-miles (Figure 11.3). We used a parametric value of the third-segment KCAS, from 260 to 320 kt (these values of the KCAS do not satisfy the condition of green dot speed at the terminal point). All of the cases refer to an initial descent from FL-350. The results show how the descent fuel increases as the descent time decreases. At this point, a proper choice must be done in terms of the cost index that relates the cost of fuel to the cost of time.

## 11.2 Final Approach

The final approach is that flight segment below a threshold altitude (1,500 feet above ground level), when the aircraft is preparing for landing. In the general case, the aircraft may have to perform complex three-dimensional trajectories to avoid obstacles and minimise noise on the ground. In this instance, we will consider only the case of a rectilinear flight path, from the threshold altitude to the start of the landing phase (~50 feet above the airfield), as sketched in Figure 11.4. A complex series of operations takes place in such manoeuvre, most of which are automated (instrument landing). With reference to this figure, a few clarifications are required.

- "O": Green Dot Speed: This is the speed of maximum L/D with engines inoperative and clean configuration.
- "S-speed": This is the minimum speed at which the slats may be retracted at take-off. On approach, it is used as a speed target when the aircraft is in CONF 1. The S-speed is equal to 1.22-1.25 $V_{stall}$ at clean configuration.
- "F-speed": This is the minimum speed at which the flaps may be retracted at take-off. On approach, it is used as a speed target when the aircraft is in CONF 2 or CONF 3. The F-speed is 1.18-1.22 $V_{stall}$ with the airplane in configuration CONF 1+F.

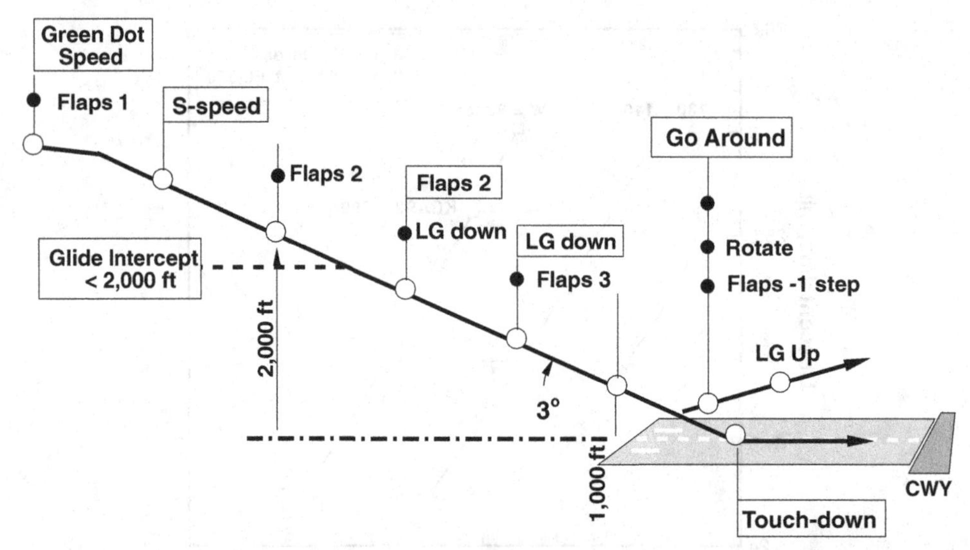


Figure 11.4. Series of events in the terminal area manoeuvre (not on scale); H/L = high-lift.

How does the air speed change during this flight segment? Data provided by Airbus indicate that typical deceleration rates are 10 to 20 kt per n-mile (10 to 20 km/h per km) on a descent slope of 3 degrees, with landing flaps and landing gear deployed. Deceleration in level flight is higher, up to 30 kt per n-mile. One recommendation is to use an approach speed $V_{app}$ that is approximately equal to

$$V_{app}[\mathrm{kt}] \simeq K_s V_{stall} + 5 + \frac{1}{3} V_w, \tag{11.2}$$

where all of the speeds are given in knots. Thus, the approach speed is equal to the stall speed (corrected for the stall margin), an additional 5 kt and one third of the headwind. We assume that the airplane starts from the green dot speed and ends with the approach speed. Such a decrease in speed leads to a lower drag and, hence, lower thrust setting and lower engine noise. Let us look first at the changes in configuration: these are associated to the deployment of the high-lift devices and the landing gear.

**HIGH-LIFT SYSTEM DEPLOYMENT.** The airplane is on a fixed glide slope, flying at altitude $h$ with true air speed $U$, with a weight $W$. The 1-g (unaccelerated) flight condition requires a lift coefficient $C_L = 2W/\rho A U^2$. The lift coefficient in nominal condition is derived from

$$C_L = C_{L_o} + C_{L_\alpha} \alpha, \tag{11.3}$$

where $C_{L_o}$ is the wing's zero-angle of attack; $C_{L_\alpha}$ is the lift-curve slope and $\alpha$ is the effective mean angle of attack. The latter parameter is unknown unless the attitude of the airplane is available. The two expressions for the lift must be the same; therefore, one could calculate the effective mean angle of attack from Equation 11.3. However, the wing might be unable to provide the lift required as the airplane descends. In other words, in the first analysis, we would allow the airplane to increase its attitude whilst decelerating. If such an increase in attitude, compatible with a condition

Table 11.1. *Flap and slat settings for the Airbus A320-200. Angles in degrees*

| CONF | Flap | Slat | Description | Flight Segment |
|---|---|---|---|---|
| 0 | 0 | 0 | Clean | Cruise |
| 1 | 0 | 18 | 1 | Hold |
| 2 | 10 | 18 | 1 + F | |
| 3 | 15 | 22 | 2 | Take-off |
| 4 | 20 | 22 | 3 | Approach |
| 5 | 40 | 27 | Full | Landing |

of flight stability, is not sufficient to produce the lift required, consider the lift deficiency

$$\Delta C_L = \frac{2W}{\rho A U^2} - [C_{L_o} + C_{L_\alpha}\alpha]. \tag{11.4}$$

Equation 11.4 expresses the difference between the lift required for a 1-g flight at speed $U$ and gross weight $W$ and the lift that can be generated at the prescribed inflow condition $\alpha$. If $\Delta C_L < 0$, the wing is unable to provide the lift required; therefore the high-lift systems must be deployed. This is done in sequence, the exact nature of which depends on the airplane. Consider, for example, the case of the Airbus A320-200. The possible slat and flap settings are given in Table 11.1, which shows that only a limited number of settings are possible.

This change in $C_L$ is associated to a combination of slat and flap deflection, according to

$$\Delta C_{L_{hl}} = \Delta C_{L_{slat}} + \Delta C_{L_{flap}} \tag{11.5}$$

so that

$$\Delta C_L = \frac{2W}{\rho A U^2} - [C_{L_o} + C_{L_\alpha}\alpha] + \Delta C_{L_{hl}} \simeq 0. \tag{11.6}$$

A numerical procedure in the flight-mechanics model is the following:

- If Equation 11.4 provides $\Delta C_L \simeq 0$, set configuration to zero (no slat/flap deployment).
- If $\Delta C_L < 0$, set configuration to 1. From Table 11.1: CONF 0; Flap 0; Slat 18. Calculate the $\Delta C_{L_{hl}}$ from Equation 11.5; if $\Delta C_L \simeq 0$ in Equation 11.6, then the correct setting is 1; otherwise, set configuration to 2 (CONF 2; Flap 10; Slat 18) and repeat the procedure.

Each time a switch is activated from the procedure described, the airplane changes configuration. This procedure generally works well, but it relies on the fact that the attitude of the airplane is a known quantity. A small $\Delta C_{L_{hl}}$ can be obtained with a small increase in airplane attitude rather than a flap setting. The FMS could be programmed to elaborate the best configuration, to control these switches and to prevent changes in flap setting above a certain altitude and TAS. The FCOM of each airplane gives information about these altitude-speed limits.

**LANDING-GEAR DEPLOYMENT.** This operation is a single switch from retracted to fully deployed and, in practice it appears instantaneous. Landing gears are prevented from

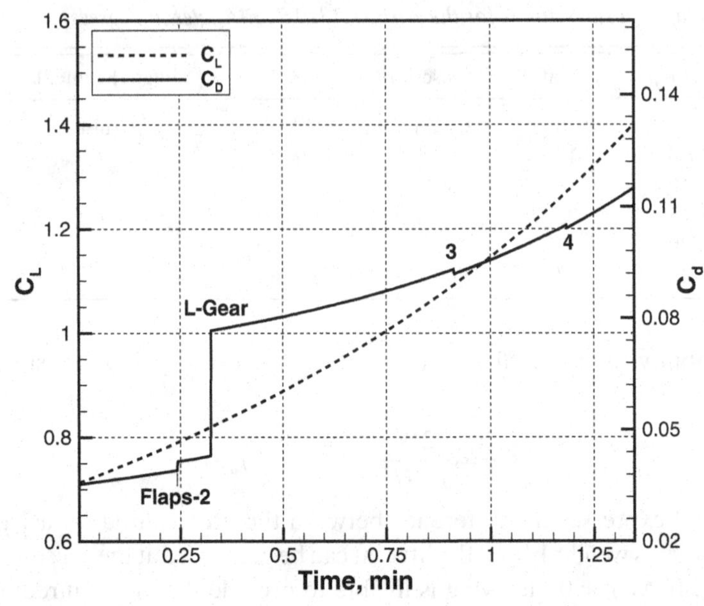


Figure 11.5. Lift and drag coefficients during the final-approach manoeuvre of the Airbus A320-200 (calculated).

deploying above a threshold altitude and air speed. Deployment occurs over a narrow range of altitudes above the airfield. The landing-gear deployment is prevented from taking place before the `Flap_2` setting (this is typically an Airbus procedure). By the time the landing gears are deployed, the aircraft must be stabilised in its glide slope.

**FLIGHT-MECHANICS MODEL.** We set the airplane over a fixed glide slope. Landing on a steeper gradient is constrained considerably by a number of factors, including stall speed, descent rates and safety concerns[1]. The differential equation is

$$\frac{dU}{dt} = \frac{1}{m}(T - D + W\sin\gamma). \tag{11.7}$$

The thrust is a free parameter and must be adjusted for minimum fuel burn, subject to the airplane maintaining an air speed sufficiently higher than the stall speed:

$$U > K_s V_{stall}, \tag{11.8}$$

where $K_s$ is the stall margin ($K_s \simeq 1.15$). There is the possibility, at high landing weights, relatively low air speed and decelerating aircraft, that the net thrust is zero in Equation 11.7. This event would require the aircraft to fly in idle mode. Although this procedure would save fuel, it is not completely safe: if an increase in thrust is required to reconfigure the aircraft or adjust the flight path, the increase (*engine spool-up*) may require several seconds, as explained in § 11.8. This delay in taking control of the aircraft would compromise safety. Therefore, a final-approach procedure should take place with at least some thrust in the final stages to allow full control of the airplane by engine power if a go-around is required.

Figure 11.5 shows the change of the aerodynamic coefficients in the approach trajectory. The $C_D$ increases as the aircraft descends. The small $C_D$ drop following the increase in flap setting is due to a slight decrease in main under-carriage drag.

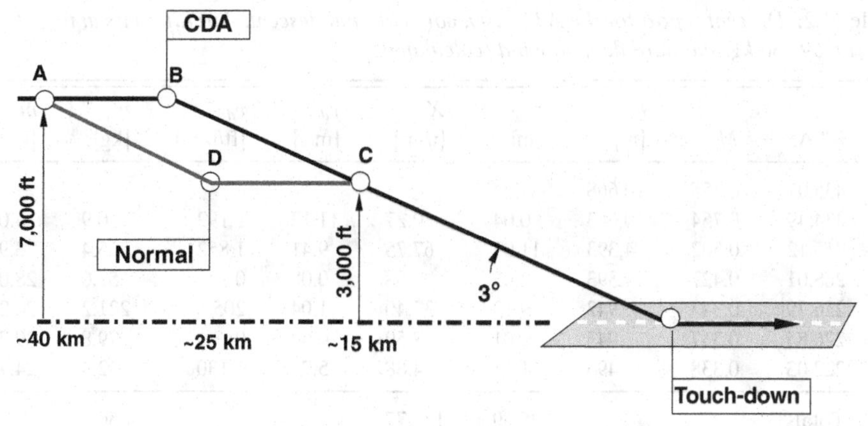


Figure 11.6. Continuous descent approach (CDA) versus conventional approach (not on scale).

This effect is due to the blocking of the flow past the under-carriage caused by the flap deflection.

## 11.3 Continuous Descent Approach

In a continuous descent approach (CDA), the aircraft descends on a fixed glide slope of about 3 degrees from an altitude of about 7,000 feet above the airfield. In this procedure, there is no level flight until the aircraft reaches the airfield. By contrast, a conventional flight path is made up of three segments, the second of which is a level flight at an altitude of 3,000 feet. The sketch in Figure 11.6 shows the differences between the conventional flight path and the CDA. A number of advantages are claimed with this new approach, including:

1. With the aircraft flying higher than normal, the distance to noise receivers on the ground is larger and the noise impact is lower.
2. With the aircraft flying higher, without a level segment, less engine thrust is required, which also contributes to lower noise emissions.
3. As a consequence of the previous point, also the fuel consumption is reduced.

First, a CDA does not affect the perceived noise EPNL (see Chapter 17) because by the time the airplane switches from conventional to CDA procedures, it is way too far to be picked up by the reference microphones (further discussion is in Chapter 18). The level flight segment at 3,000 feet, taking place over a distance of up to 10 km (~5.5 n-miles), can affect some communities below the flight path. On the CDA trajectory at ~25 km, as indicated in Figure 11.6, the airplane maintains an altitude of ~520 m (~1,700 feet) above the conventional trajectory at point D. Thus, the flight altitude point D has increased to ~4,700 feet, which is a 57% increase. Noise radiation is affected by the distance $r$ as $1/r^2$, all other parameters being the same. Hence, the CDA trajectory would have a noise signature ~4dB less than a conventional level flight at point D; this result partially justifies the claim of reduced noise. Further discussion of noise performance is available in Chapters 16 and 17.

Table 11.2. *Descent report for the A320-200, conventional descent. Aircraft mass at top of descent: 69,380 kg; standard day, no wind (calculated)*

| | KTAS | $M$ | $h$ [m] | $t$ [min] | $X$ [n-m] | $v_d$ [m/s] | $v_d$ [ft/min] | $m_f$ [kg] | $\dot{m}_f$ [kg/s] |
|---|---|---|---|---|---|---|---|---|---|
| 0 | 435.07 | 0.755 | 10,668 | | | | | | |
| 1 | 434.49 | 0.754 | 10,643 | 0.04 | 0.27 | 11.13 | 2,192 | 0.9 | 24.000 |
| 2 | 315.12 | 0.502 | 4,393 | 11.07 | 67.75 | 9.41 | 1,852 | 265.4 | 23.972 |
| 3 | 268.01 | 0.427 | 4,393 | 2.03 | 9.88 | 0.00 | 0 | 57.0 | 28.038 |
| 4 | 226.79 | 0.347 | 943 | 9.13 | 37.40 | 1.04 | 205 | 221.2 | 24.229 |
| 5 | 226.85 | 0.347 | 943 | 2.01 | 7.59 | 0.00 | 0 | 79.8 | 39.732 |
| 6 | 222.03 | 0.338 | 493 | 1.31 | 4.88 | 5.74 | 1,130 | 32.4 | 24.786 |
| | Totals | | | 25.59 | 127.77 | | | 656.7 | |

With regard to fuel consumption, point 3 is demonstrated by the results in Tables 11.2 and 11.3, which are for the Airbus A320-200 with CFM engines. These tables are summaries of aircraft performance, with the aircraft starting at the top of descent at the same conditions (altitude, Mach number, weight). The CDA procedure saves about 80 kg of fuel and takes place over a shorter distance.

A more extreme case is the one of continuous descent from the final cruise altitude. The same aircraft burns the following amounts of fuel: ~636 kg in a conventional descent; ~556 kg in a CDA descent; and ~450 kg in a continuous descent from the TOD. Thus, a saving of 180 kg of fuel could be achieved.

## 11.4 Steep Descent

The term *steep descent* refers to a glide slope greater than the conventional slope. Steep descent trajectories can reduce noise by virtue of a larger distance from the receiver on the ground but are more complicated to perform than conventional trajectories. In this section we highlight some of the problems, mostly focusing on the aerodynamics. By assuming a small angle of gliding and using the definition of the $C_L$, the descent rate is written as

$$v_s = \sqrt{\frac{2}{\rho}}\sqrt{\frac{W}{A}\frac{C_D}{C_L^{3/2}}}. \tag{11.9}$$

Table 11.3. *Descent report for the A320-200, continuous descent approach; aircraft mass at top of descent: 69,380 kg; standard day, no wind (calculated)*

| | KTAS | MM $M$ | $h$ [m] | $t$ [min] | $X$ [n-m] | $v_d$ [m/s] | $v_d$ [ft/min] | $m_f$ [kg] | $\dot{m}_f$ [kg/s] |
|---|---|---|---|---|---|---|---|---|---|
| | 435.07 | 0.755 | 10,668 | | | | | | |
| 1 | 434.49 | 0.754 | 10,643 | 0.04 | 0.27 | 11.13 | 2,192 | 0.9 | 24.000 |
| 2 | 315.12 | 0.502 | 4,393 | 11.07 | 67.75 | 9.41 | 1,852 | 265.4 | 23.972 |
| 3 | 268.01 | 0.427 | 4,393 | 2.03 | 9.88 | 0.00 | 0 | 57.0 | 28.038 |
| 4 | 222.03 | 0.338 | 493 | 10.44 | 42.28 | 1.63 | 321 | 253.6 | 24.298 |
| Totals | | | | 23.58 | 120.18 | | | 576.9 | |

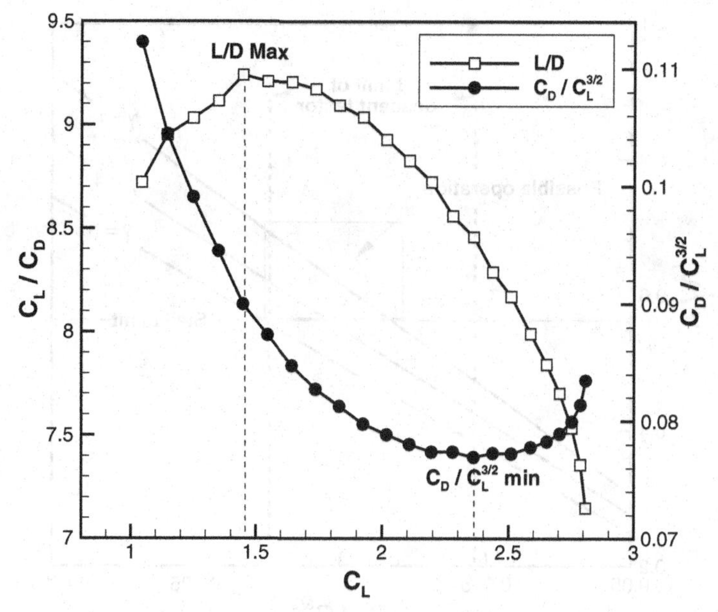


Figure 11.7. Aerodynamic parameters of the Airbus A310 scaled airplane.

The slope of the trajectory is $\gamma = v_s/U$. For a fixed weight, a change in the descent slope requires a change in the aerodynamic coefficients. If the descent is performed at the same TAS as the conventional flight, an increase in $\gamma$ requires an increase in descent rate. Alternatively, if the descent rate is kept to the value of the conventional trajectory, the TAS would have to decrease; in this case there is a risk of stalling the aircraft. The TAS must decrease according to

$$U \sin \gamma = \text{const.} \tag{11.10}$$

If the underscore "r" denotes the reference flight conditions (or conventional trajectory), the ratio between the steep-descent TAS and the reference TAS is

$$\frac{U}{U_r} \simeq \frac{\gamma_r}{\gamma} = \left(\frac{C_{Dr}}{C_D}\right)\left(\frac{C_L}{C_{Lr}}\right)^{3/2}. \tag{11.11}$$

At a constant $C_L$, the drag would have to increase as much as the flight-path angle (in relative terms). However, a slower air speed would require a higher $C_L$. There is a combination of flap and slat settings (for example, Table 11.1) to achieve this. In any case, the $C_D$ would have to grow faster than the $C_L^{3/2}$; in other words, the factor

$$f_d = \frac{C_D}{C_L^{3/2}} \tag{11.12}$$

must be larger in a steep-descent trajectory than in a conventional trajectory. A plot of such a parameter is shown in Figure 11.7, which refers to a wind-tunnel model of the Airbus A310[2]. High values of the aerodynamic factor $f_d$ are obtained with either very low or very high $C_L$. Low values of $C_L$ would increase the stall speed and thence the risk of stalling the aircraft as it slows down. Figure 11.7 also shows that the aerodynamic factor is at a minimum at relatively high $C_L$ values. Equation 11.11

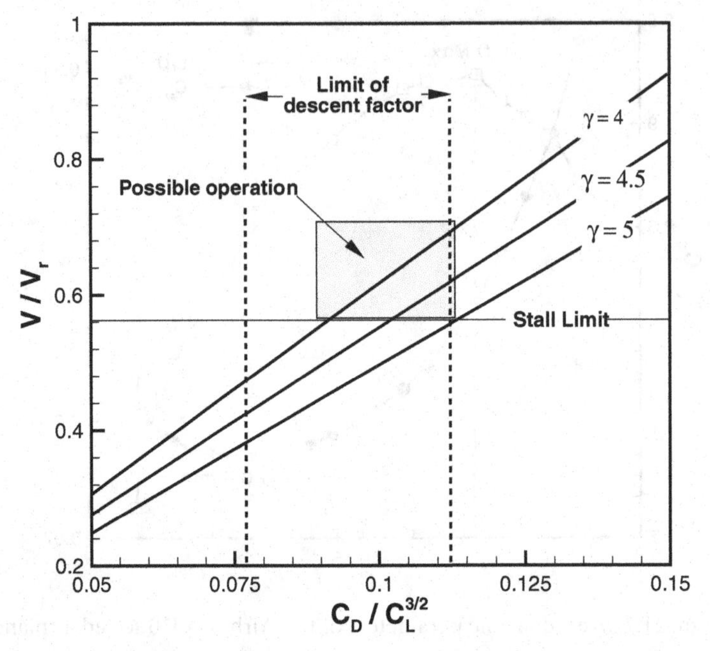


Figure 11.8. Parametric study of steep descent.

must be solved with the constraint

$$\frac{U}{U_r} \geq K_s \frac{V_{stall}}{U_r}. \tag{11.13}$$

This limit is slightly more restrictive than the one required by FAR §25.103. The stall speed is defined as

$$V_{stall} = \sqrt{\frac{2}{\rho} \frac{W}{A} \frac{1}{\sqrt{C_{L_{usable}}}}}, \tag{11.14}$$

where $C_{L_{usable}}$ denotes a fraction of the maximum lift coefficient. Next, we need to solve Equation 11.11 with the constraint Equation 11.13 and the definition Equation 11.14. With a fixed configuration, there is a small decrease in the $V_{stall}$ as the aircraft descends.

Now fix the flight-path slope. The aircraft is required to fly along this trajectory from the initial altitude so that at the final altitude the descent rate is below a threshold value considered safe:

$$v_s < v_{s_r}, \qquad \frac{v_s}{v_{s_r}} = \frac{f_d}{f_{d_r}}. \tag{11.15}$$

A parametric analysis of Equation 11.15 is shown in Figure 11.8. The horizontal line denotes the stall constraint, Equation 11.14 for this case. A relative air speed $U/U_r$ below this line is not feasible. The other constraint is the band delimited by the two vertical lines, $0.078 < f_d < 0.115$, which is the aerodynamic limit shown in Figure 11.7. The wing system cannot provide a value of $f_d$ below or above this band. It is evident that a descent with a slope $\gamma = 5.0$ degrees is not possible (see lower right corner of the constraint box). A descent at $\gamma = 4.5$ degrees is only possible if the aircraft operates at low $C_L$, as implied by Figure 11.7. However, there is a narrow band in the aerodynamic performance that allows a descent at $\gamma = 4.0$ degrees. To extend the descent envelope, the aircraft will have to provide

$$f_d = C_D/C_L^{3/2} > 0.12.$$

The aerodynamic limit will have to shift to the right of the graph for a very steep descent to be possible. The next problem is to examine how this can be achieved. There are four possibilities: 1) increase the $C_{D_o}$ at given $C_L$; 2) decrease the $C_L$ at fixed $C_{D_o}$; 3) increase the $C_{D_o}$ and decrease the $C_L$ at the same time; and 4) increase both $C_{D_o}$ and $C_L$, with the condition that $f_d$ increases. A detailed analysis is carried out in Ref.[1] that also shows some control theory on trajectories with increasing angles $\gamma$. In brief, the most effective method for increasing the glide slope is to increase the non-lifting portion of the drag. This can be achieved with a combination of technologies, such as clam shells (BAE RJ 146), conventional spoilers, or inverted spoilers[3].

## 11.5 Unpowered Descent

A fixed-wing aircraft is said to be gliding if it maintains a stable and controlled flight with minimum or no engine power. In the following sections we discuss some gliding flights and their optimal conditions.

### 11.5.1 Minimum Sinking Speed

The sinking speed is

$$v_s = U \sin\gamma = \sqrt{\frac{2}{\rho}}\sqrt{\frac{W}{A}\frac{C_D}{C_L^{3/2}}}, \tag{11.16}$$

having assumed that $C_D^2 \ll C_L^2$. The best glide ratio is in the range of 14 to 18, depending on the aircraft; hence $C_D^2$ is less than 0.5% of $C_L^2$. If we maintain the convention of positive $\gamma$ clockwise from the horizontal plane, then the sinking speed is a negative vertical velocity. From Equation 11.16 the condition of minimum sinking speed is equal to the condition of minimum power for a powered aircraft:

$$U_{mp}^2 = \left(\frac{2W}{\rho A}\right)\sqrt{\frac{k}{3C_{D_o}}}. \tag{11.17}$$

The corresponding value of the aerodynamic factor is

$$\frac{C_D}{C_L^{3/2}} = \frac{4C_{D_o}}{(3C_{D_o}/k)^{3/4}}. \tag{11.18}$$

From Equation 11.18 the minimum sinking speed is

$$v_s = \left(\frac{2}{\rho}\frac{W}{A}\right)^{1/2}\frac{4C_{D_o}}{(3C_{D_o}/k)^{3/4}}. \tag{11.19}$$

Equation 11.19 shows that the minimum $v_s$ decreases as the aircraft descends – which is a good thing. Also, the air speed decreases as the aircraft descends, which is another good thing, as long as the speed can be safely maintained above stall (for example, Equation 11.14) and the aircraft is fully controlled.

### 11.5.2 Minimum Glide Angle

The flight-path angle in absence of engine thrust is given by

$$\sin\gamma = \frac{T-D}{W} = -\frac{D}{W} \simeq -\frac{D}{L}. \tag{11.20}$$

A minimum for $\sin\gamma$ is also a minimum for $\gamma$; this occurs for an aircraft operating at maximum glide ratio. Therefore, an optimal condition at a fixed altitude is

$$\frac{\partial}{\partial U}\left(\frac{D}{L}\right) = 0. \tag{11.21}$$

This condition was found previously (Equation 8.51). We proved that this was a condition of minimum drag (or *green-dot speed*, $V_{green}$):

$$U_{md}^2 = \left(\frac{2W}{\rho A}\right)\sqrt{\frac{k}{C_{D_o}}}. \tag{11.22}$$

We now compare two different descent conditions: minimum sinking speed, as previously discussed, and minimum glide angle. The minimum sinking speed is given by Equation 11.17; therefore the ratio between the speed of minimum glide angle and the speed for minimum descent speed is

$$\frac{U_{vs}}{U_\gamma} = \frac{1}{\sqrt[4]{3}} = 0.7598 \simeq 0.76. \tag{11.23}$$

Therefore, *the speed of minimum descent rate is equal to 76% the speed of the minimum glide angle*. This ratio is the same as the ratio between the speed of minimum power and minimum drag (Equation 8.51),

$$\frac{U_{vs}}{U_\gamma} = \frac{U_{mp}}{U_{md}}. \tag{11.24}$$

The flight path can be found from the steady-state equations

$$\frac{\partial x}{\partial t} \simeq U, \qquad \frac{\partial h}{\partial t} = v_s, \tag{11.25}$$

with the additional condition

$$\frac{v_s}{U} \simeq \gamma. \tag{11.26}$$

The solution of this problem is found numerically. A typical result is shown in Figure 11.9. An aircraft of the class of the Airbus A300-600 with a mass $m = 145{,}000$ kg was simulated without any meteorological effects. The initial altitude was $h = 11{,}000$ m (~36,000 feet), with an initial Mach number $M = 0.79$.

The maximum gliding range is ~225 km (~120 n-miles); this is obtained with a minimum-glide-angle program. This is enough to establish that the aircraft without fuel should be able to glide the whole distance of 100 n-miles without any engine power. The gliding time is ~32 minutes. Head or tail winds will change the result.

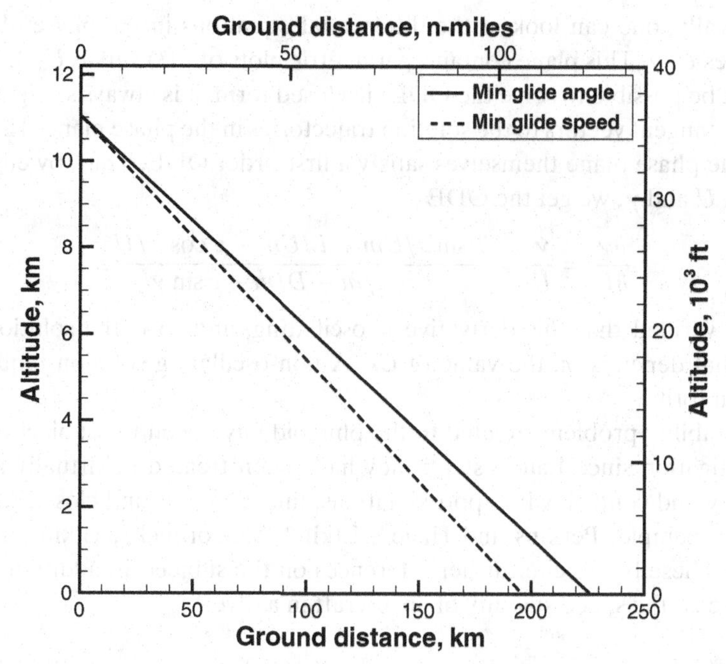


Figure 11.9. Glide range at two different flight conditions. Aircraft mass $m = 145{,}000$ kg.

### 11.5.3 General Gliding Flight

The results shown in the previous cases with the quasi-steady approximations can also be obtained by integration of the flight dynamics equations. If some thrust is available, the equations are

$$\dot{U} = \frac{T \cos \alpha}{m} - \frac{D}{m} - g \sin \gamma, \tag{11.27}$$

$$\dot{\gamma} = \frac{T \sin \alpha}{Um} + \frac{L}{Um} - g \frac{\cos \gamma}{U}, \tag{11.28}$$

having assumed that the thrust is aligned with the drag force. Because the glide slope is small, we make an approximation:

$$\sin \gamma \simeq \gamma \simeq \frac{v_s}{U}, \qquad \cos \gamma \simeq 1, \qquad U \simeq V. \tag{11.29}$$

Therefore, the drift-down equations become

$$\dot{U} = \frac{T \cos \alpha}{m} - \frac{D}{m} - g\gamma, \tag{11.30}$$

$$\dot{\gamma} = \frac{T \sin \alpha}{Um} + \frac{L}{Um} - \frac{g}{U}. \tag{11.31}$$

These equations can be integrated as an initial-value problem. The use of the full flight-path equations is required to study more general gliding problems, such as gliding at constant $C_L$, gliding with and without pitch damper. A glide at constant $C_L$ may give rise to a large-amplitude oscillation (*phugoid*), if a pitch damping is not used. The phugoid oscillation occurs when a powered or an unpowered vehicle operates at a speed and attitude away from its equilibrium and when no adjustments are made to its flight control systems.

Generally, one can look at the differential equations in the phase plane with coordinates $U, \gamma$. This plane contains parametric plots of $\gamma(t)$ versus $U(t)$. Whereas it may not be possible to solve the ODEs in closed form, it is always straightforward to plot the tangent vectors to the solution trajectories in the phase plane. The trajectories in the phase plane themselves satisfy a first-order ODE. Thus, by eliminating $t$ between $U$ and $\gamma$, we get the ODE

$$\frac{\partial \gamma}{\partial U} = \frac{\dot{\gamma}}{\dot{U}} = \frac{T \sin\alpha / Um + L/Um - g \cos\gamma / U}{T \cos\alpha / m - D/m - g \sin\gamma}. \tag{11.32}$$

It can be verified that this derivative is oscillating, and so is the solution, by an amount that depends on the value of $C_L$. A non-oscillating solution requires that $\dot{\gamma}/\dot{U}$ be uniform.

The stability problems related to the phugoid have been the subject of analytical investigation since Lanchester[4]; they have been treated in virtually all books on stability and control, with approximations, linearisations and closed-form solutions (for example, Perkins and Hage[5], Etkin[6], McCormick[7], Etkin and Reid[8], Nelson[9]). These books cite further references on the subject. In addition, Campos *et al.*[10] discuss the speed stability of an aircraft in a dive.

### 11.5.4 Maximum Glide Range with the Energy Method

We solve the maximum glide range for a supersonic jet fighter with the energy method and show how the solution to this problem can be substantially different from that of subsonic flight. Take the definition of the horizontal velocity component, $U = \dot{x}$, and divide by Equation 10.46

$$\frac{\dot{x}}{\dot{E}} = \frac{U}{(TU - DU)/m}. \tag{11.33}$$

In the absence of thrust, this equation can be simplified to

$$\frac{\dot{x}}{\dot{E}} = -\frac{m}{D} \tag{11.34}$$

or

$$\frac{\partial x}{\partial E} = -\frac{m}{D}, \qquad \frac{\partial E}{\partial x} = -\frac{D}{m} = -g\frac{D}{L}. \tag{11.35}$$

To maximise the gliding range, we must minimise $dE/dx$, subject to the constraint $L = W$. In other words, the aircraft must descend by moving along a flight path of minimum-drag or maximum-glide ratio. If we eliminate $h$ in the drag force by introducing the energy height, then the problem is a minimum drag with respect to the flight speed:

$$\min_U = -D(E, U), \tag{11.36}$$

for a fixed energy level. It is possible that the initial conditions of the aircraft (speed $U_o$ and altitude $h_o$ for given energy level $E_o$) are not on the maximum range path. This implies that the drag is not at a minimum, and the aircraft must zoom-dive or zoom-climb to the optimal energy level. Once it has reached the point on the $E = E_o$ level, the aircraft glides along the optimal flight path. The flight-path angle is found from the dynamic equation along the flight path with $T = 0$:

$$\sin\gamma = -\frac{D}{W} + \frac{1}{g}\frac{\partial U}{\partial t}. \tag{11.37}$$

Because the derivative of the speed is negative, this term decreases the drag and thus extends the glide path. The glide range is found from integrating Equation 11.35. It consists of two terms:

$$X = \int_o^2 dx = \int_{E_o}^{E_1} \frac{m}{D} dE + \int_{E_1}^{E_2} \frac{m}{D} dE, \tag{11.38}$$

where $E_o$ is the initial energy; $E_1$ is the energy at ground level; and $E_2$ is the minimum level speed at ground level. The first term in Equation 11.38 is due to the loss of energy from the starting point to ground level; the second term is due to deceleration at ground level, to a speed above stalling speed. Therefore, the aircraft can glide to nearly ground level; then it performs another stretch by increasing the angle of attack and deploying the high-lift control surfaces. If the initial speed is supersonic, the aircraft will first dissipate its kinetic energy whilst keeping nearly constant altitude; then it will glide, losing also its altitude (or potential energy).

## 11.6 Holding Procedures

It is 8:30 a.m. at London Heathrow. There are about 20 aircraft waiting to land. The sky is speckled with their lights in the far distance. Each of these airplanes is burning between 0.5 and 1.5 kg/s of fuel. The ATC is striving to manage this congestion. Unfortunately, the aircraft may be required to "hold" on a "stack" before being allowed to land, which causes considerable waste of time and fuel. The situation is similar at major airports around the world every day.

When aircraft are forced on a "hold" flight path due to local air traffic congestion, they are placed on a racetrack pattern made of two parallel segments joined by half-circles (*stacks*). A typical hold altitude is 5,000 or 7,000 feet. Thus, the aircraft must perform two U-turns for every lap. In a hold situation, range is not important but endurance is[11]. Minimum-fuel turns are discussed in § 13.2.4 and show that minimum turns are performed at the speed of minimum drag.

The minimum fuel flow corresponds approximately to the minimum drag (for jet-powered airplanes) and to the minimum power (for propeller-powered airplanes). In the former case, the best speed for a clean configuration is the speed of maximum L/D (*green dot speed*). Thus, "green dot" is selected to hold the airplane.

There are instances when the green dot speed is too high, due to obstacle proximity. If the aircraft is forced to maintain a lower speed, it must do so with a slat/flap extended by one notch (CONF1, or "S" speed on Airbus). There are also cases when the ATC requires the aircraft to maintain a speed that can be quite different from the green dot. The hold altitude is not a free parameter; this is enforced, but the rule is simple: the higher the flight altitude, the lower the fuel flow.

An example of this situation is shown in Figure 11.10, which displays the simulated hold performance of the Airbus A320-200 with CFM engines. The green dot speed is estimated at 248 kt at 10,000 feet (~3,048 m), at the weight indicated.

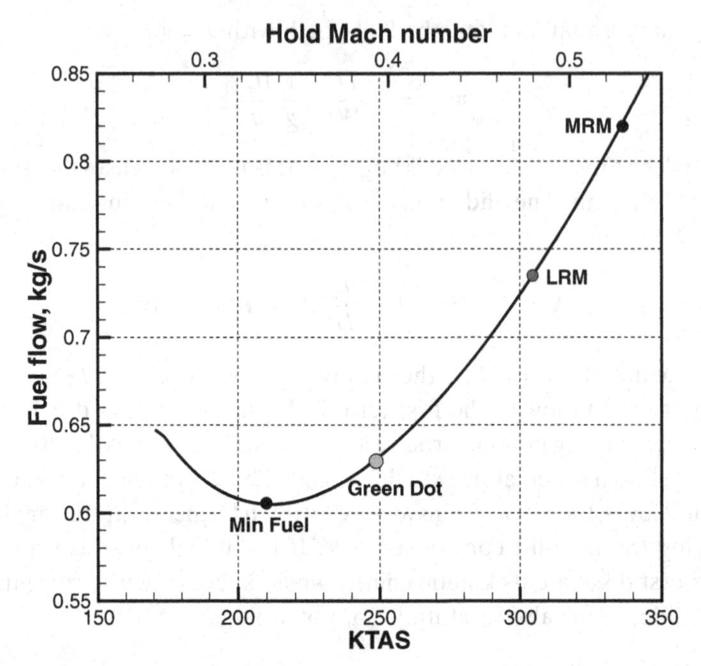


Figure 11.10. Holding speed of the Airbus A320-200 at 10,000 feet (3,048 m), $m = 69{,}380$ kg (calculated). Standard day, no wind.

The ICAO recommends that the hold speed does not exceed 230 kt at this altitude; therefore the CONF1 configuration will have to be selected (see Table 11.1). A few other points are indicated in the figure: the minimum fuel flow speed, the long-range Mach/speed and the maximum-range Mach/speed. It is worthy of note that in this case the green dot speed does not correspond to minimum fuel, a fact evidenced by Airbus on some of their airplanes †.

If conditions allow, it is preferable to perform a linear hold by extending the cruise at the green dot speed and then return to the airfield to join the stack for a more limited amount of time. Each major airport has its own hold areas, variously named for easy identification.

## 11.7 Landing Performance

Landing is the final flight segment of the aircraft when, upon reaching an altitude of about 50 feet above the airfield, the aircraft proceeds to touch-down, rotation and ground run, until the aircraft is stopped or taxied off the runway. Landing is a nearly completely automated process but one that is fairly complicated by a variety of factors due to atmospheric conditions such as precipitation, wind gusts, fog and poor visibility at night. The first instrumented landing that took place in 1929 is now an almost-forgotten piece of aeronautical history. There are two distinct phases: the airborne phase and the ground run, exactly the reverse of the take-off. A sequence of events is shown in Figure 11.11. Table 11.4 is a summary of relevant aircraft speeds

† Airbus, Flight Operations Support. Getting to Grips with Fuel Economy, Issue 3. Blagnac, France (2004).

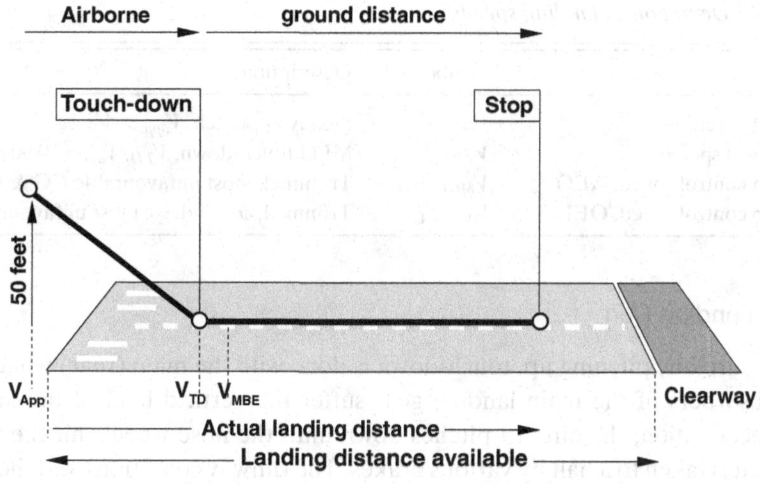


Figure 11.11. Sequence of events in a landing procedure.

at landing. The required landing distance must be less than the landing distance available[‡].

## 11.7.1 Airborne Phase

The airborne phase starts at a conventional altitude above the airfield (say, 50 feet), with the airplane coming down at a relatively high-pitch attitude. The airplane slowly pitches up as it descends. This phase ends when the main landing-gear wheels touch the ground. This point is critical for the whole under-carriage unit because it can be subject to extreme loads. Because there is no pre-spin of the tyres, at touch-down there is an impulsive load that sets the wheels in motion to follow the speed of the airplane. The impact energy $E$ between main tyres and ground can be estimated from

$$E = \frac{1}{2}\left(m - \frac{L}{g}\right)V^2 \sin^2 \gamma, \tag{11.39}$$

where $L$ is the aerodynamic lift at touch-down, $V$ is the ground speed and $\gamma$ is the descent angle. This impact energy can be decreased by increasing the lift and decreasing the approach angle.

Taking the aircraft to a halt on the runway requires finding ways to dissipate the mechanical energy of the vehicle. This energy is roughly proportional to the square of the velocity at the touch-down. Because all aircraft must take off and land on the available runway length, the landing speed is somewhat independent of the gross weight; therefore the energy that must be dissipated by the braking system is proportional to the aircraft's weight.

The length of the airborne phase is $x_1 \sim h_{ref} \cos \gamma$, where $h_{ref}$ is the altitude of the conventional screen; the value of $x_1$ is generally below the uncertainty of the length of the ground run, as discussed next.

‡ The relevant regulations for transport aircraft are the FAR § 25.125 and § 121.195.

Table 11.4. *Definition of landing speeds*

| Definition | Symbol | Description |
|---|---|---|
| Air speed at screen | $V_{app}$ | Steady approach, $V_{app} > V_S$ |
| Touch-down speed | $V_{TD}$ | MLG touch-down, $V_{TD}/V_{app} = 0.90$–$0.98$ |
| Minimum control speed, AEO | $V_{MCL}$ | Trimmed, most unfavourable CG & weight |
| Minimum control speed, OEI | $V_{MCL-1}$ | Trimmed, $\phi < 5$ degs, most unfavourable CG |

### 11.7.2 Landing Run

With the airplane pitching up, touch-down is done with the main (rear) wheels. The shock absorbers of the main landing gear suffer the vertical load of the airplane. Upon deceleration, the aircraft pitches down until the nose wheels hit the runway and then it is taken to a halt by various brakes. The runway conditions and the effects of the brakes are discussed in Chapter 9.

There are several techniques to force the airplane to reduce its pitch attitude upon landing. A combination of CG position, elevator setting and ground speed can be used. Forward CG position and deceleration due to high drag work in combination to produce a pitch-down moment.

The deceleration technique is similar to the case of aborted landing. The main difference is that the airplane is lighter and should come to a stop in a shorter distance. Brakes of a various nature can be used (thrust reversal, tyre brakes, ailerons). Thrust reverse is to be used sparingly and only when the other means of bringing the aircraft to a halt are not sufficient. At certification level, the landing distances have to be demonstrated without thrust reversal. The differential equation is

$$m\frac{\partial U}{\partial t} = D + \mu_r(W - L) + W \sin\gamma_r + B = R, \tag{11.40}$$

where the right-hand side is the total resistance, $B$ is the total braking force, and $\gamma_r$ is the slope of the runway. The $C_L$ and $C_D$ correspond to landing configuration and account for ground effects. The landing run is obtained by integration of Equation 11.40. The first step is to recast the equation as

$$U\frac{\partial U}{\partial x} = \frac{R}{m}. \tag{11.41}$$

Integration of the last equation is performed between the touch-down point (subscript "o") and the point at which the aircraft halts (subscript "1"). The result is the landing distance

$$x_l = \int_o^1 m\frac{U}{R}\,dU. \tag{11.42}$$

**CLOSED-FORM SOLUTIONS.** Assume that the runway has no inclination ($\gamma_r = 0$) and that there is no braking force. The solution is found from

$$x_l = m\int_o^1 \frac{1}{D + \mu_r(W - L)} U dU. \tag{11.43}$$

By defining the constant coefficients $c_1 = \rho A(C_D - \mu_r C_L)/2$ (with constant aerodynamic coefficients) and $c_2 = \mu_r W$, the integral assumes the form

$$\frac{1}{2}\int_o^1 \frac{dU^2}{c_1 U^2 + c_2} = \frac{1}{2c_2} \ln(c_1 U_1^2 + c_2). \tag{11.44}$$

Therefore, the landing run is

$$x_l = \frac{m}{2c_2} \ln(c_1 U^2 + c_2). \tag{11.45}$$

Equation 11.45 is a simplification because in practice some form of braking is used (aero-brakes, wheel brakes, and so on). The solution is similar to the one outlined if the braking force is either constant or proportional to $U^2$. The braking force is expressed as

$$B = c_3 T + c_4 U^2 \tag{11.46}$$

and contains the effects of the thrust reversal (via the coefficient $c_3$) and the mechanical brakes (via the coefficient $c_4$). A constant braking force $-T$ is assumed in the case of thrust reversal. The total resistance is

$$R = D + \mu_r(W - L) + c_3 T + c_4 U^2. \tag{11.47}$$

The integral of Equation 11.42 becomes

$$x_l = \frac{1}{2} m \int_o^1 \frac{1}{\rho A(C_D - \mu_r C_L + c_4)U^2/2 + \mu_r W + T} dU^2. \tag{11.48}$$

In this case, the coefficients are

$$c_1 = \rho A(C_D - \mu_r C_L + c_4)/2, \qquad c_2 = \mu_r W + T, \tag{11.49}$$

and the solution has the form of Equation 11.45.

**NUMERICAL SOLUTIONS.** Integration of Equation 11.40 is done numerically with a time-marching scheme to provide the complete trajectory of the airplane as well as the key performance parameters (airplane state, engine state). It is useful to introduce some operators that define the state of the airplane and to calculate the exact aerodynamic coefficients, which in the general case are not constant. In normal circumstances, the change in state is defined by aileron, spoiler and flap settings.

The differential Equation 11.40 is solved with the relevant equations for the rolling coefficients and normal loads on the wheels, as discussed in Chapter 9. Figure 11.12 shows the landing run of a model Airbus A320-200 with an initial mass $m = 59{,}900$ kg at sea level with a head wind $V_w = -2$ m/s.

Landing is complicated by several factors, including atmospheric conditions. Therefore, the data quoted by the manufacturers refer to a statistical analysis rather than precise measurements or other predictions. However, abstracting from some of these stochastic events, the main parameters reported in the FCOM are the landing weight, the airfield altitude, the wind speeds and the flap setting. An example of landing charts is shown in Figure 11.13. This figure displays the FAR runway length

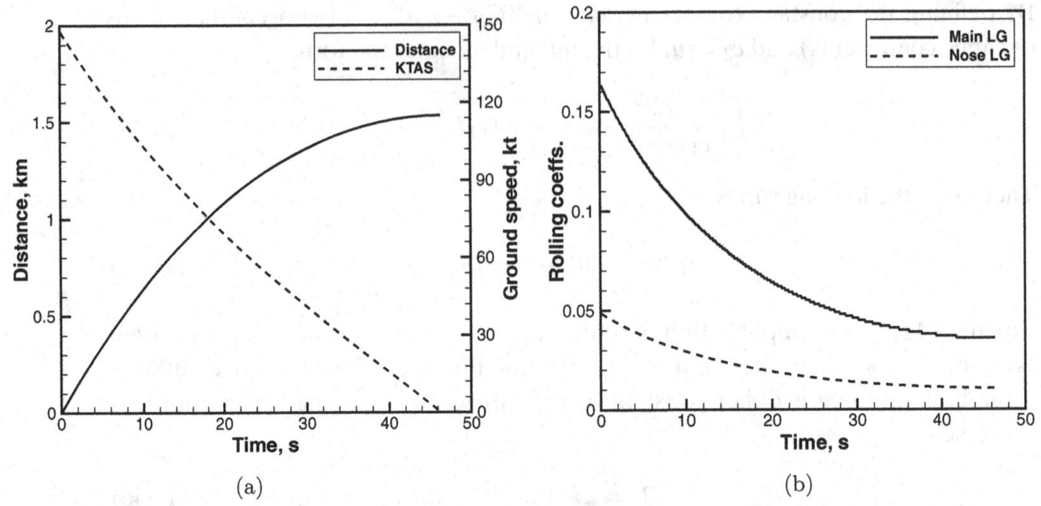


Figure 11.12. Landing run of the Airbus A320-200 model at sea level, $V_w = -2$ m/s, $m =$ 59,900 kg (calculated).

requirements for the Boeing B747 (various versions, as indicated in the graph) for a fixed-flap configuration (25 degrees), as elaborated from Boeing[12].

### 11.7.3 Crab Landing

Strong cross winds at landing are a cause of hazard. Maximum wind speeds are recommended and/or certified, along with procedures to control the aircraft. Adverse weather conditions are encountered in about one-third of all landing accidents.

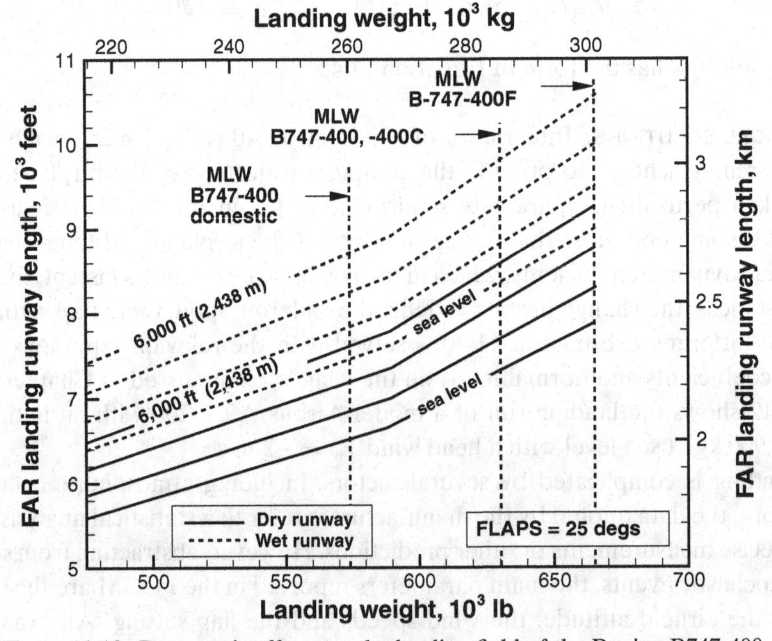


Figure 11.13. Parametric effects on the landing field of the Boeing B747-400.

Table 11.5. *Limit crosswind speeds coupled with runway conditions*

| Braking action | Runway friction | Maximum crosswind | Runway condition |
|---|---|---|---|
| Good | > 0.40 | Tested | Dry or wet; no aquaplaning |
| Good/Medium | 0.36–0.39 | 30 kt | Slush |
| Medium | 0.30–0.35 | 25 kt | Dry snow |
| Medium/Poor | 0.26–0.29 | 20 kt | Standing water, wet snow |
| Poor | < 0.25 | 15 kt | Risk of aquaplaning |
| Unreliable | — | 5 kt | |

Runway contamination adds to the hazard*. Table 11.5 is a summary of limit crosswind speeds, runway-friction characteristics and braking-action effectiveness.

The final approach must be performed with wings level and with a steady side-slip angle to align the longitudinal axis of the airplane with the centre of the runway. This operation must be done with a combination of flight controls (aileron–rudder). During the flare, any tendency of the airplane to roll downwind should be minimised with an appropriate response of the side-stick. In any case, the airplane should not have a crab angle or bank angle $\phi > 5$ degrees.

The limitations involved in crab landing (Figure 11.14) can be summarised by two conditions: 1) bank angle for a given crab angle, and 2) crab angle for a given bank angle. Figure 11.15 shows one such chart with the approach speed $V_{app}$ as the independent parameter. In Figure 11.15, point A denotes the condition of zero-crab angle (at the given $V_{app}$), which corresponds to a bank angle $\phi \simeq 3$; point B is the condition of zero-bank angle, which corresponds to a crab angle of about 4 degrees. A negative crab angle (aircraft pointing away from the wind) is a result of an excessive rudder control. The data in Figure 11.15 change with the wind speed. There are a number of limitations to be considered, including geometry (wing and tail strike) and control authority (ability to maintain a steady side-slip under crosswinds). It is possible that with high crosswinds, the aircraft is unable to perform a wing-level landing. Even if the geometrical conditions would allow a large bank angle, there is a risk of damage to the main landing gear.

Calculation of charts like the one shown in Figure 11.15 can be done with the method explained in § 7.2 (control speed in air); the aerodynamic derivatives of the airplane to side-slip and roll will have to be determined before setting up the system of equations.

**WING STRIKE.** Aileron and rudder attitude are important for lateral control of the aircraft. However, the geometric configuration is what determines the severity of ground strike when the controls are ineffective. An example of wing strike is shown in Figure 11.16. In this specific case, Figure 11.16a shows that the line between the wheel contact (W) and the wing tip (T) runs above the lowest point of the engine (E). Therefore, this would be a case of *engine strike*. If the engine's diameter is scaled down (Figure 11.16b), the engine is above the W-T line and is "protected" by

* Flight Safety Foundation, *Flight Safety Digest*, Vols. 17 (Nov. 1998) & 18 (Feb. 1999).

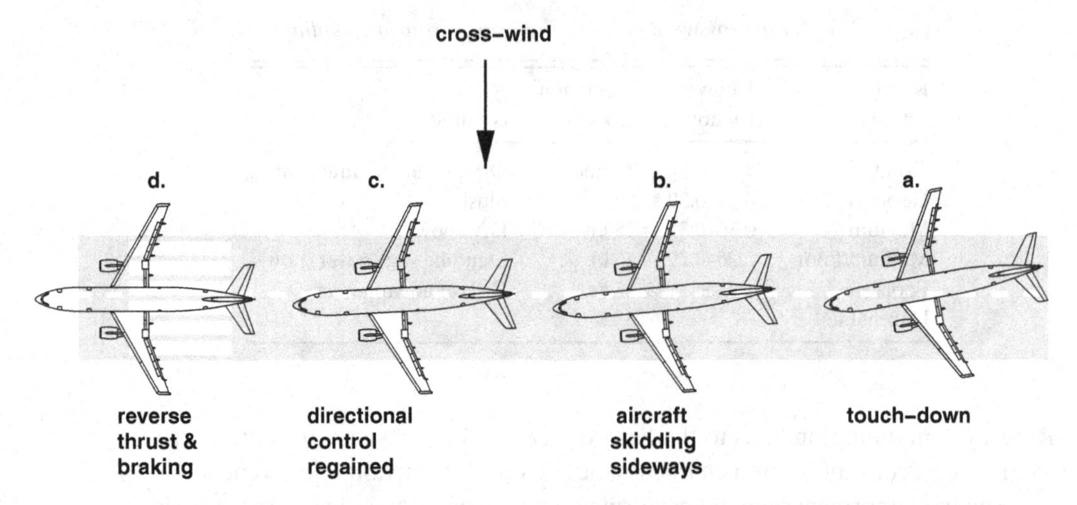


Figure 11.14. Crab landing in presence of strong side gusts.

the wing tip. In the latter case, the maximum bank angle $\phi$ that prevents wing strike can be estimated from

$$\tan\phi = \tan\varphi - \frac{2h_g}{b - b_t}\tan\theta\tan\Lambda, \tag{11.50}$$

where $\Lambda$ is the wing sweep, $b_t$ is the wheel track, $h_g$ is the height of the landing gear and $\varphi$ is the wing's dihedral angle. There are cases in which the aircraft strikes the ground with the engine instead of the wing tip.

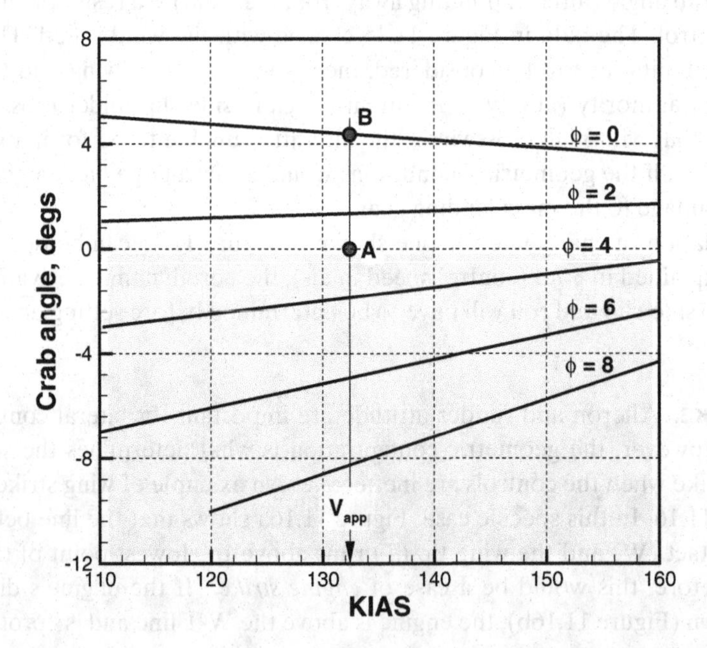


Figure 11.15. Crab angle versus bank angle for a given wind speed (extrapolated from Airbus data).

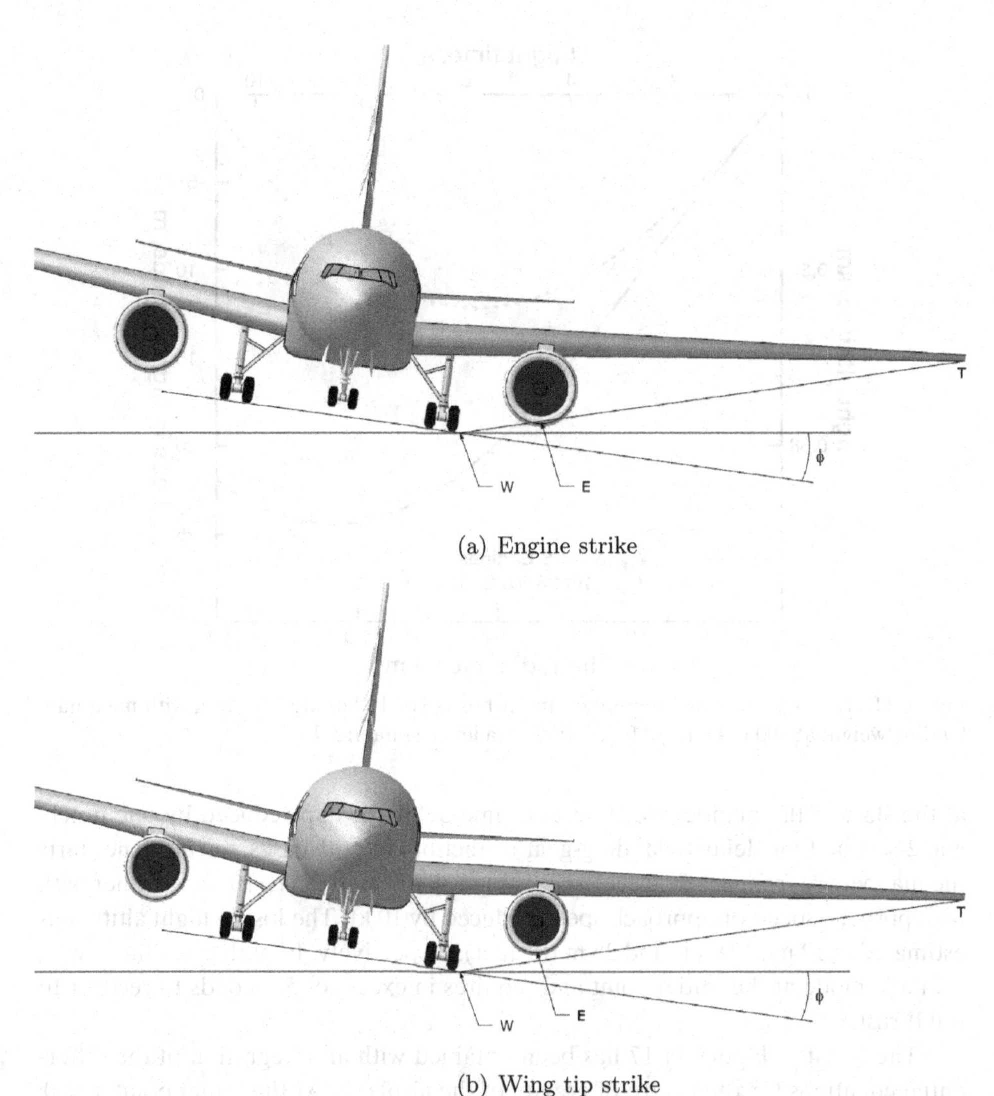


(a) Engine strike

(b) Wing tip strike

Figure 11.16. Engine and wing strike.

## 11.8 Go-Around Performance

As illustrated in Figure 11.4, there is the possibility that the aircraft is unable to land and thus it is forced to *go-around*. This means that landing is aborted, the throttle is increased, the aircraft is set onto a climb flight path (after rotation), the landing gears are retracted and the high-lift systems are retracted by one notch (or more, depending on circumstances). The engines are unable to respond immediately to the request of additional thrust (see also § 5.2.2 and Figure 5.2).

During the final approach the engine is set on idle, or nearly so. If a go-around is required, the engine must spool up quickly to provide the required thrust.

**CASE STUDY: GO-AROUND OF AN AIRBUS A319-100 MODEL.** Figure 11.17 shows two simulated trajectories of an Airbus A319-100 airplane model with CFM56-5B5 turbofan engines. The assumptions include standard day, maximum landing weight

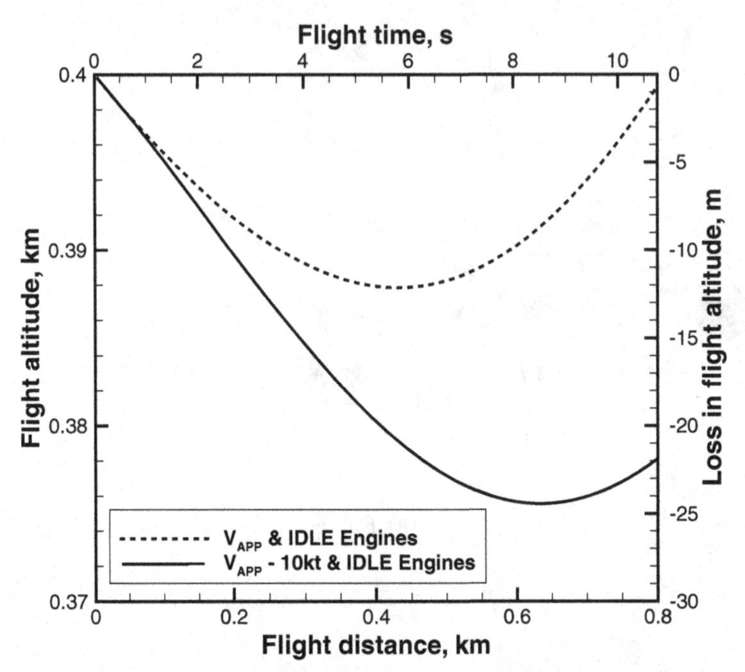

Figure 11.17. Go-around manoeuvre of an Airbus A319-100 model, starting with maximum landing weight at 400 m (~1,310 feet) above sea level, standard day.

at the start of the manoeuvre, 1-second time delay for flap reduced by one notch, and 2-second time delay to landing-gear retraction. In both cases, the airplane starts the manoeuvre from a 3-degree descent slope at 400 metres (1,310 feet), either with the approach speed or approach speed reduced by 10 kt. The loss in flight altitude is estimated at 12 m (40 feet) and 25 m (83 feet), respectively. In both cases the engine is in idle mode at the initial point and requires in excess of 5 seconds to recover to full thrust.

The result in Figure 11.17 has been obtained with an integration of the differential equations for the centre of gravity of the airplane. At the initial point, $t = 0$; $x = 0$. The loss in flight altitude can be reduced if the airplane is operating in stabilised mode, with the engine at a minimum setting (not in idle). In fact, it would take less time for the engine to respond to the increase in fuel flow commanded by the throttle. There is a combination of flight altitude and speed that will not allow the safe recovery of the airplane; in that case the airplane must proceed to landing. Parametric studies around this problem include the effects of initial altitude, weight, atmospheric condition, time lag in pilot's control actions, and engine response modes.

## Summary

In this chapter we presented the problem of powered and unpowered descent in a fairly general form. We started from the en-route descent; we showed that the aircraft is required to follow standard flight paths with two or more level flight segments, which may increase the flight time and the fuel burn. We have shown that the CDA saves fuel. A more complex manoeuvre consists in setting the aircraft on a larger glide slow (steep descent). We demonstrated that although there can be some advantages, such as reduced noise at a receiver on the ground, the procedure is seriously limited by the aerodynamics and that it is not possible to exceed 4.5 to 5 degrees. We then reviewed the holding procedures and stacking-patterns practices. Finally, we considered landing problems, such as a conventional landing, crab landing (in presence of side gusts), the risk of wing and engine strike and go-around performance. In the latter case, we demonstrated that to be able to take control of the aircraft in the case of an aborted landing, it is necessary to operate with a minimum thrust. This procedure allows a more rapid recovery of thrust, which must be used to go-around, that is, climb out of the airfield.

## Bibliography


[1] Filippone A. Steep-descent manoeuvre of transport aircraft. *J. Aircraft*, 44(5):1727–1739, Sept. 2007.

[2] Kiock R. The ALVAST model of DLR: Technical Report IB 129 96/22, DLR, Lilienthal Platz, 7, D-38018 Braunschweig, Germany, 1996.

[3] Filippone A. Inverted jet spoilers for aerodynamic control. *J. Aircraft*, 46(4):1240–1252, 2009.

[4] Lanchester FW. *Aerodonetics*. Constable, London, 1908.

[5] Perkins CD and Hage RE. *Airplane Performance Stability and Control*. John Wiley, 1949.

[6] Etkin BE. *Dynamics of Flight*. John Wiley & Sons, New York, 1959.

[7] McCormick BW. *Aerodynamics, Aeronautics and Flight Mechanics*. John Wiley, 2nd edition, 1995.

[8] Etkin BE and Reid LC. *Dynamics of Flight: Stability and Control*. John Wiley & Sons, 1996.

[9] Nelson RC. *Flight Stability and Automatic Control*. McGraw-Hill, 2nd edition, 1998.

[10] Campos A, Fonseca L, and Azinheira R. Some elementary aspects of non-linear airplane speed stability in constrained flight. *Progress Aerospace Sciences*, 31(2):137–169, 1995.

[11] Sachs G and Christodoulou T. Reducing fuel consumption of subsonic aircraft by optimal cyclic cruise. *J. Aircraft*, 24(5):616–622, 1987.

[12] Anonymous. 747-400 airplane characteristics for airport planning. Technical Report D6-58326-1, The Boeing Corporation, Dec. 2002.


## Nomenclature for Chapter 11

Symbols in Table 11.4 are not repeated here.

$A$ = wing area

$b$ = wing span

$b_t$ = wheel track

$B$ = braking force

$c_i$ = constant coefficients

$C_D$ = drag coefficient

$C_{Do}$ = profile-drag coefficient

$C_L$ = lift coefficient

$C_{L_o}$ = zero-$\alpha$ lift-coefficient
$C_{L_\alpha}$ = lift-curve slope
$D$ = drag force
$E$ = energy
$f_d$ = aerodynamic parameter, Equation 11.12
$g$ = acceleration of gravity
$h$ = altitude
$h_g$ = height of landing gear, including wheel
$k$ = induced-drag factor
$K_s$ = stall margin
$L$ = lift force
$m$ = mass
$m_f$ = fuel mass
$\dot{m}_f$ = fuel flow
$M$ = Mach number
$P$ = power
$R$ = rolling resistance
$t$ = time
$T$ = net thrust
$U$ = air speed
$v_s$ = descent rate
$V$ = ground speed
$V_{app}$ = approach speed
$V_{green}$ = green dot speed
$V_{stall}$ = stall speed
$V_{TD}$ = touch-down speed
$V_{MBE}$ = maximum brake energy speed
$V_w$ = wind speed
$x$ = distance
$x_1$ = airborne distance at landing
$x_l$ = ground run at landing
$X$ = flight distance
$W$ = weight

### Greek Symbols

$\alpha$ = angle of attack
$\gamma$ = descent angle
$\gamma_r$ = runway-inclination angle
$\Lambda$ = wing-sweep angle
$\phi$ = bank/roll angle
$\varphi$ = wing's dihedral angle
$\mu_r$ = rolling-friction coefficient
$\sigma$ = relative air density
$\rho$ = air density

### Subscripts/Superscripts

$[\cdot]_d$ = descent
$[\cdot]_{hl}$ = high lift
$[\cdot]_l$ = landing
$[\cdot]_{md}$ = minimum drag
$[\cdot]_r$ = relative value

# 12 Cruise Performance

## Overview

In this chapter we introduce the concepts of distance flown by an aircraft with and without stop for refuelling (§ 12.1). We discuss a number of cruise programs at subsonic and supersonic speeds and some optimal problems in long-range cruise, with or without constraints. First we present the analysis of the instantaneous cruise parameters (§ 12.2), including the specific range. We then provide numerical solutions of the specific range for real aircraft (§ 12.3). For the sake of generality, we also present the range equation, which is solved in closed form (§ 12.4), and deal with the separate problems of jet aircraft at subsonic speed (§ 12.5) and propeller aircraft (§ 12.6). We show a number of more advanced studies in cruise altitude selection (§ 12.7), cruise performance deterioration (§ 12.8), cost index and economic Mach number (§ 12.9) and various other effects, such as the effects of high-altitude winds and centre of gravity position (§ 12.10). We deal briefly with supersonic aircraft (§ 12.11).

**KEY CONCEPTS:** Point Performance, Specific Air Range, Cruise Altitude Selection, Range Equation, Long-Range Mach, Maximum-Range Mach, Cost Index, Economic Mach Number, Effects of Winds, Centre of Gravity Effects, Supersonic Cruise.

## 12.1 Introduction

For most commercial aircraft the fuel consumed during the cruising phase of the flight makes up the bulk of the fuel carried and is a key factor in the productivity and direct operating costs of an aircraft. Since the early 1970s, with the price of fuel soaring, both the airlines and the military have been concerned with energy-efficient operations. For this reason, several cruise conditions have been studied. The fuel costs of major international airlines are constantly reviewed. Whilst the DOC have been reduced considerably, the fuel continues to be one of the major cost items*. Jet fuel prices have grown about 350% from the year 2000 to the current day. On average, they currently make up to 30% of the total operating costs and in excess

* Fuel costs are monitored in real time by IATA and other stakeholders, who provide real-time data such as price per gallon, price variations, impact of fuel bill on commercial aviation and more.

of 40% per block hour. Considerable research now exists on issues such as costs, emissions, efficiency, fuel hedging and trends in aviation.

**OPTIMALITY ANALYSIS.** It has been demonstrated that the steady-state cruise is not always fuel-optimal. Speyer *et al.*[1;2] demonstrated that an oscillating control leads to a decrease in fuel consumption. Gilbert and Parsons[3] applied periodic cruise control on the McDonnell-Douglas F-4 interceptor, and Menon[4] showed mathematically the conditions under which oscillating solutions occur in a long-range cruise. Sachs and Christodoulou[5] analysed the benefits of cyclic flight, with dolphin-type climbs and glides, and showed that (at least theoretically) the range can be increased with such a flight program. Due to the number of free parameters involved (Mach number, altitude, lift coefficient, angle of attack, gross weight, block fuel) and number of external constraints (ATC, flight corridors, atmospheric conditions), there is a variety of optimal and sub-optimal conditions. ATC issues and terminal-area constraints are reviewed by Visser[6]. Various ESDU data items deal with cruise performance[7–9].

**DEFINITIONS.** The aircraft range is the distance that can be covered in straight flight at a suitable flight altitude. The aircraft cannot use all of the fuel, and allowance must be made to account for the terminal phases (take-off and landing), manoeuvres (loiter, holding at altitude) and fuel reserves for contingency. In practice, the different flight sections are calculated separately and hence the *range equations* are limited to steady flight at altitude.

The *block fuel* is the fuel weight required to fly a specified mission and includes the fuel to taxi at the airport. The *fuel reserve* is a contingency amount of fuel that takes into account the risk of not being able to land at the destination airport. The *mission fuel* includes 1) the fuel required to take off, accelerate and climb to the initial cruise altitude; 2) the cruise fuel; 3) the descent, terminal area approach and landing fuel; and 4) manoeuvring and reserve fuel. For the determination of the gross take-off weight (TOW), the taxi fuel at the departure airport is not included. The taxi fuel after landing is extracted from the reserve fuel.

The *endurance* is the time on station, that is, the time the aircraft can be flown without landing or in-flight refuelling. *Maximum endurance* performance applies to search and surveillance operations. Refuelling has been traditionally a problem for the military, but some commercial operations are also envisaged[10], on the grounds that the amount of fuel that has to be carried by a wide-body aircraft on a long-haul flight can be four times the useful payload. Smith[11] has published a historical account of military refuelling technology.

## 12.2 Point Performance

The instantaneous conditions in aircraft cruise are called *point performance*. The basic point parameters are the glide ratio, the specific range, the figure of merit ($FM$), the instantaneous endurance and the glide ratio. Other instantaneous parameters of interest include the product $C_L M^2$ or $W/\rho$. These parameters are important because the optimal flight path or flight program for a long-range cruise can be derived from integration of the point-performance parameters. Generalised point-performance optimisation is discussed in detail by Torenbeek and Wittenberg[12].

### 12.2.1 Specific Air Range at Subsonic Speed

The *specific range* is the distance flown by burning one unit of weight (or mass or volume) of fuel. For operational reasons, other units may be used. For example, to evaluate the productivity of an aircraft, the airlines may quote a fuel consumption in unit of volume or weight per unit of distance per passenger.

If the symbol $f_j$ denotes the thrust-specific fuel consumption (TSFC), the range $dX$ obtainable by burning a small amount of fuel $dm_f$ is

$$dX = U dt = \frac{U}{\dot{m}_f} dm. \tag{12.1}$$

The specific air range (SAR) is the derivative of the range with respect to the mass:

$$\text{SAR} = \frac{\partial X}{\partial m} = \frac{U}{\dot{m}_f} = \frac{U}{f_j T}. \tag{12.2}$$

The physical dimensions of SAR are distance × unit mass of fuel. If $\dot{m}_f$ as a function of the speed and altitude is known, the SAR can be plotted directly from the first equivalence of Equation 12.2. For a subsonic jet aircraft with a parabolic drag equation, the specific range is

$$\text{SAR} = \frac{U}{f_j D} = \frac{1}{c_1 f_j (C_{D_o} + kC_L^2) U} = f(h, U, m), \tag{12.3}$$

with $c_1 = \rho A/2$. Equation 12.3 satisfies the conditions $D = T$, $L = W$. In Equation 12.3 the speed can be replaced by $U = aM$, and the mass can be normalised with the aircraft mass at the start of the cruise ($\xi = m/m_i$). The SAR for a subsonic commercial jet with a parabolic drag is shown in Figure 12.1. It is demonstrated that 1) at a constant flight altitude and Mach number, the specific range increases with the decreasing aircraft's mass; and 2) at a constant aircraft weight and Mach number, the specific range increases with the cruise altitude.

For a flight at constant speed, one can study the effects of an aircraft's initial mass $m_i$ on the specific range. The effect of the cruise speed on the specific range can be found by studying the derivative $\partial\text{SAR}/\partial U$. The condition that gives the speed of maximum SAR is

$$\frac{\partial \text{SAR}}{\partial U} = 0. \tag{12.4}$$

Assuming that the cruise TSFC is constant, solution of this equation leads to the speed for best range

$$U = \left(\frac{2W}{\rho A}\right)^{1/2} \left(\frac{3k}{C_{D_o}}\right)^{1/4}. \tag{12.5}$$

This speed depends on the wing loading and on the flight altitude.

**SUPERSONIC FLIGHT.** A similar expression can be derived for supersonic flight after replacing the correct drag equation in Equation 12.2. The result is

$$\text{SAR} = \frac{1}{c_1 f_j a\sigma (C_{D_o} + \eta C_{L_\alpha} \alpha^2) M}, \tag{12.6}$$

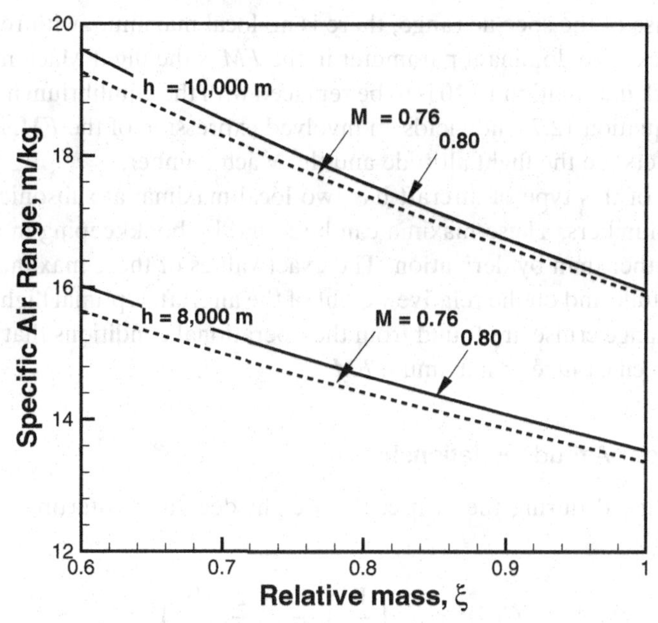


Figure 12.1. Estimated specific range for a subsonic commercial jet with parabolic drag; mass $m = 160{,}000$ kg.

with $c_1 = \rho_o A/2$. The angle of attack is calculated from the vertical equilibrium $L = W$, which requires the following equation to be satisfied:

$$\alpha = \alpha_o + \frac{2W}{\rho A a^2} \frac{1}{C_{L_\alpha} M^2}. \tag{12.7}$$

### 12.2.2 Figure of Merit

The figure of merit ($FM$) is defined as the product between the flight Mach number and the glide ratio. At subsonic speeds with Mach numbers $M < M_{dd}$, we have

$$M\left(\frac{L}{D}\right) = M\frac{C_L}{C_{D_o} + kC_L^2} = \frac{c_1 m/a^2 M\sigma}{C_{D_o} + kc_1^2 m^2/\sigma^2 a^4 M^4}, \tag{12.8}$$

where

$$c_1 = \frac{2g}{\rho_o A}. \tag{12.9}$$

Equation 12.8 allows to study the effects of flight altitude, aircraft mass and flight Mach number. The analysis is limited to a fixed Mach number. A parametric study of Equation 12.8 shows that 1) at constant Mach number, the $FM$ increases with the flight altitude; and 2) at a constant Mach number and constant altitude, the $FM$ decreases with the decreasing aircraft's mass (i.e., it decreases during the cruise).

For the transonic and supersonic cruise of a reference supersonic aircraft, we use the alternative drag Equation 4.85 that represents correctly the effects of Mach number on the aerodynamic drag of the aircraft. The $FM$ becomes

$$M\left(\frac{L}{D}\right) = M\frac{C_{L_\alpha}\alpha}{C_{D_o} + \eta C_{L_\alpha}\alpha^2}. \tag{12.10}$$

As for the case of the specific range, there is no local maximum with respect to the angle of attack. The dominant parameter in the *FM* is the flight Mach number. The angle of attack in Equation 12.10 is to be replaced with the equilibrium in the vertical direction, Equation 12.7, and yields an involved expression of the *FM*, in which the free parameters are the flight altitude and the Mach number.

The *FM* of this type of aircraft has two local maxima: at subsonic and supersonic Mach numbers. These maxima can be found by bookkeeping in a numerical procedure rather than by derivation. The exact values of these maxima depend on the flight altitude and on the relative weight of the aircraft. Optimal flight conditions over a long-range cruise are found from the operational conditions that always give maximum specific range or maximum *FM*.

### 12.2.3 Weight-Altitude Relationship

As fuel is burned during the cruise, the weight decreases. Starting from the lift equation:

$$C_L M^2 = \frac{2}{A}\left(\frac{W}{\rho}\right)\frac{1}{a^2} = \frac{2}{A}\left(\frac{W}{p}\right). \tag{12.11}$$

For a constant *FM*, the product $C_L M^2$ must be constant. In the stratosphere, where the speed of sound is constant, the aircraft must climb as it burns fuel in order to maintain a constant $W/\rho$. In the troposphere, where the speed of sound is variable, the aircraft must climb to maintain constant $W/p$.

Continuous changes in altitude and Mach number are not allowed by the ATC. By convention, two flight levels are 1,000 feet apart and are specified by the flight altitude in feet divided by 100. Thus, 31,000 feet corresponds to the flight level 310, or FL-310*.

Current regulations for commercial aviation require aircraft to maintain restricted flight corridors above a *transition altitude*. There is no universal agreement on this altitude because it varies from 18,000 feet in North America to 13,000 feet in New Zealand to 3,000 feet in Europe. Some rationalisation can be expected in the future. Anyway, above the transition altitude, the aircraft is bound to assume a conventional flight level or climb between flight levels. Over a long-range cruise there can be two or more such flight levels, depending on local regulations and computer flight procedures. Typical separations are 1,000 feet, 2,000 feet, and 4,000 feet.

## 12.3 Numerical Solution of the Specific Air Range

The solutions shown so far are simplified analyses derived with a closed-form expression of the aerodynamic drag and without proper consideration of the engine performance in terms of air speed, flight altitude and throttle setting. This section deals with a numerical solution of the SAR problem and it proposes a detailed and validated analysis method. We start with the case of a turbofan-powered aircraft.

* Additional information on UK practice is available in the CAA document CAP 410: "Manual of Flight Information Services", Part A. CAA, Gatwick, UK (2002). ISBN 0-86039 851 X.

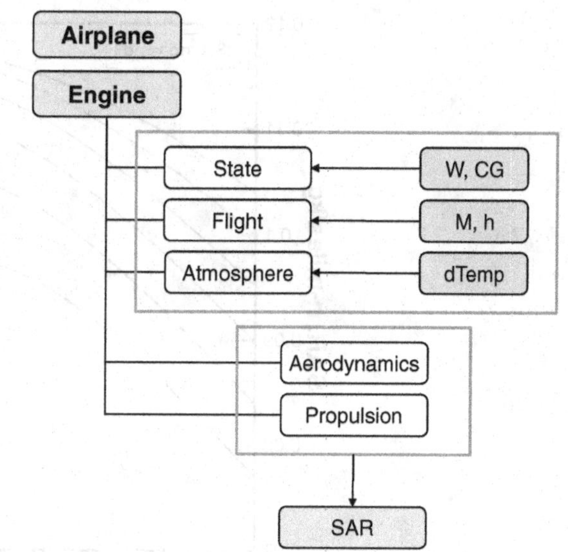


Figure 12.2. Flowchart for the numerical calculation of the SAR.

Our analysis will always refer to a unit of mass (kg) of fuel burned, so that the SAR is expressed in m/kg, km/kg or n-miles/kg. Starting from Equation 12.2, we have:

$$\mathrm{SAR} = \frac{U}{\dot{m}_f} = \frac{aM}{f_j T} = \frac{aM}{f_j D} = f(h, M, C_D, f_j, W). \tag{12.12}$$

In the presence of a wind $U_w$, we also define a specific range (SR):

$$\mathrm{SR} = \frac{U}{\dot{m}_f} = \frac{U \pm U_w}{\dot{m}_f} = \mathrm{SAR} \pm \frac{U_w}{\dot{m}_f}. \tag{12.13}$$

Note that in any case we want to maximise the ground distance flown by burning a unit of fuel. With the SAR, it will always be possible to calculate an optimal flight condition; this is defined as the operation point in the flight envelope ($M - h$) that yields the maximum SAR. The flight trajectory can be established from the relationship between SAR and all-up weight. From the relationship established by Equation 12.12, a full analysis of the SAR is multi-dimensional because the parameter depends on a fair number of factors, as illustrated in the flowchart in Figure 12.2.

Once the airplane-engine model is chosen, we need to establish the airplane state (weight and CG position), flight conditions (altitude and Mach) and atmospheric conditions (limited to a change in temperature with respect to the standard day). This is the block of input parameters which is required by the aerodynamics model and the propulsion model. The combination of these models yields the SAR. There are some more practical ways of performing an SAR analysis; these are based on:

- SAR-versus-Mach at fixed AUW and fixed altitude. This analysis leads to the calculation of the optimal Mach numbers for a given $h$ – AUW.
- SAR-versus-altitude at fixed Mach number and AUW. This analysis leads to the calculation of the optimal cruise altitude at the given $M$ – AUW.

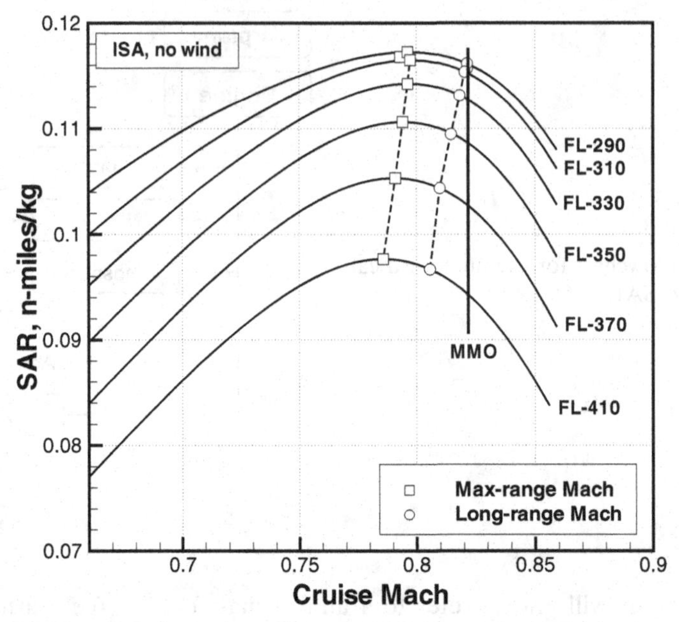


Figure 12.3. Calculated Mach number effects on SAR for a generic airplane, $m = 150{,}000$ kg; standard day.

Figure 12.3 shows the SAR versus cruise Mach number at selected flight levels. The trend shown in this figure is typical of other aircraft in the same class. In particular, at Mach numbers above the point of maximum SAR, the SAR itself decreases rapidly, mostly as a result of the transonic effects. To the left of the point of maximum SAR, the deterioration of performance is more moderate. This result has important implications in the cruise analysis. The best choice of cruise Mach number is obviously the Mach corresponding to the maximum SAR. For a given all-up weight and flight altitude, the Mach number corresponding to maximum SAR is called the *maximum-range Mach number* (MRM). Flying at this Mach number guarantees the maximum fuel efficiency. A few comments are in place here:

- For a given altitude and Mach number, the SAR decreases as the weight increases.
- The maximum SAR is virtually independent of cruise altitude – which is a considerable result. In fact, the aircraft can maintain a constant Mach number throughout the cruise.
- At altitudes below 33,000 feet (FL-330; ~10,000 m), the same Mach number leads to a higher ground speed; this would result in a shorter cruise time.
- Slowing down the airplane is not viable because the fuel consumption would increase.
- Having to choose between flying slower or faster for a given loss in SAR, it is best to fly faster to gain time.

The latter point leads to the definition of the *long-range cruise Mach number* (LRM). The long-range cruise Mach number is defined as the Mach number *higher than MRM* that corresponds to 99% of maximum SAR. In other words, flying at a Mach number equal to the LRM causes a 1% loss in specific range whilst flying reasonably faster than the MRM. This conclusion is of considerable benefit in the operation of commercial aircraft; in fact there is a widespread practice in operating the aircraft at the LRM instead of the MRM, unless external factors intervene. As in the case of the MRM, the LRM is only weakly dependent on the cruise altitude and sometimes on the all-up weight. This is another important advantage because it allows the aircraft to fly at a constant Mach over a wide range of altitudes and weights, with only a minimal loss in efficiency.

**TURBOPROP AIRCRAFT.** In this case it is necessary to trim the propeller at the prevailing flight conditions at the required thrust. The trim procedure is explained in § 6.3. This procedure provides the correct propeller settings and the shaft power, which is then used to solve the engine problem in the inverse mode (see also Chapter 5). Numerically the problem is far more involved because it requires a propeller model, an engine model, an aerodynamic model and their full integration.

### 12.3.1 Case Study: Gulfstream G550

This airplane is described in Appendix A. The verification of the computational method with the `FLIGHT` code is shown in Figure 12.4. A comparison is shown with data from the FCOM over a range of altitudes and gross weights. A number of points are made to clarify these charts:

- All weights are in thousands of pounds (for example, 50 = 50,000 lb = 22,680 kg) to comply with the way performance is shown in the FCOM.
- The SAR is given in both n-miles/lb (left scale) and n-miles/kg (right scale).
- The low end of each curve refers to the low-speed buffet Mach number (estimated); the high end of the curves denotes the high-speed buffet Mach number (also estimated).

The curves from the FCOM have been digitally interpolated up to the point where they are clearly defined. The comparison is satisfactory, particularly in consideration of the fact that the curves from the FCOM do not specify the position of the CG (in all of the calculations it was assumed a CG at 25% MAC). The sharp decrease in SAR after a relatively shallow and flat curve is due to the transonic effects arising from the fuselage fore-body at incidence and the transonic effects on the wing–body junction; neither case is modelled in the present case. Note, however, that lack of accuracy in this area is of small consequence. More important is the prediction of maximum SAR and the corresponding flight Mach number. Both quantities are overall well predicted, except in some cases when the aircraft is relatively light (50,000 lb). The Mach number corresponding to maximum SAR (MMR) is sometimes well predicted; at other times it can be as much as $M = 0.1$ off the reference value.

This analysis can be repeated for the OEI case. It is possible to show how with only one engine operating, the aircraft would have to fly at a much slower air speed ($M = 0.5$ to $0.6$) in order to maximise the SAR.

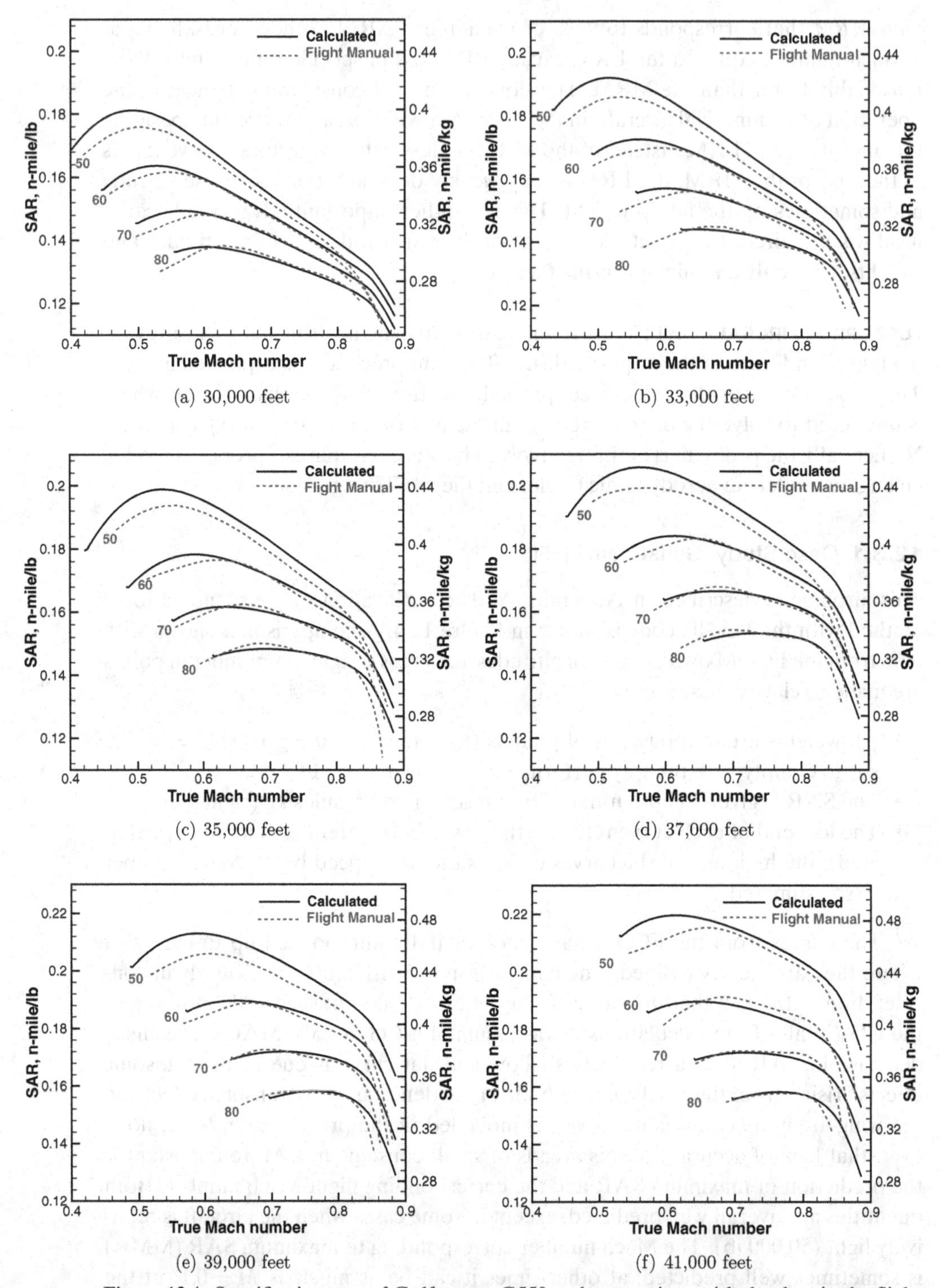


Figure 12.4. Specific air range of Gulfstream G550 and comparison with the data in the flight manual. Standard day; no wind.

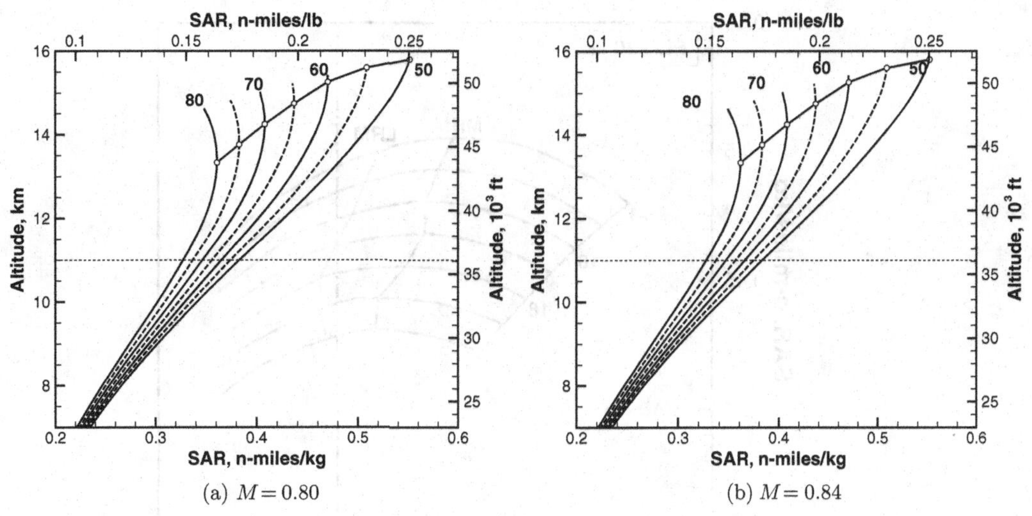


(a) $M = 0.80$ (b) $M = 0.84$

Figure 12.5. Calculated altitude performance of Gulfstream G550, standard day, no wind; weights in thousand of pounds (for example, 50 = 50,000 lb = 22,680 kg).

Figure 12.5 shows the altitude performance of the G550 at constant Mach numbers. The charts show the SAR versus the flight altitude at constant weight and constant Mach number. The results indicate that as the aircraft becomes lighter, the optimum altitude increases. In this particular instance, for a given all-up weight, the optimum SAR decreases as the Mach number is increased. The SAR growth up to the tropopause is essentially linear.

Finally, we show the figure of merit as a function of Mach number at two selected flight altitudes, Figure 12.6. The weight is a parameter of the analysis. Specifically, consider the case of the FL-350. The $FM$ increases rapidly up to $M \sim 0.82$, and then it falls off. The Mach number corresponding to maximum $FM$ varies greatly, from about 0.65 to 0.82.

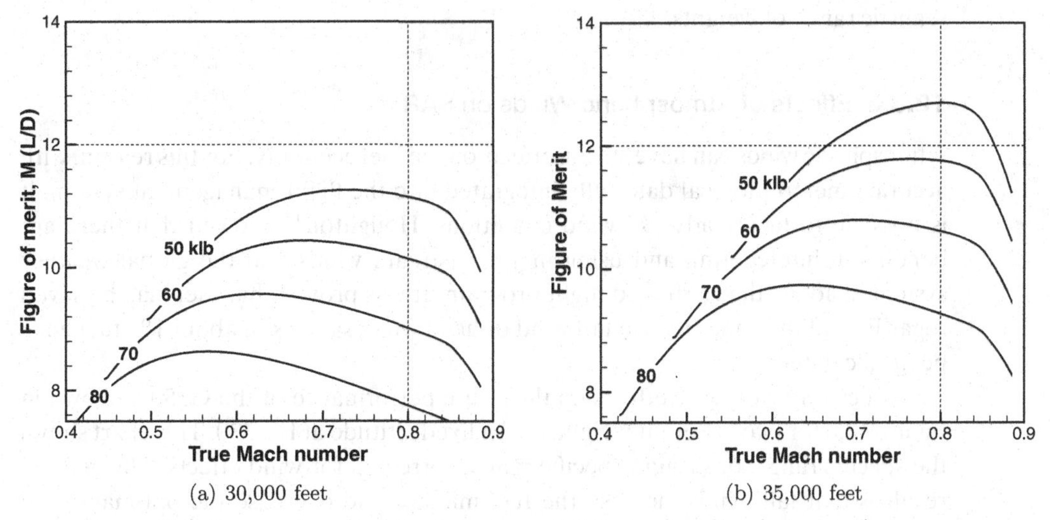


(a) 30,000 feet (b) 35,000 feet

Figure 12.6. Figure of merit of the Gulfstream G550; standard day (calculated).

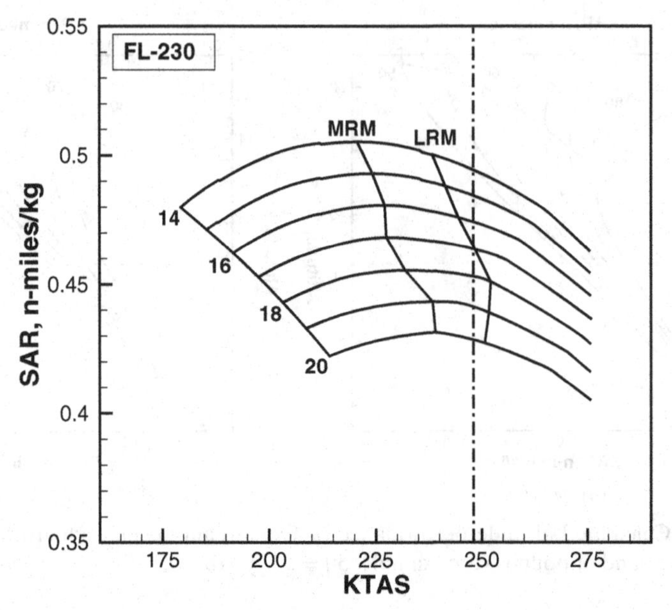


Figure 12.7. Specific air range of the ATR72-500 with PW127M turboprop engines and F568-1 propellers. The dash-dot vertical line shows the cruise speed inferred from the FCOM; standard day, no wind.

### 12.3.2 Case Study: ATR72-500

This turboprop airplane is described throughout this book: see § 6.2.4 for the propellers, § 5.4.1 for the engine model, § 14.1 for some aerodynamics. We now discuss the calculation of the point cruise performance of the airplane. Figure 12.7 shows the calculated SAR at flight level FL-230 (23,000 feet; ~7,000 m). The graph displays the estimated MRM and LRM with an AUW variable from 14 to 20 tons. The vertical dash-dot line corresponds to the typical cruise air speed and has been inferred from the FCOM. The simulation of the LRM is fairly close to the reference data for a realistic range of weights.

### 12.3.3 Effects of Atmospheric Winds on SAR

Atmospheric winds can have strong effects on the fuel economy. For this reason, with accurate meteorological data fully integrated into the flight management system, it is possible to tackle adverse wind conditions. Houghton[13] showed that there are benefits in intercepting and using major jetstream winds and the global weather system to adjust the route and flight program. It was proved that fuel may be saved regardless of the wind being a tail wind or head wind; savings of about 1% fuel have been calculated.

An example of wind effects on the cruise performance of the G550 is shown in Figure 12.8 for two different weights and a fixed altitude at FL-330. This chart shows the specific range, or ground specific range, corrected for wind effects. The general result is that tail winds increase the fuel mileage and decrease the optimal cruise Mach number. The effect of a head wind is the opposite.

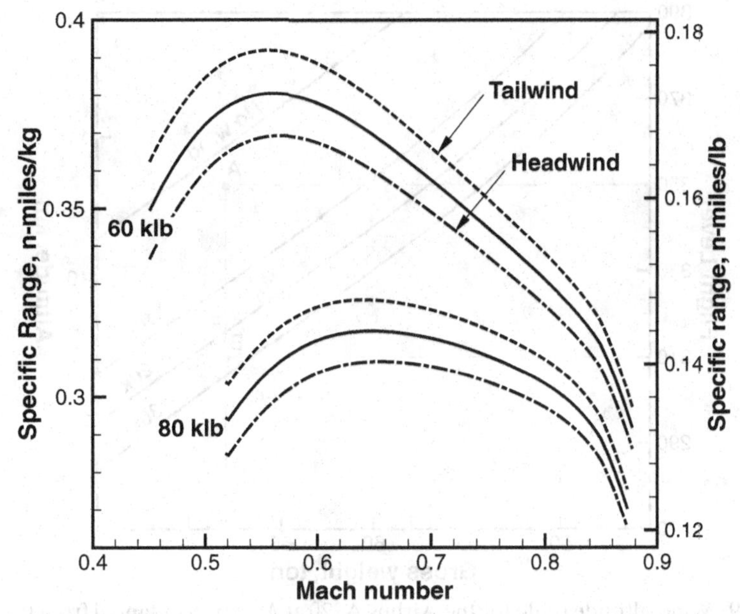


Figure 12.8. Wind effects on the specific range of Gulfstream G550, standard day, altitude = 33,000 feet; head wind = −10 kt; tail wind = +10 kt; weights as indicated.

The wind speed changes with altitude; therefore it is possible that switching altitude is more economical. For example, it is possible that at a lower altitude there is a favourable wind, which would increase the specific range. Such a decision cannot be left to the pilot's judgement; the FMS intervenes to optimise the flight conditions.

Modern airplanes have wind-altitude trade charts that allow a determination of the optimum cruise altitude. One such example is shown in Figure 12.9 for the Airbus A320. For example, assume an AUW = 68,000 kg is flying at FL-350 with a 10 kt head wind. For this aircraft, a typical altitude switch is 4,000 feet, or 40 flight levels. Thus, the airplane can switch from FL-350 down to FL-310 if the wind speed difference is about 23 kt. This is equivalent to a tail wind of at least 13 kt (13 + 10 = 23 kt).

## 12.4 The Range Equation

The previous analysis has dealt with point performance, that is, main flight parameters that depend locally on the cruise vector state $\mathcal{S} = \{h, M, W, U_w, dT\}$. The next problem is to *integrate* some of these local parameters (specifically, the SAR) in order to determine the cruise performance of the aircraft. A number of options are possible, including the following:

- aircraft range for a given fuel mass
- mass required to fly a specified range

The range equation (also known as the *Breguet* equation) is the main tool for the calculation of the cruise range of a generic aircraft, although it does not provide directly any optimal cruise conditions and does not discriminate between

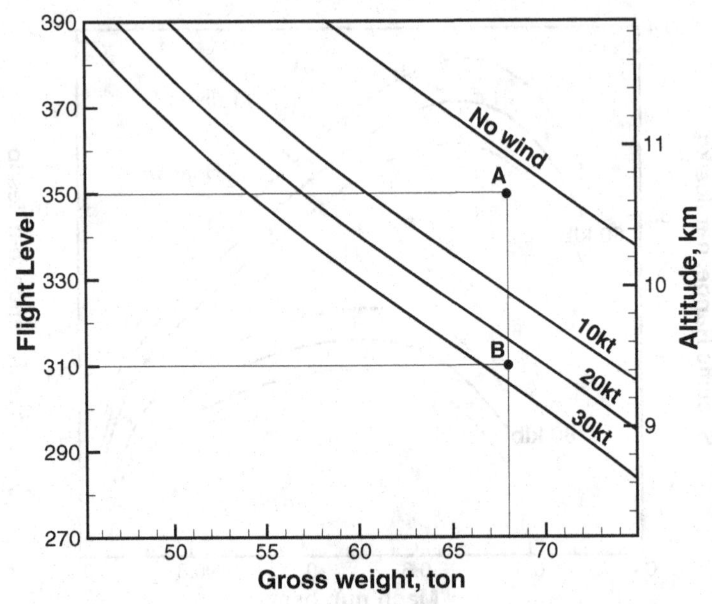


Figure 12.9. Wind-altitude trade for the Airbus A320 at $M = 0.78$ (adapted from the FCOM).

flight programs. A number of approximations are used in practice to get first-order estimates. The aircraft range is defined by the integral

$$X = \int_o^t U\,dt = \int_i^e U\frac{dm_f}{\dot{m}_f} = -\int_i^e U\frac{dm}{\dot{m}} \tag{12.14}$$

because the loss of weight is only due to fuel burn. The indexes "i" and "e" denote "initial" and "end" conditions, respectively. For a jet aircraft the fuel flow is replaced by the equation $\dot{m}_f = f_j T$. Now multiply and divide Equation 12.14 by the aircraft weight. At cruise conditions the vertical forces are in equilibrium, $L = W$. Thus, we have

$$X = \int_e^i \frac{U}{f_j T}\,dm = \int_e^i \frac{U}{f_j D}\,dm = \frac{1}{g}\int_e^i \frac{U}{f_j}\left(\frac{L}{D}\right)\frac{dm}{m}. \tag{12.15}$$

In Equation 12.15 the fuel flow $\dot{m}_f$ must be expressed in units of mass per unit of time. If other units are on input, the result would be wrong by one order of magnitude or more. The product $g f_j$ is a fuel flow expressed in units of force per unit of time.

If a first-order approximation is sufficient, all the factors within the integral can be considered constant, and the range equation is reduced to

$$X \simeq \frac{U}{g f_j}\left(\frac{L}{D}\right)\ln\left(\frac{m_i}{m_e}\right) = \frac{U}{g f_j}\left(\frac{L}{D}\right)\ln\left(\frac{1}{1-\zeta}\right), \tag{12.16}$$

where $m_i$ is the mass at the start of the cruise, $m_e$ is the mass at the end, $\zeta$ is the block fuel ratio, and $\zeta = m_f/m_i$

$$\frac{m_i}{m_e} = \frac{m_i}{m_i - m_f} = \frac{1}{1-\zeta}. \tag{12.17}$$

We conclude that at a fixed altitude the range increases with the flight speed, with the glide ratio and with the block fuel ratio. More rigorously, the calculation of the cruise range requires the solution of an integral that contains several parameters of the aircraft: the initial weight, the block fuel, the flight altitude, the air speed, the TSFC and the drag, as formulated by

$$X = f\left(h, U, f_j, \frac{L}{D}, m_i, m_f\right). \tag{12.18}$$

The speed can be replaced by the Mach number. The glide ratio is a function of the aircraft's angle of attack and the Mach number, $L/D = f(\alpha, M)$. Finally, the specific fuel consumption depends on the flight altitude and the Mach number, $f_j = f(h, M)$. The initial weight and the fuel mass can be combined into the non-dimensional block fuel ratio, Equation 12.17. Therefore, Equation 12.18 becomes

$$X = f(h, M, \alpha, \zeta). \tag{12.19}$$

The angle of attack is not a useful parameter and it is replaced with the $C_L$ because $C_L = C_{L_\alpha}\alpha$. Also, instead of the flight altitude, it is useful to have the relative air density (density altitude), that appears in the aerodynamic terms. Thus, we arrive at the expression

$$X = f(\sigma, M, C_L, \zeta), \tag{12.20}$$

that is a function of non-dimensional parameters only. Each solution of Equation 12.20 is called a *flight program*. Global optima lie in a four-dimensional space. Flight programs of interest are found with one or two parameters being constant.

### 12.4.1 Endurance

Endurance is the flight time for a unit of fuel (expressed in volume, mass or weight). If the speed and flight altitude are constant, then the endurance is the ratio between the speed and the range

$$E = \int_e^i \frac{1}{f_j}\frac{1}{D}\,dm = \frac{1}{g}\int_{m_1}^{m_2} \frac{1}{f_j}\left(\frac{L}{D}\right)\frac{dm}{m}. \tag{12.21}$$

For a propeller-driven aircraft instead, the corresponding equation is

$$E = \int_e^i \frac{1}{f_c}\frac{\eta}{DU}\,dm = \frac{1}{g}\int_e^i \frac{1}{f_c}\frac{\eta}{U}\left(\frac{L}{D}\right)\frac{dm}{m}. \tag{12.22}$$

It is useful to find a relationship between the range and the corresponding endurance because the calculation of the former performance will lead to the latter. Nevertheless, endurance problems are less interesting than range problems, with a few exceptions, such as loiter, surveillance operations, flight within prescribed area and so on. Some of these problems are analysed by Sachs[14].

## 12.5 Subsonic Cruise of Jet Aircraft

We consider three flight programs: 1) cruise at constant altitude and constant Mach number; 2) cruise at constant Mach number and constant lift coefficient; and 3) cruise at constant altitude and constant lift coefficient. All of these cases require a starting value of altitude, mass and speed $(h, m, U)$, for which an optimal solution can be found according to the methods presented in § 12.3. Another case is constant altitude and constant throttle setting. The decreasing aircraft's weight will result in an increasing Mach number. The ATC rules do not favour variable flight speed; therefore this flight condition is less interesting than the previous ones. The mass of the aircraft is variable from an initial value $m_i$ to a final value $m_e$.

### 12.5.1 Cruise at Constant Altitude and Mach Number

During the cruise, the aerodynamic drag decreases because of the decreasing weight. As the drag decreases, also the engine thrust must decrease and the throttle has to be stepped down. This is done automatically with the modern FMS. From Equation 12.15 the range becomes

$$X = \frac{aM}{gf_j}\int_e^i \left(\frac{L}{D}\right)\frac{dm}{m} \tag{12.23}$$

because the speed of sound depends only on the altitude, and the thrust-specific fuel consumption depends on both altitude and Mach number. The maximum cruise range is found from operating the aircraft at maximum glide ratio. If we use the parabolic drag equation, the glide ratio becomes a function of the aircraft mass:

$$\frac{L}{D} = \frac{c_1 m}{C_{D_o} + kc_1^2 m^2}, \tag{12.24}$$

where

$$c_1 = \frac{2g}{\rho A a^2 M^2} \tag{12.25}$$

is a coefficient depending on altitude and Mach number. The angle of attack of the aircraft does not appear explicitly in Equation 12.24. From Equation 12.23 we find

$$X = \frac{aM}{gf_j}\int_e^i \frac{c_1}{C_{D_o} + kc_1^2 m^2}\, dm. \tag{12.26}$$

The solution of the integral is from the tables of indefinite integrals[15]:

$$\int \frac{c_1}{C_{D_o} + kc_1^2 m^2}\, dm = \frac{1}{\sqrt{kC_{D_o}}}\tan^{-1}\left(c_1 m\sqrt{\frac{k}{C_{D_o}}}\right). \tag{12.27}$$

The final expression for the cruise range is

$$X = \frac{aM}{gf_j}\frac{1}{\sqrt{kC_{D_o}}}\left[\tan^{-1}\left(c_1 m_i\sqrt{\frac{k}{C_{D_o}}}\right) - \tan^{-1}\left(c_1 m_e\sqrt{\frac{k}{C_{D_o}}}\right)\right]. \tag{12.28}$$

This equation can be reduced further by using some trigonometric equivalences (see § 12.5.5). From the known values of the mass at the start and end of the cruise, it is possible to find the range from Equation 12.28. In particular, we find that at a given altitude the range increases with the flight Mach number. Finally, note that Equation 12.28 is a *direct* solution of the cruise problem, one that provides the cruise range for a given fuel load. The *inverse* problem consists in the determination of the cruise fuel corresponding to a fixed range $X$; this is a more involved problem from the point of view of the mathematical manipulation.

### 12.5.2 Cruise at Constant Altitude and Lift Coefficient

An approach to the study of this flight program is to consider the balance of forces in the vertical direction, $L = W$. With the expression of the $C_L$ we find that the air speed and the Mach number are proportional to $W^{1/2}$:

$$M = \left(\frac{2W}{a\rho A C_L}\right)^{1/2}. \tag{12.29}$$

Therefore, the Mach number must be constantly reduced as the engines burn fuel. With the substitution of Equation 12.29 the range equation is written as

$$X = \frac{a}{g f_j}\left(\frac{C_L}{C_D}\right)\left(\frac{2g}{a\rho A C_L}\right)^{1/2}\int_e^i m^{-1/2}\,dm, \tag{12.30}$$

where we have assumed that the changes in fuel consumption due to changes in Mach number can be neglected. Constant $C_L$ implies constant $C_D$ and constant $C_L/C_D$. By further manipulation, the cruise range becomes

$$X = \frac{2}{f_j}\left(\frac{C_L^{1/2}}{C_D}\right)\left(\frac{2a}{g\rho A}\right)^{1/2}\left[\sqrt{m_i} - \sqrt{m_e}\right]. \tag{12.31}$$

This flight program has a number of drawbacks, namely: 1) there is a loss in jet engine efficiency with a reduced Mach number; 2) the cruise time is increased; 3) a continuous variation of throttle setting is required; and 4) the cruise range is shorter than other flight programs.

### 12.5.3 Cruise at Constant Mach and Lift Coefficient

The cruise of a jet aircraft with a constraint on Mach number and lift coefficient is a *cruise-climb flight*. With these constraints the $C_D$, the glide ratio $C_L/C_D$ and the $FM$ are constant. The range equation becomes

$$X = \frac{1}{g}M\left(\frac{L}{D}\right)\int_e^i \frac{a}{f_j}\frac{dm}{m}. \tag{12.32}$$

The conditions for a constant $FM$ were found in § 12.2.2. We proved that a climb is necessary to compensate for the changes in mass. For flight in the stratosphere, also the speed of sound is a constant. Then the cruise range can be written as

$$X \simeq \frac{a}{g f_j}M\left(\frac{L}{D}\right)\ln\left(\frac{1}{1-\zeta}\right). \tag{12.33}$$

The TSFC is not a constant but depends on both temperature and Mach number, as explained in Chapter 5. Above 11,000 m the temperature is constant; therefore cruise at a constant Mach number in the lower stratosphere is an optimal condition.

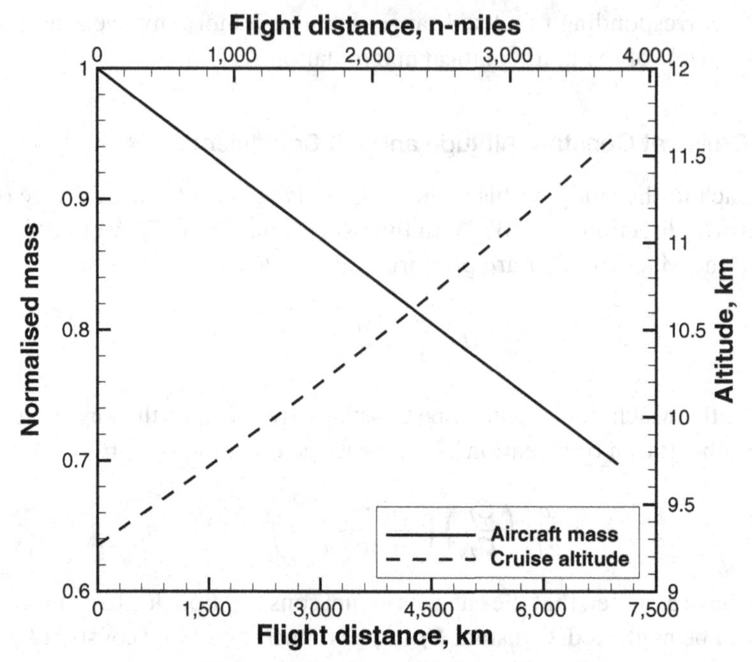

Figure 12.10. Cruise-climb profile.

The global optimum range is found from the maximum of

$$f(h, M, C_L) = \frac{1}{f_j} M \left(\frac{L}{D}\right). \tag{12.34}$$

Figure 12.10 shows how the flight altitude has to be increased as the aircraft flies. When a climb is required, the aircraft moves from one flight level to another in relatively short time. The increase in altitude shown in Figure 12.10 corresponds to several *step-ups*. The fuel required for these short segments can be calculated from energy considerations.

A further consideration is that for maximum range, $L/D$ has to be optimal at the start of the cruise. This maximum is a function of the aircraft mass; for a fixed mass (as calculated from the mission analysis), it depends on the speed and altitude.

### 12.5.4 Comparison among Cruise Programs

We are now in a position to compare the various flight programs, assuming that the aircraft has a fixed AUW, a fixed fuel mass and the same initial conditions – namely the flight altitude, the Mach number and the lift coefficient. We study the best range at the conditions summarised in Table 12.1.

Consider one aircraft model with an $m = 145{,}000$ kg and a fuel fraction $\zeta = 0.138$. The starting point is $h = 11{,}000$ m (~36,000 feet), $M = 0.80$ for all of the cruise profiles. The initial conditions require that the initial cruise lift by $C_L = 0.539$ and the glide ratio $L/D \simeq 17.98$, and $FM \simeq 14.38$. Assume an average TSFC $\simeq 1.162 \cdot 10^{-5}$ kg/Ns. The results are summarised in the right-hand column of Table 12.1. The

Table 12.1. *Summary of subsonic cruise conditions, jet aircraft; starting conditions: $h = 11{,}000$ m (~36,000 feet), $M = 0.80$*

| Flight Program | Constraints | Range Equation | Range (km) |
|---|---|---|---|
| A | $h$, m | Equation 12.28 | 5,314 |
| B | $h$, $C_L$ | Equation 12.31 | 5,328 |
| Cruise/Climb | m, $C_L$ | Equation 12.33 | 5,528 |

cruise-climb program achieves a range about 4% higher than the cruise at constant altitude and Mach number. Therefore, the cruise-climb is the best flight program among the cases considered.

### 12.5.5 Fuel Burn for Given Range

We now calculate the fuel burn of a jet aircraft for a given range (or segment) and a constant altitude and Mach number. We start from the conclusions of § 12.5.1 and write the range equation for the generic flight segment $i$-$j$.

$$X_{ij} = \frac{aM}{gf_j}\frac{1}{\sqrt{kC_{D_o}}}\left[\tan^{-1}\left(m_j c_1\sqrt{\frac{k}{C_{D_o}}}\right) - \tan^{-1}\left(m_i c_1\sqrt{\frac{k}{C_{D_o}}}\right)\right]. \qquad (12.35)$$

If the altitude and the Mach number are specified, to simplify Equation 12.35, introduce the factors

$$c_3 = \frac{aM}{gf_j}\frac{1}{\sqrt{kC_{D_o}}}, \qquad c_4 = c_1\sqrt{\frac{k}{C_{D_o}}}, \qquad (12.36)$$

so that

$$X_{ij} = c_3\left[\tan^{-1}(c_4 m_j) - \tan^{-1}(c_4 m_i)\right]. \qquad (12.37)$$

Equation 12.37 can be further reduced to

$$X_{ij} = c_3\left[\tan^{-1}\left(\frac{c_4 m_i - c_4 m_j}{1 + c_4^2 m_i m_j}\right) + \pi\right]. \qquad (12.38)$$

The term $\pi$ only appears if $c_4^2 m_i m_j > 1$. Now solve Equation 12.38 to find the value of the fuel mass during the flight segment:

$$\tan\left(\frac{X_{ij}}{c_3} - \pi\right) = \left(\frac{c_4 m_i - c_4 m_j}{1 + c_4^2 m_i m_j}\right). \qquad (12.39)$$

The unknowns are the aircraft mass at the end of the flight segment and the fuel mass. We assume that

$$m_i = m_j - m_f \qquad (12.40)$$

and solve Equation 12.39 for the unknown fuel burn. Use the known coefficient

$$c_5 = \tan\left(\frac{X_{ij}}{c_3} - \pi\right) \tag{12.41}$$

to simplify the equation. Therefore:

$$c_5 = \frac{c_4 m_i - c_4 m_j}{1 + c_4^2 m_i m_j}, \tag{12.42}$$

$$c_4 m_i - c_4 m_j = c_5\left(1 + c_4^2 m_i m_j\right). \tag{12.43}$$

Finally, the mass of fuel burned during the flight segment from $i$ to $j$ is

$$m_f = \frac{c_5 + c_4^2 c_5 m_i^2}{c_4 + c_4^2 c_5 m_i}. \tag{12.44}$$

The problem is not closed because the initial mass $m_i$ may not be known exactly. Equation 12.44 represents the fuel burn over the segment $i$-$j$ for a given altitude, Mach number and mass at the start of the segment.

## 12.6 Cruise Range of Propeller Aircraft

We now get down to more mundane speeds and altitudes, to deal with propeller-driven aircraft. These can be powered by gas turbine or reciprocating engines. The range equation (Equation 12.14) is valid in the general case. For a propeller-driven aircraft we need to replace the fuel-flow equation: $\dot{m}_f = f_c P$. The range becomes

$$X = \int \frac{1}{f_c}\frac{U}{P}\, dm. \tag{12.45}$$

We replace the power equation, $\eta P = TU$, in Equation 12.45 to obtain

$$X = \frac{1}{g}\int \frac{\eta}{f_c}\frac{L}{D}\frac{dm}{m}. \tag{12.46}$$

The latter equation contains the equilibrium conditions $L = W$, $D = T$. There is a new variable: the propeller's efficiency; this, as we have seen, depends on the advance ratio $J$ (see Chapter 6). If the engine power depends on the weight, then the propulsive efficiency is also dependent on the weight by way of the shaft power.

The types of cruise conditions for the propeller-driven aircraft are similar to the jet-engine aircraft. In particular, we note that the range is a function of this type:

$$X = f(\sigma, \eta, U, f_c, \zeta). \tag{12.47}$$

We discuss the solution of the cruise flight at constant speed and altitude. The remaining cruise programs can be found with an analysis similar to the subsonic jet aircraft.

**CRUISE AT CONSTANT ALTITUDE AND SPEED.** The range equation for this flight condition becomes

$$X = \frac{\eta(J,\theta)}{g\, f_c}\int_e^i \frac{L}{D}\frac{dm}{m}, \tag{12.48}$$

where we have assumed that the propulsive efficiency $\eta$ is a function of the propeller's advance ratio $J$ and pitch setting $\theta$. The SFC is only a function of $h$ and $U$; therefore it is taken out of the integral. If we write the glide ratio as a function of the aircraft mass, as in § 12.5.1, then

$$X = \frac{\eta(J,\theta)}{g\, f_c} \int_e^i \frac{c_1 dm}{C_{D_o} + kc_1^2 m^2}. \tag{12.49}$$

The solution of the integral is

$$X = \frac{\eta(J,\theta)}{g\, f_c} \tan\left( c_1 m \sqrt{\frac{k}{C_{D_o}}} \right)_e^i. \tag{12.50}$$

If the block fuel and the aerodynamics are fixed, then the range is maximised by the function $\eta(J,\theta)/f_c$. This factor is a combination of engine and propeller performance. The solution of Equation 12.50 inevitably needs propeller charts.

## 12.7 Cruise Altitude Selection

From an optimisation point of view, the flight levels present a constraint. Cruise optimisation is done according to a three-stage procedure. In the first stage we select the best ICA and the best Mach number (§ 12.3); in the second stage we estimate the final cruise altitude from the range equation (§ 12.4); and in the final stage we select a sequence of constant-altitude and climb flight segments to reach the end point.

Let us assume, without loss of generality, that we want to operate the aircraft at the long-range Mach number. The range equation can be used to estimate the fuel required and the altitude at the end of the cruise. The solution is

$$\rho_e = \rho_i \left( \frac{m_e}{m_i} \right). \tag{12.51}$$

The altitude can be found directly from the final air density $\rho_e$ by using the atmosphere model. Let us assume that the total climb required is $\Delta h$. However, this altitude gain will have to be adjusted to the closest flight level. An item of concern is that occasionally the aircraft is unable to reach the optimal condition, due to a number of factors, including available thrust, increased gross weight, ATC constraints and adverse atmospheric conditions. The final stage in the cruise optimisation contains a fairly large number of parameters. If $n$ is the number of constant-altitude steps, the number of climb steps is $n - 1$:

$$n - 1 = \frac{\Delta h}{\text{FLS}}, \tag{12.52}$$

where "FLS" is the flight level separation. The final segment is always a cruise. For example, if the adjusted climb is $\Delta h = 4{,}000$ feet (~1,200 m) and FLS = 100 (or 1,000 feet), then $n = 5$ (four climb segments; five cruise segments). If FLS = 200, then $n = 3$ (two climb segments; three cruise segments).

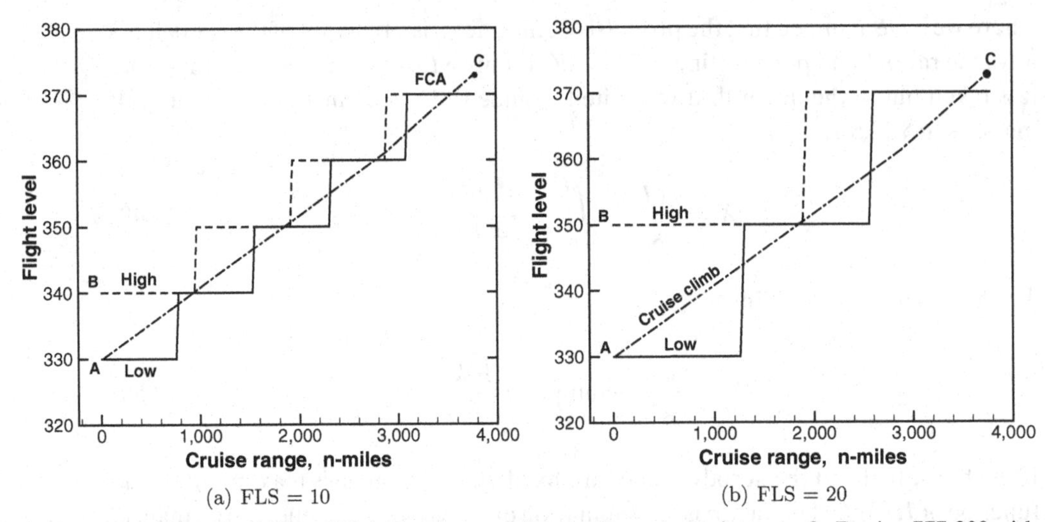


(a) FLS = 10 (b) FLS = 20

Figure 12.11. Effect of flight-level separation on the cruise trajectory of a Boeing 777-300 with GE-92 turbofan engines. Adapted from Ref.[16].

The distances flown in the cruise mode are called $X_i$ and the distances covered by the climb steps are called $X_{c_i}$. The constraints are

$$X = \sum_i^n X_i + \sum_i^{n-1} X_{c_i}, \tag{12.53}$$

$$\Delta h = (n-1)dh. \tag{12.54}$$

The total number of free parameters is $2n-1$, that is, $n$ constant-altitude stage lengths and $n-1$ climb rates between flight levels. Mathematically, such a problem is quite complex. Practical solutions deal with specified climb rates of equal magnitude and cruise stages of equal magnitude, except the last one; this is calculated from the difference between the required range and the distance covered in all of the previous stages (cruise and climb). Thus, the number of free parameters is reduced to three (two cruise stage lengths and one climb rate). An alternative cruise procedure relies on the change of flight level based on the AUW of the aircraft. There are two possible starting points of the cruise, as shown in Figure 12.11: a low profile starting at point A and a high profile starting at point B. Once the cruise has started, there are three possibilities:

1. *Low profile*: The step climb starts at the weight corresponding to the intersection between the low flight level A and the theoretical optimal altitude, Figure 12.11a. The flight profile will often be below the optimal cruise altitude; the aircraft will have better manoeuvre margin, and if the FL is below 330, the TAS will be higher. This procedure requires a limited number of intermediate calculations with the range equation, to estimate the AUW at the switch point.
2. *Middle profile*: The step climb starts when the SAR at the higher flight level is better than the SAR at the current flight level (Figure 12.11b). This procedure requires the calculation of the SAR at two flight levels at each time step; therefore, it is computationally more demanding.

Table 12.2. *SAR penalty due to non-optimal cruise altitude for some Airbus airplanes*

| Airplane | A300B4-600 | A310-324 | A320-211 | A330-203 | A340-642 |
|---|---|---|---|---|---|
| +FL-200 | 2.0% | 0.9% | – | 1.8% | 1.6% |
| −FL-200 | 1.9% | 1.4% | 1.1% | 1.3% | 0.6% |

3. *High profile*: The step climb starts at the weight (or distance) corresponding to the flight level above the theoretical optimal altitude point B. The aircraft moves to the next flight level when the flight level reaches the theoretical optimal altitude, Figure 12.11a.

Table 12.2 is a summary of cruise-performance penalties arising from cruise at the wrong flight level*. The penalty can be as much as 2%.

The minimum climb rate for transport aircraft is ~300 feet/min (~1.5 m/s). The effect of the climb rate on the cruise fuel of the reference transport airplane is shown in Figure 12.12 for a 2,000 km (1,079 n-miles) required range. The result indicates that some fuel savings can be achieved by switching the flight level as quickly as possible. The reference point for the calculation of the fuel savings is a climb with $v_c =$ 1.9 m/s (~370 feet/min).

## 12.8 Cruise Performance Deterioration

Cruise performance tends to degrade over time. This inconvenient reality can be the result of engine performance deterioration, airframe drag deterioration or a combination of both. There are various (expensive) experiments that can be carried out to verify how the deterioration affects cruise performance. For example, take as a reference a new airplane with new engines; test flights can then be carried out with old airframes and new engines, and new airframes with old engines. Four sets of SAR data ($\mathcal{S}$ = SAR) can be generated to isolate the problem:

- new airframe with new engines, $\mathcal{S}_{nn}$
- old airframe with old engines, $\mathcal{S}_{oo}$
- new airframe with old engines, $\mathcal{S}_{no}$
- old airframe with new engines, $\mathcal{S}_{on}$

The tests will have to be carried out with the same weight, with the same position of the CG, over the same route and with *sufficiently similar* atmospheric conditions. A further complication in the performance study arises from the fact that new and old airplanes are not necessarily identical. There are always minor differences in configurations, which must be carefully analysed before drawing any conclusions. With these precautions, if there is no change in SAR resulting from a change of engines, the loss in performance must be attributed to the deterioration in aerodynamic drag. The relative loss in performance can be expressed in two ways:

$$\Delta\mathcal{S}_1 = \frac{\mathcal{S}_{nn} - \mathcal{S}_{on}}{\mathcal{S}_{nn}}, \qquad \Delta\mathcal{S}_2 = \frac{\mathcal{S}_{no} - \mathcal{S}_{oo}}{\mathcal{S}_{no}} \qquad \text{[change airframe]}. \qquad (12.55)$$

* Elaborated from Airbus: *Getting to Grips with Fuel Economy*, Issue 3, July 2004, Blagnac, France.

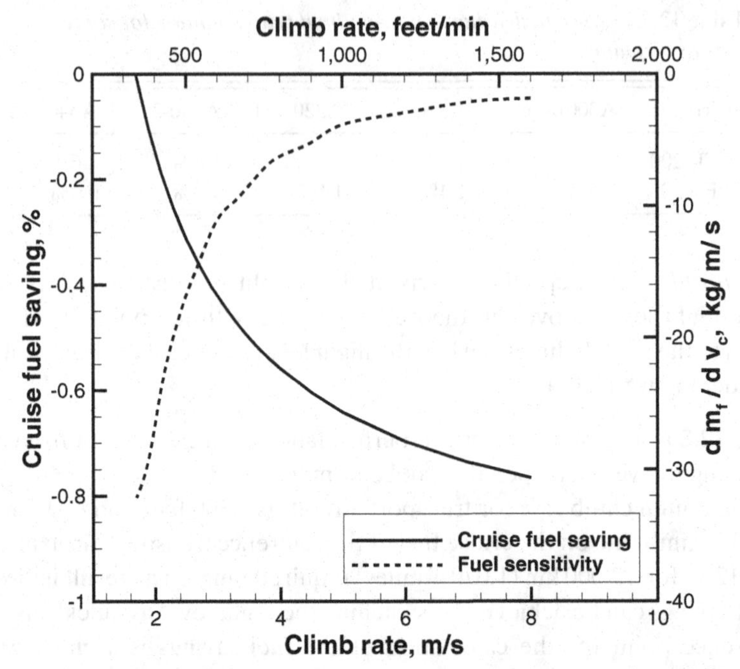


Figure 12.12. Effect of climb rate on cruise fuel for a 2,000-km (1,079 n-miles) required range; standard day; no wind.

The first and second equations denote the effects of the engine on the new and old airframes, respectively. Several causes could contribute to airframe drag deterioration, including panel mis-alignment, broken seals, leakages and surface roughness. The other extreme case is when the performance deterioration is fully attributed to the engines. This happens when the same airframe (regardless whether new or old) is equipped with new and old engines. The relative loss in performance is:

$$\Delta\mathcal{S}_3 = \frac{\mathcal{S}_{nn} - \mathcal{S}_{no}}{\mathcal{S}_{nn}}, \qquad \Delta\mathcal{S}_4 = \frac{\mathcal{S}_{on} - \mathcal{S}_{oo}}{\mathcal{S}_{on}} \qquad \text{[change engines].} \tag{12.56}$$

The deterioration in engine performance can be due to a number of causes, including mis-alignment of fan and compressor blades, blade erosion, loss in pressure ratio, over-heating, increased air bleed, loss of combustion efficiency, leakages and so on.

The loss in performance can be a fraction of a percent or as high as a few percent. Even in the most optimistic scenario, the costs arising from the operation of a commercial airplane that is not kept as new can be phenomenally high.

## 12.9 Cost Index and Economic Mach Number

The economic Mach number, $M_{econ}$, is a performance parameter that takes into account the cost of fuel and the cost of time. In fact, a higher Mach number reduces the cruise time at the expense of increased fuel consumption; vice versa for a lower Mach number. The minimum direct operating costs (DOC) are somewhere in between (see § 15.8). In quantitative terms, we have

$$\text{DOC} = c_o + c_1 m_f + c_2 t, \tag{12.57}$$

where $c_o$ is a fixed-cost term, $c_1$ is a factor depending on the cost of fuel, and $c_2$ is a factor depending on the cruise time $t$. The DOC can be plotted against the cruise Mach number.

Minimum fuel costs are obtained at the Mach number corresponding to maximum SAR. The time-related costs decrease with the increasing Mach number. These considerations lead us to conclude that the $M_{econ}$ is slightly higher than the MMR. To characterise this result, a cost index ($C_I$) is used:

$$C_I = \frac{\text{Cost of time}}{\text{Cost of fuel}} = \frac{c_2}{c_1}\frac{t}{m_f} = \frac{c_2}{c_1}\frac{1}{\dot{m}_f}. \tag{12.58}$$

The calculation of the cost index is dependent on the aircraft operator. Modern flight-management systems scale the $C_I$ between 0 and 99 (or between 0 and 999). Parametric studies of the economic Mach number can be done in terms of normalised values of $C_I$.

The physical dimensions of the $C_I$ are units of fuel per unit of time. $C_I = 0$ means that $c_1$ is large and $c_2$ is small. If $C_I$ is maximum, $c_1$ is small and $c_2$ is large; this occurs when the aircraft operates in minimum-time mode and maximum Mach number (MMO). If the $C_I$ is high, the economic Mach number is insensitive to weight. Conversely, if the $C_I$ is low, the economic Mach number is sensitive to both weight and flight levels. Tax on aviation fuel, carbon trading schemes and other environment-related costs will inevitably increase the cost of fuel compared to the cost of time; therefore the value of this index is bound to decrease in the future. To minimise the total trip cost we need to minimise the sum

$$c_1 m_f + c_2 t.$$

Now consider the cost function $\mathcal{F}_c$ per unit distance:

$$\mathcal{F}_c = \frac{c_1 m_f + c_2 t}{x} = \frac{c_1}{\text{SAR}} + \frac{c_2}{U}, \tag{12.59}$$

where $U = aM + U_w$ is the true air speed and $U_w$ is the wind speed. When the cost associated to the fuel is equal to the cost associated to the flight time, then

$$\frac{c_1}{\text{SAR}} = \frac{c_2}{U}, \tag{12.60}$$

or

$$\frac{1}{\text{SAR}} = \frac{c_2}{c_1}\frac{1}{U} \sim \frac{C_I}{U}. \tag{12.61}$$

It is not essential to know exactly the value of the factors $c_1$ and $c_2$. It will be sufficient to know that $C_I \propto c_2/c_1$. From this equivalence, we find the cost index

$$C_{I_{eq}} \sim \frac{U}{SAR}. \tag{12.62}$$

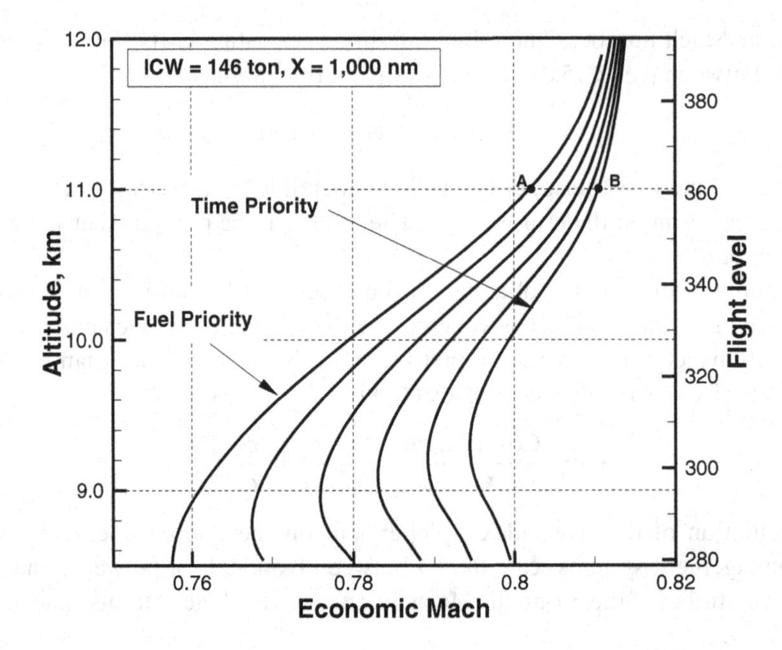


Figure 12.13. Analysis of economic Mach number.

A value twice this $C_{I_{eq}}$ corresponds to a time priority. A further scaling can be done so that $C_{I_{max}} = 99$, as in a modern flight-management system. The minimum trip cost is achieved by properly adjusting the Mach number, to take into account the atmospheric conditions, aircraft performance, the cost of fuel and the marginal costs because they are incurred by an additional flight time. An example of calculation of the economic Mach number for a transport airplane is shown in Figure 12.13 against the cruise altitude. This graph shows the difference in cruise Mach at fuel priority and time priority. This difference decreases as the aircraft flies at higher altitudes.

The application of the cost index can be expanded to other flight conditions, such as climb to initial cruise altitude (see Chapter 10) and en-route descent from cruise altitude (see Chapter 11).

## 12.10 Centre of Gravity Position

In this section we study the effects on the cruise performance due to position of the centre of gravity. We must always ensure that the loading of the aircraft gives a CG position within the certified range (i.e., within the forward and aft limits). A forward CG requires a nose-up pitching moment to establish a longitudinal trim (§ 7.1). This trim can be achieved through a lower loading of the tail plane and/or a higher loading of the wing. The result of the trim is to increase the induced drag (§ 7.1.1).

Most modern aircraft have an automatic flight-management system for the CG that is based on pumping fuel in or out of a *trim tank*. The flight system optimises the position of the CG, depending on the actual weight of the aircraft, so as to minimise the total drag and hence the cruise fuel consumption. The position of the CG can be calculated from the zero-fuel weight, the GTOW and the status of each fuel tank. The software can be quite sophisticated, but it provides the position of the CG with accuracy and celerity. For example, the Airbus A310 has CG limits between 20% and 45%, with a mid-position at 27%. Flight data for this aircraft show that flight with a CG in forward position can add up to 1.8% to the cruise fuel, compared with the mid-position; conversely, a flight with the CG in aft position can save about 1.8% of the cruise fuel, compared with the mid-position, irrespective of the flight level. However, this performance is not general and must be studied for each aircraft. Both Airbus and Boeing provide CG data and their effects on cruise on all of their aircraft. For example, for the Airbus A318/A319/A320, the manufacturer requires the following sequence of fuel use:

- additional centre tanks (ACT2, ACT1)
- centre tank, until empty
- inner wing tanks, until 750 kg are left in tanks
- outer wing tanks: fuel transferred into inner tanks
- inner wing tanks

In this sequence, only one fuel transfer is required: from the outer wing tanks to the inner wing tanks, when the inner tanks' level is below the threshold. The fuel transfer is done by gravity.

**CASE STUDY: CENTRE OF GRAVITY EFFECTS.** The effect of the CG position on the SAR has been calculated for a model Boeing B777-300[16]; this is shown in Figure 12.14. This graph displays two sets of curves, corresponding to a CG shift equal to $\pm$ 6% MAC, respectively. For each position of the CG, the change in SAR depends on both altitude and gross weight. For a fixed weight, moving the CG aft causes a loss in SAR that increases with the flight altitude; moving the CG forward causes an increase in SAR that improves with the flight altitude. Consider now the solid lines (largest weight). The airplane would operate at FL-350 (point A or B); as it burns fuel, it must climb to maintain the same SAR (or to avoid a larger SAR penalty). After burning about 20 tons of fuel, the altitude must shift to FL-370 (point C or D).

## 12.11 Supersonic Cruise

We now deal with a more exotic flight performance. We study the optimal conditions for cruise range at supersonic speeds for a given AUW and a given initial block fuel. A number of cases are possible because the cruise conditions depend on the altitude, Mach number and angle of attack. The range equation is given by Equation 12.14.

The general problem of supersonic cruise can be quite complicated. Windhorst *et al.*[17] developed guidance techniques for minimum flight time, minimum fuel consumption and minimum DOC for a fixed-range cruise. For example, a minimum fuel trajectory consists of an initial minimum-fuel climb, a cruise-climb and a maximum $L/D$ descent. The minimum DOC trajectory is nearly the same as a minimum fuel trajectory.

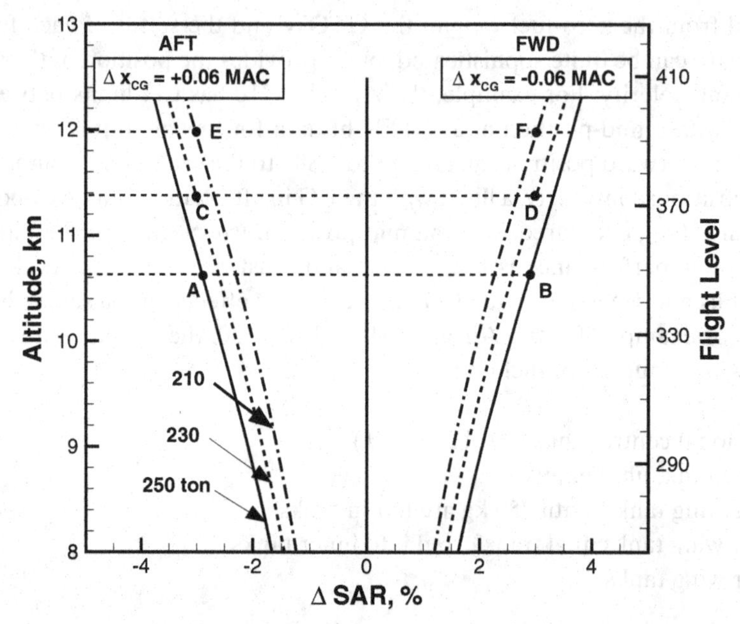


Figure 12.14. Effect of gross weight and the CG position on the optimal cruise altitude.

### 12.11.1 Cruise at Constant Altitude and Mach Number

For cruise in the stratosphere, the speed of sound is constant. In addition, if the Mach number is also constant, the range equation becomes

$$X = \frac{aM}{gf_j}\int_e^i \frac{C_L}{C_D}\frac{dm}{m} = \frac{aM}{gf_j}\int_e^i \frac{C_{L_\alpha}(\alpha-\alpha_o)}{C_{D_o}+\eta C_{L_\alpha}(\alpha-\alpha_o)^2}\frac{dm}{m}. \tag{12.63}$$

The angle of attack at cruise condition is specified by $L = W$:

$$\alpha = \alpha_o + \frac{2W}{\rho A a^2}\frac{1}{C_{L_\alpha}M^2}. \tag{12.64}$$

When we replace this condition in Equation 12.63, we find

$$\frac{L}{D} = \frac{c_o/M^2}{C_{D_o}+\eta c_o^2/M^4} = \frac{c_1(\sigma, M)m}{C_{D_o}+c_2(\sigma, M)m^2}, \tag{12.65}$$

with

$$c_o(\sigma, m) = \frac{2mg}{\rho A a^2}, \qquad c_1(\sigma, M) = \frac{2g}{\rho A a^2 M^2}, \qquad c_2(\sigma, M) = \frac{\eta}{C_{L_\alpha}}\left(\frac{2g}{\rho A a^2 M^2}\right)^2. \tag{12.66}$$

The glide ratio can be optimised with respect to the Mach number or the mass. Equation 12.65 shows the dependence of the glide ratio from the Mach number and the aircraft's mass. When we use the coefficients defined by Equation 12.66, the range becomes

$$X = \frac{aM}{gf_j}\int_e^i \frac{c_1(\sigma, M)}{C_{D_o}+c_2(\sigma, M)m^2}\,dm. \tag{12.67}$$

The integral is rather involved, but it is similar to an expression seen before (Equation 12.28):

$$\int_e^i \frac{dm}{C_{D_o} + c_2 m^2} = \frac{1}{\sqrt{C_{D_o} c_2}} \left[ \tan^{-1}\left( m \sqrt{\frac{c_2}{C_{D_o}}} \right) \right]_e^i, \tag{12.68}$$

Therefore, the solution is

$$X = \frac{aM}{gf_j} \sqrt{\frac{C_{L_\alpha}}{C_{D_o}\eta}} \left[ \tan^{-1}\left( m_i \sqrt{\frac{c_2}{C_{D_o}}} \right) - \tan^{-1}\left( m_e \sqrt{\frac{c_2}{C_{D_o}}} \right) \right]. \tag{12.69}$$

Equation 12.69 is similar to Equation 12.28, and is valid for the same flight program at subsonic speeds. A specified range can be achieved by subsonic cruise at a low altitude or a supersonic cruise in the lower stratosphere.

### 12.11.2 Cruise at Constant Mach Number and Lift Coefficient

Consider a high-speed cruise at a fixed Mach number and $C_L$. The corresponding angle of attack $\alpha$ is constant and is set at the start of the cruise, according to Equation 12.7. Thus, the figure of merit $M(C_L/C_D)$ is also a constant. To maintain a constant $C_L$ and $\alpha$ with a decreasing mass, the aircraft must climb, according to the same principle illustrated in § 12.5.3. In fact, from Equation 12.7 $\alpha$ is constant if $m/\sigma = constant$. Therefore, this flight program is a *cruise-climb*. With these considerations, the range equation becomes

$$X = \frac{a}{g} M \left( \frac{C_L}{C_D} \right) \int_e^i \frac{1}{f_j} \frac{dm}{m}. \tag{12.70}$$

The changes in $f_j$ must be considered; therefore we need to refer to an accurate aero-thermodynamic model for the engine cycle, which provides the correct behaviour of the fuel flow and hence the TSFC. For a given block fuel ratio, the range given by Equation 12.70 is maximised if

$$R(\sigma, M) = \frac{M}{f_j} \frac{C_L}{C_D} = \max. \tag{12.71}$$

If the TSFC were a constant, then the supersonic cruise would be optimal with a maximum glide ratio. The function $R(\sigma, M)$ (the *range factor*) is shown in Figure 12.15: there are two local maxima; one maximum can be at subsonic or supersonic speed, depending on the altitude. The other maximum is at about $M \simeq 1.6$ and does not vary much with the altitude. The TSFC is shown in the right scale. The two lines denote the lowest and highest flight altitudes (for clarity, the TSFC at intermediate altitudes is not plotted).

## Summary

This chapter has dealt with cruise problems. We have shown that a large portion of the fuel is burned in this flight segment, with the possible exception of very short-haul flights. Therefore, numerical optimisation methods have traditionally addressed cruise flight. In this context, the key performance parameter was the specific air

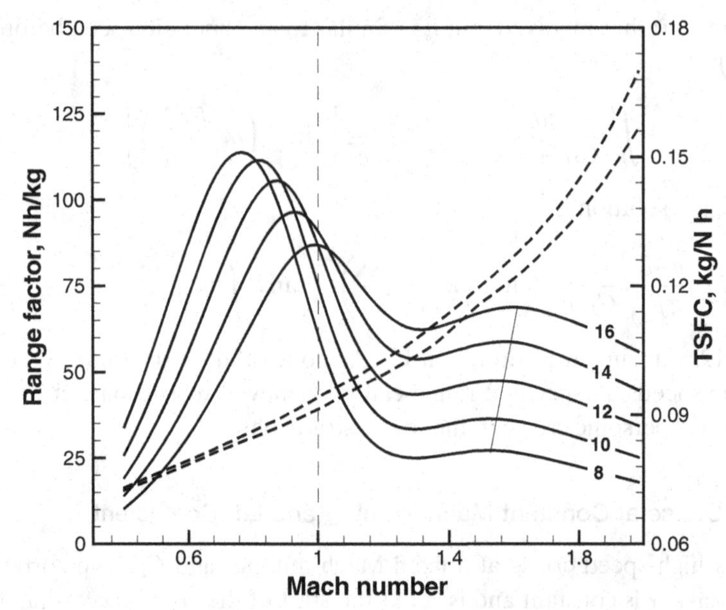


Figure 12.15. Range factor of supersonic aircraft model; flight altitudes are in [km].

range (SAR). We demonstrated that the study of the SAR is the basis of all cruise optimality conditions. From this parameter we derived the maximum-range cruise Mach number and the long-range cruise Mach number. Commercial operators use the latter speed as a compromise between faster service and increased cost of fuel. However, the problem is often more complicated and requires the determination of the cost index function, which is a balanced approach to the cost of fuel and time. On modern airplanes, the cost index is determined by the FMS.

Another key concept was the determination of the optimal cruise altitude. We have shown that current regulations only permit flight at selected levels and that a flight level switch (with a climb) is required to maintain the SAR as high as possible. The combination of long-range Mach number and initial cruise altitude is a two-parameter optimisation problem that depends on the weight. We solved a number of problems, both for subsonic and supersonic flight. We also investigated the effect of weight, CG position, winds and aerodynamic deterioration.

## Bibliography


[1] Speyer JL. Non-optimality in the steady-state cruise for aircraft. *AIAA J.*, 14(11):1604–1610, Nov. 1976.

[2] Speyer JL, Dannenmiller D, and Walker D. Periodic optimal cruise of an atmospheric vehicle. *J. Guidance*, 8(1):31–39, Jan. 1985.

[3] Gilbert EG and Parsons MG. Periodic control and the optimality of aircraft cruise. *J. Aircraft*, 13(10):828–830, Oct. 1976.

[4] Menon PKA. Study of aircraft cruise. *J. Guidance, Control and Dynamics*, 12(5):631–639, Sept. 1989.

[5] Sachs G and Christodoulou T. Reducing fuel consumption of subsonic aircraft by optimal cyclic cruise. *J. Aircraft*, 24(5):616–622, 1987.

[6] Visser HG. Terminal area traffic management. *Progress Aerospace Sciences*, 28:323–368, 1991.

[7] ESDU. *Approximate Methods for Estimation of Cruise Range and Endurance: Aeroplanes with Turbojet and Turbofan Engines.* Data Item 73019. ESDU International, London, 1982.
[8] ESDU. *Introduction to the Estimation of Range and Endurance.* Data Item 73018. ESDU International, London, 1980.
[9] ESDU. *Estimation of Cruise Range: Propeller-driven Aircraft*, volume 5, Performance of *Data Item 75018.* ESDU, London, 1975.
[10] Bennington MA and Visser KD. Aerial refueling implications for commercial aviation. *J. Aircraft*, 42(2):366–375, Mar. 2005.
[11] Smith RK. Seventy-five years of inflight refueling. Technical Report R25-GPO-070-00746-1, Air Force History and Museum Program, Washington, DC, Nov. 1998. (Available on the U.S. Government online store; document out of print as of 2005.)
[12] Torenbeek E and H Wittenberg. Generalized maximum specific range performance. *J. Aircraft*, 20(7):617–622, July 1983.
[13] Houghton RC. Aircraft fuel savings in jet streams by maximising features of flight mechanics and navigation. *J. of Navigation*, 51(3):360–367, Sept. 1998.
[14] Sachs G. Optimization of endurance performance. *Progress Aerospace Sciences*, 29(2):165–191, 1992.
[15] Abramovitz M and Stegun I. *Handbook of Mathematical Functions.* Dover, 1972.
[16] Filippone A. Comprehensive analysis of transport aircraft flight performance. *Progress Aero Sciences*, 44(3):185–197, April 2008.
[17] Windhorst R, Ardema M, and Kinney D. Fixed-range optimal trajectories of supersonic aircraft by first-order expansions. *J. Guidance, Control and Dynamics*, 24(4): 700–709, July 2001.

## Nomenclature for Chapter 12

| | | |
|---|---|---|
| $a$ | = | speed of sound |
| $A$ | = | wing area |
| $b$ | = | wing span |
| $c_i$ | = | constant factors ($i = 0, 1, 2, \dots$) |
| $C_L$ | = | drag coefficient |
| $C_{D_o}$ | = | profile-drag coefficient |
| $C_L$ | = | lift coefficient |
| $C_{L_\alpha}$ | = | lift curve slope |
| $C_I$ | = | cost index |
| $C_{I_{eq}}$ | = | equivalence cost index |
| $D$ | = | drag |
| $E$ | = | endurance |
| $f_c$ | = | specific fuel consumption (also SFC) |
| $f_j$ | = | thrust-specific fuel consumption (also TSFC) |
| $\mathcal{F}_c$ | = | cost function |
| $FM$ | = | Figure of Merit |
| $g$ | = | acceleration of gravity |
| $h$ | = | flight altitude |
| $J$ | = | propeller's advance ratio |
| $k$ | = | induced-drag factor |

$L$ = aerodynamic lift
$m$ = mass
$\dot{m}_f$ = fuel flow
$M$ = cruise Mach number
$M_{econ}$ = economic Mach number
$M_{dd}$ = divergence Mach number
$n$ = number of constant-altitude cruise stages
$p$ = pressure
$P$ = power; shaft power
$R$ = range factor, Equation 12.71
$\mathcal{S}$ = cruise vector state
$\mathcal{S}_{xy}$ = specific air range, with $x = n/o$, $y = n/o$, Equation 12.55
SR = specific (ground) range
$t$ = time
$T$ = net thrust
$\mathcal{T}$ = air temperature
$v_c$ = climb rate
$U$ = air speed
$U_w$ = wind speed
$W$ = weight
$X$ = cruise range; flight distance
$X_{ij}$ = range of flight segment $i$-$j$
$X_i$ = distance flown in cruise mode
$X_{Ci}$ = distance flown in climb mode
$x, y, z$ = Cartesian coordinates

## Greek Symbols

$\alpha$ = angle of attack
$\alpha_o$ = zero-lift $\alpha$
$\eta$ = propulsive efficiency; lift-induced factor
$\theta$ = propeller-pitch angle
$\zeta$ = fuel-block ratio, $m_f/m_i$
$\xi$ = relative change in aircraft mass; $\xi = m/m_i$ in § 12.2.1
$\rho$ = air density
$\sigma$ = relative air density

## Subscripts/Superscripts

$[.]_c$ = constant portion
$[.]_e$ = end (or final)
$[.]_f$ = fuel
$[.]_i$ = initial (or start)
$[.]_{ij}$ = generic flight segment, § 12.5.5
$[.]_o$ = reference value
$[.]_p$ = payload

| | | |
|---|---|---|
| $[.]_w$ | = | wind |
| $[.]_{nn}$ | = | new airframe and new engines |
| $[.]_{no}$ | = | new airframe and old engines |
| $[.]_{on}$ | = | old airframe and new engines |
| $[.]_{oo}$ | = | old airframe and old engines |

# 13 Manoeuvre Performance

## Overview

This chapter deals with basic airplane manoeuvres, including flight in the vertical plane and in the horizontal plane. After an introductory description of these manoeuvres (§ 13.1), we start with powered turns (§ 13.2) and unpowered turns (§ 13.3); the latter turn essentially refers to glider-type vehicles, including soaring flight. Next, we introduce the manoeuvre envelope of the airplane (§ 13.4) and discuss pilot-related issues, such as sustainable $g$-loads. Roll performance (§ 13.5) is one example of manoeuvre on a straight line by a high-performance aircraft, and pull-up (§ 13.6) is one example of flight path in the vertical plane. Finally, we discuss aircraft flight in a downburst (§ 13.7).

**KEY CONCEPTS:** Powered Turns, Banked Turns, Minimum-Fuel Turn, Turn Rates, Cornering Speeds, Unpowered Turns, Manoeuvre Envelopes, $V$-$n$ Diagram, Sustainable $g$-Loads, Roll Performance, Pull-Up Manoeuvre, Downbursts.

## 13.1 Introduction

The term *manoeuvre* refers to any change of the flight path. There are at least three basic types of manoeuvres: 1) a turn in the horizontal plane, when the aircraft changes heading and *banks* on one side (lateral manoeuvre); 2) a turn in a vertical plane, when the aircraft increases or decreases its altitude (*pull-up* and *pull-out*); and 3) a roll around the aircraft's longitudinal axis, when the aircraft rotates around itself whilst following a straight flight path (longitudinal manoeuvre). Many complex manoeuvres can be reduced to a combination of these.

During a turn on the horizontal plane, the aircraft gradually changes its banking angle from zero to a maximum and then back to zero. A pull-up occurs with a variable radius of curvature and speed. Therefore, a turn manoeuvre is always unsteady. Banking is always associated to some centrifugal accelerations that can reach relatively high levels. In addition to the cases listed, there are a large number of manoeuvres performed only by military aircraft, aerobatic aircraft and demonstrators. A concept of interest in high-performance aircraft is the *agility*, or the ability of moving rapidly from one manoeuvre to another[1].

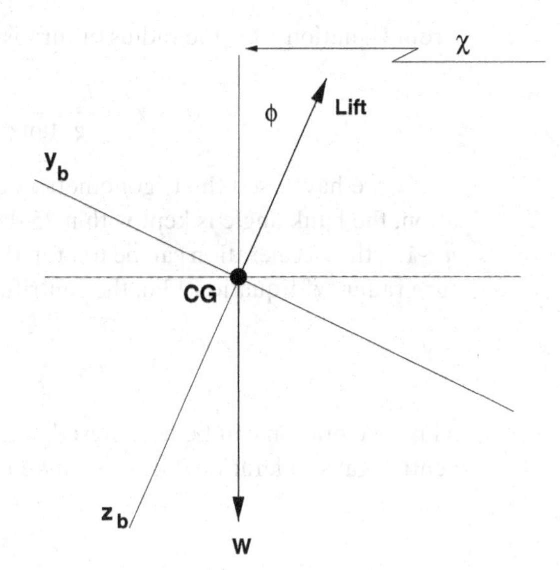


Figure 13.1. Nomenclature for a banked turn.

## 13.2 Powered Turns

A *banked level turn* (or coordinated turn) is a turn during which all of the forces on the aircraft are balanced. Therefore, the aircraft flies with a constant air speed and constant altitude. However, due to the change of heading, there is a centrifugal acceleration. The nomenclature and the reference system for a banked turn are shown in Figure 13.1. This is a view of the aircraft forces in the vertical plane $\{y, z\}$; $\phi$ is the bank angle, that is, the angle between the vertical plane and the aircraft's symmetry plane; $\chi$ is the radius of curvature of the flight trajectory.

We denote the quantities during a turn with a subscript "t". In a correctly banked turn the balance of forces along the horizontal and vertical direction are, respectively,

$$L \cos\phi - W = 0, \tag{13.1}$$

$$L \sin\phi = m\frac{U^2}{\chi}. \tag{13.2}$$

The right-hand term of Equation 13.2 denotes the force due to the centrifugal acceleration of the CG. Combination of the two equations provides the bank angle $\phi$,

$$\tan\phi = \frac{1}{g}\frac{U^2}{\chi}. \tag{13.3}$$

The bank angle increases with the speed. When the aircraft operates at the smallest values of $\chi$, it is said to perform a *tight* turn. During manoeuvres of this type, the lift is higher than the aircraft's weight. We define the normal load factor $n$ as the ratio

$$n = \frac{L}{W}. \tag{13.4}$$

In a straight level flight $n = 1$. This value, also called 1-$g$ flight, is the neutral value. Now the bank angle, Equation 13.3, can be written in terms of $n$, by using Equation 13.1:

$$n = \frac{1}{\cos\phi} = \sec\phi. \tag{13.5}$$

From Equation 13.3, the radius of turn is a function of the load factor

$$\chi = \frac{U^2}{g}\frac{1}{\tan\phi} = \frac{1}{g}\frac{U^2}{\sqrt{n^2-1}}, \tag{13.6}$$

where we have used the trigonometric equivalence $\sec^2\phi = 1 + \tan^2\phi$. In civil aviation, the bank angle is kept within 25 degrees, which leads to a normal load factor $n \sim 1.1$; this acceleration can be tolerated by most passengers. From the definition of turn radius $\chi$, Equation 13.6, the centrifugal acceleration is

$$a = \frac{U^2}{\chi} = g\sqrt{n^2-1}. \tag{13.7}$$

This acceleration can be normalised with $g$, to provide the $g$-factor, or the relative centrifugal acceleration during a banked turn

$$\frac{a}{g} = \sqrt{n^2-1}. \tag{13.8}$$

A normal load factor $n = 1$ corresponds to zero centrifugal acceleration, which proves that $n = 1$ is a flight with no acceleration.

### 13.2.1 Banked Turn at Constant Thrust

We consider a case in which the turn is performed at constant thrust. The corresponding drag coefficient is

$$C_{D_t} = C_{D_o} + kC_L^2 = C_{D_o} + k\left(\frac{2nW}{\rho AU^2}\right)^2. \tag{13.9}$$

There is an increase in lift-induced drag proportional to $n^2$ and hence an increase in overall drag, $\Delta D = D_t - D > 0$. If we write the conservation of total energy of the aircraft, then

$$\frac{\partial h_E}{\partial t} = \frac{T-D}{W}U, \tag{13.10}$$

where $h_E$ is the energy height (see Equation 10.45); the right-hand side is the specific excess power. This equation expresses the fact that an excess power can be used to increase the speed or the altitude, or both. A change in aircraft drag $dD$ can be introduced in Equation 13.10, so that

$$d\left(\frac{\partial h_E}{\partial t}\right) = \frac{T}{W}dU - \frac{dD}{W}U - \frac{D}{W}dU, \tag{13.11}$$

$$\frac{dD}{W}U = -d\left(\frac{\partial h}{\partial t} + \frac{U^2}{2g}\right) + \frac{T}{W}dU - \frac{D}{W}dU = -d\left(\frac{\partial h}{\partial t} + \frac{U^2}{2g}\right). \tag{13.12}$$

If the flight altitude is constant ($dh/dt = 0$), an increase in drag causes a decrease in speed, and vice versa:

$$\frac{dD}{W}U = -\frac{U}{2g}dU. \tag{13.13}$$

If, instead, the speed is maintained constant ($dU = 0$), then there must be a decrease in altitude

$$\frac{dD}{W}U = -d\left(\frac{\partial h}{\partial t}\right) = v_s < 0, \qquad (13.14)$$

where $v_s$ denotes the sinking speed. The magnitude of $v_s$ can be found from

$$v_s = \frac{D_t - D}{W}U = \frac{\rho A U^2}{2W}(C_{D_t} - C_D) = \frac{2W}{\rho A U^2}k(n^2 - 1). \qquad (13.15)$$

For a given speed, the rate of descent increases with $n^2$. The change in drag produces a change in the speed of minimum drag, $U_{md}$. The $C_L$ corresponding to minimum drag in a banked turn is found from the condition

$$\frac{\partial}{\partial C_L}\left(\frac{C_{L_t}}{C_{D_o} + kC_{L_t}^2}\right) = 0. \qquad (13.16)$$

The solution of Equation 13.16 is

$$C_L = \sqrt{\frac{1}{n}\frac{C_{D_o}}{k}}. \qquad (13.17)$$

The ratio of minimum drag speeds between banked turn and level flight is

$$\frac{U_{md_t}}{U_{md}} = n^{1/2}. \qquad (13.18)$$

If the air speed before the turn is $U > U_{md}$, the drag increase due to the turn will slow down the aircraft; this will contribute to a decrease in drag to the point when the forces on the aircraft are balanced, and the aircraft can sustain a turn at constant altitude, albeit at a lower speed. If, instead, the air speed before the turn is $U < U_{md}$, a decrease in speed will produce a further increase in drag. If the drag rise cannot be matched by the engine thrust, the aircraft is forced to descend.

### 13.2.2 Turn Power and High-Speed Manoeuvre

Assume that the $C_L$ during a turn is constant and equal to the value in straight level flight:

$$C_L = C_{L_t}, \quad \text{or} \quad \frac{2W}{\rho A U^2} = n\frac{2W}{\rho A U_t^2}, \qquad (13.19)$$

which leads to the equation

$$U_t = U\sqrt{n}. \qquad (13.20)$$

Equation 13.20 expresses the equivalence between aircraft speeds at *constant lift coefficient*. If the turn is done at *constant speed*, then the relationship between lift coefficients becomes

$$C_{L_t} = nC_L. \qquad (13.21)$$

Next, consider a turn at constant power, $P = P_t$; this condition requires

$$TU = T_t U_t. \qquad (13.22)$$

Equation 13.22 can be satisfied by increasing the net thrust as the air speed decreases. The power in steady-state flight is calculated from the equilibrium of the forces in the flight direction multiplied by the flight speed:

$$P = \frac{1}{2}\rho A C_D U^3. \tag{13.23}$$

Next, eliminate the velocity in Equation 13.23 by using the definition of $C_L$. Equation 13.23 becomes

$$P = \sqrt{\frac{2}{\rho A}\frac{C_D}{C_L^{3/2}}}(nW)^{3/2}. \tag{13.24}$$

We conclude that the power required for a steady banked turn grows with the factor $(nW)^{3/2}$; at constant weight, the power for a banked turn increases with $n^{3/2}$. If also the aerodynamic factor $C_D/C_L^{3/2}$ is held constant, the ratio between power required in a turn and the steady-state power is

$$\frac{P_t}{P} = n^{3/2}. \tag{13.25}$$

If the aircraft has a parabolic drag equation, Equation 13.23 becomes

$$P = \frac{1}{2}\rho C_{D_o} U^3 + k\frac{2}{\rho A}\frac{(nW)^2}{U}. \tag{13.26}$$

Equation 13.26 can be plotted versus the flight speed at different values of the load factor $n$. Matching this power with the available engine power will give the operation point. The minimum speed that the aircraft can sustain is the stall speed (within a safety margin $K_s$); hence, the lower limit of the curves will be given by the stall curve.

**SUPERSONIC FLIGHT.** We now consider the more complicated problem of high-speed turn performed by a supersonic aircraft. Consider a turn at the flight altitude $h = 8{,}000$ m ($\sim$26,245 feet), with an aircraft mass $m = 12{,}000$ kg. The problem consists in solving Equation 13.24 over a range of Mach numbers with the appropriate aerodynamics and engine at selected throttle (for example, full thrust). The results of the analysis are shown in Figure 13.2, for normal load factors up to $n = 3$. The available engine thrust, with and without after-burning, is also shown. The power required to make a supersonic turn is very large; a supersonic turn cannot be performed at high load factors.

**HIGH-SPEED MANOEUVRE.** In §10.5 we presented some concepts for aircraft climb and acceleration by using the total-energy method. Our considerations were limited to a 1$g$-flight. The same method can be applied to aircraft manoeuvring with an apparent acceleration $g$. The drag is changed as follows:

$$C_D \simeq C_{D_o} + n\left[\eta C_{L_\alpha}(\alpha - \alpha_o)\right]. \tag{13.27}$$

In Equation 13.27 the profile-drag coefficient is assumed to be unaffected by the manoeuvre (which is not quite the case), and the induced drag is proportional to the

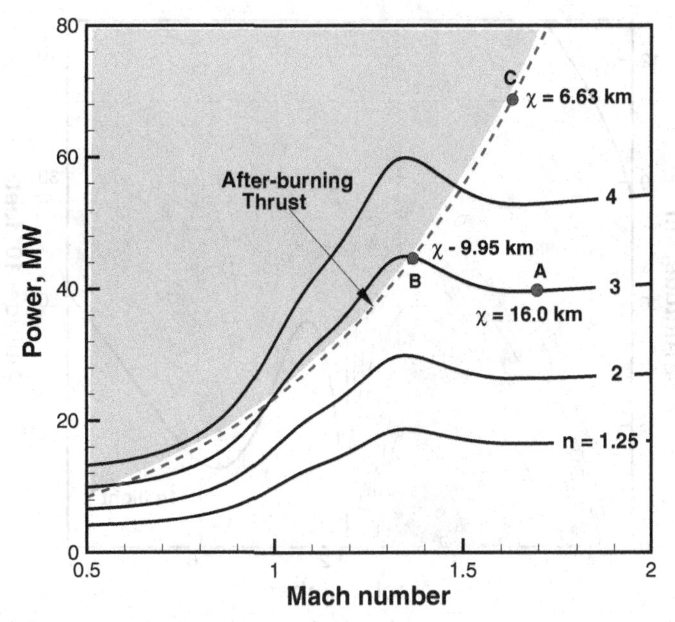


Figure 13.2. Power for supersonic turn; aircraft mass $m = 12{,}000$ kg; $\alpha = 2$ degrees (all cases); altitude $h = 8{,}000$ m (~26,245 feet).

load factor $n$. Therefore, the specific excess power is

$$\mathrm{SEP} \simeq \frac{T - nD}{W}. \tag{13.28}$$

As usual, the weight is critical and will determine whether the aircraft is capable of manoeuvring at supersonic Mach numbers. Figure 13.3 shows the predicted SEP chart at $n = 3.5$. The result shows that the aircraft is not capable of sustaining this load factor over a wide area of the envelope. The thick solid line shows the acceleration limits for the airplane at the same weight in 1$g$-flight: the only way the aircraft can move to a high supersonic manoeuvre area is to accelerate past the speed of sound at a lower load factor.

### 13.2.3 Turn Rates and Corner Speed

The *turn rate* is the angular velocity of the aircraft. Consider again the case of a coordinated turn. The turn rate is

$$\mathsf{q} = \frac{U}{\chi} = \frac{g}{U}\sqrt{n^2 - 1}. \tag{13.29}$$

If we write the turn speed in terms of the load factor $n$, we find

$$U = \sqrt{n \frac{2W}{\rho A C_L}}. \tag{13.30}$$

Thus, the turn rate becomes

$$\mathsf{q} = \sqrt{\frac{\rho g A C_L}{2W}} \sqrt{\frac{n^2 - 1}{n}}. \tag{13.31}$$

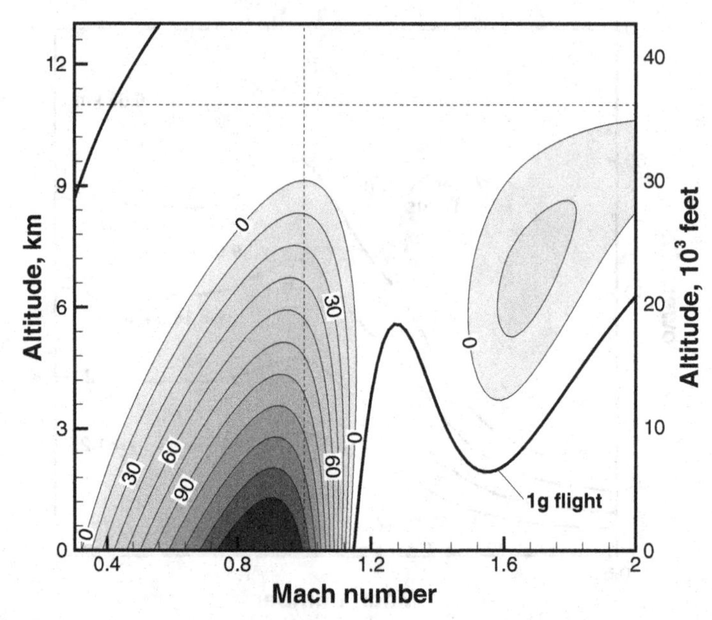


Figure 13.3. Specific excess power chart for the supersonic aircraft model in 3.5$g$ flight, mass $m = 11{,}000$ kg.

Equation 13.31 shows how the turn rate is dependent on altitude, $C_L$ and $W/A$. When maximum turn rates are quoted, it is essential to specify at least the altitude and weight.

The *corner velocity* is the minimum velocity at which the aircraft can achieve its maximum load factor. It is found at the intersection between curve of maximum constant $n$–factor with the curve corresponding to a manoeuvre at maximum $C_L$. Starting from the stall speed

$$V_S = K_s \left[ \frac{2nW}{\rho A C_{L_{max}}} \right]^{1/2}, \tag{13.32}$$

the relationship between the load factor and the Mach number is

$$n = \frac{\gamma p}{2K_s^2} \frac{1}{W/A} C_{L_{max}} M^2. \tag{13.33}$$

Equation 13.33 is obtained by replacing the equation of state of the ideal gases to remove the speed of sound. The maximum $C_L$ is a function of the Mach number; therefore Equation 13.33 is implicit. A numerical solution is shown in Figure 13.4 at selected load factors. We estimate a corner Mach number $M = 0.84$. corresponding to a turn rate $\mathsf{q} = 0.31$ rad/s (17.5 deg/s) for a $g$–limit = 9. For a given load factor, the effect of increasing flight altitude is to increase the corner Mach number. In the stratosphere, there is virtually no effect.

We conclude by recalling that general three-dimensional turn problems for minimum time and minimum fuel have been solved by Hedrick and Bryson[2;3], using the energy method. Kelley[4;5] solved the problem of differential turning with climb.

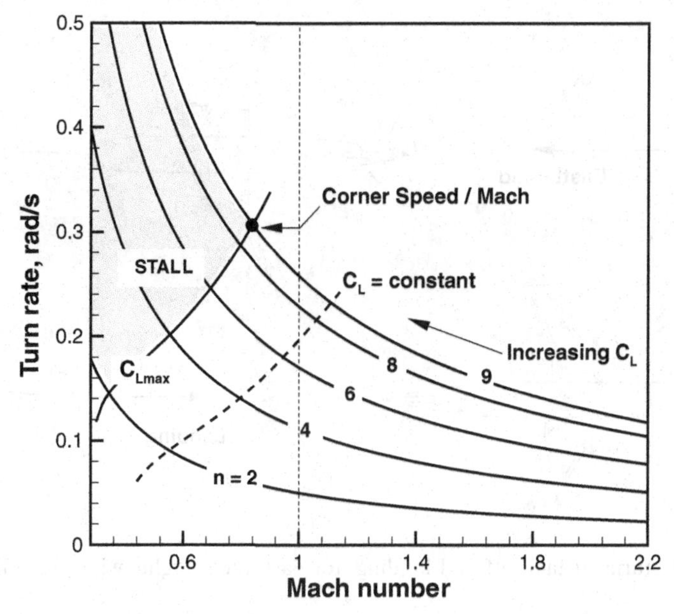


Figure 13.4. Corner speed for supersonic jet aircraft model; standard atmosphere.

### 13.2.4 Minimum-Fuel Turn

The banked turn is a common manoeuvre for a commercial aircraft. An example of this occurrence is during take-off and landing, when the runway heading is different from the assigned flight path, as shown in Figure 13.5. A simple analysis of the turn fuel is possible if the aircraft's drag is parabolic and if the turn is performed at a constant altitude. A steady-state turn is defined by the following equations:

$$D = T, \tag{13.34}$$

$$W = L \cos\phi, \tag{13.35}$$

$$mU\dot{\psi} = L \sin\phi. \tag{13.36}$$

These are the balance forces in the flight direction, in the vertical direction, and in the direction normal to the flight path, respectively.

$$dm_f = \dot{m}_f dt = f_j T dt,$$

$$T dt = T\frac{dt}{d\psi}d\psi = \frac{T}{\dot{\psi}}d\psi.$$

The fuel consumption during the turn is proportional to the time of the manoeuvre or to the angle of turn $\Delta\psi$. Therefore, the total fuel is equal to the integral

$$m_f = \int_o^{\psi} f_j T\frac{d\psi}{\dot{\psi}} \simeq f_j T\frac{\Delta\psi}{\dot{\psi}}. \tag{13.37}$$

For a *fixed* turn altitude, the problem is to minimise the ratio $T/\dot{\psi}$. The solution must be subject to the balance of forces. Divide Equation 13.34 by Equation 13.36 to find

$$\frac{T}{\dot{\psi}} = mU\frac{C_D}{C_L \sin\phi}. \tag{13.38}$$

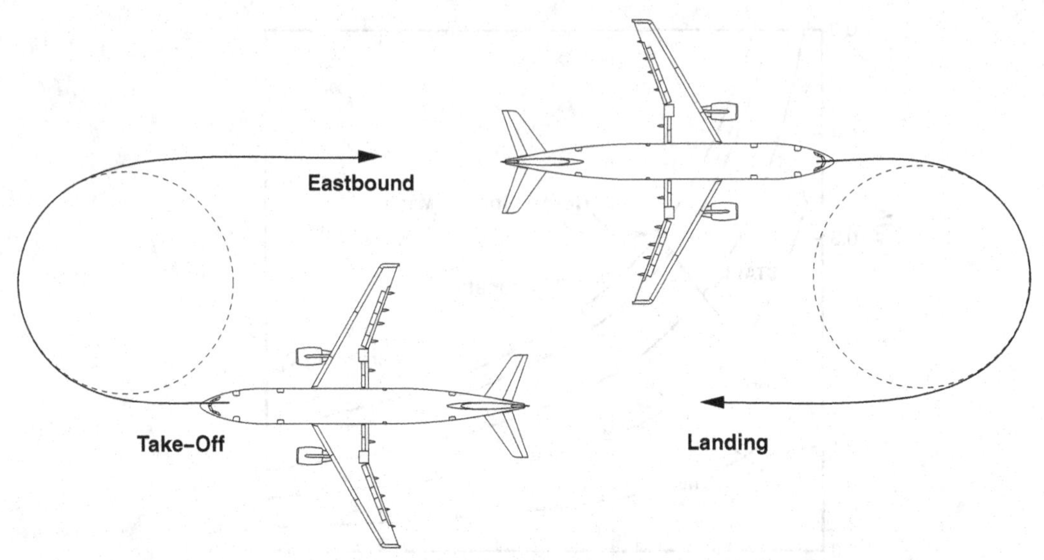


Figure 13.5. U-turns at take-off and landing, for eastbound flights with prevailing westerly winds.

We note that

$$\frac{T}{\dot{\psi}} = f(C_L, \phi, h). \tag{13.39}$$

Therefore, at a fixed bank angle and fixed air speed, the condition of minimum will be given by the condition of minimum drag, or maximum glide ratio,

$$\frac{\partial f}{\partial C_L} = \frac{mU}{\sin\phi}\frac{\partial}{\partial C_L}\left(\frac{C_D}{C_L}\right) = 0. \tag{13.40}$$

This equation shows that the condition for a minimum-fuel turn is to fly at minimum-drag speed. The corresponding lift coefficient is

$$C_L = \sqrt{\frac{C_{D_o}}{k}}. \tag{13.41}$$

The inverse of the glide ratio becomes $D/L = 2\sqrt{C_{D_o}k}$, whereas the optimal condition becomes

$$\frac{T}{\dot{\psi}} = 2\frac{mU}{\sin\phi}\sqrt{C_{D_o}k}. \tag{13.42}$$

This quantity is not dependent on the flight altitude. The fuel required to perform a U-turn at constant altitude is then

$$m_f = \int_0^{\pi} 2f_j\frac{mU}{\sin\phi}\sqrt{C_{D_o}k}\,d\psi = 2\pi f_j\frac{mU}{\sin\phi}\sqrt{C_{D_o}k}. \tag{13.43}$$

The result is that the turn fuel depends on the flight altitude. For example, if the AUW = 140 tons, a turn normal load factor $n = 1.1$ at 200 KTAS would lead to a turn fuel of the order of 120 kg. This is not a small amount of fuel: turning costs a considerable amount of fuel. The FLIGHT code has been programmed to always perform minimum-fuel turns.

## 13.3 Unpowered Turns

We present the turn performance of unpowered airplanes and gliders. One important consequence of a steady-state turn of an unpowered vehicle is a loss of altitude. The calculation of the sinking speed $v_s$, the radius of turn $\chi$ and the turn rate can be done if the bank angle $\phi$ is assigned. The dynamics equations in the direction of the flight path and in the vertical direction are, respectively:

$$D - W \sin\gamma = 0, \tag{13.44}$$

$$L\cos\phi - W\cos\gamma = 0. \tag{13.45}$$

The sinking rate is

$$v_s = \frac{\partial h}{\partial t} = -\frac{DU}{W}. \tag{13.46}$$

If we replace the parabolic drag equation in Equation 13.46, we have

$$v_s = -\frac{\rho A C_{D_o}}{2W} U^3 - \frac{\rho A k}{2W}\left(\frac{2L}{\rho A U^2}\right)^2 U^3. \tag{13.47}$$

The lift is found from Equation 13.45. Thus, we have

$$v_s = -\frac{\rho A}{2W} C_{D_o} U^3 - \frac{2W}{\rho A}\frac{k}{U}(n\cos\gamma)^2. \tag{13.48}$$

Equation 13.48 has two contributions, both negative. For its solution, we require the flight-path angle $\gamma$. The latter quantity can be calculated from the glide ratio of the unpowered airplane, $L/D$, that is found by dividing Equation 13.44 by Equation 13.45,

$$\tan\gamma = \frac{D}{L\cos\phi}, \qquad \gamma = \tan^{-1}\left(\frac{n}{L/D}\right). \tag{13.49}$$

Equation 13.49 requires the $L/D$, which is dependent on the speed. The turn rate is found from

$$\frac{U}{\chi} = \frac{g}{U}\frac{L}{W}\sin\phi = \frac{g}{U}\cos\gamma\tan\phi, \tag{13.50}$$

where we have used the combination of Equation 13.44 and Equation 13.45. The radius of turn is found from the first of the equivalences in Equation 13.50. The solution of this model is shown in Figure 13.6 for a sailplane starting a turn at an altitude of 1,000 m (3,048 feet). The data used are $C_{D_o} = 0.007$, $k = 0.022$, $m = 450$ kg, and $A = 17.0$ m$^2$. For the same case and for a load factor $n = 1.25$, we have calculated the best gliding speeds. These speeds are shown in Figure 13.7, along with the turn radius. The minimum glide angle can be found graphically, as the tangent to the sinking speed curve from the origin of the axes.

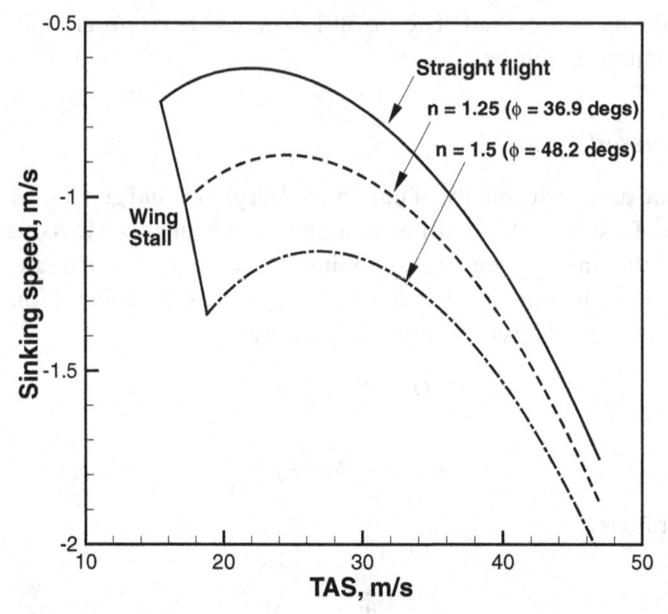


Figure 13.6. Sinking speed for glider turning at selected load factors.

## 13.4 Manoeuvre Envelope: *V*-*n* Diagram

The velocity-load-factor ($V$-$n$) diagram is a flight manoeuvre envelope involving both structural and aerodynamic limits. The speed axis can be either the Mach number or the equivalent air speed (EAS). We consider separately the cases of subsonic and supersonic flight.

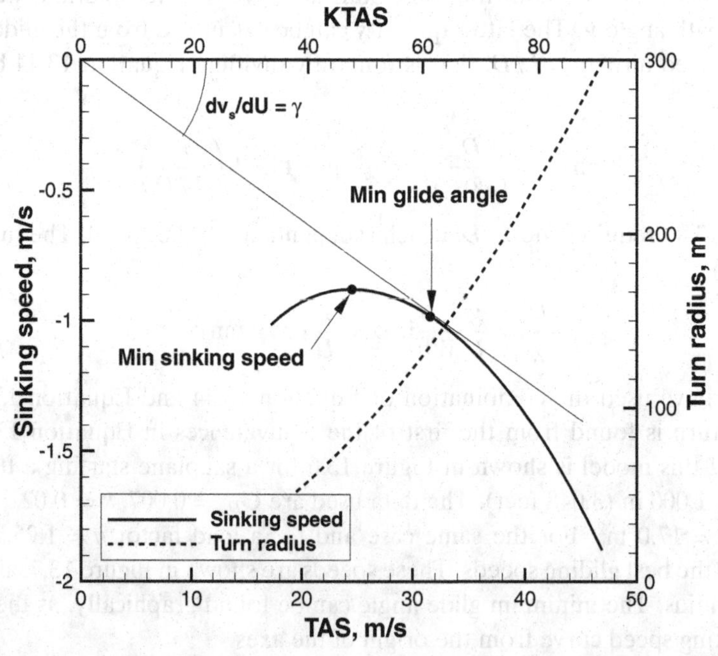


Figure 13.7. Sinking speed for the glider of Figure 13.6; load factor $n = 1.25$.

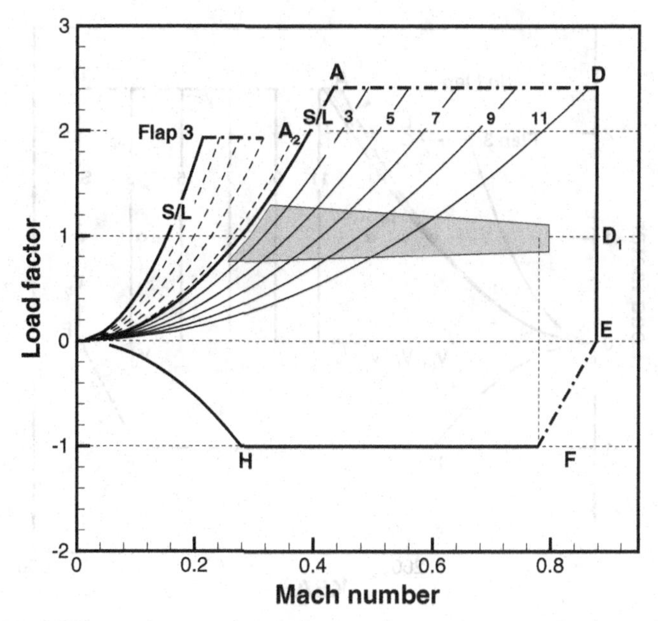


Figure 13.8. $V$-$n$ diagram for a transport aircraft model; $m=$ 58,800 kg, standard day (calculated).

**TRANSPORT AIRCRAFT.** Figures 13.8 and 13.9 show the typical manoeuvre limits of a transport aircraft (the Airbus A320). In Figure 13.8 the shaded area indicates the normal area of operation. The Federal Aviation Regulations* prescribe a positive limit load factor $2.5 \leq n \leq 3.8$, although its value is determined by the gross weight. The negative limit is $n \geq -1$.

The limit condition with flaps extended has been calculated with Flap-3 ($\delta_f = 15$ degrees; $\delta_s = 22$ degrees). The numbers 3, 5, ... indicate flight altitudes in km. The letters $A$, $B$, ... denote the standard FAR points in the $V$-$n$ diagram. In Figure 13.8 we have plotted the load factor versus the flight Mach number, from the solution obtained with the `FLIGHT` code. The Mach limit is the dive Mach number, $M_D$. For reference, the cruise Mach has been assumed as $M_D - 0.1$, although in practice there can be small variations, as discussed in Chapter 12. An increasing flight altitude pushes the left limit of the envelope to the right, both with plain and flapped configurations.

The key operating points are shown in Figure 13.9, which is a different version of the diagram, with the horizontal axis now being the equivalent air speed (KEAS) instead of the Mach number. A number of observations are reported here:

- For both plain and flapped configurations, the left limit is virtually independent of the KEAS.
- For clean configuration, the right limit depends on the flight altitude. An increasing altitude causes the KEAS to decrease.
- Point A is the design manoeuvre point with flaps retracted. When $n = 1$ (1-$g$ flight), the corresponding EAS is $V_A$.
- Point D corresponds to the design dive speed (or Mach number).

* Federal Aviation Regulations, §25.335 (Transport Category Airplanes), Subpart C (Structures).

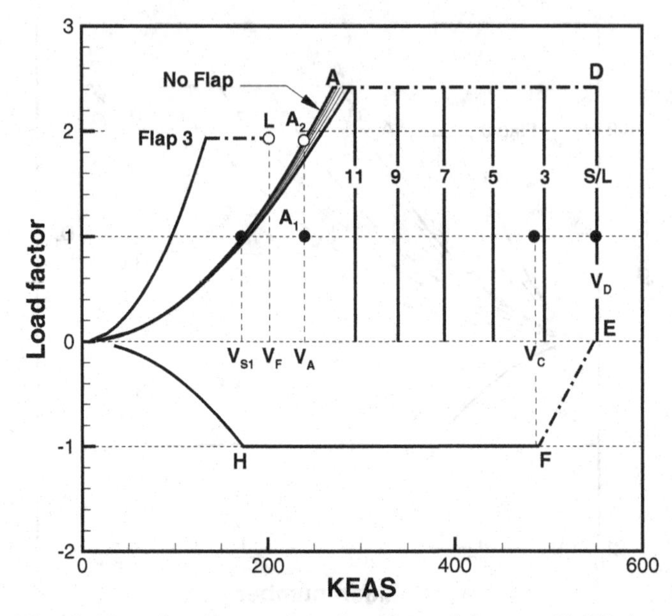


Figure 13.9. $V$-$n$ diagram for a transport aircraft model; $m =$ 58,800 kg, standard day (calculated).

- Point F corresponds to the limit $-1g$ at the design cruise speed.
- Point H corresponds to the limit $-1g$ flight with flaps retracted (or $-C_{N_{max}}$).
- Point L corresponds to the limit load factor with flaps extended.

**SUPERSONIC FLIGHT.** The manoeuvrability of an aircraft depends on its ability to perform rapid turns. This capability has some limits, which are imposed by the aerodynamic characteristics, the propulsion system and the structural dynamics, and – increasingly – on the resistance of the human body. The essential parameters are the thrust ratio $T/W$, the wing loading $W/A$ and the maximum $C_L$. There are two types of envelopes: one is the load factor versus Mach number; the other is the turn rate versus Mach number. First, consider the steady-state turn at constant altitude. From the lift equation, we have

$$C_L = \frac{2nW}{\rho A a^2 M^2}. \tag{13.51}$$

The relationship between normal load factor and Mach number becomes

$$n = \frac{1}{2}\frac{a^2 M^2 C_L}{W/A} = f(h, M, C_{L_{max}}, W/A). \tag{13.52}$$

From Equation 13.52 we draw the following conclusions:

- For a flight in the troposphere, the load factor decreases with increasing altitudes. A turn in the lower stratosphere is not affected by altitude.
- At a given altitude, the maximum load factor is limited by $C_{L_{max}}$. This parameter depends on the Mach number, as explained in Chapter 4. However, also a manoeuvre at $C_L < C_{L_{max}}$ may be of concern because of buffeting associated to the unsteady separated flow or because of control and stability problems.

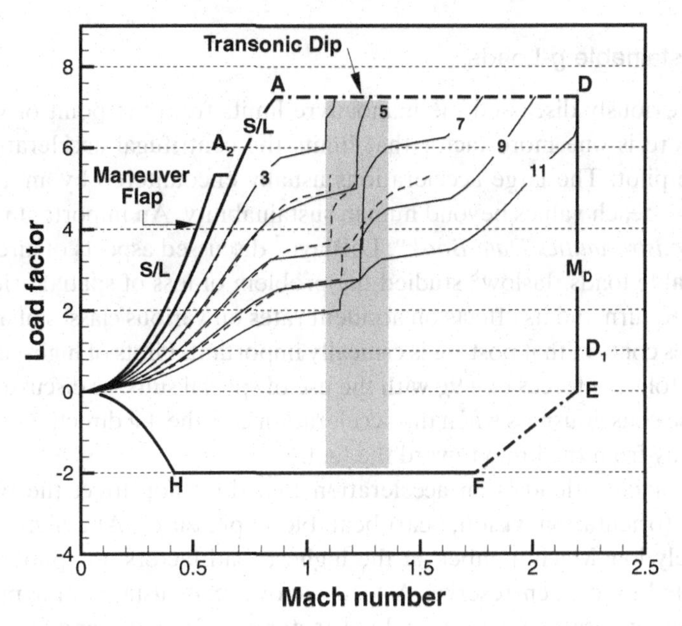


Figure 13.10. *V-n* diagram for a supersonic jet aircraft; $m = 15{,}800$ kg, standard day.

Therefore, the $C_L$ for turning must be a fraction of $C_{L_{max}}(M)$; we call this fraction the *usable lift* (see Figure 4.22 in Chapter 4).

- The third limit is due to the maximum engine thrust. One can think of moving along the line of *structural limit* and increase the Mach number as much as the engine thrust allows. The maximum Mach number during a turn also depends on the angle of attack. The governing equation is

$$nT = \frac{1}{2}\rho A\left(C_{D_o} + \eta C_{L_\alpha}\alpha^2\right) a^2 M^2, \tag{13.53}$$

or

$$M^2 = \frac{n}{c_1}\frac{T(M)}{C_{D_o} + \eta C_{L_\alpha}\alpha^2}, \tag{13.54}$$

where $c_1 = \rho A a^2/2$ is a constant coefficient at a constant turn-altitude. Equation 13.54 is an implicit relationship between the mass $m$ and the angle of attack $\alpha$.

- Finally, at a critical speed the load factor can exceed the structural limit. Therefore, if other factors do not intervene, the manoeuvre envelope has a flat top.

A qualitative example of manoeuvre envelope for a high-performance aircraft is shown in Figure 13.10, as obtained with the `SFLIGHT` program. For a given gross weight, the envelope depends on the flight altitude. The lift is limited by the subsonic buffet and the horizontal tail deflection limit. The maximum $g$−factor depends on the structural limits. The maximum level speed depends on the altitude; therefore, the speed limit in the flight envelope can move backward or forward, depending on the aircraft. The manoeuvre envelope at sea level is indicated by the shadow area.

### 13.4.1 Sustainable $g$-Loads

We have previously discussed the manoeuvre limits from the point of view of the aircraft. There is one more factor that limits the centrifugal accelerations of the aircraft: the pilot. The large accelerations usually encountered by an aircraft in a tight turn can reach values beyond human sustainability. An important reference is NASA's *Bioastronautics Data Book*[6]. LeBlaye[7] discussed aspects of aircraft agility and sustainable loads. Jaslow[8] studied the problem of loss of spatial orientation in a coordinated turn and its effects on accident rates on various classes of aircraft. All of the studies confirm that posture is critically important. Pilots of high-performance aircraft can tolerate forces of $+9g$ with the use of special suits, as discussed in Ref.[9]. Humans lose consciousness when the acceleration is in the $+g$ direction, when blood is drawn away from the brain toward the feet.

A sustainable $g$-load is an acceleration that does not affect the pilot in any serious way (orientation, vision, heart beat, blood pressure). At a high $g$-load even at a relatively low Mach number, at the highest load factors, the pilot's breathing must be aided by oxygen reserve. A load factor $n \simeq 4$ sustained for more than 5 seconds is potentially dangerous. A load factor $n \sim 2$ can be endured for a long time. Unconsciousness sets in at variable times, depending on the $g-$level (0.1 to 1 second).

## 13.5 Roll Performance

In this section we consider the one-degree of freedom roll of a high-performance aircraft. During a roll the aircraft rotates around its longitudinal axis whilst maintaining a straight course. The free parameter is the roll angle $\phi$. Roll rates and roll accelerations for these aircraft can be relatively high and are a key factor to their overall agility. Roll performance and requirements, as from U.S. Military Standards, generally refer to 1) maximum roll rate, 2) time to roll to 90 degrees, and 3) time to execute a 360-degree roll.

The roll discussed below is difficult to achieve at high speeds due to a number of unwanted effects. One such effect is the yaw response. If the aircraft is turning left in roll, it will cause a yaw in the right direction. This effect is called *adverse yaw*. Another important effect is the *inertial coupling* (or roll coupling). When an aircraft rotates around an axis that is not aligned with its longitudinal axis, the inertial forces tend to swing the aircraft out of the rotating axis, with potentially lethal consequences. The problem is compounded by low-aspect-ratio wings, high speeds, high altitudes and, consequently, high inertial forces compared to the aerodynamic forces. This problem has been treated theoretically by Phillips[10], before it was actually encountered, and is a complex subject for stability and control. Phillips pointed out that there is a precise relationship among the moments of inertia that defines the limits of inertial stability and instability in roll. The critical flight conditions for yaw divergence, pitch divergence and auto-rotational rolling have been theoretically established[11]. Seckel[12] reported that a steady-state roll is a manoeuvre that cannot be performed but must involve cyclic variations of angles of attack $\alpha$ and side-slip $\beta$.

In the present case, we consider an aircraft that does not have problems with inertial coupling. When the aircraft starts a roll ($\phi = 0$), its longitudinal axis is not

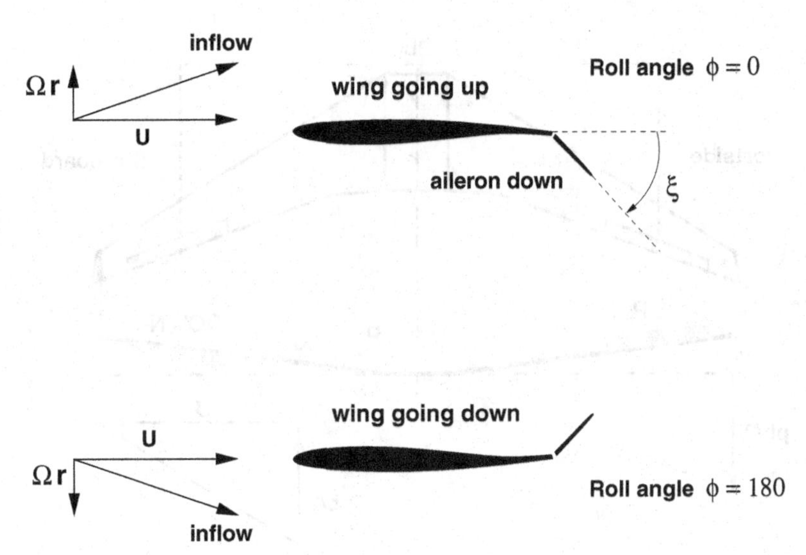


Figure 13.11. Inflow conditions on starboard wing, assuming an anti-clockwise roll in the flight direction.

aligned with the rotation axis (though by a small angle); this generates a side-slip $\beta$. When the aircraft is on its side ($\phi = 180$ degrees), side-slip force has to be generated to balance the weight. The combination of yaw and roll can lead the aircraft into inertial coupling instability.

Consider now the situation illustrated in Figure 13.11. After 180 degrees, the aircraft will find itself upside-down. The inflow on the wing is made of two components: one is due to the forward speed $U$; the other is due to the roll itself. If we assume that the former component does not change during the aircraft rotation, then the starboard wing will find itself with the same angle of attack $\alpha_o$ of the wing-level flight. This angle is generally small.

The rotational component of the speed changes is shown in Figure 13.11. The maximum angle of attack is at the tip and is estimated from

$$\tan\alpha = \frac{U}{\mathsf{p}b/2} = \frac{2U}{\mathsf{p}b}, \tag{13.55}$$

where $\mathsf{p} = \dot{\phi}$ is the roll rate. It can be verified that in many cases the forward speed is much higher than the rotational speed; therefore: $\tan\alpha \simeq \alpha$. Figure 13.12 shows the arrangement with exaggerated angles. In general, the higher the roll rate and the lower the forward speed, the higher the risk of stall. The problem is complicated by the presence of a spanwise component of the velocity. Delta wings with various aerodynamic arrangements can improve the stall behaviour at high angles of attack. Wings for high roll rates must be able to operate at high angles of attacks without stalling.

A rolling moment can be created by operating on the control surfaces. For reference, the ailerons are set at an angle $\pm\xi$ on either side of the symmetry plane ($y_b = 0$), although in practice the deflection angle can be different. In the latter case, the angle $\xi$ to consider is a mean value. A linear response cannot be guaranteed because the suction and pressure on the wing tips distort the flow. The vertical tail creates additional flow separation and damping.

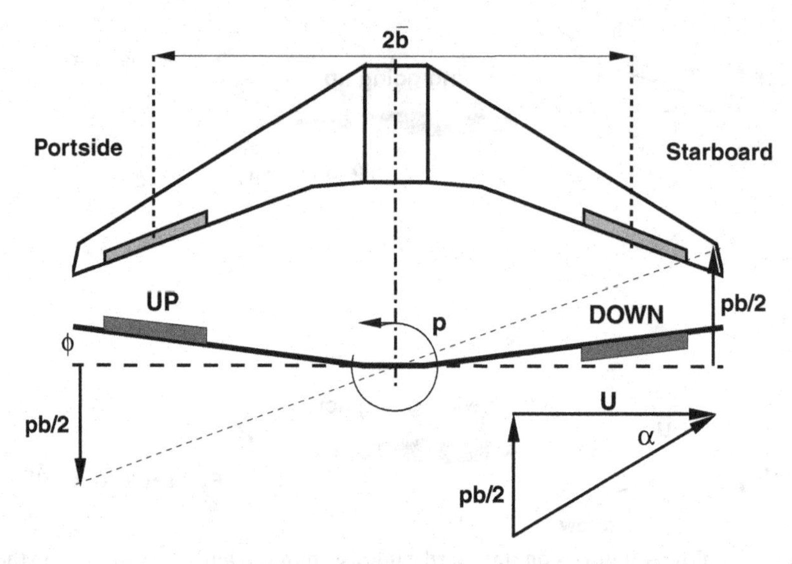


Figure 13.12. Inflow conditions on rolling wing, seen on plane normal to flight velocity. Anticlockwise roll in the flight direction.

In practice, a deflection speed of about 60 deg/s is achieved. With a maximum aileron authority of 30 degrees (up and down), about half a second is required for the aileron to reach its limit position. In most roll analyses, a step change in aileron angle is considered. To calculate the variation of angle of attack during the rotation around the $x$-axis, we have to add the effect of the rotational speed. Therefore, the inflow direction at spanwise position $y$ has the following expression:

$$\alpha(y) \simeq \tan^{-1}\left(\frac{\mathsf{p}\, y}{U}\right) \simeq \left(\frac{\mathsf{p}\, y}{U}\right). \tag{13.56}$$

The one-degree-of-freedom aileron roll is described by the differential equation

$$I_x \dot{\mathsf{p}} = \mathcal{L}_\xi + \mathcal{L}_p, \tag{13.57}$$

where $I_x$ is the principal moment of inertia around the roll axis $x$, $\dot{\mathsf{p}} = \ddot{\phi}$ is the angular acceleration, $\mathsf{p} = \dot{\phi}$ is the role rate, $\mathcal{L}_\xi$ is the rolling moment due to aileron deflection, and $\mathcal{L}_p$ is the aerodynamic damping moment due to roll rate $\mathsf{p} = \dot{\phi}$. Therefore, the one-degree-of-freedom aileron roll is governed by a forcing moment (created by the aileron) and a damping moment (created by the aerodynamics). We rewrite the moments in the following form:

$$\mathcal{L}_\xi = \left(\frac{\partial \mathcal{L}}{\partial \xi}\right)\xi, \qquad \mathcal{L}_p = \left(\frac{\partial \mathcal{L}}{\partial \mathsf{p}}\right)\mathsf{p}. \tag{13.58}$$

These derivatives express the moment curve-slopes, that is, the moment response due to a small change in aileron angle and roll speed, respectively. With these definitions the roll Equation 13.57 becomes

$$\dot{\mathsf{p}} = \frac{1}{I_x}\left(\frac{\partial \mathcal{L}}{\partial \xi}\right)\xi + \frac{1}{I_x}\left(\frac{\partial \mathcal{L}}{\partial \mathsf{p}}\right)\mathsf{p}. \tag{13.59}$$

The parameter defined by

$$L_\xi = \frac{1}{I_x}\left(\frac{\partial \mathcal{L}}{\partial \xi}\right) \tag{13.60}$$

is the *aileron effectiveness*. The effectiveness $L_\xi$ denotes the scale of lateral response to a unit input in aileron deflection. Likewise, we define the *damping in roll* parameter $L_p$:

$$L_p = \frac{1}{I_x}\left(\frac{\partial \mathcal{L}}{\partial \mathsf{p}}\right). \tag{13.61}$$

This quantity is the derivative of the rolling moment with respect to the roll rate, normalised with the moment of inertia. By using the definitions of effectiveness and damping in roll, the roll equation finally becomes

$$\dot{\mathsf{p}} = L_\xi \xi + L_p \mathsf{p}. \tag{13.62}$$

The parameters $L_\xi$ and $L_p$ are integral properties of the aircraft and depend on the Mach number, density altitude and several geometrical details. If $C_l$ is the rolling-moment coefficient, we have:

$$\frac{\partial \mathcal{L}}{\partial \xi} = \frac{\partial}{\partial \xi}\left(\frac{1}{2}\rho AbU^2 C_l\right) = \frac{\rho AU^2 b}{2}\left(\frac{\partial C_l}{\partial \xi}\right) = \frac{\rho AU^2 b}{2} C_{l_\xi}, \tag{13.63}$$

where $C_{l_\xi}$ denotes the derivative of the rolling-moment coefficient with respect to the aileron deflection. In other words, $C_{l_\xi}$ is a non-dimensional aileron effectiveness. The response to the aileron deflection $\xi$ is

$$L_\xi \xi = \frac{\rho AU^2 b}{2I_x} C_{l_\xi} \xi, \tag{13.64}$$

from which we find the relationship between $C_{l_\xi}$ and $L_\xi$:

$$C_{l_\xi} = \frac{2I_x}{\rho AU^2 b} L_\xi. \tag{13.65}$$

The damping in roll can be written as

$$L_p = \frac{1}{I_x}\left(\frac{\partial \mathcal{L}}{\partial \mathsf{p}}\right) = \frac{1}{I_x}\frac{\partial}{\partial \mathsf{p}}\left(\frac{1}{2}\rho AbU^2 C_l\right) = \frac{\rho AU^2 b}{2I_x}\left(\frac{\partial C_l}{\partial \mathsf{p}}\right) = \frac{\rho AUb^2}{4I_x}\frac{\partial C_l}{\partial(\mathsf{p}b/2U)}. \tag{13.66}$$

The damping moment is associated to the damping-in-roll coefficient

$$C_{l_p} = \frac{\partial C_l}{\partial(\mathsf{p}b/2U)} = \frac{\partial C_l}{\partial \alpha} = \frac{2U}{b}\left(\frac{\partial C_l}{\partial \mathsf{p}}\right). \tag{13.67}$$

The values of $C_{l_p}$ are dependent on the speed regime (subsonic or supersonic), on the wing geometry (taper ratio, aspect-ratio, wing sweep) and aeroelastic effects. Therefore, a generalisation cannot be made. The order of magnitude for the damping coefficient is $C_{l_p} = -0.5$ to $-0.1$. However, this value generally decreases when approaching the speed of sound and increases again at low supersonic speeds. From Equation 13.67 and Equation 13.66, we find

$$C_{l_p} = \frac{4I_x}{\rho AUb^2} L_p. \tag{13.68}$$

With substitution of Equation 13.65 and Equation 13.67 in the roll equation, we have

$$\dot{\mathsf{p}} = \frac{\rho A U b^2}{2 I_x}\left(C_{l_\xi}\frac{U}{b}\xi + \frac{1}{2}C_{l_p}\mathsf{p}\right). \tag{13.69}$$

Equation 13.69 is a differential equation that must be solved with the initial conditions

$$\mathsf{p}(t = 0) = 0. \tag{13.70}$$

The factor

$$\tau = \frac{2 I_x}{\rho A U b^2} \tag{13.71}$$

has the dimension of a time; it is a time constant depending on the aircraft. The solution of Equation 13.69 with the boundary conditions Equation 13.70 is

$$\mathsf{p} = \frac{2U}{b}\frac{C_{l_\xi}}{C_{l_p}}\xi\left(1 - e^{-t/\tau}\right). \tag{13.72}$$

The asymptotic value of the roll rate is

$$\mathsf{p}_{max} = \frac{2U}{b}\frac{C_{l_\xi}}{C_{l_p}}\xi. \tag{13.73}$$

The moment response due to aileron deflection is not always linear. The damping and aileron derivatives depend on a fairly large number of parameters, including the geometry of the aircraft (wing, tail, and aileron) and operational conditions (altitude, Mach number and aileron deflection). The functional relationships are

$$U = a(h)\,M, \qquad C_{l_\xi} = f(M, h, \text{geometry}), \qquad C_{l_p} = f(M, h, \text{geometry}). \tag{13.74}$$

These laws depend on the aircraft, as discussed next.

### 13.5.1 Mach Number Effects

The ESDU reports some useful charts for roll analysis[13;14]; in particular, it publishes charts to interpolate the values of $L_p$ and $L_\xi$ for a given wing geometry, aileron deflection and Mach number. Some experimental data are available in Sandhal[15] (delta wings of rocket-propelled vehicles at supersonic speeds up to $M = 2.0$), Myers and Kuhn[16;17] (swept-back wings at $M < 0.8$) and Anderson *et al.*[18] (straight wings at transonic speeds). The Mach number effect is marginal for most wings up to the transonic divergence, when there is a drop in the aileron effectiveness.

The effects of Mach number on the roll rate depend on the behaviour of the damping-in-roll and on the aileron effectiveness. These parameters have specific values for each aircraft but can also be extrapolated from existing general tables, such as ESDU[14;13]. Roll performance, with particular reference to damping characteristics as a function of Mach number, has been investigated by Stone[19], Sanders[20] and others. Theoretical derivations of the damping-in-roll coefficients have been done by Malvestuto *et al.*[21] and Jones and Alskne[22] at supersonic speeds.

Some roll data for early military airplanes were collected by Toll[23]. These data proved that the roll rate reaches a maximum value before decreasing more or less

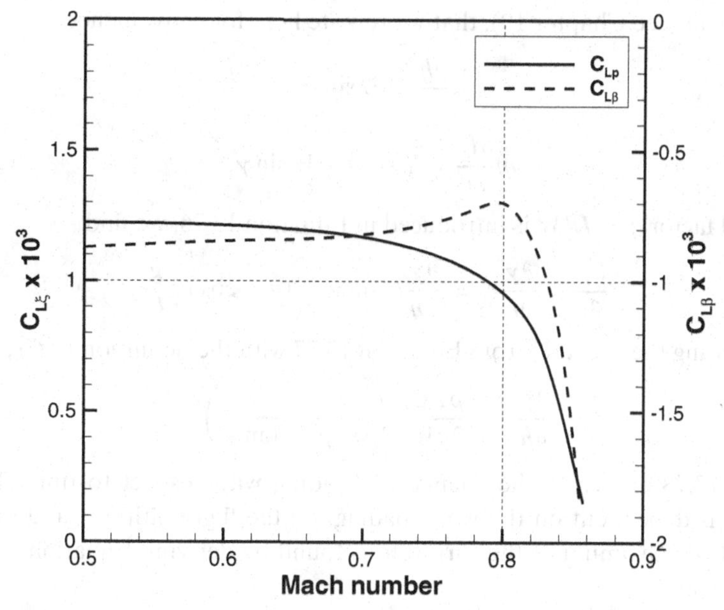


Figure 13.13. Mach number effects on aileron effectiveness and rolling-moment derivative for airplane with a straight wing; wing section; airplane Republic R-4, wing 10% thick, $\alpha = 0$ degs, twist $= -2$ degs. Elaborated from Anderson et al.[18].

sharply – depending on the airplane and with some exceptions. The best roll rates up to the end of World War II were the Spitfire with clipped wings (150 deg/s) and the Focke Wulf Fw-190 (160 deg/s). A Spitfire with clipped wings had a lower radius of gyration, a lower roll moment of inertia and hence a higher roll rate. It is reported[24] that the original version of the Spitfire-I was unable to compete with the agility of the Focke Fw-190 because this had a higher roll rate. The highest roll rates of acrobatic aircraft are in the range of 300 deg/s.

The aileron may be necessary at high speeds for roll control. In fact, as the speed increases, there is another unwanted response: an increase in effective dihedral angle (*wing dropping*, or uncommanded roll). The rolling moment starts appearing at transonic speeds in steady level flight from an asymmetry in the airplane. A right side-slip (in the flight direction) causes a left roll-off, and vice versa. "Wing drop" refers to an abrupt and irregular lateral motion of the aircraft, due ultimately to an asymmetric wing stall. If uncontrolled, it can cause an unstable roll. This is shown for a case of a straight wing in Figure 13.13, elaborated from Anderson *et al.*[18]; $C_{l_\beta}$ is the rolling-moment derivative with respect to the side-slip angle $\beta$: $C_{l_\beta} = \partial C_l / \partial \beta$. This problem appeared in the first generation of transonic aircraft in the 1950s. Wing drop has affected most high-performance military aircraft to the present day. Research on the subject is available in Rathert *et al.*[25], Chambers and Hall[26], and Hall *et al.*[27].

## 13.6 Pull-Up Manoeuvre

The pull-up manoeuvre is a turn in the vertical plane and flight path inclined by a variable angle $\gamma$. The dynamic equations of the aircraft are the same as in the case of climb flight (see Chapter 10), that we rewrite here for convenience:

$$\frac{\partial h}{\partial t} = U \sin\gamma, \tag{13.75}$$

$$m\frac{\partial U}{\partial t} = T - D - W \sin\gamma. \tag{13.76}$$

If the load factor $n = L/W$ is introduced in Equation 13.76, we find

$$\frac{\partial \gamma}{\partial t} = \frac{\partial \gamma}{\partial h} v_c = \frac{\partial \gamma}{\partial h} U \sin\gamma = (n - \cos\gamma)\frac{g}{U}. \tag{13.77}$$

By eliminating the speed $U$ from Equation 13.77 with the definition of $C_L$, we find

$$\frac{\partial \gamma}{\partial h} = \frac{g\rho A C_L}{2nW}\left(\frac{n}{\sin\gamma} - \frac{1}{\tan\gamma}\right). \tag{13.78}$$

Equation 13.78 expresses the change of heading with respect to time. This rate of change is dependent on the wing loading, on the flight altitude, and on the lift coefficient. A maximum of the turn rate is found by deriving Equation 13.78 with respect to $n$ and $C_L$.

When the aircraft pulls up, there is a change in tail-plane incidence, $\Delta\alpha_{ht}$, which depends on several factors. This change in incidence is

$$\tan \Delta\alpha_{ht} = \frac{\mathsf{q}}{U} x_{ht} = \frac{g\Delta n}{U^2} x_{ht}, \tag{13.79}$$

where $x_{ht}$ is the distance between the aerodynamic centre of the tail-plane and the CG, $\mathsf{q}$; is the turn rate. Thus, the tail-plane incidence increases with the load factor and with the distance of the tail-plane. If the resulting incidence $\alpha_{ht}$ is too high, there is a risk of tail-plane stall.

## 13.7 Flight in a Downburst

Localised thunderstorms near the ground are a serious hazard to aircraft flight, particularly in the terminal-area manoeuvre. A thunderstorm with a strong vertical component of the wind (downdraft) associated to heavy precipitations is called *downburst*. Its discovery happened only in the 1970s. Exceptionally, the downdraft has speeds of up to 30 m/s. As the winds hit the ground, they spread radially; thus they create hazards to aircraft either approaching the runway or taking off. A downburst of 5 km radius (2.7 n-miles) easily fits into an airport. Such a downburst is sometimes called a "microburst" and lasts between 2 and 5 minutes. Larger downbursts can last up to 20 minutes and are sometimes called "macrobursts."

The wind shear has two components: horizontal and vertical; the former component is basically a head or a tail wind. It can change by as much as 100 kt per n-mile; the latter component generally results in turbulence that may affect the stability of the aircraft through the wind-shear layer.

The effect of a microburst is highlighted in Figure 13.14. There are three important effects to note. First, there is the increase in IAS as the aircraft approaches the downburst and enters the outflow area. The second effect is the loss of altitude as the aircraft traverses the column of downward air. Finally, as the aircraft exits the downburst through the outflow region, the IAS drops.

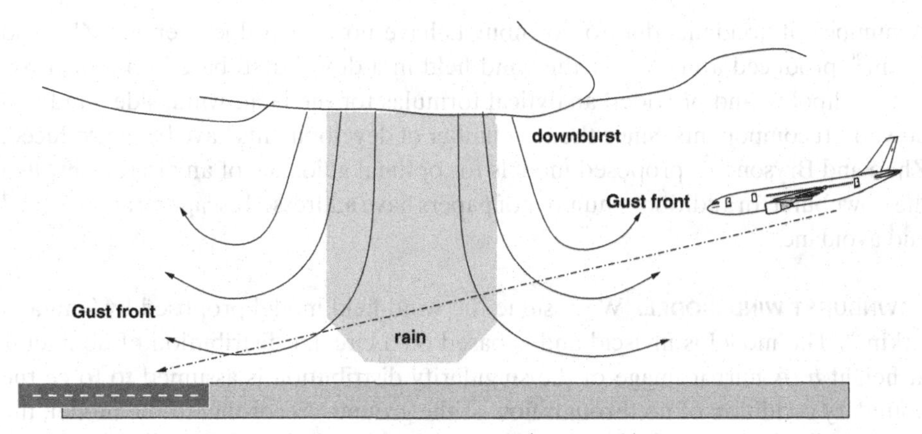


Figure 13.14. Effect of microburst on aircraft (rotorcraft) descent and landing.

**DOWNBURST PHYSICS.** For a downburst to be considered wet it must be accompanied by more than 0.25 mm of rain; in all other cases the downburst is to be considered dry. For a downdraft to be considered a downburst, the vertical wind must exceed 20 m/s (Srivastava[28]). It has also been noted that a key difference between the downburst and other downdraft phenomena is that a downburst continues to gain strength and intensity as it descends below the cloud base and that the maximum intensity is realised close to the Earth's surface. The downdraft has also been called a *reverse tornado* because rather than drawing air up through a short column into a storm cloud, the downburst creates an intense, narrow downward flow of air below the cloud which then spreads outwards radially. However, in the tornado the airflow has a large rotational component, whereas the downburst does not have a rotating wind component.

The formation of a downburst begins with a mild downdraft in a region with strong precipitation, for example in a storm cloud (rain and/or ice). The downdraft is created by an air mass having higher density (*negative buoyancy*) with respect to the surrounding atmosphere; this causes the air mass to descend. The negative buoyancy that drives the downdraft is the result of both cooling of the air mass through melting and evaporation of precipitation and the precipitation loading present in the air mass. It is possible that the downdraft's intensity increases significantly as it descends, to the point where its descent velocity exceeds ∼20 m/s; at higher descent speeds a downdraft is considered to have a magnitude sufficient to be considered a downburst[28]. The main condition required to cause this effect is a deep layer of air with a significant temperature lapse rate. There is a variety of combinations of temperature lapse rates (also referred to as temperature stratifications) and precipitation rates that can cause sufficient vertical velocity to create a downburst. A dry downburst typically has a temperature lapse rate approaching the dry adiabatic. In contrast with more stable temperature stratifications, such as approaching the wet adiabatic, heavy precipitation rates are required to drive the downdraft and increase its intensity. It is also possible for a downburst to be created and driven entirely below the cloud base. In such circumstances the precipitation loading and evaporative cooling below the cloud are sufficient to drive the downburst.

The problem of storm penetration must be addressed for safety reasons and for certification of the aircraft for operation under realistic weather conditions.

A number of accidents due to downbursts have now been documented. Zhu and Etkin[29] produced a model for the wind field in a downburst, based on a distribution of dipoles, and provided analytical formulas for the head-wind, side-wind and downdraft components. Since then, a number of developments have been produced. Zhao and Bryson[30;31] proposed models for optimal guidance of an aircraft through the downburst. In addition, a number of papers have addressed escape strategies[32–34] and avoidance[35].

**DOWNBURST WIND MODEL.** We assume the wind-field model proposed by Zhu and Etkin[29]. The model is inviscid and is based on a circular distribution of doublet $\sigma$ at height $h$. A mirror image of the singularity distribution is assumed to force the boundary condition of no through-flow at the ground. According to the model, the velocity induced at a point $x$, $y$, $z$ in a ground-based reference system by a source distribution of intensity $\sigma$ per unit area is

$$u = -\frac{3}{4\pi}\int_S \sigma(\xi,\eta)\frac{(z-h)(x-\xi)}{r^5}\,d\xi d\eta, \tag{13.80}$$

$$v = -\frac{3}{4\pi}\int_S \sigma(\xi,\eta)\frac{(z-h)(y-\eta)}{r^5}\,d\xi d\eta, \tag{13.81}$$

$$w = -\frac{1}{4\pi}\int_S \sigma(\xi,\eta)\left[\frac{3(z-h)^2}{r^5}-\frac{1}{r^3}\right]d\xi d\eta, \tag{13.82}$$

where $r$ is the distance

$$r^2 = (x-\xi)^2 + (y-\eta)^2 + (z-h)^2, \tag{13.83}$$

and $\xi, \eta$ is the coordinate of the source $\sigma$. To force the boundary condition on the ground, these velocities are augmented by the velocity induced by a source distribution $-\sigma$ at altitude $-h$. The value of these surface integrals depends on the source distribution. However, the solution is assumed to be axi-symmetric, and the source distribution $\sigma(\xi,\eta)$ is only a function of the radial distance from the centre of the downburst

$$\sigma = \sigma_m\left[1 - \sin(r/R)\right]. \tag{13.84}$$

In Equation 13.84 $\sigma_m$ is the maximum value of the momentum source. The parameters of the model are 1) the height of the source distribution, 2) the area of the area distribution, and 3) the source-strength distribution. These parameters can be chosen to fit reasonable downbursts. For the integration of Equations 13.80 to 13.82, we perform a transformation to cylindrical coordinates. The value of the source $\sigma_m$ is calculated iteratively, so that the condition of downburst velocity $\overline{w}$ is enforced at a specified altitude below the cloud base. We follow this procedure:

- Set the downwash speed $\overline{w}$ (a negative value) at the core of the downburst, at an altitude equal to half the cloud base.
- Use the bisection method with a starting guess of $\sigma_{m1} = 1$ and $\sigma_{m2} = 10^5$. These values of the source strength should bracket the solution. Upon convergence, we have forced the wind speed to be equal to $\overline{w}$ at the specified point.

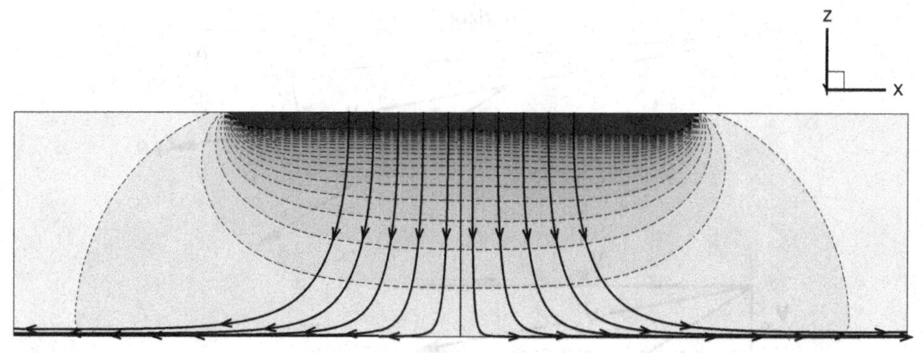


Figure 13.15. Downburst field on the plane $y = 0$; cloud base at 1,000 m, $R = 1{,}500$ m.

Figure 13.15 shows the downburst field on the vertical plane $y = 0$ through the downburst. The graph shows the streamlines from the cloud base. The contour lines represent the magnitude of the wind velocity. Note that with this wind model the velocities are unrealistically high just below the cloud base; the wind components are singular as $r \to 0$. Alternative wind models exist, including vortex rings[36] and general axi-symmetric models[37].

Another example of a field solution is shown in Figure 13.16a. This case refers to an aircraft flying level with KTAS = 250 at an altitude 1,000 m (3,048 feet) through the centre of the downburst. The figure shows both the horizontal and the vertical velocity components created by the wind model. The maximum downwash is right below the cloud base (centre of the downburst), as expected. The corresponding change in effective angle of attack is shown in Figure 13.16b.

### 13.7.1 Aircraft Manoeuvre in a Downburst

The position $\{x, y, z\}$ of the aircraft is calculated with respect to a reference system on the ground. We assume that the aircraft moves on the vertical plane and will have a position $\{x, z\}$, where the $x$-axis is in the direction of advancing flight and $z$ is a vertical

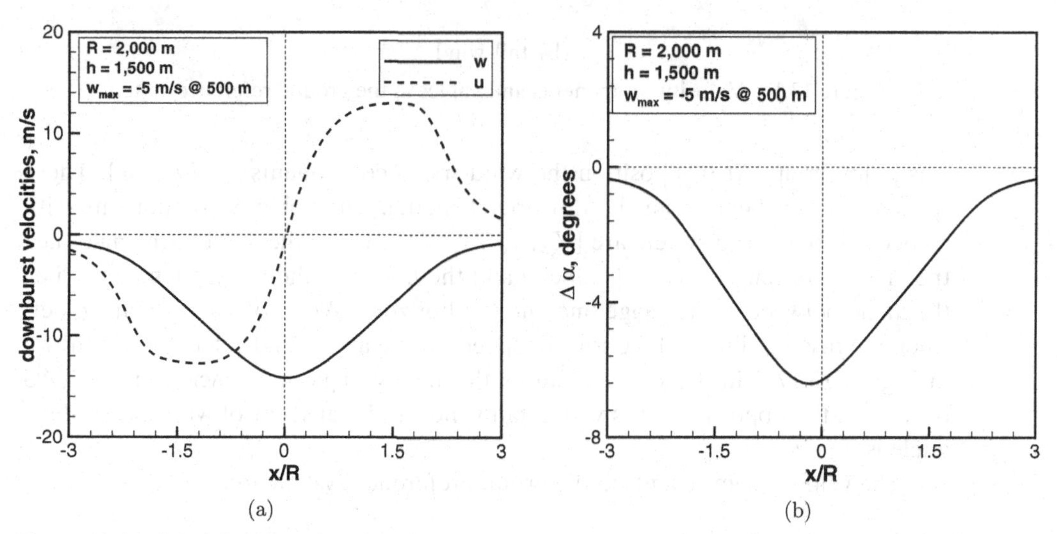


(a) (b)

Figure 13.16. Wind field in a downburst and change in wing's angle of attack in a downburst.

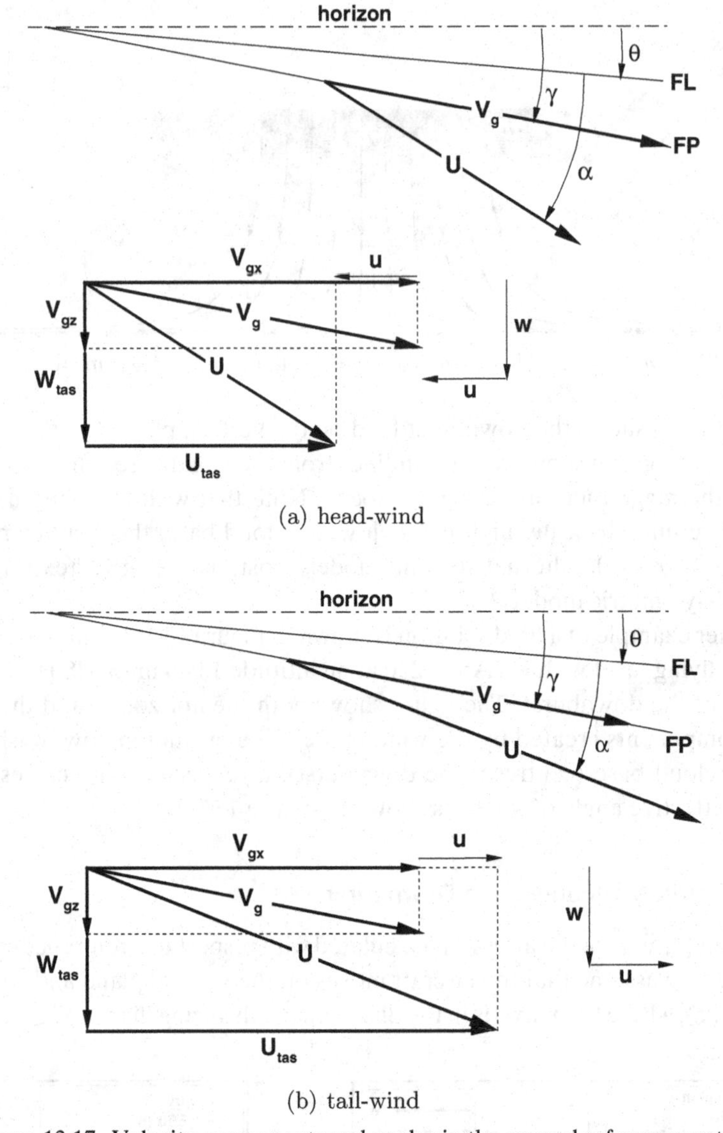


(a) head-wind

(b) tail-wind

Figure 13.17. Velocity components and angles in the ground reference system.

axis pointing up. At this position the wind-speed components are $\{u, v, w\}$. These speeds are calculated from the downburst model. The velocity components with respect to the ground system are $\{V_{gx}, V_{gy}, V_{gz}\}$. We call $\gamma$ the flight-path angle (i.e., the angle between the ground velocity and the horizon) and $\theta$ the pitch angle (i.e., the angle between the fuselage line and the horizon). We also call $\alpha$ the angle between the fuselage line and the true air speed. These angles and velocities are shown in Figure 13.17. The head wind causes the inflow angle $\alpha$ to increase; the TAS decreases; the opposite occurs with a tail wind. In the absence of winds, the inflow angle is $\alpha = \gamma - \theta$.

The velocity components in the ground reference system are:

$$V_{gx} = V_g \cos\gamma, \qquad V_{gy} = 0, \qquad V_{gz} = V_g \sin\gamma. \tag{13.85}$$

The true air speed has the following components

$$U_{tas} = V_{gx} + u, \qquad V_{tas} = v, \qquad W_{tas} = V_{gz} + w \tag{13.86}$$

with

$$\text{TAS} = U = \left[U_{tas}^2 + V_{tas}^2 + W_{tas}^2\right]^{1/2}. \tag{13.87}$$

The flight-path angle is

$$\gamma = \tan^{-1}\left(\frac{V_{gz}}{V_{gx}}\right). \tag{13.88}$$

The inflow angle $\alpha$ is

$$\alpha = \tan^{-1}\left(\frac{W_{tas}}{U_{tas}}\right) - \theta. \tag{13.89}$$

If the attitude is $\theta = 0$ in absence of atmospheric winds, the aircraft provides a lift $C_{L\theta o} > 0$, this is equivalent to having a *zero-lift* angle

$$\alpha_o = -\frac{C_{L\theta o}}{C_{L\alpha}}. \tag{13.90}$$

Thus, the lift coefficient can be written as

$$C_L = C_{L_\alpha}(\alpha - \alpha_o). \tag{13.91}$$

As the aircraft enters the downburst, there is a change in inflow angle, as illustrated in Figure 13.17, and thence a change in lift and drag. In particular, an increase in lift would cause the aircraft to decrease the descent rate and possibly even climb. If the inflow angle $\alpha$ is to be maintained, whilst maintaining the pitch angle (*pitch guidance*), a change in speed is required. From Equation 13.89, we must set the constraint

$$\tan^{-1}\left(\frac{W_{tas}}{U_{tas}}\right) = \tan^{-1}\left(\frac{V_{gz} + w}{V_{gx} + u}\right) = \tan^{-1}\left(\frac{V_{gz}}{V_{gx}}\right)_{t=0}; \tag{13.92}$$

where $t = 0$ denotes the time before entering the downburst event. Equation 13.92 contains two unknowns: $V_{gx}$ and $V_{gz}$; if we attempt to maintain the same ground speed schedule as required in a conventional approach, we only need to control the descent rate $V_{gz}$. In a typical downburst, $w > u$ (as shown by the Etkin wind model), along with $V_{gz} > V_{gx}$. Upon entering the downburst, the numerator of the argument increases in absolute value, whilst the denominator decreases. Therefore, to keep the effective inflow angle $\alpha$ constant, the true descent rate must decrease whilst the aircraft accelerates. Past the centre of the downburst, the opposite event takes place, and the aircraft must slow down whilst increasing its descent rate, although a too rapid loss in altitude would compromise safety at a very critical time.

An alternative guidance mode is a constant flight-path angle ($\gamma$-*guidance*). This mode is specified by the following equation:

$$\gamma = \tan^{-1}\left(\frac{W_{tas} - w}{U_{tas} - u}\right) = \text{const.} \tag{13.93}$$

Upon entering a downburst, the numerator of the argument decreases ($W_{tas} < 0$; $w < 0$), whilst the numerator increases (the aircraft encounters a head wind, $u < 0$).

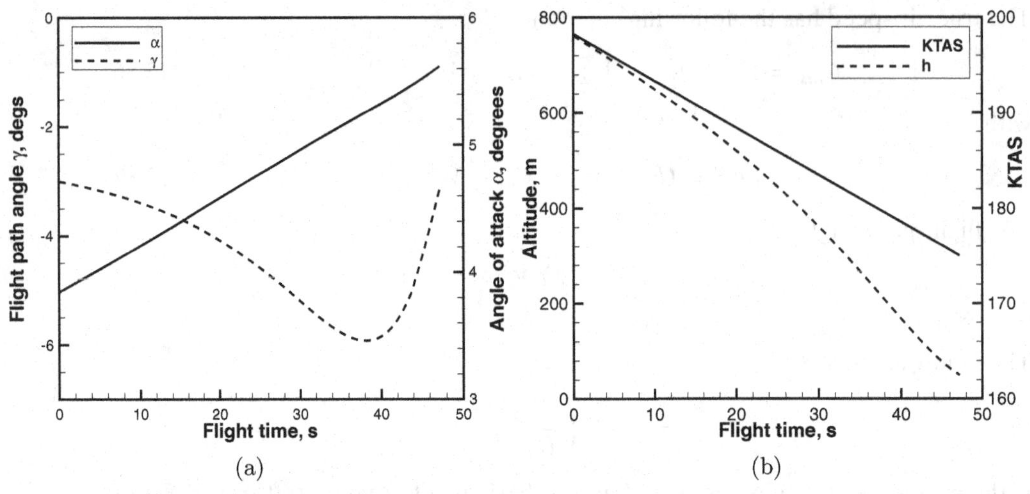


(a) (b)

Figure 13.18. Flight in a downburst with deceleration to a target approach speed.

The flight-path angle is maintained if the aircraft slows down, something that is achieved by reducing the thrust. Another form of control is based on acceleration. Figure 13.17 shows that the airspeed tends to decrease upon entering the downburst and it increases on the way out. A control law that would ensure constant TAS would require an acceleration and then a deceleration. Examples of these types of controls are found in the technical literature[38].

### 13.7.2 Case Study: Flight in a Downburst

The analysis of the downburst relies on several parameters, including the height of the cloud base, the diameter, the downdraft at the core of the downburst and the initial conditions of the aircraft (including whether the aircraft flies through the axis of the downburst or sideways). The overall performance depends on the control laws, which include at least a TAS-control, flight-path control and angle-of-attack control, with various constraints (on descent rate, true air speed, normal load factor).

We provide an example here, which was simulated for the Airbus A320-200-CFM landing in the presence of a downburst. The case refers to an initial mass $m =$ 56,400 kg, with an initial TAS $= V_{green} + 5$ kt at 2,500 m above sea level. The airplane is forced to maintain a fixed attitude, equal to the value in absence of winds. The characteristics of the downburst are 2,500 m diameter with a cloud base at 1,000 m and a maximum downdraft of $-10$ m/s on the axis at 500 m below the cloud base.

Figure 13.18 shows some characteristics of the trajectory when the airplane is guided to a target TAS, equal to an approach speed on the ground. Figure 13.19 shows a similar example, but with the airplane attempting to maintain a TAS within 95% of the initial value. In both cases, the airplane sinks below the downburst, and unless it is guided by optimal control laws, it may end up on the ground before reaching the runway. The sinking effect is shown by the flight-path angle becoming steeper as the airplane enters the core of the downburst. The results have been obtained with the flight-mechanics model of the `FLIGHT` program.

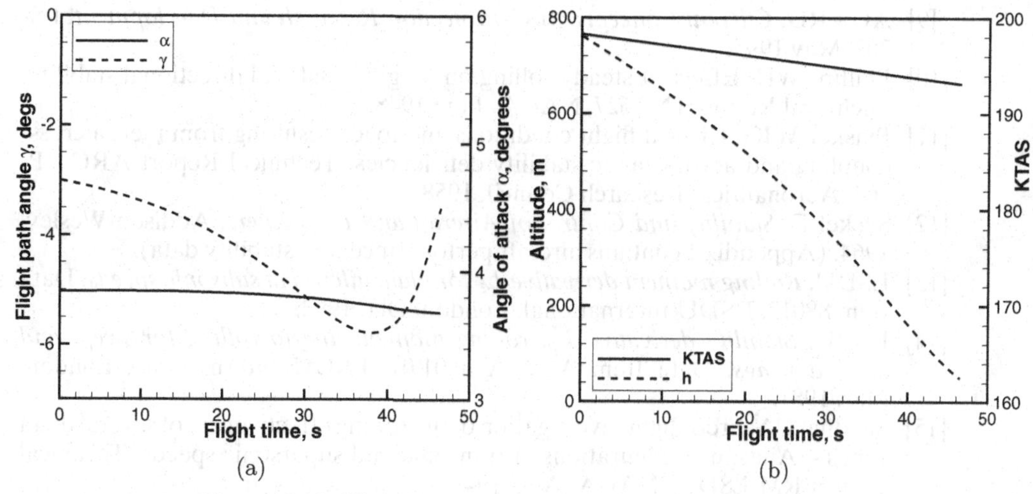


Figure 13.19. Flight in a downburst with TAS-control.

## Summary

In this chapter we considered simple aircraft manoeuvres. We discussed both powered and unpowered turns. We demonstrated that at constant thrust the aircraft would descend, due to an increase in drag and the effects of apparent centrifugal acceleration. We further demonstrated that these accelerations can exceed the resistance of the human body (sustainable *g*-loads). Thus, there are limits imposed not only by the structures and the dynamics of the airplane but also by the pilot. Clearly an unmanned vehicle would not have this problem, and it is easy to imagine new high-performance aircraft systems capable of even more rapid turns. Other high-performance manoeuvres include roll (with limitations imposed by inertial coupling). At a more mundane level, we have discussed the manoeuvre envelopes of transport-type aircraft ($V$-$n$ diagram). We concluded this chapter with a discussion of flight in a downburst, which is an adverse weather condition sometimes encountered at low altitude near an airfield.

## Bibliography

[1] AGARD WG 19. *Operational Agility*, volume AR-314. AGARD, April 1994.
[2] Bryson AE and Hedrick JK. Minimum time turns for a supersonic airplane at constant altitude. *J. Aircraft*, 8(3):182–187, 1971.
[3] Hedrick JK and Bryson AE. Three-dimensional minimum-time turns for supersonic aircraft. *J. Aircraft*, 9(2):115–121, Mar. 1972.
[4] Kelley HJ. Differential-turning optimality criteria. *J. Aircraft*, 12(1):41–44, 1975.
[5] Kelley HJ. Differential-turning tactics. *J. Aircraft*, 12(2):930–935, Feb. 1975.
[6] Webb PMD. *Bioastronautics Data Book*, volume SP-3006. NASA, 1964.
[7] LeBlaye P. *Human Consequences of Agile Aircraft*, RTO-TR-015, chapter Agility: Definitions, Basic Concepts and History. NATO, Jan. 2001.
[8] Jaslow H. Spatial disorientation during a coordinated turn. *J. Aircraft*, 39(4): 572–576, 2002.

[9] AGARD. *Current Concepts on G-Protection Research and Development*, LS-202, May 1995.
[10] Phillips WH. Effect of steady rolling on longitudinal and directional stability. Technical Report TN 1627, NACA, June 1948.
[11] Pinsker WJG. Critical flight conditions and loads resulting from inertia cross-coupling and aerodynamic stability deficiencies. Technical Report ARC CP-404, Aeronautical Research Council, 1958.
[12] Seckel E. *Stability and Control of Aircraft and Helicopters*. Addison-Wesley, 1964. (Appendix I contains aircraft performance and stability data).
[13] ESDU. *Rolling moment derivative $L_\xi$ for plain aileron at subsonic speeds*. Data Item 88013. ESDU International, London, Oct. 1992.
[14] ESDU. *Stability derivative $L_p$, rolling moment due to rolling for swept and tapered wings*. Data Item Aero A.06.01.01. ESDU International, London, Mar. 1981.
[15] Sandhal CA. Free-flight investigation of the rolling effectiveness of several delta wings – Aileron configurations at transonic and supersonic speeds. Technical Report RM-L8D16, NACA, Aug. 1948.
[16] Myers BC and Kuhn RE. High subsonic damping-in-roll characteristics of a wing with quarter-chord line swept 35° and with aspect-ratio 3 and taper ratio 0.6. Technical Report RM-L9C23, NACA, May 1949.
[17] Myers BC and Kuhn RE. Effects of Mach number and sweep on the damping-in-roll characteristics of wings of aspect-ratio 4. Technical Report RM-L9E10, NACA, June 1949.
[18] Anderson SB, EA Ernst, and RD van Dyke. Flight measurements of the wing-dropping tendency of a straight-wing jet airplane at high subsonic Mach numbers. Technical Report RM-A51B28, NACA, April 1951.
[19] Stone DG. A collection of data for zero-lift damping in roll of wing-body combinations as determined with rocket-powered models equipped with roll-torque nozzles. Technical Report TN 3955, NACA, April 1957.
[20] Sanders CE. Damping in roll of models with 45°, 60°, 70° delta wings determined at high subsonic, transonic and supersonic speeds with rocket-powered models. Technical Report RM-L52d22a, NACA, June 1952.
[21] Malvestuto F, Margolis K, and Ribner HS. Theoretical lift and damping in roll at supersonic speeds of thin sweptback tapered wings with streamwise tips, subsonic leading edges, and supersonic trailing edges. Technical Report R-970, NACA, 1950.
[22] Jones AL and Alksne A. A summary of lateral-stability derivatives calculated for wing plan forms in supersonic flow. Technical Report R-1052, NACA, 1951.
[23] Toll TA. Summary of lateral-control research. Technical Report R-868, NACA, 1947.
[24] Skow AM. Agility as a contributor to design balance. *J. Aircraft*, 29(1):34–46, Jan. 1992.
[25] Rathert G, Rolls L, Winograd L, and Cooper G. Preliminary flight investigation of the wing dropping tendency and lateral control characteristics of a 35° swept-back airplane at transonic Mach numbers. Technical Report RM-A50H03, NACA, Sept. 1950.
[26] Chambers JR and Hall RM. Historical review of uncommanded lateral-directional motions at transonic conditions. *J. Aircraft*, 41(3):436–447, May 2004.
[27] Hall RM, Woodson SH, and Chambers JR. Accomplishments of the abrupt-wing-stall program. *J. Aircraft*, 42(3):653–660, May 2005.
[28] Srivastava RC. A model of intense downdrafts driven by the melting and evaporation of precipitation. *J. Atmospheric Sciences*, 44(13):1752–1773, 1987.
[29] Zhu S and Etkin B. Model of the wind field in a downburst. *J. Aircraft*, 22(7):595–601, July 1985.

[30] Zhao Y and Bryson AE. Optimal paths through downbursts. *J. Guidance, Control and Dynamics*, 13(5):813–818, Oct. 1990.
[31] Zhao Y and Bryson AE. Control of an aircraft in a downburst. *J. Guidance, Control and Dynamics*, 13(5):819–823, Sept. 1990.
[32] Bobbitt RB and Howard RM. Escape strategies for turboprop aircraft in microburst windshear. *J. Aircraft*, 29(5):745–752, Sept. 1992.
[33] Dogan A and Kabamba PT. Escaping microburst with turbulence: Altitude, dive and guidance strategies. *J. Aircraft*, 37(3):417–426, May 2000.
[34] Mulgund SS and Stengel RF. Target pitch angle for the microburst escape maneuver. *J. Aircraft*, 30(6):826–832, Nov. 1993.
[35] de Melo DA and Hansman RJ. Analysis of aircraft performance during lateral maneuvering for microburst avoidance. *J. Aircraft*, 28(12):837–842, Dec. 1991.
[36] Ivan M. A ring-vortex downburst model for flight simulations. *J. Aircraft*, 23(3):232–236, Mar. 1986.
[37] Visser HG. Optimal lateral-escape maneuvers for microburst wind-shear encounters. *J. Guidance, Control & Dynamics*, 14(6):1234–1240, 1994.
[38] Miele A, Wang T, and Bowles RL. Acceleration, gamma and theta guidance for abort landing in a windshear. *J. Guidance, Control, & Dynamics*, 12(6):815–821, Nov. 1989.

## Nomenclature for Chapter 13

| | | |
|---|---|---|
| $a$ | = | speed of sound |
| $A$ | = | wing area |
| $\mathcal{AR}$ | = | wing aspect-ratio |
| $b$ | = | wing span |
| $b_t$ | = | thrust arm |
| $c$ | = | wing chord |
| $c_1$ | = | constant coefficient |
| $C_l$ | = | rolling-moment coefficient |
| $C_{D_o}$ | = | profile-drag coefficient |
| $C_L$ | = | lift coefficient |
| $C_{L\theta o}$ | = | lift coefficient at zero attitude, Equation 13.90 |
| $C_{l_p}$ | = | damping-in-roll coefficient |
| $C_{l_\xi}$ | = | non-dimensional aileron effectiveness |
| $C_{l_\beta}$ | = | rolling-moment derivative with respect to side-slip |
| $C_{L_\alpha}$ | = | lift-curve slope |
| $C_{L_{max}}$ | = | maximum-lift coefficient |
| $C_N$ | = | normal-force coefficient |
| $C_M$ | = | pitching moment |
| $D$ | = | drag force |
| $f$ | = | generic function, Equation 13.39 |
| $f_j$ | = | thrust-specific fuel consumption |
| $g$ | = | acceleration of gravity |
| $h$ | = | altitude |
| $h_o$ | = | arbitrary starting altitude |
| $I_x$ | = | roll moment of inertia |
| $k$ | = | lift-induced drag factor |
| $K_s$ | = | stall margin |

$L$ = lift force
$\mathcal{L}$ = rolling moment
$\mathcal{L}_p$ = aerodynamic moment due to roll rate
$\mathcal{L}_\xi$ = rolling moment due to aileron deflection
$L_p$ = damping in roll
$L_\xi$ = aileron effectiveness
$m$ = mass
$M$ = Mach number
$M_D$ = design dive Mach number
$\mathcal{M}$ = yawing moment
$n$ = normal load factor
$\mathcal{N}$ = pitching moment
$\dot{m}_f$ = fuel flow
$p$ = pressure
$\mathrm{p}$ = roll rate (or angular velocity)
$\mathrm{p}_{max}$ = maximum value of roll rate (asymptotic)
$P$ = power
$r$ = distance; distance in polar coordinates
$R$ = radius of downburst
$t$ = time
$T$ = net thrust
$u,v,w$ = wind-velocity components in downburst
$U$ = air speed
$v_c$ = climb rate
$V_D$ = design dive speed
$V_F$ = maximum design speed with flaps extended
$V_g$ = ground speed
$v_s$ = descent rate
$V_S$ = stall speed
$V_{S1}$ = stall speed, 1-*g* flight, flaps retracted
$W$ = weight
$u, v, w$ = wind-velocity components in wind model
$\overline{w}$ = average downwash
$x_{ac}$ = distance of wing's aerodynamic centre from leading edge
$x_{cg}$ = distance of centre of gravity from nose
$x, y, z$ = Cartesian coordinates (ground-based)

### Greek Symbols

$\alpha$ = angle of attack; inflow angle
$\alpha_o$ = zero-lift angle of attack/inflow
$\beta$ = side-slip angle
$\gamma$ = flight-path angle; ratio between specific heats, Equation 13.33
$\delta$ = flap or slat angle
$\eta$ = induced-drag coefficient for supersonic model
$\eta$ = coordinate in downburst-wind model
$\theta$ = pitch attitude

$\vartheta$ = angle in polar coordinates
$\xi$ = aileron deflection; coordinate in downburst-wind model
$\rho$ = air density
$\sigma$ = doublet distribution in wind model
$\sigma_m$ = maximum value of $\sigma$
$\tau$ = time constant, Equation 13.71
$\phi$ = bank angle
$\chi$ = radius of curvature
$\psi$ = turn angle

### Subscripts/Superscripts

$[\cdot]_{ail}$ = aileron
$[\cdot]_b$ = body-conformal
$[\cdot]_f$ = fuel
$[\cdot]_{md}$ = minimum drag
$[\cdot]_{ht}$ = horizontal tail
$[\cdot]_t$ = turn
$[\cdot]_{tas}$ = true air speed
$[\cdot]_w$ = wind
$[\cdot]_p$ = reference respect to roll rate
$[\cdot]_\xi$ = reference to aileron deflection

# 14 Thermo-Structural Performance

## Overview

This chapter deals with thermodynamic and structural models of various aircraft sub-systems. Two main aero-thermodynamic problems are discussed: cold-weather operation (§ 14.1), including aircraft icing (§ 14.1.1), and fuel temperature in the aircraft tanks (§ 14.3 and § 14.4). Basic characteristics of aircraft fuels are discussed in § 14.2. A lumped-mass model for thermo-structural performance of aircraft tyres is presented in § 14.5 and aims at the rapid prediction of tyre temperatures. This is useful at high speeds, heavy aircraft and rejected take-off. The final problem is that of jet blast (§ 14.6), which is of relevance during ground operations, for safety of infrastructure and ground personnel.

**KEY CONCEPTS:** Cold-Weather Operations, Aircraft Icing, Aviation Fuels, Fuel Temperature, Tyre Heating, Jet Blast.

## 14.1 Cold-Weather Operations

Cold-weather operations cause a number of problems on the ground, at take-off, at landing and in flight. There are a variety of issues of relevance, including icing, in-flight ice protection, de-icing on the ground, fuel characteristics at low temperatures and aircraft performance on contaminated runways (including breaking performance, precipitation drag, aquaplaning). We will not deal with ice protection, de-icing and the atmospheric causes of icing because they are beyond the scope of flight performance. Our presentation will focus on icing effects on performance characteristics.

The problem of aircraft icing is exemplified by a number of fatal crashes that over the years have plagued commercial aviation. Aircraft icing has its beginning in World War II, with bombers flying at high altitudes and at all-weather conditions. Ever since, a large database of icing cases has been collected. Due to the complexity of icing phenomena, it became clear that aircraft could not be certified for all icing conditions. Today, icing conditions for certification must make reference to the Federal Aviation Regulations FAR § 25, Appendix C (*Continuous Maximum Icing* and *Maximum Intermittent Icing*).

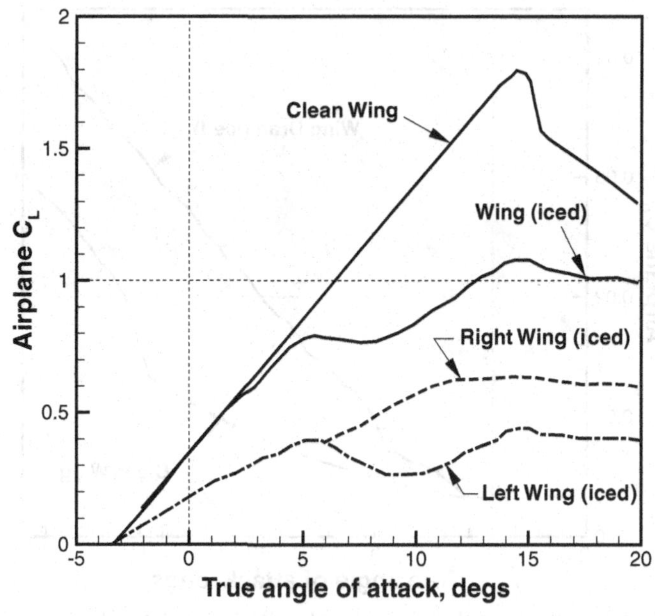


Figure 14.1. Icing effects on the wing lift of the ATR72; data adapted from Ref.[4].

Icing can occur on various parts of the vehicle, although icing on the wing, control surfaces and propeller blades (if present) is particularly severe. Typical consequences include an increase in drag and a decrease in lift – not necessarily by the same amount on right and left wings. Differential changes in aerodynamic parameters can lead to flight conditions beyond the normal flight envelope of the airplane.

Modern aircraft are certified to operate in icing conditions, within reasonable limits. Icing effects can be traced by the Flight Data Recorder (FDR), either directly or indirectly. There is abundant documentation of aircraft icing in the technical literature, from the physics of ice formation to atmospheric effects and aircraft operations. It is not within the scope of this chapter to review the whole subject, which is addressed in the appropriate literature[1;2]. Ice formation on the ground and in flight are discussed by Asselin[3].

To clarify the icing severity on airplane lift and drag, we show the recorded data for a cargo ATR72-500, adapted from Caldarelli[4], in Figure 14.1 and Figure 14.2. This case refers to a fatal crash. A peculiar aspect of this incident is the difference between left- and right-wing lift, alongside a rapid decrease in drag. For the ATR, a $\Delta C_L = -0.5$ causes an increase in stall speed by 10 kt with flaps at 15 degrees. Most critically, this loss in lift is caused by very small ice accretions.

As the aircraft begins to stall, its lateral and longitudinal stability are compromised. With regard to the former problem, wing stall occurs first at the inboard sections of the wing and proceeds outwards, thereby minimising the destabilising effects of rolling moment. However, this scenario holds only in case of symmetric ice accretion, which, as we have seen, might not be the case. With regard to the latter problem, there can be an abrupt change in pitching moment as well as tail-plane icing.

Figure 14.3 shows some typical wing conditions, with and without icing. Figure 14.3a displays the normal operation of the wing, when the airflow is smooth and the boundary layer is attached. Figure 14.3b shows a case of leading-edge icing, whose

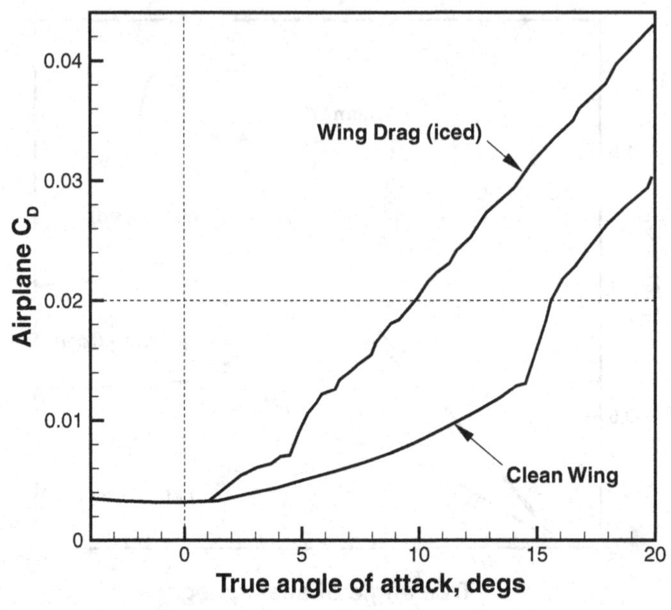


Figure 14.2. Icing effects on the wing drag of the ATR72; data adapted from Ref. [4].

effect is to change the shape of the wing section and hence to displace the air flow away from the solid wall. The boundary layer thickens and eventually separates in the aft portion of the chord. Figure 14.3c displays the case of a landing configuration in which icing is present on the leading-edge slat. Although this is potentially a dangerous condition, if there is no icing on the wing itself, it is possible that the wing boundary layer remains attached as a result of the flow through the slat-wing gap; this flow energises weak boundary layers on the suction side of the wing. Finally, Figure 14.3d shows the effect of morning frost, which typically extends through most of the suction side. The surface is modified, particularly in roughness, and boundary layer separation is likely.

The case of Figure 14.3b is used for certification purposes. The effect of aft boundary layer separation does not affect the circulation by a great deal, unless the wing is forced to operate at large angles of attack. The aircraft would be certified to land under conditions displayed in Figure 14.3c because the wing lift is not severely affected by icing on the slat. The final case is more problematic and can result in large-scale separation and consequent loss in circulation, that is, lift. This case is not certified for safe flight.

### 14.1.1 Aircraft Icing

Icing occurs with various degrees of severity, defined by the aviation authorities as *trace icing*, *light icing*, *moderate icing* and *severe icing*. Severe icing occurs when de-icing systems fail to remove the hazard. In this case, immediate flight diversion is required.

The atmosphere contains water vapour in variable proportions. At each temperature and pressure, the maximum water vapour is limited by the saturation point (or dew point) condition. At saturation, the water condenses in rain or

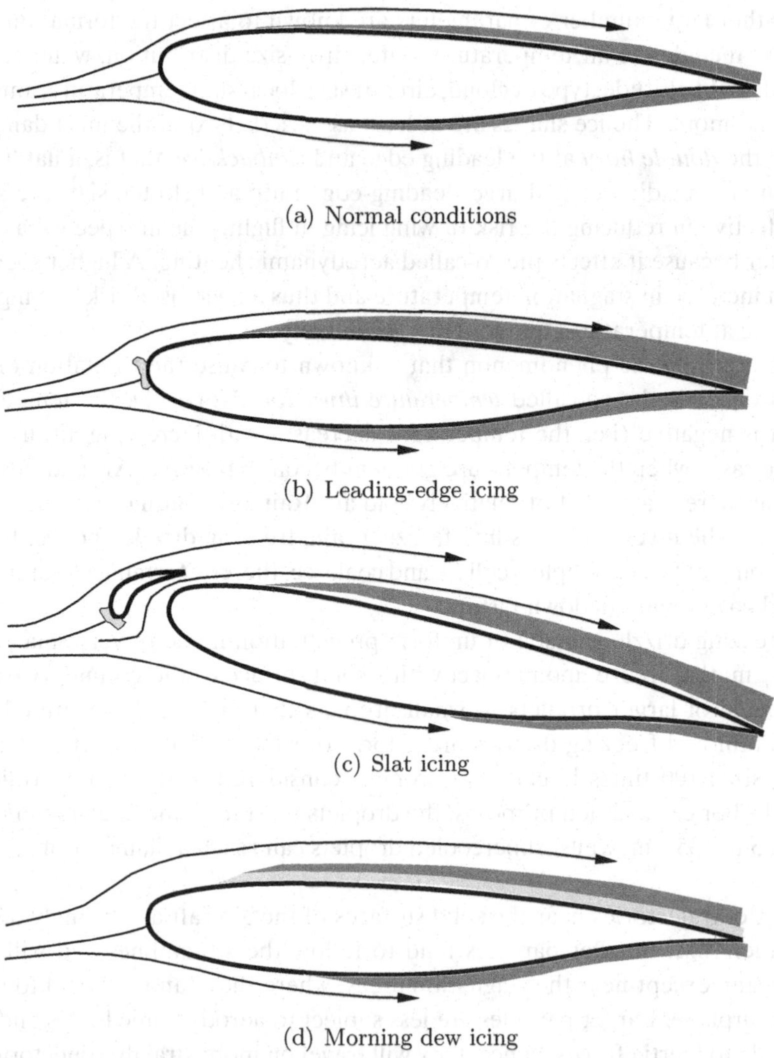


Figure 14.3. Some typical wing-icing conditions.

ice crystals. Both rain and ice form clouds. The water vapour at saturation point increases with the air temperature. Saturation is attributed to two important phenomena: lifting of warm air (due to weather instability or ground topography) and air cooling at night with a clear sky. The lower temperature causes at least part of the vapour to precipitate into water or ice. However, there are conditions when the condensation or freezing point may be exceeded without precipitation taking place. This phenomenon is called *supercooling*. This event is known to be unstable, and once a perturbation is introduced, condensation or freezing is almost instantaneous. This is the basic mechanism that causes aircraft icing. In practice, the range of air temperatures that cause icing is $-40°C < T < -10°C$, at altitudes from ground level to ~3,048 m (10,000 feet, FL-100). The parts of the aircraft most exposed to icing are the protruding parts, including lifting surfaces, leading-edge slats, hinges, engine intakes, vertical and horizontal stabilisers and so on.

A rather large number of parameters are known to affect the formation of the ice. These include the air temperature, water-drop-size distribution, water content, air speed, flight altitude, type of cloud, aircraft size, local skin temperature, anti-icing system and more. The ice shapes are at least as varied. Two of the most dangerous ones are the *double horn* at the leading edge and *runback ice*, that is, a flat lump of ice behind the leading edge. Larger leading-edge radii and slotted slats are known to be effective in reducing the risk of wing icing in flight. The air speed is a critical parameter because it affects the so-called aerodynamic heating. A higher speed can cause an increase in stagnation temperature and thus a decreased risk of wing icing. Thus, icing at temperatures above −10°C is unlikely.

One atmospheric phenomenon that is known to cause the formation of large droplets of ice is the so-called *temperature inversion*. Normally, the temperature gradient is negative (i.e., the temperature decreases with increasing altitude), but there are cases when the temperature gradient becomes positive. Around the inversion point there is a pocket of relatively cold air. Rain originating from the warmer layer above the inversion turns into freezing rain, freezing drizzle and ice. It is not uncommon that frozen droplets collide and coalesce, thereby forming larger droplets that are heavier and fall down faster.

A freezing drizzle is made of uniform precipitation made of very small drops, $d < 500\,\mu m$, that freeze upon impact with a solid surface or the ground. A freezing rain is made of larger droplets or small droplets that are widely separated. Both freezing rain and freezing drizzles are a form of supercooled droplets, which can grow to sizes 100 times larger than droplets considered in the icing certification standards. For certification purposes, the droplets must have mean diameters in the range of 5 to 135 $\mu$m, whilst supercooled droplets can reach a diameter of 2,000 $\mu$m (2 mm).

Droplet trajectories near the solid surfaces of the aircraft are strongly dependent on their size. Smaller particles tend to follow the streamlines and will travel downstream, except near the stagnation areas, where they can be forced to impact with the airplane. Larger particles are less subject to aerodynamic forces and more susceptible to inertia forces; hence, they will travel on more straight trajectories and upon impact will generate larger areas of ice coverage. Inertia forces grow with the droplet diameter $d^3$, whilst aerodynamic forces grow as $d^2$. Thus, as the droplets become larger, their dynamics is dominated by their inertia. This scenario assumes that upon impact the droplets crash on the surfaces and essentially coalesce around their impact area. The in-flight icing process itself is a result of a balance between the freezing of supercooled air on the aircraft and the removal of ice due to erosion, evaporation and sublimation.

Another icing phenomenon is the *cold soak*, in which low skin temperature comes into contact with moist air at air temperatures that would normally be above freezing. For example, it has been found in the past that the practice of tankering requires the aircraft to land at the destination airport with so much fuel that parts of it are in contact with the upper skin of the wing. Because the temperature of this fuel is below freezing at normal landing conditions, the wings become heat sinks and freeze the moisture in the air. Hence, wing icing may occur at air temperatures well above 0°C. Icing conditions are more frequent than aircraft icing. Two important

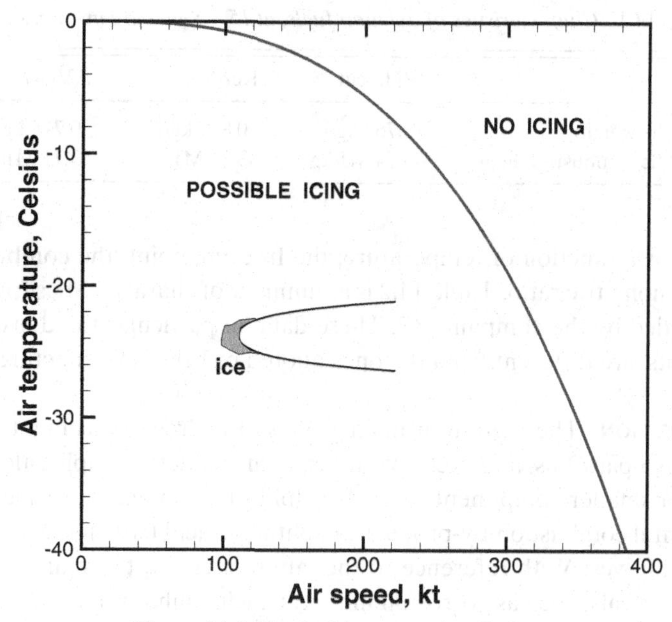


Figure 14.4. Typical-ice accretion limits.

changes in aircraft operations are increase in air speed and change of flight altitude, as shown in Figure 14.4. Ground icing, including the thin layer of morning frost, must be cleared before any aircraft operation. Even a small ice thickness can cause unwanted aerodynamic degradation during the critical ground roll, take-off and initial climb.

## 14.2 Aviation Fuels

In this section we deal briefly with the main characteristics of aviation fuel, including types of fuel, specifications, composition and fuel properties. Aviation fuel is a refined product of crude oil; about 12% of aviation fuel is derived from each barrel of crude. There are two classes of fuels within this category: 1) aviation gasoline (also called AvGas), used for reciprocating engines; and 2) turbine fuels, used for gas turbine engines. We will refer only to the latter category.

There are various commercial grades of turbine fuels, including Jet-A, Jet-A1, Jet-B, JP-4, JP-5, JP-7, JP-8 and JP-9. The most widespread fuels are Jet-A and Jet-A1; Jet-B (also called wide-cut) has a lower freezing temperature. The remaining specifications are U.S. military grades. These fuels are fully specified by international standards*, which define the full array of physical properties and characteristics (density, viscosity, volatility, composition, combustion, corrosion, additives, toxicity, and so on).

Technical information on aviation fuels is available in Goodger and Vere[5], in addition to some international standards. Table 14.1 shows some basic data of aviation fuels. The most important properties for performance calculations are the

* United States: ASTM D1655; IATA: ADD76-1; United Kingdom: DERD.2494; France: AIR 3405/C (for Jet-A1).

Table 14.1. *Characteristics of aviation fuels, at 15°C; data are averages*

| Fuel | Wide-cut | Kerosene | AVgas |
|---|---|---|---|
| Specific weight | 0.762 kg/l | 0.810 kg/l | 0.715 kg/l |
| Specific combustion heat | 43.54 MJ/kg | 43.28 MJ/kg | 43.71 MJ/kg |

fluid density as a function of temperature, the freezing point, the combustion heat and the stoichiometric ratio. Table 14.2 is a summary of characteristics for Jet-A and Jet-A1 provided by the company BP. These data, in particular the density and the freezing points, are different from the ones quoted by other manufacturers.

**FUEL COMPOSITION.** There are four main groups of hydrocarbons in a turbine fuel: 1) paraffins, iso-paraffins, and cyclo-paraffins; 2) aromatics; 3) naphthalenes; and 4) olefins. Other smaller components include sulphur (which has corrosive properties and is a harmful combustion by-product), naphthenic acid (which is also corrosive), water and additives. With reference to the latter category, there are a number of approved chemicals, such as corrosion inhibitors, icing inhibitors, static dissipators, anti-oxidants and so on.

**TEMPERATURES.** At the altitudes of normal flight operation the air temperatures can reach values below −60°C, especially on cross-polar routes via the North Pole. The aviation fuels currently used, mostly Jet A and Jet-A1, have a freezing point at a higher temperature, which poses serious safety risks. In particular, the freezing point of Jet-A1 is −47°C, the freezing point of the Jet-A is −40°C and Jet-B (*wide-cut*) has a freezing temperature of −50°C. The Russian equivalent of Jet-A1 is called TS1, although this fuel has a slightly lower freezing temperature. These are average values that do not reflect the actual freezing phenomenon. At these low temperatures there

Table 14.2. *Characteristics of turbine fuels Jet-A and Jet-A1; data are adapted from BP*

| Properties | Jet-A | Jet-A1 |
|---|---|---|
| Composition | | |
| Aromatics (volume) | 23.4% | 19.5% |
| Sulphur (mass) | 0.07% | 0.02% |
| Volatility | | |
| Boiling point | 280°C | 252°C |
| Flash point | 51.1°C | 42°C |
| Density, 15°C | 820 kg/m$^3$ | 804 kg/m$^3$ |
| Fluidity | | |
| Freezing point | −51.0°C | −50°C |
| Viscosity, −20°C | 5.2 mm$^2$/s | 3.5 mm$^2$/s |
| Combustion | | |
| Specific energy | 43.02 MJ/kg | 43.15 MJ/kg |
| Smoke point | 19.5 mm | 25.0 mm |
| Naphthalenes (volume) | 2.9% | 1.5% |

can be a precipitation of waxy material that results in blocking of the fuel feed lines and the fuel filters. As a consequence, some routes cannot be flown in very severe conditions. For example, winter flight operations across Siberia can easily encounter atmospheric temperatures of –70°C. For increased safety there is a warning system ahead of the inlet into the engine's combustors, although it is not always clear how much capability for fuel heating there is. A safety margin quoted by Airbus is that the limiting fuel temperature should be the freezing temperature, plus the engine's heating margin, plus 2°C for safety (Boeing requires 3°C above freezing point). The addition of anti-freeze additives affects the water content in the fuel. The change in temperature is more rapid toward the end of the flight, as a result of a lower heat capacity of the remaining fuel. Stirring of fuel can prevent freezing. For example, recirculating fuel with the booster pump can lower the freezing point by several degrees. Other techniques to control the fuel temperature may include waste heat from the power systems.

The properties of the fuel are somewhat variable, so we are bound to find average values of the relevant physical properties. Considerable research exists in this area, and different figures are given for certain properties. The heat capacity of the liquid fuel at temperatures between 200 and 500 K is calculated from the following semi-empirical expression (elaborated from Faith *et al.*[6]):

$$C_p\,[\mathrm{kJ/kg\ K}] = 1.4103 - 4.8877 \cdot 10^{-6} T_f + 7.1575 \cdot 10^{-6} T_f^2. \tag{14.1}$$

Simple calculations can be carried out with $C_p \simeq 2{,}000$ J/kg K. The thermal conductivity is approximated by the following polynomial expression:

$$k\,[\mathrm{W/m\ K}] = 2.6936 \cdot 10^{-1} - 4.2273 \cdot 10^{-4} T_f + 2.0251 \cdot 10^{-7} T_f^2. \tag{14.2}$$

In Equations 14.1 and 14.2 the temperature is in Kelvin.

Similar expressions could be derived for the fuel vapour. However, the data for this phase are available mostly at high temperature. Some equations exist for the fuel density of Jet-A and Jet-A1, such as:

$$\rho_f \simeq 1.0490 - 8 \cdot 10^{-4} T \ \ [\mathrm{kg/litre}], \tag{14.3}$$

where the temperature must be given in Kelvin. This equation provides a constant gradient $d\rho_f/dT_f = -8 \cdot 10^{-4}$ kg/K. A nominally empty volume is actually occupied by fuel vapour.

The vapour pressure of the jet fuels is related to the absolute temperature through the following *approximate* semi-empirical expression‡:

$$p^*\,[\mathrm{bar}] = 2.256 \cdot 10^3 \exp(-3899/T^*). \tag{14.4}$$

From Equation 14.4 we can apply the equation of state (§ 14.4.1) to calculate the corresponding vapour density $\rho_f^*$ and hence the derivative of this parameter with respect to the vapour temperature, Equation 14.21.

‡ Alternative expressions are found in the technical literature.

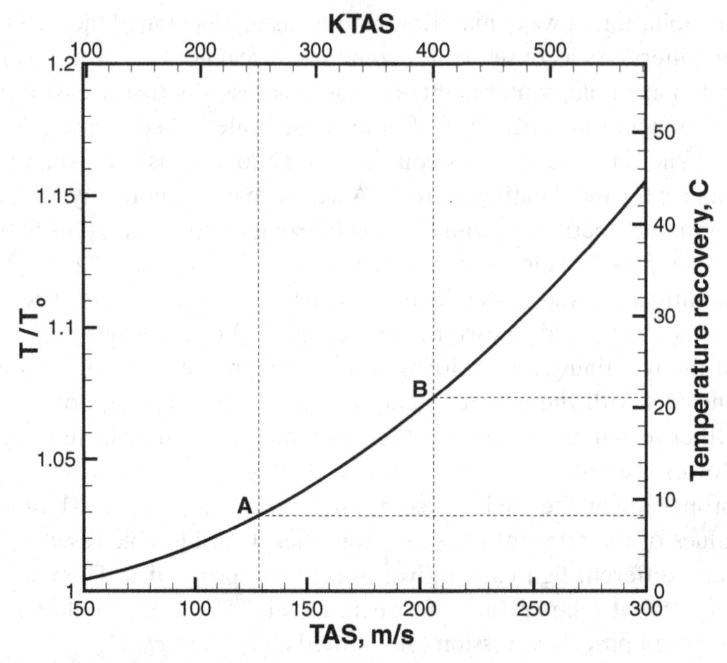


Figure 14.5. Relationship among TAT, SAT, and recovery temperature.

## 14.3 Fuel Temperature in Flight

The fuel temperature varies greatly, as the fluid is subject to heat transfer with the tanks and the atmosphere; the temperature may reach very low values, to a point at which it crystallises and freezes. However, also high temperatures are to be avoided. For example, the FCOM of the Boeing B737-300, the B747-400 and B767 specify that the maximum tank fuel temperature cannot exceed +49°C (Jet-A and Jet-A1 only). Low temperatures cause difficulty in fuel pumping, transfer between tanks and fuel moving through the engine feed lines and valves. This is particularly dangerous if freezing occurs in and around the fuel feed system, although the presence of small water particles in the fuel will certainly compound the problem. Such an event has been attributed to the crash landing of a Boeing B777 at London Heathrow in January 2008. For this reason, the fuel temperature is one of the many parameters recorded by the FDR. Data elaborated‡ on more than 140,000 flights of the Boeing B777 with three engine types (Rolls-Royce, General Electric and Pratt & Whitney) indicate that the lowest recorded fuel temperature was −39°C, although in about 17% of the cases the temperature was never below −20°C.

Extended investigations on fuel-temperature profiles were carried out in the 1970s. Ref.[7] reports fuel temperature data for all the U.S. Air Force aircraft of the time (including the B-52G, C-141, KC-135, XB70a, B-1 and others). Both Airbus and Boeing produce fuel-tank-temperature statistics for their aircraft.

Statistical data indicate that the outer wing tanks usually have a fuel temperature 3 degrees below that of the inner tanks. Hence, under severe conditions the outer tanks should be used first. The fuel temperature depends on the flight time. An example is shown in Figure 14.6, which was elaborated from Airbus's flight test data.

‡ Unofficial data of the Air Accident Investigation Branch (AAIB), 2008.

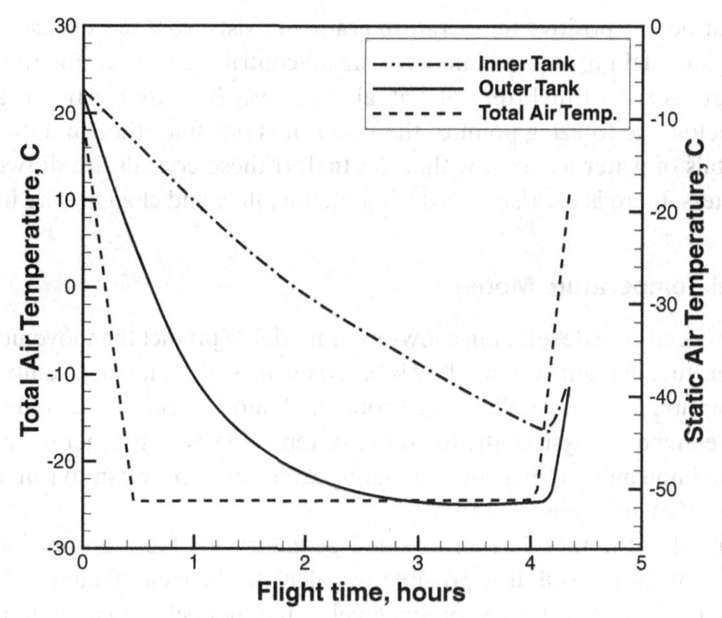


Figure 14.6. Fuel cooling during a long-range flight (elaborated from Airbus A300/A310).

The outer tanks reach a stabilisation temperature after 3 flight hours. There are a number of temperatures used in this type of analysis. The Static Air Temperature (SAT) is the temperature of still air. The Total Air Temperature (TAT) is the SAT plus a term called *kinetic heating*; the corresponding increase in temperature is called *temperature recovery*. The relationship between these temperatures is given by the energy equation

$$\mathcal{T} = \mathcal{T}_o + \frac{U^2}{2C_p}. \tag{14.5}$$

In Equation 14.5, TAT $= \mathcal{T}$, SAT $= \mathcal{T}_o$; the kinetic heating (or recovery temperature) is $U^2/2C_p$. Equation 14.5 is the compressible form of the Bernoulli equation; it expresses the fact that during an adiabatic transformation, the loss in kinetic energy is transformed in heat by an equal amount. For a flow reaching a point of stagnation, such as the leading edge of a wing, the recovery temperature is maximum. Equation 14.5 is equivalent to the following:

$$\frac{\mathcal{T}}{\mathcal{T}_o} = 1 + \frac{\gamma - 1}{2} M^2, \tag{14.6}$$

where $\gamma$ is the ratio between specific heats of the atmosphere and $M$ is the flight Mach number. The relationship among TAT, SAT and recovery temperature is shown in Figure 14.5 as a function of the air speed; we have assumed a reference SAT $= \mathcal{T}_o = 288$ K (15°C). For example, at a speed of 250 kt the temperature recovery is ~9 degrees (point A); at 400 kt it is ~21 degrees (point B). A total temperature of −30°C at $M = 0.78$ corresponds to a SAT of about −57°C.

Why is there heat transfer between the fuel and the atmosphere? The fuel tanks are not insulated for practical reasons (added structural weight and cost). Due to the size of the tanks, the minimum fuel temperature is at the contact with the tank surfaces; a positive temperature gradient exists from the surfaces inwards. Fuel pumping, fuel mixing and tank rocking all contribute to lowering the freezing temperature. A risk of fuel freezing will always exist. Because the fuel is at a temperature below the freezing point of the water, it is possible (though unlikely) that some crystals of water ice exist within the fuel. If these crystals are drawn toward the fuel filters, there is the risk of reducing the fuel flow and clogging the fuel lines.

## 14.4 Fuel-Temperature Model

We now proceed with developing a low-order model to predict the movement of the fuel temperature in flight. The method is based on an isothermal fuel (lumped mass). This assumption greatly simplifies the problem. A more accurate method must rely on a finite-element analysis of the fuel-tank system. However, the method presented here can be implemented in a general flight-performance program to run alongside other types of calculations.

Assume that the tank has an $n$–sided geometry. Its faces do not have to be necessarily flat. In general, it is possible to calculate the exact volume of the geometry and the contact surface $S$ for any level of the fuel between a minimum and a maximum. It is possible to define a numerical function

$$S = f(V) \tag{14.7}$$

that gives the total wetted (or contact) area $S$ for any fuel volume $V$. The heat transfer from the fuel to the outer system takes place through contact. The free surface of the fuel, $\overline{S}$, is shared by the liquid and vapour phases. The differential equation for this surface in terms of the fuel flow is

$$\frac{dS}{dt} = \left(\frac{dS}{dV_f}\right)\left(\frac{dV_f}{dt}\right) \simeq \left(\frac{dS}{dV_f}\right)\dot{m}_f \rho_f = \dot{m}_f \rho_f f_1(V), \tag{14.8}$$

where we have neglected the change in fuel density with respect to time; $\dot{m}_f$ is the fuel flow out of the tank; $f_1(V)$ denotes the numerical relationship between the convection surface $S$ and the fuel volume $V$. If the overline $\overline{(.)}$ is used for quantities through the free contact surface and the starred quantities $(.)^*$ denote the fuel vapour, the problem is formulated by the following ordinary differential equations:

$$\frac{dV_f}{dt} \simeq \frac{\dot{m}_f}{\rho_f}, \tag{14.9}$$

$$\frac{dS}{dt} = \rho_f \dot{m}_f f_1(V_f), \tag{14.10}$$

$$\frac{d\overline{S}}{dt} = \rho_f \dot{m}_f f_2(V_f), \tag{14.11}$$

$$\frac{dQ}{dt} = h_c (T_f - T)\, S, \tag{14.12}$$

$$\frac{d\overline{Q}}{dt} = \overline{h}_c (T_f - T_f^*)\, \overline{S}, \tag{14.13}$$

$$\frac{dT_f}{dt} = \frac{1}{C_p m_f}\left(\frac{dQ}{dt} + \frac{d\overline{Q}}{dt}\right), \tag{14.14}$$

$$\frac{d\rho_f}{dt} = \left(\frac{d\rho_f}{dT_f}\right)\left(\frac{dT_f}{dt}\right). \tag{14.15}$$

Equation 14.10 and Equation 14.11 are the rate of change of the wetted and free surface, respectively; these are functions of the actual fuel volume $V_f$; the latter quantity depends on the time via the fuel flow rate, given by Equation 14.9. Equation 14.12 expresses the heat transfer rate by contact through all of the wetted surfaces of the tank; Equation 14.13 is the heat transfer by natural convection through the free surface of the fuel; Equation 14.14 is the rate of change of the fuel temperature. Equation 14.15 is used to close the system; its right-hand side uses the chain rule of differentiation, which allows a decoupling of the thermal properties of the fuel $(dT_f/dt)$ from the time derivatives. The derivative $d\rho_f/dT_f$ is a physical property of the fuel, as discussed in § 14.2.

An improvement to the model consists in writing a similar set of equations for the vapour phase. Specifically, we would need to calculate the temperature and density (or pressure) of the fuel vapour. The coupling of the differential equations is made through the temperature $T$, that now represents the fuel-vapour temperature rather than the static atmospheric temperature. The differential equations for the vapour phase are:

$$\frac{dV^*}{dt} = -\frac{dV}{dt}, \tag{14.16}$$

$$\frac{dS^*}{dt} = -\frac{dS}{dt}, \tag{14.17}$$

$$\frac{d\overline{S}^*}{dt} = \frac{d\overline{S}}{dt}, \tag{14.18}$$

$$\frac{dQ^*}{dt} = h_c^* \left(T_f^* - T\right) S^*, \tag{14.19}$$

$$\frac{d\overline{Q}^*}{dt} = -\frac{d\overline{Q}}{dt}, \tag{14.20}$$

$$\frac{dT_f^*}{dt} = \frac{1}{C_p m_f^*}\left(\frac{dQ^*}{dt} + \frac{d\overline{Q}^*}{dt}\right), \tag{14.21}$$

$$\frac{d\rho_f^*}{dt} = \left(\frac{d\rho_f^*}{dT_f^*}\right)\left(\frac{dT_f^*}{dt}\right). \tag{14.22}$$

It can be observed that four of these equations are straightforward. In fact, Equation 14.16, Equation 14.17 and Equation 14.18 can be calculated directly from the liquid fuel and the characteristics of the tank; Equation 14.20 is also calculated by direct inference. Equation 14.21 and Equation 14.22 are more problematic. The problem is to calculate the mass of the vapour $(m_f^*)$ in thermal equilibrium with the liquid fuel. We assume that $m_f^* = \rho_f^* V^*$.

### 14.4.1 Fuel-Vapour Model

The fuel vapour is described with the non-linear gas equation of Peng and Robinson[8]. The relationship among pressure, absolute temperature and molar volume $V_m$ is

$$p = \frac{\mathcal{R}T}{V_m - b} - \frac{a\alpha}{V_m^2 + 2bV_m - b^2}, \tag{14.23}$$

where $\mathcal{R}$ is the gas constant. The other terms are:

$$a = 0.45724\mathcal{R}^2\frac{T_c^2}{p_c}, \qquad b = 0.07780\mathcal{R}\frac{T_c}{p_c}, \tag{14.24}$$

$$\alpha = \left[1 + (0.37464 + 1.54226\omega - 0.26992\omega^2)(1 - T_r^{1/2})\right]^2, \tag{14.25}$$

$$T_r = \frac{T}{T_c}. \tag{14.26}$$

The quantities denoted with the subscript "c" are *critical quantities*, whilst $\omega$ is called the *acentric factor*; $V_m$ is the molar volume. For the Jet-A fuels, on average[9] we have

$$p_c = 2,235 \text{ kPa}, \qquad v_c = 0.564 \text{ l/mol}, \qquad T_c = 537.9 \text{ K}. \tag{14.27}$$

The values of the acentric factor and the molar volume are, respectively,

$$\omega = 0.457, \qquad V_m = 0.129 \text{ kg/mol}. \tag{14.28}$$

### 14.4.2 Heat-Transfer Model

To calculate the heat-transfer rates, we need to estimate the transfer coefficients. This is done through the Nusselt number

$$\text{Nu} = \frac{h_c\mathcal{L}}{k_c}, \tag{14.29}$$

which is the ratio between the convective heat transfer and the conduction heat transfer through a solid surface. In Equation 14.29, $\mathcal{L}$ represents the characteristic length of the process, which is $\mathcal{L} \simeq S/b_{tank}$, that is, the ratio between the wetted area and the spanwise extent of the tank $b_{tank}$. The thermal conductivity $k_c$ is assigned from Equation 14.2; $h_c$ is the convection-heat-transfer coefficient. Thus, if a suitable expression for the Nusselt number is found, then the convective heat transfer is calculated from Equation 14.29.

From heat-transfer textbooks we can find various expressions for the Nusselt numbers, depending on whether the flow is laminar, transitional or turbulent; specific research addresses heat-transfer rates in jet fuels. In the present case, we use the relationship

$$\begin{aligned} \text{Nu}_L &= 0.54\text{Ra}^{1/4} \quad 10^4 < \text{Ra} < 10^7, \\ \text{Nu}_L &= 0.15\text{Ra}^{1/4} \qquad \text{Ra} > 10^7 \end{aligned} \tag{14.30}$$

where "Ra" represents the Rayleigh number; Ra = Gr Pr; Gr is the Grashof number; and $\text{Pr} = C_p/\mu/k_c$ is the Prandtl number. This equation is appropriate for a natural convection problem from a warm lower flat surface[10] and hence, with some

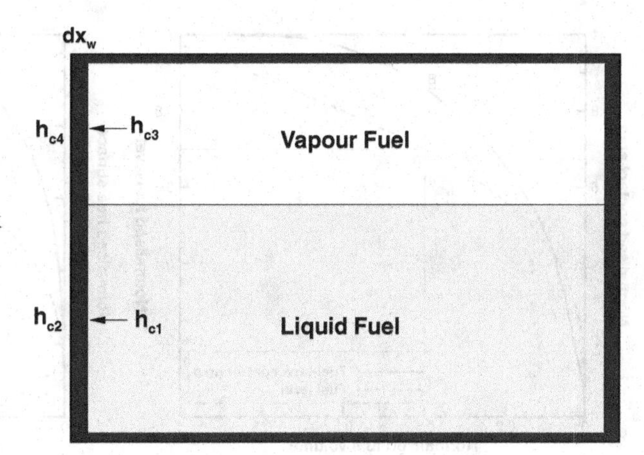


Figure 14.7. Nomenclature for fuel tank heat exchange.

extrapolation, to the cooling of the liquid fuel. The Nusselt number for the vapour, assuming that the top surface of the tank is being cooled, is nearly the same. For the calculation of the Nusselt numbers we need a reference length. This is taken as $S/b_{tank}$, although alternative choices are possible. The total heat-transfer coefficient $h_c$ depends on the tank's material construction (including its thickness) and can be written as

$$\frac{1}{h_c} = \frac{1}{h_{c1}} + \frac{dx_w}{k_c} + \frac{1}{h_{c2}}, \tag{14.31}$$

where $h_c$ is the convection-heat-transfer coefficient (fuel- and air-side), $dx_w$ is the wall thickness and as shown in Figure 14.7. The right-hand side of Equation 14.31 represents the total transmittance of the system. The heat transfer is dominated by the conduction-convection on the air-side, with some contribution from the fuel-side.

A typical value of thermal conductivity for aluminium is 220 to 250 W/m$^2$ K, whilst the convective heat transfer of the air is between 10 and 100 W/m$^2$ K. The natural convection-heat-transfer coefficient of air and fuel is another parameter that does not respond to a simple estimate.

### 14.4.3 Numerical Solution

In summary, the system is made up by 13 ordinary differential equations (Equation 14.17 is redundant). For each phase (liquid and vapour), the unknowns are the fuel temperature $T_f$; the corresponding density $\rho_f$; the contact and free-surface areas $S$ and $\overline{S}$, respectively; the fuel volume $V$; and the heat-transfer rates by contact and convection. The next step is to define the initial conditions:

- *Initial temperature.* On a hot day, $T_f$ can exceed the OAT if the aircraft spends time loading and taxiing. If the tank is partially full, there will be considerable evaporation of fuel.
- *Fuel flow.* We need to know how the aircraft has been loaded and how the fuel is to be used. Thus, we have an initial value of the fuel volume (or weight) in the reference tank and a rule for fuel used.

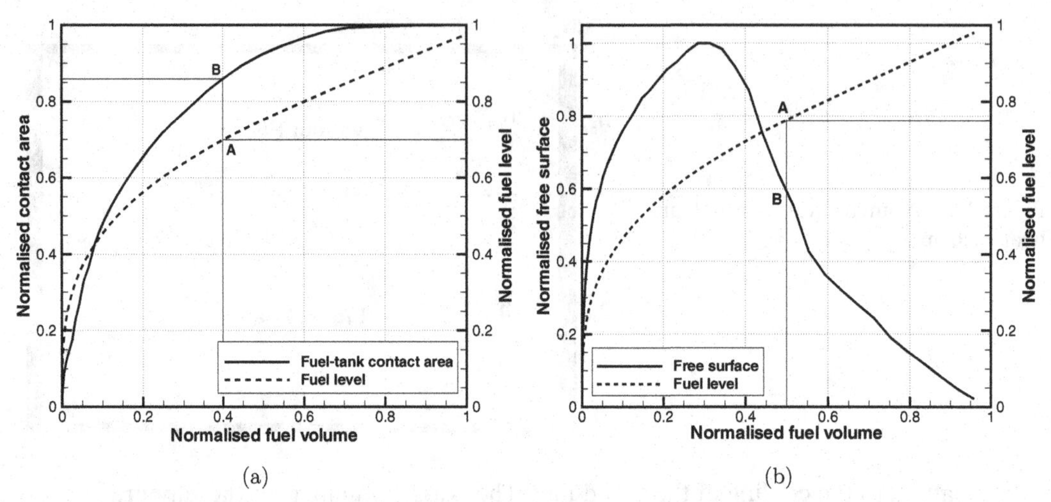


Figure 14.8. Example of wing fuel-tank calculations: normalised fuel-tank wetted areas, fuel level and fuel volume.

- *Total air temperature and the fuel flow*. The best way to acquire these data is to perform a full mission calculation and store these two quantities as a function of flight time.

An example of wing-tank characteristics is shown in Figure 14.8a; this figure displays the fuel level and fuel-tank contact area for a given fuel volume; all data are normalised. If the fuel level is 70%, the fuel volume is ~40% of the tank's capacity (point A) and the fuel-tank contact area is ~87% of the total area enveloped by the tank (point B). The free surface shown in Figure 14.8b starts from zero (with the tank empty); it reaches a maximum at a fuel level around 30% and goes down to zero in the limiting case of full tank. The lack of smoothness is due to interpolation errors.

Some simplification is required before proceeding. First, we neglect the heat transfer between the liquid and vapour through the free surface, which causes Equation 14.11, Equation 14.13, Equation 14.18 and Equation 14.20 to be redundant. A combination of the remaining equations is now possible. In particular, Equation 14.12 combined with Equation 14.14 leads to a simple equation for the liquid fuel temperature:

$$\frac{dT_f}{dt} = \frac{h_c S}{C_p m_f} (T_f - T). \tag{14.32}$$

A similar combination for the vapour fuel leads to an equation for the vapour temperature:

$$\frac{dT_f^*}{dt} = \left(\frac{h_c S}{C_p m_f}\right)^* (T_f^* - T). \tag{14.33}$$

The factor

$$\tau = \frac{C_p m_f}{h_c S} \tag{14.34}$$

represents a *time constant* in similar thermodynamic problems. The higher this factor, the more slowly the fuel responds to changes in temperature. A too rapid fuel cooling is not a good thing; therefore, one may want to find ways to increase the time constant. From the definition of Equation 14.34, it can be gleaned that for a given fuel mass, $\tau$ increases as the contact surface decreases. However, it is also clear that $\tau$ cannot be constant because fuel flows in or out of the tank, the contact surface changes and the heat-transfer coefficient changes as a result of variable outside conditions. In this context, the fuel will cool more rapidly (within a given time) when the fuel mass is low and the contact area is large. As an example, a fuel tank containing a fixed amount of 20,000 kg of fuel would have an estimated $\tau \simeq 10^5$ seconds, or about 168 minutes.

Toward the end of the flight, there is relatively little fuel left in the tanks, and the OAT increases rapidly as the aircraft descends. Because the time constant is now greatly decreased, the fuel responds more rapidly to a change of outside temperature, as demonstrated in Figure 14.6.

### 14.4.4 Numerical Solution and Verification

In this section we illustrate the steps required for a numerical solution of the fuel-temperature model. The sequence of operations is the following:

- Calculate a flight trajectory: altitude, OAT, fuel flow, as a function of flight time.
- Set the fuel-tank characteristics (fuel-tank volume, initial fuel and fuel flow).
- Set the initial conditions for the ODE: liquid/vapour fuel, geometric characteristics.
- Integrate the system of Equation 14.32 and Equation 14.32 in the time domain.

We now consider the case of a fuel tank whose characteristics are shown in Figure 14.8a. The flight trajectory is shown in Figure 14.9a, which indicates that there are four step climbs at cruise from FL-280. There is a corresponding step change in the OAT.

The system of differential equations is solved in the time domain by using a fourth-order Runge-Kutta method with the time-step set by the flight-trajectory data. Figure 14.9b displays the fuel volume and the liquid-fuel temperature during the flight. The OAT is also shown for reference. The high fuel rates during the climb cause the fuel level to go down faster (see left side of the graph). By the end of the flight, the fuel tanks are relatively empty, and the remaining fuel increases its temperature rapidly as the aircraft descends. Crucially, the method presented neglects the effects of the thermal-layer thickness near the wall, where there are considerable temperature gradients.

**VERIFICATION OF THE METHOD.** The proposed thermodynamic model is compared to the FDR data of a Boeing B777-236ER that suffered a crash upon landing at London Heathrow in January 2008[11]. This airplane, at an altitude of ~700 feet (~210 m), suffered a reduction in thrust from the right engine, due to reduction in fuel flow; a few seconds later the left engine was also reduced. The flight control system detected this loss in fuel flow and commanded a full valve opening. No change in fuel flow

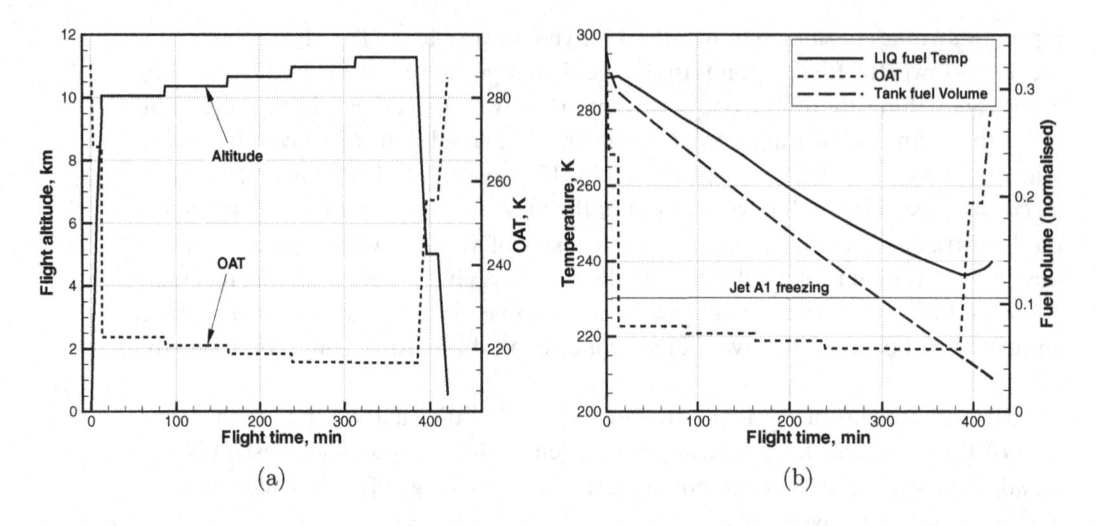


Figure 14.9. Flight trajectory used for fuel temperature calculations.

resulted, as a consequence of the fuel valves being clogged with ice within the system. The remaining thrust was not sufficient to maintain control of the airplane, which eventually crashed 1,000 feet (~300 m) from the runway.

Figure 14.10 shows the simulated fuel temperature for this case; the simulation is compared with the temperature data extracted from the FDR[12]. The graph shows the liquid and vapour fuel temperature (as calculated), the SAT (from the FDR), the tank fuel volume (from the FDR) and the fuel temperature (from the FDR).

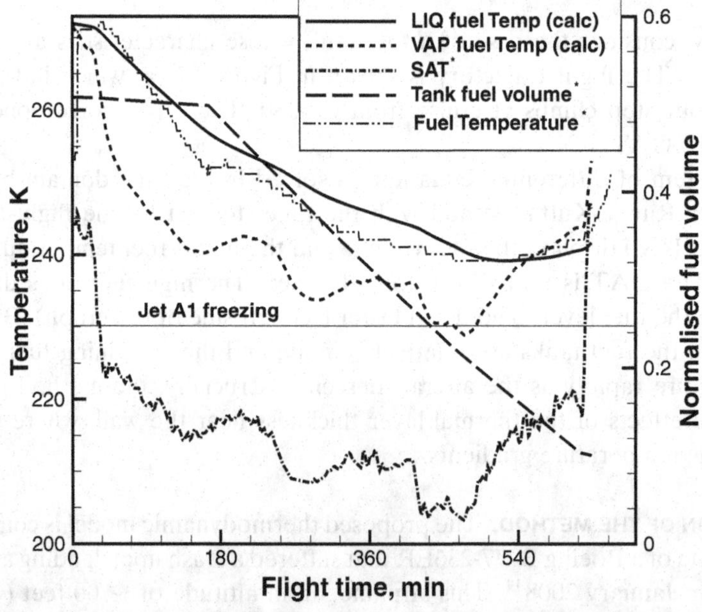


Figure 14.10. Calculated fuel temperature for the flight recorded by the AAIB[11]. The data with the asterisk ()* are FDR data.

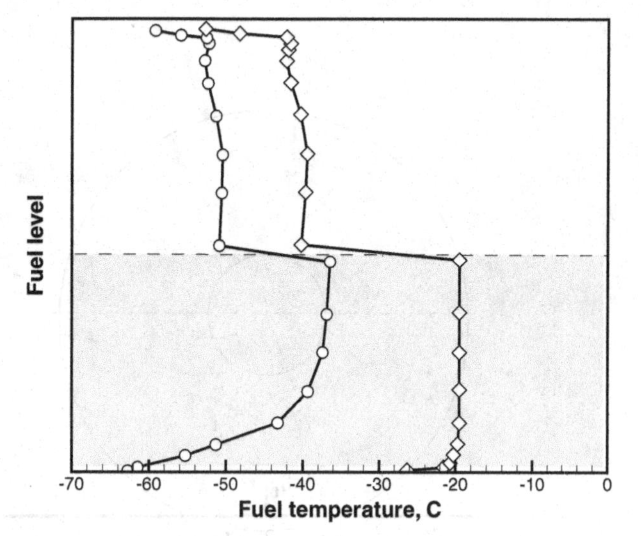


Figure 14.11. Example of cold fuel temperature distribution in box-shaped tank.

For reference we have added the conventional freezing temperature of the Jet-A1 fuel. The fuel temperature itself never reaches the freezing threshold. However, it is possible that local temperatures are lower, especially if the fuel is contaminated with water. In fact, the AAIB investigation revealed the presence of water in the fuel, whose temperature reached a minimum of −34°C; the minimum TAT was −45°C.

Because the average fuel temperature in the tank is higher than the freezing temperature, a more detailed investigation may be needed. There exist computer programs for the fuel-temperature prediction (for example, at the Boeing Company); these programs are more accurate than the model presented here and rely on field solutions of the heat-transfer problem, with a finite-element or finite-volume analysis. These programs establish that, contrary to the assumption of uniform temperature, there is a positive temperature gradient from the bottom of the tank to the free surface (the fuel at the bottom is the coldest). Furthermore, there is generally a step change in temperature at the liquid-vapour interface, the vapour phase being considerably colder, though not as cold as the fuel at the bottom of the tank. Thermal stabilisation is reached within 2 to 3 hours into the flight. Figure 14.11 shows a typical temperature distribution in a box-like fuel tank. The colder temperature is achieved after about 3 hours.

## 14.5 Tyre-Heating Model

The history of aviation is littered with catastrophic failures that in hindsight turned out to be design shortcomings†. In 1963, a Caravelle III jetliner crashed soon after take-off from Durrenasch, Switzerland, due to a tyre explosion during landing-gear retraction. Official investigations discovered that the tyre had exploded due to

† An exhaustive database of aircraft accidents around the world is available at the website of the National Transportation Safety Board, www.ntsb.gov.

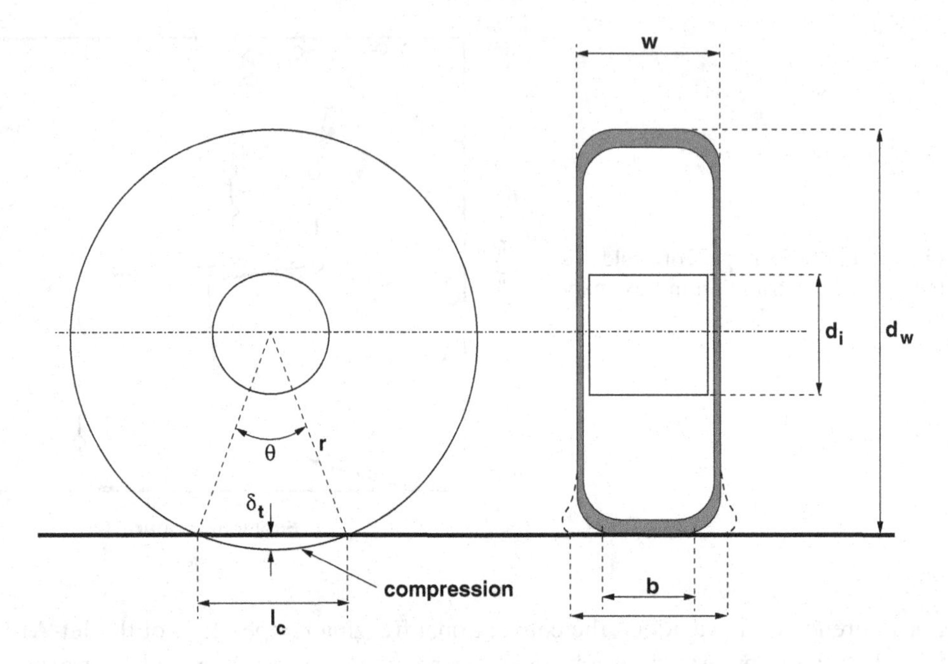


Figure 14.12. Aircraft tyre nomenclature.

over-heating during a missed take-off; tyre explosion damaged a fuel line, which in turn started an explosion; no lives could be saved. The pilot, in fact, decided to clear the morning fog with the aircraft wings, by taxiing the aircraft down to the end of the runway and back. By the time the aircraft left the ground, it had run three times on the runway. At the time there was ignorance of the abnormal tyre heating due to heavy loads, friction effects on the ground and brake operation. Nowadays, a missed take-off requires additional performance controls on the tyres and landing gear, what are commonly called *load-speed-time charts*. The relevant regulations prescribe the installation of appropriate thermal-sensitive devices. Furthermore, tyres must be stopped before assuming their folded position in the hold.

Modelling the heating of a tyre under heavy load is an exceedingly difficult problem. We propose a simple analysis that uses a limited set of data, such as those that can be found for a conventional transport aircraft. As usual some assumptions are required.

Figure 14.12 shows the nomenclature that will be used for deriving the thermodynamics and the pressure loading on an aircraft tyre. Some practical formulas for tyre deflections are available in ESDU[13]. The tyre dimensions are given according to some international standards. We will consider the specifications given as `tyre-diameter/tyre-width/tyre-rim` in *imperial units*. For example, a marking $49 \times 19.0$–$20$ indicates a tyre with a diameter $d_w = 50$ inches (1.270 m), a width $w = 19.0$ inches (0.483 m), and a rim $(d_w - d_i)/2 = 20$ inches (0.508 m). These data are converted into SI units in order to be consistent. Additional data that are required include at least the tyre pressure, the type of gas and the mass of the tyre. Aircraft tyres are generally inflated with inert gases such as helium or nitrogen to minimise the effects of expansion and contraction during extreme changes in temperature.

The front view of Figure 14.12 shows two possible deflections on the roadway, which depend on the tyre pressure. We will assume that the width of the contact $b$ is equal to the width of the tyre, $b \simeq w$. This assumption is a valid one in the context of the overall approximations that are required to estimate the heat loads. The length of the contact is $l_c$. This is conveniently found in terms of vertical deflection $\delta_t$ from the formula[13]:

$$7.2\left(\frac{\delta_t}{d_w}\right)^2 + 0.96\left(\frac{\delta_t}{d_w}\right) - \frac{F}{(p+0.08p_r)w(wd_w)^{1/2}} = 0, \tag{14.35}$$

where $F$ is the external vertical load, $p$ is the inflation pressure of the unloaded tyre and $p_r$ is the tyre rated inflation pressure. This quadratic equation is solved in terms of $\delta_t/d_w$. Only the positive solution makes sense. From the vertical compression $\delta_t$, we can calculate the compression volume $dV$, which is equal to

$$dV = \frac{1}{2}w\left[\theta r^2 - l_c\cos(\theta/2)\right], \tag{14.36}$$

where $\theta$ is the sector angle corresponding to the road contact $l_c$ (see Figure 14.12). One of the key performance parameters of a tyre is the deflection, conventionally given by

$$\delta_{tyre}\% = \frac{\delta_t}{d_w/2}\cdot 100. \tag{14.37}$$

Aircraft tyres are designed for a deflection of 32%; this value is considerably higher than tyres for other applications.

**TYRE MASS.** Often the mass of the tyre is unknown. It can be estimated from the volume equation and a weighted specific mass. The volume is:

$$V = \underbrace{\pi w d_w \tau_1}_{tread} + \underbrace{2\pi\left[\left(\frac{d_w}{2}-\tau_1\right)-\left(\frac{d_i}{2}\right)\right]\tau_2}_{sides} - \underbrace{(nwd)_g}_{grooves}. \tag{14.38}$$

In Equation 14.38 $d_i$ is the internal diameter (the rim), $\tau_1$ is the thickness of the tread, $\tau_2$ is the average thickness of the sides, $n_g$ is the number of grooves, and $w_g$ and $d_g$ are their width and depth, respectively. The resulting mass is the weighted average of the tyre components:

$$m = V\sum_i x_i\rho_i, \tag{14.39}$$

where $x_i$ is the proportion of material $i$ and $\rho_i$ is the corresponding specific mass. The volume occupied by the inflating gas is a fraction of the above, and it can be derived directly from Equation 14.38.

The rotational speed of tyre when aircraft has ground speed $U$ is $\omega = U/r$. The frequency of the loading is $f = \omega/2\pi$. With a tyre diameter of 1 m and a speed of 50 m/s, the frequency is about 16 Hz (a relatively low value), but the centrifugal acceleration is a staggering $254g$!

**NUMERICAL METHOD.** The increase in tyre pressure due to loading $F$ on the wheel is calculated iteratively. ESDU[13] suggests using the formula

$$\Delta p_t = 1.5p\left(\frac{w}{d_w}\right)\left(\frac{\delta_t}{d_w}\right)^2. \tag{14.40}$$

In our numerical method, this formula is inserted into a loop that allows the calculation of compression volume, contact length, pressure and temperature rise in static conditions:

1. Calculate unloaded tyre pressure using the current gas temperature $T_g$: $p_t = \mathcal{R}T_g\rho_g$.
2. Calculate the vertical compression $\delta_t$ from Equation 14.35.
3. Calculate the change in pressure $\Delta p_t$ from contact (Equation 14.40).
4. Calculate the compression volume $dV$ from Equation 14.36.
5. Recalculate the tyre pressure.
6. If the residual on the latest tyre pressure is below a tolerance, exit loop.

Before entering this loop, it is necessary to calculate the mass of the inflating gas. This calculation is done under unloaded conditions, by using the equation of the ideal gas. The next problem is to calculate the heat loads on both tyre and inflating gas. It is convenient to write the energy equation separately for tyre and gas.

The tyre is loaded by the external force; thus, it deforms with a frequency $f$. Part of the energy is lost due to hysteresis. The hysteresis $\mathcal{H}$ is defined as the ratio between the energy lost and the deformation energy. The hysteresis depends on tyre temperature and the frequency. Following some research published by Lin and Hwang[14], it is reasonable to assume that $\mathcal{H}$ varies between 0.1 and 0.3. The hysteresis is a local parameter that changes within the tyre. The highest values correspond to the lowest temperatures. Furthermore, the frequency effect seems to appear only at these temperatures. A secondary effect, albeit an important one at high rolling speeds, is due to the centrifugal forces. One consequence of the centrifugal loads is the generation of a standing wave that causes a distortion on the side of the tyre leaving the contact with the roadway. This condition is attributed to under-inflation and should not normally be encountered during a take-off operation. The deformation work done on the tyre during a full rotation is

$$\mathcal{W}_1 = F\delta_t\left(\frac{2\pi}{l_c}\right). \tag{14.41}$$

The work done during a time step $dt$ is taken from

$$d\mathcal{W} = \mathcal{W}_1\left(\frac{\omega dt}{2\pi}\right). \tag{14.42}$$

The work actually lost in the deformation is $d\mathcal{W}\mathcal{H}$. This work is partly dissipated by the tyre through external convection ($E_c$), through internal convection ($E_g$), through friction with the inflating gas ($E_f$) and by physical contact with the roadway ($E_r$). The remaining part ($E_t$) will serve to increase the thermal energy of the tyre itself. Combination of Equation 14.41, Equation 14.42 and an appropriate value of the hysteresis is the key to the whole method.

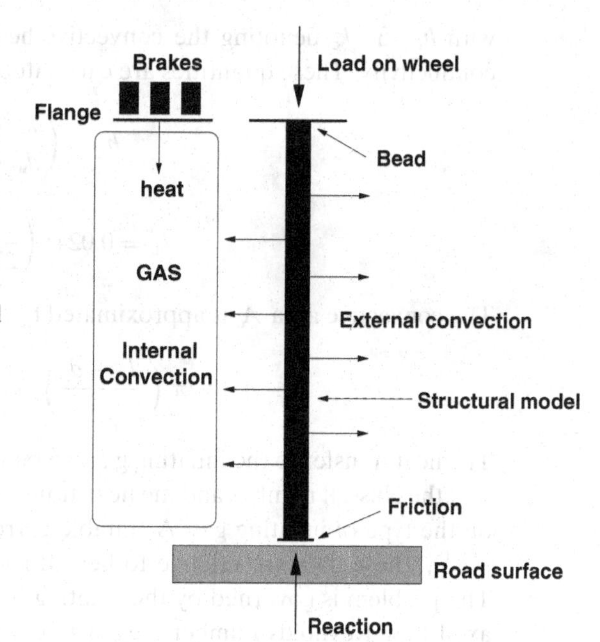


Figure 14.13. Model for thermo-structural loads on aircraft tyre.

**HEAT LOADS.** Figure 14.13 illustrates how the tyre is modelled. There is interaction between the tyre and the roadway, the inflating gas and the external environment. The gas interacts with the inner walls of the tyre. We need to find a way to estimate the heat loads, which are expressed by the equation

$$dW = dE_t - dE_c - dE_r - dE_g - dE_f. \tag{14.43}$$

Additional terms that may have to be included in Equation 14.43 arise from the internal heat that is generated during the deformation and eventually by specific cooling systems. The radiation heat transfer to the gas is about three orders of magnitude lower than the other contributions; thus, it is not considered in our analysis.

The forced convective heat transfer from the tyre to the air can be calculated by using a suitable expression of the Nusselt number as a function of the Reynolds number. This relationship can be written as

$$\text{Nu} = a\,\text{Re}^b. \tag{14.44}$$

The coefficients $a, b$ depend on the problem; the Reynolds number is calculated from the aircraft speed and the rotational speed. For a thin rotating cylinder ($w/d_w \simeq 0.017$), the equation is

$$\text{Nu} = 5.88 \cdot 10^{-3}\,\text{Re}^{0.925}. \tag{14.45}$$

There is the problem of extending this correlation to the typical widths of an aircraft tyre ($w/d \simeq 0.25$). However, once this problem is solved, the convective heat flux is

$$\dot{Q}_c = \frac{dE_c}{dt} = h_c A_c (T_t - T), \tag{14.46}$$

with $h_c$ and $k_c$ denoting the convective-heat-transfer coefficient and the thermal conductivity. These quantities are calculated with the following equations:

$$h_c = \left(\frac{k_c}{d_w}\right) \text{Nu}, \tag{14.47}$$

$$k_c = 0.0241 \left(\frac{T}{273.15}\right)^{0.9}. \tag{14.48}$$

The convective area $A_c$ is approximated by

$$A_c \approx 2\pi \left(\frac{d_w - d_i}{4}\right) + \pi w(d_w + d_i) - wl_c. \tag{14.49}$$

The heat transfer to the inflating gas can be written with a similar expression, except that the Nusselt number and the heat-transfer coefficient $k_c$ are different and depend on the type of inflating gas. A suitable correlation is found in the technical literature[15]. These data are relative to heated rotating cylinders, with an axial gas flow. The problem is governed by the rotating Reynolds number $\text{Re}_r = \omega d_w/2\nu$ and the axial flow Reynolds number $\text{Re}_a$. The data indicate that for $\text{Re}_r > 2.77 \cdot 10^5$, the Nusselt number is insensitive to the axial flow, which shows that the heat transfer is dominated by the rotation. At lower $\text{Re}_r$ the axial flow has a role that increases as the rotation decreases. Data are available for $\text{Re}_a = 0$, which corresponds broadly to the present case, although there would be the problem of accounting for the end-wall effects. The Nusselt number is calculated from

$$\text{Nu} = \begin{cases} 2.85 \cdot 10^5, & \text{Re}_r \geq 2.77 \cdot 10^5 \\ 15.77 + 6.52 \cdot 10^{-4}\text{Re}_r + 3.86 \cdot 10^{-9}\text{Re}_r, & \text{Re}_r < 2.77 \cdot 10^5. \end{cases} \tag{14.50}$$

The heat transfer with the roadway occurs through the contact surface $A_c$, which has an elliptic shape, with axes $l_c$ and $w$; the resulting formula is:

$$\dot{Q}_r = \frac{dE_r}{dt} = k_{c_t} A_c (T_t - T_{road}), \tag{14.51}$$

where $k_{c_t}$ is the thermal conductivity of the tyre. To simplify the matter, we assume that the runway is at the same temperature as the atmosphere (in the presence of direct sunlight it can be considerably higher). Finally, the energy accumulated by the tyre itself is described by

$$\dot{Q}_t = \frac{dE_t}{dt} = C_{p_t} m_t (T_t - T). \tag{14.52}$$

Equation 14.52 does not differentiate between parts of the tyre and will only provide an *average* tyre temperature. Local differences in temperature have been measured by McCart and Tanner[16], who concluded that the temperatures along the inner walls are lower than the temperatures around the outer surface. This result also implies that the road contact is a more important factor. Data from Goodyear* indicate that the difference in temperature between tread centreline and tread should be

* Manufacturers publish technical data to back up performance claims; often these are not official reports.

negligible at normal taxi speeds. The temperature of the bead is higher at taxi speeds above 30 km/h and may reach 30 degrees in excess of the tread. Thus, the method is only applicable to taxi conditions.

The heat coefficient can be estimated by the weighted average of its components. These are rubber (~80%), steel chords (~5%), and textile fabrics (~15%), but some manufacturers quote different proportions. The exact composition of a tyre is proprietary information. Average heat-transfer coefficients are rubber: $C_p \simeq 1.5$ kJ/kg K; steel: $C_p \simeq 0.46$ kJ/kg K; and nylon-based fabrics: $C_p \simeq 1.7$ kJ /kg K. The weighted average is 1.478 kJ/kg K, although there can be considerable differences with real tyres; furthermore, there can be special mechanisms that minimise tyre heating by means of cooling.

The change in energy for the gas is derived from the convective term specified by Equation 14.46 and by the friction between the inner tyre and the gas. The movement of the inflating gas inside the tyre is not amenable to simple modelling. For the purpose of the present analysis, some approximations are required. A few key physical aspects are the following:

- The inflating gas moves with the same speed as the tyre walls.
- The gas is affected by centrifugal forces that press against the outer radius.
- The gas is lagging behind the walls elsewhere.

With a leap of faith we now approximate the relative motion of the gas with respect to the tyre to a turbulent shear flow in a pipe of equivalent diameter and length. In this pipe there is no relative speed between the walls and the gas. The set of assumptions is the following:

- The tube's diameter $d_1$ is the average between the rim and the width.
- The Reynolds number of the internal flow is

$$Re_i = \frac{\rho_g u_1 d_1}{\mu_g}, \tag{14.53}$$

  where $u_1$ is the maximum rotational speed difference between the tyre and the gas (slip speed). This is a fraction of the tyre's rotational speed.

The friction coefficient $f$ is calculated from the equation

$$\frac{1}{\sqrt{c_f}} = 4\log\left(\frac{d_1}{\epsilon}\right) + 2.28 - 4\log\left(\frac{4.67(d_1/\epsilon)}{\mathrm{Re}\sqrt{c_f}+1}\right), \tag{14.54}$$

where $\epsilon$ is the average roughness of the internal walls. Equation 14.54 is in implicit form and must be solved iteratively. The average resistance between the gas and the tyre is

$$R \simeq \frac{1}{2}\rho_g\left[\frac{\pi}{4}(d_w - d_i)(d_w + d_i)\right]c_f u_1^2 \tag{14.55}$$

and the energy dissipated in the time step $dt$ is

$$E_f \simeq R u_1 dt. \tag{14.56}$$

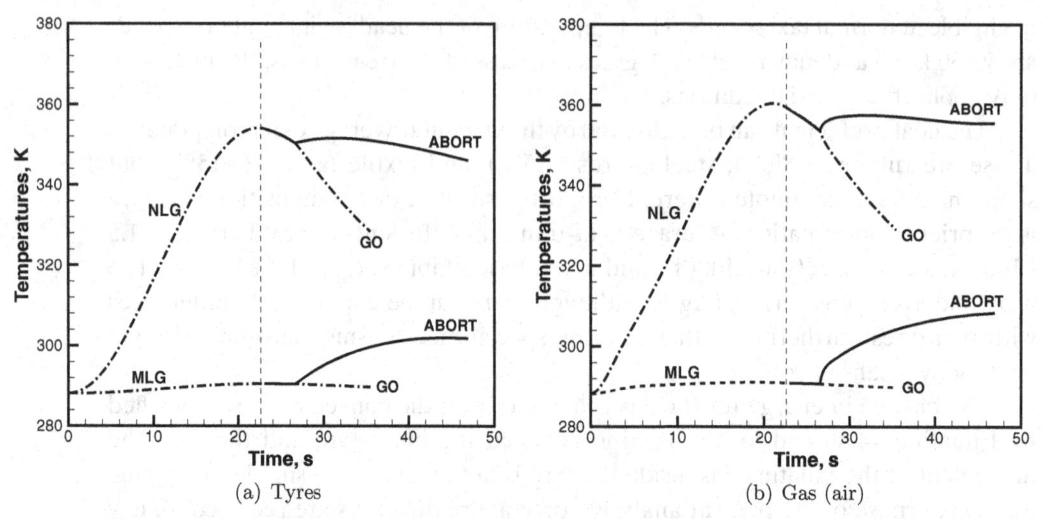


Figure 14.14. Estimated tyre and temperatures for accelerate-stop of airplane model Boeing B777-300; standard day; 2 m/s headwind; airfield at 50 m above mean sea level.

The differential equations that describe the tyre and gas temperature are, respectively:

$$\frac{d\mathcal{T}_t}{dt} = \frac{1}{C_{p_t}}\left(d\mathcal{W} - dE_c - dE_r - dE_g\right), \tag{14.57}$$

$$\frac{d\mathcal{T}_g}{dt} = \frac{dE_g}{C_{p_g} m_g}. \tag{14.58}$$

These equations are integrated with the initial conditions

$$t = 0, \quad \mathcal{T}_t = \mathcal{T}, \quad \mathcal{T}_g = \mathcal{T}, \quad \mathcal{T}_r = \mathcal{T}. \tag{14.59}$$

Equation 14.57 and Equation 14.58 can be included in the solving system for take-off and landing and solved simultaneously with the corresponding group of equations (for example, § 9.2) by using a fourth-order Runge-Kutta integration.

### 14.5.1 Numerical Simulations

The integration of the tyre-heating model with the take-off equations is shown in Figure 14.14 for a case of accelerate-go and decelerate-stop of a model Boeing B777-300 aircraft. Figure 14.14a shows the temperature distribution for the tyres and Figure 14.14b shows the corresponding gas temperature. The thermo-structural loads are higher for the nose landing gear (NLG) due to the position of the CG. These high loads on the NLG are not unusual. In any case, the temperatures decrease uniformly for the accelerate-go event because of the cooling effect of the air around the tyres. For the decelerate-stop event this cooling is absent, and the additional heating is derived from the extended tyre-runway contact, as well as internal friction with the gas and the effect of the mechanical brakes. Notice that this procedure is fully integrated with the take-off equations, which are discussed in Chapter 9.

The validation of the model proposed cannot be done directly because the few available data are not in a usable format. Some tyre-heating data for typical taxi

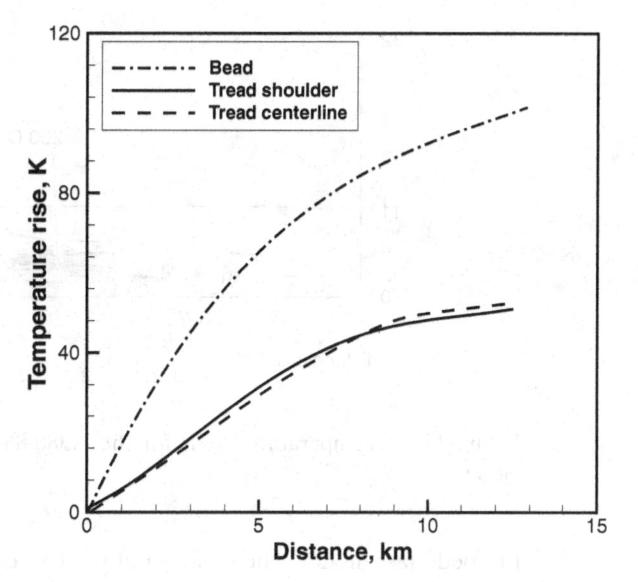


Figure 14.15. Estimation of tyre temperatures. Data adapted from Goodyear; $U = 48$ km/h (~29 kt); deformation = 0.32. Tyre dimensions and load conditions are unknown.

operations are given by Cavage[17] for the Boeing B727. The results indicate that the heating depends on the loading, the taxi speed and the type of tyre. Data are available for taxi speeds up to 70 mph (110 km/h). A qualitative comparison can be drawn with some data published by Goodyear (although the actual load conditions and tyre specifications are not disclosed), Figure 14.15. The bead temperature is always the highest. In this example, an airplane was forced to run on a long runway at a specified speed, and the temperature was measured at selected points on the tyres.

By way of comparison, Figure 14.16 shows some predicted tyre temperatures on the main under-carriage units for a model Airbus A300-600 forced to run at various speeds. The predicted temperature is an *average* within the tyre, coherently with the

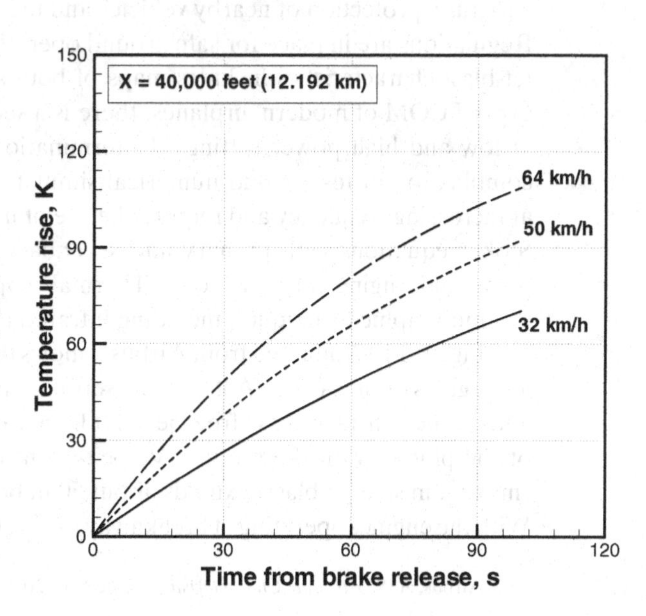


Figure 14.16. Estimation of tyre temperatures from taxiing; m = 150.0 tons; headwind = −2 m/s (−3.9 kt); altitude = 50 m; $x_{CG}$ = 30% MAC; standard day.

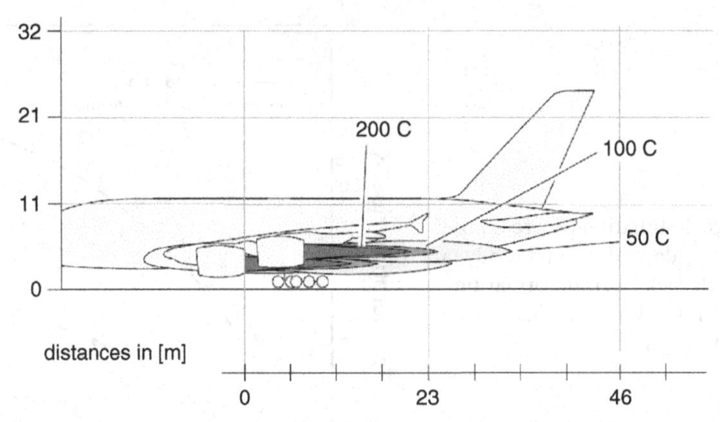


Figure 14.17. Temperature levels for the A380-861 on the ground with idle engines (side view).

lumped mass model. The position of the CG is important because it determines the load distribution on the main and nose tyres and therefore can affect the heating of both groups of tyres.

## 14.6 Jet Blast

If a jet can propel a Boeing 747 into the sky, imagine what it can do to you. It can flatten buildings, lift ground vehicles up into the air and create other damage at ramps and taxiways. A numerical study by Melber-Wilkending[18] illustrated the problem of an Airbus A380, whose jet blast hit the corner of a hangar. The wind generated around the structure creates its own weather on a length scale of at least 100 m. Most airplane manufacturers provide thermal and velocity maps of jet blast with the vehicle on the ground, with at least two operating conditions: idle and full thrust. These conditions are important in ground operations for the safeguard of infrastructure, protection of nearby vehicles and the safe operation of ground personnel. Regulations are in place for safe ground operation[19]. An accurate determination of jet-blast characteristics includes maps of both speeds and temperatures. In nearly every FCOM of modern airplanes, there is a section containing these types of data at low and high power settings. Determination of these maps can be done by a combination of testing and numerical simulation. Numerical simulations are used at increasing frequency and require the use of unsteady Reynolds-averaged Navier-Stokes equations with primary and secondary flows (by-pass), as well as ground plane and engine tilt and toe-in. Thermal maps can be determined with various thermo-graphic techniques, including infrared thermography.

Figure 14.17, adapted from Airbus*, shows the temperature distribution from the jet engines of the Airbus A380 in idle setting. The plumes are lifted from the ground and can be considered as "free" jets. At higher thrust settings there is an interaction of the plumes with the ground. Velocity contours indicate that the 56 km/h wind (maximum safe jet blast) extends about 50 m behind the airplane with idle engines. With the engines operating at breakaway thrust, the jet plume of this airplane extends

* Airbus, *A380 Airplane Characteristics*, Issue Nov. 2008, Airbus Customer Services, Blagnac, France.

to about 90 m. When the engines are at take-off thrust, the plume extends about 0.5 km behind and 50 m up in the sky.

There are no easy extrapolations that can be made from this or other cases due to the complex interaction plumes between plumes and the ground at the relevant engine settings. The scientific literature contains a number of semi-empirical relationships to predict the jet expansion and velocity distribution[20]. However, these equations are only applicable when there is no interaction with boundaries. In practical terms, because the jet-blast area can be very large, some airports are equipped with blast-deflection barriers. A recent analysis of experimental data at some major airports was published by Slaboch[21] and aimed at providing guidelines for reducing hazards on manoeuvring aircraft.

## Summary

In this chapter we presented a number of performance problems related to thermodynamics and structural mechanics. In the former class of problem, aircraft icing is considered of utmost importance, having been the source of some prominent aircraft accidents. There is a serious degradation of aerodynamic characteristics of the wing, which can even be asymmetric. Thus, aircraft control becomes impossible. Fuel-temperature problems arise both at high and low temperatures. We addressed the problem of very cold fuel and its behaviour in flight. We developed a lumped-mass model that is capable of predicting the fuel temperature change in flight. Fuel freezing is unlikely, except in the most extreme circumstances, in combination with other factors related to the fuel pipelines. In the category of structural problems, we addressed tyre loads and the response of tyre to extreme loads, which may cause abnormal heating. Also in this case we developed a lumped-mass model for tyre loads and heating, which is fully integrated with the flight mechanics of the field performance (taxi, take-off and landing).

Finally, we mentioned briefly the problem of jet blast. Jet exhausts at high power can have damaging consequences on infrastructure and ground personnel.

## Bibliography


[1] Kind RJ, Potapczuk MG, Feo A, Golia C, and Shah AD. Experimental and computational simulation of in-flight icing phenomena. *Progr. Aerospace Sciences*, 34:257–345, 1998.

[2] Lynch FT and Khodadoust A. Effects of ice accretions on aircraft aerodynamics. *Progress Aerospace Sciences*, 37(8):669–767, 2001.

[3] Asselin M. *Introduction to Aircraft Performance*. AIAA Educational Series, 1997.

[4] Caldarelli G. ATR-72 accident in Taiwan. In *SAE Aircraft & Engine Icing International Conference*, ICE 13, Sevilla, Spain, Oct. 2007.

[5] Goodger E and Vere E. *Aviation Fuels Technology*. Macmillan, London, 1985.

[6] Faith LE, Ackerman GH, and Henderson HT. Heat sink capability of a Jet-A fuel: Heat transfer and coking studies. Technical Report CR-72951, NASA, July 1971.

[7] Gray CN and Shayeson MW. Aircraft fuel heat sink utilization. Technical Report TR-73-51, US Air Force Aero Propulsion Labs, Wright-Patterson AFB, Ohio, July 1973.

[8] Peng DY and Robinson DB. A new two-constant equation of state. *Industrial & Eng. Chemistry: Fundamentals*, 15:59–64, 1976.
[9] Huber ML and Yang JC. A thermodynamic analysis of fuel vapor characteristics in an aircraft fuel tank ullage. *Fire Safety Journal*, 37:517–524, 2002.
[10] Incropera F and De Witt DP. *Introduction to Heat Transfer*. John Wiley, 1985. Chapter 9.
[11] AAIB. Accident to Boeing B-777-236ER, G-YMMM at London Heathrow Airport on 17 January 2008. Interim Report EW/C2008/01/01, Sept. 2008. Aldershot, Hampshire, UK.
[12] Filippone A. Theoretical framework for the simulation of transport aircraft flight. *J. Aircraft*, 47(5):1679–1696, 2010.
[13] ESDU. *Vertical Deflection Characteristics of Aircraft Tyres*. Data Item 86005. ESDU International, London, May 1986.
[14] Lin YJ and Hwang SJ. Temperature prediction or rolling tires by computer simulation. *Mathematics and Computers in Simulation*, 67:235–249, 2004.
[15] Seghir-Ouali S, Saury D, Harmand S, Phillipart P, and Laloy O. Convective heat transfer inside a rotating cylinder with an axial flow. *International J. Thermal Sciences*, 45:1066–1078, 2006.
[16] McCarty JL and Tanner JA. Temperature distribution in an aircraft tire at low ground speeds. Technical Report TP-2195, NASA, 1983.
[17] Cavage WM. Heating comparison of radial and bias-ply tires on a B-727 aircraft. Technical Report DOT/FAA/AR-TN97/50, US Department of Transportation, Nov. 1997.
[18] Melber-Wilkending S. Aerodynamic analysis of jet-blast using CFD considering as example a hangar and an Airbus A380 configuration. In *New Results in Numerical and Experimental Fluid Mechanics*, volume 92 of *Notes on Numerical Fluid Mechanics*. Springer, 2006.
[19] Anon. *Transport Canada Aeronautical Information Manual (TC AIM): AIR Section 1.7: Jet and Propeller Blast Danger*. Transport Canada, Oct. 2011 (continuously updated). TP 14371.
[20] Witze P. Centerline velocity decay of compressible free jets. *AIAA Journal*, 12(4):417–418, April 1974.
[21] Slaboch PE. An operational approach for the prediction of jet blast. In *AIAA Aerospace Sciences Conference*, AIAA Paper 2012-1225, Nashville, TN, Jan. 2012.

## Nomenclature for Chapter 14

| | | |
|---|---|---|
| $a, b$ | = | factors in Peng and Robinson equation of state, Equation 14.24; factors in Nusselt number, Equation 14.44 |
| $b$ | = | width of tyre contact |
| $A_c$ | = | area for convective heat transfer |
| $b_{tank}$ | = | wing-tank span |
| $C_D$ | = | airplane-drag coefficient |
| $c_f$ | = | friction coefficient, Equation 14.54 |
| $C_L$ | = | airplane-lift coefficient |
| $C_p$ | = | heat-capacity coefficient |
| $d$ | = | diameter |
| $d_i$ | = | internal diameter of the tyre (rim), Figure 14.12 |
| $d_1$ | = | tube diameter; average tyre diameter |

$d_g$ = depth of the groove
$dx_w$ = wall thickness
$d_w$ = tyre diameter
$E$ = energy
$f$ = frequency
$f_1, f_2$ = generic function
$F$ = normal load on tyre
$g$ = acceleration of gravity
Gr = Grashof number
$h_c$ = convection-heat-transfer coefficient
$\mathcal{H}$ = mechanical hysteresis
$k_c$ = thermal conductivity
$l_c$ = length of tyre-road contact
$l_{fuse}$ = fuselage length
$\mathcal{L}$ = characteristic length
$m$ = mass
$M$ = Mach number
Nu = Nusselt number
$n_g$ = number of grooves
$p$ = pressure
$p_r$ = tyre-rated pressure
Pr = Prandtl number
$Q$ = heat transfer
$R$ = rolling resistance; resistance between gas and tyre
$\mathcal{R}$ = gas constant
Re = Reynolds number
Ra = Rayleigh number
$Re_a$ = axial-flow Reynolds number
$Re_r$ = rotating-flow Reynolds number
$S$ = contact surface
$\overline{S}$ = free fuel surface
$t$ = time
$\mathcal{T}$ = temperature
$T_r$ = reduced temperature in Peng-Robinson equation, Equation 14.26
$u_1$ = maximum slip speed between tyre and gas
$U$ = ground speed or air speed
$V_c$ = critical molar volume, Equation 14.27
$V$ = volume
$V_m$ = molar volume
$V_1$ = normalised volume
$w$ = tyre width, Figure 14.12
$w_g$ = width of the groove
$\mathcal{W}$ = mechanical work
$\mathcal{W}_1$ = mechanical work in one full rotation
$x_i$ = mass fraction of component $i$

## Greek Symbols

$\alpha$ = factor in Equation 14.25
$\epsilon$ = wall roughness
$\delta_t$ = vertical deflection of the tyre
$\gamma$ = ratio between specific heats (air)
$\theta$ = sector angle for tyre-road contact
$\mu$ = dynamic viscosity
$\nu$ = kinematic viscosity
$\rho$ = density
$\rho_w$ = water density
$\tau$ = time constant
$\tau_1$ = thickness of the tread
$\tau_2$ = average thickness of the tyre sides
$\omega$ = acentric factor, Equation 14.28
$\omega$ = rotating speed of the wheels

## Subscripts/Superscripts

$[\dot{.}]$ = time derivative
$[.]_c$ = critical gas parameter or convection
$[.]_f$ = liquid fuel quantity, due to friction
$[.]_g$ = gas quantity
$[.]_i$ = internal
$[.]_L$ = with respect to reference "L"
$[.]_r$ = runway quantity; thermal quantity in Equation 14.43
$[.]_t$ = tyre quantity
$[.]^*$ = vapour fuel quantity; FDR data in Figure 14.10

# 15 Mission Analysis

## Overview

Mission analysis refers to the complete set of assumptions regarding the operation of an aircraft and the resulting weight and fuel planning. Several parametric considerations can be added, including mission optimisation, environmental emissions and various commercial or military trade-offs. This chapter illustrates how a detailed mission analysis can be carried out. We start by defining mission scenarios (§ 15.1) and the payload-range charts (§ 15.2). We then make a detailed qualitative and quantitative mission analysis (§ 15.3), including the problems of mission range and mission fuel (§ 15.4), the fuel-reserve policies (§ 15.5) and the short-range performance at MTOW (§ 15.6). In § 15.7 we discuss more advanced problems, such as service with an intermediate stop (§ 15.7.1), tankering (§ 15.7.2) and the point-of-no-return (§ 15.7.3). We proceed with the calculation of the direct operating costs (§ 15.8). We show some case studies: an aircraft selection problem (§ 15.9), a fuel mission planning for a transport aircraft (§ 15.10) and the effect of floats on the payload-range chart of a propeller airplane adapted with floats for water operations (§ 15.11). We close this chapter with an introduction to risk analysis in aircraft performance (§ 15.12).

**KEY CONCEPTS:** Mission Profiles, Flight Scheduling, Payload-Range Charts, Mission Analysis, Reserve Fuel, Fuel Planning, Tankering, En-Route Stop, Direct Operating Costs, Risk Analysis, Route Selection.

## 15.1 Mission Profiles

A mission profile is a scenario that is required to establish the weight, fuel, payload, range, speed, flight altitude, loiter and any other operations that the aircraft must be able to accomplish. The mission requirements are specific to the type of aircraft. For high-performance aircraft they get fairly complicated and require some statistical forecasting. The deterministic approach is not sufficient or even recommended.

Over the years many commercial aircraft operators have specialised in niche markets that offer prices and services to selected customers. These niches include the executive jet operators between major business centres, operators flying to particular destinations (oil and gas fields), the all-inclusive tour operators to holiday resorts

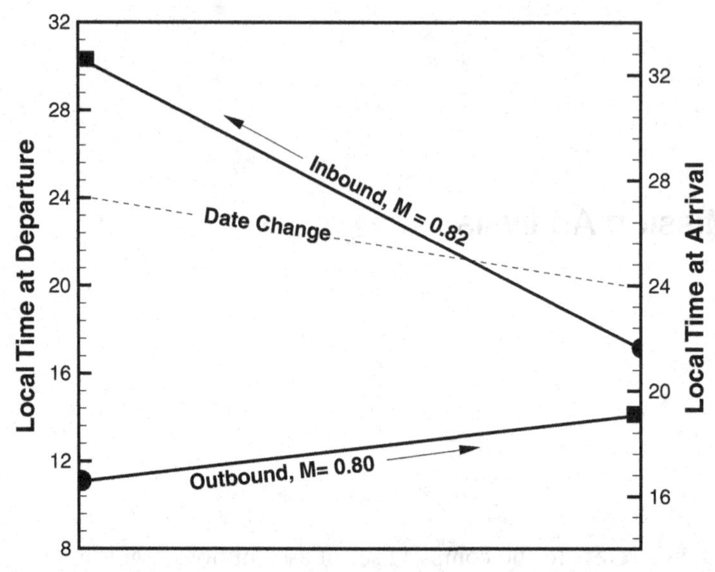


Figure 15.1. Scheduling of trans-Atlantic flight.

and the no-frills airlines, flying to minor and under-used airports. These operators have different schedules and cost structures.

Let us start with long-range passenger operations, which are serviced by subsonic commercial jets. The basic principle is that the airplane takes off from airport A and flies to airport B along a recognised flight corridor; then it returns to A. The main parameters of the mission planning are the distance between the airports, the flight time, the down time at the airport for getting the airplane ready (also called *time-on-station*), the flight speed, the local air traffic and the departure times at both ends. Back at the airport of origin, the day is not over; the operator may wish to utilise the airplane for another flight to the same destination or to another destination. The key is the departure time and the minimisation of the curfew. Figure 15.1 shows a typical schedule of such an airplane over a trans-Atlantic route from a major airport in Europe to an airport on the East Coast of the United States.

Due to the time-zone effect, a late-morning departure from Europe arrives to the United States in the mid-afternoon. An early-evening departure arrives back in Europe in the early hours of the day after. Over a 24-hour period the airplane will have done a return flight and worked about 14 to 16 hours. For a flight arriving late in the evening, a return may not be possible until early morning on the next day. This adds to the operational costs because of the need of maintaining the crew away from the home port.

The schedule shown in Figure 15.2 shows a typical schedule between a UK airport and mainland Europe. Due to the relatively short distance of 600 n-miles and one time zone, it is possible to operate two return services and one outbound service, whilst maintaining the night-time curfew (shaded area). The final outbound flight must remain at the destination airport till the following morning. This aircraft will be the first to make an inbound flight. To run efficiently, this service would require two aircraft, one at each airport. If this is not possible, then the last outbound flight must be cancelled from the schedule.

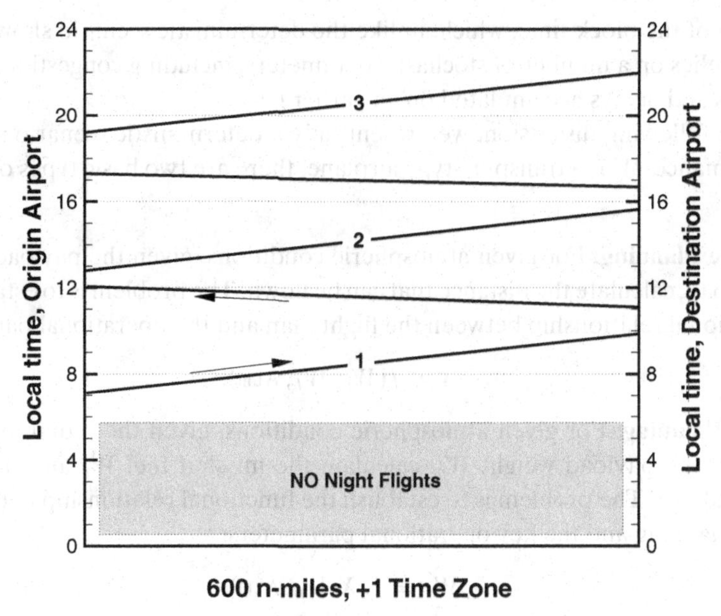


Figure 15.2. Scheduling of short–medium flight.

Another key factor in the utilisation of the airplane is the turn-around time, that is, the time required to prepare the airplane for the next flight. Considerable logistics are required to ensure all services to passengers, cargo and airplane within a target time frame. Up to six different categories of services are involved: passenger boarding/disembarking, cargo loading/unloading, refuelling, cleaning, catering, and ground handling. The time needed to get the aircraft ready for the next intercontinental flight may require up to 3 hours. However, Airbus reports that the turn-around time of the A380 is 90 minutes with standard servicing on two decks and 120 minutes with servicing from the main deck*.

### 15.1.1 Operational Parameters

*Block time* is the time required to travel from the departure gate to the destination gate. If the block time of each flight is added, at the end of one year the airplane will have a total number of block hours, which airline operators may want to maximise. There are three parts in the block time: taxi out, airborne time and taxi in. An airplane flying a day-time shorter route, and returning in the mid-afternoon, should be able to complete another return flight to the same destination or otherwise. For an airline company operating several airplanes, scheduling and optimal operation of the fleet is a complex problem. Events such as bad weather can lead to dozens of airplanes and flight crews out of position for several days. Scheduling and operation of aircraft is a subject for operations research and is addressed by specialised publications, for example Gang Yu[1], who deals with demand forecasting, network design, route planning, airline schedule planning, irregular operations, integrated scheduling, airport traffic simulation and control, and more. Coy[2] presents a method for the statistical

* Airbus, *A380 Airplane Characteristics*, Rev. Oct. 01/09 (2009), Blagnac Cedex, France.

prediction of the block time, which, unlike the deterministic scenario shown in Figure 15.1, relies on a number of stochastic parameters, including congestion, weather conditions and delays accumulated on an earlier leg.

In the following discussion, we present only a deterministic scenario and focus on performance*. For a transport-type airplane, there are two basic types of mission analysis:

- **Range Planning**: For given atmospheric conditions, given the payload and the fuel load, calculate the distance that can be flown. The problem is to establish the functional relationship between the flight plan and the operational parameters:

$$X = f(W_p, W_f, \texttt{Atm}). \tag{15.1}$$

- **Fuel Planning**: For given atmospheric conditions, given the required distance $X$ and the payload weight $W_p$, calculate the mission fuel $W_f$ and the weight breakdown. The problem is to establish the functional relationship between the fuel planning and the key operational parameters:

$$W_f = f(X, W_p, \texttt{Atm}). \tag{15.2}$$

The operator `Atm` is a vector containing all of the atmospheric conditions encountered during the flight. For convenience, these conditions are split between 1) take-off and climb-out, 2) cruise, and 3) final approach and landing. More specifically, the parameters are atmospheric temperatures, wind speed, wind direction and relative humidity. The latter parameter is only relevant for noise calculations (see Chapters 16 to 18).

## 15.2 Range-Payload Chart

The payload-range chart belongs to the range flight-performance analysis (Equation 15.1) because the distance that can be flown is unknown. The flight distance depends not only on the amount of fuel but also on the weight of the payload. In most cases the combination of maximum payload and maximum fuel load exceeds the MTOW. If we know how to calculate the flight range, then it is possible to construct a chart showing how some of the aircraft weights are related to the range. Because different cruise techniques are available, we could construct different weight-payload diagrams corresponding to each flight program. Furthermore, there are effects of atmospheric conditions, climb and descent techniques, and reserve fuel policy that can change the weight-payload performance considerably. Unless all of these conditions are specified, it is not possible to compare the range-payload performance of two aircraft.

There are several ways in which the payload-range performance can be analysed. The first approach is shown in Figure 15.3, which displays the range-payload chart for three commercial subsonic jet aircraft of the Airbus family. The manufacturer generally specifies the range at maximum passenger load (including baggage) and some value of the bulk payload; the latter item is important for most commercial airlines. However, passenger-carrying airplanes seldom operate at MTOW.

* We refer to "mass" and "weight" interchangeably, although they are physically different. The aviation jargon refers to weights rather than mass; our computer models work with "mass".

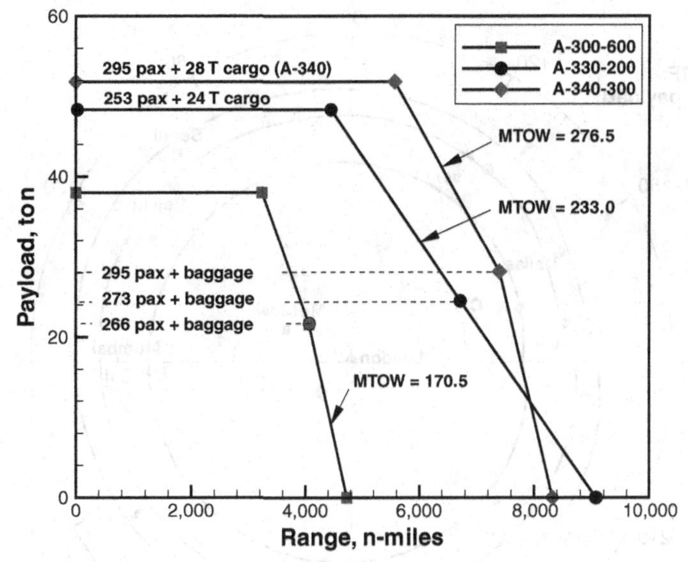


Figure 15.3. Maximum payload range for Airbus airplanes.

Another way of showing the payload-range performance is presented in Figure 15.4. The data have been elaborated from Boeing[3] and refer to the B747-400. This diagram is considerably different and more detailed than Figure 15.3. The oblique lines denote constant-BRGW values. A trade between an increase in range and the corresponding decrease in payload shall be made at constant brake-release gross weight:

$$\left(\frac{d\,\mathrm{ZFW}}{dX}\right)_{BRGW}.$$

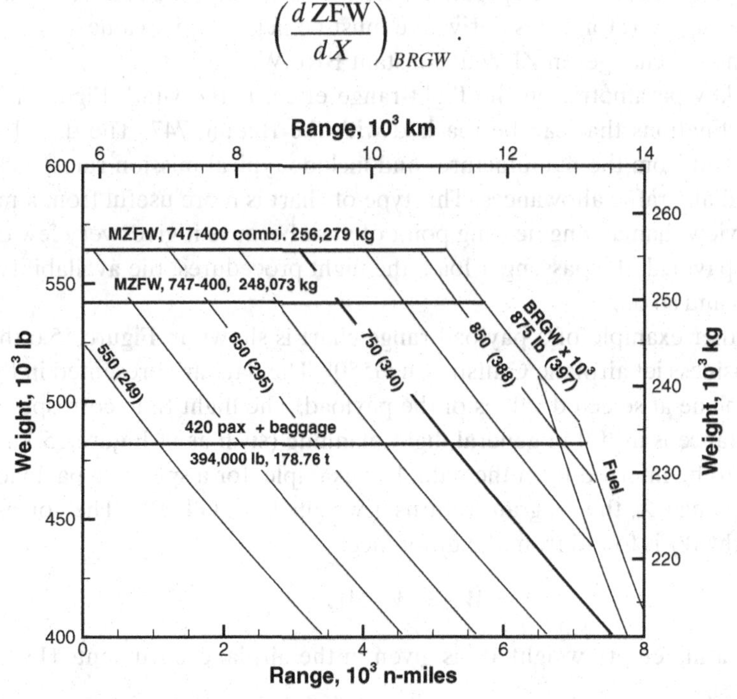


Figure 15.4. Payload-range chart for the Boeing B-747-400 and -400 Combi (CF6-80C2B1F engines); standard day, $M = 0.85$, cruise-climb profile; FAR international reserves.

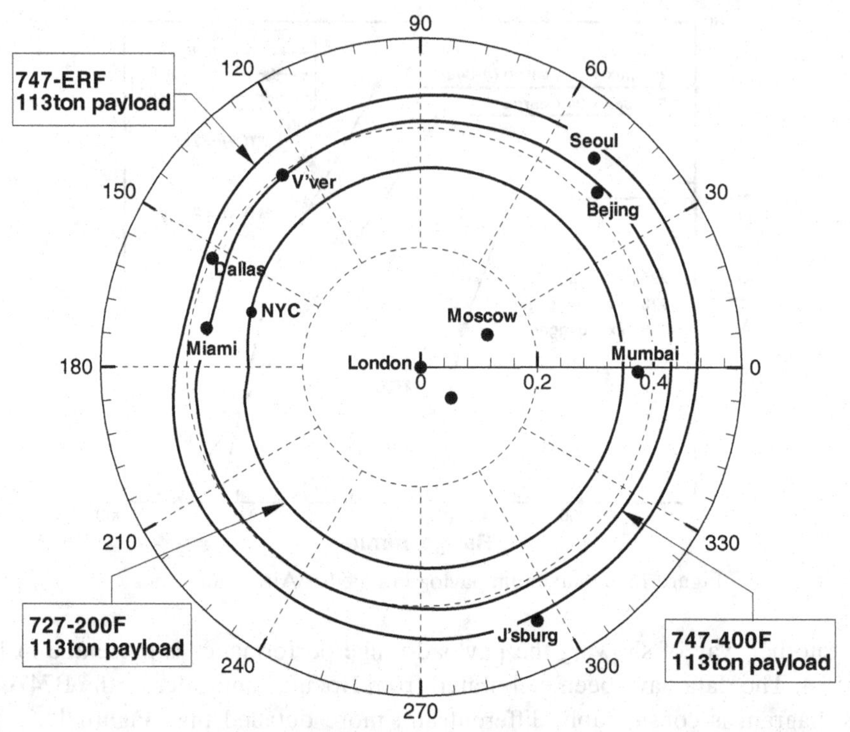


Figure 15.5. Range of Boeing 747-family of aircraft. Data elaborated from Boeing.

The ZFW is the weight of the aircraft except the loaded fuel; it includes the bulk payload, the crew and all of the operational items. At a constant BRGW, the range must decrease and, therefore, this derivative must be negative; its value is the *sensitivity* of the range to changes in ZFW at constant BRGW.

The key parametric on the flight-range effect is the wind. Figure 15.5 shows some destinations that can be reached with the Boeing 747. The data have been extrapolated from the manufacturer and include typical mission rules, 85% annual winds and air-traffic allowances. This type of chart is more useful from a marketing point of view than an engineering point of view because it gives very few details on the bulk payload, the passenger load, the flight procedures, the availability of flight corridors and so on.

Another example of a payload-range chart is shown in Figure 15.6 that refers to the business jet airplane Gulfstream G550†. The data are presented in the range-GTOW plane at selected values of the payload. The flight time corresponding to a flight distance is useful in general flight planning (such as in Figure 15.1). The fuel load has to be inferred from the data. For example, for a specified payload $W_p$ and required range $X$, the diagram returns a weight $W =$ GTOW. The corresponding fuel weight $W_f$ is found from the difference:

$$W_f = W - W_e - W_p. \tag{15.3}$$

The operating empty weight $W_e$ is given in the airplane documents (FCOM, type certificate).

† Gulfstream G550. Flight Crew Operating Manual (2005).

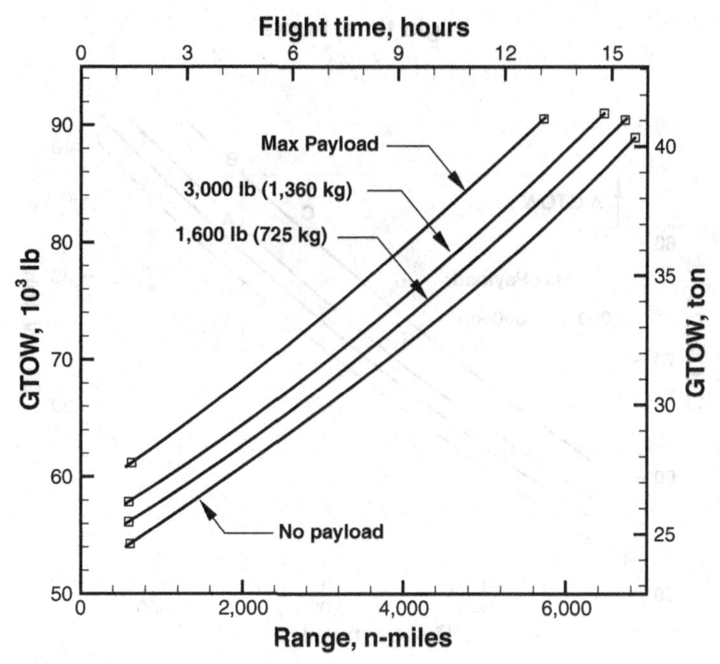


Figure 15.6. Payload-range of Gulfstream G550. Data elaborated from Gulfstream Aerospace; $M = 0.80$; 200 n-miles diversion fuel, standard day, no wind.

From Equation 15.3 the fuel fraction $\xi = W_f/W$ (fuel weight over gross weight) can be written as

$$\xi = 1 - \frac{W_e}{W} - \frac{W_p}{W}. \tag{15.4}$$

Because the empty weight is fixed, Equation 15.4 is a linear relationship between payload fraction and fuel fraction, with the gross weight $W$ being a parameter.

### 15.2.1 Case Study: Range Sensitivity Analysis

For a flight range of 5,000 n-miles (point X in Figure 15.7), the difference in GTOW between an airplane loaded to its maximum payload weight (point A) and an airplane loaded with 1,360 kg (3,000 lb) of payload (point B) is 2,052 kg (4,520 lb). The certified maximum payload is 2,812 kg (6,200 lb). This means that the difference in payload between the two cases is 760 kg (1,680 lb):

$$\Delta \text{GTOW} = 2{,}052\,\text{kg}, \qquad \Delta W_p = 760\,\text{kg}, \qquad \Delta W_f = 2{,}052 - 760 = 1{,}292\,\text{kg}.$$

This means that an increase of the payload by 760 kg requires an additional ~1,300 kg of fuel to cover the same distance, or

$$\left(\frac{dW_f}{dW_p}\right)_X = 1.7. \tag{15.5}$$

This is by any account a modest value, especially when compared to the overall operating costs of the airplane. Alternatively, consider the case of two flights at the same payload (points B and C in Figure 15.7).

$$\Delta X = 719\,\text{n-miles}, \qquad \Delta \text{GTOW} = 0, \qquad \Delta W_f = -\Delta W_p = 760\,\text{kg}.$$

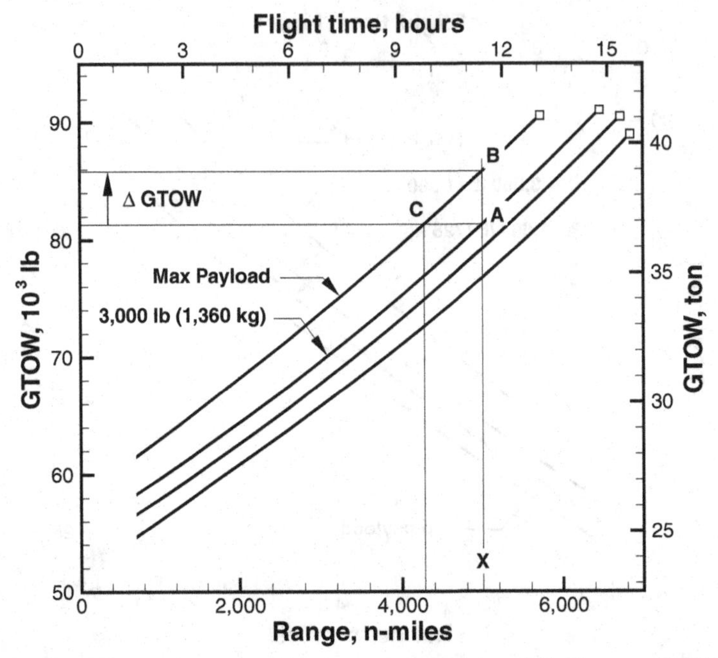


Figure 15.7. Analysis of payload-range sensitivity of Gulfstream G550.

Thus, the decrease in payload is offset by an increase in fuel, which leads to the following equivalences:

$$d\,\text{ZFW} = d\,(W_e + W_p) = dW_p = -dW_f, \tag{15.6}$$

$$\left(\frac{d\,\text{ZFW}}{dX}\right)_{GTOW} = -\frac{dW_f}{dX} = -1.057\,\text{kg/n-mile}. \tag{15.7}$$

### 15.2.2 Case Study: Payload-Range of the ATR72-500

We used the ATR72-500 powered by PW127M turboprop engines and Hamilton F568-1 propellers. This configuration has been discussed in other sections of the book. We calculated the payload-range performance for two different weights and compared our results to the "official" performance quoted by the manufacturer. These results are shown in Figure 15.8. The calculation consists of four key points: TOW-limited range, maximum-payload range, maximum-fuel range and ferry range.

### 15.2.3 Calculation of the Payload-Range Chart

We now proceed to a practical method for the calculation of the payload-range chart. To avoid computer-intensive calculations, some simplifications are required. We need to define three critical operational points: 1) the maximum-payload range $X_{p1}$; 2) the maximum-fuel range $X_{p2}$; and 3) the ferry range $X_{p3}$. The operation at ranges between $X_{p1}$ and $X_{p2}$ requires to consider the aircraft at constant BRGW. The difference is only marginal, but the additional computational burden would be considerable.

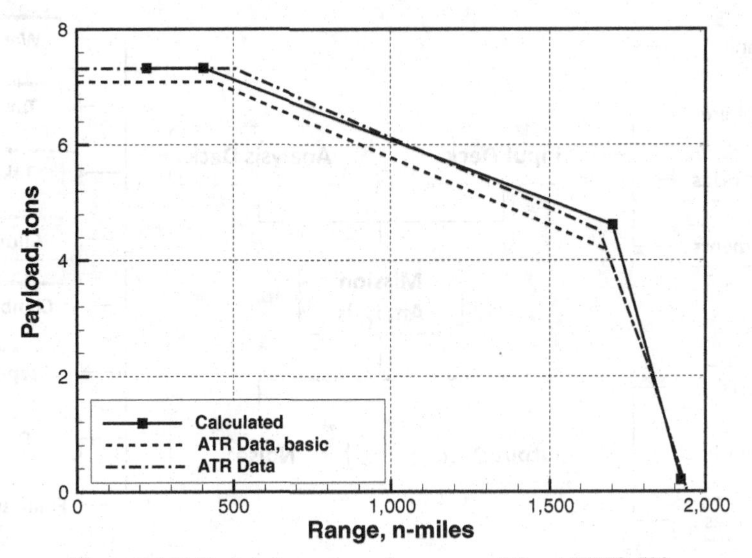

Figure 15.8. Payload-range performance of the ATR72-500.

1. **Maximum-Payload Range:** Assume that there are no passengers; assume a bulk payload equal to the MSP. Then use a bisection method to calculate the fuel planning for a tentative range $X$ with the constraint GTOW = MTOW. In other words, if the fuel planning corresponding to the tentative range $X$ leads to a GTOW < MTOW, then increase the range; otherwise decrease the range. The maximum-payload range is bracketed by $X_1 = X_{design}/2$ and $X_2 = X_{design}$. This procedure converges to $X_{p1}$ in 7 to 10 iterations, not without difficulties (it is possible that the aircraft is unable to climb to the estimated ICA, or that the GTOW corresponding to the range $X$ is considerably larger than the MTOW).
2. **Maximum-Fuel Range:** Assume GTOW = MTOW. In this case, as the range increases, payload has to be traded for fuel. If $W_{mfw}$ is maximum fuel capacity, the bulk payload must be

$$W_p = \text{GTOW} - W_e - W_{mfw}.$$

It is not useful to add passenger weight at this time because passengers require consideration of on-board services. This is a fixed quantity for the maximum-fuel range $X_{p2}$. After bracketing the range $X_{p1} < X_{p2} < X_{design}$, use the bisection method to calculate the maximum-fuel range. For a given iteration, if GTOW < MTOW, increase the tentative range, otherwise decrease the range. The procedure should converge to $X_{p2}$ in less than 10 iterations unless numerical difficulties intervene (such as at the previous point).
3. **Ferry Range:** There is no payload in this case. Thus, the take-off weight must be

$$\text{GTOW} = W_e + W_{mfw}.$$

After bracketing the solution with $X_{p2} < X_{p3} < 1.1X_{design}$, use the bisection method to calculate ferry range. For a tentative range if the mission fuel is lower than the fuel capacity, the range must increase; otherwise it must decrease.
4. **Maximum-Passenger Range:** Along with the three points specified previously, a passenger aircraft must show the design range at maximum passenger load.

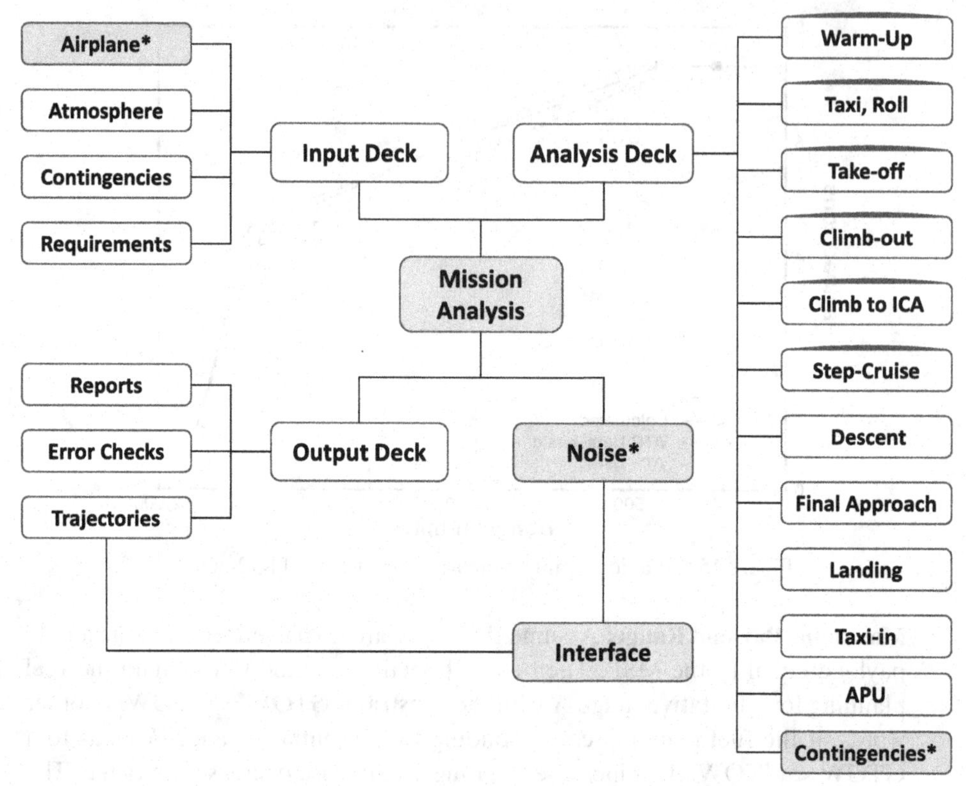


Figure 15.9. Flowchart of a mission-fuel analysis.

This range is usually between $X_{p1}$ and $X_{p2}$. For this point, the payload weight is the sum of all passengers, their baggage, average weight of service items and average number of crew members (and associated weight). After bracketing the solution with $X_{p1} < X < X_{p2}$, the solution procedure follows the same steps as in the previous scenarios.

The calculation of a payload-range must rely on a number of assumptions, some of which can be arbitrary. The fuel reserve (§15.5) and the atmospheric conditions must be specified.

## 15.3 Mission Analysis

A realistic flight planning requires an optimisation of the route (track, vertical profile), Mach number and altitude, with the constraints imposed from the air traffic control and the aviation regulators. The flight planning must be based on real-time data, such as winds, temperatures, payload and other parameters. Therefore, a detailed analysis involves several parameters. A flowchart of the computer program is shown in Figure 15.9.

The boxes with an asterisk refer to the sub-models: the airplane model is described in Chapter 2; the aircraft noise model is discussed in Chapters 16 and 17 (see also flowchart in Figure 16.3). The fuel contingencies are discussed in § 15.10. An interface is required to associate a flight trajectory to a noise calculation (§ 18.3).

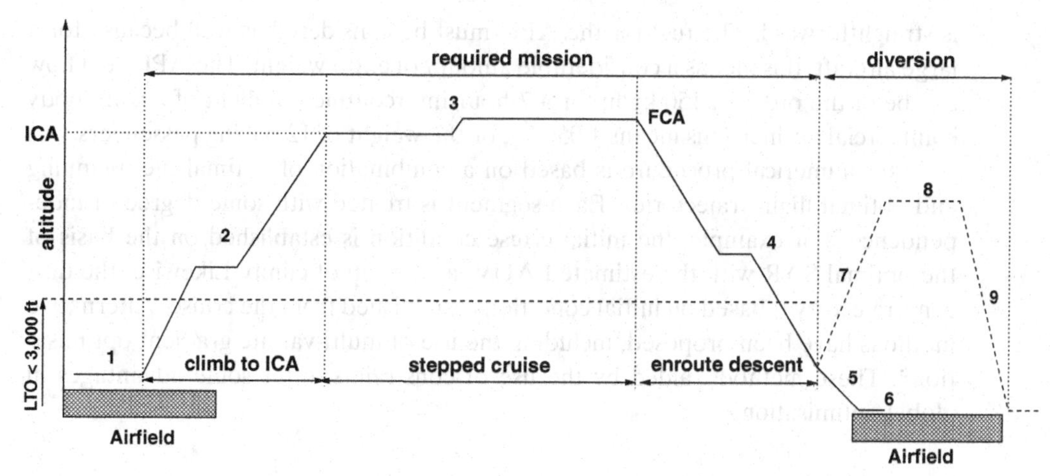


Figure 15.10. Standard mission profile of passenger aircraft, from Ref.[4].

A typical mission profile for a commercial jet is shown in Figure 15.10. It shows a climb to cruise altitude, a cruise and a descent. It also includes the case of an aborted landing and an extension of the flight at a lower altitude. Diagrams of this nature can be made more precise by considering a quantitative scale for the distance and altitude, and for each flight segment. More specifically, there are two sub-problems:

- Mission range for given fuel and payload without constraint on BRGW, § 15.3.1.
- Mission fuel for given range and payload with constraint on BRGW, § 15.4.

The weight constraint can be specified on the BRGW or the GTOW. The second case is only constrained by the fact that GTOW ≤ MTOW.

Calculation of the exact fuel required for a mission requires consideration of several objectives, some of which can be conflicting. Safety comes first. After that, the airline must consider its profitability via its own accounting system and punctuality of the service. Other items include actual payload, weather phenomena, operational constraints, route and altitude constraints. The cost of fuel is in most cases the largest cost item of a commercial airline. Price volatility makes planning difficult and risky. Fuel planning itself requires three phases:

1. Pre-flight: It refers to measures in terms of route planning, ground operations, usage of APU versus ground services, turn-around processes and other administrative tasks.
2. Flight: It refers to all of the flight procedures to be programmed in the flight computer, such as planned route, climb and descent schedules, flight level selection and so on.
3. Post-flight: It includes the collection and analysis of flight data that are used to implement strategies for reduction of fuel consumption. These measures include monitoring of the engines, the airframe and the systems from the FDR.

Calculation of the mission fuel (point 2) requires adding up the fuel required by each segment of the flight. For a passenger operation, the mission requirement is straightforward. The fuel for the APU must be considered as well because for a large aircraft, this means a considerable amount of extra weight. The APU fuel flow can be of the order of 150 kg/h; for a 7-hour intercontinental flight of a wide-body commercial airliner, this means 1,050 kg, or the weight of 12 paying passengers.

Our numerical procedure is based on a combination of optimal fuel planning and optimal flight trajectories. Each segment is treated with some degree of independence. For example, the initial cruise condition is established on the basis of the optimal SAR with the estimated AUW at the top of climb. Likewise, the descent trajectory is based on initial conditions established from the cruise. Alternative methods have been proposed, including the use of multi-variate gradient optimisation[5]. These methods, aided by the use of constraints, offer some advantages in global optimisation.

### 15.3.1 Mission Range for Given Fuel and Payload

The problem is to establish how far the aircraft can fly, with the mandatory fuel reserves and under specified atmospheric conditions. The numerical procedure is more elaborate than the calculation of a cruise range because of the implicit relationship between airplane trajectories (climb, cruise, descent) and BRGW. We propose the following algorithm:

1. Calculate the fuel from ramp up to the ICA. This operation includes taxi-out, take-off, climb to ICA and APU fuel to that point. Call this fuel $W_{f_{ICA}}$; $x_c$ is the distance to ICA.
2. Calculate the extended all-out fuel $W_{f_{ext}}$ to take into account holding and diversion procedures. This sub-task is fully specified in § 15.5.
3. Using the procedures indicated in Chapter 11, calculate the descent fuel $W_{f_d}$ and distance $x_d$ from the top of descent (final cruise altitude), with an initial weight

$$W = \text{BRGW} - \underbrace{(W_{f_{ext}} - W_{f_{res}})}_{\textit{fuel remaining}}, \tag{15.8}$$

where $W_{f_{res}}$ is the reserve fuel, which is analysed separately in § 15.5. The weight at the top of descent is equal to the BRGW minus the fuel remaining at the end of the service; this fuel includes the contribution from the diversion and the reserve. If this is the first iteration, start the descent from the ICA (the FCA is indeterminate); otherwise, estimate the FCA from the density altitude. In the latter case, the FCA is found from the condition that $W/\rho$ is a constant, or

$$\sigma_{end} = \sigma_{start} \left( W_{start} - W_{f_{cruise}} \right). \tag{15.9}$$

The step from $\sigma_{end}$ to the FCA requires the inverse solution of the atmospheric model. The FCA is then adjusted to the nearest possible flight level (§ 12.7). In any case, add the APU fuel relative to this flight segment.
4. If this is the first iteration, neglect the APU and estimate the fuel available for cruise from

$$W_{f_{cruise}} = W_{f_{usable}} - \left( W_{f_{ICA}} + W_{f_d} + W_{f_{ext}} \right). \tag{15.10}$$

In any other case, add the APU fuel for the duration of the flight.

5. If this is the first iteration, estimate the cruise range with

$$x_{cruise} \simeq \frac{U}{f_j}\left(\frac{L}{D}\right)\log\left(1+\frac{W_{f_{cruise}}}{W_f - W_{f_c}}\right). \tag{15.11}$$

In any other case, calculate the cruise fuel according to the procedures explained in Chapter 12 (stepped-climb).

6. The estimated mission range is

$$X_1 = x_c + x_d + x_{cruise}. \tag{15.12}$$

7. The calculations are repeated from point 2. The stopping criterion is based on the change of $X_1$ between iterations; when the updated mission range does not change by any meaningful amount (for example, 1 n-mile), the procedure has converged.

Note that the convergence analysis is done on the mission range. An important sub-case is when

$$W_e + W_f + W_p = \text{MRW}. \tag{15.13}$$

When $W_p$ is maximum and $W_f$ is the maximum allowable fuel weight compatible with the MRW, the corresponding range is the *maximum-payload range*. When the fuel weight is maximum and the payload is the maximum allowable weight compatible with the MRW we have the *maximum-fuel range*. Finally, if $W_p = 0$, the ramp weight is most likely below the MRW. The corresponding range is called *maximum-ferry range*.

## 15.4 Mission Fuel for Given Range and Payload

The problem is to establish the amount of fuel to load onto the aircraft in order to transport a specified payload over a specified distance $X_{req}$, under specified (or forecast) atmospheric conditions. At the end of the process, we will have a detailed weight breakdown. There are two important steps: 1) make a first estimate of the mission fuel, and 2) improve the first estimate with a suitable numerical method.

### 15.4.1 Mission-Fuel Prediction

We define the computational steps required in a procedure called `MissionFuel(..)`, which returns the fuel required for a given range, payload, atmospheric conditions and other external factors. The computational procedure is as follows:

1. Estimate the ramp mass (or weight or the BRGW) by making a guess of the fuel required by the mission, $W_f^*$. This weight is

$$W \simeq W_e + W_p + W_f^*. \tag{15.14}$$

2. Calculate the fuel from the ramp (or gate) up to the ICA. This operation includes taxi-out, take-off, climb to ICA and APU fuel to that point. Call this fuel $W_{f_{ICA}}$; $x_c$ is the distance to ICA.

3. Calculate the extended all-out fuel $W_{f_{ext}}$ to take into account holding and diversion procedures. This sub-task is fully specified in § 15.5.
4. Using the procedures indicated in Chapter 11, calculate the descent fuel $W_{f_d}$ and distance $x_d$ from the top of descent. Follow the same procedure described at point 3 on page 434.
5. Calculate the cruise distance from

$$x_{cruise} = X_{req} - x_c - x_d. \tag{15.15}$$

6. Calculate the cruise fuel required to fly the distance $x_{cruise}$ from the conditions at ICA (see point 2). This is done with step climbs, following the procedure discussed in § 12.7, by integration of the SAR. Add the APU fuel for this segment.
7. Calculate the ramp weight. This is derived from the sum of all segment fuels.

With reference to point 1 in this procedure, the initial fuel estimate is either provided externally or is specified by the following equation:

$$W_f \simeq c \left( \frac{X_{req}}{X_{design}} \right) c_{usf}, \tag{15.16}$$

where $c_{usf} < 1$ denotes the amount of usable fuel as a portion of the full tanks; the coefficient $c = 0.90$ to $0.95$ is a factor that improves the prediction and is found from extensive numerical analysis.

### 15.4.2 Mission-Fuel Iterations

There are at least three different methods for calculating iteratively the mission fuel:

a) Method based on updating the initial guess until convergence.
b) Method based on updating the latest guess with an under-relaxation.
c) Method based on a predictor-corrector of the initial guess.

These methods are listed from the less robust to the most robust. In some circumstances all of these methods converge, but there is no guarantee that they will always converge. The difficulty increases with the increasing required range and with the size of the airplane because of the increased non-linearity between the fuel weight and the mission fuel.

**A) MISSION-FUEL UPDATE.** Compare the weight $W_i$ calculated at iteration $i$ with the earlier estimate $W_{i-1}$. Now use the updated ramp weight to recalculate the mission fuel. The residual is defined as

$$E = 1 - \frac{W_i}{W_{i+1}}. \tag{15.17}$$

There is no guarantee that this procedure converges or that it converges to the correct weight. It depends both on how good is the first estimate and on the required range; it gets worse as the required range is close to the design range. In the latter case, it is possible than on the second iteration, the ramp weight exceeds the MRW and thence the climb becomes sluggish (not enough excess power is available), the

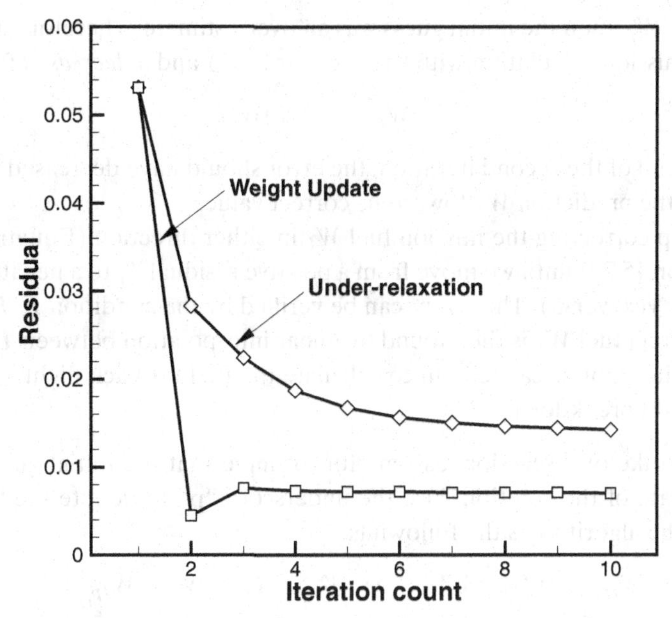


Figure 15.11. Error monitoring, Equation 15.17, in iterative mission-fuel analysis.

best ICA cannot be reached and, ultimately, the computations fail. Therefore, it may be useful to add a constraint so that in no case the ramp weight exceeds the MRW.

**B) MISSION-FUEL UPDATE WITH UNDER-RELAXATION.** A method that partially overcomes this convergence problem consists in using an under-relaxation. If $W_i$ is the initial weight estimate, and $W_{i+1}$ is the new weight estimate, then the iteration $i+2$ is started with

$$W_{i+2} = \frac{1}{2}\left(W_i + W_{i+1}\right). \tag{15.18}$$

This method does converge but it is considerably slower. A comparison between the two methods is shown in Figure 15.11. Neither option is satisfactory.

**C) PREDICTOR-CORRECTOR METHOD.** Given an estimate $W_f^*$ of the mission fuel, a mission calculation should predict that the mission fuel is $W_f = W_f^*$. However, if this were to be the case, it would be a lucky outcome. In fact, we have either $W_f > W_f^*$ or $W_f < W_f^*$. The difference $E = W_f - W_f^*$ is the error (or residual). The problem is how to improve this prediction. The algorithm is the following:

- Estimate the mission fuel $W_f^*$ and call the function `MissionFuel(..)`, which returns the fuel $W_f$. This is the first iteration, $i = 1$.
- If $W_f < W_f^*$, then the initial guess was an under-estimate. Therefore, we perform a new mission calculation with `MissionFuel(..)` and an *increased* fuel

$$W_f = (1+\epsilon)W_f^*, \tag{15.19}$$

where $\epsilon$ denotes a small fraction of $W_f^*$. Typical values are $\epsilon = 0.02$ to 0.04, although this depends on how good the first estimate $W_f^*$ is.

- If $W_f > W_f^*$, then the initial guess was an over-estimate. Therefore, we perform a new mission calculation with `MissionFuel(..)` and a *decreased* fuel

$$W_f = (1 - \epsilon) W_f^*. \tag{15.20}$$

- At the end of the second iteration, the error should have decreased because we moved the prediction $W_f$ toward its correct value.
- We keep correcting the mission fuel $W_f$ in either direction (Equation 15.19 or Equation 15.20) until we move from a positive residual $E_i$ to a negative residual $E_{i+1}$ (or vice versa). This event can be verified by the condition $E_i \; E_{i+1} < 0$.
- The correct fuel $W_f$ is then found by linear interpolation between $E_i$ and $E_{i+1}$.
- A final iteration is carried out to calculate the fuel for each flight segment and the weight breakdown.

For the sake of discussion, assume for example that the initial guess $W_f^*$ is an under-estimate of the solution. Use the underscore "p" to denote the "predicted" fuel, $W_{fp}$. The algorithm is the following:

$$W_{fp_i} = W_f^* + \epsilon \, W_f^* i, \quad \rightarrow W_{f_i}, \quad E_i = W_{f_i} - W_{fp_i} > 0 \tag{15.21}$$

$$W_{fp_{i+1}} = W_f^* + \epsilon \, W_f^*(i+1), \quad \rightarrow W_{f_{i+1}}, \quad E_{i+1} = W_{f_{i+1}} - W_{fp_{i+1}} < 0 \tag{15.22}$$

$$W_{fp_{i+2}} = \texttt{Interpolation}(E_i, E_{i+1}, W_{fp_i}, W_{fp_{i+1}}). \tag{15.23}$$

Equations 15.21 and 15.22 give a prediction of the mission fuel at iteration $i, i+1$, respectively. Equation 15.23 is the correction and the solution of the problem; the corresponding residual is $E_{i+2} \simeq 0$. Note that the residual is calculated between the fuel $W_{f_i}$ returned by the procedure `MissionFuel(..)` and the predicted fuel $W_{fp_i}$. The minimum number of iterations required for convergence is three.

Figure 15.12 shows the convergence of this method for two values of the required range. This case refers to a Boeing B747-400 with CF6 engines on a standard day, with full passenger load. In graph 15.12a convergence is achieved after four iterations. At the second iteration the residual moves below zero; therefore, the interpolation Equation 15.23 returns a corrected mission fuel. Due to the non-linearity around the solution, another iteration is required. In graph 15.12b convergence is achieved in three iterations.

## 15.5 Reserve Fuel

A fuel reserve is required by all of the regulators (FAA, ICAO, AEA, and so on). This reserve depends also on the choice of alternate airport and company policies. For example, the Association of European Airlines (AEA) specifies a 200 n-miles (370 km) diversion flight for short- and medium-range aircraft and 250 n-miles (463 km) for long-range aircraft. In addition, a 30-minute holding at 1,500 feet (457 m) altitude and a 5% mission fuel reserve for contingency are required. For domestic flights in the United States it is specified 130 n-miles (~240 km) diversion and 30 minutes holding at 1,500 feet. However, a number of alternatives are allowed, some of which are subject to prior approval. These include a 20-minute extension of the

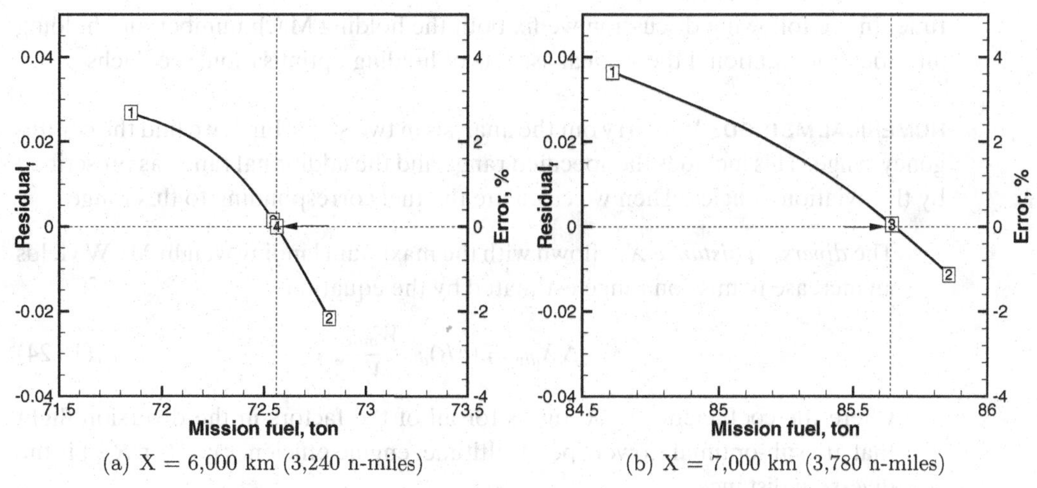


(a) X = 6,000 km (3,240 n-miles) (b) X = 7,000 km (3,780 n-miles)

Figure 15.12. Predictor-corrector analysis of mission fuel for a Boeing B747-400-CF6; standard day, no wind.

trip or a 3% trip fuel if an alternate airport can be found en route. The ICAO International* rules prescribe that the reserve fuel should take the aircraft to the alternate airport specified in the flight plan and then hold for 30 minutes at 1,500 feet before landing at standard atmospheric conditions. Upon landing, the remaining fuel must be a minimum 3% of the trip fuel.

From an operational point of view, the range may be limited by the absence of diversion airports over a predefined route. The ICAO has a particular rule (ETOPS) permitting twin-engine aircraft to fly longer routes (previously off-limits) that have no diversion airports within a 60- or 120-minute flight. This rule allows several modern twin-engine aircraft to fly over the oceans and in remote parts of the world. A discussion of ETOPS performance with OEI is available in Martinez-Val and Perez[6].

**HOLDING STACKS.** When holding is required at the destination airport, the aircraft is placed on a *racetrack pattern* that is made of two straight legs and two U-turns (§ 13.2.4). The key aspect of the holding pattern is maximum endurance rather than maximum range. The maximum endurance is achieved at minimum fuel flow or maximum glide ratio. However, at some airports the aircraft may be subject to holding at a specified air speed (or Mach number) that is sub-optimal. For the model Airbus A300-600, the holding speed is 210 kt with a partial slat extension and 240 kt in clean configuration. An alternative is to place the aircraft on a linear holding for part of the time (if the pilots are advised in time). This solution extends the cruise time and reduces the racetrack holding; potentially, it can save a considerable amount of fuel. For example, a 15-minute linear holding of the reference aircraft at $h = 35{,}000$ feet and $M = 0.80$ can save $\sim$100 kg of fuel.

Optimisation of the holding-pattern performance is critical for short-segment flights because it is not unusual to have a holding time of the same order as the cruise

* ICAO Annex 6: §4.3.6.3.

time. In the following discussion we fix both the holding Mach number and holding altitude. For additional theoretical aspects of holding optimisation, see Sachs[7].

**NUMERICAL METHOD.** We carry out the analysis in two steps. First, we find the contingency range. This includes the specified range and the additional range as prescribed by the aviation policies. Then we calculate the fuel corresponding to this range.

- The *diversion distance* $X_{div}$ flown with the maximum landing weight MLW yields an increase in mission range estimated by the equation

$$\Delta X_{div} = (cR)_{div} \frac{W_{mlw}}{W_{to}}, \tag{15.24}$$

  where the coefficient $c_{div}$ accounts for all of the factors in the diversion flight that are sub-optimal (lower speed, altitude, engine efficiency); $R$ (or $R_{div}$) is the diversion distance.
- The *holding time* (or loiter) flown at the maximum landing weight MLW yields an increase in mission range estimated by the equation

$$\Delta X_{hold} = (cUt)_{hold} \frac{W_{mlw}}{W_{to}}, \tag{15.25}$$

  where the coefficient $c_{hold}$ accounts for loss of efficiency as in the diversion flight. The holding speed is generally half of the cruising speed.
- The *extended duration* of the flight at the cruise speed for a time $t_{exd}$ yields an increase in mission range estimated by the equation

$$\Delta X = U t_{exd}. \tag{15.26}$$

- The *contingency fuel* is a percentage of the mission fuel $m_{f_{res}}/m_f = 0.05$–$0.10$, depending on actual policies.

The *equivalent all-out range* is then the sum of all contributions

$$X_{out} = X_{req}\left(1 + \frac{m_{f_{res}}}{m_f}\right) + \left[\,(cR)_{div} + (cUt)_{hold}\right]\frac{W_{mlw}}{W_{to}} + U t_{exd}, \tag{15.27}$$

where $X_{req}$ is the mission range required. The exact calculation of this parameter is essential. Torenbeek[8] proposed some practical values for the coefficients in Equation 15.27:

$$c_{div} = c_{hold} \simeq 1.1 + 0.5\,\eta_M, \tag{15.28}$$

where $\eta_M$ is the log-derivative of the propulsive efficiency with respect to the Mach number. For a modern high by-pass turbofan engine this quantity is estimated at $\eta_M \simeq 0.225$ for $M = 0.8$; it increases with the decreasing Mach number, so that $\eta_M \simeq 0.325$ at $M = 0.4$. Equation 15.27 can be further simplified if the reserve fuel is used for extension of the range at the cruise conditions.

An alternative method consists in adding separately the contributions for hold and diversion. Clearly, a number of parameters are required for the solution of this problem: the hold time, altitude and speed; the diversion distance, altitude and speed. The fuel required from these two contingency segments will be added to the mission fuel.

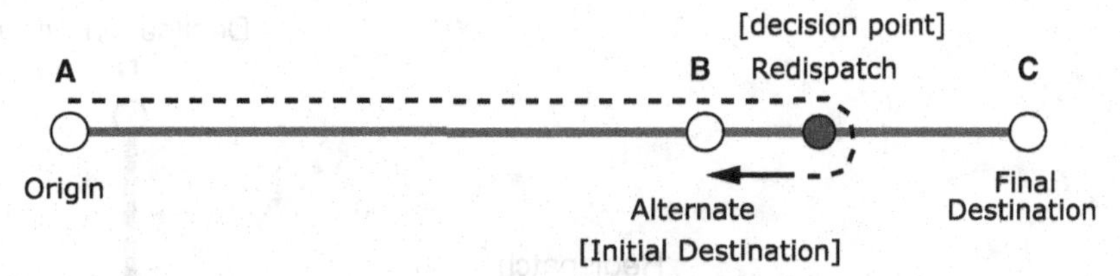


Figure 15.13. Fuel-reserve planning with redispatch.

Finally, we need to consider the *unusable fuel*, that is, the fuel that cannot be pumped into the engines. Although this is a small portion of the fuel capacity (typically 0.4 to 0.8% of the capacity), it is a constant amount of fuel that adds to the operating empty weight.

### 15.5.1 Redispatch Procedure

The reduction of the reserve fuel relies on the existence of an alternate airport, as shown in Figure 15.13. The graph indicates the departure point (*origin*), the alternate airport (*initial destination*) and the final destination. The graph shows also a waypoint (*redispatch*). The redispatch procedure is regulated by the relevant authorities*.

The reserve fuel must be sufficient to take the airplane from the origin to the initial destination B. In normal circumstances, having passed the point B, the airplane continues its flight; upon reaching the waypoint a decision must be made whether to divert to the initial destination or to continue to the final destination; this is done on the basis of a comparison between the predicted fuel burn and the available fuel. In this scenario, the reserve fuel required is limited to the flight segment AB. By the time the airplane has reached the initial destination, none of the reserve fuel is used, unless the atmospheric conditions turn out to be worse than forecast.

Now consider a more complex network of alternatives, as shown in Figure 15.14, which indicates a refuelling station, an alternative airport beyond the final destination and a second alternate after refuelling.

It is possible to show that the minimum mission fuel with the redispatch procedure is

$$m_f = m_{f_{taxi,A}} + \underbrace{m_{f_{AC}}}_{trip} + \underbrace{0.1 m_{f_{BC}}}_{reserve} + \underbrace{m_{f_{CD}}}_{altern.} + m_{f_{hold,D}} + \cdots \tag{15.29}$$

With the standard procedure, the mission fuel is

$$m_f = m_{f_{taxi,A}} + \underbrace{m_{f_{AC}}}_{trip} + \underbrace{0.1 m_{f_{AC}}}_{reserve} + \underbrace{m_{f_{CD}}}_{altern.} + m_{f_{hold,D}} + \cdots \tag{15.30}$$

The difference between the two procedures is in the reserve fuel. If the redispatch point B is chosen appropriately, a considerable amount of reserve fuel can be saved. Redispatch is not considered in the parametric fuel-reserve analysis carried out in § 15.10.

* See, for example, FAR § 121.631: Original dispatch or flight release.

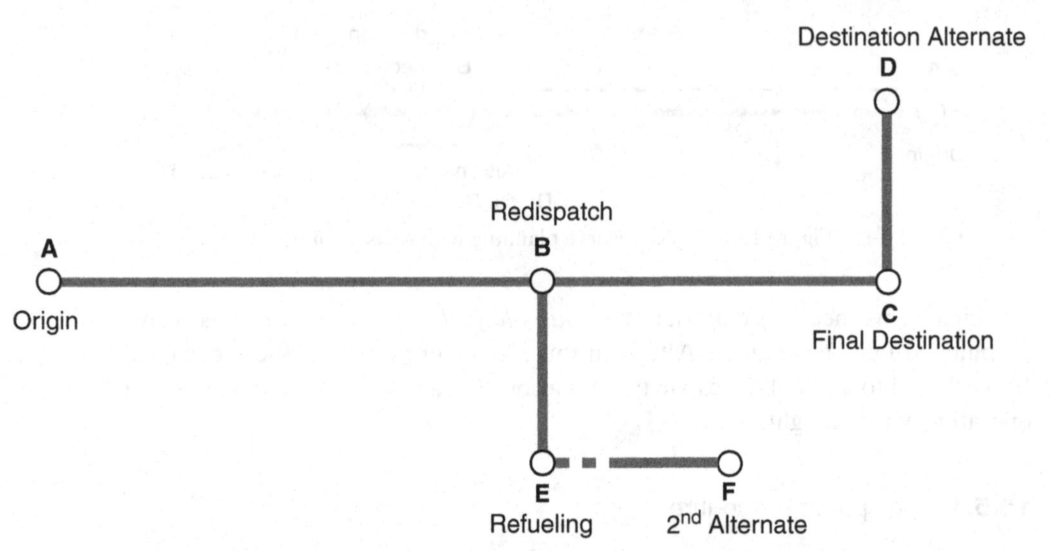


Figure 15.14. Fuel-reserve planning with redispatch for a network of alternatives.

## 15.6 Take-Off Weight Limited by MLW

Flight over short distances requires a limitation of the TOW due to the relatively low maximum landing weight (MLW). The problem is the following: calculate the distance that the airplane must fly if it is loaded with the maximum payload, takes off with the maximum take-off weight (MTOW) and lands with the MLW. This range is calculated from the following numerical procedure:

- The aircraft starts with MRW, a payload $W_{msp}$ = MSP, and whatever fuel $W_f$ can be allocated, so that

$$\mathrm{GRW} = W_e + W_{msp} + W_f. \tag{15.31}$$

- The fuel required to taxi out, take off and climb to a cruise altitude (to be determined) is calculated and added. By the time the aircraft has reached the "cruise" altitude, it will have burned a fuel $W_1$ and travelled a distance $x_1$.
- Now consider the descent phase. Estimate the weight at the top of descent from

$$W_d \simeq W_{mlw}(1+r), \tag{15.32}$$

where $r$ denotes the fuel reserve, as percent of the initial fuel. Calculate the descent fuel $W_3$ and the en-route descent distance $x_3$.
- Finally, consider the cruise range $x_2$ (unknown). The aircraft weight at the start and end of the cruise should be

$$W_{start} = \mathrm{GRW} - W_1, \qquad W_{end} = W_{start} - W_{f_c}, \tag{15.33}$$

where the cruise fuel $W_{f_c}$ is

$$W_{f_c} = W_{mtow} - W_{mlw} - W_1. \tag{15.34}$$

- Estimate the cruise range from the range equation

$$x_2 = \frac{U}{gf_j}\left(\frac{L}{D}\right)\log\left(\frac{W_{end}}{W_{start}}\right). \tag{15.35}$$

- The MLW-limited range is the following:

$$x_{mlw} \simeq x_1 + x_2 + x_3. \quad (15.36)$$

The range estimate given by Equation 15.36 can be improved by further iterations, but the result does not change a great deal.

As a practical example, consider the Airbus A-380-861 with GP-7270 engines. This aircraft has the following characteristics: MSP = 90.7 tons; OEW = 270.3 tons, MLW = 386.0 tons. If we load the airplane with the MSP, we have ZFW = 361.0 tons, which is exactly the certified MZFW. In this case, the amount of fuel required to reach the MLW is $\sim$27 tons. With such an amount of fuel, the airplane can quickly return to the airfield of origin if it was to abort the mission. Because in a typical climb the aircraft burns $\sim$9 tons of fuel, in addition to $\sim$ 2 tons for taxi and take-off, the aircraft will have to dispose of about 16 tons of fuel before landing.

## 15.7 Mission Problems

There are quite a variety of mission scenarios. Even in the relatively simple case of a transport aircraft, important decisions may have to be taken when planning a flight and during the flight. In this section we deal with a number of practical problems. We begin with the mission analysis with an intermediate stop (§ 15.7.1); in this case we want to understand whether there are benefits in terms of fuel consumption if we split the mission and call at an intermediate destination. The second problem deals with fuel tankering, which is a practice sometimes used to save money at destination airports if the fuel is more expensive there (§15.7.2). Finally, we consider a contingency scenario in which the pilot may have to abort the flight and decide whether to return to the origin airport or continue to its planned destination (§15.7.3).

### 15.7.1 Cruise with Intermediate Stop

Is it more economical to fly a long distance no-stop or with an intermediate stop? Some airlines occasionally offer cheaper seats to a destination via an off-route stop, which is counter-intuitive to both the customers and the airplane designers. The aviation industry has developed aircraft for very long range to serve distant points without en-route stops. Some of the current-generation airlines can fly, in some limited scenarios, in excess of 10,000 n-miles. This is close to the ultimate *global range*, 20,000 km ($\sim$10,790 n-miles), advocated by Küchemann[9]. The global range is half the Earth's circumference at the equator and would allow an aircraft to fly from any point to anywhere around the world – at least in principle.

The problem is one of long-range cruise in which a subsonic commercial jet has to travel from airport $A$ to a destination airport $B$. The distance along the flight corridor is within the certified maximum range of the aircraft. However, here we consider the problem of an *en-route* stop at airport $C$. The aircraft carries enough fuel to reach airport $C$, refuels at this base, takes off again and flies to its final destination $B$. We do not consider the costs of landing at $C$ or the direct operating costs incurred by the operator for increasing the time of the return journey. If $x$ is the distance to be flown in a direct flight, and $x_1$, $x_2$ are the flight segments, we assume that

$$x_1 + x_2 = x. \quad (15.37)$$

Table 15.1. *Fuel use for mixed long- and short-range service of the Boeing B777-300 (calculated)*

| $X$[n-m] | $m_f$[kg] total | $m_f$[kg] climb | $m_f$[kg] cruise | $\dot{m}_f$[kg/s] cruise |
|---|---|---|---|---|
| 300 | 6,674 | 2,553.9 | 955.0 | 2.04 |
| 5,000 | 106,287 | 3,056.3 | 94,864.6 | 2.63 |
| 5,300 | 112,396 | 3,056.6 | 100,662.7 | 2.62 |

In other words, the en-route stop $C$ is on the route to the final destination $B$. A stop with a diversion clearly increases the total range. The method consists in calculating the mission fuel for three flight segments: $x$, $x_1$, $x_2$. We then calculate the relative fuel cost. This is the ratio between the fuel required by the flight with en-route stop and the direct flight:

$$m_{fr} = \frac{m_f(x_1) + m_f(x_2)}{m_f(x)}. \tag{15.38}$$

This quantity can be plotted as a function of $x_1/x$ (the ratio between the first flight segment and the total range). Consider the case in which an airline operates an intercontinental service between A and B separated by 5,000 n-miles. Another important destination is C, which is 300 n-miles from B. The results of the calculation performed with the FLIGHT program are shown in Table 15.1. The data show that stopping at en-route airport at 5,000 n-miles and proceeding to the final destination 300 n-miles away causes an increase in fuel consumption by 0.5%.

### 15.7.2 Fuel Tankering

The aircraft has to carry the minimum required amount of fuel to perform its mission. However, this type of analysis does not take into account any contingencies arising from the costs of purchasing fuel. Some commercial operators use the practice of "tankering", that is, loading more fuel than required by a mission profile to offset the costs of purchasing fuel at a higher price at destination. As a rule, the longer the flight, the less capacity there is for tankering. Figure 15.15 shows the tankering performance of the Airbus A320-200, adapted from the manufacturer's data*. The optimum weight is not fully specified; that is, it is not clear what payload is carried.

Assume that the required range is $X_{req}$ for a total payload weight $W_p$. The fuel required for this mission is $W_f$ (including fuel reserves). To establish the optimum tankering, we need to forecast the payload and the weather for the return journey. At this point the matter can become complicated; to simplify, we assume that the aircraft performs the return journey with the same payload and atmospheric conditions. Hence, the amount of fuel required is the same.

From the point of view of the calculation, there is no difference between fuel weight and payload weight. The increase in fuel burn due to an increase in payload (or fuel itself) is the derivative

$$m = \frac{dW_f}{dW_{to}}. \tag{15.39}$$

* Airbus: *Getting to Grips with Fuel Economy*, Issue 3, Blagnac, France, July 2004.

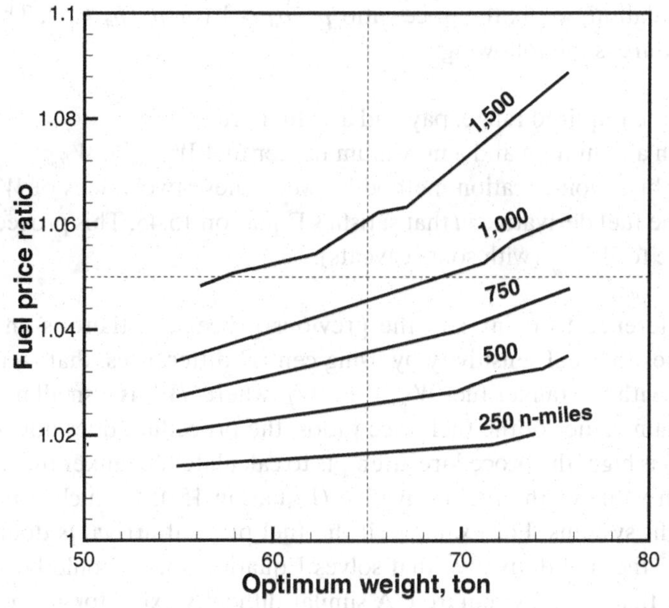


Figure 15.15. Tankering performance of the Airbus A320 relative to cruise FL-330. Data adapted from Airbus.

This derivative is always positive. Therefore, the *additional* fuel required to carry the tanker fuel $W_{f_{tanker}}$ is $mW_{f_{tanker}}$; the total increase in take-off weight (or increase in fuel weight) is

$$\Delta W_{to} = \Delta W_f = (m+1)W_{f_{tanker}}. \tag{15.40}$$

The price to be paid at departure for this additional fuel is

$$\mathcal{P}_d = mp_d W_{f_{tanker}} + p_d W_{f_{tanker}}. \tag{15.41}$$

The first contribution is due to the cost of carrying the tanker fuel; the second term is the cost of the tanker fuel itself. The cost saved at arrival is

$$\mathcal{P}_a = p_a W_{f_{tanker}}. \tag{15.42}$$

The difference $\mathcal{P}_d - \mathcal{P}_a$ must be negative (which corresponds to a saving):

$$\mathcal{P}_d - \mathcal{P}_a = mp_d W_{f_{tanker}} + p_d W_{f_{tanker}} - p_a W_{f_{tanker}} < 0, \tag{15.43}$$

which implies

$$\frac{p_a}{p_d} > 1 + m. \tag{15.44}$$

The tanker fuel in Equation 15.43 is indeterminate. Note, however, that the fuel derivative (Equation 15.39) increases with the tanker fuel. Therefore, for a given favourable fuel price ratio, it becomes progressively less convenient to tanker any fuel; alternatively, for a marginally favourable fuel price ratio, an increase in tanker fuel can be offset by the additional cost of carrying it. Mathematically, the optimum tanker fuel is $W^*_{f_{tanker}}$ such that

$$m = \frac{p_a}{p_d} - 1. \tag{15.45}$$

For our calculation, we fix the price ratio $p_a/p_d > 1$ (or $p_d/p_a < 1$). The computational procedure is the following:

1. Specify the required range, payload and fuel price ratio.
2. Establish a minimum and a maximum tanker fuel $W_{f_{tanker-1}}$, $W_{f_{tanker-2}}$.
3. Use the bisection-iteration method between these two values of $W_{f_{tanker}}$ to calculate the fuel derivative $m$ that satisfies Equation 15.45. This procedure should converge to $W^*_{f_{tanker}}$ (with some caveats).

With reference to point 3 in the previous procedure, there is an inner loop that calculates the fuel sensitivity by using central differences, that is, two mission calculations with the tanker fuel $W_{f_{tanker}} \pm dW_f$, where $dW_f$ is a small number.

For certain values of the fuel price ratios, the procedure does not converge. In fact, if $p_a/p_d$ is high, the procedure attempts to calculate the tanker fuel corresponding to a high value of the fuel derivative (Equation 15.45), which is incompatible with the flight systems. For example, if the fuel price at arrival is double the cost at departure, the fuel derivative that solves Equation 15.45 should be $m = 1$, a far higher value than actually required. A similar difficulty exists for $p_a/p_d \simeq 1$, which implies that for marginal differences in fuel costs, it is uneconomical to tanker any fuel. Realistic values of the fuel derivative are $m = 0.01$ to $0.07$, depending on aircraft, range, payload and other factors. We conclude that in these circumstances it is always convenient to use tankering.

For the Airbus A320-200, we have carried out the following analysis: $X_{req} = 750$ n-miles; bulk payload $W_p = 2{,}000$ kg; passenger load equal to 80%; standard day, with a price fuel ratio $p_a/p_d = 1.05$. The optimum tanker fuel is $W_{f_{tanker}} = 1{,}273$ kg, corresponding to a ramp weight of ~65,400 kg. This result does not match the value quoted by the manufacturer, which in any case is not fully specified. The iterative analysis carried out with the FLIGHT program is shown in Figure 15.16. The error on Equation 15.45, that is, the difference between left- and right-hand sides, is ~0.002 after four iterations.

Because fuel tankering causes additional fuel burn, it also causes additional environmental emissions. If environmental costs are added to the analysis, the incentive to tanker may disappear altogether. In any case, to be precise we would need to account for secondary costs, such as increased maintainance costs on engines, tyre and brakes wear, discounted for the shorter turn-around time at destination. Thus from an operational point of view, it may be more convenient to formulate the problem as a break-even fuel price ratio.

### 15.7.3 Equal-Time Point and Point-of-No-Return

If diversion is required due to external events, landing must be planned at an airport different from the planned destination. Such events may include engine failure, loss in cabin pressure, adverse weather, air traffic control and more. A contingency plan is formulated on the basis of the equal-time point and the point-of-no-return. The situation is shown in the sketch of Figure 15.17. These quantities are defined as follows:

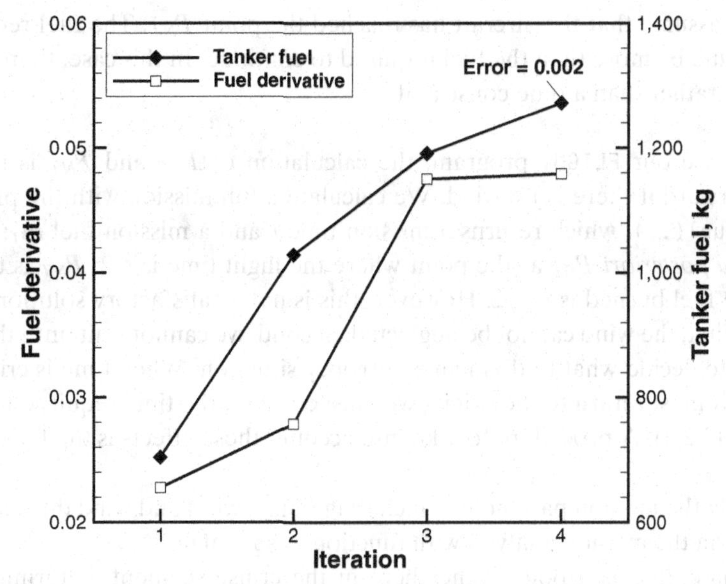


Figure 15.16. Fuel-tankering calculation for the Airbus A320-200.

- The equal-time point $P_{ET}$ is the point during the flight when the time required to return to the airport of origin and the time required to continue the flight to the planned destination are equal.
- The point-of-no-return $P_{NT}$ is the point during the flight when the fuel remaining is not sufficient to return to the airport of origin.

The point-of-no-return comes *after* the point of equal time. The wind direction is critical. These two points must be converted to a geographical position and must be assigned an estimated time of arrival.

Assume that the aircraft has reached the point $P_{ET}$. Due to external events, it must be decided whether to return or continue. At this point, the fuel required for the inbound flight (return to origin) is less than the fuel required for the outbound flight. However, the flight time is the same. In this case, there is a time constraint (or priority) rather than a fuel constraint.

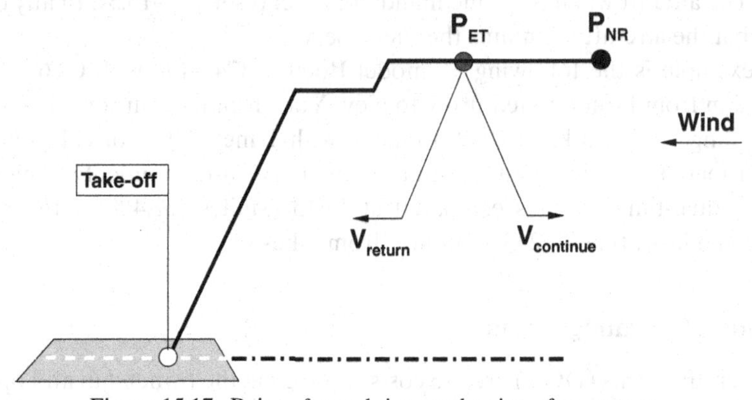


Figure 15.17. Point of equal time and point-of-no-return.

Next, assume that the aircraft has reached the point $P_{NT}$. The fuel required to return would be more than the fuel required to continue. In this case, there is a fuel constraint rather than a time constraint.

If we use our `FLIGHT` program, the calculation of $P_{ET}$ and $P_{NT}$ is relatively straightforward if there is no wind. We calculate a full mission with the procedure `MissionFuel(..)`, which returns a mission time $t$ and a mission fuel $m_f$; we then calculate *a posteriori* $P_{ET}$ as the point where the flight time is $t/2$; $P_{NT}$ is the point where the fuel burned is $m_f/2$. However, this is not a satisfactory solution for two reasons. First, the wind cannot be neglected. Second, we cannot wait until the end of the flight to decide what to do in an emergency situation. When time is critical and the aircraft must return to the origin, we must consider the time required to make a U-turn (§ 13.2.4). A procedure to take into account these effects is the following:

1. Specify the mission parameters, including range, wind and wind direction.
2. Perform the mission analysis with function `MissionFuel(..)`.
3. Estimate $P_{ET}$ as a point somewhere in the cruise segment, determined by a distance $X_{ET}$ from the origin airport and a flight time $t_{ET}$.
4. Perform the mission analysis with the function `MissionFuel(..)` up to $X_{ET}$; the distance to continue would be $X_c = X_1 - X_{ET}$; the remaining time would be $t = t_1 - t_{ET}$.
5. Perform a U-turn at the cruise altitude (§ 13.2.4).
6. Switch wind direction.
7. Continue the mission to the destination (e.g., for the remaining distance).
8. Compare the time to return $t_{ET}$ with the time to continue $t$: if $t_{ET} > t_c$, then the point $P_{ET}$ was over-estimated; otherwise, it was under-estimated.

The last point of this procedure can be implemented into a bisection method that iterates between an under-estimate and an over-estimate of the solution. The procedure can be computationally demanding because it requires two calls to `MissionFuel(..)` to bracket the solution and five to seven bisection iterations. Because each mission analysis requires about four to five iterations to converge, there are up 40 to 50 calls to `MissionFuel(..)`, unless the converge criterion is relaxed. The aircraft will have some mandatory fuel reserve (§ 15.5). In any case, we assume that the aircraft maintains the fuel reserve.

An example is the following. A model Boeing B747-400 with CF6 engines is on a mission from London Heathrow to New York Kennedy airport. The nominal required range is 5,700 km (3,078 n-miles), with 1 metric ton of cargo and 80% passenger load. It encounters headwinds of 39 kt at cruise altitude. For this specific case, the equal-time point is estimated at 1,615 n-miles (52.4% of the required distance) at a flight time of 233.5 minutes from take-off.

## 15.8 Direct Operating Costs

Direct operating costs (DOC) are the costs incurred in the ownership and operation of an aircraft. They include the cost to operate scheduled and unscheduled flights, to insure and to maintain the aircraft as airworthy. The aircraft costs money even if it stays on the ground. The parametrisation of the fuel costs is aleatory, due to the costs of aviation fuel on the international markets and to the cost of fuel at different airports around the world.

An example of DOC analysis is given by Beltramo *et al.*[10], who developed cost and weight estimating relationships for commercial and military transport aircraft. Kershner[11] has demonstrated that whilst DOC have consistently decreased over time, the impact of the fuel cost has remained high and in some historical circumstances has increased to more than 50%. The prediction of the operational costs due to fuel consumption can be calculated with the methods explained in this chapter. Other studies in this area focus on the fuel consumption for subsonic flight[12] and for supersonic transport[13].

The appearance of *budget airlines* on the major world markets has contributed to a substantial change in the structure of the DOC. It is sometimes exciting to learn that we can buy a ticket to an international destination within Europe or the continental United States at a cost comparable to the cost of this book. Cost items such as ticketing, customer service, seat allocation, baggage handling, ground services, in-flight catering, airport taxes, leasing contracts and so on have been dissected, reduced or removed altogether from the DOC. However, the cost of the fuel remains essentially the same. This cost is an essential part of the aircraft performance.

**CALCULATION METHOD.** The calculation of the DOCs of an airplane is a contentious issue. Data from major airlines are published by the IATA and are closely monitored. On the design side, there are unsubstantiated claims made by the manufacturers. On the operational side there are uncertainties arising from market conditions, financial arrangements and a long list of externalities (inflation, variable interest rates, fuel costs, tax liabilities, labour relations, adverse weather, and so on). Thus, if we want to be realistic, we need to build these uncertainties in the costs analysis. With the foregoing cautions in place, we provide a simplified analysis that is based on about half a dozen major cost items. These are:

1. *Ownership costs*, which are a combination of cash payments on acquisition, loan repayment over a fixed length of time and residual value of the airplane at the end of the financing contract.
2. *Insurance costs*, which will be considered a fraction of the value of the airplane, although this might not be the case (the third-party liabilities do not decrease over time).
3. *Fuel costs*, based on average costs between departure and arrival at *today's price, adjusted for inflation*, or an agreed deviation from it.
4. *Crew costs*, consisting of pilots and flight attendants, based on full-time rates.
5. *Spare parts costs*, consisting in parts for the propulsion system, airframe, landing gear and any other items.
6. *Labour costs*, based on recognised rates (in today's prices, adjusted for inflation) for propulsion, in-house and contracted-out operations; part of the problem is to gather reliable data for these items, the man-hours required to carry out the service and the down-time for the airplane.

7. *Other costs*, including landing fees, ground handling, pilot training, on-board service items, costs for off-station residence and so on. None of these costs can be neglected because they make up the difference between being profitable and making a loss.

We start with the ownership costs, which are the most onerous. These costs are estimated from the equation

$$C_1 = \frac{1}{n}\left[\mathcal{P} - F(1+I)^n - R_n\right], \tag{15.46}$$

where $\mathcal{P}$ is the aircraft price agreed at the acquisition date; $F$ is the financing (i.e., the fraction of the acquisition price requiring a loan); $I$ is the interest rate on the loan, assumed to be constant over the life-time $n$ (in years) of the loan itself; and $R_n$ is the residual value after $n$ years. The latter item is uncertain and must be based either on today's price or perhaps corrected for inflation at the end of the loan. To be on the safe side, we could assume that the aircraft will be written off at the end of the period, $R_n = 0$. The insurance cost is estimated from

$$C_2 = R_j \frac{I}{100}, \tag{15.47}$$

where $R_j$ is the residual value at year $j$. The fuel cost over one year is

$$C_3 = N\overline{C}_{fuel} m_f, \tag{15.48}$$

where $N$ is the number of cycles, $\overline{C}_{fuel}$ is the average fuel cost per kg, and $m_f$ is the fuel burn per cycle over the specified stage length. This quantity is calculated with a mission analysis and will depend on a realistic estimate of the passenger-load factor and gross take-off weights. Fuel-price oscillations are accounted for in the averaging. Increases in fuel consumption due to engine age are accounted for with a separate model (see, for example, § 5.3.6). The deterioration-adjusted fuel burn is

$$D_f(x) = \alpha_1\left[1 - e^{-\alpha_2 x}\right] + (n_w - 1)/c, \tag{15.49}$$

where $x = N/1{,}000$ is the number of thousand cycles since the last wash, $\alpha_2$ a deterioration lapse ($\alpha_2 \simeq 1$ to 1.5) and $\alpha_1$ is the deterioration effect on the fuel burn after 1,000 cycles ($\alpha_1 \simeq 2$ to 3); $n_w$ is the number of total engine washes and $c$ is the irretrievable loss of performance over time ($c \simeq 20$). The crew cost is

$$C_4 = n_1 S_1 \left(\frac{T_b}{T_o}\right)_1 + n_2 S_2 \left(\frac{T_b}{T_o}\right)_2 + n_1 S_3 \left(\frac{T_b}{T_o}\right)_1, \tag{15.50}$$

with $n_1$ and $n_2$ the number of pilots and average flight attendants required to operate the aircraft year round; $S_1$ and $S_2$ are the corresponding salary costs based on full-time service, $T_b$ is the total number of utilisation hours per year, and $T_o$ is the contractual number of working hours. The number of flight attendants will be estimated from the passenger load and the type of services (about 1/30 in economy and 1/10 in premium/business class). Other major crew-related costs include pilot training $S_3$, which takes place ahead of the acquisition and at scheduled intervals during the lifetime of the airplane. The costs of the spare parts is

$$C_5 = \sum_k \mathrm{SP}_k, \tag{15.51}$$

where $SP_k$ ($k = 1, \cdots$) denotes spares for the propulsion system, APU, airframe, landing gear, tyres and any other items. The maintenance schedule depends on the number of cycles and airplane age; it is divided into three broad categories: propulsion and associated systems, in-house and contracted out. Maintenance costs increase with ageing aircraft and cannot be oversimplified. As a result, the labour costs associated to maintenance are likely to go up as the number of hours required increases. The associated maintenance costs are:

$$C_6 = \sum_k \mathcal{L}_k \mathrm{MH}_k, \tag{15.52}$$

with $\mathcal{L}_k$ the labour rates (any currency per hour) for the categories listed herein and $\mathrm{MH}_k$ the number of man-hours required in each category to carry out the maintenance schedule. The number of man-hours involved depends on the airplane (gross weight, number of engines, landing gear) and the number of cycles. These costs suffer a jump as soon as the airplane comes off the warranty period. In practice, this means that part of the costs are transferred from the manufacturer to the operator. This transfer is called *burden* and is not necessarily an ageing cost.

The landing fees depend on the airport, the weight class of the aircraft and local congestion; it may be subject to additional levies, such as $CO_2$, $NO_x$ emissions and noise emissions. We assume that there are two major contributions: from the gross weight (or number of seats) and the overall emissions, with weights $c_1$ and $c_2$, respectively:

$$C_7 = 2(c_1 W + c_2 E)N. \tag{15.53}$$

As anticipated, the maintenance costs increase over time. Figure 15.18 shows such effects. These charts have been elaborated from a statistical analysis carried out by the Rand Corporation[14]. The data have been generated from a limited sample of airplanes analysed; their limitations are discussed in the report cited. The charts provide some weights to account for the costs of major items, such as airframe, engine and burden effects. In all cases, it is shown that new aircraft have rapidly increasing costs, which eventually flatten out with age. Apart from having to update the coefficients, depending on type of aircraft, technology level and other factors, the conclusions are generally valid. If we use the form

$$C_{56} = C_{5,1} + C_{6,1} \tag{15.54}$$

to denote the combined parts and man-hour costs for the engine, the function $C_{56}$ should follow the trend of Figure 15.18b. Even then, we need to establish values for labour rates $\mathcal{L}$ and man-hours MH at year zero. The DOC model, as presented, requires the evaluation of about 30 independent parameters. A summary of these parameters is in Table 15.2 for clarity.

On a final note, we report that there has been an introduction of leasing contracts for the propulsion system. These contracts allow the airline operator to simply pay a cost proportional to the number of flight hours. When this is the case, the analysis proposed here will have to be adjusted to take into account this different cost structure.

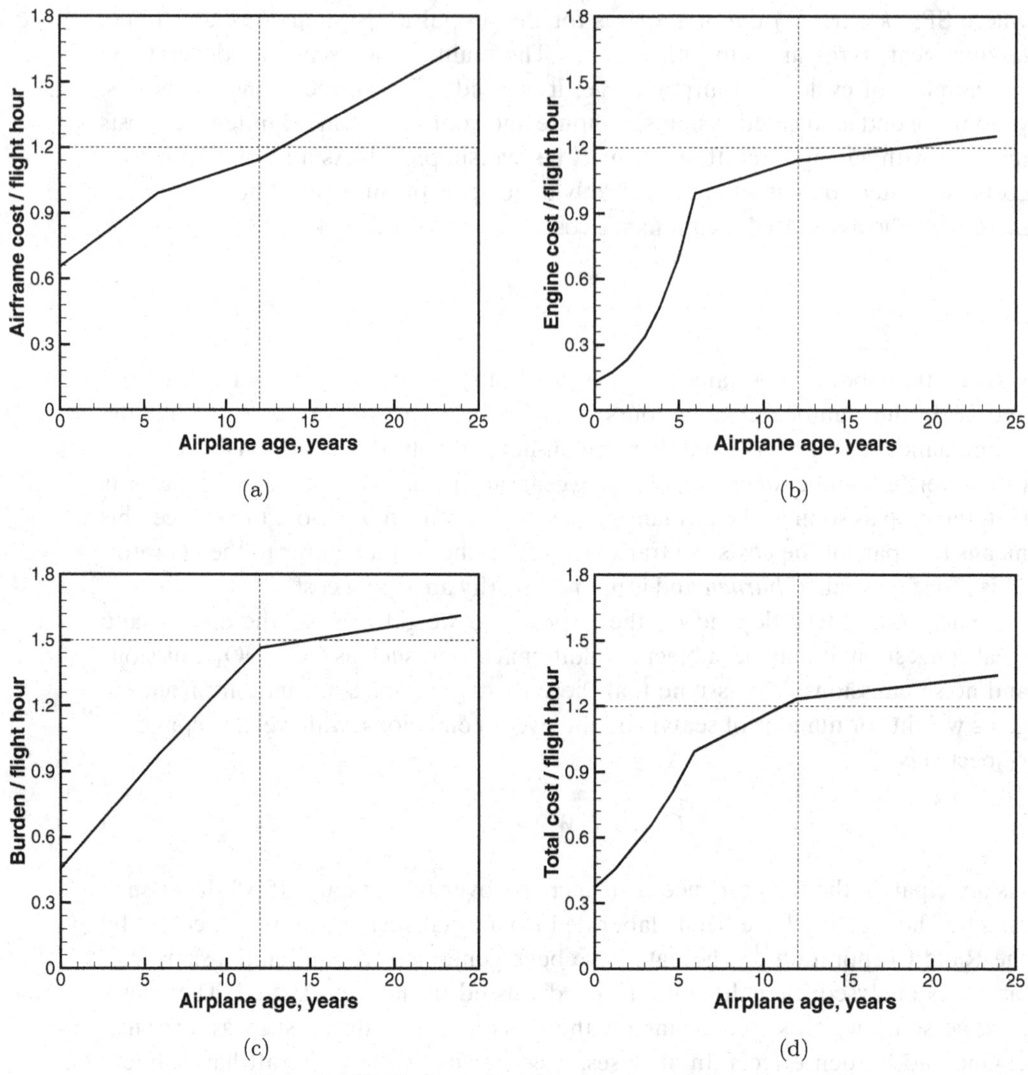


Figure 15.18. Estimated ageing effects on maintenance costs.

Table 15.2. *Summary of parameters for DOC model*

| Item | Parameters | Reference |
|---|---|---|
| Ownership | $\mathcal{P}, F, i, n, R_n$ | Equation 15.46 |
| Insurance | $I$ | Equation 15.47 |
| Fuel | $N, \overline{C}_{fuel}, m_f$ | Equation 15.48 |
| Crew | $n_1, n_2, S_1, S_2, T_o$ | Equation 15.50 |
| Spare parts | $SP_1, SP_2 \cdots$ | Equation 15.51 |
| Labour rates | $\mathcal{L}_1\ \mathcal{L}_2 \cdots\ MH_1, MH_2 \cdots$ | Equation 15.52 |
| Landing fees | $c_1, c_2, E$ | Equation 15.53 |
| Other costs | | To be specified |
| Ageing effects on fuel | $\alpha_1, \alpha_2, n_w$ | Equation 15.49 |
| Ageing effects on maintenance | | Figure 15.18 |

Table 15.3. *Calculated payload fuel efficiency for long-haul commercial flight; all flights at* $M = 0.80$*; all weights are in metric tons*

| Aircraft | OEW | MTOW | PAY | $X$ [n-m] | Fuel | GTOW |
|---|---|---|---|---|---|---|
| B | 129.9 | 275.0 | 35.0 | 7,560 | 74.62 | 244.1 |
| A | 90.1 | 165.0 | 35.0 | 3,780 | 55.69 | 156.1 |
| B | 129.9 | 275.0 | 50.0 | 7,560 | 77.94 | 263.1 |

**PRODUCTIVITY MEASURES.** There are specific parameters used by commercial operators to define their productivity. One of the most important is the Available Seat per kilometre (ASK) or nautical mile (ASM). This parameter expresses the number of seats available for sale, flown 1 km (or 1 n-mile). Then there is the Cost per available seat-km (CASK) or seat-n-mile (CASM). The CASK is equal to ASK/DOC and determines the actual cost incurred by the operator of making a seat available on a certain airplane over a certain route. Because the DOC changes as a function of the various parameters, so does the CASK. Sometimes a cost analysis is carried out by excluding the cost of fuel; this operation requires the ex-fuel CASK. Finally, there is the revenue per ASK (RASK). To make a profit, a commercial operator must have RASK > CASK.

## 15.9 Case Study: Aircraft and Route Selection

How do we select a new airplane from two or more competing airplanes? We will have to face a mountain of marketing documentation about the benefits of one airplane with respect to the other; each airplane is inevitably 20% more fuel efficient than the other one, with the costs of the competing airplane being calculated by the competitor, not the manufacturer. This is not an easy decision to make and even between two competing airplanes of one manufacturer, a business decision will have to be made on the basis of hard evidence; the process can be costly and time-consuming.

In the next analysis we will refer to the relatively simple problem of deciding which airplane to dispatch for a certain mission. We assume that we have to connect two airports separated by a distance of 14,000 km (~7,560 n-miles) and that we have to deliver a payload of 35,000 kg. We have airplanes A and B, whose OEW and MTOW are given in Table 15.3.

From the payload-range chart, we conclude that aircraft B can make a no-stop flight to destination. For this aircraft, we estimate a fuel burn of ~77.62 tons. The service can also be done with aircraft A with an en-route stop half-way (7,000 km; 3,780 n-miles). The total fuel required (including the two legs) in this case is ~55.69 tons, which is a savings of about 25%. Therefore, the second option is more economical. If, instead, we are required to deliver a 50-ton payload, aircraft A is not large enough; aircraft B instead can easily accommodate that payload, and the increase in fuel consumption over the previous case is only ~4.5%. The summary of our calculations is in Table 15.3.

Figure 15.19 shows the result of a parametric analysis carried out with a model Boeing B777-300 airplane with GE-92 engines. The purpose of this analysis was to

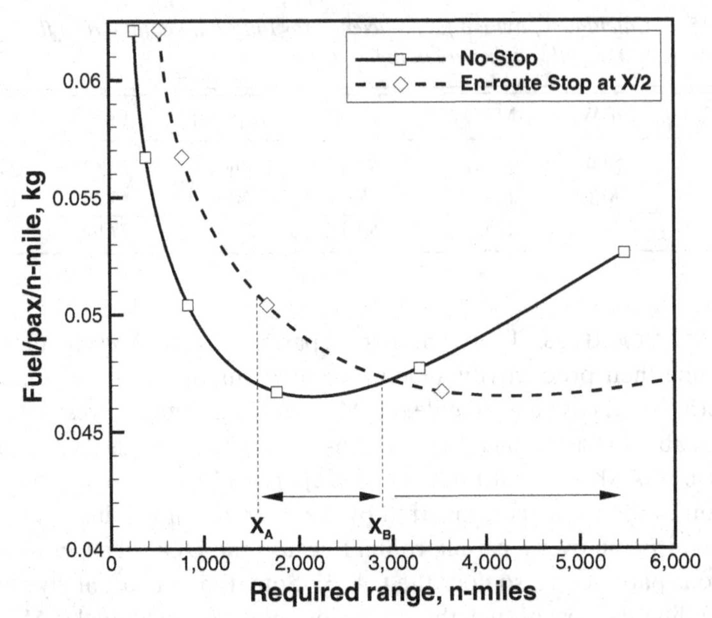


Figure 15.19. Analysis of B777-300 airplane model flight with and without en-route stop; standard day, no wind.

understand how the en-route stop affects the fuel cost. The results indicate that for a distance above ~3,000 n-miles, an en-route stop could save fuel. The reason for this is that the longer the required range, the heavier the airplane, as a result of having to carry a larger amount of fuel. The solution shown in Figure 15.19 is not as straightforward as it may look. First, the parametric analysis is carried out with the assumption that a suitable airport is available at exactly half the distance. Second, there can be considerable costs in landing at an intermediate destination. Third, for some values of the range, it would not be possible to stop anyway. For example, for a 3,000 n-miles flight it would make no sense to stop 500 n-miles to the final destination unless there is a commercial reason to do so. Refer also to the case illustrated in Table 15.1. The range $X$ between $X_A$ and $X_B$ is the most efficient and corresponds to a fuel burn (per passenger per n-mile) that is within 1% of the minimum. The results shown refer to full passenger load. Therefore, the fuel/pax/n-mile is equivalent to fuel/seat/n-mile. The result would be slightly different for a less-than-full aircraft.

The solution can be further refined to include other factors and possibly an estimate of the DOC. Green[15] showed an analysis for cargo aircraft and recommended that from the point of view of fuel efficiency, a flight segment shall not exceed 7,500 km (~4,000 n-miles) at the current level of technology. In-flight refuelling is an option that can be considered for either extending the range or increasing the payload for a fixed range. However, the calculation must include the cost of operating a tanker aircraft and to maintain a loiter trajectory for some time[16].

The next analysis focuses on the fuel use as a function of the required range. Figure 15.20 shows the fuel fraction (with respect to the fuel embarked) used for climb to the ICA and for cruise. The cruise fuel converges toward 80% of the total

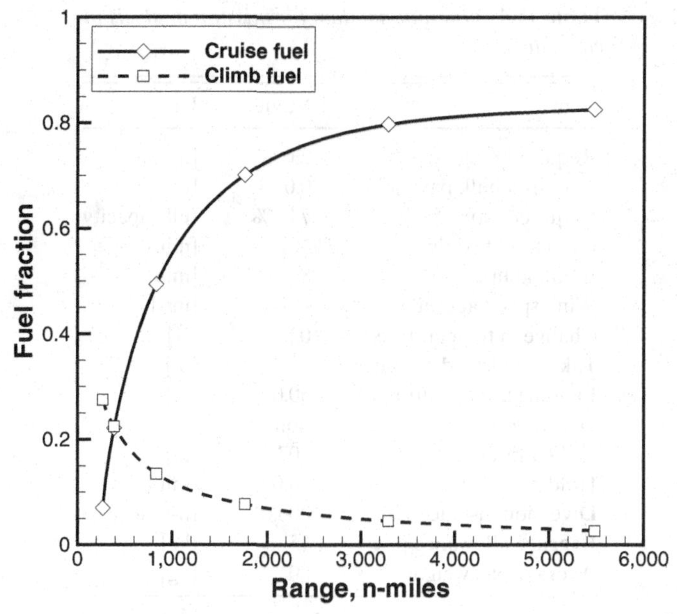


Figure 15.20. Analysis of fuel use by the Boeing B777-300 with GE engines; standard day, no wind; full passenger; no bulk cargo.

fuel embarked, whereas the climb fuel decreases to a few percent. Shorter journeys require relatively higher portions of the fuel to be used for climb. Specifically, on a 250-n-mile journey the climb fuel exceeds 25% of the total fuel embarked. The remaining fuel is used in the other flight segments and includes the reserve fuel.

## 15.10 Case Study: Fuel Planning for Specified Range, B777-300

We apply the methods presented earlier to the calculation of the mission fuel and gross take-off weight for the Boeing B777-300 with General Electric GE092 turbofan engines and 331-500 APU. A summary of mission calculations is shown here, which are computer outputs from the program `FLIGHT`. Looking at a more detailed level, we have reports for the various segments of flight. Table 15.4 shows a list of flight conditions and assumptions. Table 15.5 is a summary of fuel/weight planning resulting from the calculations. We discuss separately the different flight segments.

### Taxi-Out Report

It is not unusual that the aircraft must remain on hold waiting for a departure slot, due to local traffic or bad weather. Fuel consumption on the taxi-way can be so high that the flight must be aborted. The jet engines' performance is optimised for cruise, and considerable efficiency penalties are faced on the taxi-way. Therefore, taxi-times must be estimated with accuracy.

Some economy can be gained by taxiing the aircraft with OEI or by towing the aircraft to and out of the gate, or by parking the aircraft away from the gate. The

Table 15.4. *Operational data for mission analysis in case study*

| Item | Value | Unit |
|---|---|---|
| Required range | 2,698 | [n-miles] |
| Required bulk payload | 0.0 | [ton] |
| Required pax | 74.6% | full capacity |
| Outbound taxi time | 10 | [min] |
| Inbound taxi time | 8 | [min] |
| Wind speed at cruise | −5.0 | [m/s] |
| Change in temperature | 0.0 | [K] |
| Take-off airfield altitude | 50.0 | [m] |
| Landing airfield altitude | 50.0 | [m] |
| Derating | none | |
| Hold altitude | 507 | [m] |
| Hold time | 30.0 | [min] |
| Diversion distance | 200.0 | [n-miles] |
| Baggage allowance/pax | 15 | [kg] |
| Average pax weight | 80 | [kg] |

OEI procedure is not recommended in a number of cases, including high GTOW, slippery ground, nose-wheel steering performance, limitations on jet blasts, safety for ground personnel and other reasons. It can be more efficient to taxi at a high speed (although the limit is set to 25 to 30 kt) with a burst of engine power, and then run the engine with idle power.

A first-order estimate of the taxi fuel is found with the method shown in § 9.10 (Chapter 9). A summary of the taxi-out performance for this specific case is shown in Table 15.6. This table indicates that a considerable amount of fuel, half a ton, is required to reach the start of the runway! Thus, we face an important operational and environmental problem that must be addressed by more efficient procedures.

### Take-Off Report

This report is shown in Table 15.7. There is a large set of airplane and operational data required for the complete take-off analysis. The wind speed and runway state are specified on input. The remaining data are calculated by the computer program. The flap setting is also part of the solution, as the numerical algorithm attempts to perform a take-off with the minimum flap setting compatible with a minimum initial climb gradient. Included in the analysis is the result of the thermo-structural model for the tyres (§ 14.5) and the minimum control speed on the ground (§ 9.6).

### En-Route Climb Report

Key data are the time to climb, the distance to climb and the fuel to climb. The various climb segments are identified and report data such as average climb rate, fuel use and fuel flow. Landing-gear retraction and flaps retraction are done instantaneously. Climb is done in four segments, the last one being a constant-Mach climb to the

Table 15.5. *Summary of flight-planning analysis*

**Fuel Use Report**

| Segment | $m_f$ [kg] | $m_f$ [%] | $X$[n-m] | $t$ [min] | Notes |
|---|---|---|---|---|---|
| Climb | 4,230.2 | 8.05 | 191.76 | 28.4 | |
| Cruise | 37,415.5 | 71.16 | 2,317.72 | 299.7 | |
| Descent | 5,037.1 | 9.58 | 210.44 | 37.3 | |
| Take-off | 214.6 | 0.41 | | | |
| Taxi-out | 504.1 | 0.96 | 1.12 | 10.0 | |
| APU/ECS | 4,674.9 | 8.89 | | | |
| Taxi-in | 415.9 | 0.79 | | | from reserve |
| *Fuel burn* | 52,582.0 | | | | |

**Fuel Breakdown Report (all data [kg])**

| | | |
|---|---|---|
| Taxi-out | 504.1 | |
| Take-off | 214.6 | |
| Climb | 4,230.2 | |
| Cruise | 37,415.5 | |
| Descent | 5,037.1 | |
| Approach | 89.5 | |
| APU/ECS | 4,674.9 | |
| Diversion/Hold | 2,498.1 | |
| Reserve | 2,754.0 | |
| *Total* | 57,834.1 | 42.1% of full tanks |

**Final Weight Report (all data [kg])**

| | |
|---|---|
| OEW | 161,355 |
| Bulk payload | 0.000 |
| Passengers | 23,520 |
| Flight crew | 1,140 |
| Useful payload | 24,520 |
| Service items | 0.873 |
| Fuel | 57,834 |
| *Non-usable fuel* | 824 |
| Ramp weight | 243,658 |
| BRGW | 243,658 |
| Zero-fuel weight | 187,887 |
| Take-off weight | 249,816 |
| Final Cruise weight | 239,096 |
| Landing weight | 193,075 |

optimal (or sub-optimal) ICA. This report is shown in Table 10.3 in Chapter 10 and is not repeated here for brevity.

### Cruise Report

This report is shown in Table 15.8. The cruise flight is done in constant-altitude segments at suitable flight levels, followed by constant-climb rate between flight levels (§ 12.7). In this case, the aircraft operates between FL-330 and FL-370 and does two step climbs. Each constant-altitude cruise segment is approximately 750 n-miles. The climb rate between flight levels is specified.

Table 15.6. *Taxi-out report of fuel/weight-planning analysis*

| Parameter | Value | Unit |
|---|---|---|
| Fuel rest-to-roll speed | 14.52 | [kg] |
| Fuel in idle mode | 121.60 | [kg] |
| Fuel in roll mode | 368.02 | [kg] |
| Total taxi fuel | 504.14 | [kg] |
| Time in idle | 6.7 | [min] |
| Time in roll | 3.3 | [min] |
| Roll speed | 5.0 | [m/s] |
| Roll distance from ramp | 3.0 | [km] |
| Taxi procedure | AEO | |

Table 15.7. *Take-off report of fuel/weight-planning analysis*

| Parameter | Value | Unit | Notes |
|---|---|---|---|
| Gross take-off weight | 243.658 | [ton] | |
| Airfield altitude | 50.00 | [m] | |
| Air temperature | 0.00 | ± ISA | |
| Wind speed | −2.00 | [m/s] | −3.9 [kt] |
| Runway state | dry | | |
| Thrust angle | 0.000 | [degs] | |
| Stall margin | 1.150 | | |
| Max lift coefficient | 2.435 | | |
| Stall speed | 69.64 | [m/s] | 135.3 [kt] |
| $x_{CG}$ | 35 | [% MAC] | |
| Pitch moment of inertia | 104.70 | [$10^6$ kgm$^2$] | |
| Flap setting, $\delta_f$ | 20 | [deg] | |
| Rotation velocity, $V_R$ | 63.46 | [m/s] | 123.3 [kt] |
| Rotation distance, $x_R$ | 822.56 | [m] | |
| Rotation time, $t_R$ | 29.00 | [s] | |
| Lift-off velocity, $V_{LO}$ | 71.46 | [m/s] | 138.8 [kt] |
| Lift-off distance, $x_{LO}$ | 1,038.88 | [m] | |
| Lift-off time, $t_{LO}$ | 32.85 | [s] | |
| Velocity over screen | 71.64 | [m/s] | 139.2 [kt] |
| Mach over screen | 0.211 | | |
| Distance to screen, $x_{TO}$ | 1,150.18 | [m] | |
| Time over screen, $t_{TO}$ | 34.65 | [s] | |
| Climb angle over screen, $\gamma_{TO}$ | 12.48 | [deg] | |
| Fuel burn | 214.60 | [kg] | |
| VMCG | 55.88 | [m/s] | 108.6 [kt] |
| Max (main) tyre temperature | 293.2 | [K] | +0.6 [K] |
| Max (nose) tyre temperature | 373.7 | [K] | +26.3 [K] |
| Main tyre temp. over screen | 292.6 | [K] | |
| Nose tyre temp. over screen | 347.4 | [K] | |
| Brake-release net thrust | 877.873 | [kN] | |
| Brake-release fuel flow | 7.947 | [kg/s] | |

Table 15.8. *Cruise report of fuel/weight-planning analysis; $X_{req} = 2317.7$ [n-miles]; ICW = 239.096 [ton], M = 0.803*

| $h$ | FL | $X$ [n-m] | $t$ [min] | $m_f$ [kg] | $\dot{m}_f$ [kg/s] | $v_c$ [ft/min] |
|---|---|---|---|---|---|---|
| 10,058 | 330 | 772.57 | 99.26 | 12,837.1 | 2.156 | |
| | | 19.68 | 2.54 | 238.1 | 1.562 | 787 |
| 10,668 | 350 | 772.57 | 99.73 | 12,435.0 | 2.078 | |
| | | 19.52 | 2.54 | 230.3 | 1.510 | 787 |
| 11,277 | 370 | 733.37 | 95.58 | 11,674.9 | 2.036 | |
| Totals | | 2,317.72 | 299.65 | 37,415.5 | 2.081 | |

### Descent Report

En-route descent is performed according to the principles described in Chapter 11. From the final cruise altitude (FCA), the airplane follows some steps down to a reference altitude, for example 1,500 feet (457 m) above the airfield, with a gradual decrease in air speed. A typical descent report is shown in Table 11.2 in Chapter 11; this is not repeated here for brevity.

### Analysis of Contingency Fuel

There are various reserve fuel policies, some of which are discussed in § 15.5. A combination of various contingency policies as a function of the required range is shown in graphical form in Figure 15.21. The 20-minute extension of the flight is assumed to take place at the cruise Mach and the final cruise altitude. This fuel reserve is not dependent on the required range because the all-up mass at the start of the flight extension is virtually a constant parameter. The holding cases

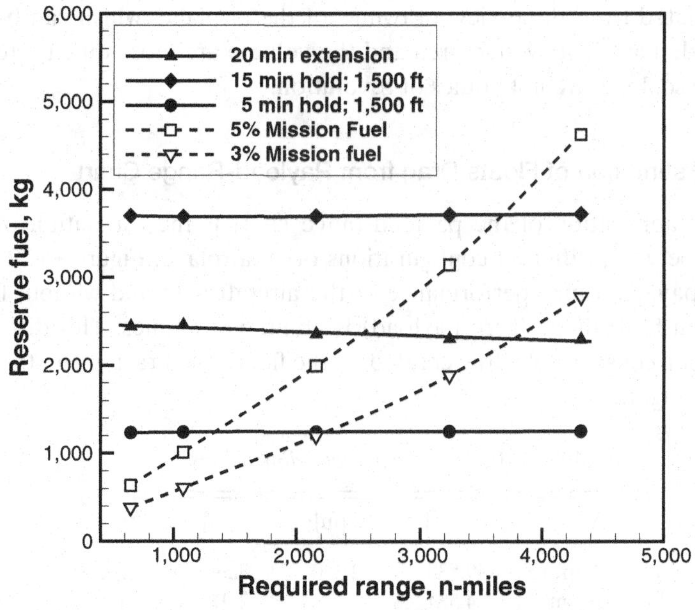


Figure 15.21. Effect of cruise range on contingency fuel; standard day, no wind; full passenger load, 300 kg of bulk cargo (calculated).

Table 15.9. *Basic performance data of model float-plane*

| Data | Value | Unit |
|---|---|---|
| Cruise TAS | 163.4 | m/s (592 km/h, 319.5 kt) |
| Cruise altitude | 6,000 | m (19,685 feet) |
| Specific fuel consumption | 0.3567 | kg/h/kW |
| Fuel capacity | 27,200 | kg |

(15 or 5 minutes) are performed at 1,500 feet above the airfield at the hold $M = 0.350$. Other flight conditions are specified in the caption. For the holding cases, we assumed that the initial mass corresponds to the all-up mass before the final approach, which yields a nearly constant reserve fuel.

If an alternate en-route airport is available, the best option is a 3% reserve fuel below 2,200 n-miles. If this is not possible, a 5% reserve fuel is less restrictive than the 20-minute cruise extension for stage lengths below 2,600 n-miles. This conclusion is quite general: each aircraft has a critical stage length beyond which a change in contingency policy may yield a weight advantage. This stage length depends on the actual conditions, in particular weather patterns and payload.

## 15.11 Case Study: Payload-Range Analysis of Float-Plane

For this study we consider the Lockheed C130J with floats. This is one of the many derivatives of the Hercules aircraft family, designed to perform special operations such as troop transport, anti-submarine warfare, mine counter-measures, search and rescue, fire-fighting, inter-island transport, oil spill response and more. Some basic performance data for the calculation of the floats' drag are given in Table 15.9. The estimated floats' dimensions are summarised in Table 15.10. These data have been calculated from three-view drawings of the airplane, which can be found in the public domain. The wetted area and the volume are more difficult to calculate because these floats are not bodies of revolution.

### 15.11.1 Estimation of Floats Drag from Payload-Range Chart

One useful application of the payload-range chart is the estimation of the drag difference between different configurations of an airplane. Figure 15.22 shows the estimated payload-range performance of the aircraft with and without floats. The difference in ferry range (zero payload) is about 570 n-miles. This difference is a reflection of a considerable drag created by the floats, the bracing and the increased structural weight.

Table 15.10. *Estimated floats' dimensions*

| View | A [m$^2$] | $x$ [m] | $y$ [m] | $z$ [m] |
|---|---|---|---|---|
| Top | 49.530 | 21.20 | 3.39 | |
| Front | 4.584 | | 2.97 | 2.00 |
| Side | 26.384 | 20.55 | | 1.89 |

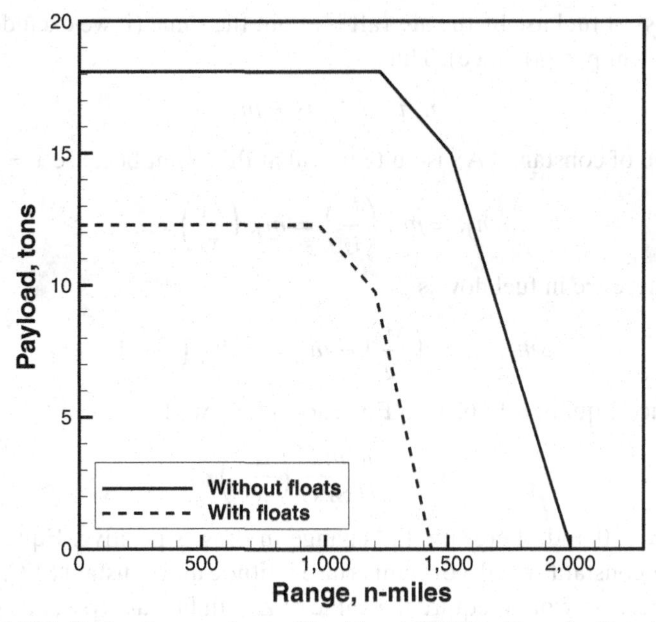

Figure 15.22. Payload range chart of Lockheed C130J, with and without floats.

A first-order estimate of the drag increase is done by assuming that 1) the structural weight of the aircraft is unchanged; 2) the cruise speed is unchanged; 3) the cruise altitude is unchanged; and 4) the drag caused by the bracing is negligible. These effects will be re-evaluated at a later time.

Assume that the nominal performance (C130J without floats) is indicated with the subscript "1", whilst the new configuration (with floats) is indicated with subscript "2". The work done to fly the airplane over the distance $x_1$ is $\mathcal{W}_1 = Tx_1$. The difference in work done to fly the airplane from $x_1$ to $x_2$ is

$$\Delta \mathcal{W} = \Delta T \Delta x = (T_1 - T_2)(x_1 - x_2). \tag{15.55}$$

In steady level flight $\Delta T = \Delta D$. The power relationship for a propeller aircraft is

$$T = \frac{\eta}{U} P, \tag{15.56}$$

where $T$ is the net propulsive thrust, $\eta$ is the propulsive efficiency, $U$ is the true air speed and $P$ is the shaft power. The relationship between shaft power and fuel flow is

$$\dot{m}_f = \text{SFC}\, P. \tag{15.57}$$

At constant TAS, constant propeller efficiency and constant SFC, we have

$$\Delta T = \Delta D = \frac{\eta}{U} \Delta P = \frac{\eta}{U} \frac{\Delta \dot{m}_f}{\text{SFC}}. \tag{15.58}$$

The difference in fuel flow required is

$$\Delta \dot{m}_f = \dot{m}_{f_2} - \dot{m}_{f_1}. \tag{15.59}$$

The total mass of fuel use by the aircraft is about the same (if we exclude the effects on climb-descent performance). Thus,

$$\dot{m}_{f_1} t_1 \simeq \dot{m}_{f_2} t_2 = m_f. \quad (15.60)$$

The condition of constant TAS is quite useful at this point because $x = Ut$.

$$\dot{m}_{f_2} = \dot{m}_{f_1} \left(\frac{t_1}{t_2}\right) = \dot{m}_{f_1} \left(\frac{x_1}{x_2}\right). \quad (15.61)$$

Thus, the difference in fuel flow is

$$\Delta \dot{m}_f = \dot{m}_{f_1} \left(\frac{x_1}{x_2}\right) - \dot{m}_{f_1} = -\dot{m}_{f_1} \left(\frac{\Delta x}{x_2}\right). \quad (15.62)$$

If we introduce Equation 15.62 into Equation 15.58, we find

$$\Delta D = \frac{\eta}{U} \frac{\dot{m}_{f_1}}{\mathrm{SFC}} \left(\frac{\Delta x}{x_2}\right). \quad (15.63)$$

Note that $\Delta x < 0$ and, therefore, the change in drag is positive. Equation 15.63 is valid only for constant speed, constant cruise altitude and constant SFC. Calculation of the drag increase would require the value of $\Delta x$. In first analysis, this value can be taken directly from the payload-range chart (for the ferry condition). However, we must also consider that the aircraft must take off, climb to cruise altitude, descend and land with a minimum of fuel reserves. Therefore, not all of the fuel capacity can be used for cruise. We assume that the fuel capacity is cleared for the mandatory reserves (say 5% of the usable fuel for a ferry operation); an additional 13% is needed for climb to altitude, descend and land. Thus, the usable fuel is about 82% of the fuel capacity.

The cruise range is lower than the mission range because the airplane performs an en-route climb and descent. In first analysis the discount is about 100 n-miles. Using the basic performance data shown in Table 15.9, we estimate that

$$m_{f_1} \simeq 1.042\,\mathrm{kg/s}, \qquad \Delta D \simeq 24{,}840\,\mathrm{kN}, \qquad \Delta C_D \simeq 0.00174. \quad (15.64)$$

In conclusion, the analysis indicates that the floats add about 174 drag counts to the cruise-drag coefficient. This is a considerable amount of drag: it corresponds to a decrease in ferry cruise range by 570 n-miles.

A more accurate calculation is possible, starting from the result just achieved. The first step forward is to consider the optimal air speed of the aircraft with floats. The optimal speed (minimum-power speed) is

$$U_{mp} = \left(\frac{2W}{\rho A} \frac{1}{C_{L_{mp}}}\right)^{1/2}, \quad (15.65)$$

with

$$C_{L_{mp}} = \left(\frac{3C_{D_o}}{k}\right)^{1/2}. \quad (15.66)$$

The difference in profile drag is $\Delta C_D = \Delta C_{D_o}$. The aircraft with the floats must fly faster, as shown by the equation

$$\frac{U_{mp2}}{U_{mp1}} = \left(\frac{C_{L_{mp2}}}{C_{L_{mp1}}}\right)^{1/2} = \left(\frac{C_{D_o} + \Delta C_{D_o}}{C_{D_o}}\right)^{1/2} = \left(1 + \frac{\Delta C_{D_o}}{C_{D_o}}\right)^{1/2} > 1. \quad (15.67)$$

For a 20% increase in profile drag, the optimal TAS would have to increase by about 10%. Thus, if we take into account these new flight conditions, the new fuel flow is

$$\dot{m}_{f_2} = \dot{m}_{f_1} \left( \frac{U_1 x_1}{U_2 x_2} \right) = \dot{m}_{f_1} \left( \frac{x_1}{x_2} \right) \left( 1 + \frac{\Delta C_{D_o}}{C_{D_o}} \right)^{-1/2}, \tag{15.68}$$

$$\Delta \dot{m}_f = \dot{m}_{f_2} - \dot{m}_{f_1} = \dot{m}_{f_1} \left[ \left( \frac{x_1}{x_2} \right) \left( 1 + \frac{\Delta C_{D_o}}{C_{D_o}} \right)^{-1/2} - 1 \right]. \tag{15.69}$$

Equation 15.69 can be used to re-iterate and calculate a new value for the increase in profile drag.

## 15.12 Risk Analysis in Aircraft Performance

Virtually all of the problems in aircraft performance presented so far are based on deterministic analysis. In other words, given a set of initial conditions and airplane state, it is possible to evaluate almost exactly how the airplane will respond. This is far from true. There are two issues at stake: 1) the risk of failure and the corresponding consequences; and 2) the uncertainty on some state parameters. An example of statistical performance analysis is available in Ref. [17].

As in any other engineering field, there is a risk associated to the operation of an airplane. Safety in air transportation is paramount but risk-free operation, like any other pursuit in life, is virtually impossible. It is perhaps an unfortunate fact that a considerable proportion of the technological progress achieved in aerospace engineering is attributed to accidents of various levels of severity. We presented some examples, but the literature on the subject is vast.

The meaning of risk itself must be clearly defined, particularly the relationship between the risk itself and its severity (or its consequences). Consider the case of ETOPS, which allows twin-engine airplanes to fly over land for a prescribed amount of time (60, 120 or 180 minutes) with one engine failure. The risk analysis associated to this type of operations has allowed twin-engine airplanes to operate across North Atlantic routes. ETOPS safety criteria correspond to less than $3 \cdot 10^{-9}$ fatalities per flight hour arising from a total loss in engine thrust from independent causes. The event of one engine failure on a twin engine is more severe than an engine failure on a three- or four-engine airplane[†].

The analysis shall consider at least the "unit of risk", which is the probability of occurrence of a certain event (incident, engine failure, flight diversion, accident) per flight hour or other relevant measure, for example number of cycles [18]. Most current regulations have been derived from the works of the ICAO and adapted by national aviation regulators, with some local variations. Figure 15.23 shows the notional relationship between the unit of risk and the severity of failure. This severity is mapped onto a failure consequence. The shaded area corresponds to an unacceptable level of risk. The ETOPS risk level is shown for reference.

[†] For operational issues, see Airbus: *Getting to Grips with ETOPS, Flight Operations Support and Line Assistance*, Issue V, Oct. 1998, Blagnac, France.

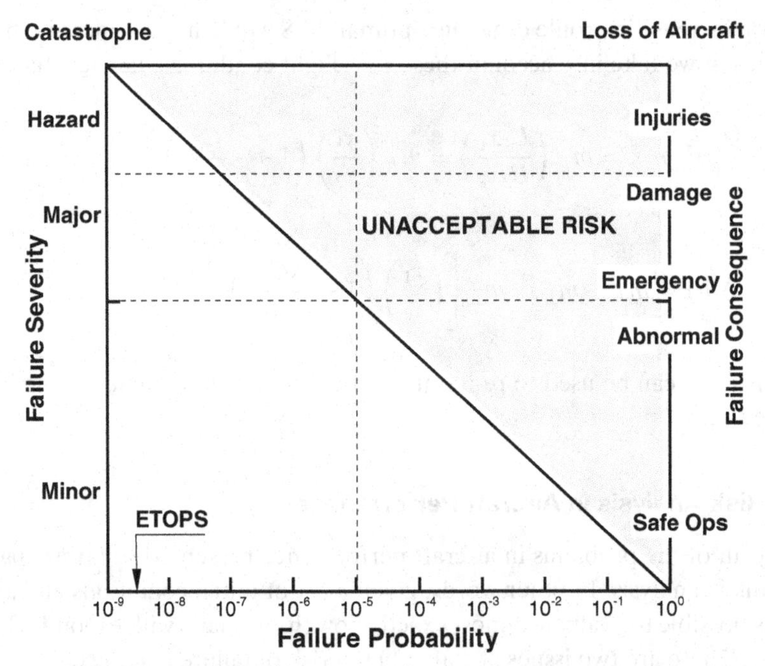

Figure 15.23. Risk analysis in aircraft performance.

Aside from the unit of risk, there is the problem of uncertain parameters (due to stochastic events) and/or imprecise evaluation of parameters. Performance prediction, then, can only be probabilistic. The performance charts produced by the manufacturers (as part of the FCOM) are a case in point. Although the graphs associated to the FCOM show clean lines, the predicted performance must be interpreted in a statistical sense with a high level of confidence.

Some of the performance calculations that are most subject to parameter uncertainty, stochastic external factors and risks of mechanical failure include take-off and landing[19;20]. ESDU has a number of data items on this subject, covering basic statistical analysis and practical methods for evaluating the performance risk[21;22].

Within the context of this discussion, we aim to carry out a "sensitivity" analysis for selected parameters. Because the number of system parameters is quite large, the investigation space is inconveniently large. Therefore, it is necessary in the first instance to isolate the dominant parameters, to evaluate their standard deviation and to establish the effects of such parameters. We give a few examples here.

**SPECIFIC AIR RANGE GUARANTEE.** The guarantee usually carries a penalty if the performance cannot be matched to the specifications. An example of variability was shown in § 12.8, where we examined the effects of some parameters on cruise performance deterioration. In this instance, we simply assume that a number of tests are carried out; the tests indicate some variability in the SAR; the result is given in the following array.

$$\Delta \text{SAR}, \% = \{1.21, 1.35, -0.59, 0.23, -2.01, 1.05, 1.45, -1.27, -0.97, 0.45\}. \tag{15.70}$$

Equation 15.70 is a 10-point array containing the percentage change in SAR calculated with 10 flight tests. The statistical properties of this array are: sample standard deviation = 1.21%. On a normal distribution, the mean value has a confidence level of 68.2% within one standard deviation (±1.21%), 95.4% confidence within two standard deviations (±2.42%). This level of confidence will have to be considered carefully before signing any delivery contract.

**TAKE-OFF FIELD LENGTH.** A preliminary analysis was carried out in Chapter 9, when we considered the effects of an engine failure at a critical point. We then defined the balanced field length and determined the actions to be taken to either stop the airplane or continue the take-off run. Take-off charts contain a number of parametric events that take into account brake-release gross weight, atmospheric temperature, longitudinal wind speed, airfield altitude, runway gradient and flap setting. This type of analysis is in fact deterministic, in the sense that a parameter variation is assumed and the response of the aircraft is measured or simulated. The actual statistical analysis is based on "risk", that is, the probability that certain combinations of winds, airfield altitude, engine failure and so on can conspire (independently or otherwise) to cause an unacceptable level of risk, as indicated in Figure 15.23. Additional considerations may be necessary if the airplane has changed configuration and its condition has deteriorated since the last inspection. The exercise of bringing all these effects together is a major project.

**NOISE PERFORMANCE GUARANTEE.** A detailed analysis of aircraft noise is carried out in Chapter 16. Noise measurements are also based on statistical values*. The effects of various sources on overall noise can be assessed by calculating the noise response to a set of random variations in parameter settings. This analysis is strongly non-linear because of the log-sum between components and because the noise metric is an integral over the full trajectory.

## Summary

We addressed several problems in aircraft-mission analysis, with focus on transport operations. We first examined issues of flight scheduling to maximise the use of the aircraft, and demonstrated that departure and arrival are time-critical. Next, we illustrated the importance of the payload-range chart, which is one of the main drivers in aircraft utilisation. We presented several numerical methods for the computation of payload-range charts and for mission planning; in particular, we demonstrated how the predictor-corrector scheme applied to the mission fuel is the most robust method for calculating the mission fuel. We presented a detailed case study of transport-mission analysis and discussed each flight segment separately. A number of other transport of practical interest include flight with en-route stop and fuel tankering. We highlighted the importance of the corresponding choices, which respond to market conditions. The DOC depend on the fuel consumption but also on a considerable number of parameters in, around and external to the aircraft.

* An example is available in the certification of the Airbus A318, A319, A320, A321. See EASA, Type Certificate for Aircraft Noise, TCDSN EASA.A.064, Issue 10, June 2011.

Calculation of the costs is made difficult by the lack of data that can be generalised. We showed a case study of a float-plane mission and the influence of the floats' drag on its payload range chart. We concluded the chapter by illustrating the importance of risk analysis in aircraft performance.

## Bibliography


[1] Gang Yu, editor. *Operations Research in the Airline Industry*. Kluwer Academic Publishers, 1998.

[2] Coy S. A global model for estimating the block time of commercial passenger aircraft. *J. Air Transport Management*, 12:300–305, 2006.

[3] Anonymous. 747-400 airplane characteristics for airport planning. Technical Report D6-58326-1, The Boeing Corporation, Dec. 2002.

[4] Filippone A. Comprehensive analysis of transport aircraft flight performance. *Progress Aero Sciences*, 44(3):185–197, April 2008.

[5] ESDU. *Examples of Flight Path Optimisation Using a Multi-Variate Gradient-Search Method*. Data Item 93021. ESDU International, London, Mar. 1995.

[6] Martinez-Val R and Pérez L. Extended range operations of two and three turbofan engined airplanes. *J. Aircraft*, 30(3):382–386, 1993.

[7] Sachs G. Optimization of endurance performance. *Progress Aerospace Sciences*, 29(2):165–191, 1992.

[8] Torenbeek E. Cruise performance and range prediction reconsidered. *Progress Aerospace Sciences*, 33(5-6):285–321, May-June 1997.

[9] Küchemann D. *The Aerodynamic Design of Aircraft*. Pergamon Press, 1978.

[10] Beltramo MN, Trapp DL, Kimoto BW, and Marsh DP. Parametric study of transport aircraft systems cost and weight. Technical Report CR 151970, NASA, April 1977.

[11] Kershner ME. Laminar flow: Challenge and potential. In *Research in Natural Laminar Flow and Laminar-Flow Control*, volume NASA CP-2487. NASA Langley, Mar. 1987.

[12] Isikveren AT. Identifying economically optimal flight techniques of transport aircraft. *J. Aircraft*, 39(4):528–544, July 2002.

[13] Windhorst R, Ardema M, and Kinney D. Fixed-range optimal trajectories of supersonic aircraft by first-order expansions. *J. Guidance, Control and Dynamics*, 24(4):700–709, July 2001.

[14] Dixon M. The maintainance costs of aging aircraft. Technical Report Rand MG456, RAND Corporation, 2006.

[15] Green JE. Greener by design – The technology challenge. *Aero J.*, 106(1056): 57–113, Feb. 2002.

[16] Bennington MA and Visser KD. Aerial refueling implications for commercial aviation. *J. Aircraft*, 42(2):366–375, Mar. 2005.

[17] Anon. Statistical loads data for the Boeing B777-200ER aircraft in commercial operations. Technical Report DOT/FAA/AR-06/11, U.S. Dept. of Transportation, Office of Aviation & Research, Washington, DC, Nov. 2006.

[18] Wagenmakers J. A review of transport airplane performance requirements might benefit safety. *Flight Safety Foundation Digest*, 19(2):1–9, 2000.

[19] ESDU. *Example of Risk Analysis for an Aircraft Subject to Performance Errors*. Data Item 08006. ESDU International, London, Aug. 2009.

[20] ESDU. *Example of Risk Analysis Applied to Aircraft Landing Distance*. Data Item 08005. ESDU International, London, June 2008.

[21] ESDU. *Statistical Methods Applicable to Analysis of Aircraft Performance Data*. Data Item 91017. ESDU International, London, Oct. 1991.

[22] ESDU. *Variability of Standard Aircraft Performance Parameters*. Data Item 91020. ESDU International, London, Oct. 1991.

## Nomenclature for Chapter 15

$A$ = wing area
$c$ = irretrievable loss of engine performance; constant
$c_{usf}$ = usable fuel ratio, Equation 15.16
$C_1$ = ownership cost
$C_2$ = interest rate
$C_3$ = fuel cost
$C_4$ = crew cost
$C_5$ = spare-parts cost
$C_6$ = maintenance cost
$C_7$ = landing fees
$C_{56}$ = combined spare-parts and man-power cost, Equation 15.54
$C_D$ = drag coefficient
$C_{D_o}$ = profile-drag coefficient
$C_{fuel}$ = fuel cost per kg
$C_{L_{mp}}$ = lift coefficient for minimum power
$D$ = aerodynamic drag
$E$ = fees due to environmental emissions; residual in fuel update
$f_j$ = thrust-specific fuel consumption
$F$ = financing
$g$ = acceleration of gravity
$i$ = iteration counter
$I$ = interest rate on loan
$k$ = induced-drag coefficient
$k_w$ = counter in DOC analysis (§ 15.8)
$L$ = aerodynamic lift
$\mathcal{L}$ = labour rate (currency units per hour)
$r$ = fuel reserve, in percent
$m$ = mass
$m_f$ = fuel burn
$m_{fr}$ = fuel-mass ratio, Equation 15.38
$\dot{m}_f$ = fuel flow
$M$ = Mach number
MH = man-hours
$n$ = lifetime of loan
$n$ = number of iterations/bisection steps
$n_w$ = total number of engine washes
$n_1, n_2$ = number of pilots/flight attendants, Equation 15.50
$N$ = number of cycles for engine wash
$p_a$, $p_d$ = fuel price (arrival, departure)
$\mathcal{P}$ = price
$P$ = power
$P_{ET}$ = point of equal time

$P_{NT}$ = point-of-no-return
$r$ = fuel-reserve ratio
$R_{div}$ = diversion distance
$R_n$ = residual value after $n$ years
$S$ = salary (wages) based on full-time rates
SP = spare parts
$t$ = time
$T$ = net thrust
$T_b$ = utilisation time (block time plus turnaround time)
$T_o$ = full-time working hours per year
$U$ = true air speed (also TAS)
$U_{mp}$ = speed of minimum power
$\mathcal{W}$ = mechanical work
$W$ = weight
$W_d$ = weight at the top of descent
$W_e$ = empty weight
$W_f$ = fuel weight
$W_{land}$ = landing weight
$W_{mfw}$ = maximum fuel weight
$W_{msp}$ = maximum structural payload weight
$W_{mlw}$ = maximum landing weight
$W_p$ = payload weight
$W_{to}$ = take-off weight
$x$ = number of cycles in thousands, $N$/1,000; distance
$x_c$ = en-route climb distance
$x_d$ = en-route descent distance
$x_i$ = flight-distance segments (climb, cruise, descent ... )
$X_{design}$ = manufacturer's design range
$X$ = stage length
$X_{div}$ = diversion distance
$X_{hold}$ = holding distance
$X_{out}$ = equivalent all-out range
$X_p$ = payload range
$X_{req}$ = required range

### Greek Symbols

$\alpha_1$ = deterioration, Equation 15.49
$\alpha_2$ = fuel-efficiency-deterioration-lapse coefficient
$\eta$ = propulsive efficiency
$\eta_M$ = log-derivative of the propulsive efficiency
$\epsilon$ = fraction of the estimated mission fuel
$\mu_r$ = rolling resistance
$\xi$ = fuel fraction, Equation 15.4
$\rho$ = air density
$\sigma$ = relative air density

### Subscripts/Superscripts

| | | |
|---|---|---|
| $[\cdot]_c$ | = | climb |
| $[\cdot]_d$ | = | descent |
| $[\cdot]_e$ | = | empty |
| $[\cdot]_{ET}$ | = | equal-time |
| $[\cdot]_{ext}$ | = | extended range |
| $[\cdot]_f$ | = | fuel |
| $[\cdot]_k$ | = | index for maintenance items or spare parts |
| $[\cdot]_i$ | = | iteration counter |
| $[\cdot]_{mlw}$ | = | maximum landing weight |
| $[\cdot]_{NT}$ | = | no return time |
| $[\cdot]_{msp}$ | = | maximum structural payload |
| $[\cdot]_p$ | = | payload; predicted value |
| $[\cdot]_{div}$ | = | diversion |
| $[\cdot]_{hold}$ | = | hold |
| $[\cdot]_{res}$ | = | reserve |
| $[\cdot]_{to}$ | = | take-off |
| $[.]^*$ | = | estimated quantity |

# 16 Aircraft Noise: Noise Sources

## Overview*

The subject of aircraft noise is too vast to be covered in a single chapter; thus, we focus on those issues that are required for a low-order simulation of aircraft noise emission and propagation. The chapter is organised as follows: We first discuss the importance of the subject (§ 16.1) and introduce basic definitions that are used to characterise aircraft noise (§ 16.2). We then establish the framework for the development of an aircraft-noise program that is synchronised with the flight mechanics (§ 16.3). The sources of noise are split into propulsive (§ 16.4) and non-propulsive components (§ 16.6). We include a discussion of the APU noise as a separate item (§ 16.5). The propeller-noise model (§ 16.7) is discussed in isolation from airframe and engines. Chapter 17 deals with propagation effects. Chapter 18 deals with noise trajectories.

**KEY CONCEPTS:** Sound, Noise Metrics, Engine Noise, Airframe Noise, Propeller Noise, APU Noise, Aircraft-Noise Model.

## 16.1 Introduction

In the beginning there was silence, or so the story goes. Then came human activities and jet engines. This is when the noise started to hurt. Low-flying aircraft cause nasty physiological effects and are bad for your well-being[1]. To be fair, noise pollution is just one of many environmental issues that face the aviation industry; it is part of the larger context of local air quality, combustion emissions, environmental compatibility, policies and regulations and public health.

In this chapter we shall identify the main causes of aircraft noise and some methods for noise reduction. Aircraft certification must meet increasingly tough international regulations (FAR Part 36 and ICAO Annex 16) and restriction of operations at several airports around the world. It will be explained how excessive noise emission can make an aircraft obsolete. In the current world of regulations, Concorde would not be allowed to fly; a first-generation jet airplane would be immediately grounded.

* Dr. Zulfaa Mohamed-Kassim contributed to this chapter.

Because a considerable number of models still rely on empirical and semi-empirical evidence, at times the physics is replaced by a direct approach to noise, which is currently the most straightforward solution. This is a rapidly expanding area of research, with new methods being proposed on a regular basis. The use of physics-based models is computationally more demanding and, for some types of aircraft performance study, it is not practical. Engineering applications include aircraft noise from conventional and unconventional configurations.

## 16.2 Definition of Sound and Noise

The essential parameters for defining noise are the sound pressure level (SPL), the acoustic power $P$, and the acoustic intensity $I$. The sound pressure $p$ is calculated with respect to a reference value $p_{ref} = 20\ \mu$Pa; this value corresponds to the sound pressure at the threshold of hearing. The sound intensity is defined as the sound pressure multiplied by the speed of propagation, or the acoustic power per unit area:

$$I = pa = p\frac{a^2}{a} = \frac{p}{a}\frac{p}{\rho} = \frac{p^2}{\rho a} = \frac{F_p}{A}\frac{r}{t} = \frac{E}{t}\frac{1}{A} = \frac{P}{A}, \tag{16.1}$$

where $F_p$ is the force created by the acoustic pressure $p$, $\rho$ is the air density, $E$ is the energy, $t$ is the time, $r$ is the distance from the source and $A$ is the propagation area. For a point source, the propagation area is spherical, so $P$ decays as $1/r^2$; for a line source, the propagation front is cylindrical, and $P \sim 1/r$. A measure of the loudness of the sound is given in *Bel*. Because this unit is generally a too large quantity, a *decibel* ($10^{-1}$ Bel $= 1$ dB) is used in practice. The sound loudness in decibels is defined from the ratio between the actual sound intensity $I$ and the sound intensity at the threshold of hearing, $I_o$

$$\text{SPL}(dB) = 10\log\left(\frac{I}{I_o}\right) = 10\log\left(\frac{p}{p_{ref}}\right)^2 = 20\log\left(\frac{p}{p_{ref}}\right), \tag{16.2}$$

where "log" is the symbol for logarithm with base 10. The threshold of hearing is set at $p_{ref} = 20\ \mu$Pa at which the corresponding SPL is defined as 0 dB. From Equation 16.2, about 12% increase/decrease in sound pressure corresponds to $\pm 1$ dB change in relative sound intensity; $\pm 1$ dB is the minimum perceptible sound difference by a sharp ear (*just noticeable difference*); 6 dB corresponds to a doubling of the sound pressure; and $+10$ dB is required before sound is perceptibly louder.

Generally, 120 dB is the threshold of pain, 110 dB is a very noisy jet aircraft, 80–90 dB is street noise in a busy city, 65 dB is a noisy office in which verbal communication is disrupted, and 50 dB is the average office environment. In a sound-proof room the SPL is of the order 10–20 dB.

The noise is also characterised by its frequency. The range of frequencies that can be heard by a human ear is between 20 and 20,000 Hz, and the range of intensity is 4 to 120 dB, although this range depends on the person. Human hearing is less sensitive at very low and high frequencies. To account for this, some weighting filters can be applied. The most common frequency weighting in current use is "*A*-weighting" that conforms approximately to the response of the human ear. The *A*-weighted sound level, called *dBA*, accounts for the fact that the middle to high frequency range (500

Table 16.1. *Summary of integral noise metrics*

| Metrics | Acronym | ANSI[3] | Full Name | Reference |
|---|---|---|---|---|
| | | **A-Weighted Noise** | | |
| Exposure-based | SEL | $L_{AE}$ | Sound exposure level | § 16.2.2 |
| | LAEQ | $L_{AeqT}$ | Equivalent sound level | Equation 16.11 |
| Maximum-level | LAMAX | $L_{ASmx}$ | Maximum sound level | |
| | | **Tone-Corrected Perceived Noise** | | |
| Exposure-based | EPNL | $L_{EPN}$ | Effective perceived noise level | § 16.2.1 |
| Maximum-level | PNLTM | $L_{PNLSmx}$ | T-corrected max perceived noise | p. 474 |

to 5,000 Hz) is more annoying to human ears than low frequencies. However, this is partly compensated by the fact that high frequencies show larger decay rates as they propagate through the atmosphere; therefore the most important frequency range from the point of view of annoyance is the 200–2,000 Hz range. A full description of the basic noise metrics is given in many acoustics textbooks, for example, Pierce[2].

The frequencies used in aircraft noise are a *1/3 octave band* series. When two frequencies $f_1$ and $f_2$ are such that $f_2 = 2f_1$, it is said that they are separated by an octave. This would normally be the size of a band. However, in aircraft noise, the band is 1/3 octave, and the audible spectrum is divided in 1/3 octave bands. Therefore, two consecutive frequencies are

$$f_2 = 2^{1/3} f_1. \tag{16.3}$$

This means, for example, that if $f_1 = 20$ Hz, then $f_2 = 25$ Hz, $f_3 = 31.5$, and so on (the frequencies are approximated to the nearest integer).

The application of these concepts to aircraft noise is somewhat more complicated than a single acoustic pressure in dB. In fact, the problem of noise is addressed in term of *annoyance* and its related causes, such as peak, duration, frequency of events, and so on. Thus, other measures of noise have been developed to characterise the complete aircraft trajectory. These new metrics are *integral*, in the sense that a segment of the flight trajectory is associated to a single noise level. Integral noise metrics are discussed next. Basically, there are three types of metrics: 1) exposure-based, 2) maximum noise level, and 3) time-based metrics. Most of these metrics have been standardised. A partial list of metrics is given in Table 16.1. A full list of terms is published by the ANSI[3].

### 16.2.1 Integral Metrics: Effective Perceived Noise

Because the noise heard on the ground is better characterised by a single scalar parameter, we need a method to transform the SPL of the individual component noise, at each frequency and time step, into such a quantity. This is currently done with a quantity called *Effective Perceived Noise Level* (EPNL). The EPNL is a weighted value that takes into account loudness and duration. The mathematical

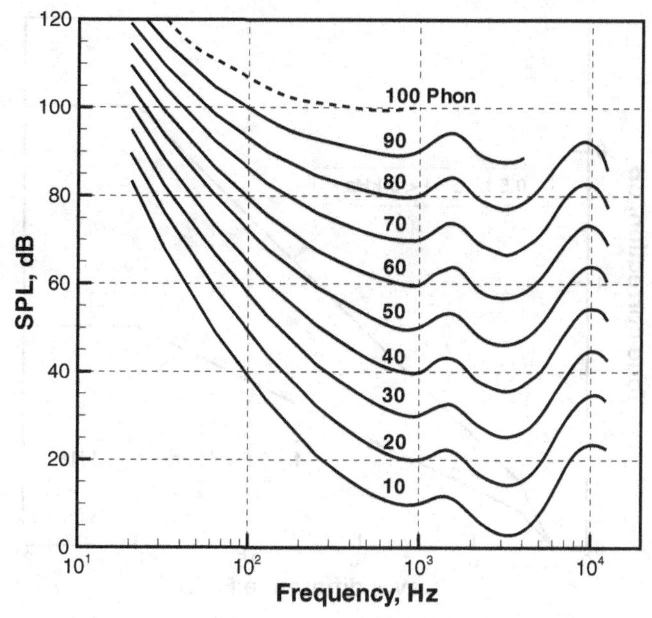


Figure 16.1. Perceived noise with selected Noy levels. Adapted from ISO 226.

definition of this quantity is

$$\mathrm{EPNL} = \mathrm{PNL}_{max} + C + D, \tag{16.4}$$

where $\mathrm{PNL}_{max}$ is the maximum perceived noise level (in dB), $C$ is a correction (in dB) for tonal noise and $D$ is a correction for duration of the noise (also expressed in dB). The main procedure to calculate the EPNL is done in a number of steps (readers are referred to Smith[4] for detailed calculations of the EPNL).

1. The SPL signal is converted into a PNL (Perceived Noise Level) through a conversion chart that is shown in Figure 16.1. This chart shows levels of constant *Noy*, which is a measure of *equal loudness*. Take, for example, the frequency $f$, corresponding to SPL($f$). The combination of an SPL($f$) for a single $f$ gives one value of the Noy (directly or by interpolation). A total $\mathrm{Noy}_{Total}$ can be derived by integrating each Noy value across the entire one-third octave-band frequency spectrum where

$$\mathrm{Noy}_{Total} = 0.85\mathrm{Noy}_{max} + 0.15 \sum_{i=1}^{24} \mathrm{Noy}_i. \tag{16.5}$$

   The relationship between $\mathrm{Noy}_{Total}$ and PNL at each time-step in the trajectory is

$$\mathrm{PNL}(dB) = 40 + 10 \log_2(\mathrm{Noy}_{Total}). \tag{16.6}$$

2. Next, we have to define a tonal correction $C$ for each time-step. This correction depends on the frequency band and most critically on a *level difference* $F$, whose calculation can be laborious (see Edge and Cawthorn[5] and Smith[4]). Once the level difference is determined, the tonal correction is given in the chart

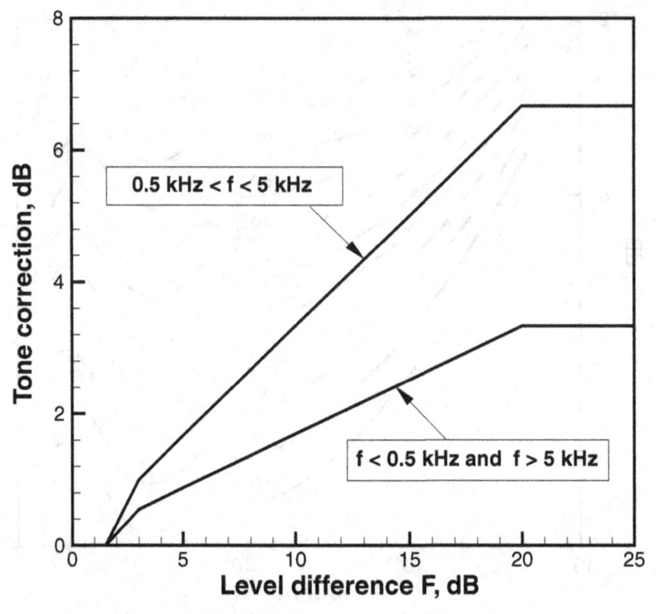


Figure 16.2. ICAO tone-correction chart.

of Figure 16.2 for various frequency bands. The maximum tone correction is 6.7 dB. Thus, the tone-corrected PNL is

$$\text{PNLT} = \text{PNL} + C \quad [dB]. \tag{16.7}$$

3. Calculate the maximum PNLT, defined as PNLTM, which is a scalar value.
4. Calculate a correction for duration of noise. If $t = t_2 - t_1$ is the time of the flight path analysed, the duration is

$$D = 10 \log_{10} \left[ \frac{1}{t} \int_{t_1}^{t_2} 10^{(\text{PNLT}/10)} \right] dt - \text{PNLTM}. \tag{16.8}$$

5. The effective perceived noise is calculated by summing the PNLTM and its duration in the flight path

$$\text{EPNL} = \text{PNLTM} + D. \tag{16.9}$$

In this sense, the signal has been cleared of frequency effects, but it takes into account other aspects, such as tone and duration.

The method described indicates that the EPNL cannot be measured directly; this parameter has to be calculated using information of the entire flight path, SPL, spectrum level and shape. The value of the EPNL can be considerably different from the loudness.

When using the EPNL metrics, noise reduction assumes a different meaning. In fact, an EPNL reduction can be achieved by a combination of changes in the flight path, changes in the acoustic signature that would limit the effects of tonal correction and other factors. Furthermore, because the EPNL metric is an integral quantity, by itself it is not representative of the accuracy of the noise-prediction method.

### 16.2.2 Integral Metrics: Sound Exposure Level

The sound exposure level (SEL) is also called $L_{AE}$ (dBA). It is calculated as follows:

- Calculate the maximum loudness along the trajectory, $L_{max}$.
- Define the reference sound level $L_{ref} = L_{max} - 10$ dB.
- Calculate the time interval $[t_1, t_2]$ of the trajectory when $L > L_{ref}$.
- Calculate the SEL from the integral:

$$L_{AE} = 10 \log_{10} \left( \int_{t_1}^{t_2} 10^{SPL(t)} \, dt. \right) \tag{16.10}$$

In the A-corrected SPL spectrum, the loudness is defined as the maximum SPL(dBA). Another integral parameter related to the SEL, is the *Equivalent Continuous Noise Level*, called $L_{AeqT}$. This is defined as

$$L_{AeqT} = L_{AE} - 10 \log_{10} t_o, \tag{16.11}$$

where $t_o$ is the time duration of interest.

## 16.3 Aircraft Noise Model

The noise produced by the aircraft is modelled with the method of components. This method consists in reconstructing the noise from the contribution of all of the subsystems. Inevitably, there will be approximations and assumptions, some of which are not completely rigorous. The second important aspect is that the noise model for each component can be approximated, and better models may be developed in the future. The advantage of this approach is that if one component provides unsatisfactory answers, then it can be improved independently from the remaining systems.

Figure 16.3 shows the breakdown in noise-source components. These are given in three broad categories: propulsion, airframe and interference. Among the propulsion sub-components, the propeller is not relevant if the airplane is powered by turbofan engines. The APU contribution is generally not considered, although its effect is substantial when the aircraft is on the ground (§ 16.5). The propeller noise will be discussed separately from the propulsive noise (§ 16.7).

Among the airframe components, the effects of the fuselage itself is not included because it is not significant. The interference contributions that are considered are the jet-by-jet noise shielding (although this is not important on turboprop engines), the fuselage scattering effects and the wing reflection of the noise sources; the latter item is particularly important if the engines are mounted above the wing. The flowchart indicates that the aircraft-noise simulation can be built in a modular fashion: it is possible to replace one sub-model with another one without having to reconsider the rest of the framework. A full review of sub-models for aircraft noise is available in Casalino *et al.*[6]

System analysis indicates that approximations of the same order are needed to either reduce noise from different systems or to calculate the noise emission with a

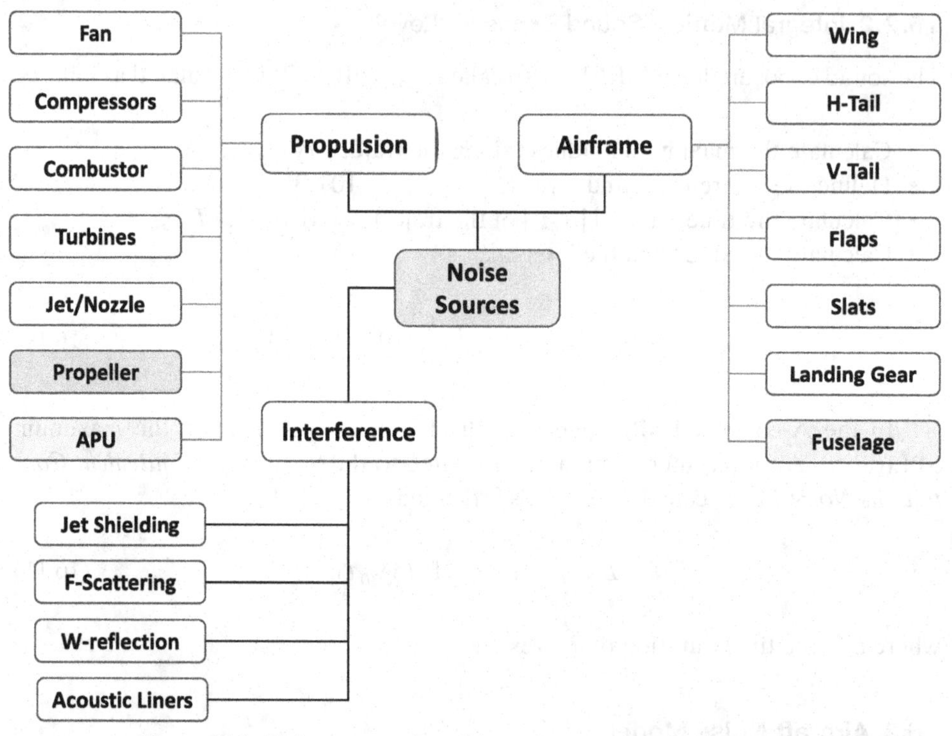


Figure 16.3. Noise-sources breakdown in a conventional airplane.

satisfactory accuracy. The far-field noise SPL is calculated from the sum:

$$\mathrm{SPL} = 10 \log_{10} \left( \sum_k 10^{SPL_k/10} \right). \tag{16.12}$$

Note that the resulting noise is calculated by first converting the various contributions into decimal scale; the final result requires a decimal logarithm. Due to the arbitrary definition of SPL such that it is 0 when $p = p_{ref}$, Equation 16.12 yields a strange result on the summation of the total SPL. For example, the sum of three contributions $\mathrm{SPL}_k = 0$ dB yields a total SPL= 4.77 dB, not zero as one would expect in a conventional sum.

An airplane that is *50% quieter* is an aircraft whose overall sound pressure is 50% lower. Because the noise metric is logarithmic, this figure translates into an OASPL that is reduced by 6 dB. This result is derived from

$$\begin{aligned}
\Delta(\mathrm{OASPL}) &= 20 \log_{10} \left( \frac{p}{2p_{ref}} \right) - 20 \log_{10} \left( \frac{p}{p_{ref}} \right) \\
&= 20 \left[ \log_{10} \left( \frac{p}{2p_{ref}} \right) - \log_{10} \left( \frac{p}{p_{ref}} \right) \right] \\
&= 20 \log_{10}(1/2) \simeq -6\,\mathrm{dB}.
\end{aligned} \tag{16.13}$$

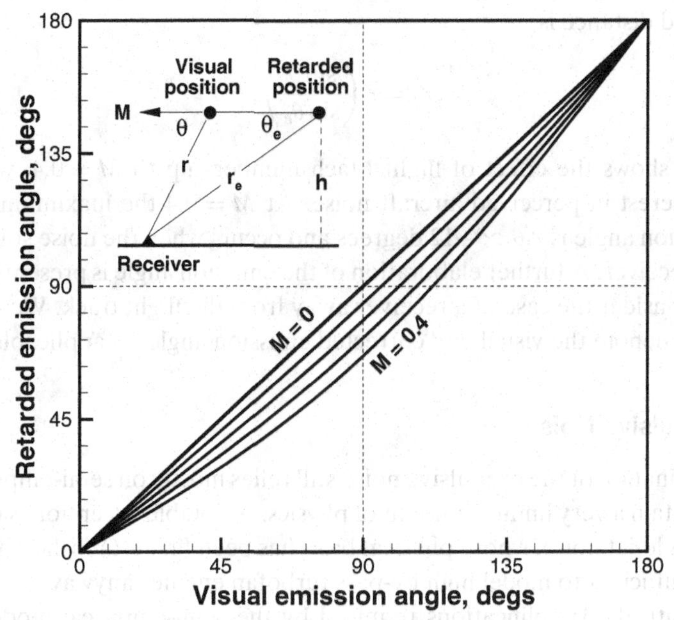


Figure 16.4. Corrected polar emission angle as function of flight Mach number. Curves are shows at intervals $\Delta M = 0.1$.

If we invert Equation 16.13, the change in acoustic pressure $\Delta p$ corresponding to a known change in OASPL is

$$\frac{\Delta p}{p} = 10^{(\Delta \mathrm{OASPL}/20)} - 1. \tag{16.14}$$

In practice, this means that at sea level and standard day, the decrease in acoustic pressure corresponding to a reduction of OASPL by 6 dB is only 480 Pa. However, if the metric used is a "weighted" one, such as the perceived noise level (§ 16.2.1), then noise reduction has a different meaning.

### 16.3.1 Polar-Emission Angle

In general, the aircraft or the noise source is found at some point above the receiver. In the most basic case, the source-receiver position is determined by the distance $r$ and the emission angle $\theta$; alternatively, it is defined by a distance and an altitude. However, most noise-radiation effects depend on the emission angle and thence on the directivity. The polar-emission angle is a function of the aircraft's speed (or Mach number). This is a result of the nature of wave propagation at a finite speed when the source itself travels at a comparable speed.

With reference to Figure 16.4, if $r$ is the "visual" distance source-to-receiver and $\theta$ is the corresponding "visual" emission (or radiation) angle of a noise source travelling with a Mach number $M$, the "corrected" (or retarded) emission angle $\theta_e$ is calculated from

$$\cos\theta_e = \cos\theta\sqrt{1 - M^2\sin^2\theta} + M\sin^2\theta. \tag{16.15}$$

The retarded distance is

$$r_e = r\left(\frac{\sin\theta}{\sin\theta_e}\right). \tag{16.16}$$

Figure 16.4 shows the effect of flight Mach numbers up to $M = 0.4$, which is the range of interest in perceived aircraft noise. At $M = 0.4$ the maximum correction in the emission angle is $\Delta\theta \simeq -12$ degrees and occurs when the noise source is right above the receiver. A further elaboration of the emission angle is presented in § 16.6, where we consider the case of a receiver away from the flight track. We will use the symbol $\theta$ to denote the visual *and* corrected emission angle, as applicable.

## 16.4 Propulsive Noise

The determination of the propulsive noise still relies mostly on semi-empirical methods that contain a very limited amount of physics. A notable exception is the calculation of a single-jet noise, whose physical basis has been first established by Lighthill, which is insufficient to model high by-pass turbofan engines anyway.

Even with the simplifications required by the semi-empirical models, a comprehensive approach to engine noise requires a considerable amount of engine data, divided into two categories: 1) geometry and configuration data, and 2) aero-thermodynamic data. In the former category we have parameters such as fan, compressors and turbine diameters, number of blades, rpm. Once the engine model has been developed, as explained in Chapter 5, the aero-thermodynamic simulation provides the operating data required.

### 16.4.1 Noise-Propulsion System Interface

The numerical determination of the propulsive noise relies on a reliable estimate of the engine state for a specified flight condition. We begin with the turbofan-powered aircraft. Our model for a turbofan engine relies on the engine architecture shown in Figure 16.5. There are two possibilities:

- The fuel flow rate is specified; the aero-thermodynamics of the engine is solved to provide net thrust, pressures, temperatures and speeds at the relevant engine sections (*direct mode*).
- The net thrust is specified; the aero-thermodynamics of the engine is solved to provide mass-flow rates, fuel-flow rates, pressures, temperatures and speeds (*inverse mode*).

In the first instance, we rely on externally generated noise trajectories such as, for example, FDR data. The fuel flow is directly measured and monitored, whilst the net thrust can only be determined through a flight-mechanics analysis.

The case of the turboshaft engine coupled to a propeller is similar, but it has some important differences, which are shown in Figure 16.6. In the inverse mode, the shaft power is determined via flight-mechanics equations. The problem is then reduced to the determination of the fuel flow corresponding to the specified shaft power. This calculation must be done iteratively to take into account the losses at

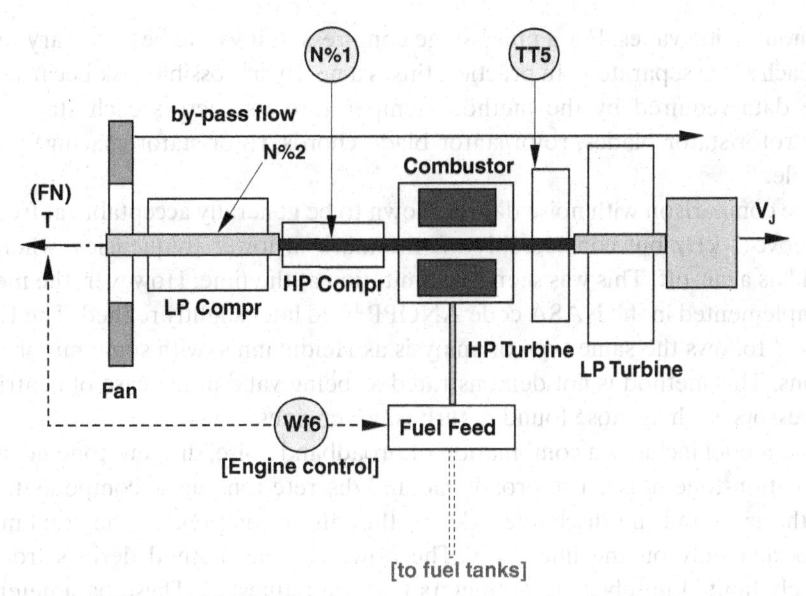

Figure 16.5. Turbofan engine model for noise simulation.

the gear-box and the effects of the residual thrust. The engine state is calculated from the fuel-flow rate.

### 16.4.2 Fan and Compressor Noise

A semi-empirical method for compressor and fan noise was developed by Heidmann[7] on the basis of an earlier method at Boeing-NASA. The method relies on an experimental database and applies to fan and single-stage compressors, with

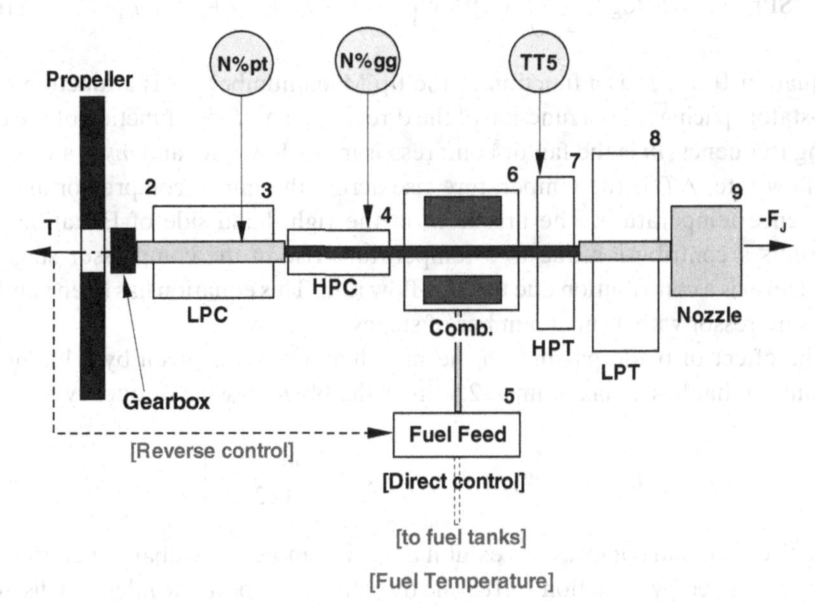

Figure 16.6. Turboshaft-propeller model for noise simulation.

or without guide vanes. For a multi-stage compressor, it would be necessary to consider each stage separately. In practice, this is a nearly impossible task because most of the data required by the method (temperature rise across each stage, number of rotor/stator blades, rotor/stator blade chords, rotor-stator spacing) are not available.

The comparison with noise data is shown to be generally acceptable at frequencies above 1 kHz but considerably off the mark at lower frequencies, where the model has a cut-off. This was seen as a limitation at the time. However, the method was implemented in the NASA code ANOPP[8] and later slightly refined. The ESDU method[9] follows the same type of analysis as Heidmann's, with some minor modifications. This method is not demonstrated as being valid in the case of centrifugal compressors, such as those found in turboshaft engines.

The model includes a combination of broadband noise, discrete-tone noise and combination-tone noise. The broadband and discrete-tone noise components arise from the inlet and the discharge sides of the fan or compressor; the combination tone occurs only on the inlet side. The power of the method derives from the relatively limited number of parameters that are requested. These parameters are mass-flow rate, temperature rise through the stage, design, and operating tip Mach numbers. Some corrections can be introduced to take into account the presence of guide vanes, rotor-stator spacing, inlet flow distortion tone cut-off and blade-row attenuation. The method involves the prediction of the spectrum shape and level and the noise directivity for each component.

**BROADBAND NOISE.** This noise component is associated to turbulence in the flow through the blade rows, arising from inlet turbulence, wall boundary layers, and vortex interaction between blades and vanes. The latter item depends on non-periodic wakes. The formula for the *broadband noise* is

$$\mathrm{SPL}(f) = 20\log\left(\frac{\Delta T}{T_o}\right) + 10\log\left(\frac{\dot{m}}{\dot{m}_o}\right) + F_1 + F_2 + F_3 + F_4. \qquad (16.17)$$

In Equation 16.17, $F_1$ is a function of the tip Mach number, $F_2$ is a function of the rotor-stator spacing, $F_3$ is a function of the directivity and $F_4$ is a function of the blade passing frequency; $\dot{m}$ is the fan or compressor mass-flow rate, and $\dot{m}_o$ is a reference mass-flow rate; $\Delta T$ is the temperature rise across the fan or compressor and $T_o$ is a reference temperature. The first term in the right-hand side of Equation 16.17 represents a contribution due to a temperature rise in the compressor stage; the second term is a contribution due to mass-flow rate. This equation has been validated for a compressor with a small number of stages.

The effect of blade passing on the broadband noise is given by a log-normal distribution that has a maximum at 2.5 times the blade passing frequency $f_b$:

$$F_4 = 10\log_{10}(e^c), \qquad c = \left[\frac{\ln(f/f_b)}{\ln 2.2}\right]^2. \qquad (16.18)$$

In practice, fans and compressors exhibit a more complex broadband spectrum than the one predicted by Equation 16.18. The directivity, for both the inlet and discharge sides, is shown in Figure 16.7.

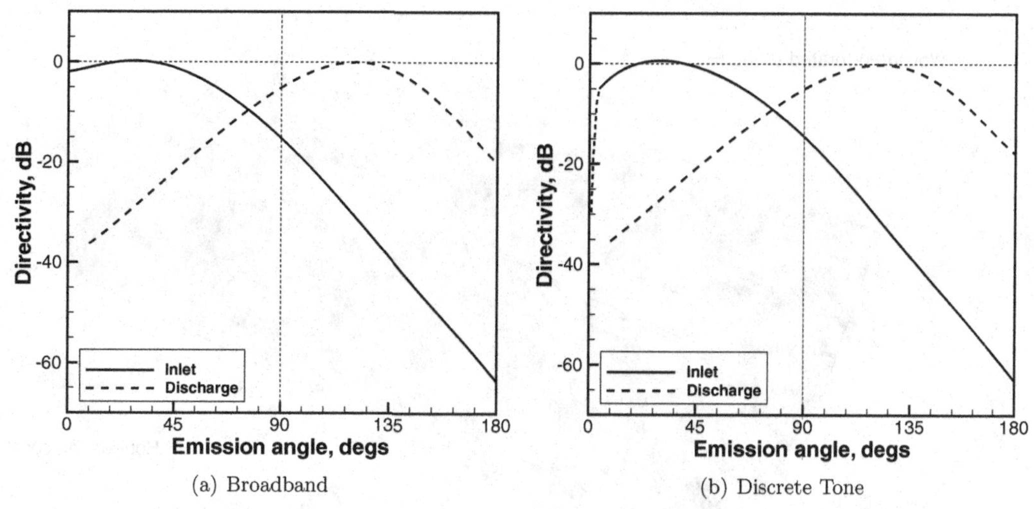


Figure 16.7. Single-stage fan and compressor noise directivities.

**DISCRETE-TONE NOISE.** The formula for the *discrete-tone noise* is

$$\mathrm{SPL}(f) = 20\log\left(\frac{\Delta T}{T_o}\right) + 10\log\left(\frac{\dot{m}}{\dot{m}_o}\right) + F_1 + F_2 + F_3. \tag{16.19}$$

In Equation 16.19, the functions $F_1$, $F_2$, and $F_3$ depend on tip Mach number, the rotor-stator spacing, and the directivity, respectively. However, these functions are different from those defining the broadband noise (see Equation 16.17).

**COMBINATION-TONE NOISE.** This is generated by supersonic tips at 1/2, 1/4 and 1/8 of the blade passing frequency and is given by a similar equation:

$$L_p = 20\log\left(\frac{\Delta T}{T_o}\right) + 10\log\left(\frac{\dot{m}}{\dot{m}_o}\right) + F_1 + F_2. \tag{16.20}$$

Again, the functions $F_1$ and $F_2$ depend on the tip Mach number and the directivity, respectively. Formulas are given for the 1/2, 1/4 and 1/8 blade passing frequencies. A *blade row attenuation* correction is applied for multiple-stage compressors but not for a single-stage fan.

The temperature rise across the stage can be estimated from the adiabatic compression

$$\frac{\Delta T}{T} = \frac{1}{\eta_c}\left[p_r^{(\gamma-1)/\gamma} - 1\right], \tag{16.21}$$

where $\eta_c$ is the thermodynamic efficiency of the stage, $p_r$ is the compression ratio of the stage, and $\gamma$ is the ratio between specific heats ($\gamma = 1.4$ will suffice for most calculations).

The method described applies to a single-stage fan or compressor, although the method has been tested successfully for two-stage compressors. The use of multi-stage compressors in modern turbofan engines complicates the analysis. One approach is to approximate the multi-stage compressor as a single stage by taking a temperature and pressure rise between the first and the last stages. Unfortunately, this approximation yields unrealistically high noise for a large number of stages.

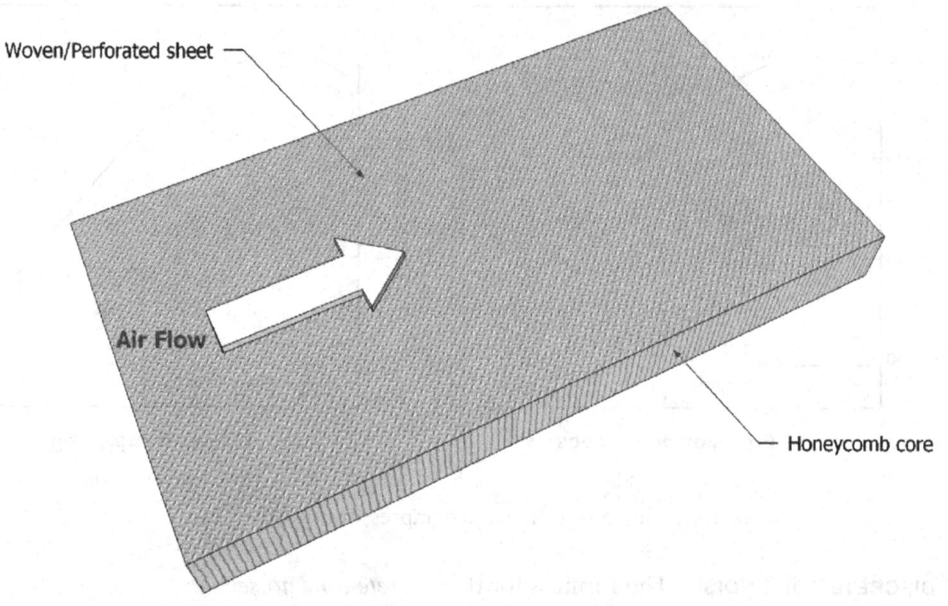


Figure 16.8. Example of acoustic liner sheet for duct-fan noise reduction.

Another option is to consider only the effects of the first stage of LP and HP compressors as representative of the far-field noise[7].

The equations providing the noise level (Equation 16.17, Equation 16.19 and Equation 16.20) are valid for an observer located 1 metre away from the source. Therefore, for an observer at a distance $r$ the result must be corrected for spherical spreading, atmospheric absorption and other dissipative and amplification factors. For the sake of generality, we assume that the noise spreads along a spherical front without encountering any obstacles. In this case, the SPL varies inversely to the square of the distance from the source (i.e., the inverse square law). Assume that at a distance $r_1$ the sound pressure level is $\text{SPL}_1$; at a distance $r_2 > r_1$, this quantity will be reduced to $\text{SPL}_2$. The relationship between the two SPLs is

$$\text{SPL}_2 = \text{SPL}_1 + 20 \log\left(\frac{r_1}{r_2}\right). \tag{16.22}$$

Note that if the distance increases, the "log" is negative; therefore, there must be a + sign to the right of $\text{SPL}_1$. Equation 16.22 shows that a doubling of the distance causes the SPL to be reduced by 6 dB.

With reference to the fan, which is one of the largest causes of engine noise, we must address some additional aspects, which involve the presence of the inlet duct and the so-called *acoustic liners*, both contributing to some noise reduction[10]. The acoustic liners are passive or adaptive systems consisting of a variety of perforated plates, honeycomb structures, and woven materials, whose purpose is to increase the noise absorption in the duct over a specified range of frequencies. An example of such liners is shown in Figure 16.8. A vast amount of research exists in this area[11–13]. ESDU has some practical methods for estimating the acoustic impedances and absorption[14;15]. The effectiveness of these systems depends on a large number of factors, including the length of the duct (long or short), the type of liner, the impedances and the directivity. The inlet ducts of modern turbofan engines are relatively large, up to one-third of the fan diameter*. This means that the fans are buried deep into the nacelles and allow the installation of a variety of technologies for further noise reduction, including the liners.

Modern turbofan engines have wide-chord fans, with a greatly reduced number of blades (typically 22 to 26 blades); this design contributes to the reduction of the blade-passing frequency. The broadband noise signature is similar to the narrow-chord fans. The reduction of the fan noise level requires appropriate strategies that address the tonal components and lower frequency bands (600 to 4,000 Hz).

In many cases it is possible that the fan blades operate in supersonic flow. This condition is typical of the tip regions at high engine speeds and moderate to high aircraft speed (especially at take-off). The resulting noise is referred to as "buzz-saw". This event is characterised by acoustic pressures at frequencies multiple of the engine rotational speed[16;17].

### 16.4.3 Combustor Noise

The semi-empirical theory proposed by ESDU[18] (derived from earlier methods by the Society of Automotive Engineers [SAE]) relies on an equation that correlates the sound pressure level, the noise power level and the directivity:

$$\text{SPL}(f,\theta) = \text{PWL}(f) + \mathcal{D}(\theta) - 20\log r + 10\log_{10}\left(\frac{\rho_o a_o W_{ref}}{4\pi p_{ref}^2}\right), \tag{16.23}$$

where PWL is the one-third octave band noise power level in dB, and $\theta$ is the angle between the engine axis and the receiver (emission angle); the subscript "o" denotes ambient conditions outside the engine. The PWL is derived from a spectral correction of the A-scale overall PWL:

$$\text{PWL} = \text{OAPWL} + \mathcal{S}(f), \tag{16.24}$$

with the overall sound power level defined by:

$$\text{OAPWL} = 10\log_{10}\left(\frac{a_o \dot{m}_2^2}{W_{ref}^2}\right) + 10\log_{10}\left[\left(\frac{\Delta\mathcal{T}}{\mathcal{T}_3}\right)^2\left(\frac{p_3}{p_o}\right)^2\left(\frac{\Delta\mathcal{T}_{turb}}{\mathcal{T}_o}\right)^{-4}\right] - C, \tag{16.25}$$

where $\dot{m}_2$ is the core-mass-flow rate, $\Delta\mathcal{T}$ is the temperature rise in the combustor, $\mathcal{T}_3$ is the combustor inlet temperature, and $p_3$ is the corresponding pressure; $\Delta\mathcal{T}_{turb}$ is the temperature drop across the whole turbine system downstream of the combustor; and $C$ is an assigned constant, which in this case is $-60.5$dB. The item provides the directivity $\mathcal{D}(\theta)$ and the spectral function $\mathcal{S}(f)$ to be used in Equation 16.23 and Equation 16.24. The SPL given by Equation 16.23 must be corrected for Doppler effects and atmospheric attenuation. The approximation of the method is estimated between $-3$ dB and $+5$ dB, although the method lacks a rational validation and verification.

* For example, the length of the inlet duct of the GE90-115B is about 1.5 m; the total liner area of the PW4048 engine powering the B777-300 is about 6.2 m$^2$.

### 16.4.4 Turbine Noise

A method to predict the tonal and broadband noise from a single axial-flow turbine is presented here following the semi-empirical approach of the General Electric Company[19;20]. The method relies on empirical relations that express the dependence of sound pressure level on frequency and polar directivity angle. Relative to other noise-producing engine components, fast first-order predictive methods and empirical data for turbine noise are still lacking and the available ones are rather crude[21]. Recent efforts focus more on developing physics-based modelling to produce robust predictions without any reliance on empiricism[22] and thus widen their implementations on various types of engines, but these are unlikely to be favourable options for fast real-time applications in the near future.

Turbine noise, particularly from low-pressure turbines, can be substantial at approach[23]. Its importance on the overall engine noise increases with increasing turbofan engine by-pass ratio[22]. The sound radiated from a turbine is highly dependent on the polar angle and peaks with an emission angle of 110 to 130 degrees relative to the intake axis[4]. This aft-radiated sound is influenced by sound refraction through the turbulent jet mixing region. Pure tones dominate the sound spectra with the fundamental harmonic set at the turbine-blade-passing frequency. The turbulent mixing layer may also influence the tonal noise by diffusing it into more broadened or "hay-stacked" signals[24].

The empirical modelling used in the *Aircraft Noise Prediction Program*[19] is described here to calculate the far-field SPL from an axial-flow turbine, which is given as

$$\mathrm{SPL} = 10\log_{10}\left(\frac{\langle p^2\rangle}{p_{ref}^2}\right). \tag{16.26}$$

The mean-square acoustic pressure $\langle p^2\rangle$ is formulated as

$$\langle p^2\rangle = \frac{\rho_\infty^2 a_\infty^4 A_i}{4\pi r^2}\frac{\Pi^* \mathcal{D}(\theta)\mathcal{S}(\eta)}{[1 - M_\infty \cos\theta]^4}, \tag{16.27}$$

with $p_{ref} = 2\times 10^{-5}$ Pa as the reference acoustic pressure. Equation 16.27 takes into account the effects of spherical spreading and forward flight via the Doppler term $[1 - M_\infty \cos\theta]$. In addition, it uses two empirical relations to quantify the effect of polar angle $\theta$ and frequency $f$ on the turbine noise. The dimensionless acoustic power parameter $\Pi^*$ is detailed in § 16.4.4.

The first empirical relation is a polar directivity function $\mathcal{D}(\theta)$ that is given as two sets of dimensionless datasets, each for the broadband and tonal noise, respectively; these are tabulated as functions of the polar directivity angle $\theta$ in Table 16.2.

The second empirical relation is a spectrum function $\mathcal{S}(\eta)$ that is given as a tabular dataset for the broadband noise and an empirical function for the tonal noise; $\mathcal{S}(\eta)$ is a function of the dimensionless frequency parameter $\eta$, defined as

$$\eta = [1 - M_\infty \cos\theta]\frac{f}{f_b}, \tag{16.28}$$

Table 16.2. *Polar directivity levels*

| $\theta$ [deg] | Broadband $\log_{10} \mathcal{D}(\theta)$ | Tonal $\log_{10} \mathcal{D}(\theta)$ |
|---|---|---|
| 0 | −0.789 | −1.911 |
| 10 | −0.689 | −1.671 |
| 20 | −0.599 | −1.471 |
| 30 | −0.509 | −1.261 |
| 40 | −0.409 | −1.061 |
| 50 | −0.319 | −0.851 |
| 60 | −0.219 | −0.641 |
| 70 | −0.129 | −0.431 |
| 80 | −0.029 | −0.231 |
| 90 | −0.071 | −0.021 |
| 100 | 0.151 | 0.189 |
| 110 | 0.221 | 0.389 |
| 120 | 0.231 | 0.589 |
| 130 | 0.211 | 0.259 |
| 140 | 0.111 | −0.191 |
| 150 | −0.029 | −0.591 |
| 160 | −0.229 | −0.931 |
| 170 | −0.549 | −1.271 |
| 180 | −0.869 | −1.611 |

where $f_b = NB$ is the turbine-blade-passing frequency, with $N$ being the turbine rotational speed (in $Hz$) and $B$ the number of rotor blades for the single-stage turbine.

For the broadband noise, the spectrum $\mathcal{S}(\eta)$ is given in Figure 16.9. This dataset is extrapolated to extend the range of the broadband spectrum levels beyond $\log_{10}(\eta) < -0.903$ and $\log_{10}(\eta) > 0.602$. With these parameters, the broadband SPL for any

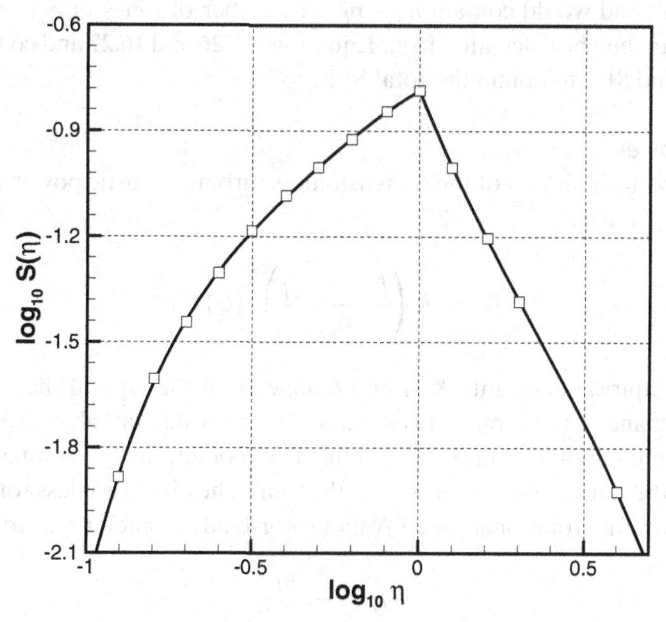


Figure 16.9. Spectrum function for broadband noise.

Table 16.3. *Empirical constants for turbine acoustic power*

| | Broadband | Pure Tones |
|---|---|---|
| $K$ | $8.589 \times 10^{-5}$ | $1.162 \times 10^{-4}$ |
| $a$ | 1.27 | 1.46 |
| $b$ | −1.27 | −4.02 |

specific frequencies can be calculated. To obtain the 1/3 octave band SPL, each band is divided into smaller sub-bands where the sub-band SPLs are computed using the sub-band centre frequencies. These are then integrated and normalised to produce the 1/3 octave band SPL for the broadband noise.

For the tonal noise, the spectrum function is given by

$$\mathcal{S}(\eta) = 0.6838 \times 10^{(1-n)/2}, \tag{16.29}$$

where $n$ is the harmonic number of each tone. The fundamental harmonic, that is, $n = 1$, is determined by the turbine-blade-passing frequency $f_b$. In this method, the tonal SPLs are combined with the 1/3 octave broadband SPL to obtain the total 1/3 octave band SPL. This is done by locating the number of harmonics as well as their harmonic numbers that are contained within each band. Given a 1/3 octave band centre frequency $f_c$, the lowest ($n_l$) and highest ($n_u$) harmonic numbers within the band can be found as follows:

$$\begin{aligned} n_l &= 1 + \texttt{int}\left[10^{-1/20}\eta\right] \\ n_u &= \texttt{int}\left[10^{1/20}\eta\right]. \end{aligned} \tag{16.30}$$

From this, a band would contain $n_u - n_l + 1$ number of tones only if $n_l \le n_u$. The tonal SPL can then be calculated from Equations 16.26 and 16.27 and combined with the broadband SPL to obtain the total SPL.

#### Acoustic Power

The empirical formulation of the dimensionless turbine acoustic power $\Pi^*$ in Equation 16.27 is given as

$$\Pi^* = K\left(\frac{h^*_{t,i} - h^*_{s,j}}{h^*_{t,i}}\right)^a (U^*_T)^b, \tag{16.31}$$

where the empirical constants $K$, $a$ and $b$ depend on the type of the noise source, either broadband or pure tones. These values are provided in Table 16.3; $h^*$ denotes the specific enthalpy: the index "i" denotes inlet condition; "j" denotes exit conditions; "s" is the static condition and "t" is the total. The dimensionless rotor tip speed $U^*_T$ depends on the rotational speed $N$ and rotor-blade diameter $d$ as follows:

$$U^*_T = \pi\frac{Nd}{a}. \tag{16.32}$$

Table 16.4. *Spectrum function for broadband noise*

| Constituent Gas | Molecular Mass |
|---|---|
| Nitrogen, $N_2$ | 28.01340 |
| Oxygen, $O_2$ | 31.99880 |
| Carbon Dioxide, $CO_2$ | 44.00995 |
| Water Vapour, $H_2O$ | 18.01534 |
| Argon, Ar | 39.94800 |

The ratio of specific enthalpies in Equation 16.31 requires the input of the total temperature at the turbine inlet ($\mathcal{T}_{t,i}$) and the static temperature at the turbine exit ($\mathcal{T}_{s,j}$). Following the documentation of Ref.[25], the specific enthalpy $h^*(\mathcal{T})$ for any arbitrary temperature $\mathcal{T}$ is defined as

$$h^*(\mathcal{T}) = \frac{1}{\mathcal{T}_\infty}\left[h_r^*\left(\frac{\mathcal{T}}{\mathcal{T}_o}\right) - h_r^*\left(\frac{\mathcal{T}_\infty}{\mathcal{T}_o}\right)\right], \tag{16.33}$$

where $\mathcal{T}_\infty$ is the ambient temperature and $\mathcal{T}_o$ is a reference temperature taken as the sea-level temperature. The reference-specific enthalpies ($h_r^*$) depend on a number of parameters for the constituent gases of air, namely nitrogen ($N_2$), oxygen ($O_2$), carbon dioxide ($CO_2$), water vapour ($H_2O$) and argon (Ar). For a given temperature ratio $\mathcal{T}/\mathcal{T}_o$, $h_r^*$ is calculated from

$$h_r^*\left(\frac{\mathcal{T}}{\mathcal{T}_o}\right) = \sum_{k=1}^{5} \mathcal{R}_k^* x_k h_{r,k}^*\left(\frac{\mathcal{T}}{\mathcal{T}_o}\right). \tag{16.34}$$

Here, $\mathcal{R}_k^*$, $x_k$, and $h_{r,k}^*$ are the dimensionless gas constant, the mass fraction and the component specific enthalpy, respectively, for the $k$-th constituent gas; $\mathcal{R}_k^*$ and $x_k^*$ are defined as follows:

$$\mathcal{R}_k^* = \frac{1}{\mathcal{R}}\left(\frac{\mathcal{R}_u}{m_k}\right)$$
$$x_k = \left(\frac{m_k}{m_T}\right) \tag{16.35}$$
$$m_T = \sum_{k=1}^{5} m_k.$$

Here, $\mathcal{R}_u = 8,314.32$ kgm$^3$/Ks$^2$ is the universal gas constant and $\mathcal{R}$ is the dry-air gas constant. The molecular masses $m_k$ of the gases are shown in Table 16.4. The masses of the component gases $x_k$, which depend on the engine fuel-to-air ratio and the absolute humidity of the ambient air, are tabulated in Ref.[25]. The component-specific enthalpies $h_{r,k}^*$ are also tabulated as a function of the temperature ratio $\mathcal{T}/\mathcal{T}_o$.

#### Turbine Noise Simulation

The turbine-noise model for a single engine is simulated to study the effect of various parameters on the turbine SPL level received 500 m away from the engine

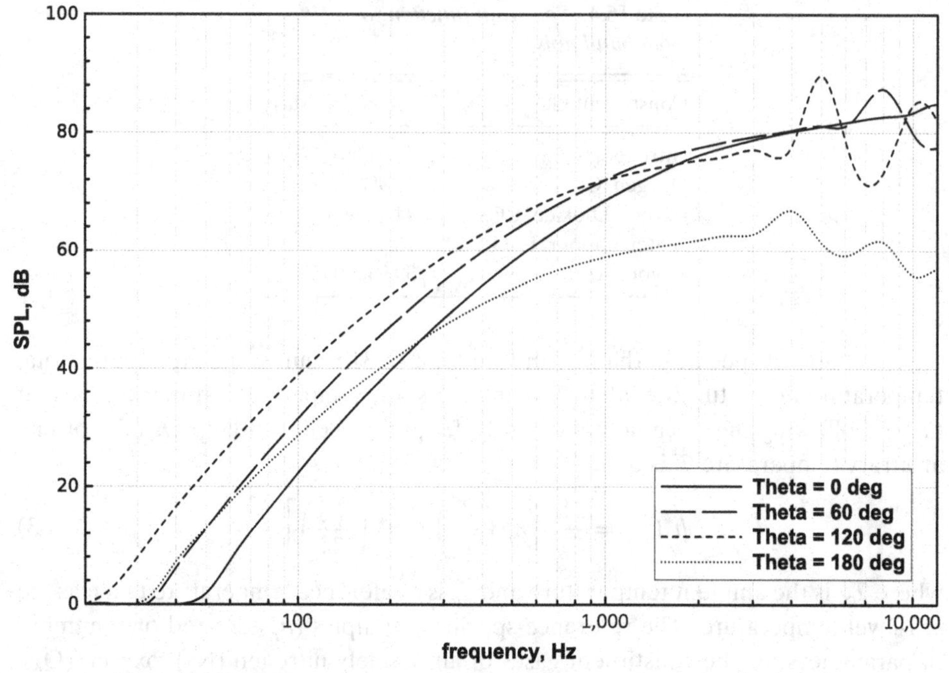


Figure 16.10. Effect of polar directivity angle $\theta$ on 1/3 octave band SPL.

with a flight Mach number $M = 0.5$. As mentioned previously, only the acoustic and attenuating effects of spherical spreading and forward flight are incorporated in the formulation; attenuation due to ground reflection, atmospheric absorption, refraction and turbulence scattering are not accounted for.

In Figure 16.10, SPL is plotted against the 1/3 octave band centre frequencies for various polar directivity angle $\theta$ ranging from 0 to 180 degrees. Generally, $\theta$ influences the SPL via the empirical *polar directivity function* $\mathcal{D}(\theta)$ and the Doppler shift term.

The shape of the SPL across the frequency spectra is governed by the spectrum function $\mathcal{S}(\eta)$. The results show that the turbine noise generally peaks at frequencies between 5 kHz and 10 kHz, depending on the shift from the effect of forward flight. The peaks are mainly due to the pure tones which have been averaged into the 1/3 octave band scale. With respect to changes in the polar angle, the noise is most attenuated aft of the engine at $\theta = 180$ degrees. The overall noise peaks for $\theta$ of 0 and 180 degrees differ by about 20 dB.

The turbine rotational speed $N$ can have a significant influence on turbine noise because it determines the turbine-blade-passing frequency $f_b$ and thus the tonal-peak frequencies. In Figure 16.11, $N$ is varied from 3,000 rpm to 12,000 rpm. In these simulations, the turbine noise is loudest for the smallest rotational speed with a significant contribution of about 24 dB from the fundamental tonal noise level relative to the highest broadband noise level. The overall 1/3 octave band SPL generally shifts downward in the dB scale across the entire frequency spectra when the rotational speed is increased. In addition, the first tonal peak shifts toward larger frequencies with increasing rotational speed. This tonal contribution toward the turbine SPL relative to those of the broadband noise also decreases with increasing

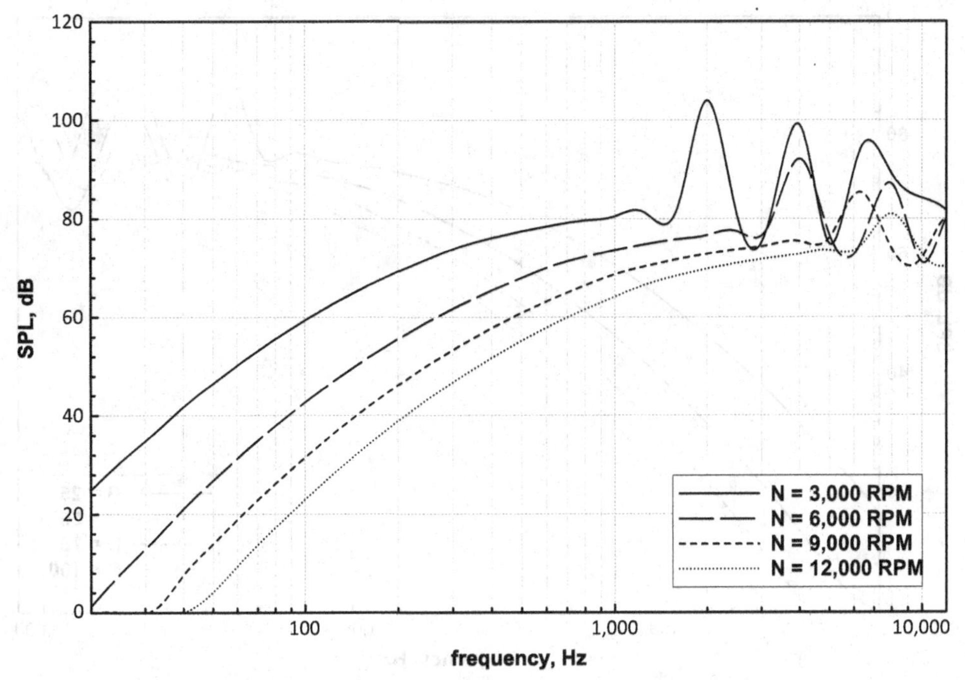


Figure 16.11. Effect of turbine rotational speed (rpm) on 1/3 octave band SPL.

rotational speed. At $N = 12{,}000$ rpm, the first tonal-peak noise level is about 8 dB above that of the highest broadband noise level.

The effects of blade number $B$ on the SPL are depicted in Figure 16.12. Because the blade number directly determines the blade-passing frequency, part of the blade-number effect on the noise level is to shift the tonal peaks in proportion with the number of blades, similar to the effect of turbine rotational speed as shown in Figure 16.11. This is clearly seen in the results as the fundamental harmonic is shifted from $f = 500$ Hz to 10 kHz when the blade number is increased from 25 to 100. The SPLs of these fundamental harmonics remain unchanged at about 90 dB. The second effect of increasing the blade number is seen in the attenuation of the broadband noise levels particularly at frequencies away from the blade-passing frequencies.

### 16.4.5 Single-Jet Noise

The single-jet noise model applies only to pure turbojet engines. This type of engines has virtually disappeared due to the high noise level and low efficiency. Fundamental studies in this field were done by Lighthill[26;27]. Lighthill's analysis shows that the acoustic power of a high-speed jet in stationary surrounding is

$$P \sim \rho_j A_j V_j^3 M^5 = \frac{\rho_j A_j V_j^8}{a_j^5}, \tag{16.36}$$

where $A_j$ is the jet area, $M = V_j/a_j$ the Mach number, and $a_j$ the average speed of sound in the jet. This equation is also known as Lighthill's eighth power law. The

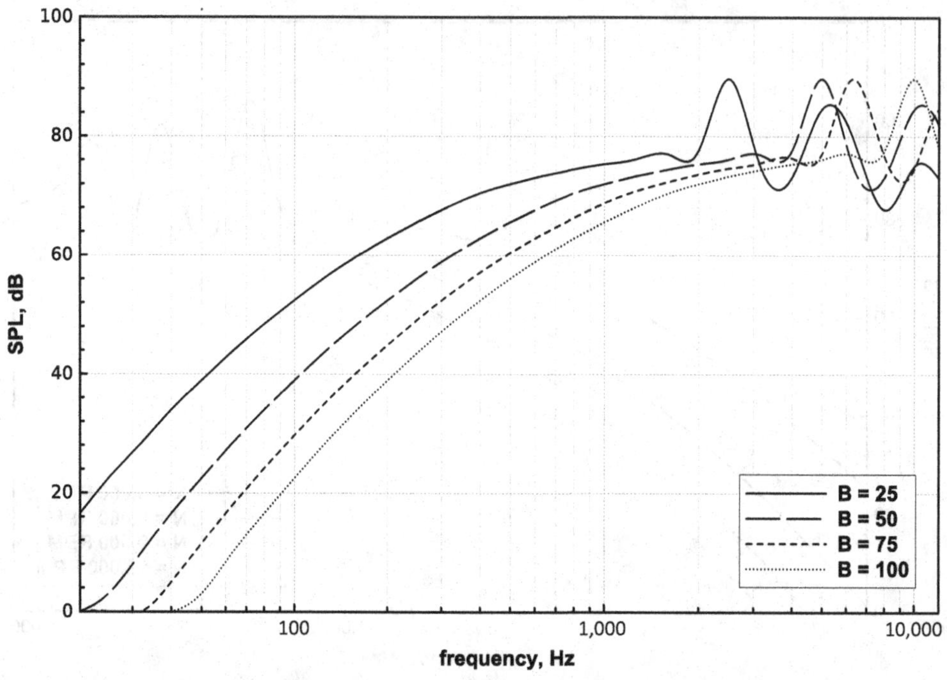


Figure 16.12. Effect of blade number $B$ on 1/3 octave band SPL.

mass flow of the jet is $\dot{m} = \rho_j A_j V_j$; therefore

$$P \sim \frac{\dot{m} V_j^7}{a_j^5}. \tag{16.37}$$

The far-field noise produced by a high-speed jet behaves as if generated by a monopole; hence, the intensity $I$ decays like the inverse of the distance, $1/r$,

$$I \sim \frac{P}{A} \sim \frac{\dot{m} V_j^7}{2\pi r a^5} \sim \frac{\dot{m} V_j^7}{a^5 r}. \tag{16.38}$$

The corresponding SPL is

$$\text{OASPL} = 10 \log \left( \frac{I}{I_o} \right) = 10 \log \left( \frac{c_1}{W_{ref}} \frac{\dot{m} V_j^7}{a_j^5 r} \right), \tag{16.39}$$

where $c_1$ is the constant of proportionality in Equation 16.38 and $W_{ref}$ is the reference acoustic power. The importance of the jet speed is evident from $V_j^7$. It is verified that if we reduce the jet speed by half, Equation 16.39 gives a reduction in OASPL of 21 dB to an observer at the same distance $r$. If, conversely, the distance of the observer is doubled but the jet speed is maintained to the original level, then the reduction in OASPL is only 6 dB. This result emphasises once again the importance of using high by-pass turbofan engines instead of pure jet engines.

If the jet is derived from an engine flying at a speed $V$, then it is convenient to introduce the engine thrust in Equation 16.38. A rough approximation for the thrust is $T \simeq \dot{m}(V_j - V)$. Thus, we have

$$I \sim \frac{T}{(V_j - V)} \frac{V_j^7}{a_j^5 r}, \qquad I \sim \frac{T V_j^6}{a_j^5 r}. \tag{16.40}$$

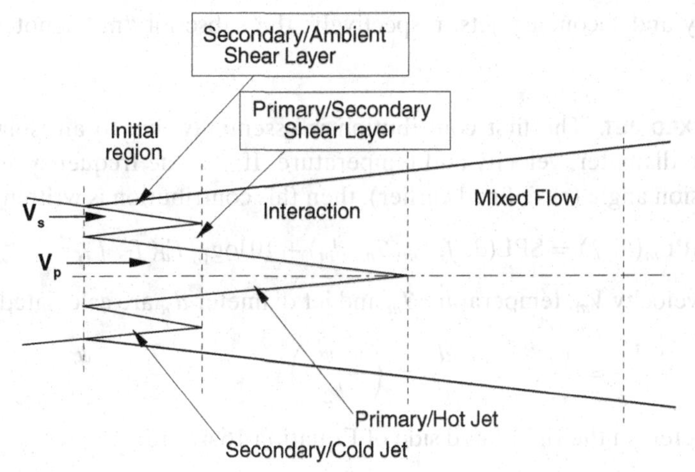


Figure 16.13. Flow features of a coaxial hot-cold jet from turbofan engine.

This equation shows the importance of cutting back the thrust, something that is routinely done in take-off operations. A simple calculation from Equation 16.40 shows that a reduction of the thrust by 50% leads to a reduction of the OASPL by ~3 dB.

### 16.4.6 Co-Axial Jet Noise

The prediction of a single-jet noise relies on the theory of Lighthill[26;27] and the subsequent developments, including the method of Balsa and Gliebe[28] and Stone *et al.*[29] for coaxial jets. In modern jet engines the jet is made of a hot core and a cold by-pass flow at a lower speed. The problem is considerably more complicated than a single-stream jet. The additional parameters are at least the velocity ratio between jets, the area ratio and the temperature ratio.

Experimental noise data exist[30;31] for a wide range of jet velocity ratios, nozzles area, Strouhal numbers and polar angles, although some limitations exist as well, including the interpolation of data. Other methods have been published by the SAE on the basis of methods developed by NASA, Boeing and others. A somewhat more comprehensive study, involving a reasonable degree of physics, is available from Fisher *et al.*[32;33]. A full description of the model is given in those references. We will limit the discussion to the acoustic-model equations and to the implementation strategy within our comprehensive model.

Figure 16.13 shows a sketch of a typical co-axial jet, made of a hot-core (primary) and a cold by-pass (secondary) jet. The initial region near the nozzle is characterised by a turbulent mixing between jets of different velocities and temperatures. Two shear layers are identified: a shear layer between the secondary jet and the atmosphere, and an internal shear layer between the primary and secondary jets. The mixing of the secondary jet with the atmosphere generates high-frequency noise.

Following Fisher *et al.*, the noise from a co-axial jet flow is calculated by summing three contributions: 1) the fully mixed jet, 2) the secondary jet, and 3) the effective jet. Some further notation is required at this point: the subscripts "p" and "s" denote the primary and secondary jets, respectively; the subscript "m" denotes the fully mixed jet.

**1. FULLY-MIXED JET.** The first contribution is essentially due to an isolated jet of appropriate diameter, velocity and temperature. If $f$ is the frequency and $\theta$ is the polar emission angle (as defined earlier), then this contribution is written as

$$\mathrm{SPL}_m(\theta, f) = \mathrm{SPL}(\theta, f, V_m, \mathcal{T}_m, d_m) + 10\log_{10} F_d(f_1, f), \tag{16.41}$$

where the velocity $V_m$, temperature $\mathcal{T}_m$ and jet diameter $d_m$ are calculated from

$$\frac{V_m}{V_p} = \frac{\psi_2}{\psi_o}, \qquad \frac{d_m}{d_p} = \left(\frac{\psi_o \psi_1}{\psi_2}\right)^{1/2}, \qquad \frac{\mathcal{T}_m}{\mathcal{T}_p} = \frac{\psi_o}{\psi_1}. \tag{16.42}$$

The parameters in the right-hand side of Equation 16.42 are

$$\psi_o = 1 + V_r A_r, \qquad \psi_1 = 1 + V_r A_r \sigma_r, \qquad \psi_2 = 1 + V_r^2 A_r \sigma_r, \tag{16.43}$$

where $V_r = V_s/V_p$ is the velocity ratio, $A_r = A_s/A_p$ is the area ratio, and $\sigma_r = \rho_s/\rho_p$ is the gas density ratio. Equation 16.41 gives the SPL of the fully mixed jet at frequencies above the cut-off frequency $f_1 = V_m/d_m$. The factor $F_d$ is a function that represents the fraction of the energy radiated from positions *downstream* of the axial coordinate $x_1$:

$$F_d(x_1, f, f_1) = \left[1 + \hat{f} + \frac{1}{2}\hat{f}^2 + \frac{1}{6}\hat{f}^3\right] \exp(\hat{f}), \tag{16.44}$$

where

$$\hat{f} = m\left(\frac{f}{f_1}\right). \tag{16.45}$$

The factor $m$ is a shape parameter (typically, $m = 4$). The frequency $f_1$ is such that the energy radiated upstream and downstream of the position $x_1$ is about the same. The first term on the right-hand side of Equation 16.42 is calculated by using the method of ESDU for subsonic circular nozzles[34]. This term is discussed separately in § 16.4.7.

**2. SECONDARY JET.** The secondary-jet contribution is

$$\mathrm{SPL}_s(\theta, f) = \mathrm{SPL}(V_s, \mathcal{T}_s, d_s, \theta, f) + 10\log_{10} F_u(f_1, f). \tag{16.46}$$

In Equation 16.46 the function $F_u$ is the complement of $F_d$ and represents the fraction of the energy radiated *upstream* of the axial position $x_1$:

$$F_u(x_1, f_1, f) = 1 - F_d(x_1, f_1, f) \tag{16.47}$$

and denotes the fraction of the energy radiated downstream of the position $x_1$.

**3. EFFECTIVE JET.** The effective-jet contribution is

$$\mathrm{SPL}_e(\theta, f) = \mathrm{SPL}(V_p, \mathcal{T}_p, d_e, \theta, f) + \Delta dB, \tag{16.48}$$

with the effective diameter $d_e$ calculated from

$$d_e = d_p(1 + A_r^2 V_r)^{1/2}. \tag{16.49}$$

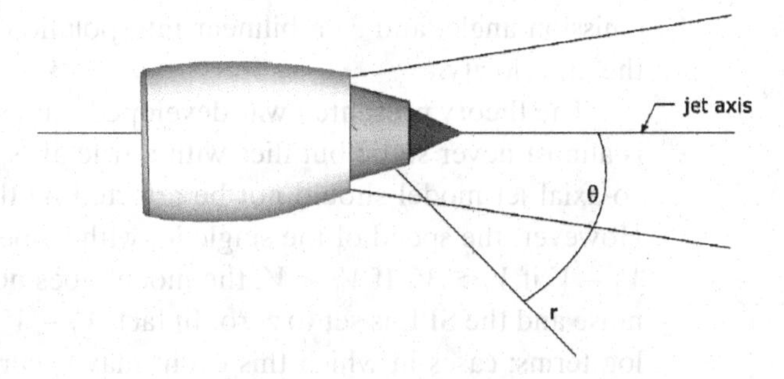


Figure 16.14. Nomenclature for single jet.

The last term on the right-hand side of Equation 16.48 is a noise reduction that depends on the level of *relative* turbulence. This is in fact the ratio between the average turbulence level in the co-axial jet and the primary jet and was found to be in the range of ~2/3. The actual SPL reduction derived in Ref.[33] is

$$\Delta \text{dB} = 10 \log_{10} \left[ 0.2 \left( \frac{1+16y}{1+7y} \right) \right], \tag{16.50}$$

with

$$y = \frac{(\tau-1)^2 M_j^2}{1+0.65(\tau-1)}, \qquad M_j = \frac{V_j}{a_o}, \qquad \tau = \frac{T_j}{T_o}. \tag{16.51}$$

## 16.4.7 Far-Field Noise from a Subsonic Circular Jet

The method of ESDU[34;35], based on semi-empirical data, is used in this context for the determination of the noise from a single-stream circular jet. This term appears in the right-hand side of Equation 16.41, Equation 16.46 and Equation 16.48. The OASPL corresponding to a frequency $f$ at a receiver position as indicated in Figure 16.14 is

$$\text{SPL} = \text{SPL}_n + 10 \log_{10} \left( \frac{\rho_j}{\rho_o} \right)^{\omega} + 10 \log_{10} \left( \frac{A_j}{r^2} \right) + 20 \log_{10} \left( \frac{p_o}{p_{ISA}} \right), \tag{16.52}$$

where $\text{SPL}_n$ denotes a normalised SPL, $A_j$ is the area of the jet, and $\omega$ is a variable density index; $p_o$ is the atmospheric pressure and $p_{ISA}$ is the atmospheric pressure under standard conditions. Thus, the second term of Equation 16.52 is a density correction, the third term is a form factor and the fourth term in the right-hand side is a correction for atmospheric conditions. The normalised SPL, the variable density index, the normalised spectra and other quantities are calculated from charts. The data are limited to polar-emission angles between 30 and 120 degrees; therefore, calculation of noise from a distant source at a small or large emission angle must be done by careful extrapolation. The reader is referred to the original documents for details.

One further important point in the formulation of Equation 16.52 is that only the SPL is given. To calculate the spectrum pressure levels, the method relies on an experimental database. This database allows the calculation of the 1/3 octave band SPL for a given set of Strouhal number, jet temperature and speed. We have used the database through a complex interpolation method consisting in three main steps: 1) a cubic spline through the Strouhal number; 2) a cubic spline through the emission angle; and 3) a bilinear interpolation through the temperature ratio and the jet velocity.

The theory presented was developed for a static engine. In our case, the engine is almost never static but flies with a true air speed $V$. The velocity ratio $V_r$ in the co-axial jet model should not be affected by the flight speed to any large degree. However, the speed of the single jet with respect to a still air must be corrected to $V_j - V$ if $V_j > V$. If $V_j < V$, the model does not provide a reliable estimate of the noise and the SPL is set to zero. In fact, $V_j - V < 0$ causes a singularity in all of the log terms; cases in which this event may occur include en-route descent with idle thrust.

The aero-thermodynamic data for the jet noise calculation are $(T, p, V)$ for both primary and secondary jets. If the area ratio is not known, use the approximation

$$\frac{d_{core}}{d_{bypass}} \simeq 0.46, \qquad \frac{d_{core}}{d_{fan}} \simeq 0.48, \tag{16.53}$$

that has been inferred from the analysis of large engines with by-pass ratios BPR > 5. The fan diameter is generally a known parameter, so it is possible to calculate the remaining quantities.

### Co-axial Jet-Noise Verification

The verification of the model is done with the data available in Fisher *et al.*[32;33]. The data were taken by arrays of microphones placed at 12 m from the nozzle and at polar-emission angles from 30 to 120 degrees. Data are available for a range of primary jet temperatures and for a range of speed ratios. These data have been interpolated from Ref.[33] and are affected by digital interpolation error of up to 0.5 dB (the original datasets were not available).

The nozzle dimensions of the test cases are $d_p = 33.2$ mm; $d_s = 58.2$ mm. We have considered the case having a primary-jet temperature $T_p = 800$ K and have further assumed that the secondary jet has an atmospheric temperature $T_s = 288$ K (this datum was not readily available from the published data). The primary and secondary jet velocities are 265 m/s and 170 m/s, respectively. The reference polar angles are 40, 90 and 120 degrees.

The predictions have been extrapolated to the lowest frequencies, where the database does not provide useful information on the jet noise. In many real cases the temperature ratio $T_p/T_s$ is greater than 2.5, which is above the limit of the temperature database. The results indicate that the extrapolation is not very good, with errors in excess of 5 dB at frequencies below 100 Hz. The comparisons shown by Fisher *et al.* have a lower cut-off frequency of about 300 Hz, depending on the test case.

Figure 16.15 shows the spectral analysis of the calculated co-axial jet noise and the comparison with the reference data. The squares have approximately the same size as the error bars on the reference SPL.

## 16.4.8 Stone Jet Noise Model

The co-axial jet noise described previously is accurate enough for implementation into a comprehensive flight-mechanics program but lacks generality due to the limitation of the experimental database on which it is founded. This database has a

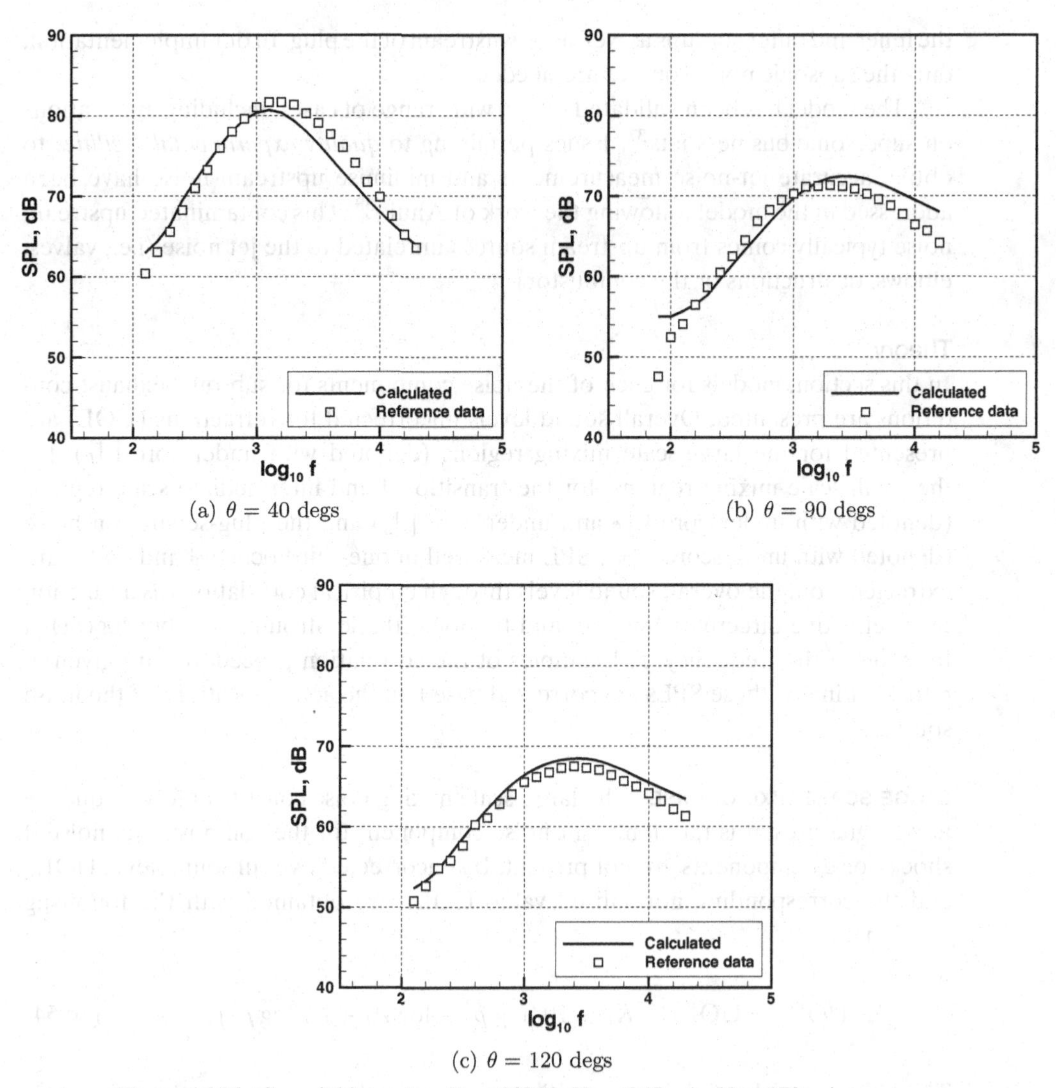


(a) $\theta = 40$ degs

(b) $\theta = 90$ degs

(c) $\theta = 120$ degs

Figure 16.15. Co-axial jet noise; $T_p = 800$ K, $V_p = 265$ m/s, $V_s = 170$ m/s.

relatively limited range of velocity ratios $V_p/V_s$ and jet velocity ratios. At take-off and high-power climb, it is likely that the jet operation point exceeds by far the matrix of the database. An alternative model is described next to overcome these limitations (Stone *et al.*[36]). The model is a semi-empirical prediction method for both single- and dual-stream nozzles derived from an analytical approach using real physical scaling laws; its empirical database covers a wide range of engine size. The dual-stream model will be used in our applications, which are on commercial transport aircraft with high-bypass-ratio turbofan engines.

In this model, jet noise is decomposed into multiple components of subsonic and supersonic noise sources based on their flow-length scales and source locations. The subsonic noise sources come from large-scale, small-scale and transitional (or intermediate) turbulent mixing structures. In addition, a jet nozzle with a plug would generate noise as the inner-stream flow separates from the plug. When at least one of the exhaust streams is supersonic, shock noises are generated from either or both of the inner and outer streams as well as downstream of the plug. In our implementation, only the subsonic noise sources are needed.

The model has been validated over a wide range of cases, including applications on supersonic business jets[37]. Issues pertaining to *quality experimental facilities* to obtain accurate jet-noise measurements and minimise upstream noise have been addressed in the model following the work of Ahuja[38]. This contaminated upstream noise typically comes from upstream sources unrelated to the jet noise (i.e., valves, elbows, obstructions or the combustor).

### Theory

In this section, models for each of the noise components for subsonic exhaust conditions are presented. Overall sound levels uncorrected for refractions (UOL) are presented for the large-scale mixing regions (denoted with underscore $[.]_L$), for the small-scale mixing regions, for the transitional and intermediate scale regions (denoted with underscore $[.]_T$ and underscore $[.]_S$) and the plug-separation noise (denoted with underscore $[.]_P$). SPL measured in one-third octave-band scales are extracted from the overall sound levels through empirical correlations as a function of an effective directivity angle $\theta'$ and the logarithmic Strouhal number log(St), a function of the frequency $f$. Examples of the correlation procedures are given in § 16.4.8. Finally, these SPLs are corrected based on the actual locations of the noise sources.

**LARGE-SCALE MIXING NOISE.** The large-scale mixing noise consists of low-frequency noise signatures; it is the dominant noise component for the composite jet noise if shock noise components are not present. Its uncorrected overall sound level $\text{UOL}_L$ and the corresponding normalised value $\text{UOL}'_L$ are obtained with the following relation:

$$\text{UOL}_L = \text{UOL}'_L - K_L + 10\left(\log \hat{p}^2 + \log \hat{A}_L + \omega_L \log \hat{\rho}_L\right). \tag{16.54}$$

The UOL normalisation is done with respect to the flow and geometry parameters and follows a lengthy formulation. The term $K_L$ is a function of a corrected Mach number and a corrected emission angle; $\hat{A}_L$ is a corrected nozzle area. These and other parameters in Equation 16.54 are defined through a dozen algebraic equations. Their derivation would require a lengthy discussion. The reader is referred to the original paper[36]; all of the terms within the parentheses in the right-hand terms are normalised quantities and are denoted with a hat $\widehat{[.]}$.

Once the $\text{UOL}_L$ is obtained, the SPLs for each one-third octave-band centre frequencies are correlated from an empirical database for the large-scale mixing noise using the following two parameters:

$$\theta'_L = \theta_{cor,L}\left(\frac{V_L}{a_\infty}\right)^{0.1}, \tag{16.55a}$$

$$\log \text{St} = \log\left[\left(\frac{f d_L}{V_{e,L}}\right)\left(\frac{T_{t,L}}{T_{t,\infty}}\right)^{0.4(1+\cos\theta_L)}\right]. \tag{16.55b}$$

Corrections to the directivity angle $\theta_{cor,L}$ and the source-to-receiver distance $r_{cor,L}$ requires the source location $X_{s,L}$ of the region generating the noise, where

$$X_{s,L} = \left[L_1 + \left(4 + \frac{\theta}{30}\right) d_L \left(PR_1 PR_2\right)^{-1/4}\right] / \text{LSF}, \tag{16.56}$$

with $\theta$ in degrees; $PR_1$ and $PR_2$ are noise-suppression parameters for the inner and outer stream nozzles, respectively; LSF is the linear scale factor (ratio of full-scale to model-scale dimensions). $L_1$ is a length scale. These parameters are defined as the ratios of the wetted perimeters for the suppressed to the unsuppressed nozzles. Jet nozzles can be modified with triangular-serrated chevrons at the trailing edges to suppress mixing noise [39;40].

**SMALL-SCALE MIXING NOISE.** The small-scale mixing noise, also termed the outer shear layer mixing noise, occurs at relatively high frequencies, generated near the nozzle exit. Its uncorrected overall sound level $\text{UOL}_S$ is given by

$$\text{UOL}_S = C_S - K_S + 75 \log \hat{V}_{e,S} + 10 \left(\log \hat{p}^2 + \log \hat{A}_S + \omega_L \log \hat{\rho}_S\right), \tag{16.57}$$

where $C_S$ is a noise-component coefficient, given in dB. The other parameters are defined in the original paper [36] by twelve algebraic equations.

Similar to the procedure for the large-mixing noise, the SPLs for the small-scale mixing noise for each one-third octave-band centre frequencies are correlated from a relevant empirical database mapped from the following two parameters:

$$\theta'_S = \theta_{cor,S} \left(\frac{V_S}{a_\infty}\right)^{0.1}, \tag{16.58a}$$

$$\log \text{St} = \log \left[\left(\frac{f d_S}{V_{e,S}}\right) \left(\frac{T_{t,S}}{T_{t,\infty}}\right)^{0.4(1+\cos\theta_S)}\right]. \tag{16.58b}$$

Corrections in the SPL, directivity angle and distance are based on the following definition of the source location:

$$X_{s,S} = \left(\frac{\theta}{45}\right) d_S / \text{LSF}. \tag{16.59}$$

**TRANSITIONAL/INTERMEDIATE-SCALE MIXING NOISE.** The transitional or intermediate-scale mixing noise is primarily composed of middle-to-high frequency noise. Unlike the large and small-scale mixing noise, the effect of the density ratio for the transitional-scale mixing noise was found to be negligible. The uncorrected sound level $\text{UOL}_T$ is defined as

$$\text{UOL}_T = C_T - K_T + 75 \log \hat{V}_{e,T} + 10 \left(\log \hat{p}^2 + \log \hat{A}_T\right), \tag{16.60}$$

where the parameters in the right-hand side of the equations are calculated through nine algebraic equations (see again the original paper).

Correlations with an empirical database to obtain the one-third octave-band SPL are mapped using the following parameters:

$$\theta'_T = \theta_{cor,1}\left(\frac{V_{e,T}}{a_\infty}\right)^{0.1}, \tag{16.61a}$$

$$\log \mathrm{St} = \log\left[\left(\frac{f d_{2,1,th}}{V_{e,T}}\right)\left(\frac{T_{t,1}}{T_{t,2}}\right)^{0.4(1+\cos\theta_T)}\right], \tag{16.61b}$$

where $d_{2,1,th}$ is the outer physical throat diameter of the inner-stream nozzle. Corrections in the SPL, the directivity angle and the source-to-receiver distance are based on the following definition of the source location:

$$X_{s,T} = \left[L_1 + \left(\frac{\theta}{45}\right) d_{h,1,th}\right]/\mathrm{LSF}, \tag{16.62}$$

where $d_{h,1,th}$ is the hydraulic diameter of the inner-stream throat.

**INNER-STREAM PLUG-SEPARATION NOISE.** A jet nozzle with a blunt-tipped plug would produce a noise when the inner-stream separates from the plug. This noise is relatively important for an engine with high by-pass ratio and low mixed-jet velocity. The following relationships determine the uncorrected sound level $\mathrm{UOL}_P$ for the plugged-tip nozzle:

$$\mathrm{UOL}_P = \mathrm{UOL}'_P - K_P + 10\left(\log\hat{p}^2 + \log\hat{A}_P + \log\hat{\rho}_P\right), \tag{16.63}$$

where the right-hand parameters are calculated via a number of algebraic equations, like in the previous cases[36]. Correlations with an empirical database to obtain the one-third octave-band SPL are mapped using the following parameters:

$$\theta'_P = \theta_{cor,P}\left(\frac{V_1}{a_\infty}\right)^{0.1}, \tag{16.64a}$$

$$\log \mathrm{St} = \log\left[\left(0.5 f d_P/\sqrt{V_1 a_\infty}\right)(1 - M\cos\theta_{cor,P})\right], \tag{16.64b}$$

where $r_P$ is the physical plug-tip radius of the inner-stream nozzle. Finally, the SPL correction is derived from the following source location:

$$X_{s,P} = \left[L_1 + L_P + 2\left(\frac{\theta}{45}\right) r_P\right]/\mathrm{LSF}. \tag{16.65}$$

### Empirical Correlations

Once the uncorrected sound levels have been determined for each of the noise components, the one-third octave-band SPL can be derived through correlations with the empirical database mapped using two parameters: the effective polar directivity angle $\theta'$ and the logarithmic Strouhal number log(St). The SPL obtained is the free-field SPL at the receiver location. The data are stored as SPL – UOL. Selected plots of the empirical data for the large-scale, small-scale, transitional and plug-separation noise are shown in Figure 16.16. In the detailed report by Stone *et al.*[36], $\theta'$ covers a range from 0 to 250 degrees only, mostly at intervals of 10 degrees except for larger angles. The logarithmic values of the Strouhal number for all of the noise components run from −3.6 to 3.6. Extrapolations are necessary for conditions beyond the recorded data.

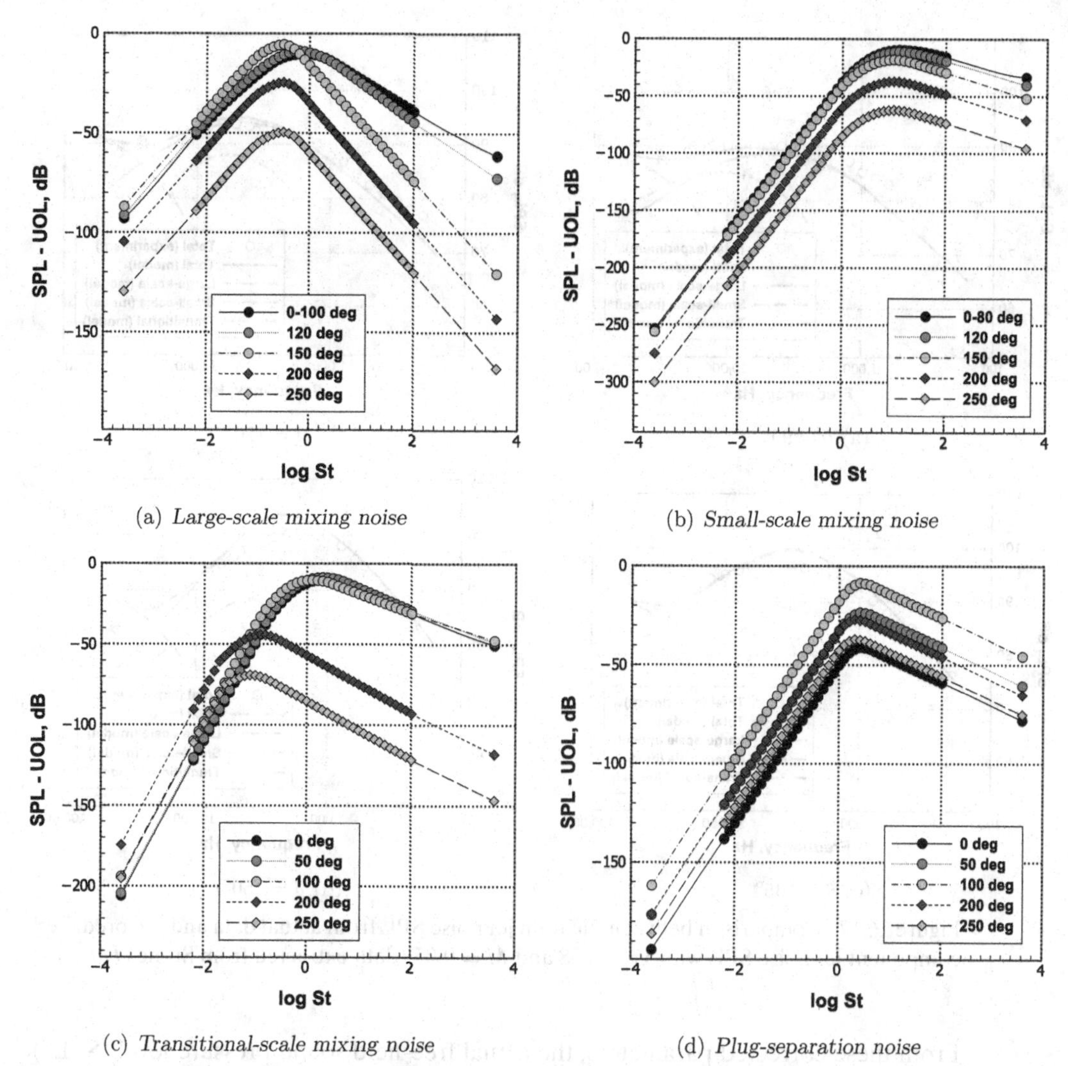


Figure 16.16. Database of SPL – UOL for selected $\theta'$ between 0 and 250 degrees, as shown in the legend (data extracted from Stone *et al.*[36]).

## Sound-Level Corrections

The SPL obtained from the empirical database must be corrected due to the varying source locations of the individual component noise. With this regard, the source-to-receiver location and the polar- directivity angle must be corrected from their initial values taken relative to the centre of the nozzle exit plane. The corrected, or actual, values are derived from the following relationships:

$$r_{cor} = r\sqrt{1 + \left(\frac{X_s}{r}\right)^2 + 2\left(\frac{X_s}{r}\right)\cos\theta}, \tag{16.66a}$$

$$\theta_{cor} = \cos^{-1}\left[\frac{X_s + r\cos\theta}{r_{cor}}\right]. \tag{16.66b}$$

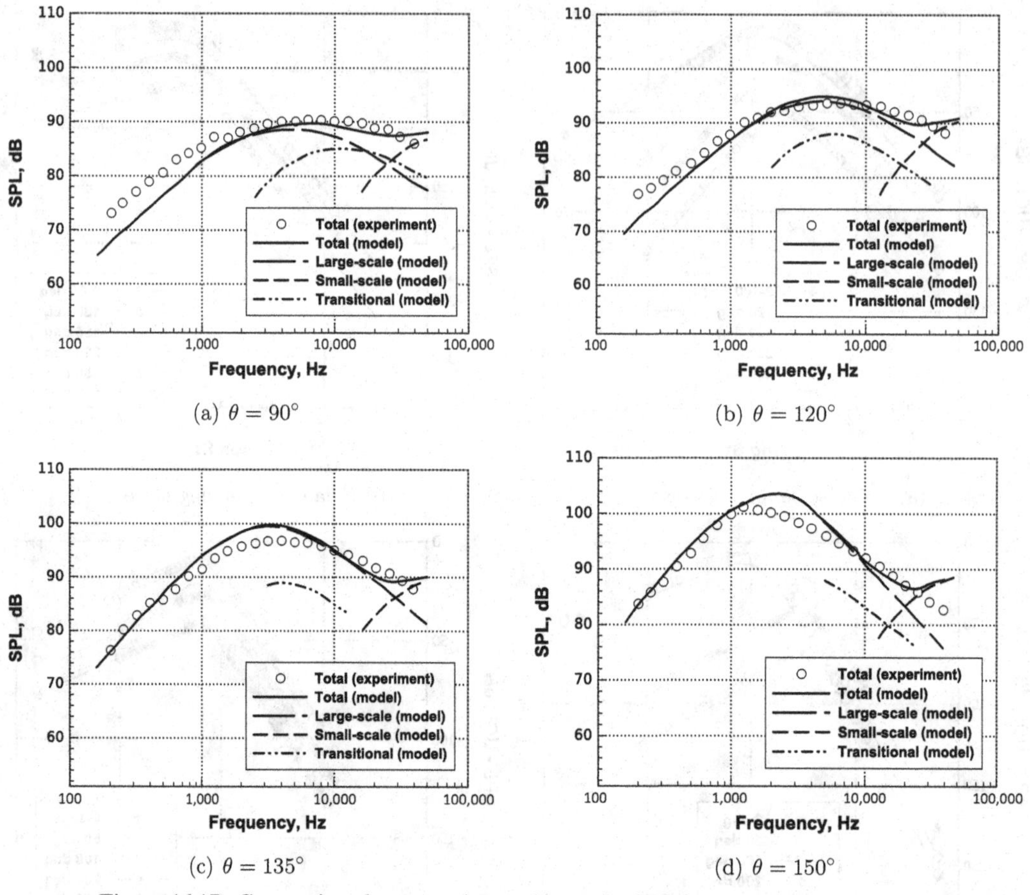


(a) $\theta = 90°$ (b) $\theta = 120°$

(c) $\theta = 135°$ (d) $\theta = 150°$

Figure 16.17. Comparison between the total jet noise SPL from actual data and the predicted component jet noise SPL with $V_1 = 0.98$ and $M = 0.05$ (data extracted from Stone *et al.*[36]).

From these corrected parameters, the actual free-field sound pressure level ($\mathrm{SPL}_a$) is computed as

$$\mathrm{SPL}_a = \mathrm{SPL} - 20\log\left(\frac{r_{cor}}{r}\right). \tag{16.67}$$

### Model Verification

Limited validations of the Stone jet model have been performed on model-scale dual-stream nozzles[36]. Important results from this study are shown in Figures 16.17 and 16.18. The experimental data highlight the dominance of the large-scale mixing noise over the other noise components when the exhaust conditions are subsonic. Higher jet velocity was found to increase the contributions of the transitional and small-scale mixing noise; in one case, the peaks of both the large-scale and the transitional-scale noise are comparable, as shown in Figure 16.18(a).

In these limited datasets, predictions from the Stone jet model are found to provide satisfactory results relative to the actual data, particularly when the jets are near the overhead locations, that is, for $\theta$ between 90 and 120 degrees. For larger values of the polar angles, the actual peak SPL are over-predicted by about 1–2 dB.

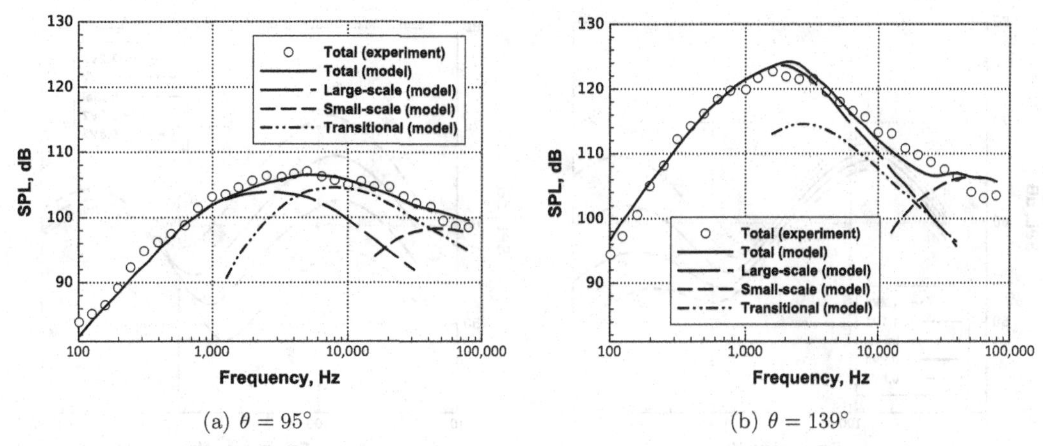


(a) $\theta = 95^\circ$ (b) $\theta = 139^\circ$

Figure 16.18. Comparison between the total jet noise SPL from actual data and the predicted component jet noise SPL with $V_1 = 1.74$ and $M = 0.0$ (data extracted from Stone *et al.*[36]).

### Sensitivity Analyses

Sensitivity analyses are performed to identify critical input parameters for the simulation of the jet noise. In these analyses, only the parameters relevant to the primary- and secondary-jet properties are studied: these are the jet velocities $V_1$ and $V_2$, the nozzle areas $A_1$ and $A_2$, the total temperatures $\mathcal{T}_{t,1}$ and $\mathcal{T}_{t,2}$, and the mass-flow rate $\dot{m}_1$ and $\dot{m}_2$. The nominal input parameters for this case study are taken from that of an Airbus A320 aircraft at its closest encounter with a receiver located at a FAR point during take-off. The secondary stream is thus the exhaust flow from the turbofan.

The overall jet noise is found to be most sensitive to variations in both the primary- and secondary-jet velocities. If the primary-jet velocity is changed by ±10% of its normal operating value, the overall sound level OASPL changes by at most 2 dB. A 20% change to $V_1$ would increase the OASPL by 4.6 dB. Most of these changes occur in the high-frequency regime of the noise spectra. Conversely, the secondary-jet stream influences the lower-frequency noise, with an increase in the OASPL of 3.3 and 6.7 dB if $V_2$ is increased by 10% and 20% of its initial value, respectively. Figure 16.19 illustrates the effect of both jet streams on the SPL in the one-third octave-band scale.

The other input parameters have only minor effects on the overall sound level of the jet noise. Of these inputs, the primary-jet nozzle has the most influence, with a change of OASPL equal to 0.4 or 0.9 dB if $A_1$ is enlarged by 10% or 20% of its original size, respectively. A closer inspection of its effect on the one-third octave-band noise spectra reveals that the changes influenced by the primary nozzle area occur mostly on the high-frequency noise. Variations in the turbofan area also affect the high-frequency noise with minor but noticeable influence on the low-frequency noise. However, the overall effect on the OASPL by the turbofan area is small, with an increase of 0.2 dB if the area is enlarged by 20%. Figure 16.20 shows the sensitivity analysis on the SPL due to the variations of the nozzle areas.

## 16.4.9 Jet-Noise Shielding

When multiple jets are present, an interference event takes place called *jet shielding*. When propagating through a high-temperature and high-speed jet, an acoustic wave

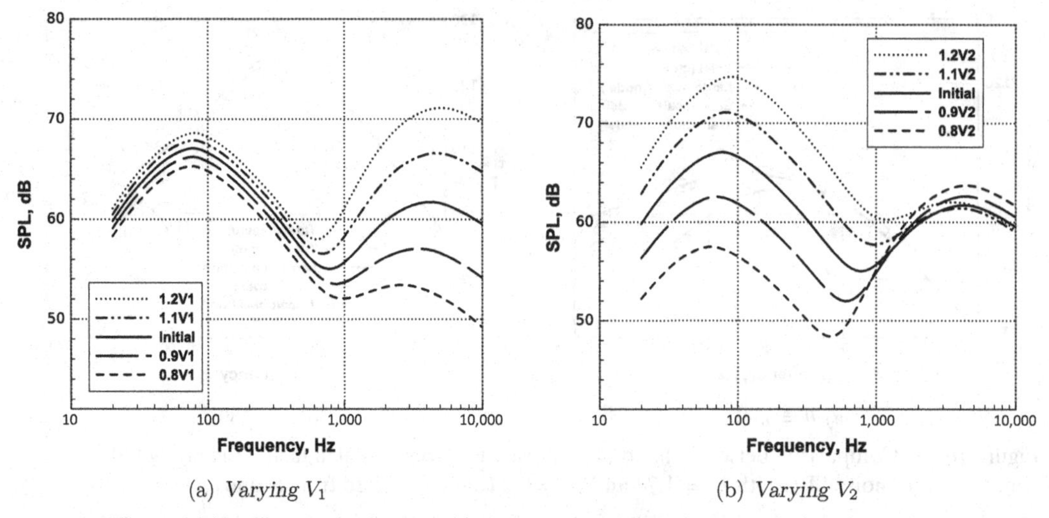


(a) *Varying* $V_1$ (b) *Varying* $V_2$

Figure 16.19. Changes in the one-third octave-band SPL due to variations in the input speeds $V_1$ and $V_2$, whose initial values are 400 m/s and 275 m/s, respectively.

encounters a different impedance that would tend to absorb, reflect and diffract the incident wave. The jet-shielding mechanisms are the following:

- Sound waves from a jet may be reflected, refracted, diffracted, scattered (due to flow turbulence) or fully transmitted upon encounter with another jet that acts as a shielding jet.
- The acoustic shielding level of the source depends on the location of the receiver with respect to the dual-jet system. Because the total acoustic power is conserved within a sphere enclosing the source, a noise reduction at one receiver location as a result of jet shielding would result in a noise increase at some other location. For example, a receiver standing in the jet axes plane might be completely or partially shielded from the noise of the jet farthest away by the adjacent jet

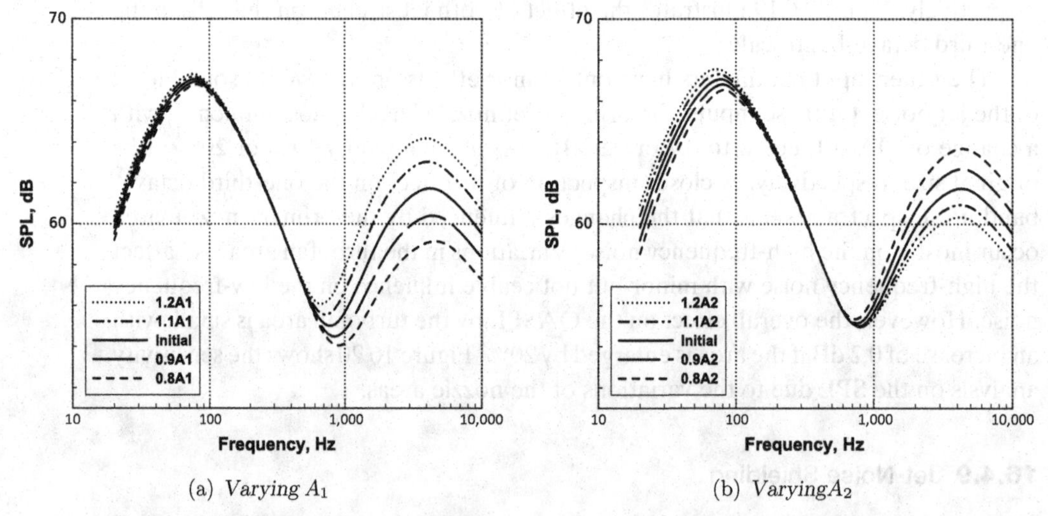


(a) *Varying* $A_1$ (b) *Varying*$A_2$

Figure 16.20. Changes in the one-third octave-band SPL due to variations in the input speeds $A_1$ and $A_2$, whose initial values are 0.6 m$^2$ and 1.3 m$^2$, respectively.

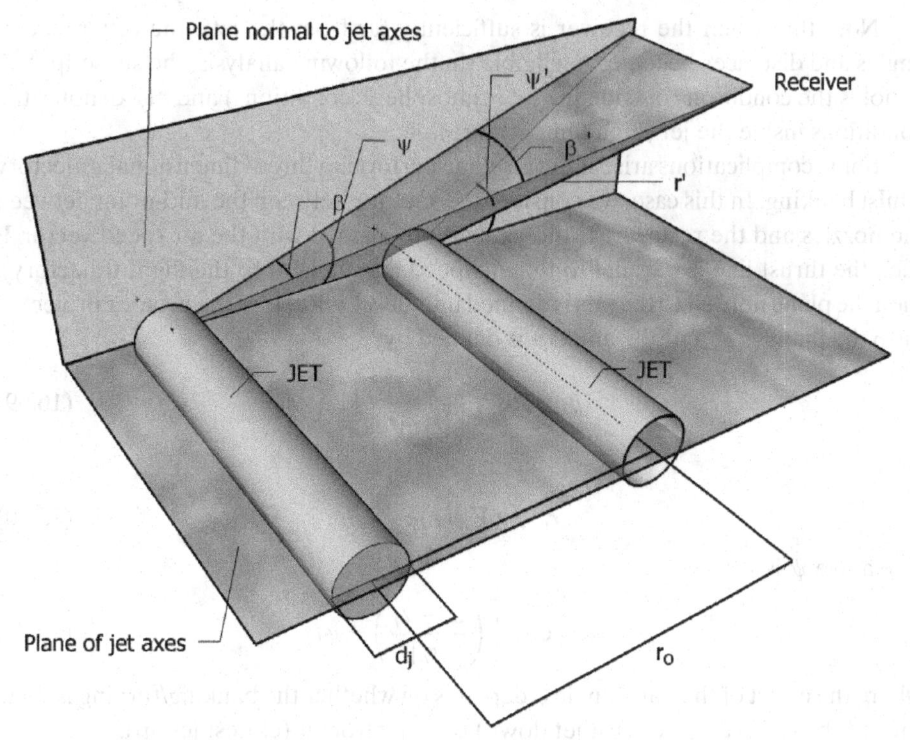


Figure 16.21. Nomenclature for jet-shielding noise model.

located in between. However, these *shielded* sound waves are actually deflected toward another location by the shielding jet; these deflected sound waves would then superimpose with other sound waves to increase the noise level at the new location.

- Constructive or destructive phasing between multiple coherent sound waves (i.e., between direct and deflected sound waves) may occur to increase or decrease the noise level at a receiver.

If $P_1$ and $P_2$ denote the sound power at a receiver in absence of noise shielding, and $P_S$ is the resulting sound power due to shielding, the difference in noise is

$$\Delta\text{SPL} = 10 \log_{10}\left(\frac{P_S^2}{P_1^2 + P_2^2}\right). \tag{16.68}$$

An analytical formulation of noise shielding is due to Gerhold[41]. Simonich *et al.*[42] performed experiments on dual jet noise and provided a refined mathematical model. ESDU[43] (and associated documents) give a practical implementation that is affected by numerical errors at large frequencies.

Consider the situation shown in Figure 16.21. The two jets are identical and have a nominal diameter $d_j$; their axes are separated by a distance $r_o$. The distance between the observer and the centre of the farthest jet at the nozzle is called $r$; $r'$ denotes the distance from the closest jet. The angle $\beta$ in this case denotes the polar angle measured from the plane normal to the jet axis; likewise, $\Psi$ denotes the azimuthal angle on a plane normal to the jet axis.

Note that when the receiver is sufficiently far from the jets, the difference in angles and distances becomes negligible. In the following analysis, the subscript "o" denotes the conditions outside the jet (atmospheric conditions) and "1" denotes the conditions inside the jet (high temperature).

Some complications arise if the airplane performs a three-dimensional trajectory whilst banking. In this case, we consider the vector $\boldsymbol{r}$ between the mid-point between the nozzles and the receiver. If the jet axes are aligned with the air speed vector $\boldsymbol{V}$ (i.e., the thrust line is parallel to the air speed and tangent to the flight trajectory), then the plane normal to the jets is defined uniquely by $\boldsymbol{V}$. Thus, the vector $\boldsymbol{r}$ projected onto the plane is $\boldsymbol{r}_n$ and the angle $\beta$ is defined by

$$\beta = \cos^{-1}\left[\frac{\boldsymbol{r}\cdot\boldsymbol{r}_1}{|\boldsymbol{r}|\cdot|\boldsymbol{r}_n|}\right], \tag{16.69}$$

with

$$\boldsymbol{r}_n = (\boldsymbol{V}\times\boldsymbol{r})\times\boldsymbol{V}. \tag{16.70}$$

The angle $\psi$ is

$$\psi = \cos^{-1}\left(\frac{z_r - z_j}{|\boldsymbol{r}_n|}\right) \pm \varphi, \tag{16.71}$$

where the effect of the bank angle $\varphi$ depends on whether the banking/turning is done toward the receiver $z_r$ (closest jet down) or away from it (closest jet up).

The key idea of the model is to express the acoustic pressure $p$ in terms of a velocity potential $\Phi$, which is done through the relationship

$$p = -\rho_o\left(\frac{d\Phi}{dt}\right). \tag{16.72}$$

Because the potential is a complex number (as discussed herein), also the acoustic pressure from this equation is a complex number. Expanding from this result, the difference in SPL is estimated from

$$\Delta\text{SPL} = 10\log_{10}\left(\frac{p}{p_o}\right)^2 = 20\log_{10}\left(\frac{|\Phi|}{|\Phi_o|}\right). \tag{16.73}$$

Equation 16.73 is only valid to compute the $\Delta$SPL of a single jet shielded by another jet with reference to the single unshielded jet. To compute the realistic effect of a dual-jet system with shielding effect on a receiver at any specific location, the sound power of each jet shielded by the adjacent jet has to be treated separately. The $\Delta$SPL is then calculated by superimposing the sound power of both jets with shielding operative and normalising it with the total power of both jets operating individually without any shielding effect. If jet "1" is the jet closest to the receiver and jet "2" is the one farther away, we call

- $p_1^2$ the power spectral density for jet 1 shielded by jet 2.
- $p_2^2$ the power spectral density for jet 2 shielded by jet 1.
- $p_{o1}^2$, $p_{o2}^2$ the total sound power of unshielded jet 1 and jet 2, respectively.
- For a discrete tone, we have:

$$\Delta\text{SPL} = 20\log\left(\frac{p_1^2 + p_2^2}{p_{o1}^2 + p_{o2}^2}\right). \tag{16.74}$$

The quantity $p^2$ is the square of the RMS pressure $\overline{p^2}$, which must be computed if one were to physically record the pressure signal of a discrete tone in real time. In a numerical computation, the ratio of the squares of these RMS pressures for discrete tones is equivalent to the ratio of the squares of the pressure magnitude $|p|^2$. In the formulation, each power spectral density term ($p^2$) is written by acknowledging that it is the square of the pressure magnitude $|p|^2$. This $\Delta$SPL is only for a single frequency because each sound power is actually the power spectral density (i.e., sound power for a single frequency, or discrete tones). To obtain the one-third-octave bands sound power or $\Delta$SPL, the results associated with the frequencies of smaller sub-bands within each one-third-octave band have to be integrated and normalised appropriately.

Because the undisturbed air density $\rho_o$ is a known quantity, the problem now focuses on the determination of a suitable potential function. The theory gives the velocity potential $\Phi$ due to a unit source at a distance $r$, corresponding to a wave number $k_o = \omega/a_o$:

$$\Phi = \frac{1}{4\pi r} e^{-i\omega t} \sum_{m=0}^{\infty} \epsilon_m \cos(m\Psi) \exp(-im\pi/2) \exp(ik_o r) F_m(\beta). \tag{16.75}$$

This is a cumbersome expression that requires further specification. First, the factor $\epsilon_m$ has the following values: $\epsilon = 1$ for $m = 0$, $\epsilon = 2$ for $m > 0$. The complex function $F_m$ of the polar angle $\beta$ is

$$F_m(\beta) = J_m(x_o) - \frac{F_{m1}}{F_{m2}} \tag{16.76}$$

with

$$\begin{cases} F_{m1}(\beta) = H_m(x_o)\left[\rho_1 a_1^2 T_1^2 \cos\beta J_m(x_1) J'_m(x_2) - \rho_o a_o^2 T_2 J_m(x_1) J'_m(x_2)\right] \\ F_{m2}(\beta) = \qquad \rho_1 a_1^2 T_1^2 \cos\beta J_m(x_1) H'_m(x_2) - \rho_o a_o^2 T_2 H_m(x_1) J'_m(x_2) \end{cases} \tag{16.77}$$

where, in order, we have:

- $J_m = m$-th order Bessel function of the first kind
- $H_m = m$-th order Hankel function of the first kind

Properties of these complex functions are given in most textbooks of mathematical special functions. Furthermore, we have the following definitions:

$$T_1 = a_o/a_1 - M \sin\beta, \quad \text{with } M = V_j/a_o \tag{16.78a}$$

$$T_2 = (T_1^2 - \sin^2\beta)^{1/2} \tag{16.78b}$$

$$x_o = k_o r_o \cos\beta \tag{16.78c}$$

$$x_1 = k_o r_j T_2 \tag{16.78d}$$

$$x_2 = k_o r_j \cos\beta \tag{16.78e}$$

$$J'_m(x_2) = \partial J_m/\partial x_2 = \partial J_m/\partial\beta \tag{16.78f}$$

$$H'_m(x_2) = \partial H_m/\partial x_2 = \partial H_m/\partial\beta \tag{16.78g}$$

The velocity potential due to a unit source in absence of the shielding jet is

$$\Phi_o = \frac{1}{4\pi r} e^{-i\omega t} \sum_{m=0}^{\infty} \epsilon_m \cos(m\Psi) \exp(-im\pi/2) \exp(ik_o r) J_m(x_o). \tag{16.79}$$

Both velocity potentials are complex numbers in the form

$$z(t) = |z| y.$$

Such complex numbers describe a signal in the time domain. Now all of the elements are in place for the calculation of the shielding effects, according to the procedure discussed here.

The jet-shielding formulation is valid for all azimuthal angles ($0 < \beta < 360$ degrees) and for all polar-emission angles $\theta$ ±20 degrees away from jet axis (i.e., $20 < \theta < 160$ or equivalently, $-70 < \beta < 70$).

### Computational Procedure

- Set jet characteristics, geometrical arrangement and position of the receiver.
- Set atmospheric conditions.
- Set frequency (or the wave number).
- Calculate the parameters given in Equation 16.78; the derivatives are calculated numerically*.
- Calculate the factor $F_m(\beta)$ from Equation 16.76.
- Calculate the shielding potential $\Phi$ from Equation 16.75 via a fast Fourier transform.
- Calculate the isolated potential $\Phi$ from Equation 16.79 via a fast Fourier transform.
- Calculate the change in SPL from Equation 16.73.
- (The calculation can be repeated for other wave numbers and receiver positions.)

There are a number of methods to simplify the integration and normalisation of the results. The exact approach would be to calculate the total sound power within each one-third-octave band by integrating the power spectral densities of its smaller sub-bands multiplied by the widths of the sub-bands. The total sound power (with shielding operative) of each band is then normalised with the bands' total sound power of both jets without shielding. The resultant ΔSPL formulation would be as follows:

$$\Delta \text{SPL} = 10 \log \left( \frac{P_s}{P_t} \right). \tag{16.80}$$

$$P_s = \sum_{i=1}^{ns} \left[ w_b \left( p_1^2 + p_2^2 \right) \right]_i , \tag{16.81}$$

$$P_t = \sum_{i=1}^{ns} \left[ w_b \left( p_{o1}^2 + p_{o2}^2 \right) \right]_i , \tag{16.82}$$

* There are a variety of open-source programs for the calculation of the Bessel functions and their derivatives.

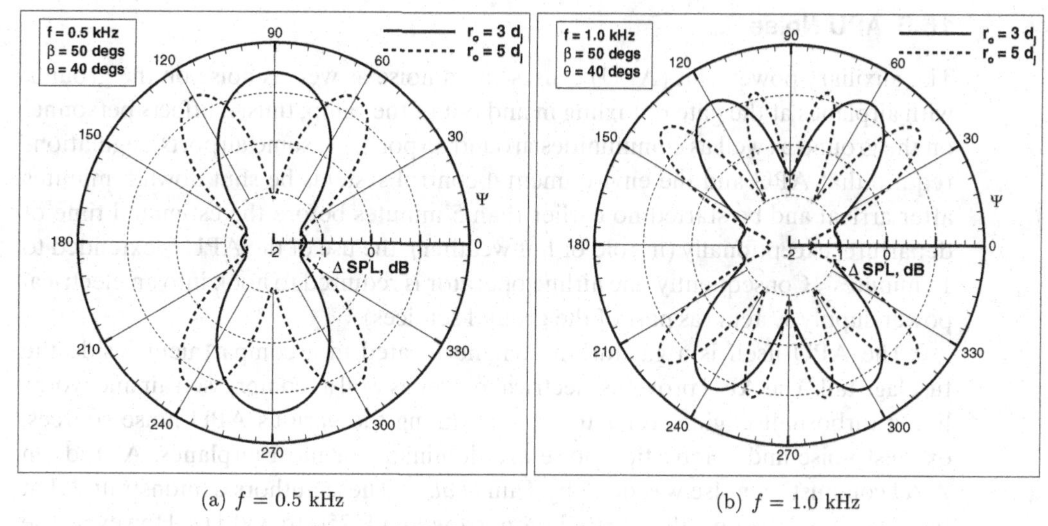


(a) $f = 0.5$ kHz (b) $f = 1.0$ kHz

Figure 16.22. Jet-shielding effects.

where $P_s$ and $P_t$ are the shielded and unshielded total sound pressure power, respectively, for each 1/3 octave band, obtained by summing the power spectral densities of both jets multiplied by their corresponding sub-bandwidths across the 1/3 octave bandwidth. Here, $ns$ is the number of sub-band partitions within each 1/3 octave band, $i$ is the $i$–th sub-band partition within a 1/3 octave band, and $w_b$ is the sub-bandwidth.

Typical values of the ratio $d_j/r_o$ are as high as 0.215 for the Boeing MD-90 and as low as 0.145 for the Boeing B777-300. Thus, it is reasonable to assume that practical values of this parameter are included within 0.15 and 0.22. Figure 16.22 shows the jet shielding effects at increasing azimuthal angles $\phi_i$, the effects of frequency, polar-emission angle and distance between jet axes are shown. Figure 16.23 shows the jet shielding as a function of the polar-emission angle.

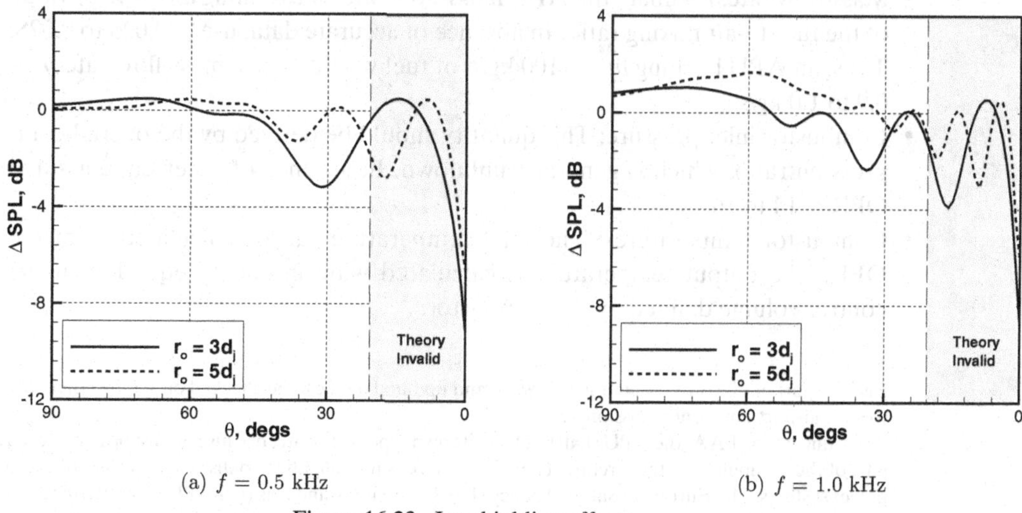


(a) $f = 0.5$ kHz (b) $f = 1.0$ kHz

Figure 16.23. Jet-shielding effects.

## 16.5 APU Noise

The auxiliary power unit (APU) causes cabin noise as well as noise on the ground, with airplanes at the gate or taxiing in and out of the ramp; thus, it affects personnel on the ground as well as communities around airports. At some airports, regulations require that APU and the environmental control systems be shut down 5 minutes after arrival and be started no earlier than 5 minutes before the estimated time of departure; exceptionally (in cold or hot weather), the use of the APU is extended to 15 minutes*. Consequently, the airline operator is required to hook into an electrical power-supply system (as part of the ground services).

The APU itself is a gas-turbine engine located in a compartment inside the fuselage tail. The APU provides electrical power as well as compressed air and works like a turboshaft engine (refer to § 5.7). Among the various APU noise sources, exhaust noise and combustion noise are dominant for most airplanes. A study in APU combustion noise was done by Tam *et al.*[44]. These authors demonstrated that the peak combustion SPL occurs in a narrow band, 250 to 350 Hz. However, the exact noise level of the measurements was unpublished.

For receivers beyond the immediate vicinity of the aircraft, the estimation of the noise signature must rely on an accurate model for noise scattering and reflection, as well as a detailed description of the horizontal and vertical obstacles between the source and the receiver. For receivers in the immediate vicinity, the use of a ground-reflection model should be sufficient.

The aero-thermodynamic data needed by the model include inlet and exit combustor temperatures, inlet pressure, mass-flow rate and atmospheric conditions. Even this limited set of data is difficult to gather. One of the few options available to realistically collect these data is to stand directly behind the airplane and take measurements†. As an alternative, we propose the following method:

- Fuel flow: To estimate the APU fuel flow, use the data in Table 5.5. If the APU is not in the table, extrapolate the fuel flow on the basis of the number of seats in the airplane.
- Mass-flow rate: Estimate the APU mass-flow rate by assuming an average value of the fuel-to-air mixing ratio. In absence of accurate data, use $\sim$ 0.025 to 0.028. Thus, an APU burning $\dot{m}_f \sim 100$ kg/h of fuel would have a mass-flow rate $\dot{m} \sim$ 0.9 to 1.0 kg/s.
- Combustor inlet pressure: This quantity should be derived by the overall compression ratio, which is generally unknown. In absence of other data, assume OPR $\sim$ 14 to 16.
- Combustor temperatures: The inlet temperature can be calculated from the OPR. The output temperature is calculated with an energy equation in the control volume defined by the combustor.

* Each airport has its own regulations. Specific and updated data can be checked on the Internet; key words "airport noise and emissions".

† According to the FAA, the APU shall not have its own type certificate, because it must be considered part of the equipment in the aircraft. Therefore, there is no obligation to disclose any data of these power systems. The European Safety Agency (EASA) issues standards (CS-APU) for "certification specifications", which is what is needed for basic engineering simulation.

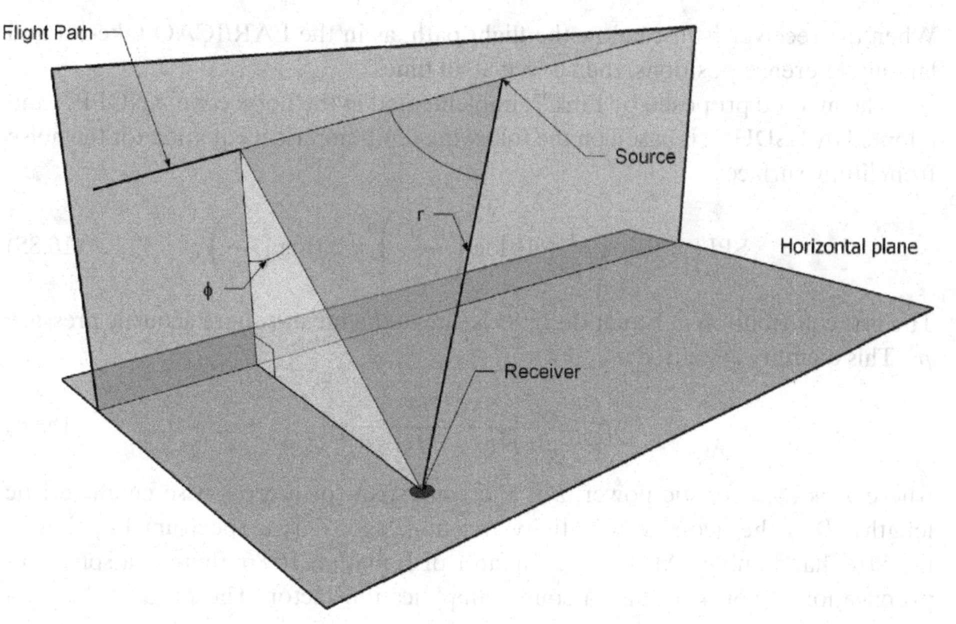


Figure 16.24. Reference systems for airframe noise models.

- Exhaust gas temperature (EGT): This is given in the Type Certificate Data Sheet. If this document is not available, use EGT $\sim 850^{\circ}$C (starting) and EGT $\sim 550^{\circ}$C (continuous).
- Exhaust gas velocity: If $d_j$ is the nozzle's diameter, then $V_j \sim 4\dot{m}/d_j^2 \rho_j$. Assume a fully expanded jet ($\rho_j = \rho$) to simplify the analysis.

With the quantities listed here, it is possible to estimate the APU noise.

## 16.6 Airframe Noise

We now proceed with the analysis of selected airframe-noise components. Some models that rely on flow physics are available to predict the noise from simple geometries. However, the problem is still dealt with by using a considerable amount of empiricism. In our case, we generally rely on approximate geometries that include few details beyond the planform of the component (by itself approximate) and the deflection angle.

The basic nomenclature is shown in Figure 16.24, which displays the position of the source (interpreted as the position of the aircraft's CG), the generic position of the receiver, the polar-emission angle $\theta$ and the azimuth angle $\phi$. The polar-emission angle is

$$\theta = \cos^{-1}\left(\frac{\boldsymbol{r} \cdot \boldsymbol{V}}{Vr}\right) \tag{16.83}$$

and must be corrected for Mach number effects as explained in § 16.3.1.

The azimuth angle $\phi$ is calculated by first taking the minimum distance $r_1$ between the flight path and the receiver. The altitude of the source at this point is $h$. The uncorrected azimuth angle is

$$\phi = \cos^{-1}\left(\frac{h}{r_1}\right). \tag{16.84}$$

When the receiver is just below the flight path, as in the FAR/ICAO take-off and landing reference positions, then $\phi = 0$ at all times.

The method proposed by Fink[45], implemented in the noise code ANOPP[46] and adopted by ESDU[47], is based on the following semi-empirical equation for the noise from lifting surfaces:

$$\mathrm{SPL} = 10 \log p^2 + 10 \log \left( \frac{\rho^2 a^4}{p_{ref}^2} \right) - 20 \log \left( \frac{p}{p_o} \right). \tag{16.85}$$

The first contribution in Equation 16.85 is due to the mean-square acoustic pressure $p^2$. This quantity is written as

$$p^2 = \frac{P b^2 \mathcal{D} \mathcal{F}}{4\pi r^2 (1 - M \cos\theta)^4}, \tag{16.86}$$

where $P$ is the acoustic power, $b$ is the wing span (or a cross-wise characteristic length), $\mathcal{D}$ is the acoustic directivity function, and $\mathcal{F}$ is a spectrum function of the Strouhal number. At the denominator of Equation 16.86, there is a spherical-propagation factor $4\pi r^2$ and a source-amplification factor. The product $Pb^2$ is a function of the flight Mach number and several empirical parameters and has the form:

$$P b^2 = k_1 k_3 M^{k_2}, \tag{16.87}$$

where the coefficient $k_2$ depends on the geometrical configuration of the item. The directivity function in Equation 16.85 depends on the polar-emission angle $\theta$ and the azimuth angle $\phi$. The Strouhal number involved in the method is in fact a "corrected" Strouhal number, given by the following expression:

$$\mathrm{St} = \frac{fl}{V}(1 - M \cos\theta), \tag{16.88}$$

where $f$ is the frequency, $l$ is a characteristic length and $V = Ma$ is the true air speed. For all the lifting components, the corrected Strouhal number is

$$\mathrm{St} = 0.37 f \left( \frac{S}{b} \right) \left( \frac{V}{\nu} \frac{S}{b} \right)^{-1/5} \frac{(1 - M \cos\theta)}{V}, \tag{16.89}$$

where $S$ and $b$ are the surface and the spanwise characteristic length of the item. Hence, the ratio $S/b$ is a streamwise length.

The third term in the formulation of Equation 16.85 is a correction due to the difference in atmospheric pressure between source and receiver locations ($p$ = pressure at source; $p_o$ = pressure at receiver).

### 16.6.1 Wing Noise

Equation 16.85, in addition to being empirical, gives the value of an SPL that is not dependent on whether or not the lifting surface produces a lift. This is in contrast with Lilley's theory[48;49] of airframe noise, which is discussed next.

In many analyses, following the models of Ffowcs-Williams and Hall[50] and Howe[51], the wing is modelled as a semi-infinite flat plate in turbulent flow. For a wing of high-aspect-ratio, the problem is associated with the scattering of turbulent kinetic energy from the trailing edge. The noise intensity just below an aircraft flying at an altitude $h$ above the ground, with a speed $V$, is found from

$$I = \frac{1.7}{2\pi^3}\frac{\rho A V^3 M^2}{h^2}\left[\left(\frac{u_o}{V}\right)^5\left(\frac{\delta}{\delta*}\right)\right]_{te} = \frac{1.7}{2\pi^3}\frac{\rho A V^5}{a^2 h^2}\left[\left(\frac{u_o}{V}\right)^5\left(\frac{\delta}{\delta*}\right)\right]_{te}, \tag{16.90}$$

where $u_o$ is the characteristic speed for the turbulent flow; $\delta$ and $\delta^*$ are the boundary layer thickness and boundary-layer displacement thickness at the trailing edge, respectively. Equation 16.90 ignores the presence of other systems (fuselage, tail plane, engines) and the sweep of the trailing-edge line, which would require a factor $\cos^3 \Lambda_{te}$. Finally, the noise intensity at aircraft positions other than the fly-over must account for a corrective factor $\sin^2(\theta/2)$. The corrected sound intensity is

$$I = \frac{1.7}{2\pi^3}\frac{\rho A V^5}{a^2 h^2}\left[\left(\frac{u_o}{V}\right)^5\left(\frac{\delta}{\delta*}\right)\right]_{te}\sin^2(\theta/2)\cos^3\Lambda_{te}. \tag{16.91}$$

The ratio $(\delta/\delta^*)_{te}$ depends on the state of the boundary layer at the trailing edge. For a turbulent boundary layer, this ratio is at least equal to 8. We take the value $(\delta/\delta^*)_{te} \simeq 10$, which should account for turbulent flow with some separation at relatively high $C_L$. This ratio, along with the turbulent characteristics $u_o/V$, can be calculated more precisely with modern computational fluid dynamics programs for the three-dimensional wing. The air speed $V$ in Equation 16.90 is found from the definition of $C_L$,

$$I = \frac{1.7}{2\pi^3}\frac{\rho A V^3}{a^2 h^2}\frac{2W}{\rho A C_L}\left[\left(\frac{u_o}{V}\right)^5\left(\frac{\delta}{\delta*}\right)\right]_{te}, \tag{16.92}$$

$$I \simeq \frac{17}{\pi^3}\frac{V^3}{a^2 h^2}\frac{W}{C_L}\left(\frac{u_o}{V}\right)^5_{te}, \tag{16.93}$$

$$I \simeq c_1\left(\frac{V^3 W}{h^2 C_L}\right), \tag{16.94}$$

with

$$c_1 = \frac{17}{\pi^3 a^2}\left(\frac{u_o}{V}\right)^5_{te}. \tag{16.95}$$

This coefficient will depend on the local atmospheric conditions (speed of sound, altitude), the Reynolds number and the geometry of the aircraft; $c_1$ has a nearly constant value that fits remarkably most experimental data collected in the past 30 years, for systems as diverse as wide-body aircrafts and birds. Typical values of the parameters in Equation 16.95 are $u_o/V \simeq 0.1$, $a \simeq 340$ m/s (sea level). With these values we find $c_1 \simeq 4 \cdot 10^{-11}$.

What happens if the $C_L$ is increased? High lift is generally associated to large suction peaks on the upper surfaces of the wings, which trigger instability in the boundary layer. This issue is a complex matter for computational aerodynamics. CFD analysis, as discussed by Lockard and Lilley[49], indicates that the average $C_L$ is related to the turbulent quantities by the approximate equation

$$\left(\frac{u_o}{V}\right)^5\left(\frac{\delta}{y_m}\right)_{te} = \left(1+\frac{1}{4}C_L^2\right)^4, \tag{16.96}$$

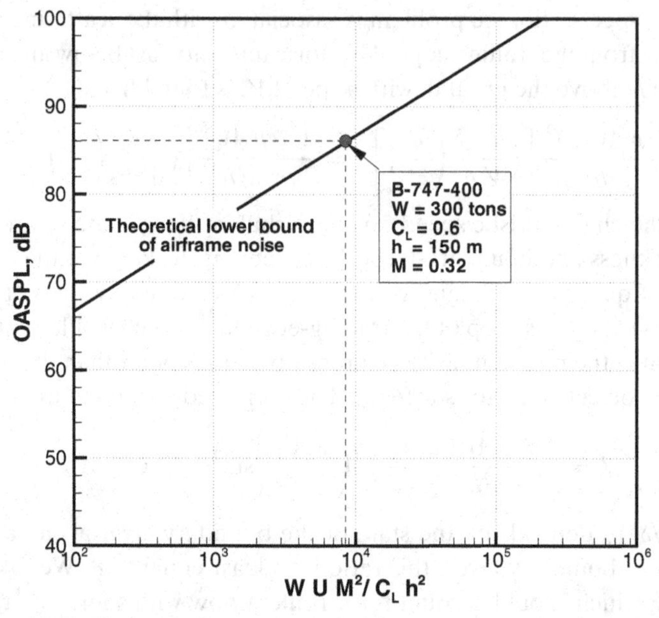


Figure 16.25. Theoretical lower bound of fly-over wing noise in level flight.

where $y_m$ is the length scale of eddies. Thus, the parameter $y_m$ replaces the displacement thickness $\delta^*$ in Equation 16.92. In areas of adverse pressure gradients $y_m/\delta^* \simeq$ 0.2 to 0.3. Operating this substitution in Equation 16.92, these authors found

$$I = \frac{1.7}{\pi^3} \frac{WVM^2}{C_L h^2} \left(1 + \frac{1}{4} C_L^2\right)^4, \tag{16.97}$$

where $C_L$ is a mean value on the wing. Equation 16.97 does not take into account the presence of partly deployed flaps. Equation 16.90 and Equation 16.97 show that the sound intensity $I$ depends on the factor

$$F = \frac{WVM^2}{C_L h^2}. \tag{16.98}$$

Equation 16.97 represents the lower bound of aircraft noise, such as if it is created by the airframe alone, without engines, under-carriage, tail plane and high-lift devices. Figure 16.25 shows the trend of this lower bound in terms of the factor $F$. This factor can be obtained for an infinite combination of aircraft weight, speed, lift coefficient and flight altitudes.

The range of the factor $F$, Equation 16.98, covers five orders of magnitude. A medium-size aircraft, such as the Airbus A300-600, with a fly-over altitude of the order of 100 m (348 feet), will have $F \simeq 6 \cdot 10^3$. A drawback of this formulation is that it does not provide the spectral content of the acoustic pressures.

### 16.6.2 Landing-Gear Noise

The determination of the landing-gear noise is one of the most difficult task in the analysis of aircraft noise because the acoustic response results from the complicated

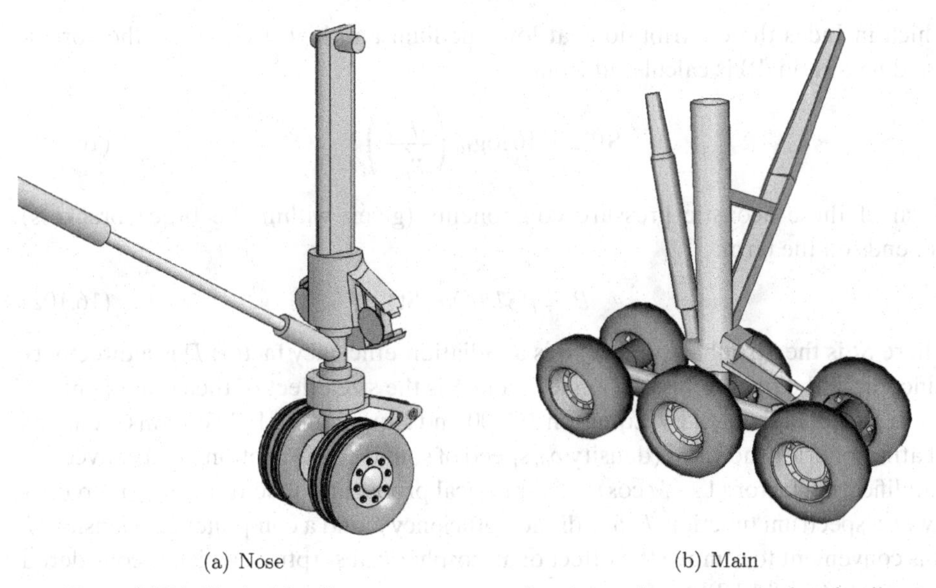


Figure 16.26. Typical landing-gear geometry showing main struts and other details.

geometry of the unit. In fact, there are various non-aerodynamic components (struts, wheels) directly exposed to the airflow, along with the under-carriage bay (a cavity with open doors). An example of landing gears is shown in Figure 16.26.

The only realistic option within a flight-mechanics model is to use validated semi-empirical models. One of these has been proposed by Guo[52] and validated with experimental data[53;54]. Experimental data on a 6.3%-scale-model landing gear of the Boeing B777 (spectra and directivity) have been published by Humphreys and Brooks[55].

Following this method, the spectral characteristics of the landing gears depend on three contributions: a low-frequency contribution, mostly due to the wheels; a mid-frequency contribution, mostly due to the main struts; and a high-frequency contribution, due to the small geometrical details of the unit. The overall acoustic emission is then calculated by summing the incoherent noise energy of the spectral components. Such a decomposition arises from a detailed analysis of experimental and flight-test data. Each of the contributions has a different spectral response and directivity. For each component, the method requires to define scaling factors, reference lengths, Strouhal numbers and normalised spectral functions.

The mean-square acoustic pressure is written as a sum of contributions at low, medium and high frequencies:

$$\langle p^2 \rangle = \langle p_L^2 \rangle + \langle p_M^2 \rangle + \langle p_H^2 \rangle. \tag{16.99}$$

The landing-gear noise is

$$\langle p^2 \rangle = \frac{\rho_o a_o^2 M^6 \mathcal{A}(r)\mathcal{D}(\theta)}{r^2(1 - M\cos\theta)^4} \left\{ P_L + P_M + P_H \right\}, \tag{16.100}$$

which includes the contributions at low, medium and high frequency; the corresponding SPL in dB is calculated from

$$\mathrm{SPL} = 10\log_{10}\left(\frac{\langle p^2\rangle}{p_{ref}^2}\right). \tag{16.101}$$

Each of these acoustic pressure components (given within the brace brackets) depends on the term:

$$P = \beta S \mathcal{D}(\theta)\mathcal{F}(\mathrm{St}), \tag{16.102}$$

where $St$ is the Strouhal number; $\beta$ is a radiation-efficiency factor; $\mathcal{D}$ is a directivity function of the polar-emission angle $\theta$; and $S$ is the size effect of the component.

The model defined by Equation 16.100 and Equation 16.102 contains elements of atmospheric conditions (density $\rho_o$, speed of sound $a_o$, absorption $\mathcal{A}$), a convective amplification factor $(1 - M\cos\theta)^4$, a spherical propagation factor $(1/r^2)$, a directivity $\mathcal{D}$, a spectrum function $\mathcal{F}$, a radiation efficiency $\beta$ and a component dimension $S$. It is convenient to remove the effect of atmospheric absorption, which is considered separately (see § 17.2).

For the low-frequency component, the size effect is due mainly to the wheels. The Strouhal number and the size effect are, respectively,

$$\mathrm{St}_L = \frac{fd}{V}, \qquad S_L = \pi N_w w d. \tag{16.103}$$

For the medium-frequency component that is due mainly to the vertical struts, the Strouhal number and the size effect are

$$\mathrm{St}_M = \frac{f\bar{s}}{V}, \qquad S_M = \sum_i s_i l_i, \tag{16.104}$$

where $s_i$ is the perimeter of the cross-section of the $i$–th strut, $l_i$ is its length, and $\bar{s}$ is the average cross-section of the struts. The sum is extended to all of the main struts of the assembly. The contribution at high frequency is

$$\mathrm{St}_H = \frac{fl}{V}, \qquad S_H = \eta l^2, \tag{16.105}$$

where $l$ is the length scale of the high-frequency noise. This length scale must be of the same order as the geometrical features in the landing-gear assembly that are responsible for turbulence noise (hydraulic hoses, wires, gaps, and so on). Clearly, this quantity depends on the landing gear. However, an average value given by Guo is $l = 0.15\bar{s}$. The factor $\eta$ is a complexity factor that is related to the high-frequency noise. This factor is defined by the following equation:

$$\eta = \left[1 + 0.028\left(\frac{N_w}{N_{ref}}\frac{l_t}{l_{ref}}\frac{W}{W_{ref}} - 1\right)\right]\left[1 + 2\left(\frac{N_w - 2}{N_{ref}}\sin 2\gamma_t\right)\right]. \tag{16.106}$$

In this equation, the reference values have been chosen as $N_w = 2$, $W_{ref} = 150{,}000$ lb (68,040 kg) and $l_{ref} = 300$ inches (7.62 m); $l_t$ is the total length of the struts of the under-carriage unit; $\gamma_t$ is the wheel-track-alignment angle. The normalised spectral functions are given by

$$\mathcal{F}(\mathrm{St}) = A\frac{\mathrm{St}^{\varsigma}}{(B + \mathrm{St}^{\mu})^{q}}, \tag{16.107}$$

where $A, B, \sigma, \mu$ and $q$ are semi-empirical coefficients. In particular, $\varsigma, \mu$ and $q$ define the shape of the normalised spectrum and are calculated by fitting experimental data; $A$ and $B$ normalise the function $\mathcal{F} = 1$ at some value of the Strouhal number, $St_o$. The directivity function is

$$\mathcal{D} = (1 + h_d \cos^2\theta)^2, \tag{16.108}$$

where $h_d$ is another empirical parameter. A summary of all the parameters for this model is given in Guo[52]. In absence of detailed geometrical data, the following expressions can be used for a large airplane (OEW > 100 ton):

$$\begin{aligned} l_{main} &= 3.71(l_{vstrut} + l_{axle})_{main}, \; \overline{s}_{main} = 0.01 l_{main} \\ l_{nose} &= 2.85(l_{vstrut} + l_{axle})_{nose}, \; \overline{s}_{nose} = 0.024 l_{nose}. \end{aligned} \tag{16.109}$$

For a smaller airplane, use

$$\begin{aligned} l_{main} &= 3.45(l_{vstrut} + l_{axle})_{main}, \; \overline{s}_{main} = 0.015 l_{main} \\ l_{nose} &= 3.71(l_{vstrut} + l_{axle})_{nose}, \; \overline{s}_{nose} = 0.027 l_{nose}. \end{aligned} \tag{16.110}$$

In all cases, following the analysis shown by Guo[52], the length scale of the small details (responsible for the broadband noise) is

$$l = 0.15\overline{w}. \tag{16.111}$$

Finally, we must account for the installation effects of the landing gear. These effects are essentially due to the reflection of acoustic waves from the wing and the fuselage. Detailed models do not exist, but Guo provides a semi-empirical estimate for some airplanes (Boeing B777) that correspond to a maximum of 0.8 dB increase in SPL when the aircraft is overhead ($\theta = 90$ degrees). This is achieved through a corrected directivity function

$$\mathcal{D} = 1.2(1 - 0.9\cos^2\theta)^2. \tag{16.112}$$

ESDU[47] uses the following expression for the mean-square acoustic pressure:

$$\langle p^2 \rangle = \frac{l^2 \mathcal{D}(\theta)\mathcal{F}(\mathrm{St})}{4\pi r^2 (1 - M\cos\theta)^4} P. \tag{16.113}$$

Apart from the inclusion of the atmospheric absorption, the two models are similar. The calculation is done by using an additional empirical equation that correlates the reference length $l$, the acoustic power $P$ and the flight Mach number:

$$Pl^2 = c_1 c_3 M^{c_2}, \tag{16.114}$$

where the coefficients $c_1$, $c_2$ and $c_3$ are dependent on the component.

The method presented is applicable to most isolated under-carriages, whether they are main or nose units. A correction for the main under-carriage is suggested, to take into account the fact that the inflow Mach number at the under-carriage location is generally lower than the flight Mach number. A value of $M = 0.75$ to 0.80 should be sufficient to take into account the effects of the flow around the airplane, in absence of more detailed data. However, more accurate information about the flow around the landing gear could further improve the noise prediction.

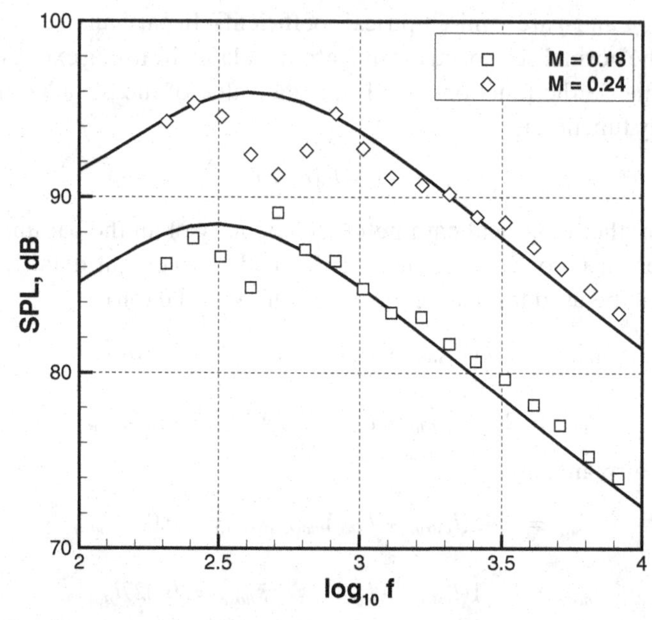


Figure 16.27. Landing-gear noise of the the Boeing B737; polar-emission angle = 90 degrees.

Another point of concern is that although the installation effects are accounted for by the term given in Equation 16.112, the model does not include the effects of the under-carriage bays and their relative cavities and cavity doors. For some modern airplanes, such as the Airbus A380, the cavities are partially closed by repositioning the doors. This technique limits the exposure of the bays to the airflow and thus reduces both aerodynamic drag and acoustic emission.

### Landing-Gear Noise Verification

The method described has been validated for a number of cases. We now refer to the Boeing B737 main landing-gear system, whose data are available in Ref.[53]. A full breakdown of systems components and dimensions is given in the aforementioned paper. We report here only the summary of data relevant for the noise calculation.

The noise measurements were taken on an isolated landing gear, with an array of microphones placed on a line at about 10 feet (~3 metres) from the centre of the unit. Data are available for a range of Mach numbers (from $M = 0.18$ to $M = 0.24$) and a range of polar-emission angles. The comparison between simulated acoustic emission and reference data is shown in Figure 16.27 for two Mach numbers and polar-emission angles equal to 90 degrees. The noise-level dip at frequencies $f \sim 500$ Hz is attributed to a spurious vortex shedding that is not captured by the theory.

Figure 16.28 shows the noise directivity for two different aircraft as predicted by this theory. The comparison with the test data[56–58] is fair; however, note that the data are somewhat scattered.

## 16.7 Propeller Noise

The propeller noise has two fundamental components: the harmonic (or rotational) noise, arising from the blade-passing frequency, and the broadband noise, arising

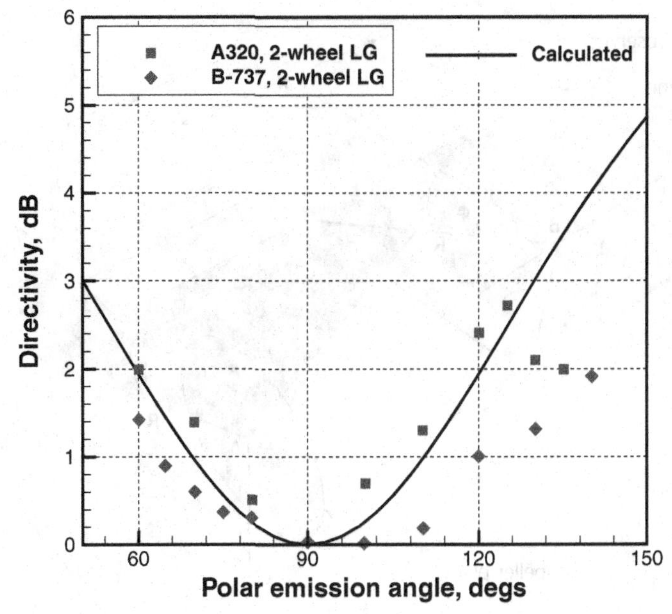


Figure 16.28. Predicted landing-gear noise directivity and comparison with data for the Airbus A320 and Boeing B737.

from turbulence-related acoustic excitation. The latter contribution is dominant at relatively high frequency and provides an acoustic signature that is continuous in the frequency spectrum. By contrast, the former contribution is a signal at discrete frequencies depending on the blade-passing frequency. It turns out that the propeller noise is dominated by the harmonic content; therefore our analysis will focus on this contribution. We will present a low-order numerical model for harmonic noise based on an implementation of the model proposed by Hanson and Parzych[59]. This method is suitable for implementation in a comprehensive flight-mechanics code.

### 16.7.1 Propeller's Harmonic Noise

Refer to Figure 16.29. Suppose that the centre of the propeller is $A$; the propeller is moving toward the observer $O$ along the $x$–direction. By the time the acoustic waves emitted at $A$ reach the observer, the propeller will be at position $B$. The initial distance to the observer is $r$; the remaining distances and angles are clearly indicated in Figure 16.29. $\{r, \theta, \phi\}$ is a spherical coordinate system when the $x$-axis is in the *flight direction*; $\{r, \theta', \phi'\}$ is a spherical coordinate system when the $x$-axis is in the *propeller* axis.

There are two distinct acoustic-power contributions, one due to lift and the other one due to volume (or thickness), described as follows:

- The lift (or loading) contribution $P_L$ is generated by steady-state blade loading that rotates with the blade. As such rotation takes place, there is a phase velocity between blade and receiver. When the flow is not axi-symmetric (a condition that arises when there is a yaw or a pitch), there is also unsteady blade loading. The lift noise is represented by a dipole distribution.
- The volume (or thickness) contribution $P_T$ arises from the displacement of the air flow when it passes through the blade. The higher the flow deflection, the

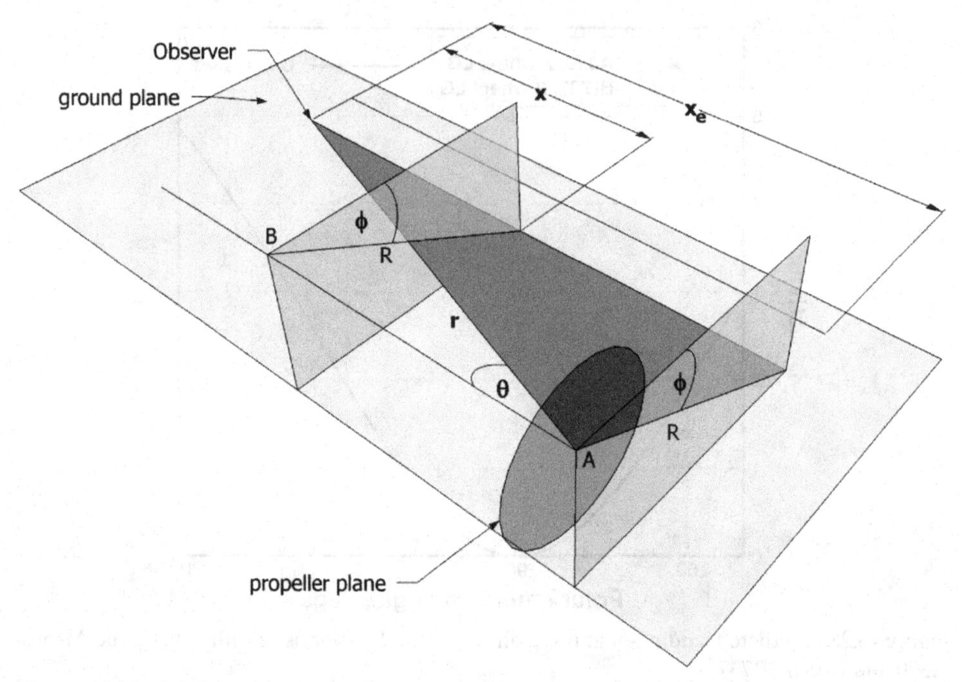


Figure 16.29. Reference system and nomenclature for propeller-noise model; $r, \theta, \phi$ is a spherical coordinate system.

higher the thickness effect. An unsteady contribution arises, as in the previous case, due to varying incidence and inflow as the blade performs a full revolution. The thickness noise is represented by a monopole distribution.

The thickness and loading noise are represented by the monopole and dipole distribution, respectively. Quadrupole noise induced by non-linear effects only arises when the propeller blades rotate at high speeds such that the blade tips' Mach number $M_{tip}$ is within the transonic or supersonic regimes, as reported by Hanson and Fink[60]. The broadband or random component is mainly induced by the turbulent flow which only becomes significant when the blades are subjected to high angles of attack.

The dominant noise for a subsonic propeller can be predicted using the frequency-domain computational approach developed by Hanson and Parzych[59]. Both noise components are computed at the blade-passing frequency as well as at its higher harmonics. Unsteady flow effects are accounted for in the model when non-co-axial, or angular, flow enters the propeller. The angular inflow produces the unsteady periodic effect because it causes the relative blade section velocity to vary within a blade revolution.

Hanson and Parzych reported that results from the model compare well with experimental propeller data for various parametric conditions and inflow angles. The model validation is not satisfactory, however, when compared to prop-fans data; they reasoned that this is probably due to the aerodynamic loading estimation for the prop-fans and not the acoustic radiation because it is based on exact equations.

The starting point for the Hanson and Parzych propeller noise model is the exact theory of Goldstein's acoustic analogy applied to moving surfaces embedded within a moving medium. It describes the acoustic pressure disturbance $p' = \rho' a_o^2$ of a source as the sum of integrals involving three source terms: the thickness noise due to displaced volumes of air, the loading noise due to surface forces and the quadrupole noise due to the non-linear effects accounted for in Lighthill's stress tensor. The model takes into account the various coordinate systems involved, that is, the propeller and flight coordinate-systems for moving and stationary observers; a careful treatment of the coordinate system conversions has been detailed elsewhere[61;59].

Of interest to the application presented here is the prediction of far-field noise radiation received by a stationary observer on the ground. This is achieved through expansion of each of the source terms in Goldstein's formula. Computationally efficient forms are obtained following normalisation into non-dimensional variables, discretisation of the blade-area elements and conversion using Fourier series into the frequency domain. Assuming that the quadrupole noise is negligible for the subsonic propeller application, the non-dimensional acoustic pressure disturbances for the thickness noise $p_{Tm}$ and the loading noise $p_{Lm}$ are shown in Equations 16.115 and 16.116, respectively. These pressures are normalised with respect to the ambient pressure $p_o$. Evaluations of the Bessel function $J_n(x)$ at various orders are required for the assessment of the non-dimensional acoustic pressures:

$$p_{Tm} = \frac{-i\gamma B k_m e^{i[k_m r + mB(\pm\phi' - \pi/2)]}}{4\pi S_o d_f} \sum_{\mu} \sum_{\nu} e^{-i[k_m S_c x_{\mu,\nu} + mB\phi_{\mu,\nu}]}$$
$$\times \left\{ V_{0\mu,\nu} J_{mB}(k_m S_s r_o) + \frac{1}{2}(V_{c\mu,\nu} - i V_{s\mu,\nu}) e^{i(\pm\phi' - \pi/2)} J_{mB+1}(k_m S_s r_o) \right.$$
$$\left. + \frac{1}{2}(V_{c\mu,\nu} + i V_{s\mu,\nu}) e^{-i(\pm\phi' - \pi/2)} J_{mB-1}(k_m S_s r_o) \right\} \tag{16.115}$$

$$p_{Lm} = \frac{-i B k_m e^{i k_m r}}{4\pi S_o} \sum_{\mu} \sum_{\nu} e^{i[(mB-k)(\pm\phi' - \pi/2 - \phi_{\mu,\nu}) + k\phi_r ef]} e^{-i k_m S_c x_{\mu,\nu}}$$
$$\times \left\{ i S_s F_{rk\mu,\nu} J'_{mB-k}(k_m S_s r_o) + \frac{mB-k}{k_m r_o} F_{\phi k\mu,\nu} J_{mB-k}(k_m S_s r_o) \right.$$
$$\left. + S_c F_{xk\mu,\nu} J_{mB-k}(k_m S_s r_o) \right\}, \tag{16.116}$$

where the counters $\mu$ and $\nu$ refer to the radial and chordwise elements; $m$ is the harmonic blade-passing frequency, $k$ is the index for the loading harmonic, and $k_m$ is the wave number in the harmonic Green function, $k_m = mBM_{tip}$; $d_f$ is the Doppler shift, $B$ is the number of blades, $r_0 = r/d_i$. The factors $S_c$ and $S_s$ are defined by:

$$S_c = \frac{\cos\theta'}{1 - M\cos\theta'} \qquad S_s = \frac{\sin\theta'}{1 - M\cos\theta}. \tag{16.117}$$

Finally, $S_o$ denotes the amplitude radius: $S_o = \sqrt{x^2 + \beta^2(y^2 + z^2)}$. Equation 16.115 describes the thickness-source terms $V_{0\mu,\nu}$, $V_{c\mu,\nu}$ and $V_{s\mu,\nu}$. The loading-source terms for the tangential and axial loading, $F_{\phi k\mu,\nu}$ and $F_{xk\mu,\nu}$ respectively, are given in Equation 16.116. The radial loading $F_{rk\mu,\nu}$ is found to have a negligible contribution of only about 0.6 dB[59] toward the loading noise and therefore is not included in the total noise prediction. The lift and drag coefficients, $C_L$ and $C_D$, both for steady

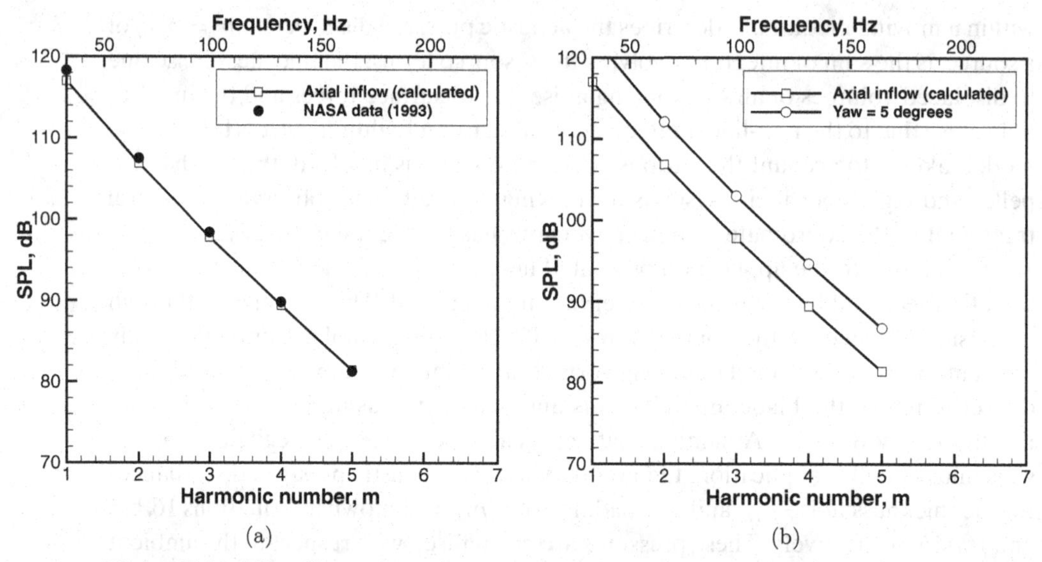


Figure 16.30. The effects of angular inflow on propeller $SPL_h$ for various harmonics of the blade-passing frequency. Left graph shows comparison with the data in Ref. 59.

and unsteady flows, are treated as inputs to the noise model. These characteristics are calculated with the propeller model discussed in Chapter 6, although alternative methods can be used, from tabulated data to computational fluid dynamics methods. The unsteady loading for the analysis presented here utilises the ESDU method developed specifically for subsonic propeller blades in non-axial inflow[62].

The parameter of interest for the total propeller noise, that is, the sound pressure level, in Equation 16.118 is calculated by summing the square of the thickness and loading acoustic pressures normalised with the reference acoustic pressure $p_{ref}$. These noise metrics are calculated at the blade-passing frequency and its higher harmonics and presented as $SPL_h$:

$$SPL_h = 10\log_{10}\left[\left(P_{Tm}\frac{p_o}{p_{ref}}\right)^2 + \left(P_{Lm}\frac{p_o}{p_{ref}}\right)^2\right]. \qquad (16.118)$$

Due to the Doppler effect, a stationary observer would receive the noise with the same $SPL_h$ but shifted to frequencies different from the source frequencies ($f_s$) on the moving aircraft. The SPLs must be shifted to these receiver frequencies via the Doppler relation $f = f_s/(1 - M\cos\theta)$, then re-scaled into the one-third octave-band scales where the SPL of the shifted harmonics that lie within the same band are integrated accordingly.

This propeller-noise model is fully coupled with the aerodynamic-performance model described in Chapter 6 to generate the correct distribution of loads. The following results illustrate the changes in the propeller SPLs when the blades are subjected to non-zero angular inflows. In Figure 16.30, the SPLs are plotted against the harmonics for the conditions without and with a yaw angular inflow of 5 degrees. The noise level is measured at a location 1.34$d$ away from the propeller hub, where $d$ is the blade diameter. The SPL decreases with increasing harmonics for both

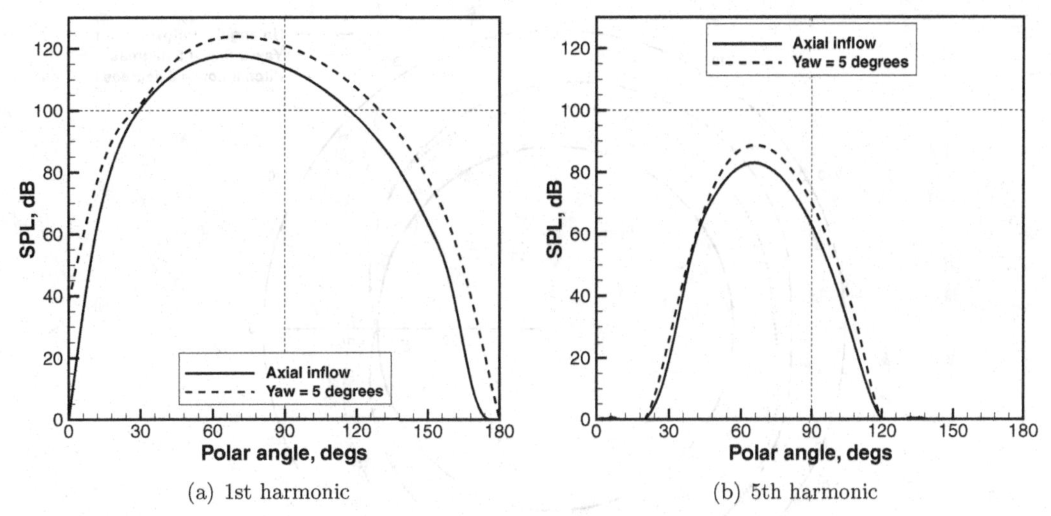


(a) 1st harmonic (b) 5th harmonic

Figure 16.31. Effect of angular inflow on propeller $SPL_h$ for the 1st and 5th harmonics.

conditions; this value increases by about 5 dB for each harmonic when the blades are subjected to unsteady flows.

The SPL variations with the polar directivity angle $\theta$ for the 1st and 5th harmonics are shown in Figure 16.31. The azimuthal directivity angle $\phi$ is set at 90 degrees, equivalent to a position below the propeller. The unsteady loads experienced at an inflow setting of 5 degrees cause the noise level to increase; the increase is both a function of the polar angle and the harmonics. For all configurations presented, the peak noise is achieved forward of the propeller at a polar angle between 60 and 80 degrees. In Figure 16.32, the azimuthal or circumferential directivity angle for the plane perpendicular to the flight path is varied with a fixed polar angle of 70 degrees. When the inflow is co-axial with the propeller axis (i.e., no angular inflow), the SPL around the propeller is uniform at about 118 dB, unaffected by any unsteady loading. Subjecting the blades to unsteady loading with a pitch or yaw angular inflow of 5 degrees generally increases the SPL but non-symmetrically, due to the blade rotational direction and the specific angular shift of the inflow.

### 16.7.2 Propeller's Broadband Noise

In many cases, the broadband noise is characterised by lower SPL than the tonal noise. However, whilst the tonal noise is limited to a few frequencies, the broadband noise contribution is distributed through the full spectrum of audible frequencies.

There are several analytical, semi-analytical and semi-empirical methods to account for the broadband noise. The methods are applied with various degrees of accuracy to a number of different systems, including propellers and helicopter rotors. Without making a full review of the subject, which is quite extensive, we consider a method described by Magliozzi *et al.* [63], where further technical details and references can be found. This method has the advantage of providing a sufficiently accurate estimate of the broadband noise due to the scattering of the turbulent boundary layer at the trailing edge.

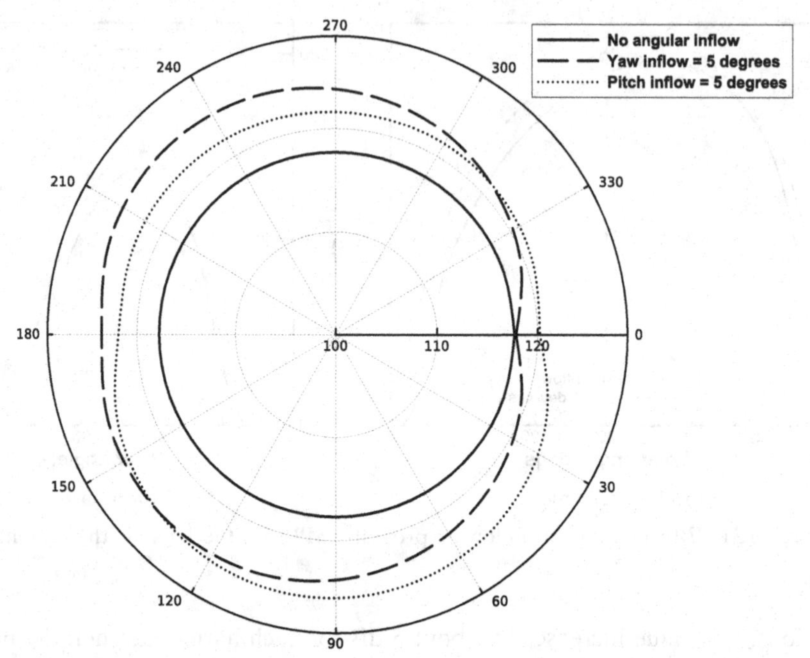


Figure 16.32. The effects of angular inflows on propeller $SPL_h$ for the first harmonic and various azimuthal directivity angles.

If $\omega$ is the angular frequency corresponding to the frequency $f$, then $\omega = 2\pi f$. This frequency is normalised with the boundary-layer thickness at the trailing edge, $\delta^*$, and the inflow velocity $U$:

$$\tilde{\omega} = \omega \left( \frac{\delta^*}{U} \right). \tag{16.119}$$

The one-third octave band SPL is then estimated from the following equation:

$$\mathrm{SPL} = C + 10 \log_{10} \left\{ 0.613 \left( \frac{\tilde{\omega}}{\tilde{\omega}_{max}} \right)^4 \left[ \left( \frac{\tilde{\omega}}{\tilde{\omega}_{max}} \right)^{3/2} + 0.5 \right]^{-4} \right\}, \tag{16.120}$$

where $\tilde{\omega}_{max}$ is the maximum value of $\tilde{\omega}$. It is reported that $\tilde{\omega}_{max} \simeq 0.1$. The log term in Equation 16.120 is always negative, with a maximum around $\tilde{\omega}/\tilde{\omega}_{max} \simeq 1$. The factor $C$ denotes an overall SPL and is given by

$$C = 10 \log_{10} \left( M^5 \frac{\delta^* b}{r^2} \mathcal{D} \right) + 141.3. \tag{16.121}$$

The remaining factors in Equation 16.121 include a lateral length scale $b$ (span or element width) and a directivity $\mathcal{D}$, which is calculated from

$$\mathcal{D} = \frac{2\cos^2(\theta/2)}{(1 - M\cos\theta)[1 - (M - M_c)\cos\theta]^2}. \tag{16.122}$$

Equation 16.122 is the only function of the corrected emission angle, the flight Mach number and the critical Mach number $M_c$; it further contains a Doppler correction. When the corrected emission angle is $\theta = \pi$, $\mathcal{D} = 0$, irrespective of Mach number. This causes Equation 16.121 to become singular; Equation 16.121 is singular also in static conditions ($M = 0$).

The displacement thickness $\delta^*$ required to normalise the angular frequency is calculated from semi-empirical formulas in turbulent flow; one such equation is

$$\frac{\delta^*}{c} = 0.047\,\mathrm{Re}_c^{-1/5}, \tag{16.123}$$

where $\mathrm{Re}_c$ is the Reynolds number based on the chord $c$. Equation 16.123 only applies to a flat plate; therefore, some additional corrections must be introduced to account for the effects of effective angle of attack $\alpha_e$, camber and thickness. From extensive aerodynamic analysis we find that a good correction is $\delta^*(\alpha) = \mathcal{F}\delta^*$. The correction factor is

$$\mathcal{F}(\alpha) = \frac{\exp(c_1\alpha_e + c_2)}{c_3}, \tag{16.124}$$

where $c_i$ are factors depending on the pressure and suction side; the parameter $c_3$ is a normalisation factor that forces $\mathcal{F} = 1$ at $\alpha_e = 0$.

This basic method is to be applied to the calculation of broadband noise from a propeller. To do so, we apply the usual blade-element theory. For a generic element, we are able to calculate the inflow conditions and the aerodynamic coefficients by following the methods presented in Chapter 6. In particular, we have the chord $c$, the element width $dy = b$, the inflow velocity $U$, the Reynolds number $\mathrm{Re}_c$ and the displacement thickness at the trailing edge. The critical Mach number is calculated as shown in § 4.2.3. If the propeller is sufficiently far from the receiver, the corrected emission angle $\theta$ and distance $r$ are virtually unaffected by the propeller's rotation. This assumption greatly simplifies the calculation. For a given frequency $f$ we need to perform an integration in the radial direction to account for the contribution of all of the blade elements. Then we need to account for all of the blades. Again, if the propeller is sufficiently far from the receiver, we assume that all of the blades contribute equally to the broadband noise.

A typical result of this numerical method is shown in Figure 16.33, which displays the noise spectrum for the Dowty R408 six-bladed propeller trimmed to provide a power $P = 1{,}328$ kW at a flight speed 115 KTAS at 200 m above sea level. The receiver is at a visual distance $r \sim 165$ m. There are only three dominant tonal contributions, but overall the broadband noise is very low.

## Summary

This chapter has reviewed some of the low-order acoustic methods that can be used for the estimation of the noise sources from a variety of aircraft systems. We have illustrated how a noise-prediction program can be configured and proceeded with splitting the noise sources between propulsive and non-propulsive sources. Among the propulsive sources there is the propeller and the APU, which is modelled in a fashion similar to a gas-turbine noise. A number of limitations are evident. First of all, the modelling is carried out on a narrow basis, with relatively few geometrical and operational parameters. Second, the models themselves rely on limited understanding of very complex phenomena. Nevertheless, the framework for noise prediction is a robust one; if more advanced submodels do become available, they can be plugged into the noise program to improve the predictions. All of the inaccuracies that may arise from any of the models presented in this chapter must be

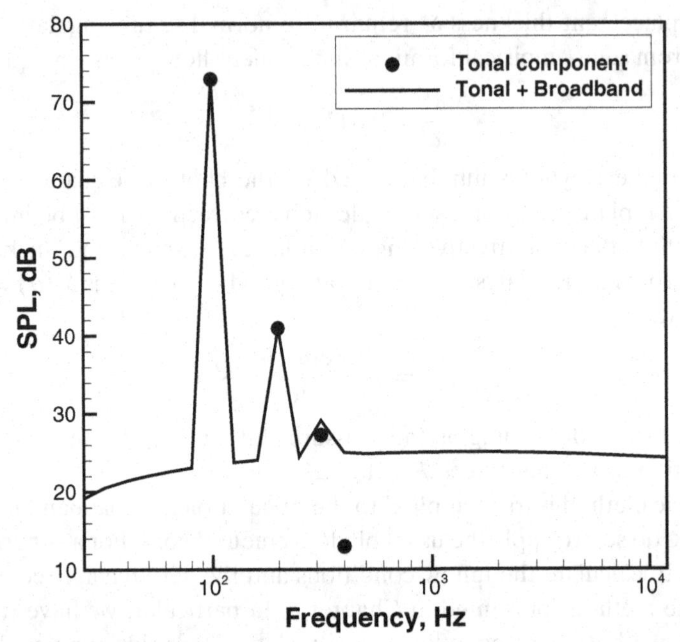


Figure 16.33. Noise spectrum for the R408 six-bladed propeller model trimmed at a specified power. Result uncorrected for atmospheric absorption.

viewed in the wider context of noise-propagation effects, which can substantially change the nature of noise, in terms of both levels and frequency content. These statements will be demonstrated in the next chapter.

## Bibliography

[1] Franssen EAM, van Wiechen CMAG, Nagelkerke NJD, and Lebret E. Aircraft noise around a large international airport and its impact on general health and medication use. *Occup. Environ. Med.*, 61(5):405–413, 2004.
[2] Pierce AD. *Acoustics: An Introduction to Its Physical Principles and Applications*. Acoust. Soc. America, 1989.
[3] Anon. Acoustical terminology. Technical Report S1.1-1994, American National Standards Institute, June 1994.
[4] Smith MTJ. *Aircraft Noise*. Cambridge University Press, 2004.
[5] Edge PM and Cawthorn JM. Selected methods for quantification of community exposure to aircraft noise. Technical Report TN-D-7977, NASA, 1976.
[6] Casalino D, Diozzi F, Sannino R, and Paonessa A. Aircraft noise reduction technologies: A bibliographic review. *Aero Sci. & Techn.*, 12(1):1–17, 2008.
[7] Heidmann MF. Interim prediction method for fan and compressor noise source. Technical Report TM X-71763, NASA, 1979.
[8] Hough JW and Weir DS. Aircraft noise prediction program (ANOPP) fan noise prediction for small engines. Technical Report CR-198300, NASA, 1996.
[9] ESDU. *Prediction of Noise Generated by Fans and Compressors in Turbojet and Turbofan Engines Combustor Noise from Gas Turbine Engines*. Data Item 05001. ESDU International, London, June 2003.
[10] Bielak GW, Premo JW, and Hersch AS. Advanced turbofan duct liner concepts. Technical Report CR-1999-209002, NASA, Feb. 1999.

[11] Ko SH. Sound attenuation of acoustically lined circular ducts in the presence of uniform flow and shear flow. *J. Sound & Vibration*, 22(22):193–210, 1972.
[12] Tam CWK, Kurbatskii KA, Ahuja KK, and Gaeta RJ. A numerical and experimental investigation of the dissipation mechanisms of resonant acoustic liners. *J. Sound & Vibration*, 245(3):545–557, 2001.
[13] Astley RJ, Sugimoto R, and Mustafi P. Computational aero-acoustics for fan duct propagation and radiation. Current status and application to turbofan liner optimisation. *J. Sound & Vibrations*, 330(16):3832–3845, 2011.
[14] ESDU. *The Acoustic Attenuation of Absorbent Linings in Cylindrical Flow Ducts*. Data Item 00012, Amend. B. ESDU International, London, June 2010.
[15] ESDU. *The Acoustic Attenuation of Absorbent Linings in Rectangular Flow Ducts with Application to Annular Flow Ducts*. Data Item 00024. ESDU International, London, Mar. 2011.
[16] McAlpine A and Fisher MJ. On the prediction of "buzz-saw" noise in aero engine inlet ducts. *J. Sound & Vibration*, 248(1):123–149, 2001.
[17] McAlpine A, Fisher MJ, and Tester BJ. "Buzz-saw" noise: A comparison of measurement with prediction. *J. Sound & Vibration*, 290(3-5):1202–1233, 2006.
[18] ESDU. *Prediction of Combustor Noise from Gas Turbine Engines*. Data Item 05001. ESDU International, London, Feb. 2005.
[19] Zorumski WE. Aircraft noise prediction program (ANOPP) theoretical manual. Technical Report TM-83199, Part 2, NASA, Feb. 1982.
[20] Matta RK, Sandusky GT, and Doyle VL. GE core engine noise investigation – Low emission engines. Technical Report FAA-RD-74, FAA, Feb. 1977.
[21] ESDU. Aircraft Noise Prediction. Technical Report Data Item 02020, ESDU International, Sept. 2009.
[22] van Zante D and Envia E. Simulation of Turbine Tone Noise Generation Using a Turbomachinery Aerodynamics Solver. Technical Report TM-2010-216230, NASA, Mar. 2010.
[23] Chien E, Ruiz M, Yu J, Morin B, Cicon D, Schweiger P, and Nark D. Comparison of predicted and measured attenuation of turbine noise from a static engine test. 13th AIAA/CEAS conference, AIAA paper 2007-3533, Rome, Italy, May 2007.
[24] ESDU. *An Introduction to Aircraft Noise*. Data Item 02020. ESDU International, London, 2002.
[25] Zorumski WE. Aircraft noise prediction program (ANOPP) theoretical manual. Technical Report TM 83199, Part 1, NASA, Feb. 1982.
[26] Lighthill MJ. On sound generated aerodynamically – Part I. *Proc. Royal Soc. London, Sect. A*, 211:564–587, 1952.
[27] Lighthill MJ. On sound generated aerodynamically – Part II. *Proc. Royal Soc. London, Sect. A*, 222:1–32, 1954.
[28] Balsa TF and Gliebe PR. Aerodynamics and noise of coaxial jets. *AIAA J.*, 15(11):1550–1558, Nov. 1977.
[29] Stone JR, Groesbeck DE, and Zola CL. Conventional profile coaxial jet noise prediction. *AIAA J.*, 21(3):336–342, Mar. 1983.
[30] Olsen W and Friedman R. Jet noise from coaxial nozzles over a wide range of geometric and flow parameters. In *AIAA Aerospace Meeting*, AIAA Paper 1974-0043, Washington, DC, Jan. 1974.
[31] ESDU. *Computer-based Estimation Procedure for Coaxial Jet Noise*. Data Item 01004. ESDU International, London, May 2001.
[32] Fisher MJ, Preston GA, and Bryce WD. A modelling of the noise from coaxial jets. Part 1: With unheated primary flow. *J. Sound & Vibration*, 209(3):385–403, Jan. 1998.
[33] Fisher MJ, Preston GA, and Mead WD. A modelling of the noise from coaxial jets. Part 2: With heated primary flow. *J. Sound & Vibration*, 209(3):405–417, Jan. 1998.

[34] ESDU. *Estimation of Subsonic Far-Field Jet-Mixing Noise from Single-Stream Circular Nozzles*. Data Item 89041. ESDU International, London, Feb. 1990.
[35] ESDU. *Computer-based Estimation Procedure for Single-Stream Jet Noise*. Data Item 98019. ESDU International, London, Nov. 1998.
[36] Stone JR, Krejsa EA, Clark BJ, and Berton JJ. Jet Noise Modeling for Suppressed and Unsuppressed Aircraft in Simulated Flight. Technical Report TM-2009-215524, NASA, Mar. 2009.
[37] Stone JR, Krejsa EA, and Clark BJ. Jet Noise Modeling for Supersonic Business Jet Application. Technical Report TM-2004-212984, NASA, Mar. 2004.
[38] Ahuja KK. Designing Clean Jet Noise Facilities and Making Accurate Jet Noise Measurements. Technical Report AIAA-2003-0706, AIAA, Jan. 2003.
[39] Bridges J and Brown CA. Parametric testing of chevrons on single flow hot jets. In *10th AIAA/CEAS Conference*, Manchester, UK, Sept. 2004.
[40] Rask O, Kastner J, and Gutmark E. Understanding how chevrons modify noise in a supersonic jet with flight effects. *AIAA J.*, 49(8):1569–1576, Aug. 2011.
[41] Gerhold CH. Analytical model of jet shielding. *AIAA J.*, 21(5):694–698, May 1983.
[42] Simonich JC, Amiet RK, and Schlinker RH. Jet shielding of jet noise. Technical Report CR-3966, NASA, 1986.
[43] ESDU. *Jet-by-Jet Shielding of Noise*. Data Item 88023. ESDU International, London, Mar. 1992.
[44] Tam C, Pastouchenko N, Mendoza J, and Brown D. Combustion noise of auxiliary power units. AIAA Paper 2005-2829, May 2005.
[45] Fink MR. Noise component method for airframe noise. *J. Aircraft*, 16(10):659–665, 1979.
[46] Zorumski WE. Aircraft noise prediction program theoretical manual, Part 1. Technical Report TM-83199, NASA, Feb. 1982.
[47] ESDU. *Airframe Noise Prediction*. Data Item 90023. ESDU International, London, June 2003.
[48] Lilley GM. The prediction of airframe noise and comparison with experiment. *J. Sound & Vibration*, 239(4):849–859, 2001.
[49] Lockard DP and Lilley GM. The airframe noise reduction challenge. Technical Report TM-213013, NASA, 2004.
[50] Ffowcs-Williams JE and Hall LH. Aerodynamic sound generation by turbulent flow in the vicinity of a scattered half plane. *J. Fluid Mech*, 40:657–670, 1970.
[51] Howe MS. A review of the theory of trailing edge noise. *J. Sound & Vibration*, 61(3):437–466, 1978.
[52] Guo YP. Empirical prediction of aircraft landing gear noise. Technical Report CR-2005-213780, NASA, 2005.
[53] Guo YP, Yamamoto KJ, and Stoker RW. Experimental study on aircraft landing gear noise. *J. Aircraft*, 43(2):306–317, Mar. 2006.
[54] Guo YP. A component-based model for aircraft landing gear noise prediction. *J. Sound & Vibration*, 312:801–820, 2008.
[55] Humphreys WM and Brooks TF. Noise spectra and directivity for a scale-model landing gear. In *13th AIAA/CEAS Aeroacoustics Conference*, AIAA 2007-3458, Rome, Italy, May 2007.
[56] Guo YP, Yamamoto KJ, and Stoker RW. Component-based empirical model for high-lift system noise prediction. *J. Aircraft*, 40(5):914–922, Sept. 2003.
[57] Stoker RW. Landing gear noise report. Technical report, NASA Report NAS1-97040, 1997.
[58] Dobrynski W and Buchholtz H. Full-scale noise testing on Airbus landing gears in the German-Dutch wind tunnel. AIAA paper 1997-1597, May 1999.

[59] Hanson DB and Parzych DJ. Theory for noise of propellers in angular inflow with parametric studies and experimental verification. Technical Report CR-4499, NASA, Mar. 1993.
[60] Hanson DB and Fink MR. The importance of quadrupole sources in prediction of transonic tip speed propeller noise. *Journal of Sound and Vibration*, 62(1):19–38, Jan. 1979.
[61] ESDU. *Prediction of Near-Field and Far-Field Harmonic Noise from Subsonic Propellers with Non-Axial Inflow*. Data Item 95029. ESDU International, 1996.
[62] ESDU. *Estimation of the Unsteady Lift Coefficient of Subsonic Propeller Blades in Non-Axial Flow*. Data Item 96027. ESDU International, London, 1996.
[63] Magliozzi B, Hanson DB, and Amiet RK. Propeller and propfan noise. In Hubbard HH, editor, Aeroacoustics of Flight Vehicles: Theory and Practice, RP 1258. NASA, Aug. 1991.

## Nomenclature for Chapter 16

| | | |
|---|---|---|
| $a$ | = | speed of sound |
| $a,b$ | = | parameters in Equation 16.31; Table 16.3 |
| $A$ | = | area |
| $\mathcal{A}$ | = | attenuation function |
| $A_r$ | = | area ratio in co-axial jet noise |
| $A, B$ | = | parameters in landing-gear model, Equation 16.107 |
| $b$ | = | wing span in Equation 16.86; spanwise reference length |
| $B$ | = | number of propeller blades |
| $c$ | = | exponent defined by the second Equation 16.18 |
| $c_1, c_2, c_3 \cdots$ | = | constant factors |
| $C$ | = | factor in broadband noise, Equation 16.121 |
| $C$ | = | correction of SPL for tonal noise in Equation 16.4; constant in Equation 16.25 |
| $C_D$ | = | drag coefficient |
| $C_S, C_T$ | = | parameters in UOL equations for small-scale, transitional regions |
| $C_L$ | = | lift coefficient |
| $d$ | = | wheel diameter; nozzle diameter; other diameter |
| $d_e$ | = | effective jet diameter |
| $d_f$ | = | Doppler factor |
| $d_{h,1,th}$ | = | hydraulic diameter of the inner-stream throat, Equation 16.62 |
| $d_{2,1,th}$ | = | outer physical throat diameter of the inner-stream nozzle |
| $\mathcal{D}$ | = | directivity function |
| $D$ | = | correction of SPL for duration, Equations 16.4 and 16.8 |
| $E$ | = | acoustic energy |
| $G$ | = | Green function |
| $\mathcal{F}$ | = | correction factor for displacement thickness, Equation 16.124 |
| $f$ | = | frequency |
| $f_1$ | = | frequency corresponding to equal downstream/upstream propagation |

$\hat{f}$ = corrected frequency in co-axial jet noise, Equation 16.45
$f_b$ = blade-passing frequency
$f_c$ = centre frequency in a one-third octave band
$F$ = noise-level difference; factor in wing-noise model, Equation 16.98
$F_d$ = fraction of energy radiated downstream, Equation 16.44
$F_m$ = complex function in shielding model, Equation 16.76
$F_p$ = force created by acoustic pressure
$F_u$ = fraction of energy radiated upstream, Equation 16.47
$\mathcal{F}$ = normalised spectrum, Equation 16.107
$F_d$ = downstream spectral-energy function, Equation 16.44
$F_i$ = sound-pressure correction in engine noise, $i = 1, \ldots, 5$
$F_u$ = upstream spectral-energy function, Equation 16.47
$F_1$ = a function of the tip Mach number
$F_2$ = function of the rotor-stator spacing
$F_3$ = function of the noise directivity
$F_4$ = function of the blade-passing frequency
$F_{\phi k\mu,\nu}$, $F_{xk\mu,\nu}$, $F_{rk\mu,\nu}$ = tangential, axial, radial loading in propeller noise
$h$ = flight altitude
$h_d$ = parameter in directivity function, Equation 16.108
$h^*$ = specific enthalpy, § 16.4.4
$h^*_r$ = reference specific enthalpy
$h^*_{r,k}$ = component-gas specific enthalpy
$\mathcal{H}$ = relative humidity
$H_m$ = $m$-th order Hankel function of the first kind
$i$ = imaginary unit
$I$ = noise intensity (dB)
$J_m$ = $m$-th order Bessel function of the first kind
$k_1, k_2, k_3$ = factors in airframe noise model, Equation 16.87
$k_m$ = wave number in harmonic Green function, $k_m = mBM_{tip}$, Equations 16.115 and 16.116
$k_p$ = sinusoidal period of oscillation for the pressure transform
$k, k_o$ = wave numbers
$K$ = factor in Equation 16.31, Table 16.3
$K_L, K_S, L_T$ = parameters in UOL equations for large-scale, small-scale, transitional regions
$l$ = reference length
$l_i$ = length of $i$–th landing-gear strut
$l_t$ = total length of struts in landing gear
$L$ = sound pressure level; turbulence length scale
$L_{AE}$ = sound exposure level
$L_{AeqT}$ = equivalent continuous sound level
$L_p$ = peak level of fundamental tone in fan noise, Equation 16.20
$L_1$ = length scale in jet-noise model

| | | |
|---|---|---|
| LSF | = | linear scale factor (ratio of full-scale to model-scale dimensions) |
| $m$ | = | mass or molecular mass |
| $m$ | = | jet-shape parameter, Equation 16.44; wave number in Equation 16.75; harmonic number in propeller noise |
| $m_T$ | = | total molecular mass of a gas |
| $\dot{m}$ | = | mass-flow rate |
| $M$ | = | Mach number |
| $M_c$ | = | critical Mach number |
| $M_{tip}$ | = | tip Mach number |
| $n$ | = | wave number ratio in ground reflection |
| $n$ | = | harmonic number in spectral function, Equation 16.29 |
| $n_k$ | = | number of computational points |
| $n_l, n_u$ | = | lowest and highest harmonic numbers, Equation 16.30 |
| $ns$ | = | number of sub-band partitions, Equation 16.81 |
| $N$ | = | rotational speed of turbine stage |
| $N_\nu$ | = | number of propeller elements in chordwise direction |
| $N_w$ | = | number of wheels |
| $p$ | = | atmospheric or acoustic pressure |
| $p_r$ | = | pressure ratio |
| $p_{ref}$ | = | reference acoustic pressure, $20 \cdot 10^{-6}$ Pa |
| $p'$ | = | acoustic pressure disturbance |
| $P$ | = | acoustic power |
| $\langle p^2 \rangle$ | = | mean square acoustic pressure |
| $p_L, p_T$ | = | acoustic pressure due to loading/thickness in propeller noise |
| $P_s, P_t$ | = | shielded and unshielded sound pressure level, Equation 16.81 and Equation 16.82 |
| $P_S$ | = | acoustic power resulting from jet shielding, Equation 16.68 |
| $PR$ | = | suppressed-to-unsuppressed wetted perimeter ratio |
| $q$ | = | semi-empirical parameter in landing-gear noise |
| $Q$ | = | image source strength |
| $r$ | = | distance source-to-receiver |
| $\boldsymbol{r}$ | = | vector distance source-to-receiver |
| $\boldsymbol{r}_n$ | = | projection of $\boldsymbol{r}$ onto plane normal to jet |
| $r_j$ | = | jet radius |
| $r_o$ | = | distance between jets; radial coordinate normalised with blade diameter (propeller noise) |
| $r'$ | = | distance between receiver and closest jet, Figure 16.21 |
| $r_1$ | = | minimum distance, Equation 16.82 |
| $\mathcal{R}_k^*$ | = | constituent gas dimensionless gas constant |
| $\mathrm{Re}_c$ | = | Reynolds number based on chord $c$ |
| $\mathcal{R}_u$ | = | universal gas constant |
| $\mathcal{R}$ | = | dry-air gas constant |
| $s$ | = | perimeter in Equation 16.104 |

$\bar{s}$ = average diameter of landing-gear struts
$S_o$ = amplitude radius
$S$ = reference area for airframe noise components, Equation 16.89
$\mathcal{S}$ = spectral function
$S_c$, $S_s$ = factors in Equations 16.113 and 16.114
St = Strouhal number
$t$ = time
$t_o$ = reference time
$\mathcal{T}$ = atmospheric temperature
$\mathcal{T}_r$ = reference temperature
$\mathcal{T}_{i,i}$ = total inlet-turbine temperature
$\mathcal{T}_{s,j}$ = static exit-turbine temperature
$\mathcal{T}_o$ = water triple-point temperature
$T$ = thrust in jet-noise model, Equation 16.38
$T_1$, $T_2$ = factors in jet-shielding noise, Equations 16.78a and 16.78b
$u_o$ = isotropic turbulence level, Equation 16.90
$U$ = inflow velocity
$U_T^*$ = dimensionless tip speed, Equation 16.32
UOL = overall sound level, uncorrected for refraction, dB
UOL' = normalised UOL
$V$ = true air speed
$V_r$ = velocity ratio in co-axial jet noise
$\mathbf{V}$ = vector velocity with components $u$, $v$, $w$
$V_{0\mu,\nu}$, $V_{c\mu,\nu}$, $V_{s\mu,\nu}$ = thickness-source terms in propeller noise
$w$ = wheel width
$w_b$ = sub-band width in jet-shielding noise
$W$ = weight
$W_{ref}$ = reference acoustic power, $10^{-12}$ W in § 16.4.3
$W_{ref}$ = reference weight in landing-gear model, Equation 16.106
$W_F$ = complex error function
$x_1$ = reference position for co-axial jet noise
$x_1$, $x_2$, $x_3$ = factors in jet-shielding model, Equations 16.78c, 16.78d, 16.78e
$x_k$ = constituent-gas mass fraction
$X_s$ = axial source location relative to outer stream nozzle exit
$y$ = co-axial jet parameter defined by Equation 16.51
$y_m$ = length scale of turbulent eddies in Equation 16.96
$z$ = height from the ground

## Greek Symbols

$\alpha_e$ = effective inflow angle
$\beta$ = radiation efficiency factor, Equation 16.102
$\beta$ = polar angle between receiver and plane normal to jet axis, Equation 16.67

| | | |
|---|---|---|
| $\delta$ | = | boundary-layer thickness; relative pressure, $p/p_o$ |
| $\delta^*$ | = | boundary-layer displacement thickness |
| $\epsilon$ | = | factor of the wave number, Equation 16.75 and Equation 16.79 |
| $\eta$ | = | landing-gear complexity function, Equation 16.106 |
| $\eta$ | = | dimensionless frequency, Equation 16.28 |
| $\eta_c$ | = | thermodynamic efficiency of a compressor stage |
| $\theta, \theta'$ | = | polar-emission angle, polar angle in spherical coordinates (propeller noise), see Figure 16.28 |
| $\gamma$ | = | ratio between specific heats, $\gamma = C_p/C_v$ |
| $\gamma_t$ | = | wheel-track-alignment angle, Equation 16.106 |
| $\epsilon_m$ | = | factor in potential, Equation 16.75 |
| $\lambda$ | = | wavelength |
| $\Lambda$ | = | sweep angle of a lifting surface |
| $\mu$ | = | dynamic viscosity |
| $\mu$ | = | semi-empirical parameter in landing-gear noise |
| $\nu$ | = | kinematic viscosity in Equation 16.89 |
| $\mu, \nu$ | = | index for radial/chordwise elements in propeller noise, Equations 16.115 and 16.116 |
| $\Pi^*$ | = | acoustic power parameter, Equation 16.31 |
| $\rho$ | = | air density |
| $\tau$ | = | temperature ratio, last of Equation 16.51 |
| $\sigma$ | = | semi-empirical parameter in landing-gear-noise model |
| $\sigma_r$ | = | gas-density ratio in co-axial jet |
| $\varsigma$ | = | power coefficient in normalised spectra, Equation 16.107 |
| $\varphi$ | = | bank angle, Equation 16.71 |
| $\phi_o$ | = | shielding velocity potential due to a unit source |
| $\phi$ | = | azimuth angle in flight path, Figure 16.24 |
| $\phi, \phi'$ | = | azimuthal angle in spherical coordinate system (propeller noise) |
| $\Phi$ | = | velocity potential |
| $\psi$ | = | angle between source-to-receiver and plane normal to jet, Figure 16.21, Equation 16.71 |
| $\psi_o, \psi_1, \psi_2$ | = | parameters in co-axial jet noise, defined by Equation 16.43 |
| $\Psi$ | = | azimuthal angle on plane normal to jets |
| $\omega$ | = | variable density index, Equation 16.52 |
| $\omega$ | = | angular frequency |

## Subscripts/Superscripts

| | | |
|---|---|---|
| $[\cdot]_a$ | = | referred to airframe or atmosphere |
| $[\cdot]_{cor}$ | = | corrected for source location |
| $[\cdot]_e$ | = | referred to engines |
| $[\cdot]_e$ | = | corrected or retarded value |
| $[\cdot]_h$ | = | harmonics in propeller noise |
| $[\cdot]_H$ | = | high frequency |
| $[\cdot]_i$ | = | inlet condition |
| $[\cdot]_{ISA}$ | = | International Standard Atmosphere |
| $[\cdot]_j$ | = | jet quantity; exit condition in turbine-noise model |
| $[\cdot]_L$ | = | low frequency; large-scale mixing noise in § 16.4.8 |

| | | |
|---|---|---|
| $[\cdot]_M$ | = | medium frequency |
| $[\cdot]_m$ | = | mixed jet |
| $[\cdot]_{max}$ | = | maximum value |
| $[\cdot]_{main}$ | = | relative to main landing gear |
| $[\cdot]_n$ | = | projection of vector to a plane |
| $[\cdot]_{nose}$ | = | relative to nose landing gear |
| $[\cdot]_p$ | = | primary jet flow |
| $[\cdot]_P$ | = | inner-stream plug-separation noise in § 16.4.8 |
| $[\cdot]_o$ | = | standard atmospheric conditions; other reference conditions |
| $[\cdot]_r$ | = | primary/secondary ratio in co-axial jet noise |
| $[\cdot]_{ref}$ | = | reference quantity |
| $[\cdot]_s$ | = | secondary-jet flow; static or source |
| $[\cdot]_S$ | = | small-scale mixing scale, in § 16.4.8 |
| $[\cdot]_t$ | = | total or stagnation condition in co-axial jet noise, § 16.4.8 |
| $[\cdot]_{te}$ | = | trailing edge |
| $[\cdot]_T$ | = | transitional separation noise in § 16.4.8 |
| $[\cdot]'$ | = | derivative in jet-shielding model |
| $\overline{[\cdot]}$ | = | average value |
| $\hat{[\cdot]}$ | = | normalised quantity in jet model |
| $\langle[\cdot]\rangle$ | = | mean-square value |
| $[\cdot]_\infty$ | = | far-field conditions |
| $[\cdot]_1$ | = | inner stream (i.e., the core jet flow) |
| $[\cdot]_2$ | = | outer stream (i.e., the by-pass flow or the ambient) |
| $[\cdot]_1, [\cdot]_2$ | = | jet number in jet-shielding model (1 is closest jet) |
| $\tilde{[\cdot]}$ | = | normalised value |

# 17 Aircraft Noise: Propagation

## Overview*

Noise propagation involves all of the physical events that take place between the noise source and the receiver. Due to the relatively large distance between source and receiver (from a minimum of ~100 m to several kilometers), atmospheric absorption is particularly important – no less important than the determination of the noise sources themselves. If propagation takes place over an unbounded medium, the use of the atmospheric absorption model would be sufficient to estimate the effect at the receiver. However, there are cases where the topography of the terrain and the short distance between the airplane and the ground cause effects such as reflection and scattering to become important. We begin by considering briefly the airframe noise shielding (§ 17.1); we continue with the standard absorption model in the atmosphere (§ 17.2). We describe ground-reflection problems in § 17.3. We finally discuss the combination of wind shear and temperature gradients in the propagation of aircraft noise over long distances (§ 17.4).

**KEY CONCEPTS:** Airframe Noise Shielding, Wing Scattering, Ground Effect, Atmospheric Absorption, Wind Effects, Turbulence Effects.

## 17.1 Airframe Noise Shielding

A number of interference effects take place around the airframe due to the reciprocal position between noise sources and solid surfaces. The effect is frequency-dependent. The investigation of new blended airframes, which actually benefit from the noise shielding[1–3], has increased the interest in this acoustic effect.

One method is called the *equivalent source method*[4;5], which uses a finite number of sources to satisfy the boundary condition on the airframe to simulate the acoustic scattering. Based on this method, a fast-scattering model was developed to calculate the sound reflected and scattered by the airframe, as shown by Dunn and Tinetti[6]. Although some researchers pointed out that a small number of sources may cause a

* Dr. Zulfaa Mohamed-Kassim contributed to this chapter.

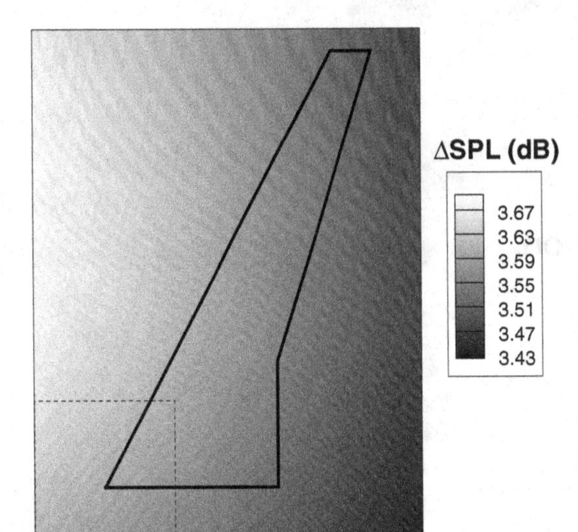


Figure 17.1. Simulated scattering noise on the Airbus A320 wing; level flight, one span below the wing.

poor accuracy, a series of investigations have shown that this accuracy is acceptable when the number of sources is carefully chosen[7].

In the equivalent scattering method, we replace a vibrating body with a system of sources (equivalent sources) in the interior of the body. Hence, the sound field is expressed by a superposition of an incident sound field (caused by the real acoustic source) and a scattering sound field (caused by the equivalent source).

An example of calculation is shown in Figure 17.1, which displays the change in SPL due to engine noise in the presence of wing scattering; the field plot refers to a plane parallel to the wing, placed one span below.

Although the equivalent source is a good choice for airframe design and noise diagnosis at specific frequencies, it still needs considerable time to compute. The most serious problem is that the method needs the numbers of equivalent sources and points on the boundary to be larger than a certain value to keep the accuracy.

Methods based on ray-tracing theory[8] have been developed to solve the sound propagation in a non-uniform atmosphere. With the ray-tracing method the sound wave front emitted from a source is divided into many small pieces. The reflection from wing and ground can be calculated if the wing and ground are treated as boundaries. The smaller the sound piece and the time interval, the more accurate the result. The number of time steps depends on the source-to-receiver distance.

To overcome the computational difficulty and satisfy the requirements of a fast comprehensive noise model, we propose a simplified method, which assumes that the reflection from the wing obeys the rule of reflection from a flat surface. This assumption is acceptable when 1) the lower wing surface has a small curvature, and 2) the receiver has a distance larger than the curve's length scale. Based on these assumptions, only one or two reflections need to be calculated for each source. We call this method the *reflection-only method*. The computational procedure is the following:

1. Divide the wing surface into quadrilateral elements.
2. Calculate the normal vector $\boldsymbol{n}$ of each element.
3. Compare these normal vectors with the source and observer positions; check if they satisfy the reflection condition: incident angle = reflected angle.

4. If the reflection condition is satisfied, a reflected wave will be calculated and superposed to the direct acoustic wave at the observer position. Otherwise, the observer will be marked as non-reflection.

The scattering from the fuselage is modelled only when the aircraft is close to the ground. In this instance, we use the model of a sufficiently long cylinder over an impedance boundary[9]. However, this model can only be applied when the receiver point is sufficiently far from the axis of the airframe; otherwise it would require a fictitious extension of the fuselage in order to account for this interference. The basic idea of this model is to express the total sound field as a sum of four components: the incident field, the reflected wave, and the scattered fields from the cylinder and its images. Theoretical details can be found in the references cited.

## 17.2 Atmospheric Absorption of Noise

Sound propagating through the atmosphere is absorbed by an amount depending on frequency, distance and atmospheric conditions. Atmospheric absorption essentially means that the noise heard by the receiver is lower than the noise due to distance propagation effects. The key parameters in the atmosphere are the changes in its temperature, pressure and relative humidity. With so many parameters at play, the result tends to be complicated. In aircraft certification, the knowledge of the atmospheric attenuation is used to derive the noise at the source from measurements on the ground. In the present case, where we attempt to predict the noise at the ground, we need the atmospheric attenuation as a correction of the noise propagating through an ideal medium in thermodynamic equilibrium (isothermal and isobaric).

If we abstract for a moment from atmospheric winds, then we define the absorption function

$$\mathcal{A} = \mathcal{F}(h, r, d\mathcal{T}, \mathcal{H}, f),$$

where $h$ is the altitude of the source with respect to the receiver, $r$ is the distance from the receiver, $d\mathcal{T}$ is the change in temperature around the standard value, $\mathcal{H}$ is the relative humidity and $f$ is the frequency of the acoustic emission. The attenuation function is generally given in dB per meter. For computational reasons, it is convenient to have an explicit relationship between the attenuation rate and its physical parameters. Therefore, the attenuation at a given frequency $f$ over a distance $r$ due to an aircraft flying at an altitude $h$ is found from the integral

$$\mathcal{A}(f, h, \mathcal{H}, r) = \int_o^r \mathcal{F}(d\mathcal{T}, \mathcal{H}, r)\, dr. \tag{17.1}$$

Values of the absorption have been found experimentally over several years (for example, Zuckerwar and Meredith[10]). The American Institute of Physics (AIP) published a standard approved by ANSI[11]. Another recognised standard is the ISO 9613-1[12], which is substantially equivalent to the ANSI standard. ESDU[13] has published a model that includes the effects of ground refraction, atmospheric turbulence and wind shear. Morfey and Howell[14] review the theory of acoustic wave propagation from aircraft.

From the ANSI model, the sound attenuation rate at a frequency $f$ is

$$\mathcal{A}[dB/m] = 8.686 f^2 \left\{ \left[ 1.84 \cdot 10^{-11} \left(\frac{p_r}{p_a}\right) \left(\frac{T}{T_r}\right)^{1/2} \right] + \right.$$

$$+ \left[ 0.01275 \exp\left(\frac{-2.2391 \cdot 10^3}{T}\right) \frac{f_{ro}}{f_{ro}^2 + f^2} \right] \left(\frac{T}{T_r}\right)^{-5/2} + \tag{17.2}$$

$$\left. + \left[ 0.10680 \exp\left(\frac{-3.3520 \cdot 10^3}{T}\right) \frac{f_{rn}}{f_{rn}^2 + f^2} \right] \left(\frac{T}{T_r}\right)^{-5/2} \right\}.$$

In this equation $p_a$ is the atmospheric pressure at the reference point, $p_r$ is the reference pressure (101,325 Pa), and $T_r$ is the reference temperature (293.15 K, or 20 °C); the terms $f_{ro}$ and $f_{rn}$ are given by

$$f_{ro} = \frac{p_a}{p_r} \left[ 24 + 4.04 \cdot 10^4 h_c \left(\frac{0.020 + h_c}{0.391 + h_c}\right) \right], \tag{17.3}$$

$$f_{rn} = \left(\frac{p_a}{p_r}\right) \left(\frac{T}{T_r}\right)^{-1/2} \left\{ 9 + 280 h_c \exp\left[ -4.170 \left( \left(\frac{T}{T_r}\right)^{-1/3} - 1 \right) \right] \right\}, \tag{17.4}$$

where $h_c$ is the molar concentration of water vapour in the atmosphere. If the relative humidity $\mathcal{H}$ is given in percent, the molar concentration is given by

$$h_c = \mathcal{H} \left(\frac{p_{sat}}{p_r}\right) \left(\frac{p_r}{p_a}\right), \tag{17.5}$$

and the saturation pressure $p_{sat}$ is

$$p_{sat} = p_r 10^V. \tag{17.6}$$

In Equation 17.6, the exponent $V$ is calculated from

$$V = a_1 \left(1 - \frac{T_{o1}}{T}\right) - b_1 \log_{10}\left(\frac{T}{T_{o1}}\right) + \tag{17.7}$$

$$c_1 \left[ 1 - 10^{c_2(T/T_{o1} - 1)} \right] + d_1 \left[ -1 + 10^{d_2(1 - T_{o1}/T)} \right] - e_1.$$

In Equation 17.7, $T_{o1}$ is the triple-point isotherm temperature (273.15 K). The numerical coefficients for this equation are given in Table 17.1. At a given position in space $(h, T, p)$, it is possible to calculate the saturation pressure $p_{sat}$ from Equation 17.6 (through the power $V$ given by Equation 17.7). The saturation pressure is used to calculate the molar concentration of water at the same point from Equation 17.5. The latter parameter is used by the terms $f_{ro}$ and $f_{rn}$ and, finally, the attenuation $\mathcal{A}$ from Equation 17.2. The cumulative effect of attenuation over a long distance requires the integration of Equation 17.1. The calculation must be repeated over the full spectrum of noise frequencies. In practice, it is convenient to use the same reference frequencies as the dB-A scale. The main difficulty consists in accounting for inhomogeneous humidity across different layers of the atmosphere. This effect is sometimes evident to the receiver when the aircraft crosses areas of different cloud thickness.

Table 17.1. *Numerical coefficients for Equation 17.7*

| $a_1$ | $b_1$ | $c_1$ | $d_1$ | $e_1$ | $c_2$ | $d_2$ |
|---|---|---|---|---|---|---|
| 10.79586 | 5.02809 | $1.50474 \cdot 10^{-4}$ | $0.42873 \cdot 10^{-3}$ | 2.21960 | 8.29692 | 4.76955 |

FAR Chapter 36, "Noise Standards: Aircraft Type and Airworthiness Certification", requires that the noise be corrected for a fixed humidity of 70%, standard conditions on the ground, no winds and for the spectrum from 50 Hz to 10 kHz.

The attenuation function calculated with this model is greatly dependent on the frequency. In fact, a 10-kHz frequency is absorbed $10^4$ times more than a 10-Hz frequency. Most of the attenuation effects occur above 1 kHz. Consequently, high-frequency noise is less likely to be heard by the receiver at a long distance, whilst low-frequency noise is persistent and is affected more by distance than atmospheric conditions. The atmosphere acts like a low-pass filter. However, unless the distance of the aircraft is considerable, the SPL corrected for atmospheric absorption is generally lower than the accuracy of most aircraft-noise predictions.

Sample calculations of the atmospheric absorption are shown in Figure 17.2 for variable altitude and relative humidity, as indicated. The data show that the relative humidity is a stronger factor than the flight altitude.

The values of the absorption $\mathcal{A}$, integrated over a long distance, may lead to a large reduction in noise level, as shown in Figure 17.3. The reduction in noise level is evident after a few hundred metres, with the high frequencies being cut first.

Because in practice the aircraft noise propagates through layers of the atmosphere that are not isobars, not isothermal and not even isohumidity, we need to carry out the integration of Equation 17.2. Several methods, of variable accuracy, have been proposed. However, due to the uncertainty on the relative humidity, we consider the propagation through piecewise segments of constant atmospheric conditions.

There are various other aspects of sound attenuation that we may need to consider, such as the geometric divergence, the effects of winds and the ground

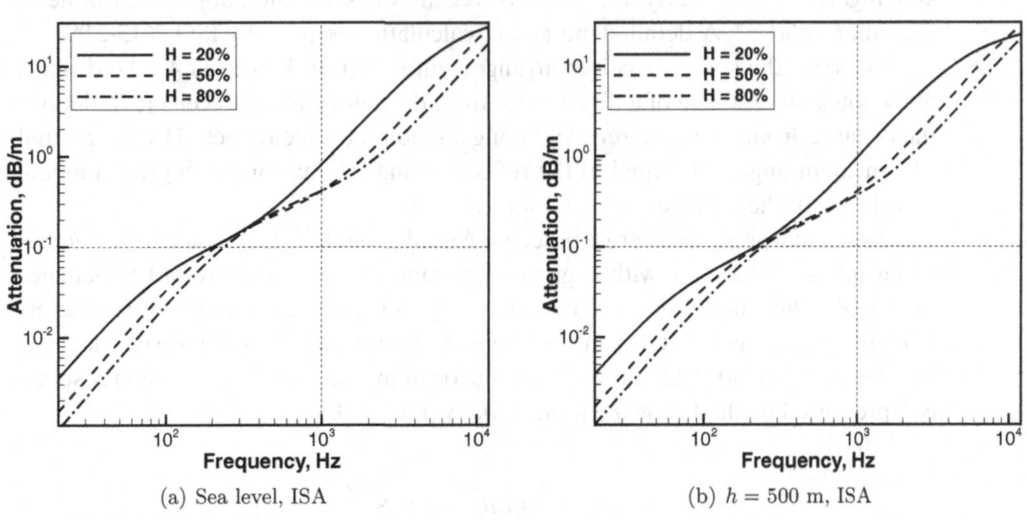


(a) Sea level, ISA (b) $h = 500$ m, ISA

Figure 17.2. Atmospheric absorption model, according to ISO 9613-1.

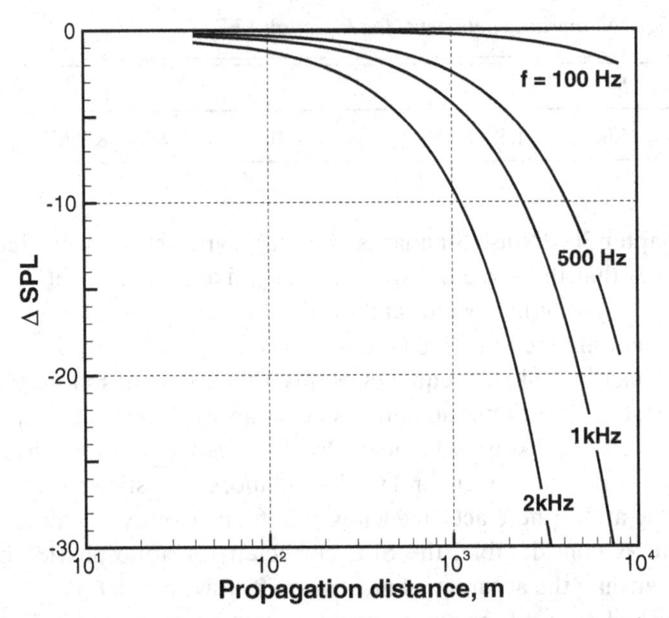


Figure 17.3. Atmospheric absorption versus distance at selected frequencies, $\mathcal{H} = 70\%$.

reflection. The latter case is important, when noise reflected from the ground propagates between the source and the receiver.

## 17.3 Ground Reflection

The analysis carried out so far assumes that the acoustic waves propagate through an unbounded medium – an assumption that is reasonable in free flight. However, when the aircraft is relatively close to the ground, the acoustic-wave propagation is *impeded* by the presence of the ground. Typical effects are refraction and absorption of acoustic waves.

There exist theoretical models that allow the prediction of such effects. Theoretical studies on this problem are well established starting with the work of Rudnick [15] and Ingard [16] in the early 1950s. More recent work on the subject was done by Attenborough [17;18]. A detailed method of calculation is given by ISO 9613-2 [19].

Consider the source-receiver arrangement shown in Figure 17.4. The distance $R_1$ denotes the distance of a direct wave from the source to the receiver; $R_2$ denotes the distance from source to receiver along an indirect wave reflected by the ground. The incident angle $\phi$ is equal to the reflection angle. This condition gives a unique travel path of the reflected acoustic wave.

The model presented here is based on Attenborough [20]. The ground is considered as an infinite boundary with a given impedance. If $Z$ is the ground impedance, $H = 1/Z$ is the admittance. The impedance is a function of a number of parameters, namely the flow resistivity $\sigma_e$ and the inverse effective depth of the surface layer, $\alpha$. The resultant sound field at the observer's location due to a single-frequency source is represented by the following total velocity potential:

$$\Phi_{tot} = \frac{e^{ik_1 R_1}}{4\pi R_1} + \frac{e^{ik_1 R_2}}{4\pi R_2} Q, \qquad (17.8)$$

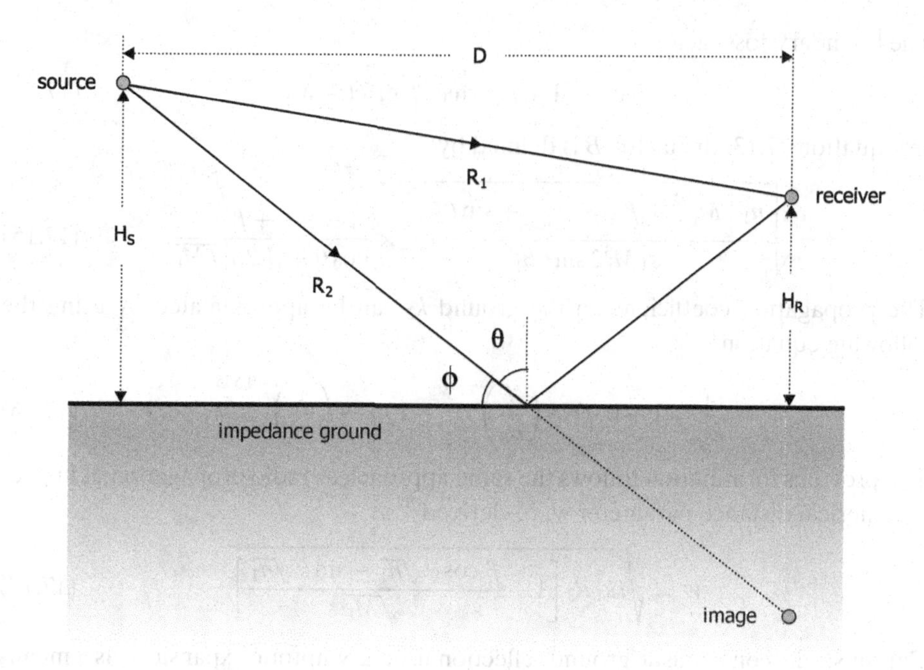


Figure 17.4. Ground reflection of acoustic waves.

where $k_1$ denotes the acoustic propagation coefficient in the air. The second term in Equation 17.8 includes the relative image source strength $Q$, which is a proportionality factor of the reflected wave being superposed with the direct wave. This factor can be interpreted as the spherical-wave-reflection coefficient. From Equation 17.8, the resulting pressure field is

$$p_{tot} = -\rho \left( \frac{\partial \Phi_{tot}}{\partial t} \right). \tag{17.9}$$

Assume that $\beta$ denotes the normalised specific admittance, equivalent to the inverse of the specific impedance $Z$; $M$ is the density ratio, $n$ is the propagation coefficient (or wave number) ratio, and $w$ is a *numerical distance*. If subscripts "1" and "2" denote the upper medium (air) and the lower medium (ground), respectively, these quantities are defined as

$$M = \frac{\rho_1}{\rho_2}, \qquad n = \frac{k_1}{k_2}, \qquad \beta = nM. \tag{17.10}$$

Furthermore, to simplify the algebra, set the following equivalences:

$$n_1 = 1 - \frac{1}{n^2}, \qquad s_1 = \left[ 1 - \left( \frac{\sin\theta}{n} \right)^2 \right], \qquad \beta_1 = \sqrt{1 - \beta^2}, \qquad M_1 = 1 - M^2. \tag{17.11}$$

With this nomenclature, the reflection coefficient is

$$R_p = \frac{\cos\theta - \beta(1 - s_1)^{1/2}}{\cos\theta + \beta(1 - s_1)^{1/2}}. \tag{17.12}$$

The image source strength is

$$Q = [R_p + B(1 - R_p)]\ F(w). \tag{17.13}$$

The boundary-loss factor is

$$F(w) = 1 + i\sqrt{\pi}we^{-w^2}\mathrm{erfc}(-iw). \tag{17.14}$$

In Equation 17.13, the factor $B$ is defined by

$$B = \sqrt{\frac{n_1\left[M_1^{1/2} + \beta\cos\theta s_1^{1/2} + \sin\theta\beta_1\right]}{s_1 M_1^3 2\sin\theta\beta_1}} \times \frac{\cos\theta + \beta s_1}{\cos\theta + \beta\sqrt{n_1/M_1}}. \tag{17.15}$$

The propagation coefficient on the ground $k_2$ can be approximated by using the following equation[18]:

$$\frac{k_1}{k_2} = 1 + 0.0978\left(\frac{f}{\sigma_e}\right)^{-0.7} + i0.189\left(\frac{f}{\sigma_e}\right)^{-0.595}. \tag{17.16}$$

The previous formulation follows the same approach as radio propagation; it utilises a numerical distance parameter $w$, re-derived[20] as

$$w = \sqrt{ik_1 R_2\left[1 + \frac{\beta\cos\theta\sqrt{n_1} - \sin\theta\sqrt{n_1}}{\sqrt{M_1}}\right]}. \tag{17.17}$$

Earlier studies on acoustic ground reflection used asymptotic expansions as a means to rapidly evaluate $F(w)$. Given that $w' = w^2$, two different series are used for different values of $w'$ divided at $|w'| = 10$:

$$F(w') = \begin{cases} 1 + i\sqrt{\pi w'}\,e^{-w'} - 2w'\left[1 + \sum_n \frac{w'^n}{(2n+1)(n-1)!}\right]e^{-w'} & \text{if } |w'| < 10 \\ -\left[\frac{1}{2w'} + \frac{1\times 3}{(2w')^2} + \frac{1\times 3\times 5}{(2w')^3} + \ldots\right] & \text{if } |w'| \geq 10 \end{cases} \tag{17.18}$$

This approach has been shown to produce errors at small values of $w'$ and is unstable particularly with $|w'|$ close to $10^6$. To overcome this problem, part of the boundary-loss factor is evaluated instead as the complex-error function, $W_F(w)$, defined as

$$W_F(w) = e^{-w^2}\mathrm{erfc}(-iw). \tag{17.19}$$

The complex-error function, also known as the *Faddeeva function*, can be efficiently computed to within machine precision using an algorithm developed by Poppe and Wijers[21]. The effect of air turbulence is incorporated into the model using a coherence parameter $\Lambda$. It takes into account the effect of turbulence on the coherence between the direct wave and the reflected wave:

$$\Lambda = e^{-0.25\beta_\Lambda P}, \tag{17.20}$$

where

$$\beta_\Lambda = \begin{cases} 0.5 & \text{if} \sqrt{D/k_1} > \ell \\ 1.0 & \text{if} \sqrt{D/k_1} < \ell \end{cases} \tag{17.21}$$

and

$$P = <\mu^2> k_1^2 D\ell\sqrt{\pi}. \tag{17.22}$$

The factor $< \mu^2 >$ is the fluctuating index of refraction, and $\ell$ is the turbulent length scale[22].

Following Chessel[23], the difference between the sound pressure levels for a single wave frequency measured with the presence of the ground relative to that obtained without the ground (i.e., the free-field SPL) is given by

$$\Delta \text{SPL} = 10 \log_{10} \left[ 1 + \left( \frac{R_1}{R_2} |Q| \right)^2 + 2 \left( \frac{R_1}{R_2} \right) |Q| \Lambda \cos \left( \frac{2R_2 - R_1}{k_1} + \alpha \right) \right], \quad (17.23)$$

with $Q = |Q| e^{i\alpha}$. A positive $\Delta$SPL value indicates that the sound due to the direct wave is enhanced due to the presence of the ground, whereas a negative value means that the actual SPL is lower than that of the free-field.

**SIMPLIFIED MODEL.** Most other first-order models typically assume a number of simplifications to arrive at more compact solutions that are relatively easier to arrive at in the derivation and are attractive for fast numerical computation. Typical simplifying assumptions are $|n| > 1$, $|M| << 1$, $|\beta| < 1$ (hard-boundary case), and $\phi \simeq 0$ (small grazing angles). In addition to these, the ground impedance can be further simplified by treating the ground as a locally reacting boundary, where the wave propagation in the ground is independent of the incident angle that leads to $|n| >> 1$. Following these simplifications, one would arrive at the reduced definitions for some of the parameters above, with $B \simeq 1$:

$$R_p = \frac{\cos\theta - \beta}{\cos\theta + \beta}, \quad (17.24)$$

$$w = \sqrt{\frac{i k_1 k_2}{2}} (\cos\theta + \beta). \quad (17.25)$$

This solution predicts the ground-reflection correction, $\Delta$SPL, for a single-frequency sound wave. In practice, most measurements are recorded in the 1/3 octave band levels, which integrate the single-frequency sound levels within each band. To obtain an octave band reading from the $\Delta$SPL calculated herein, each band is subdivided into smaller bands where the $\Delta$SPL value for each sub-band centre frequency is calculated. These sub-band $\Delta$SPL are then multiplied with their sub-band widths and integrated within each 1/3 octave band. The integrated values are then normalised with the corresponding 1/3 octave band-widths.

### 17.3.1 Ground Properties

The acoustic properties of the ground are represented by the complex impedance. Earlier research in this field modeled the ground impedance with a single parameter, the specific flow resistance per unit thickness, based on the work of Delaney and Bazley[24]. The model relies on curve fitting of various experimental measurements of actual ground impedances. Attenborough[25] shows that the single-parameter model is unreliable, particularly outside the range of the fitted data; he then introduced a more accurate model using two parameters: the effective flow resistivity, $\sigma_e$, and the rate of exponential decrease of porosity with depth, $\alpha$. The specific normalised

Table 17.2. *Typical values for flow resistivity and inverse effective depth*

| Ground | $\sigma_e$ [kNs/m$^4$] | $\alpha$ [m$^{-1}$] |
|---|---|---|
| Snow | 5 – 20 | 0 |
| Short grass | 30 – 50 | 20 – 45 |
| Sandy soil | 60 – 100 | 0 |
| Dirt road | 40 | 10 |
| Wet compact soil | 4,000 | 0 |
| Tarmac | 4,500 | 0 |

impedance $Z$ based on the two-parameter model is given by the following equation:

$$Z = \left(\frac{\sigma_e}{\gamma\pi\rho f}\right)^{1/2} + i\left[\left(\frac{\sigma_e}{\gamma\pi\rho f}\right)^{1/2} + \frac{a\alpha}{4\gamma\pi f}\right]. \qquad (17.26)$$

Practical values for $\sigma_e$ and $\alpha$ are given in Table 17.2.

Comparisons between the two approaches with experimental data are shown in Figure 17.5. Evidence from these comparisons shows that the two-parameter model is better at accurately predicting the corrected sound levels. A weakness of the single-parameter model is that it cannot accurately predict the primary-dip amplitude; it is, however, able to match the primary-dip frequency. The predictions are better with the two-parameter model given that the correct values of both $\sigma_e$ and $\alpha$ are used.

### 17.3.2 Turbulence Effects

Most noise calculations rely on the assumption of still atmosphere (no winds, no turbulence). This is in fact far from true and, to be realistic, even a simple turbulence model would be of considerable benefit. There are several models, but the simplest

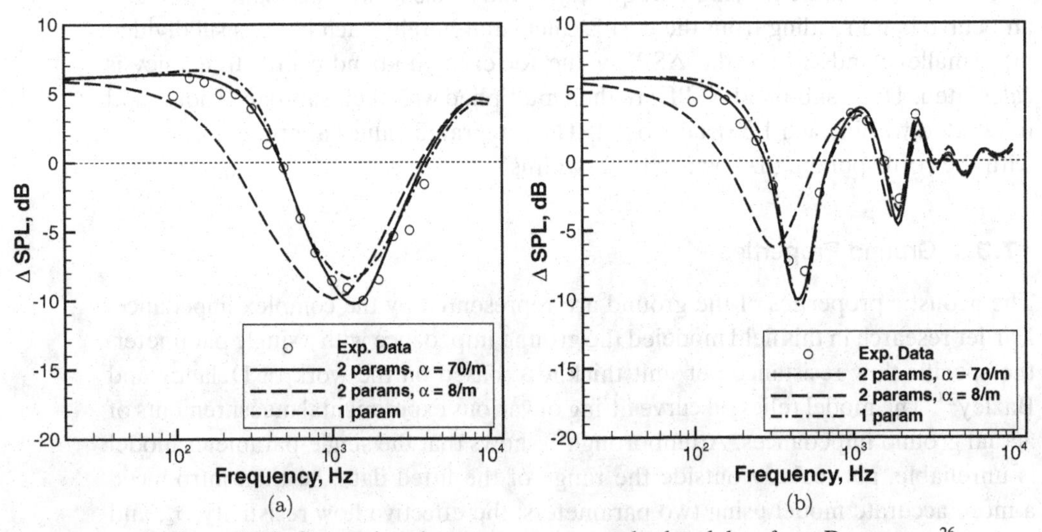


Figure 17.5. Comparison between present method and data from Rasmussen[26].

one is based on the evaluation of the *fluctuating index of refraction* $<\mu^2>$:

$$<\mu^2>=\frac{\sigma_v^2}{a^2}+\frac{\sigma_t^2}{4T^2}, \tag{17.27}$$

where $\sigma_v$ is the root mean square (RMS) of the wind-velocity fluctuation and $\sigma_t$ is the RMS of the temperature fluctuation. The effect of turbulence on the acoustic wave refraction is more pronounced at frequencies below 5kHz.

## 17.4 Wind and Temperature Gradient Effects

The effects of atmospheric winds on the propagation of aircraft noise are essential. For example, receiver positions downwind experience more annoyance than receiver positions upwind. The differences can be several dB. There is a combination between wind and ground effects, which cannot be separated. The first step in a practical analysis is to assume an average wind-shear profile, in absence of turbulence and local topographical effects, which is

$$\frac{V_w(z)}{V_{wo}}=\left(\frac{z}{z_o}\right)^{\zeta}, \tag{17.28}$$

where $V_w(z)$ denotes the horizontal wind speed at height $z$, $V_{wo}$ is the wind speed at height $z_o$, and $\zeta$ is the wind-shear exponent; this exponent depends on the type of terrain, with $0.1<\alpha<0.25$. In practice, $z_o$ will be an altitude above the ground at which wind speeds are available. A typical reference altitude is $z_o=10$ m.

The degree of the frequency-dependent attenuation or enhancement depends on a number of factors, that is, horizontal propagation distance, wind speed and direction ($V_w$ and $\theta_w$), temperature variation with height ($dT/dh$), and source and receiver heights ($H_s$ and $H_r$). The presence of temperature variations produces a speed-of-sound gradient $\gamma_c$. An increase in the speed of sound from source to receiver would refract the sound downward and generally enhance the sound received. This is due to a positive sound-speed gradient resulting from downwind propagation or temperature inversion (i.e., an increase of temperature with height). An upwind propagation or a normal decrease of temperature with height (i.e., negative $\gamma_c$) results in upward-refraction that produces a shadow region downstream; sound levels are mostly attenuated in this case. Figure 17.6 illustrates these sound-refraction effects on sound rays emanating from a single source.

A theory on the combination of wind, temperature gradient and ground reflection has been developed by Rasmussen[27], who expressed the acoustic pressure field with a Hankel transform

$$p=-2\int_o^{\infty} J_o(\kappa D)P(z,\kappa)\,\kappa d\kappa, \tag{17.29}$$

where $J_o$ is the Bessel function of the first kind of zeroth order, and $P(z,\kappa)$ is the transform of $p$, $D$ is the horizontal distance, $r$ is the propagation distance (see also Figure 17.4). In terms of pressure ratio relative to the free stream, the acoustic pressure is

$$\frac{p}{p_o}=-2r\int_o^{\infty} J_o(\kappa D)P(z,\kappa)\,\kappa d\kappa. \tag{17.30}$$

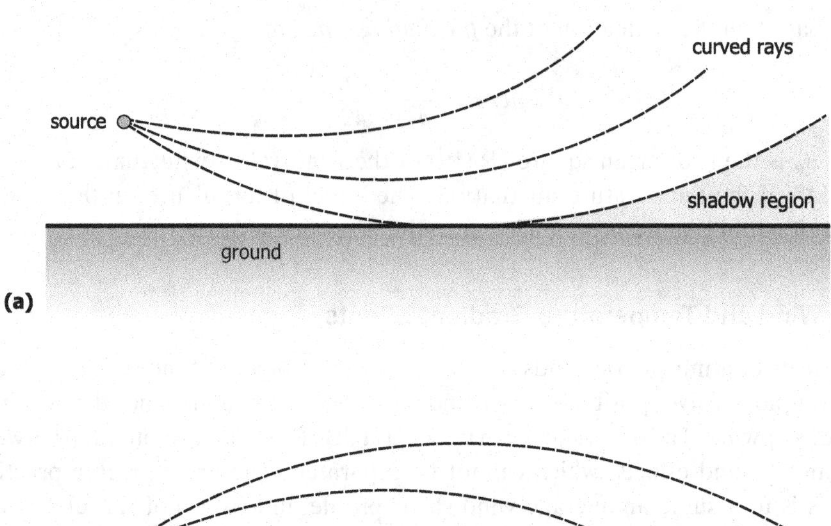


Figure 17.6. (a) Upward refraction due to upwind sound propagation or temperature inversion ($\gamma_c < 0$), and (b) downward refraction due to downwind propagation or normal temperature variation ($\gamma_c > 0$).

Following derivations by Pierce[28] and Rasmussen, with the assumptions that the sound speed varies linearly with the height and that the speed-of-sound gradient $\gamma_c z << 1$, two expressions for the pressure transform are obtained, depending on the sound-speed gradient. The vertical-density variation is assumed to be negligible in the derivation; for $\gamma_c > 0$ and $\gamma_c < 0$, we have respectively:

$$P_+(z,\kappa) = -v(\tau + y_L)\left[w(\tau + y_S) - \frac{w'(\tau) + qw(\tau)}{v'(\tau) + qv'(\tau)} w(\tau - y_S)\right]\mathcal{L} \tag{17.31}$$

$$P_+(z,\kappa) = -w(\tau - y_L) - \left[w(\tau - y_S) - \frac{w'(\tau) - qw(\tau)}{v'(\tau) - qv'(\tau)} w(\tau - y_S)\right]\mathcal{L}, \tag{17.32}$$

where

$$\mathcal{L} = \left[2|\gamma_c|k_o^2\right]^{-1/3}, \qquad \tau = \left[k^2 - k_o^2\right]\mathcal{L}, \tag{17.33}$$

$$y_S = \frac{\min(H_s, H_r)}{\mathcal{L}} \qquad y_L = \frac{\max(H_s, H_r)}{\mathcal{L}} \qquad q = ik_o\beta\mathcal{L}, \tag{17.34}$$

and $k_o$ is the wave number of the sound propagation on the ground; i.e., $k_o = 2\pi f/a$. The pressure-transform solutions utilise the Airy-Fock functions[29], $v(\tau)$, $w(\tau)$ and

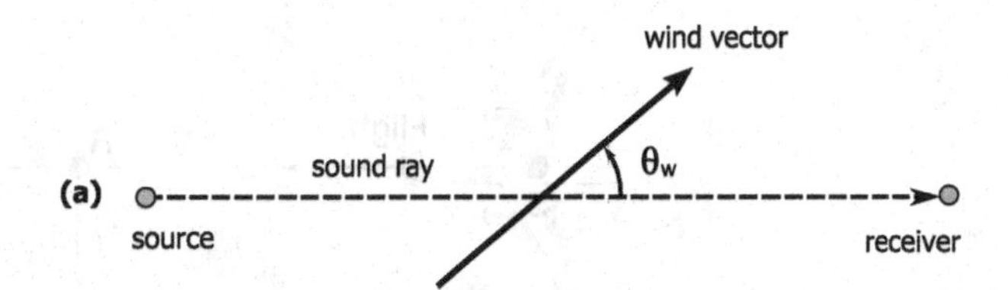


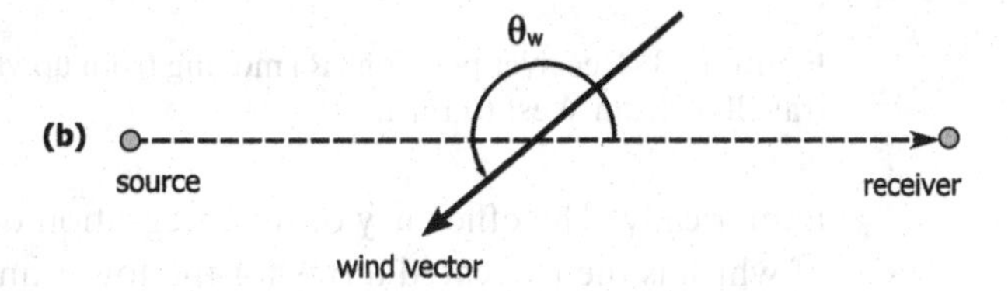


Figure 17.7. Downwind and upwind sound propagation.

$Ai(\tau)$. These functions are expressed by

$$v(\tau) = \sqrt{\pi}\,\mathrm{Ai}(\tau), \tag{17.35}$$

$$w(\tau) = 2\sqrt{\pi}\exp\left(\frac{i\pi}{6}\right)\mathrm{Ai}\left(\tau\exp\left[i\frac{2\pi}{3}\right]\right). \tag{17.36}$$

For values of $|\tau| > 3$, the Airy-Fock functions can be substituted by some approximations[2], restricted to small speed-of-sound gradients – a condition easily met in outdoor noise propagation. The gradient itself is calculated from a set of temperature, wind speed and wind direction measurements on the ground at heights $z_1$ and $z_2$, respectively.

Figure 17.7 illustrates the wind vectors relative to propagating rays at the measuring height; the wind speed on the ground is zero. The linear sound-speed gradient is

$$\gamma_c = \frac{a(z_2) - a_o}{a_o z_2}, \tag{17.37}$$

where $a_o$ is the speed of sound at ground temperature and

$$a(z_2) = a(z_2) + V_w(z_2)\sin\theta_w. \tag{17.38}$$

For a given wind direction, during a typical trajectory an aircraft can change its position from downwind to upwind (or vice versa) with respect to a fixed receiver on the ground. How this may take place is best explained with the aid of Figure 17.8. The aircraft is travelling from left to right (or West to East) with a head wind. In the first instance, the aircraft is to the West of the receiver; thus, the receiver is placed upwind. When the aircraft has moved to the East of the receiver, the receiver itself is placed downwind.

### 17.4.1 Numerical Solution

To obtain the difference in sound pressure levels ΔSPL relative to the free-field, a modified expression for pressure ratio in Equation 17.30 can be integrated

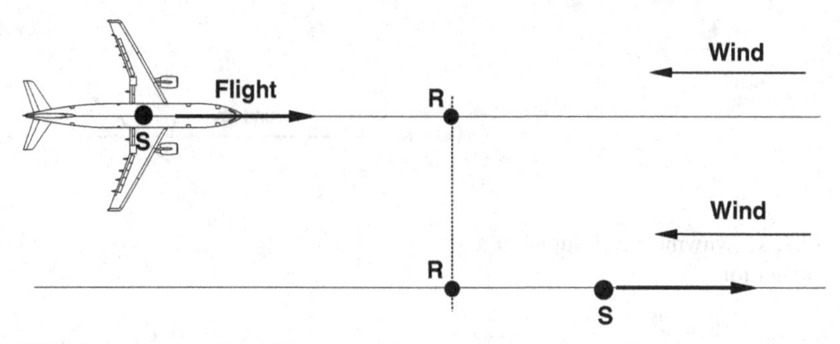


Figure 17.8. Receiver position (R) moving from upwind to downwind at fly-over. Aircraft (S) travelling from West to East.

numerically. The efficiency of the integration depends on a number of factors, one of which is the truncated terms for the lower and upper limits of the integral, $a$ and $b$, such that

$$\Delta\text{SPL} = 20\log_{10}\left(\frac{p}{p_o}\right) = 20\log_{10}\left[-2r\int_a^b J_o(\kappa D)P(z,\kappa)\,\kappa d\kappa\right]. \tag{17.39}$$

Rasmussen provided a recommendation on the limits of integration as well as the discretisation size $d\kappa$ for use on his datasets whose frequencies range from 100 Hz to 2 kHz only. The rules break down for wider and higher frequency ranges. It is found that a new set of rules can be applied to optimise the integration parameters by analysing the behaviours of the Bessel function $J_o$ and the pressure transform $P(z, \kappa)$; $J_o$ governs the oscillations of the functions to be integrated for the upwind propagation. In the downwind propagation, the oscillations are dominated by the pressure transform $P(z, \kappa)$. The discretisation size is optimised by setting

$$d\kappa = k_p/3D \tag{17.40}$$

with

$$\kappa_p = \begin{cases} 1 & \text{if } \gamma_c < 0 \\ \sqrt{k_o^2 + (6\pi|\gamma_c|k_o^2)^{2/3}} - k_o & \text{if } \gamma_c > 0 \end{cases}. \tag{17.41}$$

The lower limit of integration is $a = 0$ for all values of $\gamma_c$; the upper limit is

$$b = \sqrt{k_o^2 + 1/P_{ave}^2} \tag{17.42}$$

with $P_{ave} = 0.5$ for $\gamma_c > 0$ and $P_{ave} = 0.04$ for $\gamma_c < 0$. The number of computational points is set to $n_k = b/d\kappa$. Following this approach, the computational cost increases with the horizontal propagation distance and the frequency of propagation, particularly for the downward-refracting case where the discretisation size is much smaller as imposed by $P(z, \kappa)$. The computational efficiency can be improved significantly by reducing the terms to be computed through algebraic simplification of the Airy-Fock functions and various approximations.

This solution describes the sound correction for a single frequency. To obtain a 1/3 octave band reading of the ΔSPL, each band is subdivided into smaller bands where the ΔSPL value for each sub-band centre frequency is calculated. These sub-band ΔSPLs are then multiplied with their sub-band widths and integrated

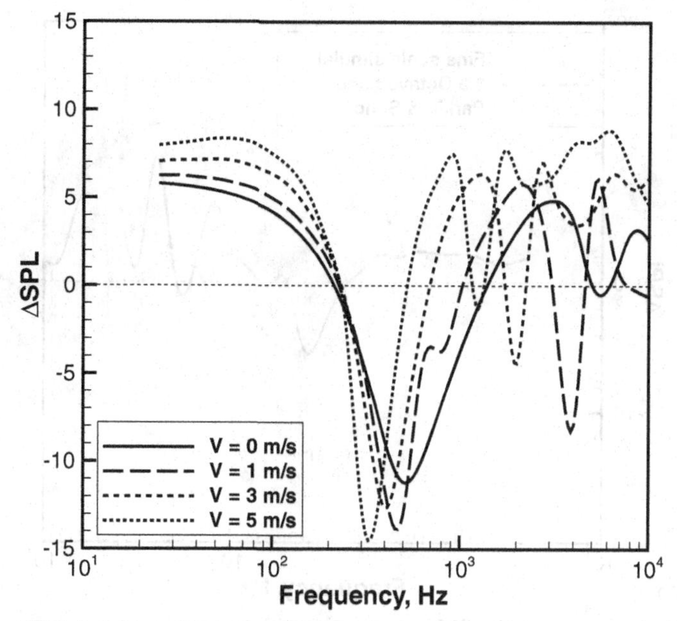


Figure 17.9. ΔSPL for downwind propagation over grassy field with $\sigma_e = 200 \cdot 10^3$ Nsm$^{-4}$. Input data are equivalent to those in Rasmussen[27]; wind speeds as indicated.

within each 1/3 octave band. The integrated values are then normalised with the corresponding 1/3 octave bandwidths.

**MODEL VERIFICATION.** In Figure 17.9, the ΔSPL received at a location 120 m downstream of a source for various wind speeds are plotted for a wide frequency range of 10 Hz $< f <$ 10 kHz. The input data are identical to those used by Rasmussen[27]. The corresponding results below match his numerical solutions, which are presented within a narrower frequency band of 100 Hz to 2 kHz, with the exception for slightly different primary-dip magnitudes. These small variations can be attributed to the different approaches of setting the integration parameters, as described in § 17.4.1.

Comparisons with experimental data in Figures 17.10 and 17.11 show that the trends of the simulated attenuation curves are similar to those measured. However, their magnitudes vary considerably to within 7 dB for the downward-refracting case and within 8 dB for the upward-refracting case. This discrepancy can be largely attributed to two factors. First, there is a possible mismatch between the simulated ground, wind, and atmospheric parameters and the actual parameters at the time the real data were recorded. For example, only the temperature-gradient regime (i.e., lapse, neutral or inversion) and the actual temperatures were reported by Parkin and Scholes[30] but not the actual temperatures at various heights. In addition, the data were recorded in various seasons of the year. The ground parameters, which may vary across the field, were not reported either. Reported uncertainties of the measurements are within the range of the differences observed in the simulated values. The uncertainty on the data varies from ±0.6 dB to ±10.3 dB, depending on the atmospheric conditions, measurement sites, time of the year and receiver distance.

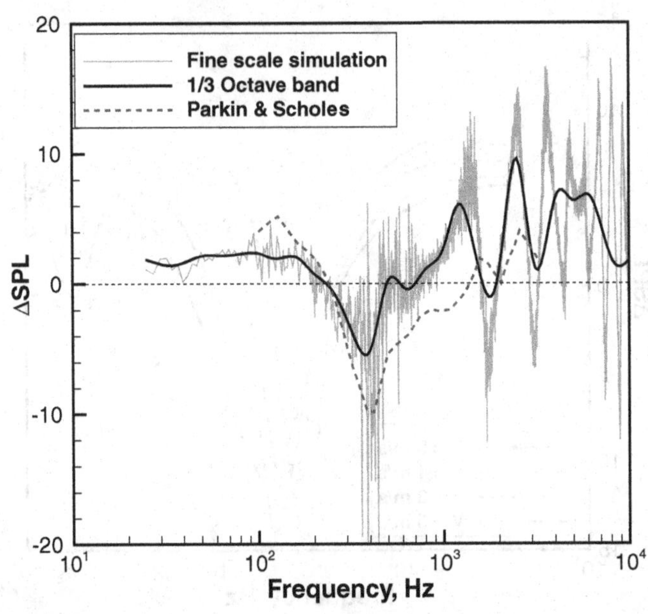


Figure 17.10. Comparison of the SPL attenuation with the experimental data of Parkin and Scholes[30]. The main inputs are $D = 109.73$ m (360 feet), $H_s = 1.83$, $H_r = 1.52$ m, $V_w = 4.57$ m/s downstream, and $\sigma_e = 200 \cdot 10^3$ Nsm$^{-4}$.

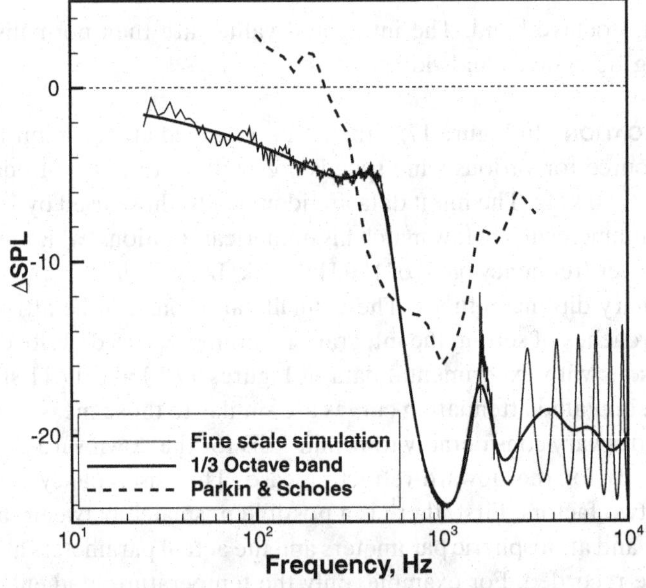


Figure 17.11. Comparison of the SPL attenuation with the experimental data of Parkin and Scholes[30]. The main inputs are identical to those in Figure 17.10; wind speed $V_w = -4.57$ m/s.

## Summary

This chapter has introduced some numerical models for the prediction of important effects in sound propagation. Because the distance source-to-receiver can be of the order of a few kilometers (or n-miles), the noise recorded at the receiver can have a completely different acoustic content from the noise sources. The most prominent effect is the atmospheric absorption in still and unbounded atmosphere with a constant level of humidity. The standard model indicates that high frequencies are cut more rapidly than low frequencies. A distant airplane can be identified by the acoustic pressures at low frequencies. Yet, this model is not sufficient to take into account all of the effects, which include ground reflection (or absorption), temperature differentials and most critically the wind. A model that combines wind and temperature gradients has been implemented in a numerical form and extended to the prediction of far-field noise. This model is computationally more expensive than the noise sources themselves and requires serious computer hardware even with the approximation of wind shear not dependent on the topography.

## Bibliography

[1] Rawlins AD. The engine over wing noise problem. *J. Sound & Vibration*, 50(4):553–569, 1977.
[2] Gerhold CH. Investigation of acoustical shielding by a wedge-shaped airframe. *J. Sound & Vibration*, 294(1):49–63, 2006.
[3] Agrawal A and Dowling A. The calculation of acoustic shielding of engine noise by the silent aircraft airframe. In *11th AIAA/CEAS Aeroacoustics Conf.*, Monterey, CA, May 2005.
[4] Ochmann M. The source simulation technique for acoustic radiation problems. *Acustica*, 81:512–527, 1995.
[5] Dunn MH and Tinetti AF. Aeroacoustic scattering via the equivalent source method. In *10th AIAA/CEAS Aeroacoustics Conference*, AIAA Paper 2004-2937, May 2004.
[6] Dunn MH and Tinetti AF. Application of fast multipole methods to the NASAx fast scattering code. In *14th AIAA/CEAS Aeroacoustics Conference*, AIAA Paper 2008-2875, Vancouver, BC, May 2008.
[7] Pinho MEV. On the use of the equivalent source method for nearfield acoustic holography. *ABCM Symposium Series in Mechatronics*, pages 590–599, 2004.
[8] Schroeder H, Gabillet Y, Daigle GA, and L'Experance A. Application of the Gaussian beam approach to sound propagation in the atmosphere: Theory and experiments. *J. Acoust. Soc. Am.*, 93(6):3105–3116, 1993.
[9] Lui WK and Li KM. The scattering of sound by a long cylinder above an impedance boundary. *J. Acoust. Soc. Am.*, 127(2):664–674, 2010.
[10] Zuckerwar AJ and Meredith RW. Low-frequency sound absorption measurements in air. Technical Report R-1128, NASA, Nov. 1984.
[11] Anon. Method for the calculation of the absorption of sound by the atmosphere. Technical Report S1.26-1978, American National Standards Institute, June 1978.
[12] Anon. Attenuation of sound during propagation outdoors – Part I: Calculation of the absorption of sound by the atmosphere. Technical Report ISO-9613-1, International Standards Organisation, Geneve, CH, 1993.
[13] ESDU. *Prediction of Sound Attenuation in a Refracting Turbulent Atmosphere with a Fast Field Program*. Data item 04011. ESDU International, London, May 2004.
[14] Morfey CL and Howell GP. Nonlinear propagation of aircraft noise in the atmosphere. *AIAA J.*, 19(8):986–992, Aug. 1981.
[15] Rudnick I. The propagation of an acoustic wave along a boundary. *J. Acoust. Soc. Am.*, 19:348–356, 1947.
[16] Ingard U. On the reflecton of a spherical sound wave from an infinite plane. *J. Acoust. Soc. Am.*, 23:329–335, 1951.
[17] Attenborough K. Review of ground effects on outdoor sound propagation from continuous broadband sources. *Applied Acoustics*, 24:289–319, 1988.

[18] Attenborough K. Sound propagation close to the ground. *Ann. Rev. Fluid Mech.*, 34:51–82, Jan. 2002.
[19] Anon. Attenuation of sound during propagation outdoors – Part II: General method of calculation. Technical Report ISO-9613-2, International Standards Organisation, Geneve, CH, 1996.
[20] Attenborough K. Propagation of sound above a porous half-space. *J. Acoust. Soc. Am.*, 68:1493, 1980.
[21] Poppe GPM and Wijers CMJ. More efficient computation of the complex error function. *ACM Trans Math. Software*, 16, 1990.
[22] Daigle GA, Piercy JE, and Embleton TFW. Line-of-sight propagation through atmospheric turbulence near the ground. *J. Acoust. Soc. Am.*, 74:1505–1513, 1983.
[23] Chessel CI. Propagation of noise along a finite impedance boundary. *J. Acoust. Soc. Am.*, 62:825–834, 1977.
[24] Delaney M and Bazley E. Acoustical properties of fibrous absorbent materials. *Appl. Acoustics*, 3:105–116, 1970.
[25] Attenborough K. Ground parameter information for propagation modeling. *J. Acoust. Soc. Am.*, 92(1):418–427, Jan. 1992.
[26] Rasmussen KB. Sound propagation over grass covered ground. *J. Sound & Vibration*, 78(2):247–255, 1981.
[27] Rasmussen KB. Outdoor sound propagation under the influence of wind and temperature gradients. *J. Sound & Vibration*, 104:321–335, 1986.
[28] Pierce AD. *Acoustics: An Introduction to Its Physical Principles and Applications*. Acoust. Soc. America, 1989.
[29] Hazewinkiel M. *Encyclopaedia of Mathematics*, chapter Airy Functions, page 65. Kluwer Academic, 1995.
[30] Parkin PH and Scholes WE. The horizontal propagation of sound from a jet engine close to the ground, at Hatfield. *J. Sound & Vibration*, 2, 1965.

## Nomenclature for Chapter 17

Symbols listed on Table 17.1 are not repeated here.

| | | |
|---|---|---|
| $a$ | = | speed of sound; lower limit of integration in Equation 17.39 |
| $\mathcal{A}$ | = | atmospheric absorption |
| $Ai(\tau)$ | = | Airy function |
| $b$ | = | upper limit of integration, Equation 17.42 |
| $B$ | = | parameter in ground-reflection model, Equation 17.15 |
| $D$ | = | ground distance between source and receiver in ground reflection |
| $f$ | = | frequency |
| $f_{rn}$ | = | factor given by Equation 17.4 |
| $f_{ro}$ | = | factor given by Equation 17.3 |
| $\mathcal{F}$ | = | absorption function |
| $F(w)$ | = | boundary-loss factor, Equation 17.18 |
| $k$ | = | wave number |
| $k_1, k_2$ | = | acoustic-propagation coefficient of air/ground |
| $h$ | = | flight altitude |
| $h_c$ | = | molar concentration |
| $\mathcal{H}$ | = | relative humidity |
| $H$ | = | ground admittance, $H = 1/Z$ |
| $H_r, H_s$ | = | height of receiver/source above ground |

| | | |
|---|---|---|
| $i$ | = | imaginary unit |
| $J_o$ | = | Bessel function of first kind, zero-th order |
| $k$ | = | wave number |
| $k_o$ | = | wave number of the sound propagation on the ground |
| $\ell$ | = | turbulence length scale |
| $\mathcal{L}$ | = | parameter defined in the first Equation 17.33 |
| $M$ | = | ratio between air densities, $M = \rho_1/\rho_2$ |
| $M_1$ | = | parameter defined in Equation 17.11 |
| $n$ | = | propagation coefficient |
| $n_1$ | = | parameter defined in Equation 17.11 |
| $n_k$ | = | number of computational points |
| $p$ | = | pressure; acoustic pressure |
| $p_a$ | = | atmospheric pressure |
| $p_r$ | = | reference pressure |
| $p_{sat}$ | = | saturation pressure |
| $p_{tot}$ | = | total acoustic pressure |
| $P$ | = | acoustic power |
| $P_{ave}$ | = | average values of the pressure transform for convergence |
| $P(z, \kappa)$ | = | pressure transform in wind-propagation effects |
| $q$ | = | function of the wave number $k_o$, Equation 17.34 |
| $Q$ | = | velocity potential source strength |
| $r$ | = | distance |
| $\mathcal{R}$ | = | gas constant |
| $R_p$ | = | reflection coefficient, Equation 17.12 |
| $R_1, R_2$ | = | distances in ground-reflection model, Figure 17.4 |
| $s_1$ | = | parameter defined in Equation 17.11 |
| $t$ | = | time |
| $\mathcal{T}$ | = | temperature |
| $\mathcal{T}_{01}$ | = | triple-point temperature Equation 17.7 |
| $V$ | = | exponent in Equation 17.6, given by Equation 17.7 |
| $V_w$ | = | wind speed |
| $V_{wo}$ | = | wind speed at altitude $z = z_o$ |
| $w$ | = | numerical distance parameter, Equation 17.25 |
| $w(\tau)$ | = | Airy-Fock function |
| $w'$ | = | square of the numerical distance, $w' = w^2$ |
| $W_F$ | = | complex-error function |
| $x$ | = | sampling point on surface boundary |
| $y_L, y_S$ | = | functions of receiver/source height, Equation 17.34 |
| $z$ | = | vertical coordinate |
| $z_o$ | = | reference altitude in wind-shear model, Equation 17.28 |
| $Z$ | = | ground impedance |

## Greek Symbols

| | | |
|---|---|---|
| $\alpha$ | = | inverse effective depth, Equation 17.26 |
| $\beta$ | = | specific normalised admittance of the ground |

$\beta_1$ = parameter defined in Equation 17.11
$\beta_\Lambda$ = function of coherence parameter $\Lambda$, Equation 17.21
$\theta$ = polar-emission angle
$\theta$ = angle between incident wave and normal to ground, $\theta = 1 - \phi$
$\zeta$ = wind-shear exponent, Equation 17.28
$\gamma$ = ratio between specific heats, $\gamma = C_p/C_v$
$\gamma_c$ = speed-of-sound gradient
$\kappa$ = variable of integration, Equation 17.29
$\kappa_p$ = parameter defined by Equation 17.41
$d\kappa$ = discretisation size
$\Lambda$ = coherence parameter, Equation 17.20
$\mu$ = semi-empirical parameter in landing-gear noise
$\langle\mu^2\rangle$ = fluctuating index of refraction, Equation 17.27
$\nu(\tau)$ = Airy-Fock function
$\rho$ = air density
$\phi$ = angle between indirect acoustic wave and the ground, $\phi = 1 - \theta$
$\Phi$ = total velocity potential, Equation 17.8
$\sigma_e$ = effective flow resistivity
$\sigma_t$ = root mean square of temperature fluctuation
$\sigma_v$ = root mean square of wind-velocity fluctuation
$\tau$ = argument of Airy-Fock functions, $\tau = (k^2 - k_o^2)\mathcal{L}$
$\chi$ = angle between wind vector and source-receiver line
$\omega$ = angular frequency

## Subscripts/Superscripts

$[.]_r$ = receiver
$[\cdot]_s$ = source
$[\cdot]_{sat}$ = saturation
$[\cdot]_{tot}$ = total
$[\cdot]_w$ = wind
$[\cdot]_o$ = reference or ground conditions

# 18 Aircraft Noise: Flight Trajectories

## Overview

In Chapter 16 we discussed various methods that can be used for the rapid prediction of aircraft-noise sources. In Chapter 17 we described the effects of atmospheric propagation, interference and scattering. In this chapter we deal with aircraft-noise trajectories. In particular, we discuss noise certification (§ 18.1) and noise-abatement procedures (§ 18.2). The implementation of the various sub-models requires a flight-mechanics integration (§ 18.3) and a data-handling structure, to process the large matrices produced by several noise sources. A noise sensitivity analysis is presented (§ 18.4) to show the relative importance of the various noise sources to explain the most useful strategies for noise reduction. We discuss two case studies: the ICAO/FAR noise trajectories of a jet-powered aircraft (§ 18.5) and the ICAO/FAR noise trajectories of a turboprop aircraft (§ 18.6). We also present a number of noise applications, such as steep descent and wind effects (§ 18.7). To complete our discussion of the noise modelling, we provide an example of verification with flight data (§ 18.8). The last item of investigation is the noise footprint around the airfield (§ 18.9).

**KEY CONCEPTS:** Aircraft Noise Certification, Noise-Abatement Procedures, Flight-Mechanics Integration, Noise Sensitivity, Noise Trajectories, Wind Effects, Footprints, Sonic Boom.

## 18.1 Aircraft Noise Certification

Unlike a Beethoven symphony, aircraft noise is not welcome music to the ear. The second movement of the Ninth Symphony does not start with a landing-gear deployment. Communities that live near large airports know this fact rather well. In fact, they have been making noise on their own by voicing their concern.

Aircraft noise has been increasing over time, but the really loud noise started with the jet-powered aircraft in the 1950s. Their presence prompted noise limitations at Heathrow Airport in London as early as 1959. London Gatwick followed suit in 1968, and the restrictions expanded worldwide, as shown in Figure 18.1. Today commercial aircraft operations include limits on night flights and restrictions on

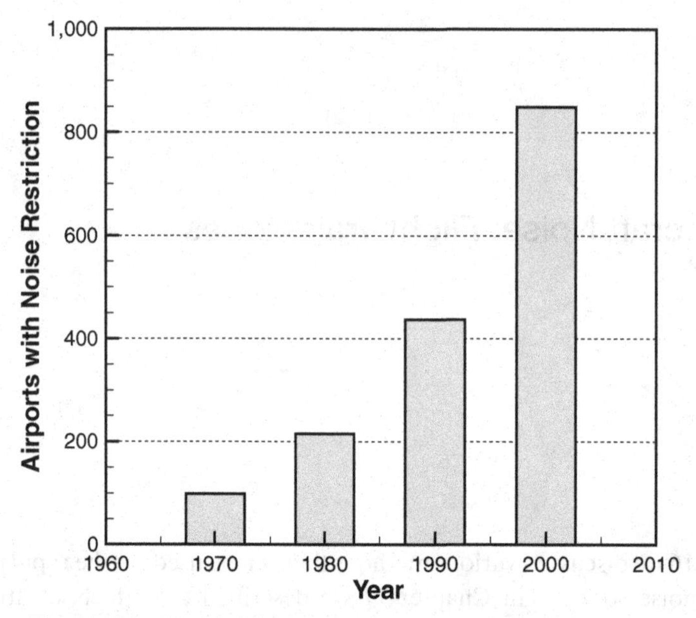


Figure 18.1. Worldwide trend in airports with noise restriction.

flight paths. Considerable research is available on health effects[1]. As a result, taxes are levied on the noisiest aircraft on most airports; charge models have been in place since the 1980s[2] to compensate for marginal damage.

Research into aircraft noise started in earnest in the late 1960s. In 1965, Gebhardt[3] reported on the noise problems of the Boeing B727 that had entered service one year earlier. Crighton[4] estimated that the noise level of a first-generation Boeing B737 (1965) was the same as the world population shouting at once, whilst a second-generation aircraft would only produce the same noise as the city of New York shouting at once. Concorde was the loudest of all aircraft; its noise performance did not comply with even the early noise regulations; exceptions were made for its case, and the aircraft was eventually prevented to fly at supersonic speeds over land. Its flight operations sparked mass protests in the 1970s[5]. For military vehicles, where detection is a critical issue, lower frequencies are also important because low frequencies have lower attenuation rates (§ 17.2) and therefore can be heard over long distances. There is also concern about operations from military airports[6].

The technical literature is rich in information regarding noise-certification issues, regulations, data and procedures, for example Refs.[7;8]. Various national and international organisations are involved in one form or another in setting up working groups, regulations and strategies. It is not possible to review these issues in any detail at this point. It suffices to mention that the problem of aircraft noise involves a number of factors beyond the aircraft itself, such as the level of traffic at an airfield. Smith[5] is a fairly good source of historical information. It includes some discussion on noise sources (power plants, fuselage, propellers, gas turbines), data acquisition and performance prediction, sonic boom, relevant historical notes and a very extensive bibliography. Several aviation organisations have considerable amounts of technical publications (continuously updated) that cover any aspect of aircraft noise.

It has now been established that the best approach to aircraft-noise reduction must be achieved through a *balanced approach*[*] that takes into account all of the aspects of the operation of the aircraft, including:

- noise reduction at the source (or aircraft design)
- improvements in land use and urban developments
- establishment of optimal flight procedures
- restriction of operations (as shown in Figure 18.1)

At the operational level, there is an increasing need for assessing noise performance for new and extended runways, changes in operational procedures, increase in traffic at airports and changes in air-space allocations.

The certification of aircraft noise relies on measurements at recognised points around the airfield. Microphones record the noise signals over the complete flight trajectory and a software-hardware interface calculates the effective perceived noise EPNL. Specifically, the ICAO has established three measuring points, which provide, respectively, the take-off noise, the lateral noise, and the final approach (landing) noise. For this purpose, the standard position of the microphones is defined as follows:

- **Take-Off Noise**: The microphone is placed at the centre of the runway, 6,500 m down from the brake-release point. The aircraft must have the retracted landing-gear configuration, with a speed greater than $V_2 + 10$ kt. Furthermore, thrust cut-back must take place at altitudes above 210 m (~690 feet); after a thrust cut-back, the aircraft must maintain a flight-path angle greater than 4 degrees (Figure 18.2a).
- **Lateral Noise**: The microphone is placed 450 m (~1,480 feet) off the centre of the runway, where the aircraft noise at take-off is maximum. The aircraft must have configuration with landing gear retracted, with a speed greater than $V_2 + 10$ kt (Figure 18.2b) and maximum thrust. Note that in this case several microphones parallel to the runway are required to establish the maximum noise level.
- **Landing (Approach) Noise**: The microphone is placed 2,000 m (~6,560 feet) upstream of the runway threshold, with the aircraft on a −3 degree slope flight path ending 300 metres beyond the threshold (Figure 18.2c). On a level ground, the microphone will be 120 metres below the aircraft. The aircraft must have its landing gear out, the flaps deployed and an air speed of $V_S + 10$ kt.

The case of lateral noise requires further discussion. In fact, the regulations only specify the lateral position of the microphone; the longitudinal position is indeterminate. A set of microphones is placed along a line parallel to the runway, as shown in Figure 18.2b. The noise level is recorded at all of the microphones, and the resulting sideline noise level is the maximum value among the recordings (interpolation of data may be required).

Table 18.1 shows the position of the microphones for community noise at London Heathrow. The data are, respectively, runway number, reference point,

[*] ICAO: Assembly Resolution A35-5, Appendix C, Doc 9848 (2001).

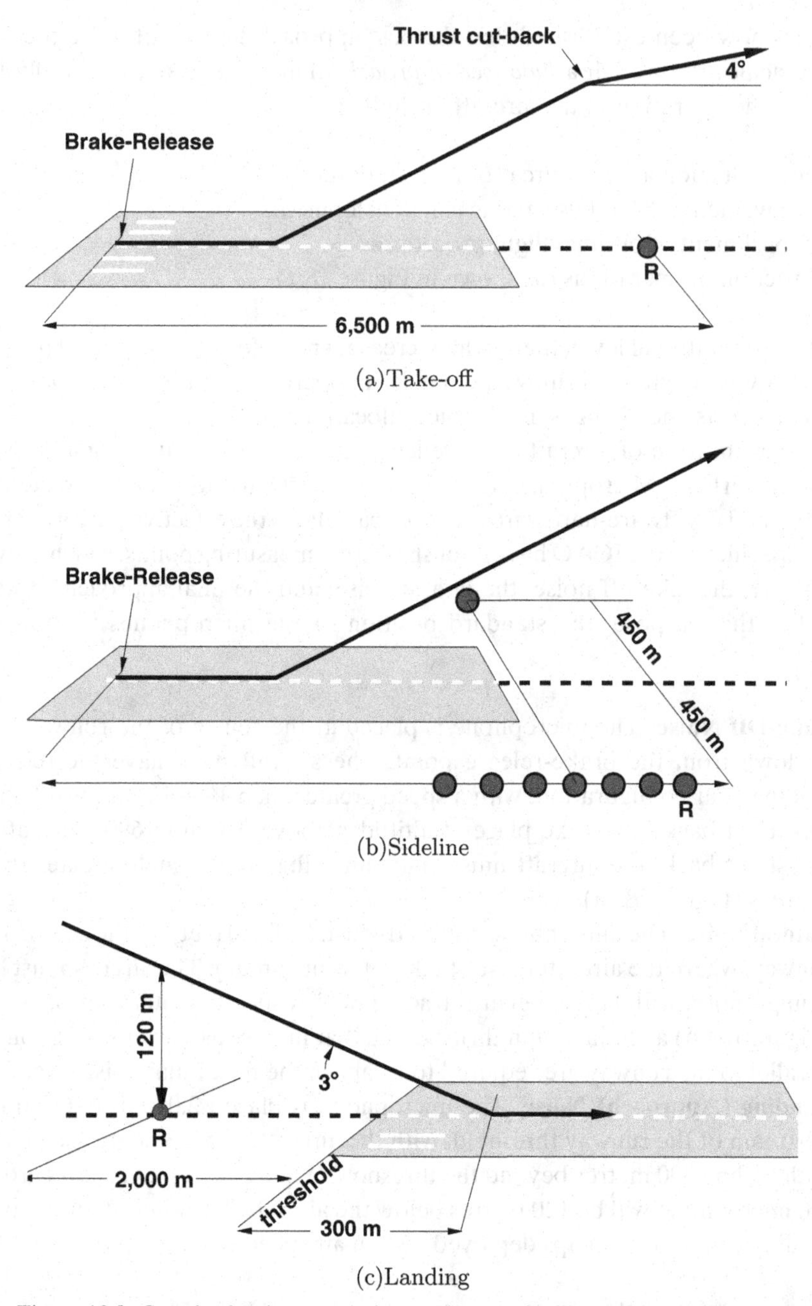


Figure 18.2. Standard noise-measuring stations according to ICAO Chapter 3.

elevation above sea level, elevation above airfield (EAA), latitude, longitude, distance from brake-release, limit correction, maximum day-time (07:00-23:00) and maximum night-time (23:00-07:00) noise.

The actual flight paths are affected by atmospheric conditions, gross weight and local air traffic procedures. Therefore, there is always a dispersion in the flight paths, both vertically and on the ground tracks. More realistically, when assessing real flight trajectories, we must consider flight-corridor boundaries.

Table 18.1. *Microphone positions for aircraft-noise measurements at London Heathrow*

| RWY | Ref. | Elev | EAA | Lat | Long | Dist | Limit | Day | Night |
|---|---|---|---|---|---|---|---|---|---|
| | | [m] | [m] | (N) | (W) | [km] | [dB] | [dB] | [dB] |
| | 6 | 18 | −6 | 51 27 56 | 00 03 15 7 | 6.58 | 0.3 | 94.4 | 87.4 |
| | A | 20 | −4 | 51 28 57 | 00 03 12 4 | 6.25 | 2.3 | 97.0 | 90.0 |
| | B | 20 | −4 | 51 28 40 | 00 03 11 7 | 6.00 | 4.8 | 99.5 | 92.5 |
| 27L/R | C | 18 | −6 | 51 28 14 | 00 03 14 8 | 6.58 | −0.3 | 94.4 | 87.4 |
| | D | 17 | −7 | 51 27 45 | 00 03 20 1 | 6.83 | −0.6 | 94.1 | 87.1 |
| | E | 17 | −7 | 51 27 24 | 00 03 21 6 | 7.20 | −1.0 | 93.7 | 86.7 |
| | F | 21 | −3 | 51 28 21 | 00 02 34 5 | 6.40 | 0.9 | 95.6 | 88.6 |
| | G | 21 | −3 | 51 28 06 | 00 02 33 8 | 6.50 | −0.1 | 94.6 | 87.6 |
| 09R | H | 21 | −3 | 51 27 45 | 00 02 34 0 | 6.37 | 1.2 | 95.9 | 88.9 |
| | I | 20 | −4 | 51 27 14 | 00 02 34 1 | 6.60 | −0.3 | 94.4 | 87.4 |

The reference atmospheric conditions are pressure $p_o = 101{,}325$ Pa; air temperature 25°C; relative humidity $\mathcal{H} = 70\%$; zero wind. Actual tests must be carried out under the following conditions: no precipitation; air temperature between −10 and +35°C; relative humidity $20\% < \mathcal{H} < 95\%$; wind speed less than 12 kt; cross-wind speed less than 7 kt (measured 10 m above the ground); no anomalous conditions that may affect the measurements.

There are a variety of operational conditions that can affect the measurements. For this reason, the certification requires several sets of measurements, data reduction and corrections.

Sound measurements are taken with microphones placed 1.2 m above the ground. Comparisons between measured and estimated data are made difficult by the great variability in atmospheric conditions as well as conditions on the ground (terrain shape, airfield altitude, presence of large obstacles). Some of these difficulties can be overcome if the microphones are placed on the ground because the ground reflection is then removed.

The maximum noise levels allowed by the ICAO depend on the number of engines and on the maximum take-off weight. The rules for Chapter 3 (valid until 2006) for take-off noise were as follows:

$$\text{EPNL}(dB) = \begin{cases} \begin{cases} 89 & m < 48.1 \\ 66.65 + 13.29 \log m & 48.1 < m < 385 \\ 101 & m > 385 \end{cases} & \text{2 engines} \\ \begin{cases} 89 & m < 28.6 \\ 69.65 + 13.29 \log m & 28.6 < m < 385 \\ 104 & m > 385 \end{cases} & \text{3 engines} \\ \begin{cases} 89 & m < 20.2 \\ 71.65 + 13.29 \log m & 20.2 < m < 385 \\ 106 & m > 385 \end{cases} & \text{4 engines} \end{cases} \qquad (18.1)$$

The maximum noise is 89 dB at weights below the thresholds (all engines); the maximum noise is 101 dB (2 engines), 104 dB (3 engines), and 106 dB (4 engines) at

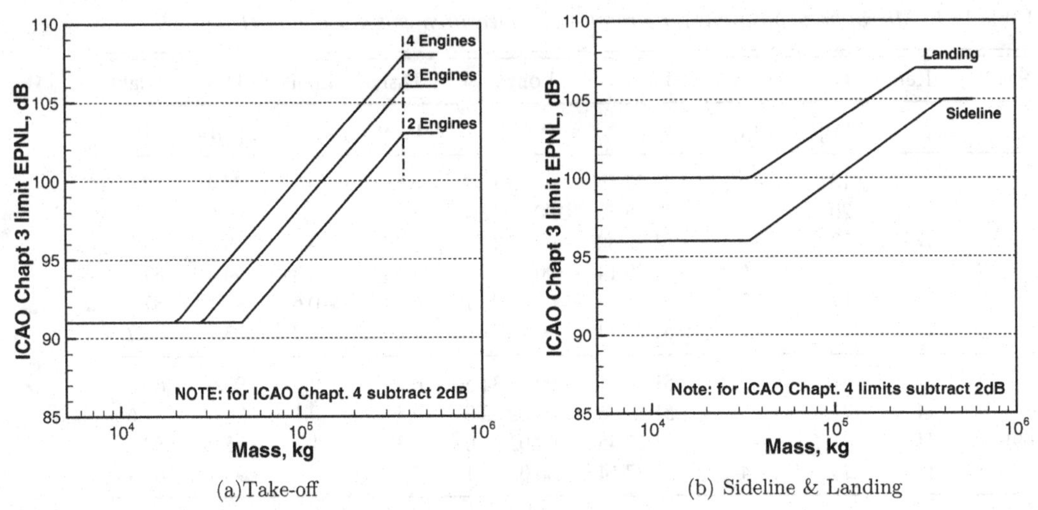


(a)Take-off (b) Sideline & Landing

Figure 18.3. ICAO Chapter 3, noise limits at take-off, sideline and landing/approach.

weights above the thresholds (385 tons). The sideline noise limits are:

$$\text{EPNL}(dB) = \begin{cases} 94 & m \leq 35 \\ 80.87 + 8.51 \log m & 35 < m < 400 \\ 103 & m \geq 400 \end{cases} \tag{18.2}$$

The landing (approach) noise limits are:

$$\text{EPNL}(dB) = \begin{cases} 98 & m \leq 35 \\ 86.03 + 7.75 \log m & 35 < m < 280 \\ 105 & m \geq 280 \end{cases} \tag{18.3}$$

In Equation 18.1, Equation 18.2 and Equation 18.3, the mass $m$ is expressed in 1,000 kg. A graphic form of these equations is shown in Figure 18.3.

Chapter 3 of the ICAO regulations (now expired) allowed a trade-off in case of non-compliance at one reference point. The more stringent regulations now enforced with the ICAO Chapter 4 require that none of the noise levels can exceed the levels defined in Chapter 3. Furthermore, the aircraft must demonstrate a cumulative margin of at least 10 dB at all reference points. This means that for a given weight, the sum of the noise levels (take-off, sideline, approach) should be at least 10 EPNL (dB) below the sum of the maximum noise levels calculated with Equation 18.1, Equation 18.2, and Equation 18.3. The margin on any single point (take-off, sideline, approach) over Chapter 3 regulations must be at least 2 EPNL (dB).

We could argue whether the regulations have fostered progress in noise-abatement technology or progress in aircraft engineering has led to noise abatement which was eventually endorsed by the regulations[5]. However, if we look at a time-line showing the progress in noise reduction and plot the data against the regulations, we find a striking correlation, which is shown in Figure 18.4. This figure shows a trend of the certified noise levels on average for selected commercial subsonic jets. The graph shows how, as a combination of more efficient engine and airframe design, the noise levels have been reduced considerably from the first generation of jet-powered aircraft. By the time the ICAO Chapter 4 was introduced, most aircraft

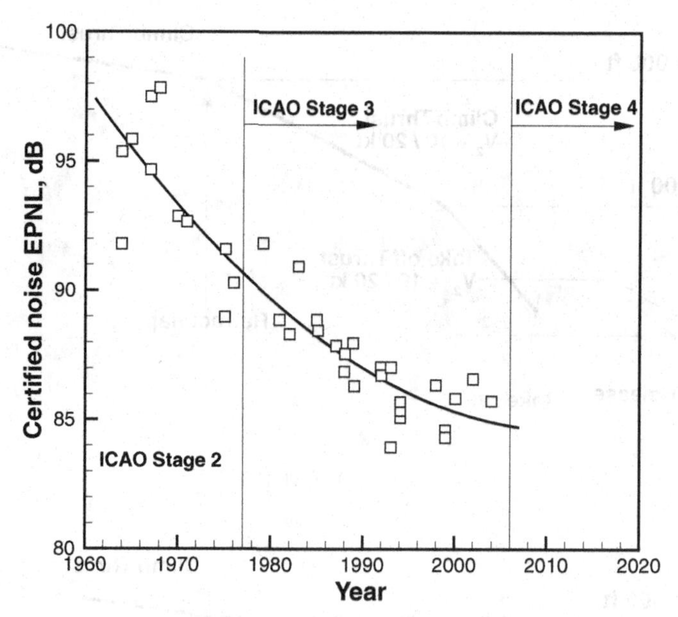


Figure 18.4. Trends in certified noise of commercial aircraft at year of introduction.

were already compliant with the more stringent regulations. As a result, one may argue that even more restrictive limits will be introduced (at least another 3 dB cumulative cut) to drive noise emissions down. Further data and comparison are shown by Waitz *et al.*[9].

**A Case for a Noise Tax.** In the past there has been considerable debate on forcing compliance with noise regulations via taxation. Non-compliance with noise regulation can be structural or accidental. Structural causes are the result of more stringent regulations that make existing aircraft obsolete, as shown in Figure 18.4 (technology obsolescence). Accidental events are those that occur infrequently due to bad manoeuvre, late departure or other reasons.

Consider the case of the technology obsolescence. One of the options proposed includes a tax per landing (or movement), which can be offset by retrofitting newer technologies. Assume that there exists a noise-reduction technology capable of bringing down the noise metrics below a threshold. Assume that this technology can be retrofitted on the aircraft at a cost $C$. This cost is an acquisition item that can be described with Equation 15.46. The cost per movement would be

$$C = \frac{1}{n}\left[\mathcal{P} - F(1+i)^n - R_n\right]\frac{1}{M_n}, \tag{18.4}$$

where $M_n$ is the number of movements in the $n$–th year of operation; $R_n$ is the residual value in the $n$–th year. To simplify, assume that no financing is required ($F = 0$). The retrofitting cost can be offset if the tax is such that

$$\text{tax} > \frac{1}{n}\frac{1}{M_n}\left(\mathcal{P} - R_n\right). \tag{18.5}$$

For example, if the price is $\mathcal{P} = 1$ (in whatever currency), $M_n$ = 2,000 and the expected life-time of the technology is 20 years, then the noise tax that would offset

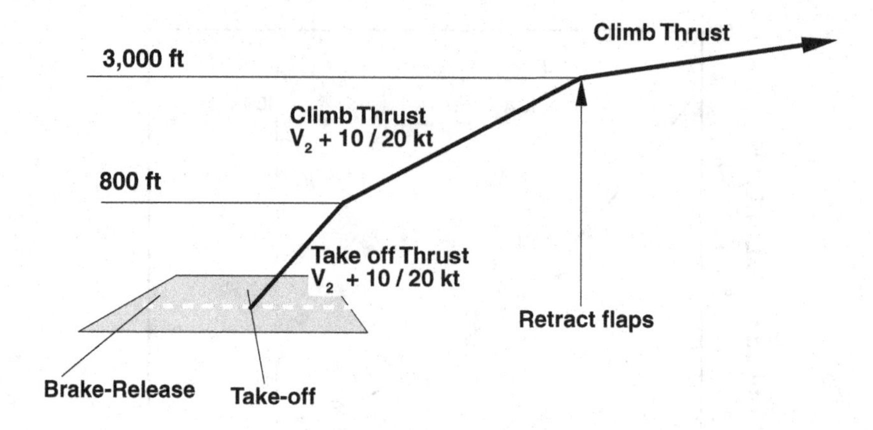


(a) NADP 1

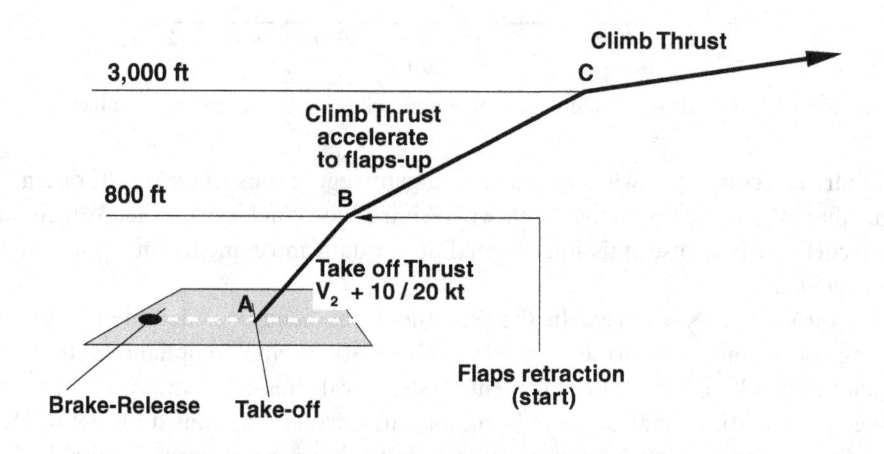


(b) NADP 2

Figure 18.5. ICAO noise-abatement procedures at departure.

the new technology is

$$\text{tax} > 0.25 \cdot 10^{-4} [1 - R_n] . \tag{18.6}$$

## 18.2 Noise-Abatement Procedures

For both departure (take-off) and arrival (approach and landing), there are some recommended flight procedures aimed at minimising noise. We limit our presentation to a few representative cases. For the departure case, ICAO recommends two procedures for noise abatement, one called *close-in* (or NADP 1) and the other called *distant* (or NADP 2). These trajectories are shown in Figure 18.5.

The noise-abatement departure procedure NADP1 (see Figure 18.5) requires a power reduction at or above a prescribed minimum altitude above the airfield (in this case, 800 feet). The retraction of the flaps is delayed until a prescribed maximum altitude is reached (in this case, 3,000 feet). Above this altitude, accelerate to normal en-route climb speed. The noise-abatement procedure is started at 800 feet. On reaching this altitude, the engine thrust is adjusted whilst maintaining a climb speed $V_2 + 10$ kt $< V < V_2 + 20$ kt, with flaps and slats in the take-off configuration. The landing gear is retracted.

The noise-abatement departure procedure NADP2 (Figure 18.5b) requires a flap/slat retraction on reaching a minimum altitude (in this case, 800 feet), whilst maintaining a minimum climb rate. As in the previous case, the noise-abatement procedure is started at 800 feet, with a climb speed $V_2 + 10$ kt $< V < V_2 + 20$ kt. Starting at this altitude, the aircraft must accelerate to the speed for flaps/slats retraction; the attitude is reduced; thrust is reduced as soon as the flaps are retracted. In this phase, a positive climb rate must be maintained with a speed $V_2 + 10$ kt $< V < V_2 + 20$ kt. Upon reaching 3,000 feet, the aircraft accelerates to normal en-route climb speed (as specified by the FCOM).

Similar procedures exist for approach and landing trajectories. For a commercial airliner, the standard procedure is a 3-degree descent trajectory in the terminal areas, although in some cases the aircraft is forced to perform complicated three-dimensional manoeuvres to go around topography and ATM constraints. The engine power setting is minimal, although there is scope for tuning on the throttle. Instead, special care must be placed toward the operation of the high-lift devices and the landing gear, which are fully deployed. Two things can be done: 1) reduce the air speed, and 2) increase the distance between the aircraft and the noise receiver on the ground. When possible, other procedures can be attempted, such as best aerodynamic configuration (optimal flap setting), continuous descent and steep descent. The latter procedures are neither standard nor possible with some aircraft[10]. The term *steep approach* is used to denote any trajectory steeper than 3 degrees, although in practice it is not possible to exceed 4 degrees. A noise reduction of up to 2 dB has been reported.

A continuous descent requires no level flight below 7,000 feet. This procedure is applied at some airports (for example, London City) with some aircraft (BAE 146). A reduction of up to 5 EPNL(dB) is claimed with this procedure.

### 18.2.1 Cumulative Noise Index

One of the issues with aircraft noise is its persistence due to *continuous* movements, that is, take-off and landings at short time intervals during the day and night. Thus, there is a *cumulative effect* on annoyance, which is described by further metrics. If there are $n$ movements in the reference time and each movement causes a noise level $L_{AE}$, then the *equivalent continuous sound level* $L_{EQ}$ is given by equations such as:

$$L_{EQ}[24\text{h}] = 10 \log \left[ \frac{1}{24 \times 3,600} \sum_{i=1}^{n} 10^{L_{AE}(i)/10} \right] \tag{18.7}$$

$$L_{EQ}[\text{day}] = 10 \log \left[ \frac{1}{(23-7) \times 3,600} \sum_{i=1}^{n} 10^{L_{AE}(i)/10} \right] \tag{18.8}$$

for the full day and day-time only, respectively. Additional metrics are available from the technical standards[7]. In any case, these cumulative metrics are used to

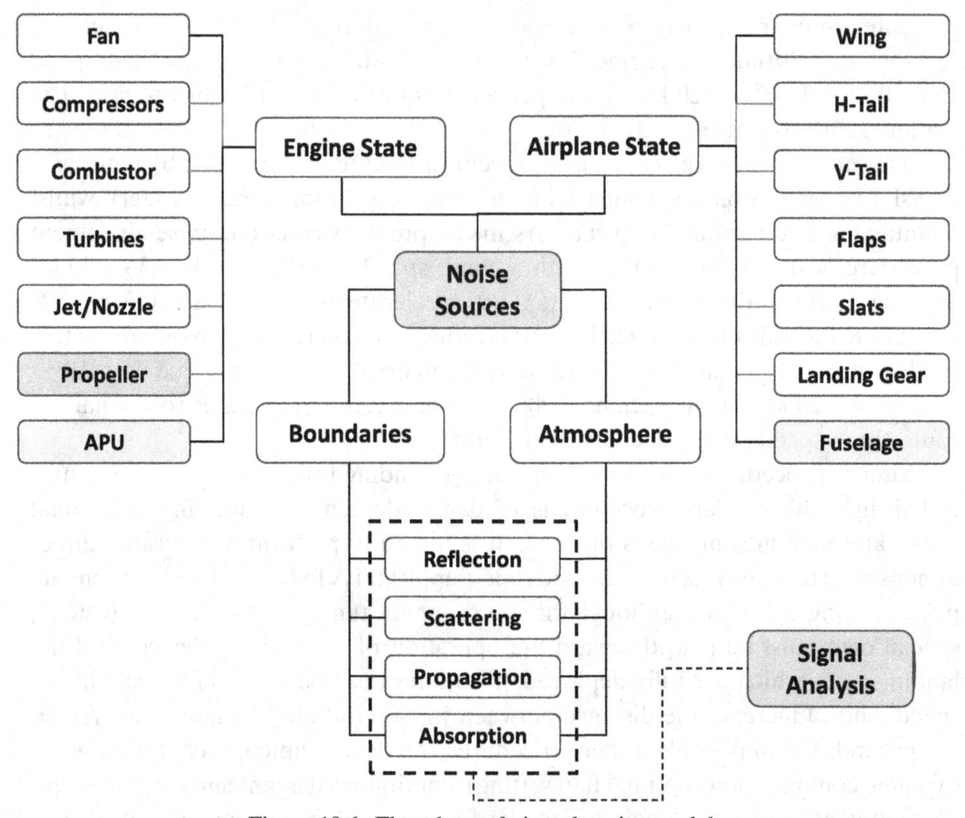

Figure 18.6. Flowchart of aircraft-noise model.

determine noise maps around airports, typically with noise levels between 57 and 72 $L_{EQ}$(dBA).

### 18.2.2 Noise-Program Flowchart

The analysis of about a dozen independent components is required to establish a rational basis for aircraft noise. A flowchart of the model is shown in Figure 18.6. The sources are split in two categories: 1) propulsive noise, arising from the power plants; and 2) non-propulsive noise (or airframe noise).

In principle, we should add also the noise generated by the APU, which can be a large gas-turbine engine on its own. However, this is generally not done. One justification for this shortcoming is that the mechanical details of the APU are undisclosed. However, this contribution is of considerable importance when the aircraft is on the ground with the APU running. To understand this effect, we can perform a sensitivity analysis on the same vein as described by Equation 16.13. Another noise component not included in the analysis is the tyre noise, which occurs during the ground roll: this is of no relevance for certification purposes. Viewed in a larger context, this example indicates that the detailed knowledge of one noise component is of limited use if other components are ignored.

Once the noise sources are identified and modelled (following the examples in Chapter 16), the acoustic perturbations have to be tracked in the travel from the source to the receiver. For this purpose there is a propagation module that takes

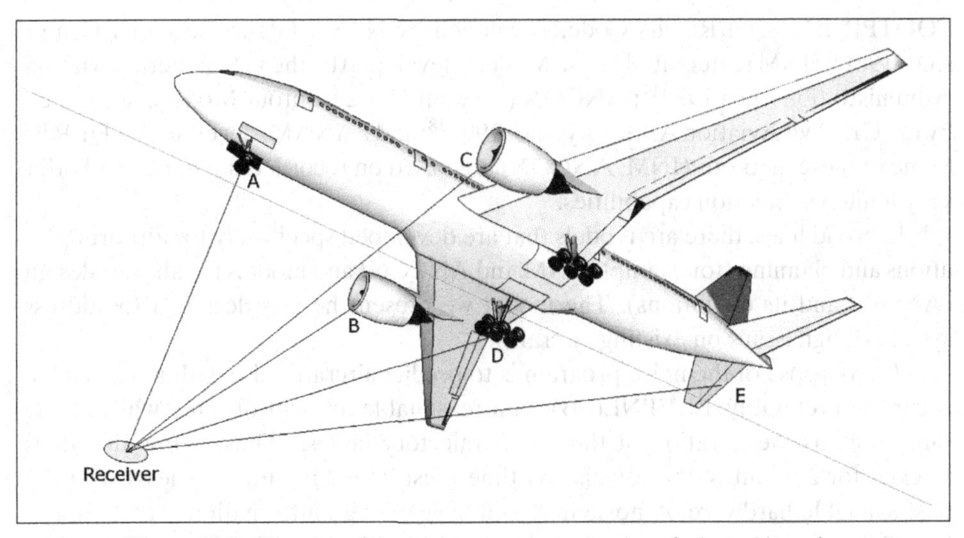


Figure 18.7. Distances from noise sources to receiver (selected items).

into account the effects of the atmosphere and the presence of boundaries (the ground or other obstacles). The combination of reflection, scattering, propagation and absorption is sent to another module called "Signal Analysis". The function of this module is to elaborate the SPL data and provide a suitable quantitative measure of the aircraft noise.

The aircraft is usually tracked over a long trajectory. During landing and take-off the aircraft changes configuration. For example, in the terminal-area manoeuvre, the flaps are extended at about 750 m (~2,500 feet) above the airfield; the landing gears are deployed at about 500 m (~1,600 feet) above the airfield; the flaps are at landing setting at about 450 m (~1,500 feet); and so on.

The noise program operates on the basis of a known flight trajectory, previously calculated with the flight-mechanics program. These trajectories can be defined as flight data (from radar tracking or flight recorder data systems), information on the state of the aircraft and the engine, and a detailed knowledge of the atmospheric conditions around the aircraft and the airfield.

The distance between the receiver and the noise source depends on the source itself, as indicated in Figure 18.7, although when the aircraft is far enough, the noise model is unable to discriminate between points A, B, C. ... It is interesting to note that the effect of such components as the landing gear is affected by their relative distance to the receiver. For a Boeing B777-300 on approach, a signal from the main landing gear reaches the receiver below the flight track about 0.09 second later than the nose landing gear. In any case, the exact position of the noise source can be determined from the aircraft geometry model (see Chapter 2). If $\boldsymbol{r}$ is the vector position of the CG with respect to the receiver at the airfield, and $\boldsymbol{r}_a$ is the position of the noise source A with respect to the CG, then the correct source-receiver distance $\tilde{\boldsymbol{r}}$ is

$$\tilde{\boldsymbol{r}} = \boldsymbol{r} + \boldsymbol{r}_a. \tag{18.9}$$

A number of aircraft-noise-simulation methods have been developed over the years, most notably NASA's code ANOPP[11–14] and its successor ANOPP2[15]; the

FOOTPR Footprint/Radius Code, developed by NASA Glenn Research Center since 1981[16]; INM (Integrated Noise Model), developed by the U.S. Federal Aviation Administration since 1978[17]; ANCON (Aircraft Noise Contour Model), developed by the UK Civil Aviation Authority since 1992[18]; and PANAM, developed at DLR[19]. Some of these methods (INM, ANCON) are based on recorded databases and offer very limited simulation capabilities.

In broad lines, there are models that are developed specifically for airport operations and planning (for example, INM and ANCON) and models for aircraft design (ANOPP and its evolutions). The model we present here is designed to address system-design issues on existing aircraft.

The purpose of the noise program is to predict aircraft noise with a reasonable accuracy, preferably ±1 EPNL(dB), in a reasonable amount of time, which is the same time as the duration of the flight trajectory or less. Thus, if the aircraft is tracked for 2 minutes, the calculation time must be ~2 minutes (or less) with the best available hardware. A noise map will have enough information if there are at least 200 receivers on the ground over at least 10 km$^2$ (10 km × 1km). To calculate such a map, we would need ~400 minutes, or 6.7 CPU hours. The calculation of a noise map due to 100 aircraft movements (50 take-off and 50 landings) will require weeks of calculation, and it would clearly be unmanageable unless the numerical methods are improved. If atmospheric winds are included, the computational times will increase by at least a factor 10. Thus, it is clear that as the complexity of the problem increases, accuracy and details must be sacrificed in order to obtain a solution. Therefore, the value of integrated methods such as INM becomes evident when complete noise maps around airports have to be established.

## 18.3 Flight-Mechanics Integration

Integration between the flight-mechanics and the noise sources array is done through an operator that provides the vector of parameters necessary to define the state of the aircraft:

$$\mathcal{S} = f\left(\underbrace{t,\ \mathtt{Lat},\ \mathtt{Lon},\ z_g}_{position},\ \underbrace{\theta, \phi, \psi,\ \text{IAS},\ V,\ V_g}_{mechanics},\ \underbrace{N_1, \dot{m}_f}_{engine},\ \underbrace{\text{LG, SF},\ W}_{config},\ \underbrace{\mathcal{T}, \mathcal{H},\ V_w,\ \Psi_w}_{atmosphere}\right) \tag{18.10}$$

Equation 18.10 defines the state of the aircraft state by a set of parameters in the categories of position, flight mechanics, engine state, configuration and atmospheric conditions. This is a standard format that can be used in other applications such as flight simulation.

The parameters $\mathtt{Lat}$ and $\mathtt{Lon}$ are the GPS longitude and latitude, respectively; $z_g$ is the geometrical altitude; $\theta, \phi$, and $\psi$ are pitch, bank and yaw angles, respectively; IAS is the indicated air speed; $V_g$ is the ground speed; $N_1$ is the engine's rotational speed; $\dot{m}_f$ is the fuel flow; $V$ is the true air speed; LG denotes the state of the landing gear (0 = retracted; 1 = deployed); SF denotes the state of the high-lift systems (see, for example, Table 11.1); $\mathcal{T}$ is the outside air temperature, $\mathcal{H}$ is the relative air humidity, $V_w$ is the magnitude of the wind speed and $\Psi_w$ is the wind direction.

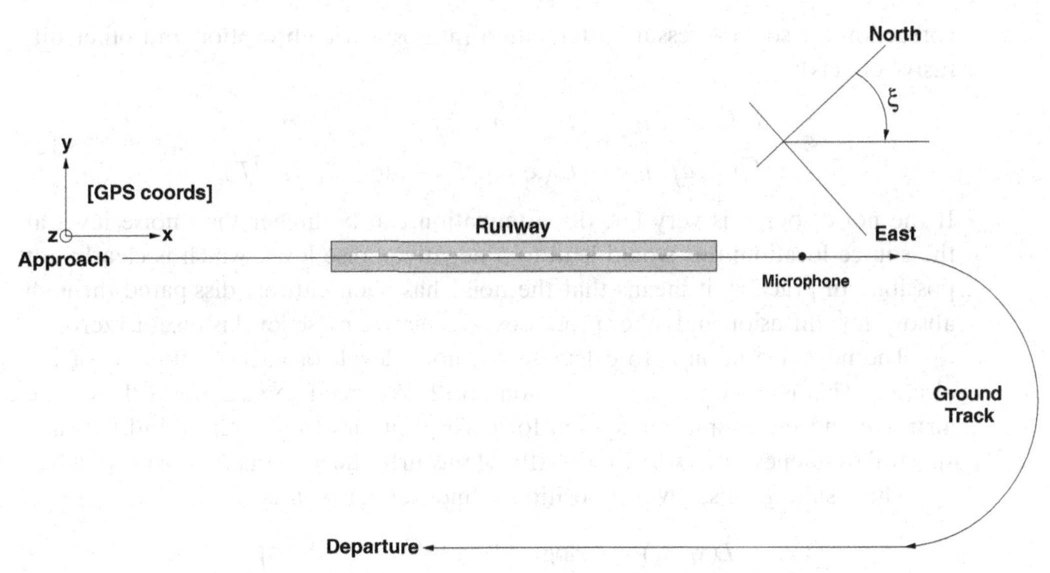


Figure 18.8. Typical departure trajectory, with a ground track and a local reference system.

Figure 18.8 shows a typical situation at departure. In our convention, there is a reference system on the ground, whose longitudinal axis is aligned with the runway. If the trajectory is taken from FDR data, the coordinates are GPS. A transformation is carried out, so that all of the GPS coordinates are transferred to the conventional reference system on the ground. The transformation is done with the Vincenty inverse distance method[20]. This method only gives the distance $r_o$ between two points on the ground. The distance is corrected to take into account the altitude of the aircraft, according to

$$r \simeq \sqrt{r_o^2 + z_g^2}. \tag{18.11}$$

### 18.3.1 Noise Data Handling

For a specified noise trajectory, such as Equation 18.10, referred to a single receiver on the ground, there will be a number of sampling steps $x_1, x_2, \cdots$. The aircraft's noise is produced by a number of airframe components $a_1, a_2, \cdots$. For each component, there will be a spectrum of frequencies $f_1, f_2, \cdots$, up to a cut-off frequency. The airframe noise is characterised by a noise level $L_a$. A similar matrix exists for the propulsive noise components. Thus, we have the matrices

$$L_a(x_i, a_j, f_k), \qquad L_e(x_i, a_j, f_k).$$

The way these data are handled is key to the determination of the noise level at the receiver position. It is also important to remember that each component is contributed by more than one unit (for example: engines, flaps, landing gear).

Let us assume that the SPL has already been corrected for Doppler effects, directivity and spectral effects because these corrections are generally already part of the noise models, as discussed in Chapter 16. The first data handling is the correction for sound-pressure attenuation (atmospheric absorption and other diffusive effects):

$$\begin{aligned} L_a(x_i, a_j, f_k) &\rightarrow L_a(x_i, a_j, f_k) - \mathcal{A}(x_i, f_k, \mathcal{H}, d\mathcal{T}) \\ L_e(x_i, a_j, f_k) &\rightarrow L_e(x_i, a_j, f_k) - \mathcal{A}(x_i, f_k, \mathcal{H}, d\mathcal{T}). \end{aligned} \tag{18.12}$$

If the noise source is very far, the attenuation can be higher than noise level at the source location; this would lead to a negative noise level, which is clearly not possible. In practice, it means that the noise has been entirely dissipated through absorption, diffusion and other processes. A negative noise level is reset to zero.

The next operation is to calculate the noise level for each position $x_i$ of the aircraft. This is done through Equation 16.12. We need to sum the SPL for the airframe and the propulsion system for each position of the aircraft and for each spectral frequency; then we sum the SPL of the airframe with that of the propulsion.

The resulting noise level at position $x_i$ and frequency $f_k$ is

$$L(x_i, f_k) = 10 \log_{10}\left(10^{L_a(x_i, f_k)} + 10^{L_e(x_i, f_k)}\right).$$

From this equation, we calculate the OASPL, the PNL and other metrics, following the standard methods of acoustics.

**FLIGHT TIME AT RECEIVER.** Another important correction arises from the distance source-to-receiver, which can be considerably higher than $r/a$. For example, in isothermal atmosphere, a signal travels at a constant speed and the distance covered after a time $t$ is $r = at$; in other words it will take $x/a$ seconds to hear the sound emitted at the source; thus, there is a time delay equal to $r/a$. The problem is more complicated in a non-isothermal atmosphere. In this case, the acoustic signal emitted at time $t$ by the source is heard at the receiver at a retarded time $t_r$:

$$t_r = t + \int \frac{dr}{a}. \tag{18.13}$$

Assuming that there is no boundary effect, the direct acoustic wave from the source reaches the receiver at time

$$t_r = t + \int_{h_r}^{h_s} \frac{dz}{a \tan\varphi} \simeq \frac{r}{\overline{a}}, \tag{18.14}$$

where $\varphi$ is the angle between the acoustic wave and the ground and $\overline{a}$ is the average speed of sound. There is no guarantee that the receiver time $t_r$ is a non-decreasing function. If the atmospheric temperature varies with altitude according to $\mathcal{T} = \mathcal{T}_o - \lambda h$, then there is a closed-form solution of Equation 18.14. The time delay is

$$\Delta t = \frac{2}{\lambda \tan\varphi \sqrt{\gamma \mathcal{R}}}\left[(\mathcal{T}_o - \lambda h_r)^{1/2} - (\mathcal{T}_o - \lambda h_s)^{1/2}\right]. \tag{18.15}$$

## 18.4 Noise Sensitivity Analysis

A noise sensitivity analysis is important for understanding which noise sources are dominant and where modelling and simulation efforts should be concentrated. As

Table 18.2. *Noise sensitivity matrix for a Boeing 777-300 for ± 2 dB on take-off and landing trajectories (simulated data); APUJ = APU jet; APUC = APU compressor; the receivers are at standard ICAO/FAR position (§ 18.1)*

| | Landing | | | | Take-Off | | | |
|---|---|---|---|---|---|---|---|---|
| Rank | Item | $EPNL_c$ | +dB | −dB | Item | $EPNL_c$ | +dB | −dB |
| 1 | MLG | 97.37 | 1.508 | −1.354 | FAN | 99.55 | 1.685 | −1.590 |
| 2 | SLAT | 88.80 | 0.291 | −0.221 | JET | 88.18 | 0.404 | −0.325 |
| 3 | FAN | 84.25 | 0.072 | −0.087 | LPT | 83.30 | 0.056 | −0.034 |
| 4 | NLG | 83.38 | 0.020 | −0.053 | COMB | 78.42 | 0.030 | −0.012 |
| 5 | COMB | 82.47 | 0.019 | −0.050 | Wing | 63.52 | 0.005 | −0.003 |
| 6 | FLAP | 77.50 | −0.005 | −0.039 | MLG | 63.30 | 0.000 | 0.000 |
| 7 | HPC | 77.18 | 0.086 | −0.123 | HSTAB | 62.81 | 0.002 | −0.001 |
| 8 | Wing | 74.29 | −0.037 | −0.015 | LPC | 61.16 | 0.001 | |
| 9 | HSTAB | 74.24 | −0.047 | −0.009 | HPC | 58.46 | 0.001 | |
| 10 | LPC | 64.65 | −0.060 | −0.001 | NLG | 48.17 | | |
| 11 | JET | 46.01 | −0.062 | | SLAT | 47.09 | | |
| 12 | VSTAB | 44.82 | −0.062 | | APUC | 19.53 | | |
| 13 | APUC | 38.37 | −0.062 | | APUJ | | | |
| 14 | LPT | | −0.062 | | VSTAB | | | |
| 15 | APUJ | | −0.062 | | HPT | | | |
| 16 | HPT | | −0.062 | | FLAP | | | |

discussed next, it is futile to embark into precise simulation of noise sources that have little impact; such is the case of a high-lift system or a landing gear in a standard take-off trajectory. On approach and landing these specific noise sources are dominant.

A noise sensitivity analysis is shown in the data of Table 18.2. $EPNL_c$ is the perceived noise level of the noise source "c" (as named in the table). The column denoted by $+dB$ shows the change in *total* EPNL due to an increase of 2 dB in the spectrum of that component at all points in the trajectory. In other words, at each position in the aircraft trajectory, the SPL of the component "c" is increased by 2 dB across the spectrum, from 20Hz to 10 kHz. In Table 18.2, on the take-off column, an increase by 2 dB in the dominant component (FAN, rank 1) causes an increase in total EPNL by about 1.69 dB. Likewise, a decrease by 2 dB in the same component causes a decrease in aircraft EPNL by ~1.59 dB. As we move down the table, to follow noise sources that are less and less dominant, the contribution of any change to the aircraft SPL becomes negligible. In the empty boxes, this change is effectively zero. At landing the contribution of the main landing gear (MLG) is dominant. This result explains why landing-gear noise reduction is so critical in this phase of flight. An increase of 2 dB in this component causes an ~1.5 dB increase in the overall noise level. However, a decrease by a similar amount causes a decrease by ~1.35 dB. In both cases, the role of the APU in generating noise is negligible. We conclude that for a take-off trajectory we should focus on the accurate prediction of the jet noise and the fan noise. Landing is more complicated by the fact that many source components are contributing, coupled with the fact that the aircraft is closer to the receiver. The role of each source changes if we change the position of the receiver.

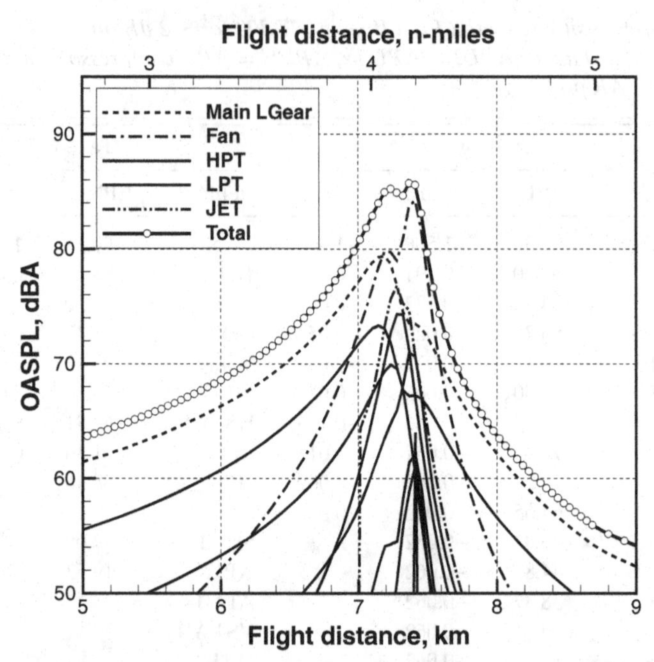


Figure 18.9. Calculated noise trajectory of the A320-200 at landing; ICAO/FAR reference.

## 18.5 Case Study: Noise Trajectories of Jet Aircraft

We consider the case of a model Airbus A320-200 with CFM56-5C4 turbofan engines. The trajectories are generated by the flight-mechanics model. The interface is given by the vector state in Equation 18.10. We consider the cases of ICAO reference points for take-off and approach noise. The reference conditions are sea-level airfield, standard day, no wind and a relative humidity level of 70%.

**LANDING.** Figure 18.9 shows the calculated noise trajectory at approach/landing. The graph shows the time-dependent overall noise level (in dB) at the receiver. Selected contributions are named in the graph, such as the main landing gear, the fan, the HPT, the LPT and jet noise. The largest single contribution is the fan, with the landing-gear noise being relatively high as well. The symbols on the OASPL indicate the time sampling rate of this trajectory.

A more detailed analysis of the contribution is done with the component bar chart, which is shown in Figure 18.10. Only the largest contributions are named. The result shows that landing-noise prediction must rely on accurate methods for several components. The bar charts indicate the change in overall EPNL due to a change by ±2 dB in each of the components. The sensitivity analysis highlights the importance of a correct prediction of the fan noise.

**TAKE-OFF.** Figure 18.11 shows the calculated noise trajectory at take-off. Graph 18.11a tracks the first three minutes of the flight from the brake-release. The shaded area at the bottom-left corner represents the first 75 seconds, which are shown in graph 18.11b. Only selected noise sources are indicated in the caption. Specifically, graph 18.11a shows the four largest contributions, which are jet, fan,

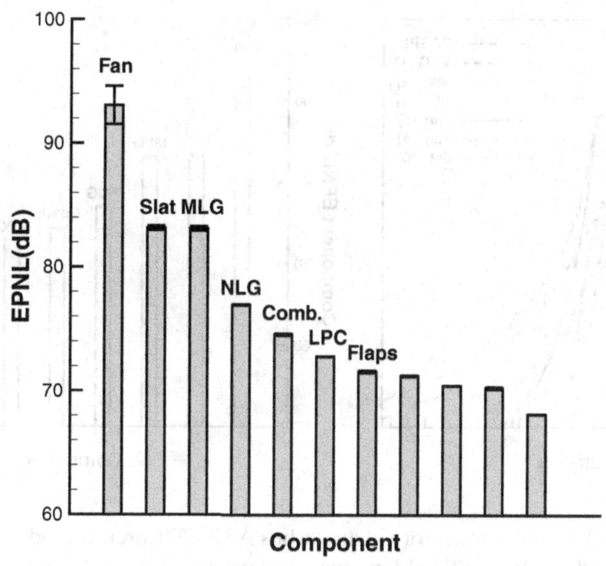


Figure 18.10. Calculated bar chart contribution of the noise trajectory shown in Figure 18.9.

combustor and LPT. The airframe contributions are smaller, for reasons evident in graph 18.11b: at the flight time indicated by A, there is a small thrust reduction, from take-off level to initial climb. At the flight time indicated by B, the landing gears are retracted; thus, the MLG and NLG contributions are zero from this point onwards.

At point C, the slat is retracted, and its contribution is discarded. At the flight time D, there is a thrust cut-back, which contributes to a large reduction in fan and jet noise.

The noise trajectories discussed herein refer to the case of ICAO/FAR certification points. These receiver points must be verified first to secure noise certification by the relevant authorities. However, the community is interested in the noise levels at points well outside the airfield, where the effects can be just as annoying

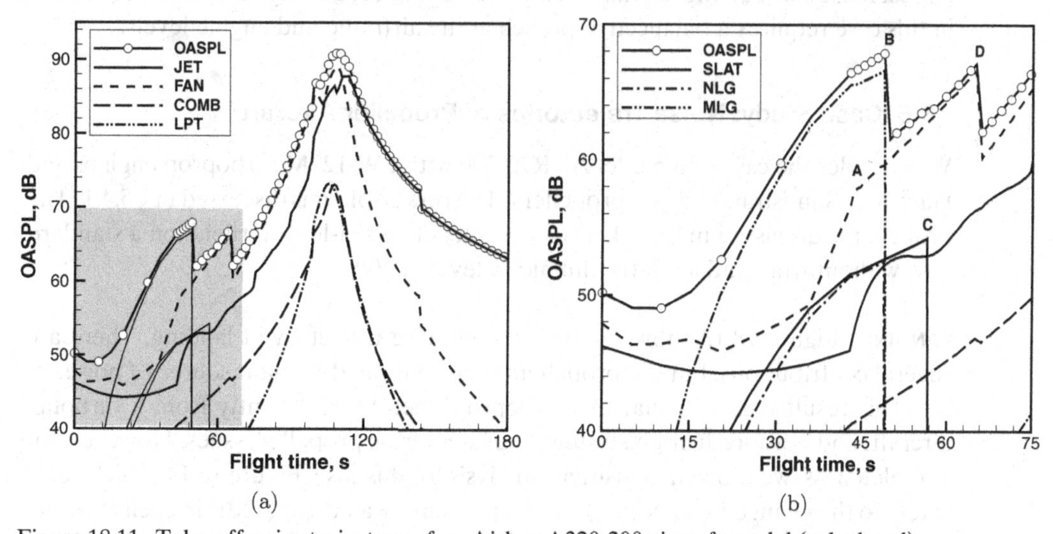


Figure 18.11. Take-off noise trajectory of an Airbus A320-200 aircraft model (calculated).

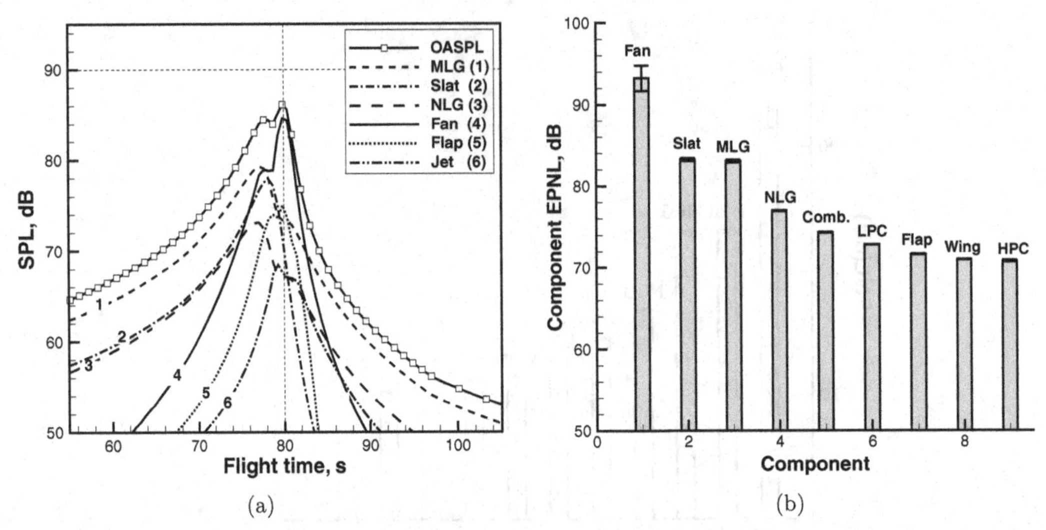


Figure 18.12. Landing-noise trajectory of an Airbus A320-200 aircraft model; reference point 1,000 m upstream of the ICAO/FAR landing; standard day.

as at the runway threshold. We now consider the case of a receiver point placed 1,000 metres upstream of the ICAO/FAR landing point and right below the flight path of the aircraft. The analysis of the noise signature of the aircraft is shown in Figure 18.12.

This analysis shows that there is a relatively long exposure to noise, with several components contributing on a comparable level. Graph 18.12a shows selected contributions (numbered 1 to 6 and named in the legend). The maximum EPNL occurs at about 80 seconds in the aircraft flight time. The fan is the largest source of concern, but the landing gear and the slat are also of considerable importance. Graph 18.12b shows the bar chart of the component contribution to the EPNL. The error bars indicate the noise sensitivity to a ±2 dB change in the noise level of each component, as previously discussed. Aside from the fan's contribution (easily the largest noise source), the airframe components are particularly high. Noise reduction in this case requires a balanced approach at the airframe and engine levels.

## 18.6 Case Study: Noise Trajectories of Propeller Aircraft

We consider the case of a model ATR72-500 with PW-127M turboprop engines and Hamilton-Sundstrand F568-1 propellers. The power-plant is discussed in § 5.4.1. The propeller is discussed in § 6.2.4. The case refers to a sea-level airfield, on a standard day, without wind and a relative humidity level of 70%.

**LANDING.** Figure 18.13 shows the calculated noise trajectory at landing. There are several contributions, but the propellers represent the dominant source of noise.

This result indicates that the turboprop operates differently from a turbofan aircraft, and noise-reduction strategies must address propeller issues. However, for completeness, we show the sensitivity analysis for this case, Figure 18.14, which again refers to the change in EPNL(dB) corresponding to an error ±2dB in each separate noise-source component. The sensitivity is shown by the error bars. The amplitude

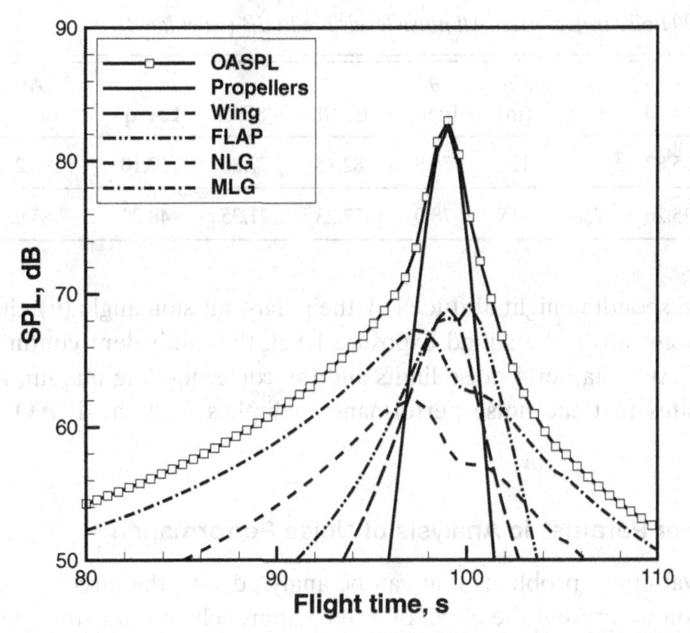


Figure 18.13. Calculated noise trajectory of the ATR72-500 at landing; ICAO/FAR reference; selected noise components shown.

of the error bars denotes the change in total EPNL(dB). In other words, an error of $\pm 2$ dB on the propeller noise causes an error on the EPNL that has an amplitude equal to $+1.551 - (-1.428) = 2.979$ dB. The contribution of the other components to the overall error decreases with the component's EPNL.

**TAKE-OFF.** Figure 18.15 shows a similar case for the take-off trajectory. Again, the propeller noise is the dominant contribution – by far.

A summary of numerical results is in Table 18.3. The data include maximum loudness [dBA], the distance to the receiver corresponding to maximum noise ($r$)

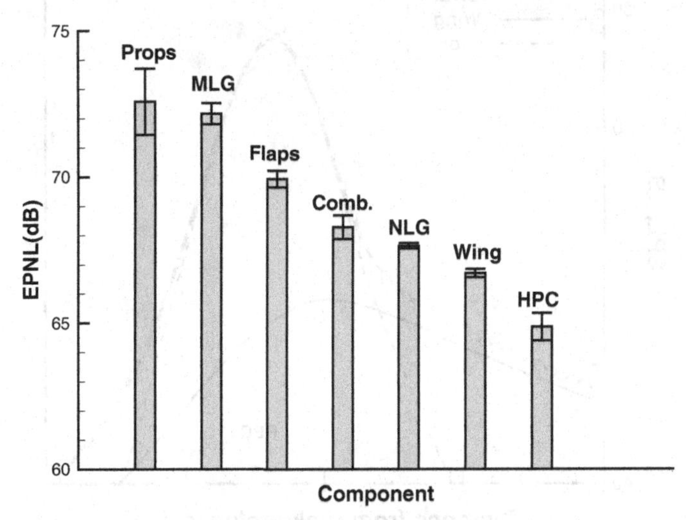


Figure 18.14. Calculated landing-noise sensitivity, case of Figure 18.13.

Table 18.3. *ATR72-500 noise trajectories; All noise levels are in dB (calculated)*

| | m [ton] | Loud | $r$ [m] | $h$ [m] | $\theta$ [deg] | EPNL | SEL | LAeqT | ICAO4 Limit | ICAO4 Margin |
|---|---|---|---|---|---|---|---|---|---|---|
| Landing | 22.67 | 85.87 | 124 | 121 | 76.8 | 82.18 | 77.20 | 57.10 | 96.2 | −14.0 |
| Take-off | 20.61 | 75.26 | 730 | 715 | 78.0 | 77.23 | 71.35 | 48.23 | 85.0 | −7.8 |

and the corresponding flight altitude ($h$), the polar-emission angle ($\theta$), the effective perceived noise level, the sound exposure level, the equivalent continuous noise level, the ICAO Chapter 4 noise limits and the corresponding margin. A negative value indicates that the noise performance complies with the ICAO Chapter 4 regulations.

## 18.7 Further Parametric Analysis of Noise Performance

There is a variety of problems that can be analysed with the aircraft noise model. In this section we present the effect of a steep approach and the atmospheric winds on approach and take-off. All of the calculations have been done with the `FLIGHT` code.

**STEEP DESCENT.** The flight mechanics of steep descent was discussed in § 11.4. We now consider the consequences of a steep glide slope on the noise metrics. The example deals again with the Airbus A320-200 with CFM engines. We have considered a conventional trajectory with $\gamma = 3.01$ degrees and a *steep* trajectory with $\gamma = 3.775$ degrees. The flight takes place in standard atmosphere with negligible wind speeds (the wind is included in the trajectory simulation but not in the noise

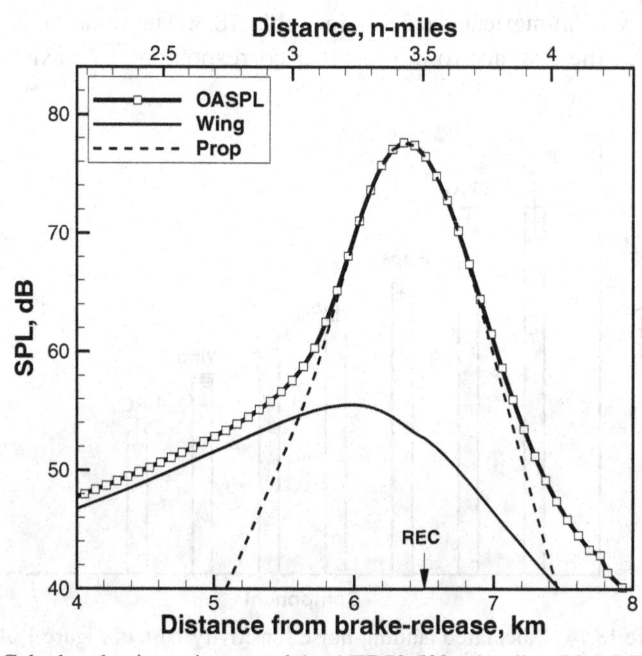


Figure 18.15. Calculated noise trajectory of the ATR72-500 at landing; ICAO/FAR reference.

Table 18.4. *Calculated noise metrics (in dB) over a conventional and steep landing trajectory at a ICAO/FAR landing point and point 1,000 m upstream (3,300 m from touch-down)*

| Metrics | ICAO/FAR | | FAR – 1,000 m | |
|---|---|---|---|---|
| | Normal | Steep | Normal | Steep |
| EPNL | 95.2 | 92.4 | 91.9 | 89.3 |
| SEL | 91.3 | 89.5 | 89.3 | 87.5 |
| LAeqT | 73.0 | 71.9 | 70.7 | 69.5 |

propagation). The resulting noise level is in shown Figure 18.16. Note that the flight times of these trajectories are not synchronised; therefore, the shift in the peak noise level does not correspond to actual conditions. However, the peak is reduced by ~2 dB, and the consequences on the integral noise metrics are of a similar magnitude. A summary of calculated performance is in Table 18.4, which also contains the results for a microphone placed 1,000 m upstream of the ICAO/FAR reference point. We conclude that some noise reduction can be achieved by flying a steep trajectory.

**WIND EFFECTS.** Two sub-cases are considered: 1) ICAO/FAR microphone at landing with head wind or tail wind, and 2) sideline microphone at landing with a 90-degree wind.

Figure 18.17a shows the calculated noise trajectories with a tail-wind, a head-wind and a no-wind condition. The wind speed is $V_w = \pm 4$ m/s, measured 1.2 m from the ground. The difference in EPNL is contained within 0.7 dB, which is due to a combination of increased OASPL peak and duration. However, there is

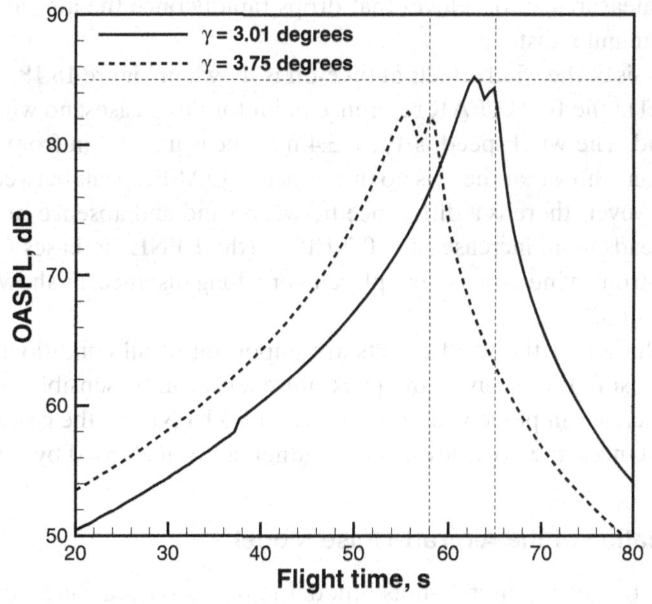


Figure 18.16. Simulated landing-noise trajectories of an Airbus A320-200, conventional and steep approach. The noise metrics at the ICAO/FAR point are given in Table 18.4.

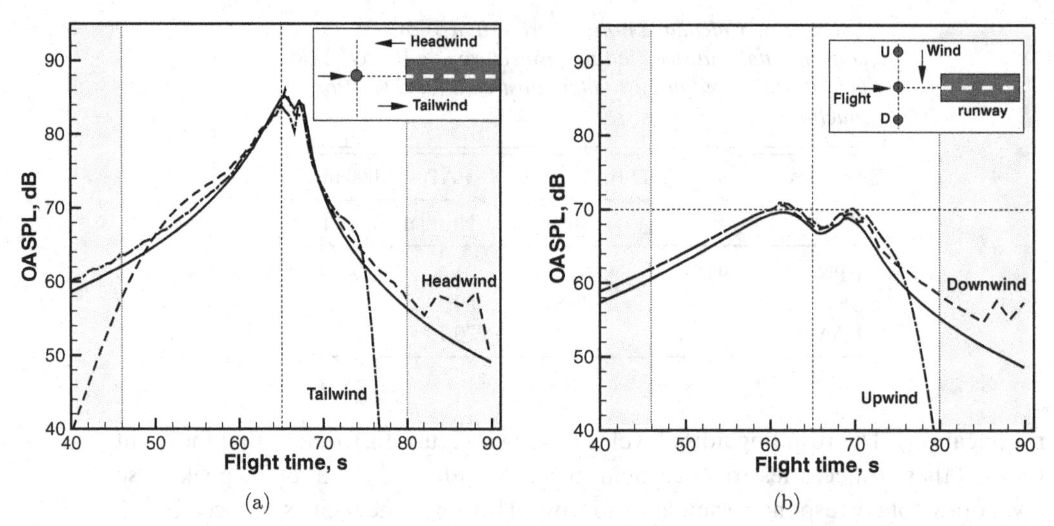


Figure 18.17. Wind effects on perceived noise at landing; aircraft model is the A320-200-CFM (calculated).

a considerable wind effect over long distances, which cuts the noise level at the receiver. One point worthy of note is that the notion of head and tail wind depends on the position of the microphone with respect to the aircraft (see also Figure 17.8).

Figure 18.17b shows the calculated noise trajectories with a fixed wind at 90-degrees, as indicated in the top-right box. The central point is the ICAO/FAR reference point; the side-line points are placed 450 metres off the runway centreline. With a fixed wind direction, one sideline microphone is upwind (U) and the other is downwind (D). The calculations show an upward shift in the OASPL, which contributes to an increase in perceived noise by about 1.5 EPNL. The downwind microphone measures an increased noise level over the full trajectory. The upwind microphone measures a noise level that drops rapidly once the aircraft has passed through its minimum distance.

The analysis of the wind effects at take-off is shown in Figure 18.18, which shows the noise level at the ICAO/FAR reference point for three cases: no wind, tail wind, and head wind. The wind speed is $V_w = \pm 4$ m/s, measured 1.2 m from the ground. The calculations show that there is no difference in OASPL peak between head and tail wind. However, there is a difference between wind and absence of it (top right box), which leads to an increase of $\sim 0.7$ EPNL (the EPNL increases from 90.9 dB to 91.6 dB). Strong wind effects take place over a long distance, as shown to the left of the main graph.

We conclude that the wind effects are important at all conditions. When the aircraft is closest to the receiver, the peak noise level can be sensibly affected so as to cause an increase in perceived noise by 0.7 to 1.5 EPNL for the cases calculated. Over long distances, the noise level can be either cut or increased by several dB.

## 18.8 Verification of the Aircraft-Noise Model

The ultimate test of an aircraft-noise model is based on real flight data that are synchronised with noise measurements on the ground. Data have been gathered

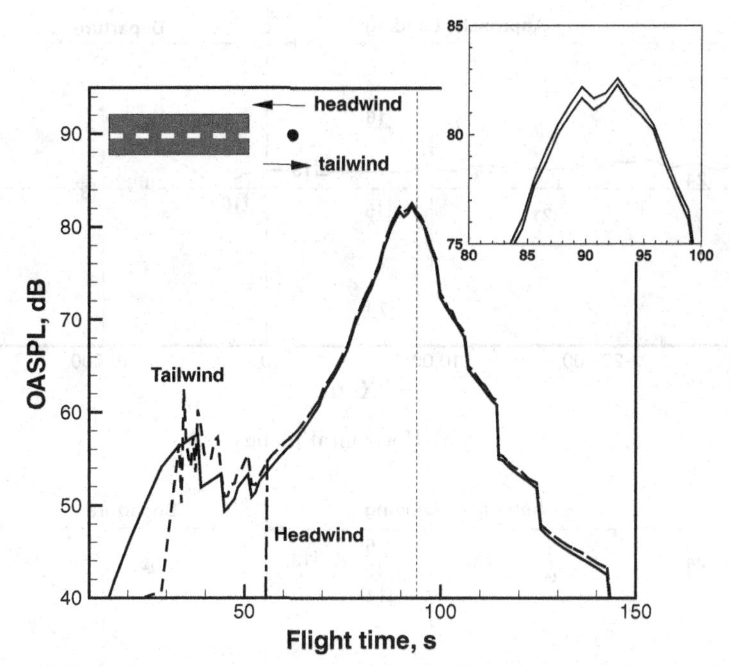


Figure 18.18. Wind effects on perceived noise at take-off; aircraft model is the A320-200-CFM (calculated).

from a campaign of measurements[21;22;23] carried out on the Airbus A319-100 flown by Lufthansa*.

Figure 18.19 shows two views of selected flight trajectories and the relative position of the microphones on the ground. With reference to graph 18.19a, the vertical scale has been expanded and shows lateral movements of the order of one wing span. The change in direction at the start or end of the trajectory is the effect of the aircraft moving along the taxi-way.

Two sets of microphones are available: one set is at 1.2 metres from the ground; the other set is on the ground and is used to remove the effects of ground reflection and diffraction. Microphones 1–12 are on the take-off/departure side; microphones 13–25 are on the approach/landing side. Only selected microphones are shown in Figure 18.19. Measurements were taken at the Schwerin-Parchim airfield (IATA code: SZW; ICAO code: EDOP). The ground track shows an apparently rough flight trajectory, with lateral oscillations of the order of one wing span. None of the microphones is directly below the airplane except microphone 13. Graph 18.19b shows the position of the landing gear (LG) deployment and retraction.

Selected results obtained with the `FLIGHT` program are shown in Figure 18.20 and Figure 18.21. In both cases we have the microphones as named in Figure 18.19. The flight data that have been used are a sub-set of the noise-state vector given by Equation 18.10. The detailed atmospheric conditions were not available. Standard values were assumed.

Overall, the results show a good comparison with the flight data. The peaks are generally well predicted, although there are cases where the peak OASPL are missed.

* Selected raw data have been kindly provided by the Institute of Aerodynamics and Flow Technology of DLR; Braunschweig, Germany, Nov. 2011.

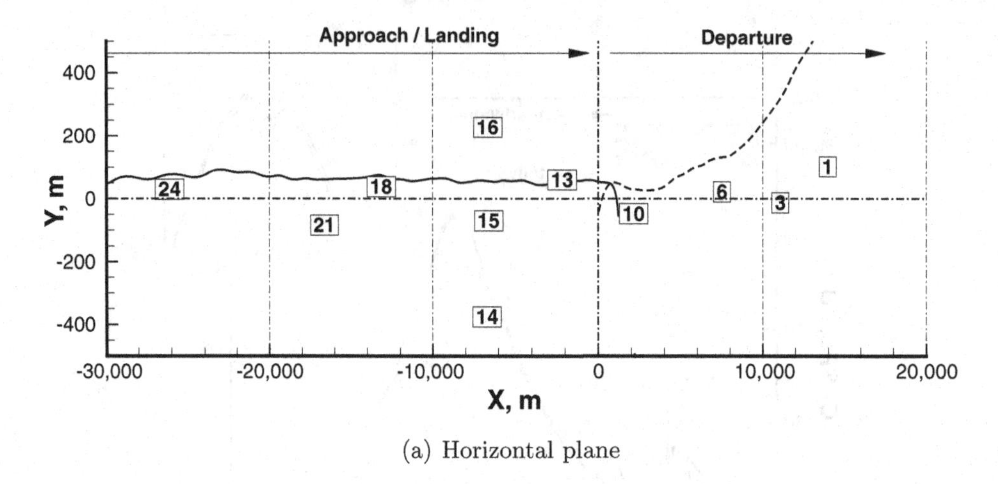


(a) Horizontal plane

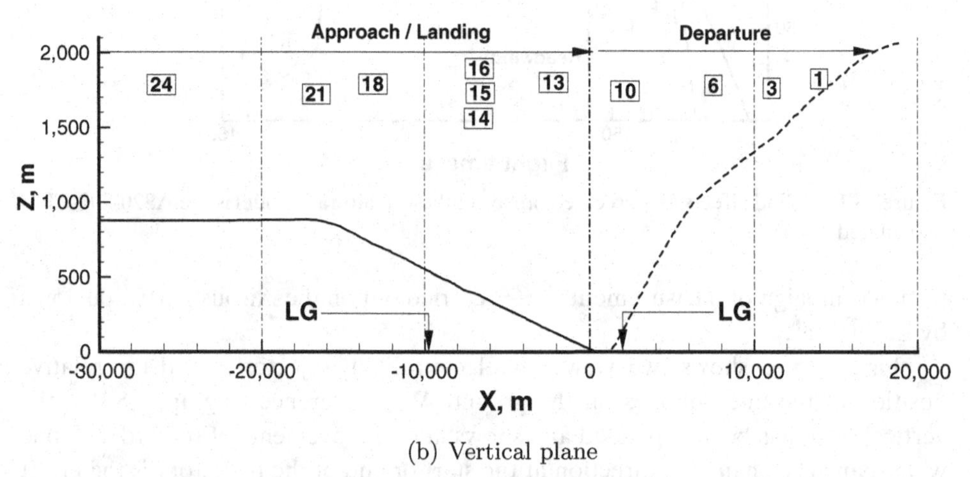


(b) Vertical plane

Figure 18.19. Landing and take-off flight trajectories with microphone positions on the ground.

There are also cases where there is a large difference in OASPL upstream of the peak. However, note that the flight data are affected by high-frequency fluctuations, possibly due to localised gusts and turbulence. Although small wind speeds at the microphone location do not make a large difference (see also Figure 18.17), sudden gusts at altitude affect the propagation characteristics and result in sudden changes in acoustic pressure; these changes can be of the order of $\pm 10$ dB.

**QUESTIONS ON NOISE-PREDICTION ACCURACY.** The complexities of the noise signal point to an important question: What can be achieved with a low-order aircraft noise model? The answer lies in the type of metrics we wish to examine. Because the key metrics for certification, trajectory optimisation and land planning are of the integral type, then we could restrict the accuracy requirements to those characteristics that allow an accurate prediction of these metrics. With reference to the EPNL, we need to be able to predict at least the following quantities: 1) peak noise level; 2) flight time corresponding to the peak level; and 3) noise level over the 10 dB-down time. Additionally, we would need some accuracy in the spectral content of the noise,

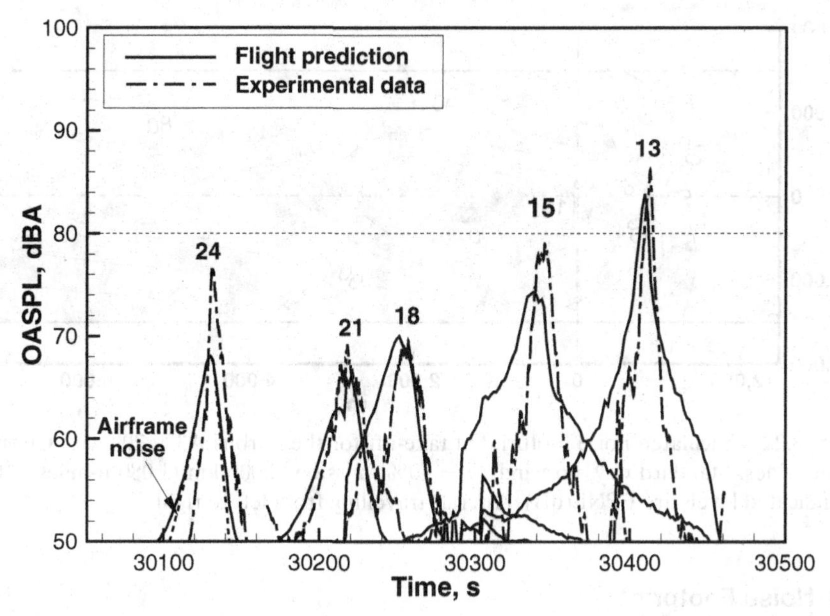


Figure 18.20. Comparison between prediction and experimental data for an approach/landing trajectory. Microphone numbers are the same as in Figure 18.19; microphones on the ground.

which is more difficult to establish because its nature depends on the combination of the various contributions.

The sensitivity analysis has highlighted the components that need addressing for the noise reduction. Therefore, the parameters listed here would certainly apply to those components.

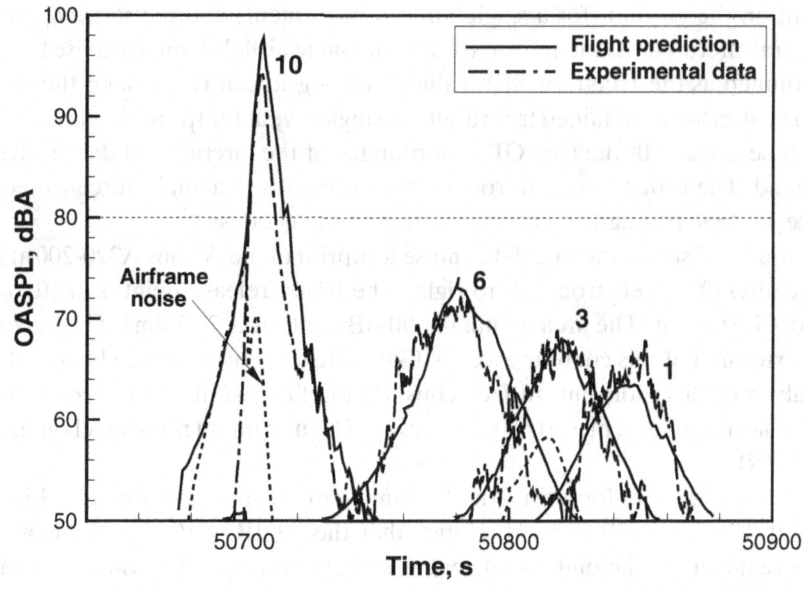


Figure 18.21. Comparison between prediction and experimental data for a take-off/departure trajectory. Microphone numbers are the same as in Figure 18.19; microphones on the ground.

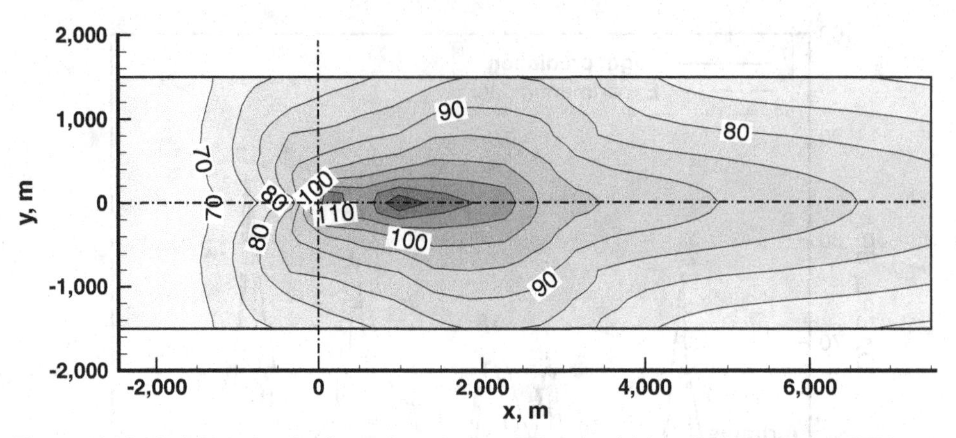


Figure 18.22. Calculated noise footprint at take-off for the Airbus A320-200 with CFM turbofan engines. Standard day, no wind, $\mathcal{H} = 70\%$; mission: 2,000 km (1,080 n-miles), 100% pax. Indicated levels are EPNL(dB). Aircraft travelling from left to right.

## 18.9 Noise Footprint

Once the relevant noise metrics have been defined, several zones around the airfield are defined for the purpose of land use and noise management, including Zone A, which is in the proximity of the runways and subject to the most disruptive noise; and Zone B, subject to more moderate noise exposure. Other zones (such as C and D) are sometimes defined. These areas can be several square kilometres around the airport, although their extent has shrunk over the years. The 90-dB-level noise footprint of the first jet was ~30 km$^2$, whilst a modern regional jet has a 90-dB-level footprint of ~5 km$^2$.

A straightforward extrapolation of the noise trajectory analysis is the noise footprint on the ground (for a single aircraft movement) and the *stacking patterns* (for several aircraft movements in and out of the same airfield). First, a noise footprint is determined as the EPNL or SEL values over a grid centred around the airfield. Stacking patterns are obtained from a sum of single-event footprints. Analyses of this type can be done with the true GPS coordinates of the aircraft and the receiver on the ground. The transformation from GPS coordinates to actual source-to-receiver distances is done through an inverse solution of the geodesics[20].

Figure 18.22 shows the calculated noise footprint of the Airbus A320-200 at take-off. The aircraft travels from left to right. The brake-release point is at (0,0). The trajectory is straight. The area within the 90-dB contour is 7.12 km$^2$. The maximum lateral extension of this contour exceeds 1 km. The sideline noise can be found from the analysis of this footprint. In fact, consider the line parallel to the centre of the runway (dash-dot line), placed 450 m laterally. The maximum noise level on this line is ~97 EPNL.

The landing-noise footprint for the same case is shown in Figure 18.23. The aircraft travels from left to right. Notice that the 90-dB contour extends about 3 km upstream of the landing point, but it is very narrow, ~0.1 km in the lateral direction. This result is very important because it shows that receivers below the flight path are subject to high noise levels, whilst receivers away from the flight path are comparatively unaffected. If atmospheric winds are taken into account,

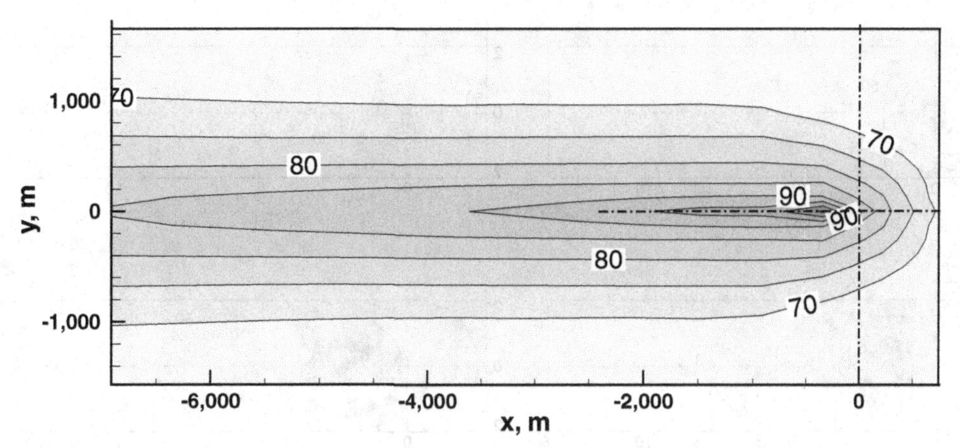


Figure 18.23. Calculated noise footprint at landing for the Airbus A320-200 with CFM turbofan engines. Standard day, no wind, $\mathcal{H} = 70\%$. Mission: 2,000 km (1,080 n-miles), 100% pax. Indicated levels are EPNL(dB). Aircraft travelling from left to right.

receivers upwind of the flight path will be affected by lower noise levels and receivers downstream will suffer higher noise.

The noise footprint of a propeller aircraft such as the ATR72-500 shows considerable differences from the turbofan engines discussed so far. Figure 18.24 shows the calculated footprint at landing. The equal-noise curves envelop a much smaller area. To be fair, we would need to consider the weight effects.

**REAL-TIME METRICS.** The previous examples of noise footprint illustrate the "steady-state" noise performance of the aircraft because we have used the EPNL. However, if we want to look at the real-time effect of a passing aircraft (approaching or taking off), the EPNL cannot be used. We then consider an instantaneous noise level described by the OASPL. One such example is shown in Figure 18.25. The noise carpet spreads far and wide, particularly after take-off. Whilst on the ground, the noise is strongly affected by the ground reflection.

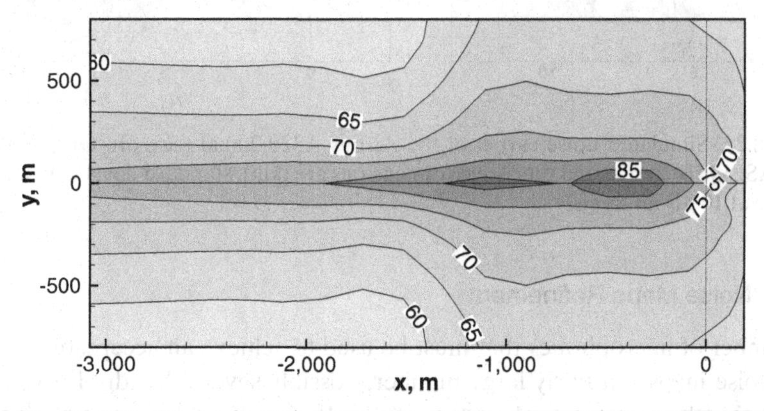


Figure 18.24. Calculated noise footprint at landing for the ATR72-500 with PW127M turboprop engines; standard day, no wind, $\mathcal{H} = 70\%$. Mission: 1,000 km (540 n-miles), 100% pax. Indicated levels are EPNL(dB). Aircraft travelling from left to right.

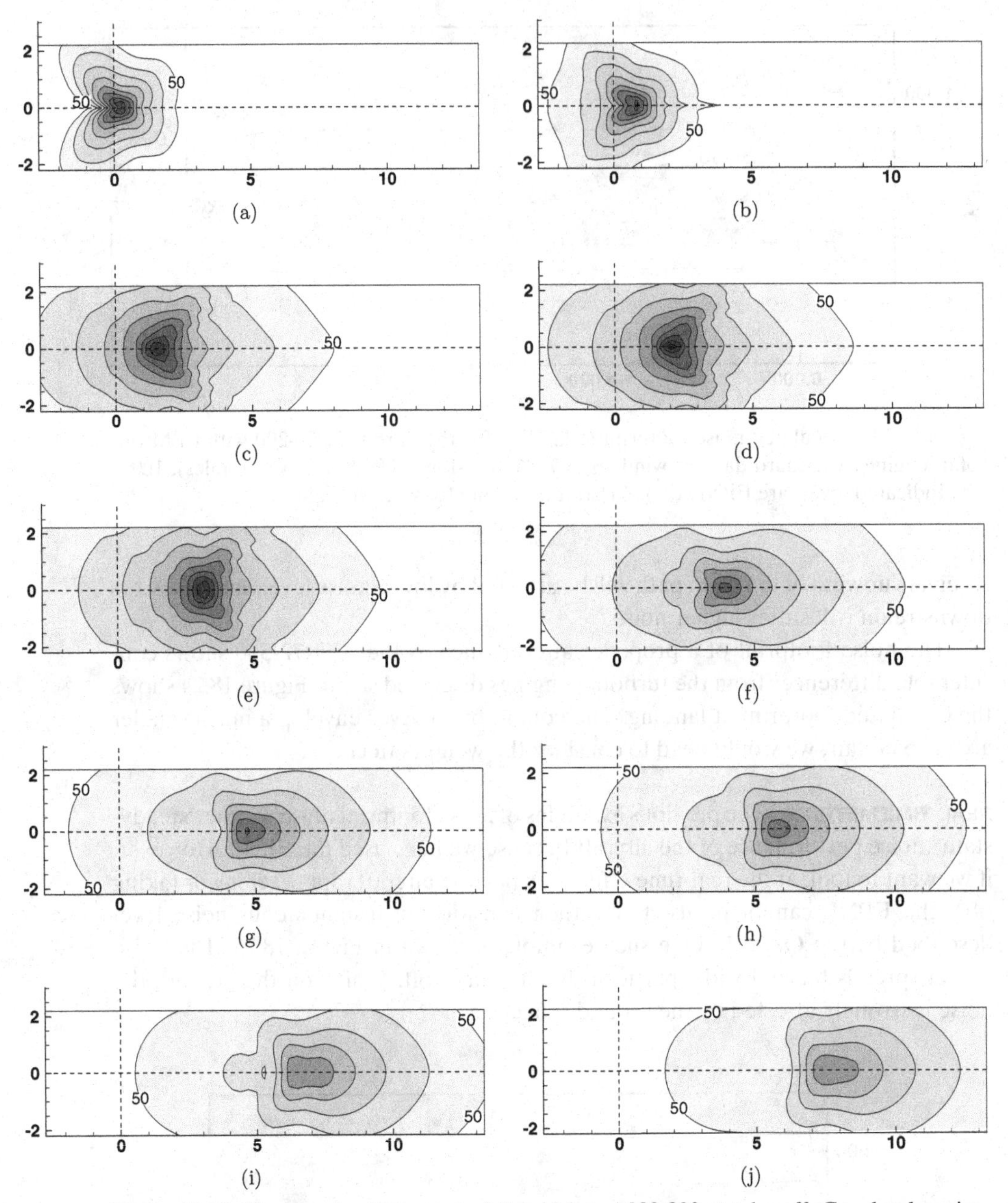


Figure 18.25. Simulated noise carpet of the Airbus A320-200 at take-off. Graphs show iso-level OASPL[dB] at selected time steps; dimensions are [km]; standard day; no wind; OASPL levels are at 10 dB intervals.

### 18.9.1 Noise Maps Refinement

The number of microphones that must be used to achieve an acceptable resolution of the noise maps is a fairly large number, possibly several hundred. Even from a computational point of view, the use of hundreds of microphones may require several hours of calculation. Therefore, we propose to refine the noise map by increasing the resolution of the grid *after* the noise footprint is calculated.

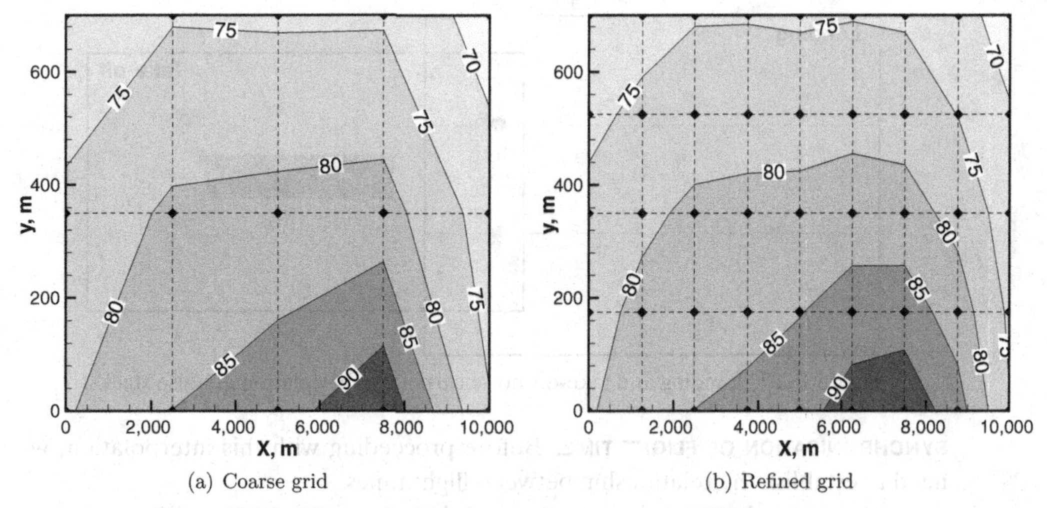


(a) Coarse grid (b) Refined grid

Figure 18.26. Effects of grid refinement on a generic EPNL map.

If the grid is defined by the arrays $x = x(1, \cdots, n_x)$ and $y = y(1, \cdots, n_y)$, we double the size of these arrays, e.g. we halve the resolution of the grid. The transformed grid will be defined by arrays $x_2 = x_2(1, \cdots, 2n_x - 1)$ and $y_2 = y_2(1, \cdots, 2n_y - 1)$. Once the grid has been refined, we apply a cubic spline interpolation in two directions to calculate the function value at the new grid points[24]. A typical result of this operation is shown in Figure 18.26. The refined map gives a more accurate estimation of the area affected by noise.

## 18.10 Noise from Multiple Aircraft Movements

We discuss the calculation of the noise carpet due to the operation of one airplane type from one airfield (*noise stacks*). This is a simplified case because, in reality, a commercial airfield will operate with several aircraft types. We have two noise carpets: approach/landing and take-off/climb-out. The reason for two separate carpets is that the area affected is relatively large.

**OVERLAP OF LANDING AND TAKE-OFF GRIDS.** The arrays $x_L(..)$, $y_L(..)$ represent the landing-carpet grid coordinates; the arrays $x_T(..)$, $y_T(..)$ represent the take-off/climb-out noise carpet. These carpets are partially overlapping, as shown in Figure 18.27. The black dots are landing grid points which fall within the take-off carpet. At these points a sum between the noise level $L_L$ and $L_T$ is required. Because the noise $L_T$ is not available at this black grid point, an interpolation is carried out to determine its value. We use a bilinear interpolation within the top cell (in dashed line). The white dots denote grid points in the take-off carpet which overlap the landing carpet. Again, a bilinear interpolation within the relevant grid cell is used to calculate the noise level $L_L$ at this point.

For the overlapping grid points the noise level is

$$L = 10 \log_{10} \left( 10^{L_T/10} + 10^{L_L/10} \right). \tag{18.16}$$

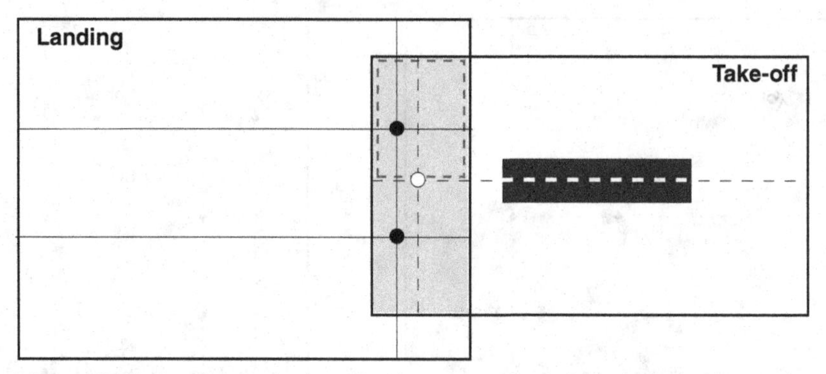


Figure 18.27. Landing and take-off noise carpets for calculation of noise stacks.

**SYNCHRONISATION OF FLIGHT TIME.** Before proceeding with this interpolation, we need to establish the relationship between flight times.

Let us consider first all operations in the same category (approach or take-off). At a given time, there can be an arbitrary number of aircraft, depending on separation and carpet size. If the flight trajectories have $n$ time steps, which are separated from each other by $n_s$ time steps, then the correlation between the flight movement $j$ and its time step $i$ is found from the following relationships:

$$t(i, j) = n, \qquad n = i_1, \cdots, i_2, \qquad i_1 = 1 + n_s(j-1), \qquad i_2 = i_1 + n - 1. \quad (18.17)$$

In other words, Equation 18.17 is a matrix with sub-vectors $\{1, 2, \cdots, n\}$ shifted by $n_s$ steps; all other entries in the matrix are zero. The number of airplanes within the carpet is determined by the relationship between $i$, $n_s$ and $n$; if $n_s > n$, there will be one airplane at the time.

We use a Universal Time (UT). The raw trajectory of an approaching flight $i_L$ has a time $t_{i_L}$, with $t_{i_L} = 0$ at the start. Likewise, the trajectory of a take-off flight $i_T$ has a time $t_{i_T}$, with $t_{i_T} = 0$ at the start (brake-release point). There is a specified separation between landing flights. To synchronise these flights, the time sequences are transferred into a universal time.

Next, consider the approach/landing carpet as a reference; if $t_{i_L}$ is the current time step, the take-off carpet must be calculated exactly at this time. The sequence of time steps $t_{i_T}$ generally does not coincide with the sequence $t_{i_L}$, but it should be possible to find an index $i_T$ such that $t_{i_T} < t_{i_L} < t_{i_T+1}$. If this is the case, we carry out an interpolation of the SPL in time, between $i_T$ and $i_T + 1$.

Figure 18.28 shows the noise signature predicted with the program `FLIGHT` by a sequence of Airbus A320-200s landing and taking off at one airfield, along a conventional straight trajectory. The aircraft movements are separated by 75 seconds.

The touch-down point is at $x = 0$; the aircraft travels from the left to the right of the page. For the case shown in Figure 18.28 the calculation of the cumulative metrics (§ 18.2.1) is straightforward because the noise level is the same for all flights. The symbols L1, L2 denote approach/landing flight numbers 1 and 2, respectively; the symbols T1, T2 denote take-off flight numbers 1 and 2, respectively. In frame 18.28h there are two approaching flights and in frame 18.28i there are two take-off flights. The noise pattern on the ground is strongly dependent on the separation time.

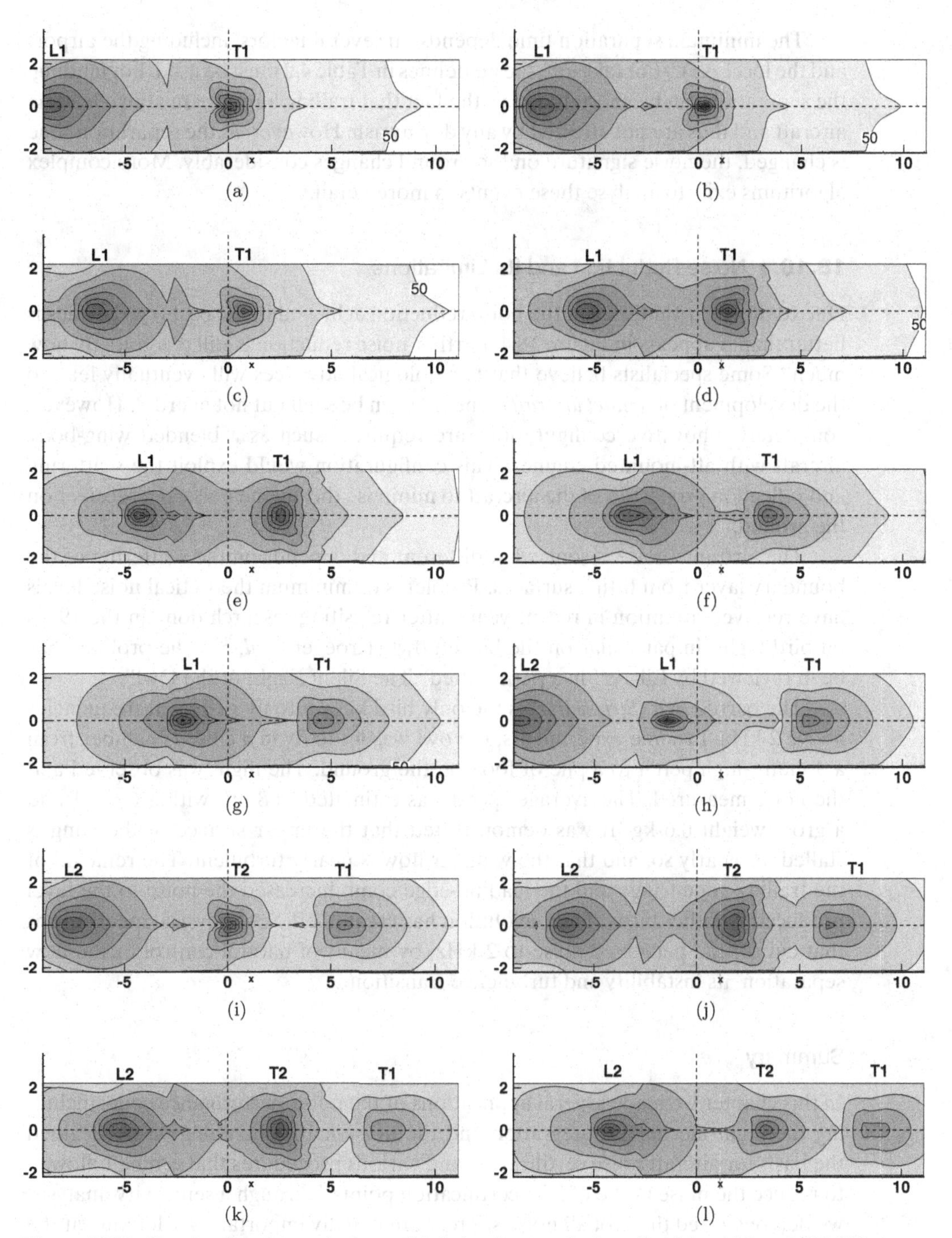


Figure 18.28. Simulated noise stacks for multiple landings and take-offs of the Airbus A320-200; $\Delta t_s = 75$ s. Graphs show iso-level OASPL[dB] at arbitrary time steps; OASPL intervals are 5 dB; carpet dimensions are [km].

The minimum separation time depends on several factors, including the airport and the local ATC. For take-off, the guidelines in Table 4.2 must be used. For landing, the separation can be shorter, due to the fact that trailing aircraft are above leading aircraft and thus are not affected by any downwash. However, as the separation time is changed, the noise signature on the ground changes considerably. More complex algorithms exist to analyse these events in more detail.

### 18.10.1 Noise Reduction and Its Limitations

Due to the logarithmic scale, the noise reduction achieved in the past few decades is better than it appears in Figure 18.4. Further noise reduction is still possible. By how much? Some specialists believe that technological advances will eventually lead to the development of a *quiet aircraft* – one that can be seen but not heard[25]. However, completely innovative configurations are required, such as a blended wing-body aircraft with aft-mounted engines. This configuration would exploit the scattering and reflection properties of the aircraft to minimise the engine noise at a receiver on the ground.

The airframe noise is somewhat different and depends on the scattering of the boundary layer from lifting surfaces. Problems of minimum theoretical noise levels have received attention in recent years, after revisiting research done in the 1970s on bird flight, in particular on the barred owl (Kroeger *et al.*[26]). The problem has been reviewed by Lilley[27] in a paper titled "The Silent Flight of the Owl".

The barred owl (*Strix varia*) is the only bird known to fly *silently* at frequencies above 2 kHz. In some experiments, the owl was let to fly in a closed chamber from a 3-metre-high perch to a pile of food on the ground. The flight was observed and the noise measured. The average speed was estimated as 8 m/s with a $C_L \simeq 1$ and a gross weight 0.6 kg. It was demonstrated that the upper surface of the wing is stalled, or nearly so, and that the wing air flow is nearly turbulent. The removal of the trailing-edge fringe and the leading-edge comb increased the noise to the level produced by other birds. Thus, evolution has created efficient physical mechanisms that cut off the perceived noise to 2 kHz, by means of natural control of the flow separation, its instability and turbulence transition.

## Summary

In this chapter we made several applications of the noise-performance code, including the flight-mechanics integration and the noise-matrix handling. We explained the basis for aircraft noise certification and various procedures that can be followed to reduce the noise at the official certification points. Through a sensitivity analysis, we demonstrated that not all noise sources are equally important. By looking at the EPNL(dB) integral metrics, we made a detailed analysis of the noise-level contributions at take-off and landing. The latter case is demonstrated as more complicated because there are several noise sources that contribute to the overall EPNL at a similar level. At take-off, the most important noise components are the engine fan and jet/nozzle. For a propeller aircraft we demonstrated that the dominant acoustic effect is due to the propeller itself. Wind effects are always important. Over short distances, they may contribute to a change in EPNL of the order of 1.5 dB. Over long distances, they can affect the sound level by several dB. A comparison with flight-noise data has shown the complexities of the acoustic signals that must be simulated.

The analysis of the noise trajectories, especially approach and landing, has demonstrated that the highest noise level is concentrated below the flight path. It is safe to assert that in absence of atmospheric winds, at a lateral distance of ~1.5 km from the flight path, the noise level is relatively low (with respect to other sources of noise). In the presence of atmospheric winds, it is best to be placed at an upwind position with respect to the flight path. Having the option between living in a mansion below/downwind the flight path or in a modest dwelling, the choice should be the latter – every time.

## Bibliography


[1] Franssen EAM, van Wiechen CMAG, Nagelkerke NJD, and Lebret E. Aircraft noise around a large international airport and its impact on general health and medication use. *Occup. Environ. Med.*, 61(5):405–413, 2004.

[2] Alexandre A, Barde J-Ph, and Pearce DW. The practical determination of a charge for noise pollution. *J. Transport Econ. & Policy*, 14(2):205–220, May 1980.

[3] Gebhardt GT. Acoustical design features of Boeing Model 727. *J. Aircraft*, 2(4):272–277, 1965.

[4] Crighton DG. Model equations of nonlinear acoustics. *Ann. Rev. Fluid Mech.*, 11:11–33, 1979.

[5] Smith MTJ. *Aircraft Noise*. Cambridge University Press, 2004.

[6] Shahadi PA. Military aircraft noise. *J. Aircraft*, 12(8):653–657, 1975.

[7] ECAC. Report on standard method of computing noise contours around civil airports. Vol. 1: Applications guide. Technical Report Doc. 29, European Civil Aviation Conference, Dec. 2005.

[8] ECAC. Report on standard method of computing noise contours around civil airports. Vol. 2: Technical guide. Technical Report Doc. 29, European Civil Aviation Conference, Dec. 2005.

[9] Waitz I, Kukachko S, and Lee J. Military aviation and the environment: Historical trends and comparison to civil aviation. *J. Aircraft*, 42(2):329–339, 2005.

[10] Filippone A. Steep-descent manoeuvre of transport aircraft. *J. Aircraft*, 44(5):1727–1739, Sept. 2007.

[11] Fink MR. Noise component method for airframe noise. *J. Aircraft*, 16(10):659–665, 1979.

[12] Fink MR and Schlinke RH. Airframe noise component interaction studies. *J. Aircraft*, 17(2):99–105, 1980.

[13] Zorumski WE. Aircraft noise prediction program theoretical manual, Part 1. Technical Report TM-83199, NASA, Feb. 1982.

[14] Kontos K, Janardan B, and Gliebe P. Improved NASA-ANOPP noise prediction computer code for advanced subsonic propulsion systems – Volume 1. Technical Report CR-195480, NASA, Aug. 1996.

[15] Lopes LV and Burley CL. Design of the next generation aircraft noise prediction program: ANOPP2. In *17th AIAA/CEAS Aeroacoustics Conference*, AIAA 2011-2854, Portland, June 2011.

[16] Clark BJ. Computer program to predict aircraft noise levels. Technical Report TP-1913, NASA, 1981.

[17] Boeker ER, Dinges E, He B, Fleming G, Roof CJ, Gerbi PJ, Rapoza AS, and Hemann J. Integrated noise model (INM) Version 7.0. Technical Report FAA-AEE-08-01, Federal Aviation Administration, Jan. 2008.

[18] Ollerhead JB, Rhodes DP, Viinikainen MS, Monkman DJ, and Woodley AC. The UK civil aircraft noise contour model ANCON: Improvements in Version 2. Technical Report R&D 9842, Environmental Research and Consultancy Dept., Civil Aviation Authority (CAA), June 1999.
[19] Bertsch L, Dobrzynski W, and Guerin S. Tool development for low-noise aircraft design. *J. Aircraft*, 47(2):694–699, Mar. 2010.
[20] Vincenty M. Direct and inverse solutions of geodesics on the ellipsoid with application of nested equations. Technical Report XXIII, No. 176, Survey Review, 1975.
[21] Pott-Pollenske M, Dobrzynski W, Buchholz H, and Guerin S. Airframe noise characteristics from flyover measurements and predictions. In *12th AIAA/CEAS Aeroacoustics Conference*, AIAA 2006-2567, Cambridge, May 2006.
[22] Guerin S and Michel U. Prediction of aero-engine noise: comparison with A319 flyover measurements. In *Tech. Rep., DLR*, DLR IB 92517-04/B3, 2007.
[23] Bertsch L, Guérin A, Looye G, and Pott-Pollenske M. The parametric aircraft noise analysis module – status overview and recent applications. In *17th AIAA/CEAS Aeroacoustics Conference*, AIAA-2011-2855, Portland, Oregon, USA, 5–8 June 2011.
[24] Press WH, Teukolsky SA, Vetterling WT, and Flannery BP. *Numerical Recipes*. Cambridge University Press, 2nd edition, 1992.
[25] Crichton D, de la Rosa Blanco E, Law T, and Hileman J. Design and operation for ultra low noise take-off. In *45th AIAA Aerospace Sciences Meeting*, AIAA 2007-0456, Reno, NV, Jan. 2007.
[26] Kroeger RA, Gruska HD, and Helvey TC. Low speed aerodynamics for ultra quiet flight. Technical Report TR-971-75, US Air Force Flight Directorate Laboratory, 1971.
[27] Lilley GM. A study of the silent flight of the owl. AIAA Paper 1998-2340, 1998.

## Nomenclature for Chapter 18

$a$ = speed of sound
$A$ = wing area
$\mathcal{A}$ = atmospheric attenuation of sound
$C$ = cost, Equation 18.4
$C_L$ = lift coefficient
$d$ = diameter
$f$ = frequency
$F$ = financing
$h$ = flight altitude
$\mathcal{H}$ = relative humidity
$i, j$ = counters for grid arrays $(x, y)$
$i_1, i_2$ = index for start/finish of noise trajectory in noise stacks, Equation 18.17
$l$ = reference length
$L$ = noise level in dB
$L_{AE}$ = sound exposure level
$L_{EQ}$ = equivalent continuous sound level (prescribed time)
LG = binary value for landing-gear state
$m$ = mass
$\dot{m}_f$ = fuel flow

$M$ = Mach number
$M_n$ = number of aircraft movements on $n-$th year of service
$n$ = number of steps in flight trajectory, Equation 18.17; number of years; number of aircraft movements in Eq. 18.7 and 18.8
$n_s$ = number of separation time steps, Equation 18.17
$n_x, n_y$ = number of grid points in footprint calculation
$N_1$ = engine's rpm
$p$ = pressure; acoustic pressure
$p_a$ = atmospheric pressure
$\mathcal{P}$ = price
$r$ = distance
$\boldsymbol{r}$ = position vector
$\mathcal{R}$ = gas constant
$R_n$ = residual value of noise technology at $n-$th year, Equation 18.5
$\mathcal{S}$ = vector operator, Equation 18.10
SF = binary value for flaps/slats state
$t$ = time
$t_r$ = time at receiver location
$\mathcal{T}$ = atmospheric temperature
$V$ = true air speed
$V_g$ = ground speed
$V_S$ = stall speed
$V_w$ = wind speed
$V_2$ = take-off speed
$W$ = aircraft weight
$x$ = distance on the ground
$x, y$ = grid coordinates for footprint calculation
$z$ = vertical coordinate or altitude
$z_g$ = ground elevation

## Greek Symbols

$\gamma$ = ratio between specific heats
$\delta$ = relative pressure
$\theta$ = pitch angle/attitude
$\lambda$ = temperature-lapse rate
$\rho$ = air density
$\phi$ = bank angle
$\varphi$ = angle of incidence of acoustic wave
$\Psi_w$ = wind direction
$\psi$ = heading angle, Equation 18.10

## Subscripts/Superscripts

$[.]_a$ = airframe
$[.]_c$ = component
$[.]_e$ = engine

| | | |
|---|---|---|
| $[.]_i$ | = | position index; array index |
| $[.]_j$ | = | component index; array index |
| $[.]_L$ | = | landing, in § 18.10 |
| $[.]_k$ | = | frequency index |
| $[.]_o$ | = | reference or sea-level conditions |
| $[.]_r$ | = | receiver |
| $[.]_s$ | = | source |
| $[.]_T$ | = | take-off, in § 18.10 |
| $\tilde{[.]}$ | = | corrected value |
| $\overline{[.]}$ | = | average value |

# 19 Environmental Performance

## Overview

The previous chapters have shown several aspects of environmental performance, including aircraft noise and carbon emissions. This chapter elaborates further on the impact of aviation. We discuss the effects of commercial aviation on the formation of condensation trails (§ 19.1) and possible methods for mitigation, including altitude flexibility. We discuss briefly the controversial issue of radiative forcing of various forms of pollution (§ 19.2) and a method for calculating the landing and take-off emissions (§ 19.3). We show a case study for a transport aircraft to illustrate key parametric effects on carbon-dioxide emissions (§ 19.4). We then propose an example of "perfect flight" (§ 19.5), that is, a flight that is not constrained by air traffic regulations. One of the key aspects that is due to dominate emissions (the trading scheme) is briefly reviewed in § 19.6.

This chapter does not attempt to enter the debate of environmental performance and its relationship with the climate change, as many stakeholders and interdependencies are involved. Furthermore, these deeper issues now overlap those of atmospheric physics, biochemistry and policy-making[1], which are beyond our context.

**KEY CONCEPTS:** Aircraft Contrails, Contrail Factor, Radiative Forcing, Altitude Flexibility, Carbon Emissions, LTO Emissions, Emissions Trading, Perfect Flight.

## 19.1 Aircraft Contrails

The blue skies! – What a rare pleasure at our latitudes. One cold afternoon, a clearing in the clouds brings crispy skies. The air traffic, which is normally invisible above the clouds, reveals the full scale of its impact, as shown in Figure 19.1. The photo shows at least 14 distinct vapour trails, from less than 1 minute old to about 30 minutes. The oldest contrails evolve into cirrus clouds.

The contrail cover is a well-known side effect of commercial flights. Thus, it is natural to ask: What can we do to prevent them from appearing? Several aspects must be considered. First is the understanding of the atmospheric conditions that lead to extensive contrail formation. Second is the understanding of how the gas-turbine

Figure 19.1. Contrail cover at 16:00 GMT, 22nd January 2011 (sun-down); geo-coordinates 53° 37′ N; 2° 36′ W (approximate), looking SW.

engines operate and also how their exhaust gases contribute to these artificial clouds. There is the question of whether a change in operational parameters (flight altitude, speed and routing) can minimise the impact. Finally, to be taken seriously, we should also ask whether the elimination of the contrails addresses more important environmental aspects. With regard to the latter concern, there are academic studies that have highlighted considerable effects, beyond the pollution of the blue skies. For example, a study by Travis *et al.* [2;3] showed that during the unprecedented four-day halt in commercial flights above North America following the 9/11 events, the absence of contrails caused an increase in daily temperature change by more than 1°C. This result is attributed to persisting contrails, which can reduce the radiative heat transfer between the ground and the higher atmosphere, thus trapping heat at lower altitudes at night time and reducing radiative heat from the sun during day time. Such an experiment could have been repeated over Northern Europe in Spring 2010, when most of the air traffic was shut down due to unprecedented (again) ash-cloud fallout from a volcanic eruption in Iceland.

Not all contrails are clearly visible with a naked eye. For a given viewing angle, the visibility is determined by light scattering in the direction of the observer. If the scattering generates a luminance contrast above a threshold value, the contrail is visible with a naked eye. The threshold value is determined by the optical thickness (or optical depth), defined as[4]

$$\tau = \frac{3}{2}\frac{D}{d}\left(\frac{\rho}{\rho_p}\right) m_l Q_{ext}, \tag{19.1}$$

where $D$ is the geometrical thickness of the contrail in the direction of the observer, $d$ is the average diameter of the particles in the contrail, $\rho_p$ is the mean particles density, $m_l$ is the mass fraction of liquid or iced water, and $Q_{ext}$ is a parameter called *extinction efficiency*, which depends on the wave-length of light in the visible spectrum. A detailed analysis of Equation 19.1 indicates that the optical depth is dependent on the emission index for ice particles and varies from zero to approximately 3.84 for a particle diameter $d \sim 6\ \mu$m. Review of data in the technical literature indicates that a contrail having a physical depth of 50 m is visible if made of water droplets with $d \sim 2\ \mu$m and water content $\sim 10^2$ kg/m$^3$. As far as viewing angles are concerned, the most favourable conditions are when the contrails are at a small angle with respect to the sun.

The speculation that contrails affect the climate is not new. A short article on contrails appeared in the magazine *Popular Mechanics* in March 1943. At that time there was only a limited effect, due to constraint on both altitude and speeds; airplanes of that time were powered by internal-combustion engines.

A study published in 1970[5] indicated that the ice crystals in the contrail caused an inadvertent cloud seeding, thus leading to increased cloud cover and modification of precipitation conditions. The research on contrail formation is now mature and extensive. The research on mitigating options is still lagging behind. A work published by Fichter *et al.*[6] addressed the impact of the flight altitude and the possibility of changing flight levels. This paper demonstrated that a descent in 2,000 feet ($\sim$ 610 m) steps decreases the likelihood of contrail formation. A descent by 6,000 feet ($\sim$1,830 m) would decrease the contrail coverage by 45%. Ström and Gierens[7] carried out a simulation of a liquid-hydrogen–powered airplane and found that although the amount of water vapour in the exhausts is considerably larger than in the exhaust plumes of a conventional airplane, the absence of aerosol particles (soot) decreases the formation of ice crystals; hence the contrails are optically thinner.

Consideration of alternative fuels such as hydrogen is important, but it does not consider the real constraints of the aviation industry. The only available power plants are gas-turbine engines running on aviation kerosene. This is not going to change anytime soon. Commercial airplanes of the current generation will still be in service 25 years thence, with some airplanes operating for as many as 40 years. New airplanes being rolled out now (Airbus A380, Boeing 787, Airbus A350) will be operating beyond 2050. No one will scrap these airplanes with the argument that a new, more environmental airplane is on the way. Although power-plant re-engineering is not uncommon, as discussed in Chapter 1, the question is to what extent a revolutionary power plant (that yet does not exist) can be integrated into an existing airframe and ground infrastructure.

### 19.1.1 Cirrus Clouds

Aside from the visual aspects of the contrails moments after formation, there are several – more serious – aspects that must be considered. We are specifically concerned with the evolution of the contrails into *cirrus* clouds and the effects of the radiation balance within the lower atmosphere[8;9].

Cirrus clouds are established at relatively high altitudes, between the high troposphere and the lower stratosphere. They are composed exclusively of ice particles. These types of cirrus clouds have a long life in supersaturated air. The cirrus clouds expand over a period of several minutes to a few hours, aided by high-altitude winds; eventually they transform into cirrus clouds that are almost natural, although their origin is not. One such example is shown in Figure 19.2 where a cirrus cloud, several minutes old (top right), is seen near a fresh contrail (left).

Figure 19.2. Example of degeneration of a contrail into a cirrus cloud.

The origin of the contrail takes place within a fraction of a second from the emission. In some instances it forms rings that become unstable and eventually break up and disappear, thanks to the evaporation of the ice crystals. One such example is shown in Figure 19.3.

The technical literature reports also cases of aircraft dissipation trails named *distrails*[10]. These events are much less common, less visible and of shorter duration. They occur when an aircraft passes through a supercooled *altocumulus* cloud. Because these clouds have altitudes below the normal cruise altitude (below 20,000 feet), the aircraft is likely to encounter this situation when changing altitude.

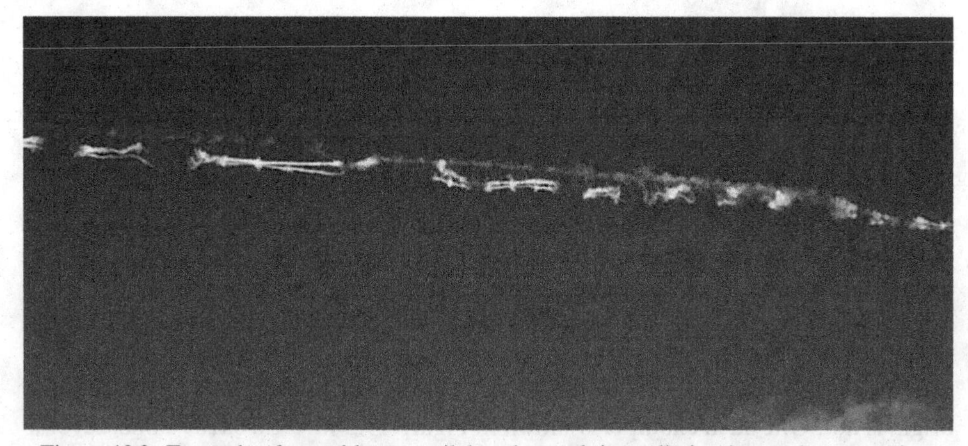

Figure 19.3. Example of unstable contrail; break-up of rings, dissipation and evaporation.

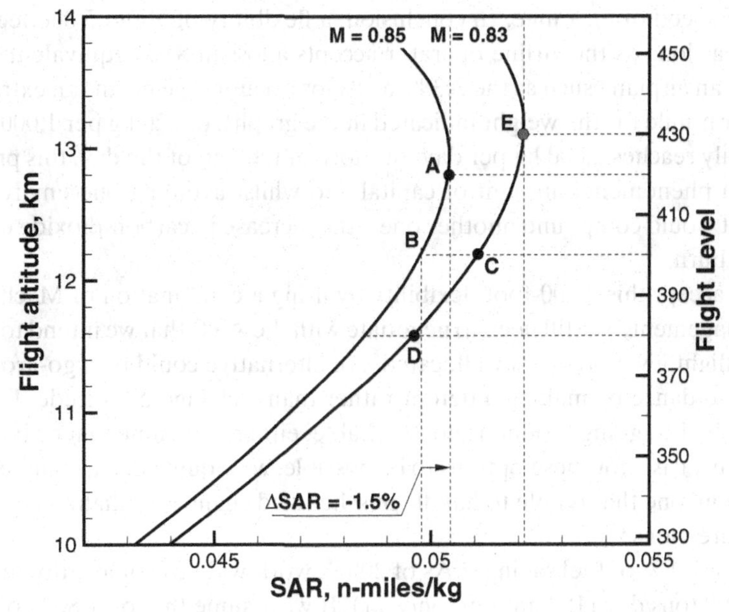


Figure 19.4. Altitude flexibility analysis for the Airbus A380-861; W = 323.70 tons; standard day, no wind.

## 19.1.2 Cruise Altitude Flexibility

We investigate the changes in operational conditions *at the current technology level* that would lead to a decrease in contrail impact. When we look at these options, we soon find out that they must be discarded. For example, a trade in flight altitude cannot be done at the expense of fuel consumption. As we demonstrated in Chapter 12, commercial airliners operate at maximum SAR over a restricted range of altitudes. The optimal cruise conditions for a given all-up-weight are established by a unique combination of Mach-altitude parameters. One way of overcoming this limitation is to increase the *altitude flexibility* so that the optimal cruise conditions become *independent* of altitude. We have demonstrated[11] that this is not possible in the general case because of limits in the flight physics; however, a ±2,000-foot altitude flexibility can be obtained by trading in the cruise Mach number.

**ALTITUDE-SAR RELATIONSHIP.** One example is shown in Figure 19.4, which shows the simulated altitude performance of the Airbus A380-861 with GP-7270 engines. There are two curves showing the SAR at $M = 0.85$ and $M = 0.83$. Assume the airplane is cruising at $M = 0.85$, which is close to the long-range Mach number (LRM). The best flight level would be FL-420 (point A), unless the aircraft is forced to a lower level by the ATC. A descent by 2,000 feet to FL-400 (point B) would cause a loss in SAR equal to about −1.5%, which is relatively high. Alternatively, the airplane could fly at FL-400 (point C) and at the lower Mach $M = 0.83$ (close to the maximum-range Mach number). This procedure has the advantage of actually increasing the SAR, although the airplane would have to sacrifice some speed. It can do even better by descending to FL-380 (point D) at $M = 0.83$. This option requires again a loss of about 1.5% SAR. One more option is to move higher to FL-430, either at $M = 0.85$ or $M = 0.83$, which can perhaps avoid some high-altitude weather conditions without sacrificing speed or air range. In conclusion, a flexibility of about 5,000 feet can be achieved, as long as the airline operator accepts a loss in SAR equivalent to about 1.5%. For an airplane such as the A380, a 1% loss in air range means an extra 0.22 kg of fuel per n-mile (at the weight indicated in the graph), or 220 kg per 1,000 n-miles, which easily reaches 2,000 kg per day. In short, at the end of the day, this procedure will cost a phenomenal amount of capital and whilst avoiding one environmental problem it would compound another one – the increase in carbon-dioxide emissions from fuel burn.

If we accept this 2,000-foot flexibility by using a combination of Mach-altitude flight management, we still need to negotiate with the ATC that we intend to descend from the flight level previously allocated. An alternative could be a go-around, i.e., contrail avoidance by making a detour rather than a change of altitude. Extending the cruise by just going around is not workable, either in commercial or in practical terms. In any case, for these options to be possible, we would need a completely new ATC system, one that is able to handle unscheduled changes of flight level, possibly from all directions.

What is a 1% of fuel savings? As of 2007*, worldwide use of aviation fuel stood at ~515,000 tons/day (188 million ton/year). If we assume that only 80% of the fuel is burned at cruise (on average), then a 1% savings would lead to a savings of ~4,100 ton/day of aviation fuel, or ~13 million kg of $CO_2$/day.

**DESIGN ISSUES.** Assume that an airplane is to operate at constant TAS over the altitude range suitable for cruise. To make the SAR insensitive to altitude, we would need to make the drag insensitive to altitude. The TSFC is an engine-performance indicator and is temporarily excluded from the analysis. The drag analysis shown in § 4.2.12 indicates that the dominant effect is that of the induced drag, which grows rapidly with the altitude; this effect is ultimately attributed to the air density. Let us set $dD/dh = 0$. This condition determines the optimum altitude with respect to aerodynamic configuration alone[11]. After carrying out the algebra, this equation takes the following form:

$$\frac{d\rho}{dh} + c\frac{d}{dh}\left(\frac{1}{\rho}\right) = 0, \tag{19.2}$$

where the factor $c$ is defined

$$c = \frac{k}{C_{D_o}}\left(\frac{2W}{AV}\right)^2 \frac{1}{V^2} < 1. \tag{19.3}$$

A quick analysis of Equation 19.2 shows that the optimal cruise altitude decreases when the factor $c$ increases. Such an effect can be achieved with a number of options, most of which are unreasonable. One reasonable option, albeit a challenging one, is the reduction in profile drag.

If we look closer at the TSFC, we find that it is not constant; in fact, it may suffer large variations at cruise altitude as a result of increasing or decreasing fuel flow (i.e., increasing or decreasing thrust or drag). If the drag increases, so does the TSFC. In other words, the TSFC and the drag move in the same direction as the

* ICAO Environmental Report 2010.

altitude increases or decreases. Therefore, the optimum design problem involves aerodynamics as well as propulsion.

A number of studies[12;13] proposed a cruise altitude shift. This shift was found to be dependent on the latitude and the time of year. These papers also refer to a number of studies that address the consequences of shifting cruise altitude above and below the current levels. In particular, Mannstein *et al.*[13], following extensive measurements, contend that a ± 2,000-foot shift in altitude would be sufficient to avoid 50% of the contrails. Measurements shown by Sussmann[14] behind a Boeing B747 show that the vertical dispersion of the vortex wake determines the vertical dispersion of the contrail. An increased diffusion would limit the effectiveness of any cruise-level shift.

### 19.1.3 The Contrail Factor

Having excluded large changes in cruise levels, reduction in cruise Mach number, changes of aircraft routings and major upgrades in the air traffic control, we are left with the engine technology. What can be done? First of all, let us examine the likelihood of contrail formation. There is a relatively simple rule that can be used: this is the Appleman-Schmidt criterion, which is generally written as

$$G = \frac{\Delta e}{\Delta T} = p \frac{\Delta q}{\Delta h_j} \frac{C_p}{\epsilon}, \tag{19.4}$$

where $G$ is the ratio between changes in water vapour pressure ($e$) and the corresponding change in temperature during mixing; $p$ is the atmospheric pressure at the flight level; $C_p$ is the specific heat capacity of air; $\Delta q$ is the change in water content in the plume; and $\Delta h_p$ is the corresponding change in enthalpy. The *contrail parameter* is

$$C = \frac{\Delta q}{\Delta h_j} \frac{C_p}{\epsilon}. \tag{19.5}$$

The contrail parameter is proportional to the contrail factor: $G = p\,C$. The function $e(T)$ is called *mixing line*; thus, Equation 19.4 gives the slope $G$ of the mixing line.

The threshold temperature $\mathcal{T}_M$ for $\mathcal{H} = e/e_{sat} = 100\%$ (saturation) can be fitted by the following equation[4]:

$$\mathcal{T}_M = -46.46 + 9.43 \ln(G - 0.053) + 0.720 \Big[ \ln(G - 0.053) \Big]^2, \tag{19.6}$$

with $\mathcal{T}_M$ in °C and $G$ in Pa/K. The threshold temperature corresponding to a relative humidity $\mathcal{H} < 100\%$ is

$$\mathcal{T}_C = \mathcal{T}_M - \frac{1}{G} \left[ e_{sat}(\mathcal{T}_M) - \mathcal{H} e_{sat}(\mathcal{T}_C) \right]. \tag{19.7}$$

Equation 19.4 can be written in a form that contains the propulsive efficiency of the engines. This form of the equation was further demonstrated to be extremely important because it links the probability of contrail formation to the engine performance.

**THE APPLEMAN-SCHMIDT CRITERION.** An important consequence of the criterion is the relationship among flight altitude, humidity level, and contrail existence. An example is shown in Figure 19.5a. This graph shows levels of constant relative humidity in the temperature-altitude space. In particular, it shows levels corresponding to

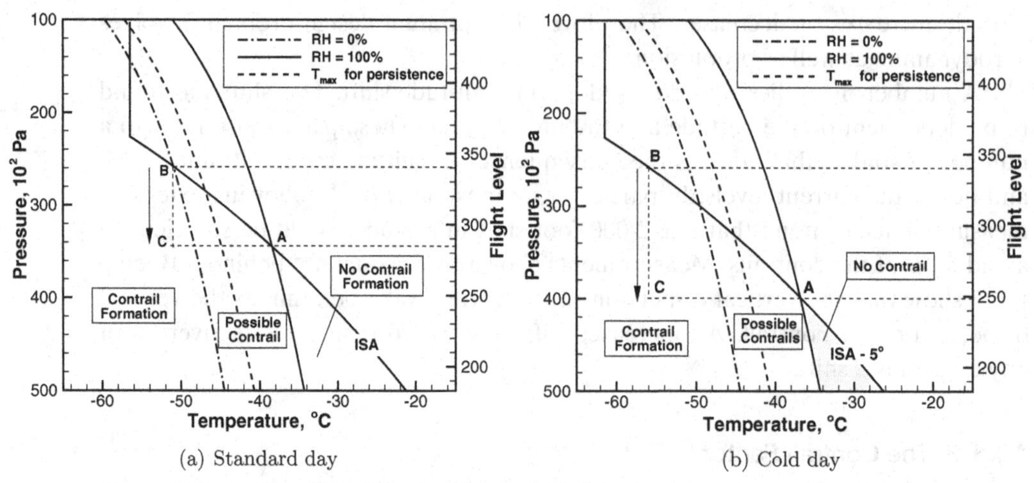


Figure 19.5. Application of the Appleman-Schmidt criterion to contrail prediction; threshold temperature at selected levels of relative humidity.

dry air ($\mathcal{H} = 0\%$) and saturated air ($\mathcal{H} = 100\%$). Assume that an aircraft is flying at ISA conditions at FL-340. Its operation point is indicated by point B in Figure 19.5a. At this point the atmosphere is dry. Therefore, there is a high probability of creating a persistent contrail. To avoid this situation, the aircraft should descend to FL-290 to point A. At this point, in fact, the air is saturated with water vapour and the aircraft will not create a contrail. A descent to an intermediate altitude cannot guarantee that a contrail will be avoided. Thus, the descent required is 50 flight levels, that is, 5,000 feet or more. In a standard atmosphere, contrails will form at altitudes above 8.2 km.

If the atmosphere is colder, for example ISA −5 °C, then the situation is shown in Figure 19.5b. Assume again that the aircraft is flying at FL-340. Its operational point is B. This point is in the region of dry air; therefore, a contrail will certainly form. To avoid the contrail, the aircraft would have to descend to FL-250.

### 19.1.4 Effects of Propulsive Efficiency

Let us begin from the propulsive efficiency of the jet engine, which is

$$\eta = \frac{F_N V}{\dot{m}_f Q}, \tag{19.8}$$

where $F_N$ is the net thrust, $V$ is the true air speed, $\dot{m}_f$ is the fuel flow, and $Q$ is the combustion heat of the aviation fuel ($Q \sim 43.5$ MJ/kg; see § 14.2). At cruise conditions this efficiency is between 0.27 and 0.33; it decreases slightly as the aircraft burns fuel. The enthalpy of the plume* is

$$h_p = h + \frac{1}{2}(V_j - V)^2, \tag{19.9}$$

where $V_j - V$ is the relative speed of the plume with respect the true flight speed $V$. We now write the energy balance in a frame of reference moving with the aircraft.

* There is a difference between jet and plume: a turbulent jet is a flow produced by a pressure drop through a nozzle. A turbulent plume is a fluid motion whose primary source of energy and momentum is body forces derived from density variations.

The change in total energy (kinetic and thermal) is equal to the flow of combustion energy due to fuel burn. If fuel is burned at a rate $\dot{m}_f$, the input energy rate is $\dot{m}_f Q$. The change in total energy of the plume is the product between the total mass-flow rate and the change in enthalpy

$$\dot{m}_f Q = \dot{m}_j \left[ h_j - h_e + \frac{1}{2}(V_j^2 - V^2) \right], \tag{19.10}$$

where $h_e$ denotes the enthalpy of the external gas. The same energy-budget equation written in a reference fixed on the ground is

$$\dot{m}_f Q - F_N V = \dot{m}_j \left[ h_j - h_e + \frac{1}{2}(V_j^2 - V^2) \right]. \tag{19.11}$$

The product $F_N V$ between the net thrust and the true air speed $V$ is the useful work done on the aircraft, which is moving by *reaction*. Thus, this quantity must be subtracted from the inflow of energy. Insert the definition of propulsive efficiency (Equation 19.8) into Equation 19.11 to find

$$(1-\eta)\dot{m}_f Q = \dot{m}_j \left[ h_j - h_e + \frac{1}{2}(V_j^2 - V^2) \right] \tag{19.12}$$

$$(1-\eta) Q = \frac{\dot{m}_j}{\dot{m}_f} \Delta h_p, \tag{19.13}$$

where

$$\Delta h_p = \left[ h_j - h_e + \frac{1}{2}(V_j^2 - V^2) \right] \tag{19.14}$$

denotes the change in enthalpy of the plume. The mass budget for water vapour is

$$\Delta q = q_j - q_e = EI_{H20} \frac{\dot{m}_f}{\dot{m}_j}, \tag{19.15}$$

where $EI_{H20}$ is the emission index of water, that is, the amount of $H_2O$ produced by the jet for a unit mass of fuel burned; $\dot{m}_f$ is the fuel flow, and $\dot{m}_j$ is the mass-flow rate of the jet. If we insert Equation 19.15 and Equation 19.13 into the contrail parameter (Equation 19.5), we finally get the following expression

$$G = EI_{H2O} \frac{p C_p}{\epsilon\, Q(1-\eta)}. \tag{19.16}$$

The relationship between the contrail threshold temperature and the overall propulsive efficiency is shown in Figure 19.6 for selected values of the relative humidity. The shaded area indicates the range of propulsive efficiency of commercial aircraft; the highest values of $\eta$ correspond to the most modern engines. This is an inconvenient result. In fact, as engines become more efficient, they become more likely to create the conditions for contrail formation. This result was demonstrated by Schumann *et al.* [15] on the basis of flight tests with an Airbus A340 and a Boeing B707 flying side by side: the A340 turned out to produce contrails, whereas the B707 did not, under comparable conditions (same altitude, same air speed).

Graph 19.6b shows that the vertical corridor for contrail formation increases with the increasing efficiency. This result would point to the option of flying higher rather than lower.

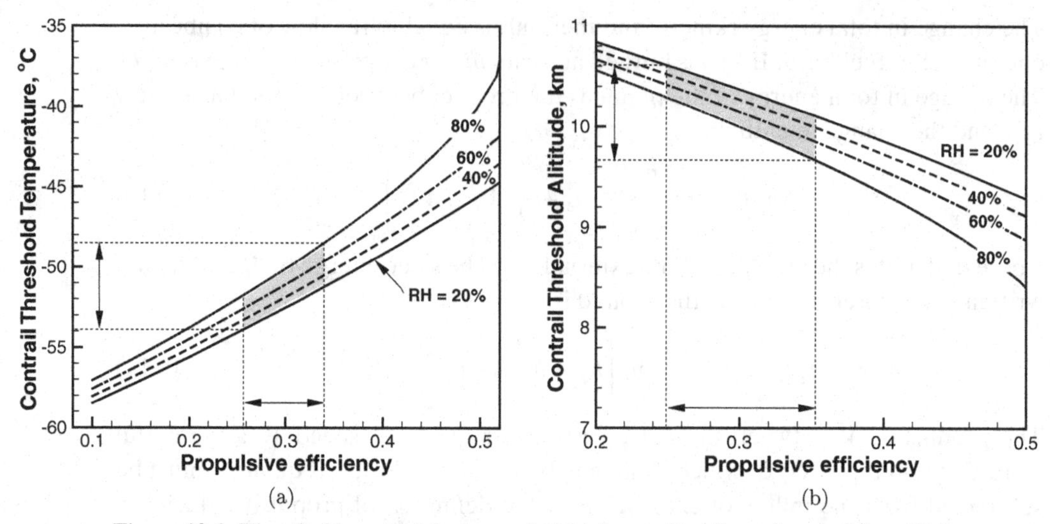


Figure 19.6. Threshold contrail factor and altitude as a function of propulsive efficiency.

A further application of the altitude flexibility concept is shown in Figure 19.7. The shaded area indicates the vertical displacement of the contrail and must be interpreted as a region where there is a high contrail factor. The aircraft travels from the left of the graph to the right. The normal procedure indicates that at some point the aircraft is going to enter contrail conditions at FL-330. However, if sufficient information is available well ahead of time, the flight control system can be programmed to avoid the grey area. This is done with two stepped descents from the initial FL-310 and one stepped climb until the altitude stabilises at FL-290. Aside from ATC issues, there is the problem in gathering enough atmospheric information ahead of the aircraft. The scenario shown in Figure 19.7 cannot be generalised.

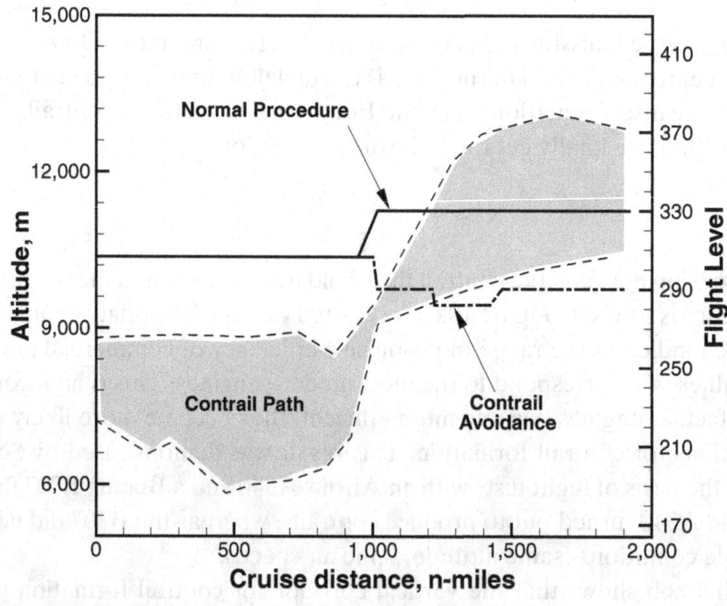


Figure 19.7. Example of contrail-avoidance strategy with shift in cruise altitude.

However, there are software models capable of modelling the contrail frequency over a large region from a mix of weather forecast, atmospheric parameters and aircraft characteristics [16]. The work of Irvine *et al.* [17] shows an analysis of the effect of weather patterns, including winds and contrail factors, on the optimal flight altitude of a trans-Atlantic flight: the weather-optimal trajectory is often outside the acceptable flight levels (FL-290 to FL-410).

### 19.1.5 Heat Released in High Atmosphere

We consider the problem of combustion heat released in the high atmosphere during a conventional cruise. The heat released is calculated from

$$Q_c \sim (1-\eta) Q \dot{m}_f t_c n_e, \tag{19.17}$$

where $\eta$ is the average propulsive efficiency, $Q$ is the specific combustion heat of the aviation fuel, $t_c$ is the total cruise time, and $n_e$ is the number of engines. We have verified that the propulsive efficiency changes very little during the cruise; therefore, the approximation is acceptable. This means that a commercial airplane in the same class as the Airbus A320-200 releases about 140 GJ of heat during the cruise segment of a 2,000 km (1,089 n-miles) flight. This is equivalent to burning $\sim$ 3,200 kg of aviation fuel in the open air at the flight altitude, corresponding to an average propulsive efficiency $\eta \simeq 0.28$. If we expand from this result and apply the calculation to the whole of commercial aviation, we conclude that heating of the upper atmosphere is not only a result of the persistence of greenhouse gases and persistent contrails but also a direct heating of the atmosphere. It has been demonstrated that the diffusion of the exhaust plume has a complex behaviour [18;19]. There are several implications, including, for example, the dilution of the plume and the vertical size of the plume.

## 19.2 Radiative Forcing of Exhaust Emissions

There are several contributions to the exhaust emissions, some of which are direct and others are indirect, that is, by-product of a pollutant with the atmosphere. In addition to contrails and cirrus clouds, as previously discussed, we have:

- *Carbon-dioxide, $CO_2$.* This is by far the largest contribution because it is approximately equal to 3.1 times the weight of the fuel burn; $CO_2$ has long been considered the main culprit of environmental pollution.
- *Nitrogen oxides, $NO_x$.* This contribution includes both NO and $NO_2$; they are a by-product of the combustion of aviation fuel; and they affect the production of ozone $O_3$ and methane $CH_4$. Their radiative forcing is investigated by Gardner *et al.* [20] and Stevenson *et al.* [21]. The problem is compounded by high combustor temperatures in modern gas-turbine engines.
- *Ozone, $O_3$.* At flight altitudes below 13 km (~42,650 feet), the emission of $NO_x$ causes an increase in $O_3$, whilst at higher altitude (16 to 20 km), it causes the destruction of $O_3$ [22]. Aircraft-induced ozone can have a long-term contribution to the atmospheric temperature.

- *Methane, $CH_4$.* Methane is a powerful greenhouse gas. This contribution is not by itself relevant because it is not a by-product of gas-turbine combustion, but $NO_x$ emissions could contribute to the reduction of $CH_4$, which would have a cooling effect.
- *Water vapour, $H_2O$.* Emission of water vapour from gas-turbine engines is relatively small, compared to the weather effects in the troposphere, and is removed by precipitation. However, at very high altitudes, where the atmosphere is dry (see also Figure 8.2), an increase in water content acts as a greenhouse effect. Due to the limited air circulation at stratospheric altitudes, the water content has a long residence time. A side effect of stratospheric vapour is to create stratospheric clouds that may have direct and indirect effects on the ozone.
- *Soot and Sulphates.* Soot is made of extremely small particles, less than 30 $\mu m$ in diameter. Even smaller unburnt liquid particles are removed from the exhaust plumes; these particles combine with the atmospheric components to produce sulphuric acid, hydrocarbons, and water vapour, as well as electrically charged molecules.

One of the key aspects of the entire pollution problem is the long-term effect of the various forms of pollutants. This effect is measured by a parameter called *radiative forcing*. In simple words, the radiative forcing is a means of estimating by how much the atmospheric temperature increases as a result of a pollutant, assuming that the pollutant effect is similar to a greenhouse. A mathematical expression often given for the radiative forcing is

$$RF = \frac{\Delta T_s}{\lambda}, \tag{19.18}$$

where $\Delta T_s$ is the change in *mean global surface temperature* and $\lambda$ is a sensitivity parameter. The radiative forcing is measured in [W/m$^2$] (or more practically in [mW/m$^2$]); therefore, the sensitivity parameter $\lambda$ is given in [K m$^2$/W]. The problem with Equation 19.18 is that the sensitivity parameter varies greatly depending on the pollutant. Despite these caveats, Equation 19.18 is used to compare different emissions.

The value of the radiative forcing is still a matter of scientific debate. We report data based on the study of Sausen *et al.* [23], which gives an indication of the various effects and their level of uncertainty. We draw the reader's attention to the fact that the warming effects caused by carbon dioxide ($CO_2$), ozone ($O_3$) and contrails are of similar magnitude (RF $\sim$ 20 mW/m$^2$). Soot and water vapour are also of the same magnitude but considerably smaller (RF $\sim$ 2 mW/m$^2$). The radiative factor of the cirrus clouds is presently indeterminate. Various estimates indicate this contribution to be higher than any other contribution, with values of ~25 mW/m$^2$, Minnis *et al.* [8]. The upper bound of the uncertainty is considerably higher, up to ~80 mW/m$^2$. Finally, the radiative factor of methane ($CH_4$) is negative (RF $\sim -14$ mW/m$^2$), which would contribute to a cooling rather than a warming.

## 19.3 Landing and Take-Off Emissions

The landing and take-off (LTO) emissions are used to estimate the environmental impact of commercial aircraft at altitudes below 3,000 feet (~915 m). This altitude

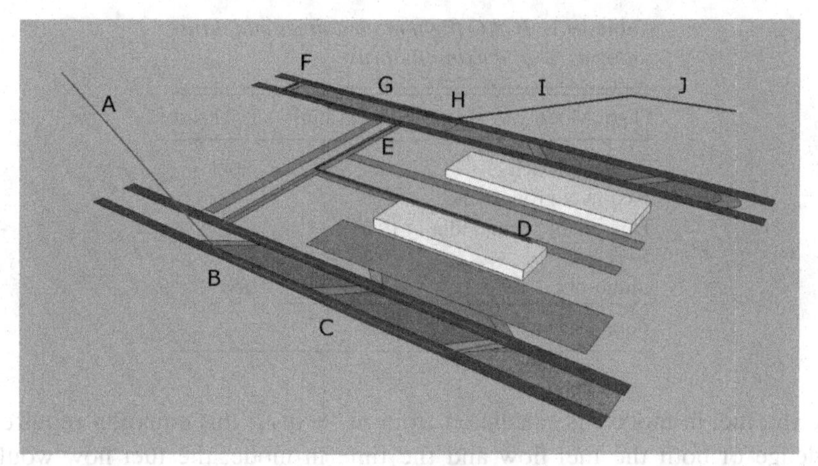


Figure 19.8. Landing and take-off operations of a commercial flight.

corresponds approximately to the mixing layer: pollutants emitted below this *average* altitude can have effects on air quality at lower altitudes. Emissions at ground level will have a shorter-term impact than emissions at higher altitudes.

Figure 19.8 shows the typical flight modes of a commercial flight at a large airport. The airplane is on a final approach (A), proceeds to landing (B) and then performs a roll to the gate (C), eventually stopping for other traffic, turning at several points, before eventually reaching the gate (D). When it is ready for departure, it rolls out (E) and reaches the start of the runway to a brake-release point (F). It then waits for the go-ahead, performs a ground run (G), a take-off (H), an initial climb (I) and reduces the thrust (J). All of these phases are included in the LTO emissions, as long as the airplane is below 3,000 feet.

The ICAO defines five different modes for the purpose of certification and for dispersion calculations. These modes are 1) final approach, 2) taxi-in and ground idle, 3) taxi-out and ground idle, 4) take-off, and 5) climb-out. The taxi-in and out are essentially the same mode. ICAO goes on to propose an average time in each mode and an average thrust setting on each mode, as given in Table 19.1. During an actual flight, neither the times nor the thrust settings are the same as those in Table 19.1. The pollutants considered in the analysis are $NO_x$ (nitrogen oxides), HC (hydrocarbons), CO (carbon-monoxide), $SO_x$ (sulfate oxides), particulate matter (or soot), in addition to carbon-dioxide $CO_2$, which is proportional to the fuel burned. With the exception of the latter component, the other pollutants depend *non-linearly* on the thrust rating. To avoid complicated analysis that involves combustion modelling, these emission indexes can be inferred from a databank published by the ICAO itself*. The emission indexes are generally given as grams of pollutant per kilogram of fuel burn. An example is shown in Figure 19.9 for the CO, $NO_x$ and HC. Thus, the emission $E_j$ of pollutant $j$ would be

$$E_j = \sum_i EI_i \, m_{f_i}, \tag{19.19}$$

* ICAO Engine Emissions Data Bank. Periodically updated. Available from the ICAO and other aviation regulators. This databank does not contain information on turboprop engines and APU.

Table 19.1. *ICAO flight modes, times and thrust rating as % of maximum thrust*

| Flight Mode | Time [min] | % Thrust |
|---|---|---|
| Final approach | 4.0 | 30 |
| Taxi-in & ground idle | 7.0 | 7 |
| Taxi-out & ground idle | 19.0 | 7 |
| Take-off | 0.7 | 100 |
| Climb-out | 2.2 | 85 |
| Total | 32.9 | |

where the fuel-in-mode $i$ is calculated from $m_f = \dot{m}_f t$; this equation requires the knowledge of both the fuel flow and the time in mode; the fuel flow would be proportional to the thrust-in-mode. However, there is a problem here because this assumption implies that the TSFC is a constant. This assumption is only approximately true at high thrust settings (take-off and climb-out), as illustrated previously (see Chapter 5).

A more accurate estimation of the LTO emissions relies on the actual fuel burn for all the flight modes below 3,000 feet. The emission indexes are dependent on the thrust rating, as shown in Figure 19.9. Therefore, we still need to approximate the ICAO emission indexes with interpolation functions. The emission of species $j$ during the time step $dt$ is calculated from

$$dm_j = \left[\dot{m}_f(N_1)dt\right] \mathrm{EI}_j(N_1). \tag{19.20}$$

This equation expresses the fact that the emission $dm_j$ of species $j$ is proportional to the fuel flow $\dot{m}_f$ and the emission index $\mathrm{EI}_j$. Both the fuel flow and the emission

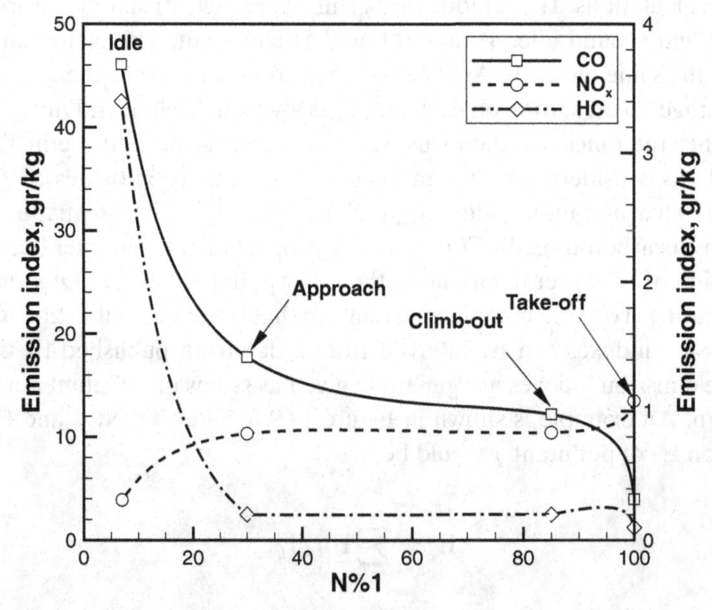


Figure 19.9. Emission indexes for the CFM56-5C4 as a function of the engine rpm. Data elaborated from the ICAO databank.

Table 19.2. *LTO emissions summary for Airbus A320-200 with CFM56 engines; standard day, no wind, 2,000 km (1,079 n-miles) mission; all data in [kg], as calculated*

| | Segment | CO | $NO_x$ | HC | t[min] | Notes |
|---|---|---|---|---|---|---|
| 1 | Taxi-out, roll | 0.942 | 0.358 | 0.065 | 6.7 | |
| 2 | Taxi-out, idle | 1.885 | 0.716 | 0.131 | 3.3 | |
| 3 | Takeoff | 4.288 | 1.727 | 0.195 | 0.4 | |
| 4 | Climb-out | 5.022 | 2.929 | 0.057 | 1.5 | to 3,000 feet |
| 5 | Approach | 1.337 | 2.391 | 0.026 | 3.1 | from 3,000 feet |
| 6 | Landing | 0.150 | 0.149 | 0.008 | 0.8 | from 35 feet |
| 7 | Taxi-in, roll | 0.410 | 0.109 | 0.029 | 6.7 | |
| 8 | Taxi-in, idle | 2.049 | 0.546 | 0.146 | 1.3 | |
| 9 | APU | 0.018 | 0.025 | 0.001 | 23.9 | total |
| | Total LTO | 16.084 | 8.925 | 0.658 | 23.9 | |

index are a function of the actual engine speed $N_1$ (or `N%1`). The actual emission index is *interpolated* from the reference values in the ICAO database. The LTO emissions are calculated by integration of Equation 19.20, that is, by integration of the fuel flow when the airplane is below 3,000 feet:

$$E_j = n_e \int_a^b dm_j = n_e \int_a^b \dot{m}_f(N_1)\mathrm{EI}_j(N_1)dt, \tag{19.21}$$

where $n_e$ is the number of operating engines. In Equation 19.21 "$a$" denotes the moment when the airplane leaves the gate; "$b$" denotes the time when the airplane reaches the gate at destination. In summary, the computational procedure is the following:

1. At the current flight condition, solve the engine problem: `Wf6`, `N%1`.
2. Calculate the emission index corresponding to `N%1` by interpolation of the ICAO data (see Figure 19.9).
3. Calculate the emission $dm_i$ within the time step $dt$ from Equation 19.20.
4. Advance the flight time by $dt$ and proceed from point 1, as long as $h < 3{,}000$ feet.

An example of calculated LTO emissions is given in Table 19.2, which refers to the Airbus A320-200 with CFM56-5C4 turbofan engines; the flight planning was done for a 2,000-km (1,079-n-mile) mission, on a standard day, without winds, full passenger load and 300 kg of bulk cargo. The flight scenario involves a 10-minute taxi-out and an 8-minute taxi-in. The engine warm-up is not included in the table. The LTO emissions for this engine quoted by the ICAO databank can be quite different from the calculated values. In fact, considerable variations can be detected among different types of flights, and caution must be exercised when taking averages. Actual data can differ by a factor 2. Note that the LTO time is 23.9 minutes instead of 32.9 minutes prescribed in Table 19.1. The main difference comes from the average taxi-out roll.

A sub-case of the LTO emissions is the ground-operations segment. At congested airports, some aircraft spend up to one hour between pushing back from the gate and starting the take-off run (segment D to F in Figure 19.8); for most of this

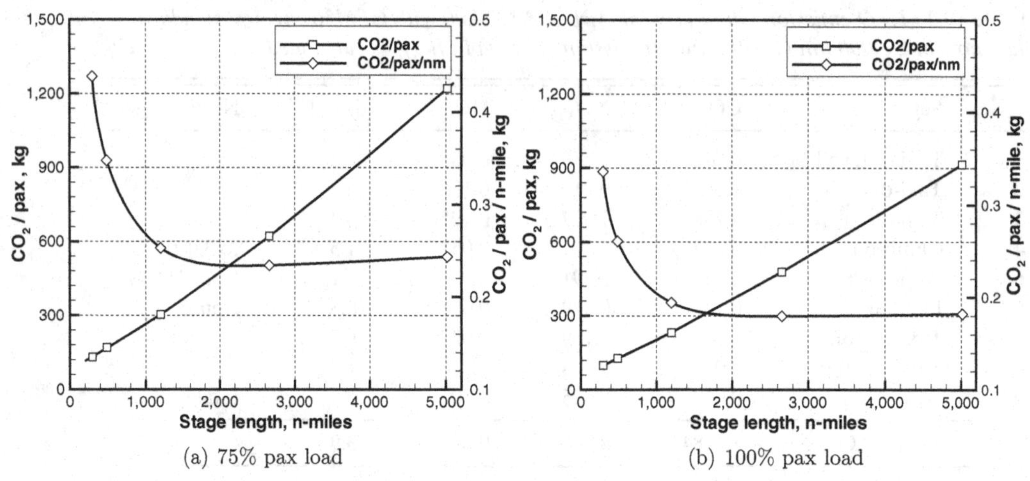


(a) 75% pax load

(b) 100% pax load

Figure 19.10. Calculated $CO_2$ emissions of the Boeing B777-300-GE, standard day, 9 kt head wind.

time, the engines run in idle mode and consume large amounts of fuel whilst emitting harmful exhaust gases. Reduction of these emissions is beyond the domain of a single airplane and must be examined in the context of airport operations, queueing strategies and flight distributions.

## 19.4 Case Study: Carbon-Dioxide Emissions

We now set the task of studying the carbon emissions of a commercial airplane, using as a parameter the stage length and the passenger load. The latter parameter is expressed as a percent of filled seats; 100% means full aircraft. These airplanes generally carry also some form of cargo. However, we have excluded this case in order to highlight the emissions as a function of passenger load only.

Figure 19.10 shows the calculated $CO_2$ emissions over a range of required stage lengths. The data are shown in terms of emissions per passenger and per n-mile. The results illustrate that the emissions per travelling passenger increase almost linearly at intermediate stage lengths; over long-haul distances the increase is super-linear, due to the fact that a considerable portion of the aircraft's weight is made up of fuel. The fuel-use efficiency is better shown by the quantity $CO_2$/pax/n-mile.

A similar analysis is carried out with the Airbus A320-200, as shown in Figure 19.11. Again, it appears that the emissions per passenger are considerably higher when the aircraft operates at only three quarters of full capacity. Additional analyses are presented in Ref.[24].

The final case refers to a parametric analysis of a typical transport mission. We have included the effects of nine different parameters, as listed in Figure 19.12. The sensitivity analysis consists in changing the parameters listed over a nominal value, which correspond to a specified mission. The mission is a London–New York flight (5,570 km; 3,078 n-miles), with 80% passenger load, 15 tons of bulk cargo, 18 minutes of taxi-out time at London, 8 minutes taxi-in at New York, and 20 kg/pax baggage allowance, on a standard day, with an average cruise wind speed equal to an 18-kt head wind.

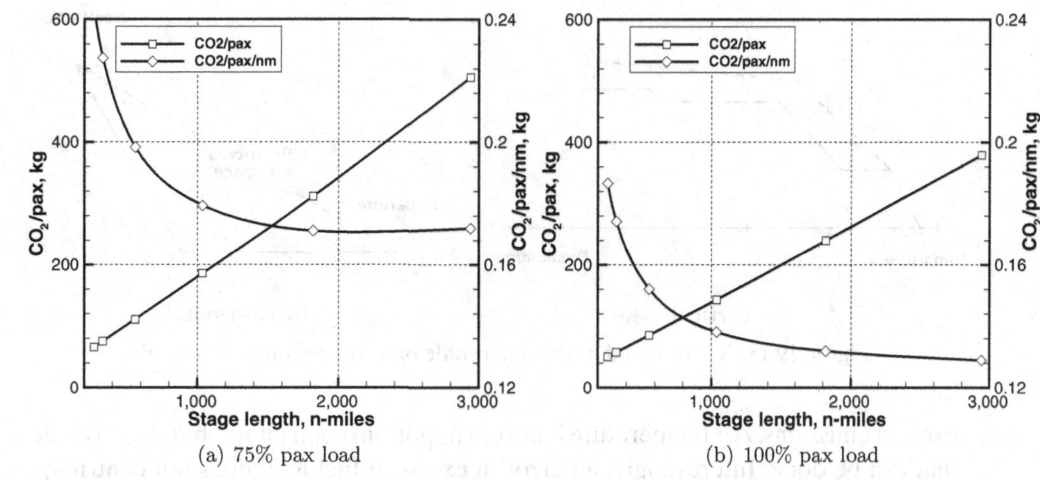


(a) 75% pax load (b) 100% pax load

Figure 19.11. Calculated $CO_2$ emissions of the Boeing Airbus A320-200-CFM, standard day, 9-kt head wind.

Variation of the nominal parameters is as follows: bulk cargo ± 5 tons (effect of commercial payload); passenger load ±10% (effect of passenger revenue); flight route −25 n-miles and +75 n-miles (effect of routing); extra fuel load +1 or +2 tons (error in fuel loading); taxi-out time ±6 minutes (effect of airport congestion); outside air temperature ±20 °C (weather effects); and wind speed of 9 kt or 36 head wind (weather effects). The final parameter is a continuous descent approach (CDA) at the final destination; this parameter can be true or false; it denotes an accurate flight trajectory, although it is subject to certification. The data in Figure 19.12 show the response to a change of a single parameter. By far the most important parameter is the wind, which can add or subtract several percentage points in fuel burn and

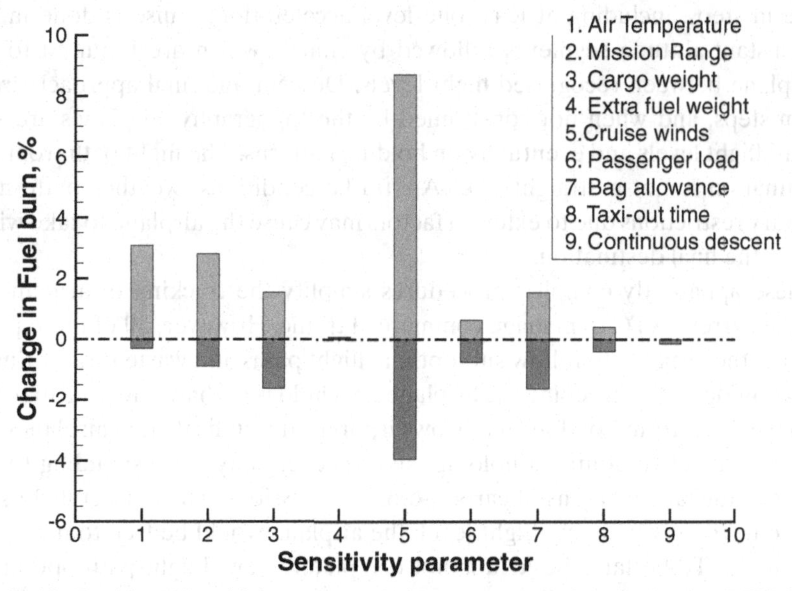


Figure 19.12. Fuel-burn sensitivity to key mission parameters for the Boeing B777-300-GE (simulated).

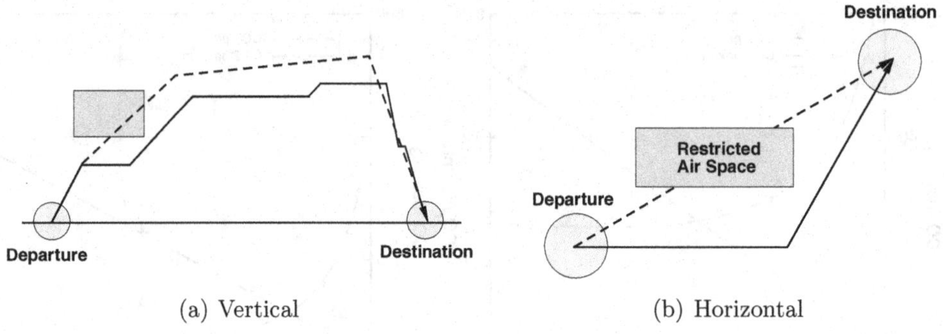


(a) Vertical (b) Horizontal

Figure 19.13. Vertical and horizontal profile of a typical commercial flight.

exhaust emissions. Air temperature is also an important contributor, but there is little that can be done. Interestingly, an error in excess in fuel load does not contribute much to the total fuel consumption, whilst baggage allowance and taxi-time are operational items that can be controlled. A CDA can have some benefits, as also shown in § 11.3.

The issues highlighted in this sections are applied to a single aircraft flight. It is possible to scale up the investigation to include one type of aircraft, one route, and one airport pair. Thence, it is possible to extrapolate to a larger or even global air transportation system, to include all types of emissions, which can then be classified by region, by season, by aircraft movements and other parameters. One such framework is available with the SAGE initiative[25].

## 19.5 The Perfect Flight

The previous chapters have explained how commercial flights operate under ATC and a considerable number of constraints, Figure 19.13. Climb to initial cruise altitude is done in steps, including at least one level acceleration; cruise is done in steps, with constant-altitude segments followed by climbs, which are required to move the airplane between recognised flight levels. Descent and final approach are also done in steps, and when not constrained by the topography, airplanes are set on standard flight levels and eventually on holding patterns. The flight path from origin to destination is not a straight one. Air traffic conditions, weather patterns and temporary restrictions due to external factors may cause the airplane to take winding routes to the final destination.

These apparently complex procedures simplify the tracking of airplanes and allow the current ATC to manage commercial traffic. However, all of these procedures force the airplane to follow sub-optimal flight paths and waste further time and fuel at terminal areas. Keeping the airplane on a hold is perhaps the principal source of wastage. Data from London Heathrow airport (2011) indicate that airplanes spend about 55 cumulative hours on holding patterns every day, corresponding to about 190 tons of fuel and 600 tons of carbon-dioxide emissions. The perfect flight should not have to comply with any flight level; the airplane would be free to move to the best altitude. To be fair, the idea is not completely new. Flight-path optimisation was applied as early as the 1970s to the DC-10 and Boeing B727 models[26–28]; it was demonstrated that a fuel savings of up to 10% could be achieved.

If we exclude considerations on taxi times, on take-off run (which is essentially unaffected by the perfect flight requirements) we would have the following flight segments.

**CONTINUOUS CLIMB.** The most economic climb will be based on the cost index, which is a weighted parameter that takes into account the cost of fuel as well as the cost of time (§ 12.9).

**CRUISE-CLIMB.** We have demonstrated that the optimal cruise profile is a continuous climb (§ 12.5.3). This strategy corresponds to minimum fuel burn for a given initial weight. We establish the cruise condition from the optimal ICA-Mach at the estimated initial cruise weight.

**CONTINUOUS DESCENT.** We assume that the airplane can perform a continuous descent from the end of the cruise (*top of descent*) to the airfield (§ 11.3). Most of this descent can be done with a nearly idle engine.

**THE OPTIMAL TRAJECTORY.** Following the previous discussion, there will be no holding on a stack and the airplane can proceed directly to landing. We assume the standard international reserves (§ 15.5). We consider a "nearly" perfect flight, to allow for the practical constraints in the real world: the flight will be as conventional, except the cruise (which will be a continuous climb) and the descent, which will be a continuous descent to 3,000 feet, with a short level flight, and descent on a 3-degree slope. The climb is constrained by two constant-CAS segments, with a level acceleration in between.

The flight-performance model has been run with the Airbus A320-211 powered by two CFM56-5C4 turbofan engines. The summary of flight conditions is:

- standard day; no wind
- required range: 2,500 km; 1,349 n-miles
- passenger load: 100%
- cargo load: 300 kg
- taxi-out time: 10 minutes
- taxi-in time: 8 minutes
- departure/destination airports at sea level
- average passenger mass: 75 kg; average baggage mass: 15 kg
- climb schedule: conventional
- diversion flight: 2,000 feet below the FCA
- APU fuel: included in the analysis (ECS mode)
- fuel reserve: minimum between 20-minute extension and 200 n-mile diversion
- holding: 30 minutes at 1,500 feet
- U-turn at 5,000 feet
- engine state: new
- aerodynamics: clean

Note that the fuel reserve in these cases is always the 20-minute extension of the cruise because the combination of 200 n-mile diversion and holding on a stack requires more fuel.

Table 19.3. *Analysis of a perfect flight with an Airbus A320-200 model; the range is 2,500 km (1,349 n-miles)*

| | Constraints | Total Fuel | Climb Fuel | Climb Fuel | Descent Fuel | $\Delta m_f$ % |
|---|---|---|---|---|---|---|
| Conventional | | 8,804 | 951 | 5,917 | 657 | 0.8 |
| Scenario I | Cruise-Climb | 8,697 | 951 | 5,875 | 598 | −1.20 |
| Scenario II | & No Restriction on ICA | 8,351 | 1,062 | 5,350 | 655 | −5.14 |

A summary of simulations is in Table 19.3; we also show the contributions of the climb to ICA, cruise and descent; $\Delta m_f$ is the change in fuel burn, in percent, compared to the conventional flight.

Figure 19.14 shows the trajectory of our nearly perfect flight. Note that there are two flight segments, at climb and descent, that are level flights (A and B, respectively). Flying at such high altitudes may have a strong side effect: this is the increased environmental impact of emissions at altitude, including the risk of generating more contrails and the effects on atmospheric ozone.

In Scenario I the aircraft is forced to set its initial cruise altitude to a recognised flight level (FL-350), but then it is free to perform a continuous climb, followed by a continuous descent from the top of climb. These conditions on cruise and descent are maintained in Scenario II.

In Scenario II the aircraft is not constrained by a flight level and climbs to a considerably higher altitude before starting its cruise (11,820 m; 38,870 feet). Consequently, the aircraft burns more fuel to climb and to descend. However, the reduced cruise fuel at a higher altitude (up to 40,000 feet) more than compensates for these segment fuels. The fuel saving is estimated at about 430 kg, or 5% of the fuel burned in a conventional trajectory; the "perfect" trajectory produces a savings of about 1,300 kg of carbon dioxide.

## 19.6 Emissions Trading

The European Emission Trading Scheme (ETS) will require the industry to purchase certificates that allow the emission of a specified amount of $CO_2$ (measured in tons). Due to the price volatility of both aviation fuel and certificates (which are traded like commodities), it is not possible to make an accurate assessment of the additional cost that airlines will incur, although it is likely to exceed 10% of the cost of fuel. Emissions beyond the level specified in the certificates will incur a fine. Thence, the argument goes, airlines will have to drive their emissions down. Any emissions savings, with respect to the certificates, can be sold in the open market at commercial rates.

The ETS by itself does not contribute to cutting emissions and may contribute to diverting profits away from the industry. However, because a price tag is associated to these emissions[29], according to the free market, there should be financial incentives in keeping the emissions as low as possible. Initially, these certificates are designed to apply to all flights to and from European airports. However, unless there is worldwide agreement for this implementation, we will

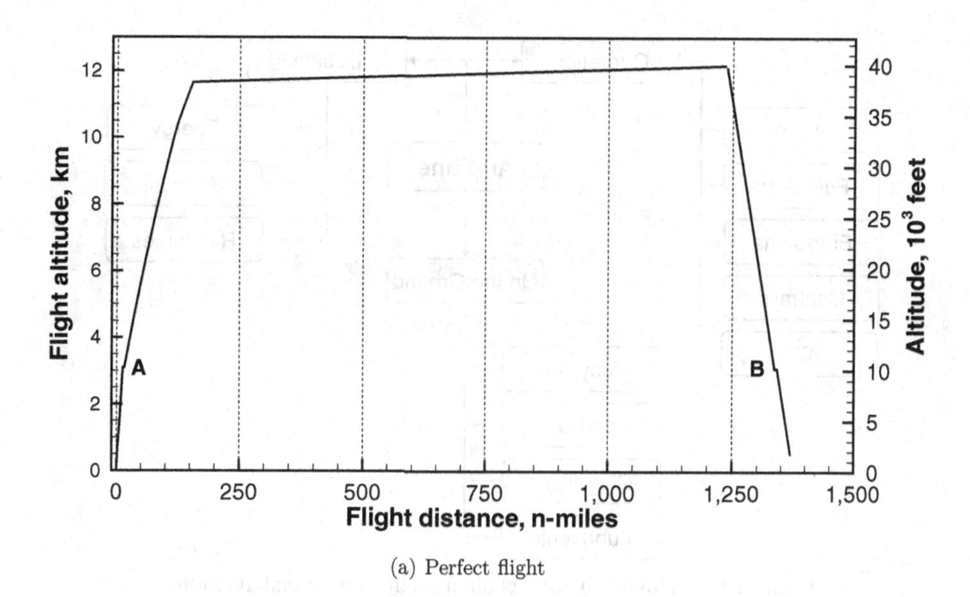


(a) Perfect flight

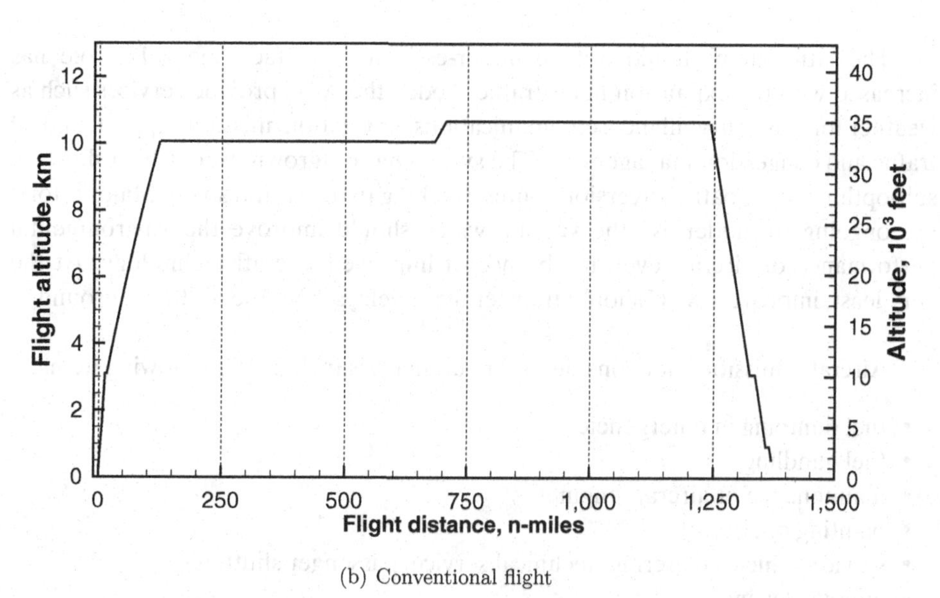


(b) Conventional flight

Figure 19.14. "Nearly" perfect trajectory (a) and conventional trajectory (b) for a commercial flight.

see a distortion of the market that will actually cause an *increase* in aviation emissions.

## 19.7 Other Aspects of Emissions

The environmental aspects of commercial aircraft can be summarised with the headings shown in the flowchart of Figure 19.15. On the operation side, we have the aspects discussed in this book: noise, fuel burn, exhaust emissions and contrails (and contributions to the climate over short, medium and long terms) and the role of the ATC.

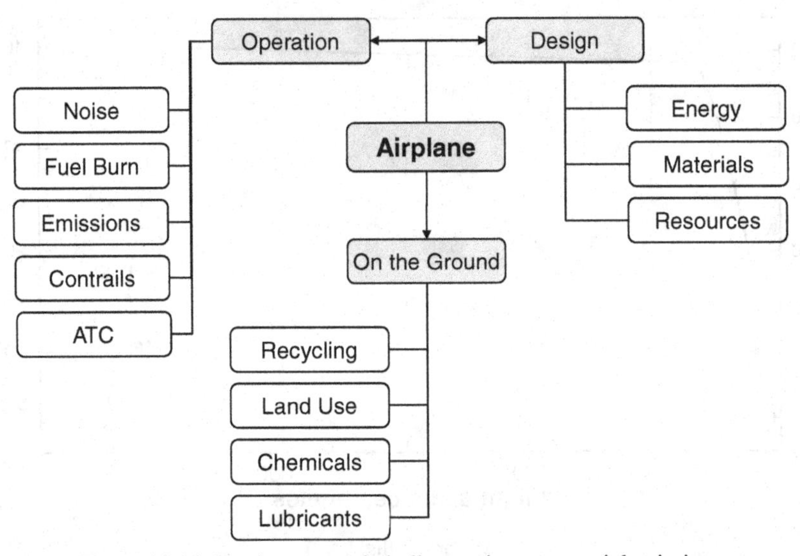


Figure 19.15. Environmental challenges in commercial aviation.

The latter item should not be under-estimated; in fact, the ATC role has increased with the expansion in air traffic. Today the ATC provide services such as weather forecast, surveillance, communications, navigation, mission support, and all traffic and congestion management. The system has outgrown itself, which shows in sub-optimal flight paths, diversion routes, stacking patterns and so on. Major efforts are ongoing to modernise the system, which should improve the environmental performance of aircraft even in absence of improved aircraft technology. At the very least, improved ATC should transfer flight delays from the air to the ground[30].

Aircraft emissions not considered in our analysis include the following items:

- fuel dumping in emergencies
- fuel handling
- maintenance of aircraft engines
- painting of aircraft
- service vehicles (catering, technical service, passenger shuttles)
- aircraft deicing.

On the design side (which is beyond our scope) there are aspects of energy use, raw materials and general resources, including land use and airport expansions. Finally, on the ground side there are issues of recycling at the end-of-life and the use of chemicals and lubricants (acquisition and disposal). What shall we do with about 300 aircraft due to retire from service every year?

## Summary

We have introduced a number of environmental issues in aircraft flight performance. We started with the problem of aircraft contrails and cirrus clouds, which are expanding rapidly and may cause further damage than the emission of $CO_2$ from gas-turbine-engine combustion. We analysed the problem of altitude flexibility and identified a potential source of improvement by using appropriate operational changes. Among the major contributors of the exhaust gases, $NO_x$, CO and hydrocarbons HC have been discussed in the context of landing-and-take-off emissions. We introduced a calculation method, which must rely on a reliable database. The database is of limited use if the airframe and engines are poorly maintained. To offset part of the emissions specifically associated to $CO_2$, we illustrated the problem of a "perfect flight", that is, a flight along the shortest trajectory that is not constrained by a flight level or by climb and descent procedures. This procedure could lead to considerable fuel savings, provided that the air traffic control will be able to cope with the changes required. Finally, we briefly reviewed some elements of environmental pollution around the aircraft itself.

## Bibliography


[1] Penney J, Lister D, Griggs D, Dokken D, and McFarland M, editors. *Aviation and the Global Atmosphere*. Cambridge University Press, 1999.

[2] Travis DJ, Carleton AM, and Lauritsen RG. Contrails reduce daily temperature range. *Nature*, 418:601, 8 August 2002.

[3] Travis DJ, Carleton AM, and Lauritsen RG. Regional variations in U.S. diurnal temperature range for the 11–14 September 2001 aircraft groundings: Evidence on jet contrail influence on climate. *J. Climate*, 17:1123–1134, 2004.

[4] Schumann U. On conditions for contrail formation from aircraft exhausts. *Meteorologische Zeitschrift*, 5(1):4–23, Jan. 1996.

[5] Murcray WB. On the possibility of weather modification by aircraft contrails. *Monthly Weather Review*, 98(10):746–748, Oct. 1970.

[6] Fichter C, Marquat S, Sausen R, and Lee DS. The impact of cruise altitude on contrails and relative radiative forcing. *Meteorologische Zeitschrift*, 14(4):563–572, Aug. 2005.

[7] Ström L and Gierens K. First simulation of cryoplane contrails. *J. Geophys. Res.*, 107(D18):4346, Sept. 2002.

[8] Minnis P, Ayers PK, Palikonda R, and Phan D. Contrails, cirrus trends, and climate. *J. Climate*, 17(8):1671–1685, April 2004.

[9] Spichtinger P and Gierens KM. Modelling of cirrus clouds – Part 1: Model description and validation. *Atm. Chem & Physics*, 9(2):685–706, 2009.

[10] Duda DP and Minnis P. Observations of aircraft dissipation trails from GOES. *Monthly Weather Review*, 130:398–406, Feb. 2002.

[11] Filippone A. Cruise altitude flexibility of jet transport aircraft. *Aero. Science & Technology*, 14:283–294, 2010.

[12] Gierens K, Lim L, and Eleftheratos K. A review of various strategies for contrail avoidance. *Open Atm. Science J.*, 2:1–7, Jan. 2008.

[13] Mannstein H, Spichtinger P, and Gierens K. A note on how to avoid contrail cirrus. *Transp. Res. D*, 10:421–426, Sept. 2005.

[14] Sussmann R. Vertical dispersion of an aircraft wake: Aerosol lidar analysis of entrainment and detrainment in the vortex regime. *J. Geophys. Res.*, 104(D2), 1999.

[15] Schumann U, Busen R, and Plohr M. Experimental test of the influence of propulsion efficiency on contrail formation. *J. Aircraft*, 37(6):1083–1087, Nov. 2000.

[16] Schumann U. A contrail cirrus prediction tool. In *Int. Conf. on Transport, Atmosphere and Climate*, Maastricht, the Netherlands, June 2009.

[17] Irvine E, Hoskins B, Shine K, Lunnon R, and Froemming C. Characterizing North Atlantic weather patterns for climate-optimal aircraft routing. *Meteorological Applications*, 2012.

[18] Schumann U, Konopka P, Baumann R, Busen R, Gerz T, Schlager H, Schulte P, and Volkert H. Estimate of diffusion parameters of aircraft exhaust plumes near the tropopause from nitric oxide and turbulence measurements. *J. Geophysical Res.*, 100(D7):147–162, July 1995.
[19] Schumann U, Schlager H, Arnold F, Baumann R, Haschberger P, and Klemm O. Dilution of aircraft exhaust plumes at cruise altitudes. *Atmospheric Environment*, 32(18):3097–3103, 1998.
[20] Gardner R, Adams K, Cook T, Deidewig F, Ernedal D, Falk R, Fleuti E, Herms E, Johnson C, Lecht M, Lee D, Leech M, Lister D, Mass B, Metcalfe M, Newton P, Schmitt A, Vandenbergh C, and van Drimmelen R. The ancat/ec global inventory of NOx emissions from aircraft. *Atmospheric Environment*, 31(12):1751–1766, 1997.
[21] Stevenson D, Dohery R, Sanderson M, Collins W, Johnson W, and Derwent R. Radiative forcing from aircraft $NO_x$ emissions: Mechanisms and seasonal dependence. *J. Geophysical Res.*, 109:D17307, 2004.
[22] Wuebbles DJ and Kinnison DE. Sensitivity of stratospheric ozone and possible to present and future aircraft emissions. Technical Report JCRL-JC-104730, Lawrence Livermore National Labs, Aug. 1990 (presented at DLR Seminar on Air Traffic and Environment, Bonn, Nov. 1990).
[23] Sausen R, Isaksen I, Grewe V, Hauglustaine D, Lee D, Myhre G, Kohler M, Pitari G, Schumann U, Frode S, and Zeferos C. Aviation radiative forcing in 2000: An update on IPCC (1999). *Meteorologische Zeitschrift*, 14(4):555–561, Aug. 2005.
[24] Filippone A. Analysis of carbon-dioxide emissions from transport aircraft. *J. Aircraft*, 45(1):183–195, Jan. 2008.
[25] Kim B, Fleming G, Balasubramanian S, Malwitz A, Kee J, Ruggiero J, Waitz I, Klima K, Stouffer V, Long D, Kostiuk P, Locke M, Holcslaw C, Morales A, McQueen E, and Gillette W. *SAGE: System for Assessing Global Aviation Emissions*, Sept. 2005. FAA-EE-2005-01.
[26] Lee HQ and Erzberger H. Algorithm for fixed-range optimal trajectories. Technical Report TP-1565, NASA, July 1980.
[27] Erzberger H and Lee HQ. Constrained optimum trajectories with specified range. *J. Guidance, Navigation & Control*, 3(1):78–85, Jan.-Feb. 1980.
[28] Ashley H. On making things the best – Aeronautical uses of optimization. *J. Aircraft*, 19(1):5–28, 1982.
[29] Scheelhaase JD and Grimme WG. Emissions trading for international aviation – An estimation of the economic impact on selected European airlines. *J. Air Transport Management*, 13:253–263, 2007.
[30] Vranas DJ, Bertsimas PB, and Odoni AR. The multi-airport ground-holding problem in air traffic control. *Operations Research*, 42:249–261, Mar. 1994.

## Nomenclature for Chapter 19

$a, b$ = limits of integration, Equation 19.21
$A$ = wing area
$c$ = factor defined by Equation 19.3
$C$ = contrail parameter, Equation 19.5; concentration
$C_{D_o}$ = profile-drag coefficient
$C_p$ = specific heat capacity of air
$d$ = particle diameter
$D$ = contrail thickness; aerodynamic drag
$e$ = vapour pressure
$e_{sat}$ = vapour-saturation pressure
$E_j$ = emission of pollutant species $j$

| | | |
|---|---|---|
| EI | = | emission index |
| $EI_{H_2O}$ | = | emission index of water |
| $F_N$ | = | net thrust |
| $G$ | = | contrail function defined by Equation 19.4 |
| $h$ | = | flight altitude |
| $h, h_j$ | = | enthalpy; enthalpy in the jet |
| $h_e$ | = | enthalpy of external flow |
| $\mathcal{H}$ | = | relative humidity |
| $k$ | = | induced-drag factor |
| $m$ | = | mass |
| $M$ | = | Mach number |
| $m_l$ | = | mass fraction of liquid, Equation 19.1 |
| $\dot{m}_f$ | = | fuel-flow rate (also `Wf6`) |
| $n_e$ | = | number of engines |
| $N_1$ | = | engine rotational speed (also `N1`) |
| $p$ | = | atmospheric pressure |
| $q$ | = | specific water-vapour mass concentration |
| $Q$ | = | combustion heat content of aviation fuel |
| $Q_c$ | = | heat released at high altitude, Equation 19.17 |
| $Q_{ext}$ | = | extinction efficiency |
| RF | = | radiative forcing |
| $\mathcal{R}$ | = | gas constant |
| $t$ | = | time |
| $t_c$ | = | cruise time |
| $\mathcal{T}$ | = | temperature |
| $\mathcal{T}_C$ | = | threshold temperature for humid air |
| $\mathcal{T}_M$ | = | threshold temperature for saturated air |
| $\mathcal{T}_s$ | = | average global surface temperature |
| $V$ | = | true air speed |
| $V_j$ | = | speed of the jet |
| $W$ | = | weight |

## Greek Symbols

| | | |
|---|---|---|
| $\epsilon$ | = | ratio between gas constants water vapour and air |
| $\eta$ | = | propulsive efficiency |
| $\lambda$ | = | sensitivity parameter in radiative forcing |
| $\rho$ | = | air density |
| $\rho_p$ | = | particle density |
| $\tau$ | = | optical depth, Equation 19.1 |

## Subscripts/Superscripts

| | | |
|---|---|---|
| $[.]_e$ | = | external (outside the jet/plume) |
| $[.]_f$ | = | fuel |
| $[.]_j$ | = | jet; counter for emission calculations |
| $[.]_p$ | = | plume |
| $[.]_{sat}$ | = | vapour-saturation conditions |

# 20 Epilogue

Commercial aviation has matured rapidly since the introduction of gas-turbine-engine technology. The investments required to develop new aircraft have increased exponentially. These investments include capital, human resources across a wide spectrum of competences, infrastructure and time. In this framework, the importance of operational aircraft performance is crucial because the existing technology is locked into a large fleet that is due to operate for several decades into the future. Airplane evolution is therefore inevitable and is proved in several case studies, both in the commercial and in the military world. The issues raised in this book will continue to dominate the debate; we expect considerable changes in the future, from the ATC, to aero-engine technology, to aircraft design, operations and demand management.

Each chapter has illustrated major issues in aircraft flight performance; conclusions have been drawn in each separate instance. To conclude, we wish to mention some key aspects of wider interest.

**ROLE OF COMPREHENSIVE ANALYSIS IN AIRCRAFT PERFORMANCE.** Performance is what sells an airplane; the ability to predict and verify performance parameters is key to the whole aircraft industry. We have emphasised the fact that the field of comprehensive analysis is not well developed (unlike rotorcraft engineering); the amount of reference data is relatively modest, as reported in the flight manual and in some type certificate documents. However, these data are often insufficient for an accurate engineering analysis. In this book we have attempted to make advancements in this area and have proposed several numerical methods, validation strategies and sensitivity analyses that can improve the prediction of aircraft performance.

**THE KNOWLEDGE BASE.** As technology progresses, the knowledge base in aircraft engineering is bound to increase – at least theoretically. The knowledge base is interpreted as the ensemble of written communications and personal experience. The written communications include papers, technical reports, technical specifications, drawings, graphs, databases, experimental data, flight data, regulations, engineering standards and so on. All of this written communication has to be studied and interpreted. Increasingly, the answer is given by the computer or the Internet. The computer calculates, on the basis of a software system that is taken for granted, and produces results that often are not critically assessed. However, decisions cannot be delegated to computers, particularly when they involve human factors, safety, compliance with regulations and large procurement contracts. With this in mind, industrial and professional expertise is crucial; maintaining expertise over time has become increasingly challenging.

**IMPACT OF AVIATION.** With regard to the commercial aspects of air transport, we have hit various limits in terms of airport capacity, air traffic control, airport security, consequences of local disruption, ground transport integration, and so on. In this book, we have focused on the airplane itself, which is at the heart of this infrastructure. The airplane may be small in comparison to an international airport or the sky, but it has become increasingly entangled in a complex web of ground services expanding way beyond the runway; in the sky, there are enough airplanes to potentially cause collisions on a regular basis. Thus, they cannot navigate on their own. Traffic congestion is no longer a term used to denote ground vehicles.

**AVIATION EMISSIONS FORECAST.** Considerable research already exists on the effects of aviation on atmospheric physics, as briefly mentioned in this book. A vast amount of technical work is available in the field of aggregate aviation emissions*.

The emissions from commercial aviation are indeed high. However, their contribution must be viewed in the larger context of emissions from other sources, in particular the effects of deforestation, the loss of natural habitats, the increase in world population and other threats of various magnitude. As we write this section, the world's population has reached the 7-billion mark; surely, it will continue to rise. The effect of population increase alone (about 1.2% per year minimum since 1965) leads to an increase of about 84 million people per year†. Each of these new citizens of the world will need to use energy. If they were to emit only 1 ton of carbon-dioxide per year (about 1/10 of the UK; 1/20 of the United States and Canada), the total *additional worldwide emissions* would be about 84 million tons/year. This effect would be compounded in the following year by an additional 85 million tons, and so on. To this effect, we would need to add the loss of natural environment to make room for more urban infrastructure. The total amount of direct carbon-dioxide emissions from aviation is estimated at about 500 million tons per year; thus, the effect of population increase in one year alone corresponds to one-sixth of these emissions, at the conservative rates that we have assumed. At the current growth rate, the population increase offsets any realistic cut in aviation emissions.

The "one $CO_2$-ton society" that has been suggested in some quarters will not be possible if we keep flying. However, we cannot imagine a modern society without air transportation. Furthermore, if we stop deforestation for one day only, we save enough $CO_2$ to transport about 4 million passengers from London to New York. This estimate is based on average $CO_2$ emissions of 810 million metric tons‡ per year (based on 2000–2005 rates), and 550 kg $CO_2$/passenger over a 3,100-n-mile stage length with a 394-seat Boeing B777-300.

* See, for example, EMEP/CORINAIR Emission Inventory Guidebook; available on the Internet, periodically updated.

† Source: United Nations; World Bank.

‡ Harris NL, Brown S, Hagen SC, Saatchi SS, Petrova S, Salas W, Hansen MC, Potapov PV, and Lotsch A. Baseline map of carbon emissions from deforestation in tropical regions. 336(6088):1573–1576, 2012.



APPENDIX A

# Gulfstream G-550

The Gulfstream G-550 is a long-range business/executive jet powered by two Rolls-Royce BR710 turbofan engines. It has advanced cockpit avionics (see Figure 1.1) and a variety of seating configurations. The G-550 is a transport-type airplane, certified for day and night, extended over-water and polar navigation. A few examples of calculations are shown throughout the book (see, for example, the SAR analysis in §12.3.1 and the payload-range in § 15.2).

A summary of weights, capacity and limitations is given in Table A.1. A summary of geometrical dimensions is given in Table A.2. The wing is swept back 27 degrees at the quarter-chord location, with a 3-degree dihedral, and with a tip-mounted upper winglet. The wing does not have a leading-edge slat. The main flight controls are a single Fowler-type flap, ailerons and trim tabs. There are three spoilers per side. There are some drag-reduction devices, such as vortex generators that wrap around the winglet junction, vortex generators on the upper side of the wing and blunt trailing edges. Furthermore, there are seals installed around rudder, elevator, and thrust reverser and fairings around antennas. A summary of operational limitation is given in Table A.3.

## Power Plant

The engine is a two-shaft high-by-pass ratio Rolls-Royce BR710 C4-11 turbofan engine with a single-stage low-pressure compressor, a ten-stage high-pressure compressor, a two-stage high-pressure turbine and a two-stage low-pressure turbine. The engine is equipped with FADEC. Basic data are shown in Table A.4. The engines are mounted high in the aft fuselage through horizontal pylons.

## Airplane Dimensions

Table A.2 shows the main dimensions of the airplane. Where data are indicated as *n.a.*, it means that they cannot be conventionally defined. For example, the vertical stabiliser has several sides, and although a root chord could be defined, it is more difficult to decide what is the leading-edge sweep.

Table A.1. *Weights and capacities of the G-550*

| Parameter | Value | Unit |
|---|---|---|
| **Weights & Capacity** | | |
| Maximum take-off weight | 41,277 | kg |
| Maximum ramp weight | 41,458 | kg |
| Operating empty weight | 20,960 | kg |
| Maximum zero-fuel weight | 27,721 | kg |
| Maximum landing weight | 34,155 | kg |
| Maximum payload | 2,812 | kg |
| Maximum fuel capacity (1 left, 1 right tank) | 18,734 | kg |
| Maximum allowable fuel asymmetry (take-off) | 454 | kg |
| Maximum allowable fuel asymmetry (flight) | 908 | kg |
| Seating capacity | 14–18 | |

The flight controls include the ailerons, Fowler flaps (one per side) and spoilers (two flight spoilers per side and one ground spoiler per side). The tail surfaces consist of a fixed-geometry vertical stabiliser with a rudder and a trimmable horizontal stabiliser with flight controls (right/left elevators). The elevators incorporate adjustable trim tabs.

Table A.5 is a summary of landing-gear characteristics. The noise gear is steerable and the main landing gear has an anti-skid system. The main landing gear has a main strut with a shock absorber, a structural post at the back, an axle fitting assy

Table A.2. *Basic dimensions of the G-550*

| Parameter | Value | Unit | Parameter | Value | Unit |
|---|---|---|---|---|---|
| **Airplane** | | | **Horizontal Tail Plane** | | |
| Overall length | 29.38 | m | Wing span | 10.72 | m |
| Overall height | 7.90 | m | Wing sweep at L.E. | 33.0 | degs |
| Fuselage length | 26.16 | m | Wing area | 23.24 | $m^2$ |
| Cabin length | 15.27 | m | Aspect-ratio | 4.96 | |
| Tail upsweep angle | 13.8 | degs | Root chord | 3.084 | m |
| Fuselage height | 2.44 | m | Taper ratio | 0.456 | |
| Fuselage width | 1.88 | m | Dihedral angle | 0.3 | degs |
| **Main Wing** | | | **Vertical Tail Plane** | | |
| Wing span (reference) | 28.48 | m | Height | 3.58 | m |
| Tip chord | 1.620 | m | Wing sweep at L.E. | n.a. | |
| Root chord | 6.130 | m | Wing area | 15.1 | $m^2$ |
| Taper ratio | 0.265 | | Aspect-ratio | 0.85 | |
| Wing sweep at L.E. | 27.0 | degs | Root chord | n.a. | m |
| Wing sweep at Q.C. | 30.5 | degs | Taper ratio | n.a. | |
| Wing area (reference) | 90.89 | $m^2$ | | | |
| Mean aerodynamic chord | 3.795 | m | | | |
| Dihedral angle | 3.0 | degs | | | |

Table A.3. *Operational limits of the G550*

| Parameter | Value | Unit |
|---|---|---|
| **Speeds** | | |
| Long-range Mach number | 0.80 | |
| Maximum operating Mach number | 0.885 | |
| Minimum control speed, landing, S/L | 110 | KCAS |
| Minimum control speed, take-off, S/L | 107 | KCAS |
| Minimum control speed, air | 112 | KCAS |
| Tail wind limit, take-off | 10 | kt |
| Tail wind limit, landing | 10 | kt |
| **Aerodynamic Controls** | | |
| Maximum UP/DOWN aileron deflection | 11 | degs |
| Maximum UP/DOWN aileron trim tab deflection | 15 | degs |
| Maximum UP/DOWN elevators deflection | n.a. | degs |
| Maximum spoiler deflection | | |
| clean wing | +30 | degs |
| full aileron deflection | +55 | degs |
| Maximum right/left rudder deflection | 22 | degs |
| Maximum flap deflection | 39 | degs |
| **Operating Limitations** | | |
| Maximum cruise altitude (level) | FL-510 | |
| AEO service ceiling | 13,015 | m |
| OEI service ceiling | 7,870 | m |
| Maximum approved take-off altitude | 3,048 | m |
| Maximum runway slope | ±2 | degs |
| Maximum payload | 5,700 | n-miles |
| 3,000 lb (1,300 kg) payload | 6,450 | n-miles |
| 1,600 lb (0.725 kg) payload | 6,700 | n-miles |
| ferry range | 6,900 | n-miles |
| BFL at MTOW, ISA, sea level | 1,800 | m |
| Final approach speed at MLW, sea level | 105 | kt |
| Maximum fuel temperature | 54 | °C |
| Minimum fuel temperature (red alert) | −37 | °C |
| Minimum turning radius | 34.14 | m |

Table A.4. *Selected data of the Rolls-Royce BR710 C4-11 gas turbine engine*

| Item | Value | Unit |
|---|---|---|
| Dry weight | 1,818 | kg |
| Overall length | 4.660 | m |
| Fan diameter | 1.785 | m |
| Number of fan blades | 24 | |
| | | |
| Maximum continuous thrust (S/L, ISA) | 64.3 | kN |
| Maximum take-off thrust (S/L, ISA) | 68.4 | kN |
| Turbine gas temperature | 860 | |
| maximum continuous thrust | 860 | °C |
| take-off (5 minutes) | 900 | °C |
| By-pass ratio | 4.2 | |
| Compressor ratio | 24 | |
| Minimum fuel temperature | −40 | °C |
| Maximum oil temperature | +160 | °C |

Table A.5. *Landing gear of the G550*

| Parameter | Value | Unit |
|---|---|---|
| **Dimensions** | | |
| Number of wheels | 6 | |
| Landing-gear groups | 3 | |
| Main gear wheels | 2 × 2 | |
| Nose gear wheels | 6 × 1 | |
| Wheel base | 13.72 | m |
| Wheel track | 4.37 | m |
| Main gear tyre | H35 × 11.0-18 | |
| Nose gear tyre | 21 × 7.25-10 | |
| Main tyre pressure | 12.8 | bar |
| Nose tyre pressure | 15.2 | bar |
| Distance of nose gear from nose | 2.1 | m |
| Distance of main gear from nose | 15.8 | m |
| Total length of main struts | 2.70 | m |
| Average diameter of main struts | 0.16 | m |
| Tyre speed limit | 195 | kt |
| Fuse plugs on each wheel | 4 | |
| Brake warning temperature | 650 | °C |

and a side-brace actuator. The nose landing gear has a main strut with a shock absorber, an inclined drag brace, a downlock actuator and two head lamps.

## Performance

Figure A.1 shows the flight envelope of the airplane, as elaborated from the FCOM. This envelope is also discussed in § 8.6.2.

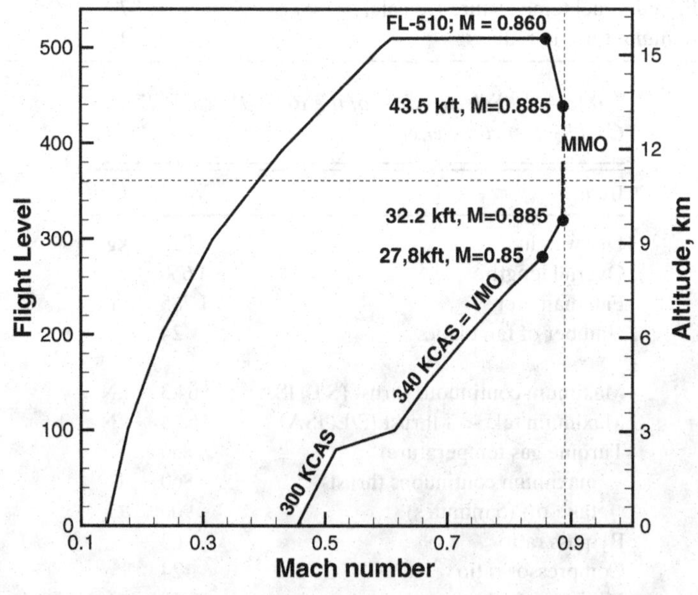


Figure A.1. Flight envelope of the Gulfstream G550, standard day, no wind.

### A.0.1 Geometry Model

The following table is an example of our control-points model for this aircraft. See Chapter 2 for details.

```
"Gulfstream G550"                  airplane name
  "1.0.0"                          airplane model version (can be updated!)
   0.5d0                           accuracy of data (percent, variable, ** estimated **)

"TOP"                              view

  "wing"          5                notes
    10.9137        1.1867          wing attachment (leading edge), wrt NOSE point
    18.1895       13.4489          wing tip (leading edge)
    19.8107       13.4489          wing tip (trailing edge)
    16.9241        2.2151          wing break (trailing edge), close to fuselage
    17.0428        1.1867          wing attachment (trailing edge)

  "flaps"         4                COMPONENT, number of control points
    16.9241        2.2547          generic points
    16.3706        2.4524
    18.5058        9.7306
    18.8617        9.7306
 "aileron"        4                COMPONENT, number of control points
    18.8617        9.7700          generic points
    18.1895        9.7700
    18.9803       12.1435
    19.4549       12.1435
  "tailplane"     5                COMPONENT, number of control points
    24.5558        0.1582          inboard leading edge
    27.8378        5.1027
    28.2728        5.3600
    29.3405        5.3600
    27.6401        0.0800          inboard trailing edge

    .......
```

APPENDIX B

# Certified Aircraft Noise Data

Table B.1 and Table B.2 summarise the certified noise performance of selected aircraft, in various categories. A full database of certification data is published by the Federal Aviation Administration and is periodically updated. The FAA database provides noise levels of hundreds of aircraft configurations, including the effects of different engine installation. Military aircraft are excluded. The tables also report the MTOW, the by-pass ratio of the engines and the position of the flaps, both at take-off (TO) and approach/landing (AP). The noise data are provided at take-off, sideline (SL) and approach.

Table B.1. *Certified noise levels for commercial aircraft*

| Aircraft | MTOW | Engines | BPR | Flap (TO) | Flap (AP) | TO | SL | AP |
|---|---|---|---|---|---|---|---|---|
| ATR72-500 | 22.50 | 2 × PW-127M | | 15 | 25 | 79.0 | 80.7 | 92.3 |
| A-300B4-622R | 149.70 | 2 × PW-4158 | 4.85 | 0 | 40 | 88.0 | 98.3 | 101.3 |
| A-310-324 | 150.00 | 2 × PW-4152 | 4.85 | 15 | 40 | 90.6 | 97.2 | 100.2 |
| A-320-211 | 68.00 | 2 × CFM56-5A1 | 6.00 | 10 | 35 | 85.3 | 94.4 | 96.4 |
| A-330-301 | 180.00 | 2 × CF6-80E1A2 | 5.05 | 14 | 32 | 87.0 | 97.9 | 98.5 |
| A-330-321 | 230.00 | 2 × PW4164 | 4.85 | 8 | 32 | 95.6 | 97.5 | 98.0 |
| A-340-312 | 220.00 | 4 × CFM56-5C3 | 6.60 | 17 | 32 | 88.0 | 95.8 | 97.3 |
| A-340-312 | 270.00 | 4 × CFM56-5C3 | 6.60 | 17 | 32 | 96.2 | 95.3 | 97.2 |
| A-380-842 | 391.00 | 4 × Trent 972 | | | | 94.6 | 94.5 | 98.0 |
| A-380-861 | 391.00 | 4 × GP7270 | 8.30 | | | 94.8 | 94.5 | 97.1 |
| BAE 146-RJ100 | 46.00 | 4 × LF 507-1F | 5.10 | 18 | 33 | 86.1 | 88.1 | 97.6 |
| B-717-200 | 64.58 | 2 × BR700-715A1 | 4.66 | 5 | 40 | 84.0 | 89.0 | 91.6 |
| B-737-300 | 56.47 | 2 × CFM56-3 | 5.00 | 1 | 40 | 82.4 | 89.7 | 98.5 |
| B-737-300 | 63.28 | 2 × CFM56-3 | 5.00 | 1 | 40 | 83.9 | 90.9 | 97.6 |

Table B.2. *Certified noise levels for commercial aircraft (part 2)*

| Aircraft | MTOW | Engines | BPR | Flap (TO) | Flap (AP) | TO | SL | AP |
|---|---|---|---|---|---|---|---|---|
| B-737-500 | 49.00 | 2 × CFM56-3 | 5.00 | 5 | 40 | 81.0 | 89.3 | 98.4 |
| B-737-500 | 63.05 | 2 × CFM56-3-B1 | 5.00 | 5 | 40 | 87.3 | 90.0 | 100.0 |
| B-737-700 | 60.33 | 2 × CFM56-7B22 | 5.40 | 1 | 40 | 82.6 | 92.5 | 95.8 |
| B-737-700 | 70.08 | 2 × CFM56-7B26 | 5.10 | 1 | 40 | 84.6 | 94.7 | 95.9 |
| B-737-700/IGW | 72.12 | 2 × CFM56-7B24 | 5.30 | 1 | 40 | 86.6 | 92.9 | 96.1 |
| B-737-700/IGW | 77.57 | 2 × CFM56-7B27 | 5.10 | 1 | 40 | 86.6 | 95.2 | 96.1 |
| B-737-800 | 79.01 | 2 × CFM56-7B27 | 5.10 | 1 | 40 | 87.0 | 94.7 | 96.5 |
| B-737-900 | 74.39 | 2 × CFM56-7B24 | 5.30 | 1 | 40 | 86.6 | 92.0 | 96.4 |
| B-737-900 | 79.01 | 2 × CFM56-7B27 | 5.10 | 1 | 40 | 86.7 | 94.2 | 96.4 |
| B-747-100 | 322.06 | 4 × JT9D-3A 4 | 5.10 | 10 | 30 | 105.4 | 102.1 | 104.6 |
| B-747-100 | 332.94 | 4 × JT9D-3A (*) | 5.10 | 10 | 30 | 109.4 | 99.6 | 107.2 |
| B-747-400 | 394.60 | 4 × CF6-80C2B1F | 5.20 |  | 25 | 99.7 | 98.3 | 101.4 |
| B-747-400 | 394.60 | 4 × PW 4056 | 4.80 | 10 | 30 | 101.5 | 99.7 | 104.7 |
| B-747-400 | 396.90 | 4 × RB211-524G | 4.30 | 10 | 30 | 99.2 | 98.0 | 103.8 |
| B-757-300 | 124.74 | 2 × RB211-535E4B | 4.10 | 5 | 30 | 88.4 | 94.8 | 95.4 |
| B-757-300 | 107.05 | 2 × RB211-535-E4 | 4.10 | 5 | 30 | 84.8 | 93.9 | 95.2 |
| B-767-200 | 160.06 | 2 × CF6-80C2-B2 | 5.00 | 1 | 30 | 89.5 | 93.7 | 96.4 |
| B-767-200 | 175.54 | 2 × CF6-80C2-B4 | 5.00 | 1 | 30 | 90.6 | 95.0 | 96.4 |
| B-767-200/ER | 136.08 | 2 × CF6-80C2B2F | 5.00 | 1 | 30 | 85.1 | 93.8 | 95.8 |
| B-767-300 | 184.61 | 2 × PW 4056 | 4.80 | 5 | 30 | 94.2 | 95.7 | 100.2 |

APPENDIX C

# Options for the FLIGHT Program

The key performance options of the FLIGHT program described in this book are given here. After loading the airplane, the following user-menus are available:

1. **Performance Charts**
   (a) Aerodynamics (Chapter 4)
   (b) Specific Air Range (Chapter 12)
   (c) Engine Charts (Chapter 5)
   (d) Flight Envelopes (Chapters 8 and 13)
   (e) Propeller (Chapter 6)
   (f) WAT (AEO Take-Off) (Chapter 9)
   (g) Balanced Field Length (Chapter 9)
   (h) Payload-Range (Chapter 15)
   (i) Economic Mach Number (Chapter 12)
   (j) CG Effects (Chapter 12)
   (k) Buffet Boundary (Chapter 4)
   (l) Specific Excess Power (Chapter 13)
   (m) Jet Thermo Charts (Chapter 14)
   (n) Atm-Speed Charts (Chapter 8)
   (o) Holding Speeds (Chapter 15)
   (p) V-n diagram (Chapter 13)
2. **Mission Analysis** (Chapter 15)
   (a) Fuel Planning
   (b) Aircraft Range
   (c) Matrix-Fuel-Plan
   (d) Equal-Time Point
3. **Aircraft Noise** (Chapters 16–18)
   (a) Take-Off at FAR/ICAO Point
   (b) Landing at FAR/ICAO Point
   (c) Sideline at FAR/ICAO Point
   (d) Arbitrary Trajectory
   (e) Noise Footprints
   (f) Stacking Patterns

4. **Exhaust Emissions**
   (a) Exhaust Emissions versus Range (Chapter 15)
   (b) Contrail Analysis (Chapter 19)
5. **Flight Optimisation**
   (a) Minimum Climb-Fuel (Chapter 10)
   (b) Optimum Climb between Flight Levels (Chapter 12)
   (c) Fuel-Tankering Analysis (Chapter 15)
6. **Manoeuvre Analysis** (Chapter 13)
   (a) Landing in Downburst
   (b) Take-Off in Downburst
7. **Trim Analysis**
   (a) Minimum Control Speed on Ground, VMCG (Chapter 9)
   (b) Minimum Control Speed in Air, VMCA (Chapter 7)
8. **Direct Operating Costs** (Chapter 15)

Each sub-option requires entering of a set of operational data. All of the geometry, mass and inertia properties (Chapter 2 and Chapter 3) are given automatically on output.

The propeller option points to another program, with user-input as herein. After loading the propeller model, the following sub-menus are available:

1. **Propeller Geometry**
2. **Performance at Design Point**
3. **Performance in Oblique Flight**
4. **Performance Charts**
5. **Propeller Design**
6. **Propeller Trim**
7. **Propeller Noise**
8. **Performance of Ducted Propeller**

# Index

Note: "t" after page number denotes a table; "f" after page number denotes a figure